AF327477

LIGHT ELEMENTS IN THE UNIVERSE

IAU SYMPOSIUM No. 268

COVER ILLUSTRATION: JET D'EAU de Genève

Designed by Cécile H.

INTERNATIONAL ASTRONOMICAL UNION

UNION ASTRONOMIQUE INTERNATIONALE

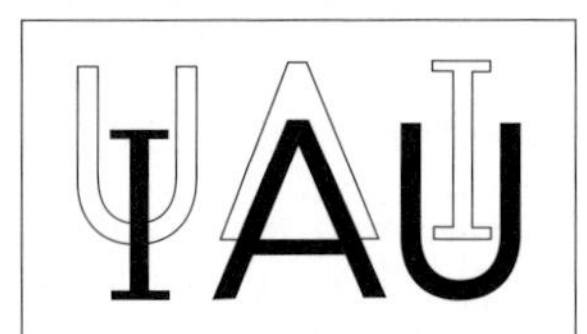

LIGHT ELEMENTS IN THE UNIVERSE

PROCEEDINGS OF THE 268th SYMPOSIUM OF THE INTERNATIONAL ASTRONOMICAL UNION HELD IN GENEVA, SWITZERLAND NOVEMBER 9–13, 2009

Edited by

CORINNE CHARBONNEL

*GENEVA OBSERVATORY, GENEVA UNIVERSITY, SWITZERLAND &
CNRS, FRANCE*

MONICA TOSI

INAF-OSSERVATORIO ASTRONOMICO DI BOLOGNA, ITALY

FRANCESCA PRIMAS

EUROPEAN SOUTHERN OBSERVATORY-ESO, GARCHING, GERMANY

and

CRISTINA CHIAPPINI

*GENEVA OBSERVATORY, GENEVA UNIVERSITY, SWITZERLAND &
INAF-OSSERVATORIO ASTRONOMICO DI TRIESTE, ITALY*

CAMBRIDGE
UNIVERSITY PRESS

CAMBRIDGE UNIVERSITY PRESS
The Edinburgh Building, Cambridge CB2 8RU, United Kingdom
32 Avenue of the Americas, New York, NY 10013-2473, USA
477 Williamstown Road, Port Melbourne, VIC 3207, Australia
Ruiz de Alarcón 13, 28014 Madrid, Spain
Dock house, The Waterfront, Cape Town 8001, South Africa

First published 2010

Printed in the United Kingdom at the University Press, Cambridge

Typeset in System LaTeX 2_ε

A catalogue record for this book is available from the British Library

Library of Congress Cataloguing in Publication data

This book has been printed on FSC-certified paper and cover board. FSC is an independent,
non-governmental, not-for-profit organization established to promote the responsible
management of the world's forests. Please see www.fsc.org for information.

ISBN 9780521765060 hardback
ISSN 1743-9213

Table of Contents

Opening Session

Session I. Production of the light elements in the first minutes of the Universe
Chair: Suzanne Talon

Session II. Abundances of D, ^{3}He and ^{4}He: observations
Chairs: Robert Rood, Monica Tosi

Session III. Abundances of LiBeB: observations
Chairs: Beatriz Barbuy, Yuri Izotov, Paolo Molaro & Francesca Primas

Session IV. Sources and sinks of light elements
Chairs: Francesca Primas, David Lambert

Session V. Evolution of the light elements in the Universe
Chairs: David Lambert, Corinne Charbonnel

Preface

The light elements (H, He, Li, Be, B and their isotopes) deserve special attention because of their relationship with several important astrophysical domains: they provide key clues for stellar and ISM structure and evolution, galaxy formation and evolution, Big Bang nucleosynthesis and cosmology. They are one of the few bridges connecting several different astrophysical communities.

The previous IAU Symposium on the light elements was held in 1999 (in Natal, Brasil, IAU Symp. 198), and other, non-IAU supported, related meetings also took place about a decade ago. Since then, there have been many significant developments both on the observational and theoretical sides. Striking progress was achieved thanks to the accurate determination of the baryon density of the Universe by recent cosmic microwave background experiments. This allowed an unprecedented precision on the determination of the yields of Standard Big Bang Nucleosynthesis and a different perspective in the comparison with the D, ^{3}He, ^{4}He and ^{7}Li abundances measured from observations of low-metallicity environments.

In parallel, the advent of new generation ground and space based telescopes allowed the observation of light elements in objects previously unreachable, with new intriguing results on the present and past abundances of D, He (both ^{3}He and ^{4}He), Li and its isotopic ratio, Be and B and their isotopic ratios. Thanks to multi-fiber instruments, a wealth of data could be gathered in a consistent way which still await to be fully understood and interpreted. Also, realistic 3D, time-dependent, hydrodynamical model atmospheres strongly helped in providing more reliable abundance determinations.

On the theoretical side, we entered a golden age for the description of stellar interiors and evolution thanks to improved treatments of stellar rotation, magnetic fields, internal gravity waves, atomic diffusion, and thermohaline instability in new generation stellar models. Most of these improvements were achieved thanks to constraints coming from combined observations of light element abundances in various types of stars. Last but not least, a wealth of independent consistent chemical evolution models able to fit the general trends of the vast majority of Galaxy physical and chemical features were developed that must be tested with respect to the evolution of light elements.

Despite all these achievements, we are far from understanding and reproducing in detail all the light element patterns, their local, short term variations, as well as their global evolution in the Universe. The complete understanding of the evolution of the light elements in the Universe is a challenging task that requires the exchange of ideas and the collaboration of astrophysicists with observational and theoretical expertise in stellar hydrodynamics and evolution, Galactic and extra-galactic astronomy, and cosmology.

We felt that the route to this goal could be found only by gathering specialists in all the different relevant fields in an IAU Symposium. The various IAU Commissions and Divisions corresponding to these research fields have favourably received and endorsed our request, thus making this conference possible. The Symposium brought together 118 participants from 23 countries around the world to discuss the achievements as well as the problems concerning the light elements. We had 25 invited reviews, 53 oral contributions and 35 posters distributed over five sessions.

Several important issues, still highly controversial, have been tackled during the meeting thanks to the collective inputs from observers, stellar and galactic physicists, and cosmologists. To let the participants have time to discuss at length the hottest topics identified by the SOC, we also organized four discussion sessions: one on the dispersion of the observational D abundances, one on the problems affecting the derivation of ^{4}He from HII regions, one on the problems affecting the derivation and interpretation of Li, Be and B abundances, and one on the stellar yields of ^{3}He, ^{4}He and ^{7}Li.

In addition to the scientific program, we enjoyed a public conference attended by more than 600 people at the Geneva University. Four speakers (the astrophysicists H.Reeves and J.Geiss, the physicist G.Laval, and the psychiatrist J.M.Aubry) presented to the public different aspects related to "Deuterium, helium, lithium: From the Big Bang to the contemporary civilisation".

Finally, with a 25% of the Symposium speakers being women, we considered it appropriate to organize a Women Networking Lunch, to let the participants from different countries exchange their experiences. The event turned out to be quite successful and we all enjoyed the opportunity to interact with other women astronomers coming from different environments. In addition F.Primas, who is co-chairing the IAU Working Group Women in Astronomy, gave a talk on Women Career Advancements for the PhD students and postdocs at the Geneva University. We are grateful to her for having had the idea of these activities, for her help in organizing them, and, most importantly, for always being ready to take initiatives against gender and minority discriminations.

It is a great pleasure to thank the Geneva Observatory, the University of Geneva, and the Museum of Natural History of Geneva for hospitality and support. We gratefully acknowledge the financial support of our sponsors, listed on the next page of these Proceedings, and the active contribution of the members of the SOC and the LOC. We are particularly grateful to Chantal Taçoy and Michel Grenon, who both did real miracles for the logistic organization of the Symposium, and to Bob Rood who took most of the pictures included in this volume. Finally, we warmly thank Cristina Chiappini and Francesca Primas for having co-edited these proceedings and shared with us their responsibility.

Corinne Charbonnel and Monica Tosi, co-chairs SOC,
Geneva, Bologna, February 4, 2010

THE ORGANIZING COMMITTEE

Scientific

C. Charbonnel (Switzerland/France, chair)
B. Barbuy (Brazil)
T. Kajino (Japan)
J. Lattanzio (Australia)
W. Moos (USA)
R.T. Rood (USA)

M. Tosi (Italy, co-chair)
Y. Izotov (Ukraine)
D. Lambert (USA)
P. Molaro (Italy)
F. Primas (Germany)

Local

C. Charbonnel (chair)
M. Dessauges-Zavadsky
M. Grenon
C. Taçoy

C. Chiappini
S. Ekström
N. Lagarde
G. Simond

Acknowledgements

The symposium is sponsored and supported by the IAU Divisions IV (Stars), VI (Interstellar Matter), VIII (Galaxies and the Universe); and by the IAU Commissions No. 28 (Galaxies), No. 29 (Stellar Spectra), No. 35 (Stellar Constitution), No. 36 (Theory of Stellar Atmospheres), No. 37 (Star Clusters and Associations).

The Local Organizing Committee operated under the auspices of the
Geneva Observatory, University of Geneva (Switzerland).

Funding by the
International Astronomical Union,
Swiss National Science Foundation,
Swiss Academy of Sciences,
Museum of Natural History of Geneva City,
University of Geneva,
Geneva Observatory,
Faculty of Sciences of the University of Geneva,
Société de Physique et d'Histoire Naturelle,
and
Swiss Society for Astrophysics and Astronomy,
is gratefully acknowledged.

Conference photograph

Participants

George **Angelou**, Monash University, CSPA, Australia — george.angelou@sci.monash.edu.au
Martin **Asplund**, Max Planck Institute for Astrophysics, Germany — asplund@mpa-garching.mpg.de
Dana **Balser**, NRAO, USA — dbalser@nrao.edu
Thomas **Bania**, Department of Astronomy, Boston University, USA — bania@bu.edu
Fabio **Barblan**, Geneva Observatory, Geneva University, Switzerland — fabio.barblan@unige.ch
Beatriz **Barbuy** , Universidade de São Paulo , Brazil — barbuy@astro.iag.usp.br
Yerra **Bharat Kumar**, Indian Institute of Astrophysics, India — bharat@iiap.res.in
Marek **Biesiada**, Dept. of Astrophysics and Cosmology, University of Silesia, Poland — marek.biesiada@us.edu.pl
Ann Merchant **Boesgaard**, Institute for Astronomy, University of Hawaii, USA — boes@ifa.hawaii.edu
Piercarlo **Bonifacio**,GEPI, Observatoire de Paris, France — Piercarlo.Bonifacio@obspm.fr
Angela **Bragaglia**, INAF - Osservatorio Astronomico di Bologna, Italy — angela.bragaglia@oabo.inaf.it
Nahuel **Cabral** , Physikalisches Institut Universität Bern, Switzerland — nahuel.cabral@space.unibe.ch
Elisabetta **Caffau**, GEPI, Observatoire de Paris, France — Elisabetta.Caffau@obspm.fr
Luca **Casagrande**, Max Planck Institute for Astrophysics, Germany — luca@MPA-Garching.MPG.DE
Oscar **Cavichia**, University of São Paulo , Brazil — cavichia@astro.iag.usp.br
Corinne **Charbonnel**, Geneva Observatory , Switzerland and CNRS, France — corinne.charbonnel@unige.ch
Sujit Kumar **Chatterjee**, IGNOU, New Alipore College Centre, India — chat_sujit1@yahoo.com
Cristina **Chiappini**, Geneva Observatory , Switzerland and INAF/OATS, Italy — Cristina.Chiappini@unige.ch
Roberto **Costa**, Universidade de São Paulo, Brazil — roberto@astro.iag.usp.br
Katia **Cunha**, National Optical Astronomy Observatory, USA — kcunha@noao.edu
Francesca **D'Antona**, INAF – Osservatorio di Roma, Italy — dantona@mporzio.astro.it
Valentina **D'Orazi**, INAF - Osservatorio Astronomico di Padova, Italy — valentina.dorazi@oapd.inaf.it
Patrick **de Laverny**, Observatoire de la Côte d'Azur, France — laverny@oca.eu
Thibaut **Decressin** , Argelander-Institut für Astronomie, Germany — decressin@astro.uni-bonn.de
Elisa **Delgado Mena**, Instituto de Astrofisica de Canarias, Spain — edm@iac.es
Miroslava **Dessauges**, Geneva Observatory, Geneva University, Switzerland — miroslava.dessauges@unige.ch
Marcella **Di Criscienzo**, INAF - Osservatorio Astronomico di Roma, Italy — dicrisci@gmail.com
Adam **Dobrzycki** , European Southern Observatory, Germany — adobrzyc@eso.org
Joanna **Dunkley**, Oxford University, United Kingdom — j.dunkley@physics.ox.ac.uk
Patrick **Eggenberger**, Geneva Observatory, Geneva University, Switzerland — patrick.eggenberger@unige.ch
Sylvia **Ekström**, Geneva Observatory, Switzerland — sylvia.ekstrom@unige.ch
Lisa **Elliott**, Monash University, Australia — Lisa.Elliott@sci.monash.edu.au
Gary **Ferland**, University of Kentucky, USA — gary@pa.uky.edu
Patrick **François**, Paris Observatory, France — patrick.francois@obspm.fr
Urs **Frischknecht**, University of Basel, Switzerland — urs.frischknecht@unibas.ch
Johannes **Geiss**, International Space Science Institute, Switzerland — johannes.geiss@issibern.ch
Cyril **Georgy**, Geneva Observatory, Geneva University, Switzerland — cyril.georgy@unige.ch
Oscar **Gonzalez**, European Southern Observatory, Germany — oagonzal@uc.cl
Jonay I. **Gonzalez Hernandez**, Universidad Complutense de Madrid, Spain — jonay@astrax.fis.ucm.es
Claudia **Greco**, Geneva Observatory, Geneva University, Switzerland — claudia.greco@unige.ch
Michel **Grenon**, Geneva Observatory, Geneva University, Switzerland — Michel.Grenon@unige.ch
Roald **Guandalini**, University of Perugia, Italy — guandalini@fisica.unipg.it
Guillaume **Hébrard**, Institut d'Astrophysique de Paris, France — hebrard@iap.fr
J. Christopher **Howk**, University of Notre Dame, USA — jhowk@nd.edu
Garik **Israelian**, Instituto de Astrofisica de Canarias, Spain — gil@iac.es
Hiroko **Ito**, The Graduate University of Advanced Studies, NAOJ, Japan — hiroko.ito@nao.ac.jp
Yuri **Izotov**, Main Astronomical Observatory, Ukraine — izotov@mao.kiev.ua
Karsten **Jedamzik**, University of Montpellier II, France — jedamzik@lpta.univ-montp2.fr
Colin **Jones**, Simon Fraser University, Canada — colin_jones@sfu.ca
Andreas **Kaufer**, European Southern Observatory, Chile — akaufer@eso.org
Andreas **Korn** , Uppsala Astronomical Observatory, Sweden — andreas.korn@fysast.uu.se
Motohiko **Kusakabe** , Inst. for Cosmic Ray Research, University of Tokyo, Japan — kusakabe@icrr.u-tokyo.ac.jp
Nadége **Lagarde**, Geneva Observatory, Geneva University, Switzerland — Nadege.Lagarde@unige.ch
David **Lambert**, The University of Texas at Austin, McDonald Observatory, USA — director@astro.as.utexas.edu
Livio **Lamia**, INFN–LNS Catania, Italy — llamia@lns.infn.it
Norbert **Langer**, Argelander-Institut fuer Astronomie, Germany — nlanger@astro.uni-bonn.de
John **Lattanzio**, Monash University, Australia — john.lattanzio@sci.monash.edu.au
Karin **Lind**, European Southern Observatory, Germany — klind@eso.org
Jeffrey **Linsky**, JILA, University of Colorado and NIST, USA — jlinsky@jila.colorado.edu
Donald **Lubowich** , Hofstra University, USA — Donald.Lubowich@Hofstra.edu
Walter **Maciel**, University of São Paulo, Brazil — maciel@astro.iag.usp.br
André **Maeder**, Geneva Observatory, Geneva University, Switzerland — andre.maeder@unige.ch
Enrico **Maiorca**, University of Perugia, Italy — emaiorca@gmail.com
Sushma **Mallik**, Indian Institute of Astrophysics, India — sgvmlk@iiap.res.in
Anna **Marino**, Universitá di Padova, Italy — anna.marino@unipd.it
Francesca **Matteucci**, Trieste University, Italy — matteucc@oats.inaf.it
Jorge **Meléndez**, Centro de Astrofisica da Universidade do Porto, Portugal — jorge@astro.up.pt
Georges **Meynet**, Geneva Observatory, Geneva University, Switzerland — georges.meynet@unige.ch
Antonino **Milone**, Universitá di Padova, Italy — antonino.milone@unipd.it
Tamara **Mishenina**, Astronomical Observatory, Odessa National University, Ukraine — tamar@deneb1.odessa.ua
Paolo **Molaro**, INAF - Osservatorio Astronomico di Trieste, Italy — molaro@oats.inaf.it
Thierry **Montmerle**, Laboratoire d'Astrophysique de Grenoble , France — montmerle@obs.ujf-grenoble.fr
Saumitra **Mukherjee**, Jawaharlal Nehru University, India — saumitramukherjee3@gmail.com
Ko **Nakamura**, National Astronomical Observatory of Japan, Japan — nakamura.ko@nao.ac.jp
Poul Erik **Nissen**, Department of Physics and Astronomy, University of Aarhus, Denmark — pen@phys.au.dk
Pierre **North**, Laboratoire d'Astrophysique – EPFL, Switzerland — pierre.north@epfl.ch
Giancarlo **Pace** , Centro de Astrofisica da Universidade do Porto, Portugal — gpace@astro.up.pt
Ana **Palacios**, GRAAL, CNRS , France — ana.palacios@univ-montp2.fr
Sara **Palmerin** , Universitá degli Studi di Perugia - - INFN, Italy — sara.palmerini@fisica.unipg.it
Igor **Panov**, Alikahnov Institute for Theoretical and Experimental Physics, Russia — Igor.Panov@itep.ru
Manuel **Peimbert**, Universidad Nacional Autonoma de Mexico, Mexico — peimbert@astroscu.unam.mx
Antonio **Peimbert**, Universidad Nacional Autonoma de Mexico, Mexico — antonio@astroscu.unam.mx
Ruth **Peterson**, Lick Observatory, University of California , USA
Marco **Pignatari**, University of Victoria , Canada — mpignatari@gmail.com
Marc **Pinsonneault**, Ohio State University, USA — pinsonneault.1@osu.edu
Nina **Polosukhina-Chuvaeva**, Crimean Astrophysical Observatory, Ukraine — polo@crao.crimea.ua
Nikos **Prantzos**, Institut d'Astrophysique de Paris, France — prantzos@iap.fr
Francesca **Primas**, European Southern Observatory, Germany — fprimas@eso.org
Tijana **Prodanovic**, Department of Physics, University of Novi Sad, Serbia — prodanvc@df.uns.ac.rs

Sofia **Randich**, INAF – Osservatorio Astrofisico di Arcetri, Italy — randich@arcetri.astro.it
Alejandra **Recio-Blanco**, Observatoire de la Côte d'Azur, France — arecio@oca.eu
Hubert **Reeves**, CNRS, France — hreeves@club-internet.fr
Adam **Ritchey**, University of Toledo, USA — adam.ritchey@utoledo.edu
Donatella **Romano**, Dipartimento di Astronomia, Universitá di Bologna , Italy — donatella.romano@oabo.inaf.it
Robert **Rood**, University of Virginia , USA — rtr@virginia.edu
Nuno **Santos**, Centro de Astrofisica, Universidade do Porto, Portugal — nuno@astro.up.pt
Luca **Sbordone**, Max Planck Institute for Astrophysics, Germany — lsbordone@mpa-garching.mpg.de
Daniel **Schaerer**, Geneva Observatory, Geneva University , Switzerland — Daniel.Schaerer@unige.ch
Kenneth **Sembach**, Space Telescope Science Institute, USA — sembach@stsci.edu
Evan **Skillman**, University of Minnesota, USA — skillman@astro.umn.edu
Rodolfo **Smiljanic**, University of São Paulo, Brazil and ESO, Germany — rsmiljanic@gmail.com
Verne **Smith**, National Optical Astronomy Observatory, USA — vsmith@noao.edu
David **Soderblom**, Space Telescope Science Institute, USA — drs@stsci.edu
Francois **Spite** , Observatoire de Paris, France — francois.spite@obspm.fr
Monique **Spite** , Observatoire de Paris, France — monique.spite@obspm.fr
Richard **Stancliffe**, Monash University, CSPA, Australia — Richard.Stancliffe@sci.monash.edu.au
Matthias **Steffen**, Astrophysikalisches Institut Potsdam, Germany — msteffen@aip.de
Gary **Steigman**, The Ohio State University, USA — steigman@mps.ohio-state.edu
Suzanne **Talon**, Université de Montréal, Canada — suzanne.talon@umontreal.ca
Friedel **Thielemann**, University of Basel , Switzerland — f-k.thielemann@unibas.ch
Monica **Tosi**, INAF – Osservatorio Astronomico di Bologna, Italy — monica.tosi@oabo.inaf.it
Takuji **Tsujimoto**, NAOJ, Japan — taku.tsujimoto@nao.ac.jp
Sylvie **Vauclair**, LATT, Université de Toulouse, France — sylvie.vauclair@ast.obs-mip.fr
Paolo **Ventura**, INAF – Observatory of Rome, Italy — ventura@oa-roma.inaf.it
Mark **Walker**, Maw Technology Manly, Australia
George **Wallerstein**, University of Washington, USA — wall@astro.washington.edu
Laimons **Zacs** , University of Latvia, Latvia — zacs@latnet.lv
Jean-Paul **Zahn** , Observatoire de Paris, France — jean-paul.zahn@obspm.fr

Opening Session

David Lambert

Gary Steigman

Light Elements in the Universe
Proceedings IAU Symposium No. 268, 2009
C. Charbonnel, M. Tosi, F. Primas & C. Chiappini, eds.

© International Astronomical Union 2010
doi:10.1017/S1743921310003790

Light elements - one observer's historical perspective

David L. Lambert

The W. J. McDonald Observatory,
The University of Texas at Austin
1 University Station, C1400, Austin, Texas, USA, 78712-0259
email: dll@astro.as.utexas.edu

Abstract. This essay attempts to provide a historical perspective on some of the key questions that engaged the attention of participants at the symposium. In particular, the writer offers and comments on a personal list of milestones in the literature published between 1957 and 1982.

Keywords. nuclear reactions, nucleosynthesis, abundances

1. Introduction

In today's rush to publish new results, one may overlook and even dismiss as unimportant the context in which new observational and theoretical findings should be placed. This essay is an imperfect attempt to provide one individual's perspective on the historical development of the roles played by the light elements in contemporary astrophysics. A warning to the reader: the perspective is biassed towards observations and, perhaps more importantly, is surely incomplete with respect to even the central events. Nonetheless, I hope I achieve my primary goal which is to identify when the seeds were planted that blossomed into the flowers whose scents we are chasing and trying to interpret at this symposium.

Here, this symposium considers the class of light elements to encompass just the first five elements: H, He, Li, Be, and B. In this quintet, the stable nuclides are ^{1}H, ^{2}H, ^{3}He, ^{4}He, ^{6}Li, ^{7}Li, ^{9}Be, ^{10}B, and ^{11}B. Each one of these will feature in talks at this symposium. Certain unstable light nuclides – ^{7}Be, ^{10}Be, and ^{8}B – also play a role in observational astrophysics and deserve a mention. ^{7}Be is important because of its role in the synthesis of ^{7}Li in stars, a topic of several presentations here. Electron capture on ^{7}Be results not only in ^{7}Li but also in a γ-ray and a neutrino. In stable stars, the γ-rays are locally absorbed deep in the interior and so lost to observers; searches for ^{7}Be γ-rays emitted by novae have been attempted but proven unsuccessful. The neutrinos are freed to roam the Universe; ^{7}Be neutrinos account for a fraction of the Sun's neutrino output. ^{10}Be with a half-life of 1.5 Myr is present in cosmic rays where it serves as a clock, a subject notably not aired at this meeting. ^{8}B's notoriety arises from a contribution to the solar neutrino problem with its ultimate resolution ending in the discovery of neutrino oscillations.

A crucial tool from the perspective of an observer is the inventory of spectral lines providing evidence for the light elements. Detection of these lines in absorption or emission is the first step in a determination of the abundance of a light element or of an isotopic ratio. Hydrogen and helium, when detectable, offer several series of lines. On the other hand, Li, Be, and B present - except in very rare circumstances - in trace amounts are detectable almost always only through the resonance lines of neutral atoms or ions. Lithium atoms are detectable in cool gas via the well known resonance doublet at 6707Å and, if Li is abundant, excited lines at 6104Å and 8126Å are also detectable. The He-like Li^+ ion and H-like Li^{++} ions are not providers of detectable lines at the abundances expected for Li. Perhaps, a great disappointment is that Be atoms are not detectable; the Be I resonance line at 2348.6Å has to my knowledge not been detected in a stellar spectrum. Beryllium abundances are based exclusively on the Be II resonance doublet at 3130Å. Higher Be ions of the H-like and He-like isoelectronic sequences are not on an observer's list for detection. With one more electron, the species of boron potentially accessible to the observer are the neutral atom and the ions B^+ and B^{++} with the He-like ion B^{3+} and the H-like ion B^{4+} far beyond accessibility. Indeed, resonance lines of B I at 2496.8Å, B II at 1362.4Å, and B III at 2065.8Å have featured prominently in discussions of the boron abundance in, as appropriate, cool and hot stars and the interstellar medium. In addition to providing estimates of the elemental abundance of a light element, this array of lines for the light elements has under appropriate circumstances provided isotopic abundances for stellar atmospheres, interstellar and extragalactic gas: D/H, ^{3}He/^{4}He, ^{6}Li/^{7}Li, and ^{10}B/^{11}B.

2. Hydrogen's title role

To stellar spectroscopists, measurements of absorption lines of element X in the spectrum of a normal (i.e. H-rich) star yield the abundance of the element X with respect to H, i.e., the ratio X/H. As shown by years of oral examinations of students, there may not always be a clear understanding of how the ratio X/H is the *natural* outcome of measurements of lines of element X. This lack of clarity reflects, in part, a failure of the student's teachers to explain a fundamental piece of stellar atmospheric physics and, in part, a lack of curiosity by the student who, nonetheless, may be adept at running the appropriate computer programmes. The crucial link between the observation of (say) X I lines and the abundance X/H is, of course, that the strength of the line is set not by the line opacity alone but by both the line and the continuous opacity.

Hydrogen's dominant role was first suggested by Cecilia Payne in her Harvard dissertation published as *Stellar Atmospheres* (Payne 1925). Not only did the dissertation show that similar compositions account for the diverse spectra of stars – spectroscopic variations are primarily due to differences in atmospheric temperature and pressure – but also that hydrogen and helium are the two most abundant elements by far in stellar atmospheres. However, as Longair (2006) reminds us in his majesterial *The Cosmic Century*, Payne qualified her conclusion by writing 'Although hydrogen and helium are manifestly very abundant in stellar atmospheres, the actual values derived from estimates of their marginal appearances are regarded as spurious', Longair adds judiciously 'The conclusion simply reflected the prevailing prejudice' but notes that the dominant role of H was proven a little later by Unsöld (1928) from an analysis of solar absorption lines and by McCrea (1929) from an analysis of the solar chromospheric (flash) spectrum.

Gaseous nebulae to invoke an old term – H II regions, planetary nebulae, for example – are also a major source of information on the abundances of the elements. In the context of the light elements, information is obtainable on H and He and their minor isotopes D and ^{3}He. Inspection of the optical emission line spectrum of a nebula shows H I, He I, and [O III] (say) lines to be of comparable intensity. Yet, the H and He are much more abundant than O. This seeming paradox is today readily understood as arising from the different excitation mechanisms for the permitted H and He lines and the forbidden O and heavy element lines. An early understanding of the paradox's resolution - perhaps, the first - was provided by Bowen (1935), the identifier of the forbidden lines (Bowen 1927, 1928). In his abstract, Bowen wrote 'A study of nebular line intensities in light of the foregoing processes indicates that H is the most abundant element and He is the second, N, O, Ne, S – and possibly C and A – are present but are very much rarer. The lines of these heavier elements are strong, not because the elements are very abundant but because they are able to make use of large sources of energy that are not available to the predominent [*sic*] H and He lines'. Bowen's attempt to provide quantitative abundance estimates for – say – O/H was frustrated, in part, by his inability to detect recombination lines of oxygen that might be compared simply and directly with the recombination lines of hydrogen. Amusingly, this comparison may now be made and, as I recall, provides an oxygen abundance at odds with that derived from oxygen forbidden lines. The first paper to provide quantitative abundance estimates may be that by Aller & Menzel (1945).

3. Helium's dark role

Every astronomical spectroscopist has surely experienced the frustration over an inability to determine the abundance of a trace element of distinctive astrophysical significance – say, Li, Be or B – but just as frustrating but more rarely expressed must be the inability to determine the He abundance of a cool star. The He abundance of the Sun, for example, is one of the least securely established elemental abundances. This important quantity is not determinable from solar absorption lines but from measurements (e.g., sampling the solar wind) that are impossible to secure on other stars.

Fortunately, all normal Galactic stars are anticipated to have a similar He abundance. The floor is set by the primordial abundance and the likely ceiling by the He abundance of local H II regions and B stars. The error affecting derived elemental abundances resulting from adopting a 'standard' He/H ratio between the floor and the ceiling is almost certainly less than that incurred from the multiplicity of other sources of error, recognized and unrecognized. Yet, there are reasons to wish to ferret out bounds on the atmospheric He/H ratio.

One recalls at once that the dredge-ups experienced by giants on the first ascent and particularly on the asymptotic giant branch are predicted to change the He content of the atmosphere along with the changes to the light elements, C, N, and O as well as the s-process in the case of AGB stars. A recent discovery has been that of multiple sequences in the colour-magnitude diagrams of globular clusters. These sequences have generally beeen attributed to differences in He abundance. A lack of spectroscopic confirmation of such differences may be inevitable but is certainly frustrating and, therefore, stands as a challenge for observers.

Helium-rich and hydrogen-poor stars are known, if rare. The prototypes might be said to be (the warm) R Coronae Borealis and (the hot) Popper's star HD124448, the latter discovered at the McDonald Observatory (Popper 1942). These stars are obviously very H-poor because at their effective temperatures the Balmer lines are either absent or very weakly present. At lower temperatures, the H-deficiency may be judged by the weak or absent CH bands.

In light of the impossibility of direct detection of He lines in photospheric spectra of cool stars, one is led to wonder if mildly He-rich cool stars exist, how they might be detected, and how they might arise (diffusion, internal nucleosynthesis and mixing, binary interactions, etc.?).

4. Milestones - a personal selection

Writing the definitive history of the light elements requires talents beyond those at my command. Here, I offer a personal selection of papers that occurred to me as I prepared my talk and which I consider to be milestones marking the road from 1957 to 1982 and, in particular, those that anticipated the hot topics featured at this symposium, and, in particular, the issues and questions providing the four discussion sessions. My selection begins in 1957, a date recognized by all (I hope!) as marking the publication of 'Nucleosynthesis of elements in stars' by Burbidge, Burbidge, Fowler & Hoyle (1957, here B^2FH, of course) and 'Nuclear reactions in stars and nucleogenesis' by Cameron (1957). The final milestone from 1982a is the seminal paper 'Lithium abundance at the formation of the Galaxy' (Spite & Spite 1982) marking the beginning of continuing studies of the Li abundances in halo stars and their relation to the primordial Li yield from the Big Bang.

4.1. *Light elements in 1957*

Hydrogen was considered the basic raw material for all element synthesis by B^2FH and Cameron. The former wrote 'It seems probable that the elements all evolved from hydrogen, since the proton is stable while the neutron is not'. Fair comment but this suppresses the issue of baryogenesis, a subject oddly not raised at this meeting.

Helium was presumed to originate from H-burning in stars. For example, Cameron in his summary table listed He along with C, N, O, Fe, and Ne as products of 'hydrogen and helium thermonuclear reactions in orderly evolution of stellar interiors.

Trace light nuclides of Li, Be, and B (and D) were noted by both B^2FH and Cameron as 'not formed in stellar interiors'. B^2FH were (reluctantly, I conjecture) forced to admit 'We have made some attempt to explain possible modes of production of deuterium, lithium, beryllium, and boron, but at present must conclude that these are little more than qualitative suggestions'. The label x-process was introduced by B^2FH to cover the 'possible modes'. This use of x presumably sprang from the authors' initiation into the symbolic language of algebra in their early schooling in different counties and countries. Cameron's says of D, Li, Be, and B 'possibly made by nuclear reactions in stellar atmospheres after the acceleration of charged particles in changing magnetic fields', a component of B^2FH's x-process.

4.2. *Cosmic He puzzle*

In my tally of milestones, the earliest recognition that He was likely not the product of H-burning in the course of orderly stellar evolution came from Hoyle & Tayler (1964) with observational evidence assembled and theoretical arguments marshalled in a short article in *Nature* with the title 'The mystery of the cosmic helium abundance'. Genesis of this milestone in a lecture course by Hoyle is described by Longair (2006) and in a delightful reminiscence by Faulkner (2009) who recalls lectures, Hoyle the man, and helium the element. Four decades after attending Hoyle's lectures as a first-year research student at Cambridge, Longair would write 'Fred Hoyle gave a course of lectures on the problems of extragalactic research. He would arrive with, at best, a scrap of paper with some notes and expound an area of current research. One week the topic was the problem of the cosmic helium abundance.' The problem and its resolution are succinctly set out in the concluding paragraph of the *Nature* article: 'There has been difficulty in explaining the high helium content of cosmic material in terms of ordinary stellar processes. The mean luminosity of galaxies come out appreciably too high on such a hypothesis. The arguments presented here make it clear, we believe, that the helium was produced in a far more dramatic way. Either the Universe has had at least one high-temperature, high-density phase, or massive objects must play (or have played) a larger part in astrophysical evolution than has hitherto been supposed.' Calculations involving He synthesis in a high-temperature high-density phase were, as Faulkner engagingly retells, very quickly completed by him with the principal result relayed to Hoyle on the Sunday morning following the key lecture. As Longair notes 'the audience had the privilege of being present as a key piece of modern astrophysics was created in real time in a graduate lecture course.'

4.3. *Stellar evolution - dredge-ups*

In the 1960s, papers first appeared describing computer calculations of low mass stars from the main sequence through the first giant branch and then the asymptotic giant branch. Dilution of the surface abundances of the light elements Li, Be, and B (with emphasis sensibly on Li) was noted early in the series of papers. Perhaps, the first paper pointing out how observations might provide quantitative verification of the calculations is Iben's 'The surface ratio of N^{14} to C^{12} during helium burning' (Iben 1964). Iben's purpose was to draw attention to a prediction that 'the ratio of N^{14} to C^{12} at the surface of a star undergoes a significant increase during the rise into the red giant region immediately preceding the phase of helium burning in the core' and to call for spectroscopic verification of this prediction. Of course, N surface enrichment implies reduction of Li, Be, and B surface abundances. Iben was one of the prime movers in this endeavour to model stellar evolution and his reviews 'Stellar evolution: Comparison of theory with observation' and 'Stellar within and off the main sequence' remain valuable reading for the student (Iben 1967a, b). Discovery of the He-shell flashes (thermal pulses) in theoretical models of post-He core burning low mass (i.e., AGB) stars by Schwarzschild & Härm (1965) marked the birth of a vigourous area still of theoretical and observational study. Subsequent and continuing observational and theoretical work on surface abundance changes arising from the first and second dredge-ups in early red giants and the third dredge-up in asymptotic giant branch stars spring from these exploratory calculations.

4.4. *Discovery of the Cosmic Microwave Background*

Discovery of the cosmic microwave background radiation has had fundamental implications for our understanding of the origins of the light elements, as is evidenced here by several papers discussing contemporary strains between prediction and observation. Perhaps, *the* discovery of the CMB was announced by Penzias & Wilson (1965) in a short note with the title 'A measurement of excess antennna temperature at 4080 Mc/s' whose abstract noted the excess temperature was 3.5K higher than expected and 'isotropic, unpolarized, and free from seasonal variations' with a 'possible explanation' being relic radiation from the Big Bang.

This discovery quickly focussed theoretical attention anew on possible observable consequences of the Big Bang, with areas of particular relevance for the light elements being related issues of the primordial nucleosynthesis and the power spectrum of the CMB's very small anisotropy which is fundamental to determining the baryon density, the value of which sets the yields from the nucleosynthesis. First theoretical explorations of the acoustic oscillations in this regard were reported by Zel'dovich & Sunyaev (1969) and Peebles & Yu (1970).

4.5. *Pioneering calculations of Big Bang nucleosynthesis*

Faulkner's calculations inspired by Hoyle's 1964 lecture course were expanded thoroughly by Wagoner, Fowler, & Hoyle (1967) in their paper 'On the synthesis of elements at very high temperatures'. Their study which went beyond the synthesis of He discussed by Faulkner and also by Peebles (1966a,b) provided 'a detailed calculation of element production in the early stages of a homogeneous and isotropic expanding universe as well as within imploding-exploding supermassive stas' with a conclusion 'if the recently measured microwave background radiation is due to primeval photons, then significant quantities of only D, He^3, He^4, and Li^7 can be produced in the universal fireball.' The insertion of 'if' presumably reflected not only natural professional caution about the reality of the microwave radiation background but also possibly tension among the authors over acceptance of a hot Big Bang. Hoyle, of course, never accepted the idea of the Big Bang, a term coined long before by him in a prestigious series of broadcasts on the BBC (Kynaston 2007).

4.6. *Back to the x-process*

Recognition that 'significant quantities' of D and ^{7}Li could be produced in the Big Bang did not address fully the identity of the x-process. What were the origins of ^{6}Li, ^{9}Be, ^{10}B and ^{11}B? This question was very largely answered when the role of Galactic cosmic rays in the spallation of abundant C, N, and O nuclei was appreciated. Reeves (1993) has provided a fascinating account of how this appreciation occurred when he caught Hoyle 'talking in class': 'In 1969, I presented these ideas in a seminar at the former IOTA (Institute of Theoretical Astronomy) in Cambridge (UK). During my seminar, Fred Hoyle kept on talking to Willie Fowler. I could overhear some of his words "I have been repeating that to you for many years. You should have listened to me." Later on, he told me that he had considered this scenario for a long time. We published a paper together on this subject.'

The paper was 'Galactic cosmic ray origins of Li, Be, and B in stars' (Reeves, Hoyle, & Fowler 1970). This and subsequent work with Reeves as an active player showed that spallation by (high energy) Galactic cosmic rays could account satisfactorily for ^{6}Li, ^{9}Be, and ^{10}B as the products of collisions between protons and α particles on the one hand and C, N, and O nuclei on the other hand; the solar system ratios ^{6}Li/^{9}Be and ^{10}B/^{9}Be were well reproduced by the calculations. But the ratios ^{7}Li/^{9}Be and ^{11}B/^{9}Be were greater than could be accounted for by high-energy spallation. ^{9}Be is taken to be a benchmark because no process of stellar Be nucleosynthesis has been identified. Spallation by low energy cosmic rays may account for some of the required extra ^{7}Li and ^{11}B, as first suggested by Meneguzzi, Audouze, & Reeves (1971). In low metallicity gas, collisions between cosmic ray and interstellar He nuclei – α on α collisions – will produce ^{6}Li and ^{7}Li but not Be and B. Neutrino-induced spallation in Type II supernovae may produce ^{7}Li and ^{11}B - the so-called ν-process (Woosley 1977; Woosley *et al.* 1990). Of course, a major contributor to ^{7}Li is the Big Bang.

4.7. *A role for diffusion?*

Discussion of diffusive separation of elements, even isotopes, in a modern form is traceable to Georges Michaud (1970) and the paper 'Diffusion processes in peculiar A stars'. Proposals generally by Michaud and his disciples concerning observable effects of diffusion – principally describing a competition between gravitational settling and radiative levitation – have since 1970 encompassed many kinds on stars across the H-R diagram.

Of relevance here, the principal appearance of diffusion is surely the role attributed by some to it in accounting for the disagreement between the Li abundance of the Spite plateau and the several times higher abundance predicted from Big Bang nucleosynthesis according the baryon density required by the fit to the cosmic microwace background fluctuations.

4.8. *Li synthesis in stars*

Speculations about the x-process included Li (and Be, B) synthesis by spallation on a stellar surface. The stellar interior was known to be highly unfavourable to Li survival;

the very low solar Li abundance (relative to that in chondritic meteorites) was early attributed to convection resulting in Li destruction. Seemingly, the first proposal for Li production in stellar interiors with a consequence for high atmospheric abundances of Li was by Cameron (1955), a proposal further developed by Cameron & Fowler (1971). Today, this idea is widely referred to as the 'Cameron-Fowler mechanism'.

In 1971, the mechanism was presented as an explanation for the high Li abundances seen in a few cool carbon and S stars. The prototype of the group is WZ Cas (McKellar 1940). 'It is tempting to believe that the lithium has been produced by some internal process. However, since the majority of the carbon and S giants do not possess nearly so much lithium, it is necessary to postulate that the production process involves some unusual events' (Cameron & Fowler 1971). The 'unusual events' were suggested by Cameron & Fowler to be the He-burning shell flashes predicted by Schwarzschild & Härm (1965) to occur in stars on the asymptotic giant branch. These flashes had already been suggested by others to be a site for operation of a s-process and thus to account for the heavy element enrichment in carbon and S stars and notably the presence of Tc in selected S stars. The paper proposed that the neutron source was the now familiar $^{13}\mathrm{C}(\alpha, n)^{16}\mathrm{O}$ with fresh $^{13}\mathrm{C}$ provided by the mixing of protons into the He-shell.

Lithium synthesis involved the sequence $^3\mathrm{He}(\alpha, \gamma)^7\mathrm{Be}(e^-\nu)^7\mathrm{Li}$ (Cameron 1955). To achieve efficient conversion of $^3\mathrm{He}$ to $^7\mathrm{Li}$, the $^7\mathrm{Be}$ must be swept out of the hot layers where it is produced to cooler layers where it and $^7\mathrm{Li}$ are immune to destruction by protons. Cameron & Fowler sited initiation of the sequence at the base of the H-rich convective envelope with $^7\mathrm{Be}$ and $^7\mathrm{Li}$ convected outward to safety. Today, we would refer to the situation as a hot-bottomed convective envelope. In addition to its applicability to AGB stars with such an envelope, the Cameron-Fowler conversion of $^3\mathrm{He}$ to $^7\mathrm{Li}$ has been applied to account for Li-rich first ascent red giants, stars with internal structures very different from that of an AGB star, as several papers at this symposium discuss.

A star's internal reservoir of $^3\mathrm{He}$ is built up from two sources. First, there is the $^3\mathrm{He}$ and $^2\mathrm{H}$ present at its birth. The $^2\mathrm{H}$, surviving Big Bang deuterium, is burnt to $^3\mathrm{He}$ in the pre-main sequence phase. The initial $^3\mathrm{He}$ is a consequence of Galactic chemical evolution but likely dominated by the Big Bang's $^3\mathrm{He}$, and any $^3\mathrm{He}$ synthesised by stars during the Galaxy's evolution. A contribution from stellar nucleosynthesis, apart from the D to $^3\mathrm{He}$ conversion, is significant only in low mass stars where their long main sequence life allows the initial reaction of the pp-chain with its very small cross-section governed by the weak interaction to provide $^3\mathrm{He}$ exterior to the energy-energy generating core. A corollary of the last point is that $^3\mathrm{He}$ is *not* synthesised by stars once they have evolved off the main sequence. Initiation of the Cameron-Fowler mechanism may begin by increasing the Li abundance in the stellar atmosphere but extended operation will eventually lead to a reduction of the Li abundance as the $^7\mathrm{Li}$ is recycled to high temperatures and destroyed and the $^3\mathrm{He}$ supply is run to exhaustion.

4.9. *Helium is primarily primordial*

In their 1964 paper, Hoyle & Tayler presented a convicing theoretical argument that the reported He/H ratio of about 0.1 in several Galactic and extragalactic objects could not be accounted for by stellar nucleosynthesis but had been produced 'in a far more dramatic way'. Apart from a measurement of the He/H ratio in a planetary nebula belonging to a globular cluster (O'Dell, Peimbert, & Kinman 1964), observational evidence that He might have a primordial origin with a minor supplement from stellar nucleosynthesis was scant in the extreme. Furthermore, the stellar ejecta comprising a planetary nebula may have been enriched in He in the course of stellar evolution.

Spectroscopy of halo stars (i.e., old and expected to be He-poor if stellar nucleosynthesis is the dominant origin of He) sufficiently hot to show He lines had displayed a tendency to present low He/H ratios but with compositions that 'are quite exotic and hard to accept as being primordial ' (Searle & Sargent 1972). Now, we identify 'exotic' as consequences of diffusion that has distorted the 'true' composition of the stars.

Definitive observational spectroscopic evidence of the primarily primordial origin of He was first provided by Searle & Sargent (1972) in their paper 'Inferences from the composition of two dwarf blue galaxies'. Emission lines in the galaxies I Zw 18 and II Zw 40 showed that O/H and Ne/H were lower than local Galactic values ($[O/H] \simeq -1$ in the case of I Zw 18) but He has a normal abundance. 'These galaxies are the first metal-poor systems of Population I to be discovered: the normal helium abundance is taken as evidence that this abundance is primordial' (Searle & Sargent). I Zw 18 remains one of a handful of O-poor galaxies anchoring the low O/H end of the relation between He/H and O/H that provides our estimates of the Big Bang's He yield (see papers at this symposium).

4.10. *Deuterium as a baryometer*

The realization that the deuterium abundance - D/H - in appropriate locales might serve as a baryometer for the Big Bang was boosted with Reeves *et al.*'s (1973) paper 'On the origin of the light elements'. While the paper offered a comprehensive survey of the origins of all the light nuclides, perhaps its most telling and lasting point were the assertions that 'The deuterium can *only* be produced pregalactically either by the big bang or in some pregalactic event' [emphasis added] and 'the most plausible origin of D is big-bang nucleosynthesis'. Another way to express this result is to say that all modes of stellar nucleosynthesis destroy ('astrate') D and possible production paths involving energetic particles in diverse places (interstellar medium, stellar atmospheres, for example) cannot produce D without overproducing other light elements. The primordial origin of D was amplified by Epstein, Lattimer, & Schramm (1976) with the caveat that 'other, more speculative, sources are not ruled out'. To date, no such 'other' sources have been ruled in.

Detection of D in the interstellar medium was first reported by Rogerson & York (1973) from *Copernicus* spectra of the hot star β Cen showing Lyman lines of H I and D I - the isotopic wavelength shift of 81 km s^{-1} provided well resolved lines. Measurements of the D/H ratio along several lines of sight followed this pioneering determination for β Cen.

Of course, the D/H measurement might reasonably be taken to be a lower limit to the Big Bang D/H ratio; cycling of gas through stars astrates D. Various attempts were made (and continue to be made) to devise the correction for the cumulative astration. Substantial additions to the library of interstellar D/H measurements, especially from *FUSE* spectra, have added challenges to modelling of GCE of D.

Adams (1976) proposed that the difficulties of correcting the interstellar D/H ratios for astration in order to obtain the pregalactic/primordial ratio could be alleviated, if not avoided, by detecting the Lyman D I lines in absorption from intergalactic clouds along the lines of sight to distant quasars. Such clouds representing almost pristine primordial gas should have essentially their primordial D abundance. The degree to which the clouds have been contaminated by products of stellar nucleosynthesis is assessable from absorption lines of ions of C, O, Si etc. Adams' proposal was ahead of its time. It took 8-meter class telescopes with high-resolution spectrographs to obtain suitable spectra. Even today, as discussed at this symposium, useful D/H ratios have been published for just a handful of quasars; many quasars turn out to have ill-suited spectra; the D I lines of a strong cloud are blended with H I lines of a weaker cloud at a different velocity. A definitive set of D/H determinations continues to elude us.

4.11. *The hyperfine line 8.7 GHz line of $^3He^+$*

The ground state of the ion ^{3}He$^+$ with a nuclear spin of 1/2 has a hyperfine line at 8.7 GHz, the analogue of the H atom's 21 cm line. Rood, Wilson, & Steigman (1979) reported on 'The probable detection of interstellar ^{3}He$^+$ and its significance'. The 'tentative detection' referred to a giant H II region. Since 1979, Rood and colleagues have doggedly pursued the 8.7 GHZ line in Galactic H II regions and planetary nebulae, as described by Bania at this symposium. Abundances of ^{3}He$^+$ in these sources have led to what has been dubbed 'the ^{3}He problem' (Galli *et al.* 1997).

The isotope ^{3}He with D and ^{7}Li is a minor product of the standard Big Bang. Additionally, main sequence stars are predicted to synthesise ^{3}He as part of the *pp*-chain and from D-burning (as outlined above in Section 4.8). If one assumes that the synthesised ^{3}He is not subsequently destroyed (astrated) and returned to the interstellar medium, the ^{3}He abundance in the interstellar medium might be expected to increase with time, i.e., the abundance in local H II regions might be higher than the protosolar value, as apparently first pointed out by Rood, Steigman, & Tinsley (1976). This possibility and the lure of pinning down the Big Bang's ^{3}He yield prompted the now 30-year old pursuit of the 8.7 GHz line.

In the 1979 paper, the single probable detection of ^{3}He in an H II region led to the determination of a ^{3}He/H ratio 'comparable to the protosolar number' and the conclusion that there is 'no evidence for the production of ^{3}He predicted for low mass stars'. Knowing the ingenuity of students of Galactic chemical evolution, the conclusion may not necessarily follow from the determination of the similar ^{3}He/H abundances. However, the '^{3}He problem' was starkly evident after the 8.6 GHz line was detected from Galactic planetary nebulae. Balser *et al.* (1997) report the first successful observations of the line with the result that the ^{3}He/H abundances 'are more than an order of magnitude larger than those found in any H II region, the local interstellar medium, or the proto-solar system'. Clearly, 'there is *some* stellar production of ^{3}He' [emphasis in the 1979 abstract]. The ^{3}He problem is the subject of several papers here.

4.12. *The Spite plateau*

My final milestone in this chronological listing is the pair of 1982 papers by Spite & Spite: The *Nature* Letter 'Lithium abundance at the formation of the Galaxy' (Spite & Spite 1982a) and the *A&A* article 'Abundance of lithium in unevolved halo stars and old disk stars: interpretation and consequences' (Spite & Spite 1982b).

The landscape in which this milestone was uncovered may be appreciated by recalling the summary of Li observations assembled by Reeves *et al.* (1973). The local Li abundance was put at $\log \epsilon \simeq 3$ with Li-richer stars (e.g., WZ Cas) attributed to internal Li synthesis and Li-poorer stars (e.g., Sun) attributed to internal Li astration. In their presentation of the possible Galactic evolution of the ^{7}Li abundance (Fig. 6 of Reeves *et al.*), two extremes were presented: (i) the assumption that ^{7}Li was a Big Bang product led to the suggestion of a factor of several *increase* in the Li/H ratio from the present ratio back to the formation of the Galaxy; (ii) an assumption that ^{7}Li came only from enrichment of the interstellar medium by Li-rich gas from red giants led, of course, to Li-poor gas at early times. A third possibility was that the local Li abundance seen in stars with ages spanning at least 5 Gyr represents the pre-Galactic abundance.

Spite & Spite exploiting the spectroscopic capabilities of the (then) new CFHT telescope observed a small sample of halo dwarfs, stars recognized as very old and known to be very metal-poor, i.e., the gas from which these stars had formed should be a fair approximation to pre-Galactic gas. To their (I presume) surprise and delight, the Li

12   D. L. Lambert

abundance of the warmest of these stars was uniform (hence, the Spite plateau) and at
$\log \epsilon(\mathrm{Li}) = 2.05$ much lower than the local value. 'After discussion, it is suggested that
the abundance in halo dwarfs is therefore representative of the interstellar matter which
formed the stars and also this matter itself retains the lithium abundance of the Big
Bang, hardly altered' (Spite & Spite 1982b).

The existence of a plateau, the Li abundance of the plateau, and the relation between
the Li abundance and the abundance provided by the Big Bang have been debated from
1982 to this day, and, indeed, these questions were among the reasons advanced for
holding this symposium.

5. Final thoughts

The four discussion sessions at the symposium were intended to focus on the 'hot'
questions involving light elements in contemporary astrophysics. My contribution sheds
light on the historical perspective behind most of these questions. Omissions will be
excused, I hope, on the grounds of lack of time and space.

In closing, I offer two rather different historical thoughts: the first on the publication of
symposium proceedings and the second on the sophistication of our scientific endeavours.

As a research student, I recall the particular joy I experienced in reading the extended
discussions that were reported at length in the two IAU symposia (#12 and #28) on
'Aerodynamic Phenomena in Stellar Atmospheres' edited by R.N. Thomas (1960, 1965).
These were not meetings I had attended but the discussions almost brought me into the
meeting room as an engaged listener, if one too shy and ignorant to participate. Why do
organisers of meetings not aspire to emulate these two volumes?

Astronomical spectra are beautiful objects, especially when a striking conclusion may
be drawn by inspection. In Geneva, I drew attention once again to Spite & Spite's (1982b)
Figure 1 showing spectra around the Li I 6707Å doublet for four metal-poor dwarfs.
Inspection shows Ca I and Fe I lines of differing strength but Li I lines of similar strength
in all but the coolest dwarf HD 103095 (alias Groombridge 1830). Thus, the Spite plateau
is there in front of ones eyes. (Li is strongly depleted in HD 103095.)

Of course, conversion of spectra to a quantitative abundance estimate cannot be done
by visual inspection alone. Tools for the conversion have certainly become more sophisticated and refined over the years spanned by milestones and certainly since 1982.
Had I not concentrated on results exclusively, I would have entertained including milestones relating to telescopes, instruments, and analytical tools. Currently, the apex of
analytical sophistication surely involves the application of 3D or hydrodynamical model
atmospheres with the selective addition of non-LTE considerations in certain cases.

In the area of light elements, 3D atmospheres are being applied in particular to analyses
of the Li I 6707Å line in metal-poor dwarf stars to address key questions from cosmology,
early Galactic evolution, and stellar physics. If I were to limit the questions to two,
my choices would be 'Is the Li abundance of Spite plateau stars reconcilable with that
predicted from the Big Bang with the baryon density set by the CMB fluctuations? and
'Is there $^{6}\mathrm{Li}$ mixed in with the $^{7}\mathrm{Li}$ in these stars?

To the first question, there appear to be two diametrically different answers to account
for the result obtained (even with 3D atmospheres) that the stellar Li abundance is a
factor of several less than the Big Bang prediction. One, diffusion is invoked to reduce
the atmospheric abundance but to account for the plateau's observed height and shape
over several dex in metallicity turbulent diffusion is introduced without thorough detailed physical understanding. Two, the standard model of the Big Bang is modified by

introducing new physics including modifications to the standard model of particle physics. One or two? Or both?

Obviously, observational tests are sought. In the case of one, there is now evidence that diffusion whose effects should be partially undone in evolution off the main sequence has occurred - see the beautiful work reported here on the composition of globular cluster stars. In the case of two, an indicator of some failures of the standard Big Bang model is the presence of ^{6}Li in significant amounts. Measurements of small amounts of ^{6}Li in the presence of ^{7}Li demand a sophistication of line profile analysis rarely required, and possibly not yet achievable.

One looks forward to the next workshop, conference, or symposium on light elements in the Universe. Perhaps, no other set of five elements brings together in such an interlocked fashion matters of cosmology, galaxy formation and evolution, and an array of matters in stellar astrophysics.

In preparing this essay for publication, I took the occasion to explore the literature somewhat more thoroughly than I did prior to the oral presentation in Geneva. This led to some modification of the presentation but, more fundamentally, I was humbled afresh by how much of the history I had forgotten or had never ever appreciated. Thus, I apologise to those whose contributions I have still overlooked or underappreciated. I thank my Austin friends who answered several questions at short notice, especially Volker Bromm and Paul Shapiro. As has been the case for more than 30 years, I are indebted to the Robert A. Welch Foundation of Houston, Texas for grant F-634 which has supported my research into the chemical compositions of stars.

References

Adams, T. F. 1976, *A&A*, 50, 461

Aller, L. H. & Menzel, D. H. (1945), *ApJ*, 102, 239

Balser, D. S., Bania, T. M., Rood, R. T., & Wilson, T. L. 1997, *ApJ*, 483,320

Bowen, I. S. (1927), *PASP*, 39, 295

Bowen, I. S. (1928), *ApJ*, 67, 1

Bowen, I. S. 1935, *ApJ*, 81, 1

Burbidge, E. M., Burbidge, G. R., Fowler, W. A., & Hoyle, F. 1957, *Rev. Mod. Phys.*, 29, 547

Cameron, A. G. W. 1955, *ApJ*, 121, 144

Cameron, A. G. W. 1957, *PASP*, 69, 201

Cameron, A. G. W. & Fowler, W. A. 1971, *ApJ*, 164, 111

Epstein, R. I., Lattimer, J. M., & Schramm, D. N. 1976, *Nature*, 263, 198

Faulkner, J. 2009, in *Finding the Big Bang*, ed. P. J. E. Peebles, L. A. Page, Jr., & R. B. Partridge, Cambridge:Cambridge University Press, p. 244

Hoyle, F. & Tayler, R. J. 1964, *Nature*, 203, 1108

Galli, D., Stanghellini, L., Tosi, M., & Palla, F. 1997, *ApJ*, 477, 218

Iben, I., Jr. 1964, *ApJ*, 140, 163

Iben, I., Jr. 1967a, *Science*, 155, 785

Iben, I., Jr. 1967b, *ARAA*, 5, 571

Kynaston, D. 2007, *Austerity Britain 1945-51*, London:Bloomsbury Publishing plc, p. 387

Longair, M. S. 2006, *The Cosmic Century:A History of Astrophysics and Cosmology*, Cambridge: Cambridge University Press

McCrea, W. H. 1929, *MNRAS*, 89, 483

McKellar, A. 1940, *PASP*, 52, 407

Meneguzzi, M., Audouze, J., & Reeves, H. 1971, *A&A*, 15, 337

Michaud, G. 1970, *ApJ*, 160, 641

O'Dell, C. R., Peimbert, M., & Kinman, T. D. 1964, *ApJ*, 140, 119

Payne, C. H. 1925, *Stellar Atmospheres:Harvard College Observatory Monographs, No. 1*, Cambridge, MA:Harvard University Press

Peebles, P. J. E. 1966a, *Phys. Rev. Lett.*, 16, 410

Peebles, P. J. E. 1966b, *ApJ*, 146, 542

Peebles, P. J. E. & Yu, J. T. 1970, *ApJ*, 162, 815

Penzias, A. A. & Wilson, R. W. 1965, *ApJ*, 142, 419

Popper, D. M. (1942), *PASP*, 54, 160

Reeves, H. 1993, in *Origin and Evolution of the Elements*, ed. N. Prantzos, E. Vangioni-Flam, & M. Cassé, Cambridge:Cambridge University Press

Reeves, H., Fowler, W. A., & Hoyle, F. 1970, *Nature*, 226, 727

Reeves, H., Audouze, J, Fowler, W. A., & Schramm, D. N. 1973, *ApJ*, 179, 909

Rogerson, J. B. & York, D. G. 1973, *ApJ*, 186, L95

Rood, R. T., Steigman, G., & Tinsley, B. M. 1976, *ApJ*, 207, L57

Rood, R. T., Wilson, T. L., & Steigman, G. 1979, *ApJ*, 227, L97

Schwarzschild, M. & Härm, R. 1965, *ApJ*, 142, 855

Searle, L. & Sargent, W. L. W. 1972, *ApJ*, 173, 25

Spite, M. & Spite, F. 1982a, *Nature*, 297, 483

Spite, F. & Spite, M. 1982b, *A&A*, 115, 357

Thomas, R. N. 1960, editor of *Aerodynamic Phenomena in Stellar Atmospheres*, Proc. IAU Symposium No. 12, Nuovo Cimento Suppl., 22

Thomas, R. N. 1965, editor of *Aerodynamic Phenomena in Stellar Atmospheres*, Proc. IAU Symposium No. 28, London & New York:Academic Press

Unsöld, A. 1928, *Z. f. Phys.*, 46, 765

Wagoner, R. V., Fowler, W. A., & Hoyle, F. 1967, *ApJ*, 148, 3

Woosley, S. E. 1977, *Nature*, 269, 42

Woosley, S. E., Hartmann, D. H., Hoffman, R. D., & Haxton, W. C. 1990, *ApJ*, 356, 272

Zel'dovich, Ya. B. & Sunyaev, R. A. 1969, *ApSS*, 4, 301

Session I

Production of the light elements
in the first minutes of the Universe

Monica Tosi, Bob Rood, Tom Bania

Richard Stancliffe, Sara Palmerini, Thibaut Decressin

Light Elements in the Universe
Proceedings IAU Symposium No. 268, 2009
C. Charbonnel, M. Tosi, F. Primas & C. Chiappini, eds.

© International Astronomical Union 2010
doi:10.1017/S1743921310003807

Constraints from the cosmic microwave background experiments

Joanna Dunkley[1]

[1] Oxford University - United Kingdom
email: j.dunkley@physics.ox.ac.uk

Abstract. I will give a review of the current constrains on light element abundances from cosmic microwave background experiments, focusing on results from WMAP and discussing prospects from upcoming data from Planck and ground-based experiments. I will describe how the production of light elements affects the CMB anisotropies, and how we use the data to extract cosmological information that includes constraints on the baryon density, and primordial abundances.

Keywords. Cosmology: cosmic microwave background

A couple of months after the Symposium, Joanna Dunkley informed the editors of these proceedings that *"The WMAP collaboration's policy is not to write conference proceedings if they contain results published in WMAP papers. Since this review has been almost entirely focused on results from the 5-year WMAP analysis it cannot be reported here. The results can be found in Komatsu et al. (2009) and Dunkley et al. (2009)."*

Given the relevance of her review to the discussions and the results of the whole Symposium we felt it would not be appropriate to simply skip her contribution. We thus report here the abstract that she originally sent and the references where the reader can find the WMAP results that she presented to the audience.

The responsibility of this decision is fully of the editors and should not be ascribed to Dr. Dunkley.

References

Dunkley, J., Komatsu, E., Nolta, M. R., Spergel, D. N., Larson, D. *et al.* 2009, *ApJS*, 180, 306
Komatsu, E., Dunkley, J., Nolta, M. R., Bennett, C. L., Gold, B. *et al.* 2009, *ApJS*, 180, 330

Michel Grenon & Sushma Mallik

Chantal Taçoy & Nadège Lagarde

Light Elements in the Universe
Proceedings IAU Symposium No. 268, 2009
C. Charbonnel, M. Tosi, F. Primas & C. Chiappini, eds.
© International Astronomical Union 2010
doi:10.1017/S1743921310003819

Primordial nucleosynthesis: A cosmological probe

Gary Steigman

Departments of Physics and Astronomy
Center for Cosmology and Astro-Particle Physics
The Ohio State University
192 West Woodruff Avenue
Columbus, OH 43210 USA
email: steigman@mps.ohio-state.edu

Abstract. During its early evolution the Universe provided a laboratory to probe fundamental physics at high energies. Relics from those early epochs, such as the light elements synthesized during primordial nucleosynthesis when the Universe was only a few minutes old, and the cosmic background photons, last scattered when the protons (and alphas) and electrons (re)combined some 400 thousand years later, may be used to probe the standard models of cosmology and of particle physics. The internal consistency of primordial nucleosynthesis is tested by comparing the predicted and observed abundances of the light elements, and the consistency of the standard models is explored by comparing the values of the cosmological parameters inferred from primordial nucleosynthesis with those determined by studying the cosmic background radiation.

Keywords. cosmology: early universe, nucleosynthesis, cosmological parameters

1. Introduction

Primordial nucleosynthesis provides a key probe of the physics and early evolution of the Universe. Big Bang Nucleosynthesis (BBN; ~ 20 minutes) and the cosmic microwave background (CMB) photons, last scattered at recombination (~ 400 kyr), provide complementary probes of the physics of the early evolution of the Universe.

For a brief period during its early evolution the hot, dense Universe is a cosmic nuclear reactor. Since the Universe is expanding and cooling rapidly, there is only time to synthesize in astrophysically interesting abundances the very lightest nuclides (D, ^{3}He, ^{4}He, and ^{7}Li). In the standard models of cosmology and particle physics described by General Relativity, the universal expansion rate, the Hubble parameter, H, is determined by the total mass/energy density: $H^2 \propto G\rho$, where $H = H(z)$, z is the redshift, G is the gravitational constant, and ρ is the energy density. During such early epochs, the Universe, which is filled with relativistic particles including three flavors of light neutrinos ($N_\nu = 3$), is "radiation dominated", and the abundances of the nuclides synthesized during BBN depend on only one cosmological parameter, $\eta_{\rm B}$, which provides a measure of the universal density of baryons.

$$\eta_{\rm B} \equiv n_{\rm B}/n_\gamma \equiv 10^{-10}\eta_{10}. \qquad (1.1)$$

In eq. 1.1, $n_{\rm B}$ is the number density of baryons and n_γ is the number density of cosmic background photons. The only baryons present at BBN are nucleons, *i.e.*, protons and neutrons. In contrast to the standard model of cosmology, there is a class of non-standard cosmological (and/or particle physics) models in which the expansion rate may differ from its standard model value, $H' \neq H$. In these non-standard models the expansion rate can be parameterized by an "expansion rate parameter", S, or equivalently, by an "effective number of neutrinos", $N_\nu \neq 3$, where

$$S^2 \equiv (H'/H)^2 \equiv G'\rho'/G\rho \equiv 1 + 7\Delta N_\nu/43. \qquad (1.2)$$

More generally, the effective number of "extra" neutrinos, $\Delta N_\nu \equiv N_\nu - 3$, parameterizes any non-standard energy density ($\rho' \neq \rho$), normalized to the contribution from one standard model neutrino by,

$$\Delta N_\nu \equiv (\rho' - \rho)/\rho_\nu. \qquad (1.3)$$

However, even if $\rho' = \rho$, it could be that $N_\nu \neq 3$ ($\Delta N_\nu \neq 0$) if the early-Universe gravitational constant differs from its current value, $G' \neq G$,

$$G'/G = S^2 = 1 + 7\Delta N_\nu/43. \qquad (1.4)$$

As will be seen below, in this class of non-standard models the BBN-predicted (nSBBN) primordial abundance of deuterium depends largely on the baryon density parameter, η_B (deuterium is a cosmological baryometer), while that of helium-4 is sensitive to the early Universe expansion rate, S (^{4}He is an early universe chronometer).

In order to test the standard models of cosmology and particle physics, two key questions are addressed:

1. Do the light element abundances predicted by BBN agree with the primordial abundances inferred from observations?

2. Do the BBN values of η_B and S (N_ν) agree with those inferred from independent, non-BBN observations (e.g., from the CMB)?

2. Standard Big Bang Nucleosynthesis (SBBN)

For SBBN ($N_\nu = 3$), the light element relic abundances are only a function of the baryon density parameter, η_B. Among the light nuclides, deuterium is the baryometer of choice. There are several reasons why D occupies this special place. One is that the post-BBN evolution of deuterium is simple and monotonic: as gas is cycled through stars (producing the heavy elements), D is only destroyed (Reeves et al. 1973, Epstein, Lattimer, & Schramm 1976). As a result, if deuterium is observed anywhere in the Universe, at any time in its evolution, the observed abundance will be no larger than the primordial value: $(D/H)_{OBS} \leqslant (D/H)_P$. In addition, for systems of low metallicity, a sign that very little of their gas has been cycled through stars which destroy deuterium, the observed D abundance should approach the primordial value: $(D/H)_{OBS} \rightarrow (D/H)_P$ (the "Deuterium Plateau"). Another reason to prefer D is that its predicted primordial abundance is sensitive to the baryon density parameter; since $(D/H)_P \propto \eta_B^{-1.6}$, a $\sim 10\%$ determination of $(D/H)_P$ results in a $\sim 6\%$ determination of η_B.

The deuterium abundance is determined by comparing the H I and D I column densities inferred from observations of absorption of radiation from background UV sources by intervening gas. In searching for the Deuterium Plateau the relelvant data is provided by observations of high-redshift, low-metallicity, QSO Absorption Line Systems (QSOALS). Unfortunately, at present there are only seven, relatively reliable D abundance determinations (Pettini et al. 2008), which are shown in Figure 1.

The weighted mean of the seven D abundances is $\log(y_{DP}) = 0.45$. However, as may be seen from the Figure 1, only three of the seven abundances lie within 1σ of the mean. Indeed, the fit to the weighted mean of these seven data points has a $\chi^2 = 18$ ($\chi^2/dof = 3$). Either the quoted errors are too small or, one (or more) of the determinations is wrong, perhaps contaminated by unidentified (and, therefore, uncorrected) systematic errors. In the absence of evidence identifying the reason(s) for such a large dispersion, the best that can be done at present is to adopt the mean D abundance and to inflate the error in the mean in an attempt to account for the unexpectedly large dispersion among the D abundances (Steigman 2007).

$$\log(y_{\mathrm{DP}}) \equiv 0.45 \pm 0.03. \tag{2.1}$$

For this relic D abundance, SBBN predicts that the baryon density parameter is

$$\eta_{10}(\mathrm{SBBN}) = 5.80 \pm 0.28, \tag{2.2}$$

corresponding to a baryon mass density $\Omega_{\mathrm{B}} h^2 = 0.0212 \pm 0.0010$.

For $\eta_{10}(\mathrm{SBBN})$, the SBBN-predicted abundances of the remaining light nuclides are

$$y_{3\mathrm{P}} \equiv 10^5 \, (^3\mathrm{He}/\mathrm{H})_{\mathrm{P}} = 1.07 \pm 0.04, \tag{2.3}$$

$$\mathrm{Y_P} = 0.2482 \pm 0.0007, \tag{2.4}$$

$$[\mathrm{Li}]_{\mathrm{P}} \equiv 12 + \log(\mathrm{Li}/\mathrm{H})_{\mathrm{P}} = 2.67^{+0.06}_{-0.07}. \tag{2.5}$$

2.1. *Consistency of SBBN?*

Having used the deuterium observations along with the predictions of SBBN to determine the baryon density parameter, we may now ask if the observed abundances of $^3\mathrm{He}$, $^4\mathrm{He}$, and $^7\mathrm{Li}$ are consistent with their SBBN-predicted primordial values.

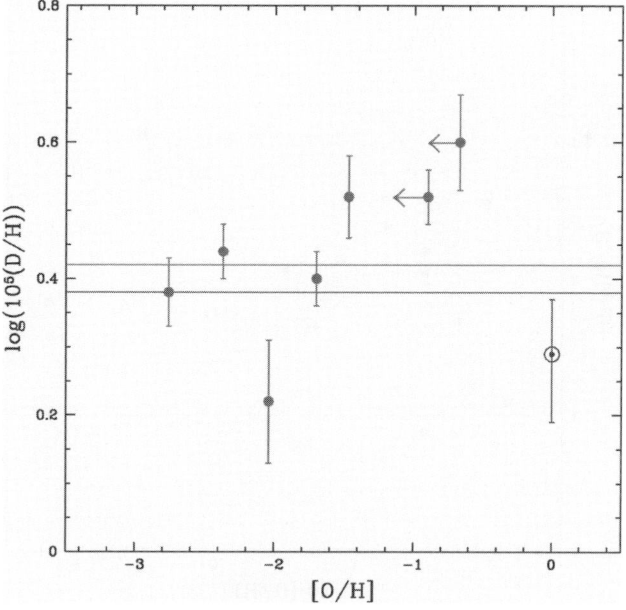

Figure 1. The logs of the deuterium abundances, $y_{\mathrm{D}} \equiv 10^5 (\mathrm{D}/\mathrm{H})$, observed in high-$z$, low-Z QSO Absorption Line Systems (Pettini *et al.* 2008), as a function of the corresponding oxygen abundances. For comparison, the solar deuterium and oxygen abundances are shown (Geiss & Gloeckler 1998). The band indicated by the solid lines is the 68% range of the SBBN-predicted primordial D abundance using the CMB-determined baryon density parameter (see §3).

Helium - 3.

The post-BBN evolution of ^{3}He is model dependent and, considerably more complicated than that of D. Overall, the ^{3}He abundance is expected to increase during Galactic chemical evolution (see, *e.g.*, Rood 1972, Rood, Steigman & Tinsey 1976). Observations of ^{3}He are limited to the relatively evolved H II regions in the Galaxy (Bania, Rood, & Balser 2002). The data are shown in Figure 2, where the observed ^{3}He abundances are plotted versus the corresponding H II region oxygen abundances. The data reveal a *minimum* (primordial?) ^{3}He abundance which is consistent with the SBBN prediction. While the higher observed abundances support the expectation of net post-BBN production of ^{3}He, the absence of a correlation with the oxygen abundances is puzzling.

Helium - 4.

The primordial abundance (mass fraction) of ^{4}He is inferred from observations helium and hydrogen recombination lines from metal-poor, extragalactic H II regions (Blue Compact Galaxies: BCDs). In using these data to determine the primordial helium abundance, the systematic errors (underlying stellar absorption, temperature and density inhomogeneities, ionization corrections, atomic emissivities, etc.) dominate over the statistical errors and the uncertain extrapolation to zero metallicity. In my opinion, the uncertainty in Y_P is $sigma(Y_P) \approx 0.006$ and **not** $\sigma(Y_P) < 0.001$, as claimed in some published papers. Therefore, rather than show the helium abundances inferred from observations of hundreds of BCDs, in Figure 3 are shown the handful of helium abundances determined from careful observations of a few H II regions where attention has been paid to some but, even here, not all, sources of systematic uncertainties (Olive & Skillman 2004, Peimbert, Luridiana & Peimbert 2007). The seven Olive & Skillman (2004) H II regions are

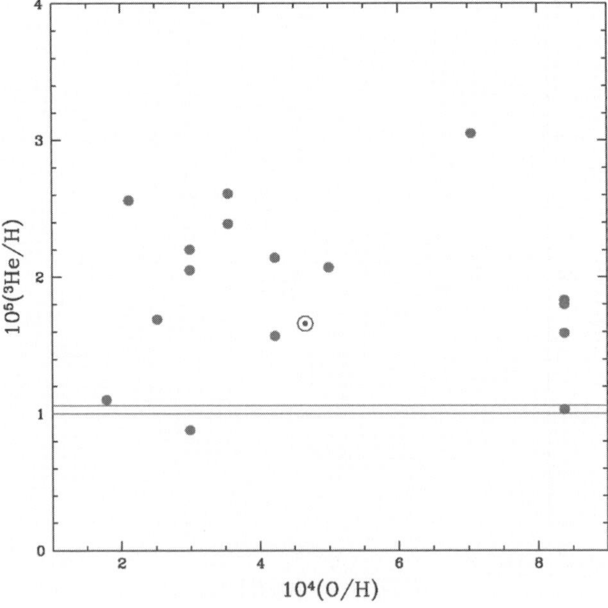

Figure 2. The ^{3}He abundances, $y_3 \equiv 10^5 (^3\mathrm{He}/\mathrm{H})$, observed in Galactic H II regions from Bania, Rood, & Balser (2002) are shown as a function of the corresponding oxygen abundances. For comparison, the solar helium-3 and oxygen abundances are shown. The band indicated by the solid lines is the 68% range of the SBBN prediction for the primordial ^{3}He abundance using the CMB-determined baryon density parameter (see §3).

consistent with **no** correlation between the helium and oxygen abundances, leading to a weighted mean, $Y_{OS} = 0.2500 \pm 0.0030$. The same is true for the five H II regions studied by Peimbert, Luridiana, & Peimbert (2007), with $Y_{PLP} = 0.2517 \pm 0.0043$, where the PLP statistical and systematic errors have been added linearly. The independent analyses of OS and PLP agree. The surprising absence of evidence for statistically significant slopes in their Y versus O/H relations prevents an extrapolation to zero metallicity in order to find Y_P. However, the weighted means do provide *upper bounds* to Y_P: $Y_P < \langle Y \rangle$. As may be seen by comparison with eq. 2.4, these data are consistent with the SBBN prediction.

Lithium - 7.

Like deuterium, lithium (^{6}Li and ^{7}Li) is fragile. In contrast to deuterium, post-BBN lithium is produced via Cosmic Ray Nucleosynthesis and by some stars (see these Proceedings). This is confirmed by Galactic observations of lithium as a function of metallicity. It is therefore expected that in the limit of low metallicity the lithium abundance should approach a plateau, the "Spite Plateau". However, while the primordial abundances of ^{3}He and ^{4}He inferred from the observational data are consistent with the SBBN predicted abundances based on the deuterium abundance, ^{7}Li poses a severe problem. As may be seen from Figure 4, the lithium abundances derived from observations of the most metal-poor halo and globular cluster stars in the Galaxy lie well below the SBBN-predicted value (see eq. 2.5). The discrepancy between the prediction and the observations is a factor of $\sim 3 - 5$. In addition, at the lowest iron abundances, the lithium

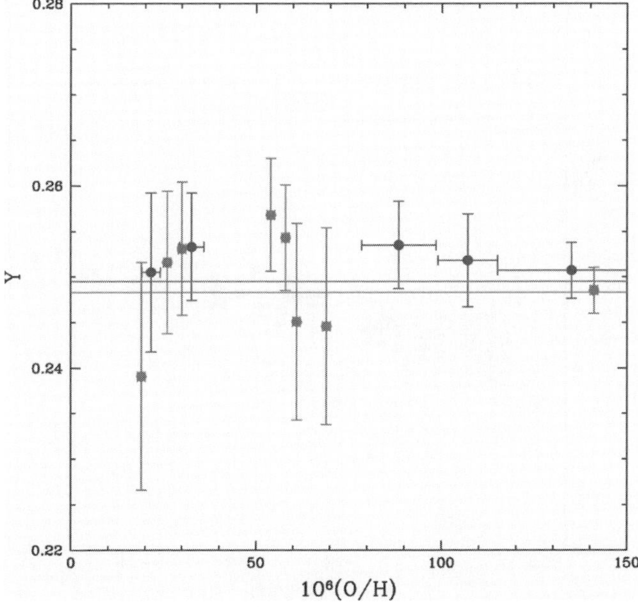

Figure 3. The ^{4}He mass fractions, Y, derived from a selected sample (see the text) of low metallicity, extragalactic H II regions, as a function of the corresponding H II region oxygen abundances. The blue filled circles are from Peimbert, Luridiana, & Peimbert (2007) and the red filled squares are from Olive & Skillman (2004). The band indicated by the solid lines is the 68% range of the SBBN prediction for the primordial ^{4}He mass fraction using the CMB-determined baryon density parameter (see §3).

abundances appear to be decreasing with metallicity. Where is the Spite Plateau? What is the value of $[Li]_P$?

Thus, although the predictions and observations of D, ^{3}He, and ^{4}He are consistent with SBBN, lithium is a problem. Setting lithium aside, we may ask if SBBN is consistent with the CMB?

3. SBBN with the CMB-inferred baryon density parameter

The CMB temperature anisotropy spectrum depends on the baryon density parameter η_B (see the contribution by Dunkley). From the WMAP data Dunkley *et al.* (2009) find $\Omega_B h^2 = 0.02273 \pm 0.00062$, which corresponds to (Steigman 2006)

$$\eta_{10}(\text{CMB}) = 6.226 \pm 0.170. \tag{3.1}$$

The baryon density parameters inferred from deuterium and SBBN, when the Universe was ~ 20 minutes old, and from the CMB, last scattered some 400 thousand years later, agree within $\sim 1.5\sigma$ (the glass is half full). SBBN and the CMB are consistent (modulo the lithium problem).

It is interesting to check the consistency of SBBN and the CMB by comparing the SBBN-predicted primordial light nuclide abundances determined using the CMB value of baryon density parameter to the observations. These comparisons are shown by the

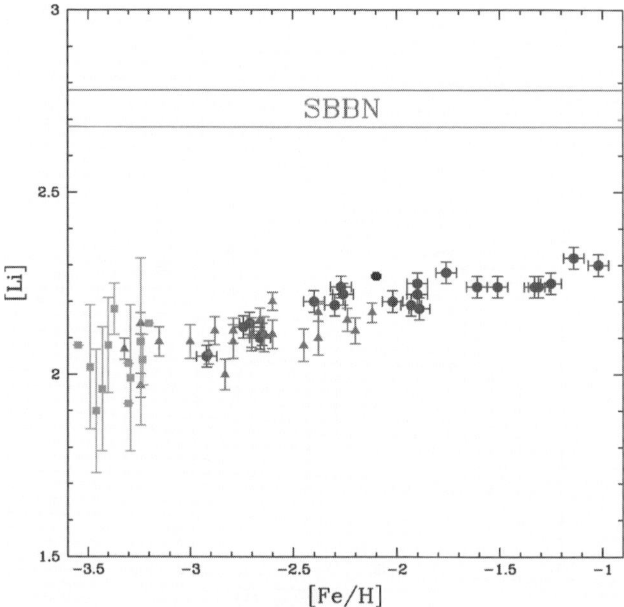

Figure 4. The lithium abundances, $[Li] \equiv 12+\log(Li/H)$, derived from observations of low metallicity Galactic halo and globular cluster stars as a function of the iron abundance (relative to solar). Blue filled circles (Asplund *et al.* 2006), red, filled triangles (Boesgard *et al.* 2005), green filled squares (Aoki *et al.* 2009). The black filled circle (Lind *et al.* 2009) is for the globular cluster NGC6397. The band indicated by the solid lines is the 68% range of the SBBN prediction for the primordial ^{7}Li abundance using the CMB-determined baryon density parameter (see §3).

horizontal bands in Figures 1-4. For SBBN with η_B(CMB),

$$\log(y_{DP}) = 0.40 \pm 0.02 \quad (y_{DP} = 2.52 \pm 0.13), \tag{3.2}$$

$$y_{3P} = 1.03 \pm 0.03, \tag{3.3}$$

$$Y_P = 0.2489 \pm 0.0006, \tag{3.4}$$

$$[\mathrm{Li}]_P = 2.74 \pm 0.05. \tag{3.5}$$

The SBBN/CMB-predicted abundances of D, ^{3}He, and ^{4}He are consistent with their observationally-inferred primordial values, but the lithium discrepancy is exacerbated. SBBN ($N_\nu = 3$) and the CMB are consistent (but lithium is a problem!).

4. Non-Standard Big Bang Nucleosynthesis (nSBBN): $N_\nu \neq 3$

For non-standard BBN (nSBBN) with $N_\nu \neq 3$, the relic abundances of the light nuclides are functions of two parameters, η_B and N_ν. First, consider deuterium. The nSBBN primordial abundance is predicted to vary as $y_{DP} \propto \eta_D^{-1.6}$, where $\eta_D = \eta_D(\eta_{10}, N_\nu)$ (see, e.g., Kneller & Steigman 2004, Steigman 2007). It is interesting to explore the consequences of using the CMB (Dunkley *et al.* 2009) to fix η_{10} and the observed primordial D abundance to determine η_D. This leads to a combined BBN and CMB prediction for N_ν. For $\log(y_D) = 0.45 \pm 0.03$ and η_{10}(CMB) $= 6.23 \pm 0.17$, $N_\nu = 4.0 \pm 0.7$. Here, the relative insensitivity of y_{DP} to N_ν has amplified the small difference between η_{10} and η_D into a relatively large value (and uncertainty) of $\Delta N_\nu = 1.0 \pm 0.7$. Although the central value of the effective number of neutrinos determined this way is $N_\nu \neq 3$, this result is consistent with $N_\nu = 3$ at $\sim 1.4\sigma$. Using this value of N_ν along with η_{10}(CMB), how do the predicted BBN abundances of the remaining light nuclides compare with their observationally inferred primordial values? Here, I concentrate on the two key elements, ^{4}He and ^{7}Li (by construction, D is de facto consistent).

For this combination of η_{10} and N_ν, the primordial ^{4}He mass fraction is $Y_P = 0.261 \pm 0.009$. Here, the sensitivity of Y_P to N_ν has amplified the small difference between η_{10} and η_D into a relatively large value (and uncertainty) of Y_P. As may be seen from Figure 3, within the large uncertainty of this prediction, the very high central value is consistent with the data.

The BBN-predicted abundances of D and ^{7}Li are very tightly correlated, both for $N_\nu = 3$ and $N_\nu \neq 3$ (Kneller & Steigman 2004, Steigman 2007). As a result, even for this example of nSBBN, the predicted primordial lithium abundance is very similar to its SBBN value, $[\mathrm{Li}]_P = 2.70^{+0.05}_{-0.06}$, in conflict with the observational data in Figure 4.

Thus, although this variant of nSBBN is consistent with D, ^{3}He, and ^{4}He, the lithium problem persists! Nonetheless, this example illustrates the potential value of combining BBN and the CMB to constrain and test non-standard models of particle physics and cosmology. A slightly different variant of this approach is presented in the next section.

5. Using ^{4}He and the CMB to constrain N_ν

Of the light nuclides synthesized during BBN, the ^{4}He mass fraction is most sensitive to non-standard physics ($N_\nu \neq 3$). Indeed, for $|\Delta N_\nu| \lesssim 1$, $\Delta Y_P \approx 0.013 \Delta N_\nu$, so that a good bound (small uncertainty) to Y_P would result in a tight constraint on N_ν. According to Kneller & Steigman (2004) and Steigman (2007), using η_{10}(CMB)$= 6.23 \pm 0.17$, a very good fit to Y_P, is

$$Y_P \approx 0.2486 \pm 0.0007 + 0.013 \Delta N_\nu. \tag{5.1}$$

As mentioned earlier, the present uncertainty in the observationally inferred value of Y_P is dominated by systematic errors. To illustrate the potential value of an accurate determination of Y_P, let's adopt the weighted mean of the Olive & Skillman (2004) helium abundances as an *upper bound* to the primordial ^{4}He mass fraction: $Y_P < \langle Y \rangle_{OS} = 0.2500 \pm 0.0030$. Comparing this BBN prediction of Y_P with the upper bound inferred from the data leads to an *upper bound* on the effective number of neutrinos,

$$\Delta N_\nu < 0.11 \pm 0.24 \quad (N_\nu < 3.11 \pm 0.24). \tag{5.2}$$

If, instead, we had adopted the weighted mean of the Peimbert, Luridiana, & Peimbert (2007) helium abundances, we would have found $\Delta N_\nu < 0.24 \pm 0.33$ ($N_\nu = 3.24 \pm 0.33$). Just a few good H II region ^{4}He abundances are all that is needed to obtain a very strong constraint on N_ν. For these combinations of η_{10}(CMB) and N_ν, the bounds to the D and ^{3}He abundances are consistent with the data, while lithium remains a problem.

6. Challenges

In answer to the first question posed in §1, yes, SBBN ($N_\nu = 3$) is consistent with the observationally-inferred primordial abundances of D, ^{3}He, and ^{4}He, but ^{7}Li poses a problem. The answer to the second question posed in §1 is also yes, the CMB and the BBN values of η_B and S agree. SBBN and the CMB in combination allow, but also constrain, some models of non-standard physics. The challenges facing BBN, largely observational, makes the timing of this meeting ideal. Having had the luxury of being one of the first speakers, I will end by presenting my list of challenges to those who follow.
1. Why is the spread in the observed deuterium abundances so large?
2. Why are the observed ^{3}He abundances uncorrelated with either the oxygen abundances or with the distance from the center of the Galaxy?
3. What are the sources (and the magnitudes) of the systematic errors in Y_P and, are there observing strategies to reduce them?
4. What is the primordial abundance of ^{7}Li (and of ^{6}Li)?

References

Aoki, W. *et al.* 2009, *ApJ*, 698, 1803.
Asplund, M. *et al.* 2006, *ApJ*, 644, 229.
Bania, T. M., Rood, R. T., & Balser, D. S. 2002, *Nature*, 415, 54.
Boesgaard, A. M., Stephens, A., & Deliyannis, C. P. 2005, *ApJ*, 633, 398.
Dunkley, J. *et al.* 2009, *ApJS*, 180, 306.
Epstein, R. I., Lattimer, J. M., & Schramm, D. N. 1976, *Nature*, 265, 219.
Geiss, J. & Gloeckler, J. G. 1998, *Space Sci. Rev.* 84, 239.
Kneller, J. P. & Steigman, G. 2004, *New J. Phys.*, 6, 117.
Lind, K. *et al.* 2009, *A&A*, 503, 545.
Olive, K. A. & Skillman, E. D. 2004, *ApJ*, 617, 29
Peimbert, M., Luridiana, V., & Peimbert, A. 2007, *ApJ*, 666, 634
Pettini, M., Zych, B. J., Murphy, M. T., Lewis, A., & Steidel, C. C. 2008, *MNRAS*, 391, 1499.
Reeves, H., Audouze, J., Fowler, W. A., & Schramm, D. N. 1973, *ApJ*, 179, 909.
Rood, R. T. 1972, *ApJ*, 177, 681.
Rood, R. T., Steigman, G., & Tinsley, B. M. 1976, *ApJL*, 207, L57.
Steigman, G. 2006, *JCAP*, 10, 016.
Steigman, G. 2007, *Ann. Rev. Nucl. Part. Sci.*, 57, 463.

Light Elements in the Universe
Proceedings IAU Symposium No. 268, 2009
C. Charbonnel, M. Tosi, F. Primas & C. Chiappini, eds.

© International Astronomical Union 2010
doi:10.1017/S1743921310003820

The cosmic lithium problem and physics beyond the Standard Model

Karsten Jedamzik

Laboratoire de Physique Théorique et Astroparticules, Université Montpellier II, place Eugne
Bataillon 34095 Montpellier, FRANCE
email: `jedamzik@lpta.univ-montp2.fr`

Abstract. In this proceeding I briefly discuss the possibility of relic decaying or annihilating particles to explain the cosmological ^{7}Li anomaly and/or to be the source of significant amounts of pre-galactic ^{6}Li. The effect of relic massive charged particles through catalysis of nuclear reactions is also discussed. The possibility of a connection of the ^{7}Li problem to the cosmic dark matter and physics beyond the standard model of particle physics, such as supersymmetry, is noted.

Keywords. Cosmology: theory, early universe, cosmological parameters

---

The standard scenario of Big Bang nucleosynthesis (BBN), with its minimalistic assumptions about the composition and state of the Universe at around one second, predicts light-element abundances generally in approximate agreement with those observed. However, after an independent precise observational determination of the baryon density through observations of the cosmic microwave background radiation by the WMAP satellite, a significant discrepancy between the predicted and observationally inferred ^{7}Li/H ratio has emerged (Cyburt *et al.* 2008). This discrepancy may not be due to seriously underestimated stellar surface temperatures and only very unlikely due to nuclear reaction rate uncertainties (Cyburt & Pospelov 2009). The mismatch may, however, be due to ^{7}Li depletion in low-metallicity halo stars (Richard *et al.* 2005, Korn *et al.* 2006), though the details of such depletion remain uncertain. Alternatively, the discrepancy may be due to physics operating during the BBN epoch itself. Such non-standard BBN scenarios also often lead to the production of considerable amounts of ^{6}Li. Interestingly, the existence of substantial ^{6}Li in low-metallicity stars has been claimed (Asplund *et al.* 2006), though due to the difficulty of these observations is currently controversial (Cayrel *et al.* 2007). A brief summary of BBN scenarios leading to a reduction of ^{7}Li and the production of ^{6}Li will be the main subject of this note.

^{7}Li is only produced in trace amounts during BBN the main reason being that its synthesis requires the presence of ^{4}He and by the time that neutrons are incorporated into ^{4}He the Coulomb barriers for the two main ^{7}Li producing reactions, i.e ^{3}H $+^4$He $\rightarrow^7$Li$+\gamma$ and ^{3}He $+^4$He$\rightarrow^7$Be$+\gamma$ are already substantial. Production of ^{6}Li in standard BBN is even further supressed compared to ^{7}Li by an additional factor $\sim 10^{-4}$ due to the weakness of the quadrupole reaction ^{2}H $+^4$He $\rightarrow^6$Li$+\gamma$. Most alternative BBN scenarios, such as inhomogeneous or with antimatter domains or with lepton chemical potentials, etc., do not lead to a reduction of the ^{7}Li abundance, rather usually the opposite happens. Attempting to reduce the ^{7}Li by later photodisintegration is highly problematic (Ellis *et al.* 2005) due to the concommitant ^{2}H photo-disintegration or ^{3}He/^{2}H overproduction due to ^{4}He photodisintegration. One known way of reducing ^{7}Li is by the assumption of varying fundamental constants, since ^{7}Li production is very sensitive to the ^{2}H and ^{7}Be

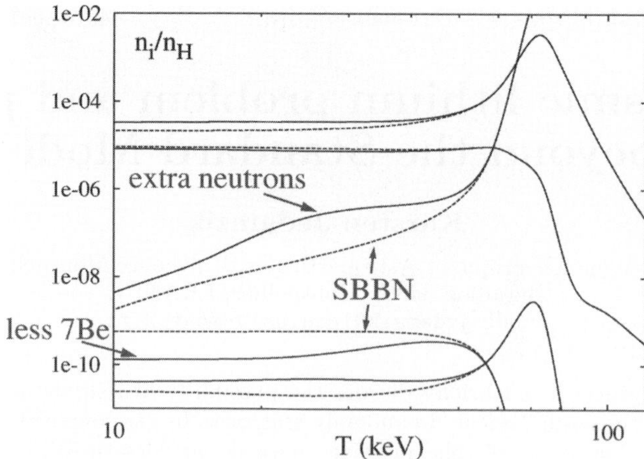

Figure 1. Evolution of light element abundances during BBN in a standard BBN scenario (solid) and in a BBN scenario with $\sim 10^{-5}$ additional neutrons/baryon injected at time $\tau \approx 1000\,\mathrm{sec}$ (dashed). From top to bottom (at low temperatures), the lines show ^{2}H/H, ^{3}He/H, n/p, ^{7}Be/H, and ^{7}Li/H, respectively. Figure taken from Jedamzik (2004a).

binding energies (Dmitriev *et al.* 2004). It is nevertheless very difficult to test for such a hypothesis by other means.

There exists, however, a very simple way of reducing the ^{7}Li yield (Jedamzik 2004a). At a baryon-to-photon ratio of $\eta \approx 6.2 \times 10^{-10}$ around 90% of all ^{7}Li is synthesized in form of ^{7}Be which after BBN electron captures to form ^{7}Li. In case this ^{7}Be is converted prematurely, already during BBN, to ^{7}Li, the resultant ^{7}Li will be destroyed by the reaction $^7\mathrm{Li}+p \rightarrow^4\mathrm{He} + {}^4\mathrm{He}$. The main reaction which converts ^{7}Be to ^{7}Li is $^7\mathrm{Be}+n \rightarrow^7\mathrm{Li}+p$. The efficiency of such conversion depends on the ambient neutron density. Due to the large cross section for this reaction already a minute amount of extra neutrons $n/p \sim 10^{-5}$ injected at temperature $T \approx 30\,\mathrm{keV}$ may severely deplete the ^{7}Li. This is seen in Fig. 1 where the evolution of abundance yields of such a model with extra neutron injection is compared to standard BBN. Such neutron injection also leads to some additional ^{2}H.

^{6}Li is a sensitive probe of non-standard BBN scenarios and the evolution of the early Universe (Jedamzik 2000), since in a large number of non-standard scenarios, already small deviations from standard BBN (SBBN) lead to ^{6}Li synthesis beyond that in SBBN. All scenarios which include the injection of energetic hadrons or photons into the plasma, lead to either photodisintegration of ^{4}He or nuclear spallation of ^{4}He. The resultant energetic ^{3}H and ^{3}He nuclei may participate in the non-thermal nuclear reaction ^{3}H $(^3\mathrm{He}) + {}^4\mathrm{He} \rightarrow {}^6\mathrm{Li} + n\,(p)$ (Dimopulos *et al.* 1988) to form ^{6}Li. Note that these reactions are not available to thermal BBN, due to the existence of energy thresholds.

When searching for scenarios which may lead to an injection of extra neutrons during BBN to solve the ^{7}Li problem, one immediate idea is the possibility of residual annihilation of cosmological particle dark matter during BBN (Jedamzik 2004b). It is well know that the correct cosmological dark matter density results when assuming a thermal freeze-out from equilibrium of a self-annihilating stable particle (e.g. a supersymmetric neutralino) with annihilation rate $< \sigma v > \approx 3 \times 10^{-26}\mathrm{cm}^3\mathrm{s}^{-1}$. Since such annihilation rates are typical for weak mass scale particles, an attractive possibility for the cosmological dark matter is a new stable particle of mass $m_\chi \sim 100\,\mathrm{GeV}$, potentially to be discovered at the LHC. Such particles would also be present during BBN with residual annihilations taking place. Since in many extensions of the standard model their annihilation products are quarks, the concommitant formation of baryons and anti-baryons

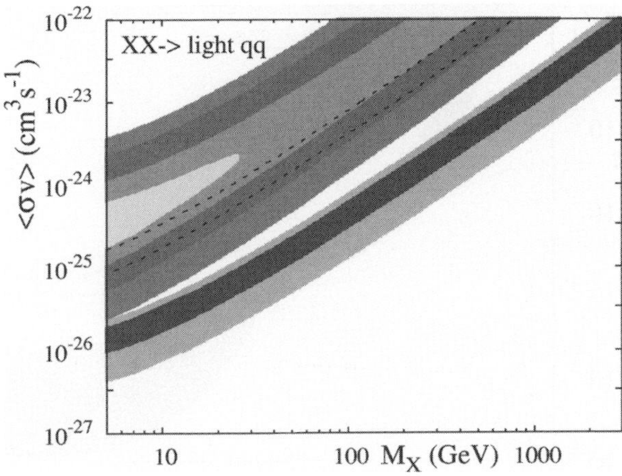

Figure 2. Dark matter annihilation rate versus dark matter mass. The blue band shows parameters where ^{6}Li due to residual dark matter annihilation may account for the ^{6}Li abundance as inferred in HD84937 (^{6}Li/^{7}Li$\approx 0.014 - 0.09$ at 2-σ), whereas the orange-red-green-yellow region shows where ^{7}Li is efficiently destroyed (e.g. green band - ^{7}Li/H$< 2 \times 10^{-10}$). From Jedamzik & Pospelov (2009).

would lead to the injection of extra neutrons. In Fig. 2 the parameter space in the annihilation rate $< \sigma v >$ and particle mass m_χ is shown where annihilations significantly affect the ^{6}Li and ^{7}Li abundances. It is seen, however, that residual dark matter annihilation can only solve the ^{7}Li problem (the orange-red-green band) at the expense of overproducing ^{6}Li (the blue band). Annihilations are therefore most likely not at the heart of the ^{7}Li problem. However, it is intruiging that a dark matter particle of $m_\chi \sim 20 - 80\,\mathrm{GeV}$ annihilating into quarks at the required rate for the dark matter density may produce *all* the ^{6}Li as claimed to exist in the star HD84937. Note, that this already includes a factor ~ 3 stellar depletion of ^{7}Li (and ^{6}Li) to solve the ^{7}Li problem. Here it has been assumed that both isotopes are depleted by the same amount.

Whereas dark matter annihilation can not solve the ^{7}Li problem, the decay of metastable particles during BBN can. This may be seen in Fig. 3, which shows the parameter space in the $\Omega_X B_h$ and τ_X plane for which the ^{7}Li problem may be solved and/or significant ^{6}Li is synthesized. Here Ω_X would be the contribution of the X-particle to the present critical density if it wouldn't have decayed. It is seen that for decay times $\tau_X \sim 100 - 2000\,\mathrm{sec}$ the ^{7}Li abundance may be significantly reduced due to the injection of extra neutrons from the decay. For larger $\tau_X \gtrsim 10^3\,\mathrm{sec}$ large amounts, i.e. ^{6}Li/^{7}Li$\approx$ $0.015 - 0.3$, of ^{6}Li are synthesized. At $\tau_X \approx 10^3$, both, ^{7}Li and ^{6}Li, are in agreement with observations. In this area, additional ^{2}H production results as well.

It has been recently realized that significant amounts of ^{6}Li may also result simply due to the presence of negatively charged weak mass scale particles during BBN (Pospelov 2007). This is due to such particles entering into bound states with nuclei and thereby leading to catalysis of nuclear reactions. The most important of these reactions is (^{4}He-X^-)+^{2}H$\rightarrow^6$Li + X^-. Nevertheless, though interesting in its physics, and having received wide-spread attention, concerning ^{6}Li production, catalysis is only more important that the ^{6}Li production due to hadronic decay products, when the particle has a hadronic branching ratio smaller than $B_h \lesssim 10^{-2}$. Though this may occur (e.g. the supersymmetric partner of the stau lepton) it is often not the case. Catalysis may not present the solution to the ^{7}Li problem (Bird *et al.* 2007, Kanimura *et al.* 2009), unless the hadronic branching ratio of the charged weak mass scale particles is very small $B_h \lesssim 10^{-5}$ and its abundance

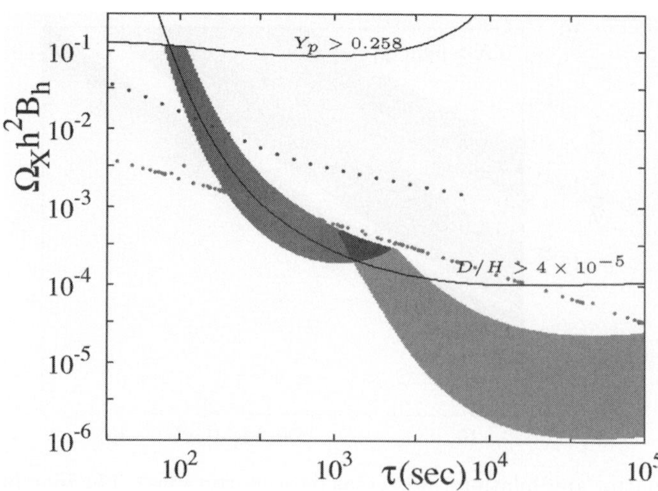

Figure 3. Parameter space in the relic decaying particle abundance times hadronic branching ratio B_h, i.e. $\Omega_X h^2 B_h$, and life time τ_X plane, where ^{7}Li is significantly reduced (red and blue) and ^{6}Li is efficiently produced (green and blue). See text for further details. From Bailly *et al.* (2009).

is very large, i.e. $\Omega_X \gtrsim 10$. However, though currently not certain, catalysis could lead to the formation of ^{9}Be in BBN (Pospelov 2008), something impossible to achieve with neutral decaying particles.

In summary, the decay of metastable relic particles during BBN may easily reduce the standard BBN predicted ^{7}Li/H$\sim 5 \times 10^{-10}$ to its observationally inferred value ^{7}Li/H$\sim 1 - 2 \times 10^{-10}$. When such decays occur around $\gtrsim 10^3$ seconds they are also associated with considerable ^{6}Li production. In contrast, residual annihilation of dark matter during BBN may account for all the ^{6}Li as claimed to exist in the star HD84937, but not solve the ^{7}Li problem without significant ^{6}Li overproduction. Catalysis of nuclear reactions due to electrically charged weak scale particles may lead to additional ^{6}Li, potentially produce ^{9}Be, but only very unlikely has impact on the cosmological ^{7}Li problem. It is the hope that many of these scenarios are testable at the LHC accelerator, either in form of production of light dark matter particles (e.g. a supersymmetric neutralinos) or metastable particles being stopped and decaying in the detector (e.g. a supersymmetric staus or gluinos). However, some particularly attractive scenarios are unfortunately not reachable in energy at the LHC (Jedamzik *et al.* 2006).

References

Asplund, M., Lambert, D. L., Nissen, P. E., Primas, F., & Smith, V. V. 2006, *ApJ*, 644, 229
Bailly, S., Jedamzik, K., & Moultaka, G. 2009, *Phys. Rev. D*, 80, 63509
Bird, C., Koopmans, K., & Pospelov, M. 2008, *Phys. Rev. D*, 78, 83010
Cayrel, R. *et al.*, 2007, arXiv:0708.3819
Cyburt, R. H., Fields, B. D., & Olive, K. A. 2008 *JCAP*, 811, 12
Cyburt, R. H. & Pospelov, M. 2009, arXiv:0906.4373
Dimopoulos, S., Esmailzadeh, R., Hall, L. J., & G. D. Starkman, G. D. 1988, *ApJ*, 330, 545
Dmitriev, V. F., Flambaum, V. V., & Webb, J. K. 2004 *Phys. Rev. D*, 69, 63506
Ellis, J. R., Olive, K. A., & Vangioni, E. 2005 *Phys.Lett. B*, 619, 30
Jedamzik, K. 2000, *Phys.Rev. Lett.*, 84, 3248
Jedamzik, K. 2004, *Phys.Rev. D*, 70, 63524

Jedamzik, K. 2004, *Phys.Rev. D*, 70, 83510
Jedamzik, K., Choi, K. Y., Roszkowski. L., & Ruiz de Austri, R. 2006, *JCAP*, 607, 7
Jedamzik, K. & Pospelov, M. 2009 *New J. Phys.*, 11, 105028
Kamimura, M., Kino, Y., & Hiyama, E. 2009, *Prog. Theor. Phys.*, 121, 1059
Korn, A. J. *et al.*, 2006, *Nature* 442, 657
Pospelov, M. 2007, *Phys. Rev. Lett.*, 98, 231301
Pospelov, M. 2007, arXiv:0712.0647
Richard, O., Michaud, G., & Richer, J. 2005 *ApJ*, 619, 538

Guillaume Hébrard

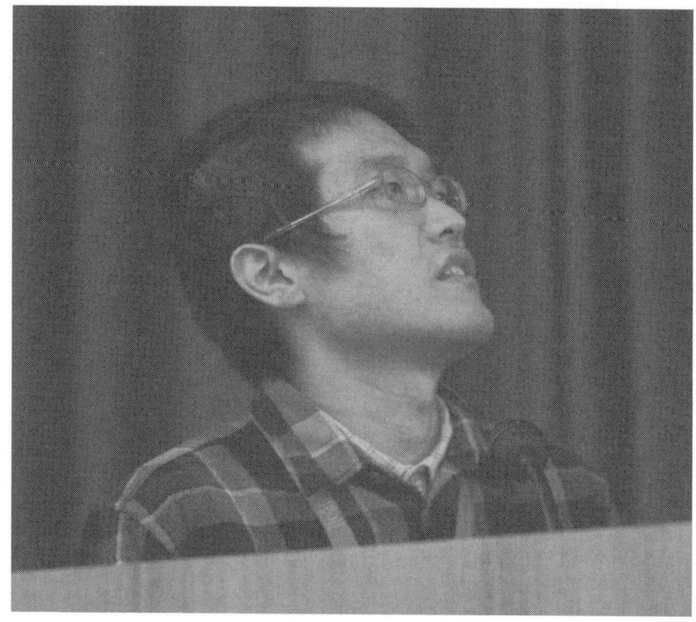

Motohiko Kusakabe

Light Elements in the Universe
Proceedings IAU Symposium No. 268, 2009
C. Charbonnel, M. Tosi, F. Primas & C. Chiappini, eds.

© International Astronomical Union 2010
doi:10.1017/S1743921310003832

Big Bang nucleosynthesis with long-lived strongly interacting relic particles

Motohiko Kusakabe[1]†, **Toshitaka Kajino**[2,3,4], **Takashi Yoshida**[2] and **Grant J. Mathews**[5]

[1]Institute for Cosmic Ray Research, University of Tokyo, Kashiwa, Chiba 277-8582, Japan
email: **kusakabe@icrr.u-tokyo.ac.jp**

[2]Department of Astronomy, Graduate School of Science, University of Tokyo,
Bunkyo-ku, Tokyo 113-0033, Japan
[3]Division of Theoretical Astronomy, National Astronomical Observatory of Japan,
Mitaka, Tokyo 181-8588, Japan
[4]Department of Astronomical Science, The Graduate University for Advanced Studies,
Mitaka, Tokyo 181-8588, Japan
[5]Department of Physics, Center for Astrophysics, University of Notre Dame,
Notre Dame, IN 46556, USA

Abstract. We study effects of relic long-lived strongly interacting massive particles (X particles) on big bang nucleosynthesis (BBN). The X particle is assumed to have existed during the BBN epoch, but decayed long before detected. The interaction strength between an X and a nucleon is assumed to be similar to that between nucleons. Rates of nuclear reactions and beta decay of X-nuclei are calculated, and the BBN in the presence of neutral charged X^0 particles is calculated taking account of captures of X^0 by nuclei. As a result, the X^0 particles form bound states with normal nuclei during a relatively early epoch of BBN leading to the production of heavy elements. Constraints on the abundance of X^0 are derived from observations of primordial light element abundances. Particle models which predict long-lived colored particles with lifetimes longer than ~ 200 s are rejected. This scenario prefers the production of ^{9}Be and ^{10}B. There might, therefore, remain a signature of the X particle on primordial abundances of those elements. Possible signatures left on light element abundances expected in four different models are summarized.

Keywords. elementary particles; nuclear reactions; nucleosynthesis; abundances; stars: abundances; cosmology: early universe

1. Introduction

Primordial lithium abundances are inferred from measurements in metal-poor stars (MPSs). Observed abundances are roughly constant as a function of metallicity (Spite & Spite 1982, Ryan *et al.* 2000, Meléndez & Ramírez 2004, Asplund *et al.* 2006, Bonifacio *et al.* 2007, Shi *et al.* 2007, Aoki *et al.* 2009) at ^{7}Li/H $= (1 - 2) \times 10^{-10}$. The theoretical prediction by the standard big bang nucleosynthesis (BBN) model, however, is a factor of $2 - 4$ higher when its parameter, the baryon-to-photon ratio, is fixed to the value deduced from the observation with Wilkinson Microwave Anisotropy Probe of the cosmic microwave background (CMB) radiation (Dunkley *et al.* 2009). A recent study including a reevaluation of nuclear reaction rate of ^{3}He$(\alpha,\gamma)^7$Be shows the standard BBN (SBBN) prediction of ^{7}Li/H $= (5.24^{+0.71}_{-0.67}) \times 10^{-10}$ (Cyburt *et al.* 2008). The discrepancy indicates some mechanism of ^{7}Li reduction having operated in some epoch from the BBN to this day. One possible astrophysical process to reduce ^{7}Li abundances in stellar surfaces is the combination of the atomic and turbulent diffusion (Richard *et al.* 2005, Korn *et al.* 2006,

† Research Fellow of the Japan Society for the Promotion of Science (JSPS).

2007). The precise trend of Li abundance as a function of effective temperature found in the metal-poor globular cluster NGC 6397 is, however, not reproduced yet (Lind *et al.* 2009).

^{6}Li/^{7}Li isotopic ratios of MPSs have also been measured spectroscopically. A ^{6}Li abundance as high as ^{6}Li/H$\sim 6 \times 10^{-12}$ was suggested (Asplund *et al.* 2006), which is about 1000 times higher than the SBBN prediction. Convective motions in the atmospheres of MPSs could cause systematic asymmetries in the observed line profiles and mimic the presence of ^{6}Li (Cayrel *et al.* 2007). Nevertheless, there still remain a few or several MPSs with certain detections of high ^{6}Li abundances after estimations of convection-triggered line asymmetries (García Pérez *et al.* 2009, Steffen *et al.* 2009). This high ^{6}Li abundance is a problem since the standard Galactic cosmic ray (CR) nucleosynthesis models predict negligible amounts of ^{6}Li yields compared to the observed level in the epoch corresponding to the metallicity of [Fe/H] < -2 (Prantzos 2006).

Be and B abundances are also observed in MPSs. ^{9}Be abundances increase linearly as iron abundance when the Galaxy evolves chemically (Boesgaard *et al.* 1999, Primas *et al.* 2000a, Tan *et al.* 2009, Smiljanic *et al.* 2009, Ito *et al.* 2009, Rich & Boesgaard 2009). In a region of very low metallicities of [Fe/H]< -3, a dispersion in Be abundances is indicated (Primas *et al.* 2000b, Boesgaard & Novicki 2006). B abundances increase linearly as iron abundance (Duncan *et al.* 1997, Garcia Lopez *et al.* 1998, Primas *et al.* 1999, Cunha *et al.* 2000). So far no primordial plateau abundances of Be and B are found.

As a cosmological solution to the Li problems, BBN models including exotic decaying particles have been studied (see contribution by Jedamzik, this volume). Nonthermal nuclear reactions triggered by the radiative decay of long-lived particles can produce ^{6}Li nuclides in amounts greater than observed in MPSs and at most ~ 10 times as much as the level without causing discrepancies in abundances of other light elements or the CMB energy spectrum (Kusakabe *et al.* 2006, 2009a). If negatively-charged leptonic X^- particles exist in the BBN epoch, they affect the nucleosynthesis (see contribution by Jedamzik, this volume). The X^- particles get bound to positively charged nuclides with binding energies of $\sim O(0.1 - 1)$ MeV. Since the binding energies are low, the bound states between the X^- and nuclides form late in BBN epoch. Nuclear reactions at this temperature are no longer efficient so that the effect of negatively-charged particles is rather small. Interestingly the X^- particle can catalyze a preferential production of ^{6}Li (Pospelov 2007) and weak destruction of ^{7}Be (Bird *et al.* 2008, Kusakabe *et al.* 2007). Non-equilibrium nuclear network calculation of this model (Kusakabe *et al.* 2008) with realistic cross sections derived from a quantum mechanical calculation (Kamimura *et al.* 2009) shows that ^{6}Li production and ^{7}Li reduction can simultaneously occur and that there are no likely signatures in primordial abundances of Be and heavier nuclei.

As an astrophysical solution to the ^{6}Li problem, the cosmological CR nucleosynthesis associated with a possible activity of supernova explosions in the early epoch of structure formation has been suggested (Rollinde *et al.* 2005, 2006). This ^{6}Li production mechanism is likely realized with coproduction of ^{9}Be and ^{10,11}B nuclides in abundances probably within reach of future observations of MPSs (Kusakabe 2008, Rollinde *et al.* 2008). The ^{6}Li production is, however, not a certain possibility since a calculation in a similar scenario but using a different star formation and chemical evolution history fails to reproduce the ^{6}Li plateau level (Evoli *et al.* 2008).

In this paper we show a new BBN scenario which includes long-lived strongly interacting relic particles which appear in some particle models beyond the standard. Constraints on their abundance and lifetime is derived. Signatures of such relic particles are found to be possibly left on the primordial abundances of Be and B. A theoretical prediction of primordial light element abundances are given in the limit of long lifetime.

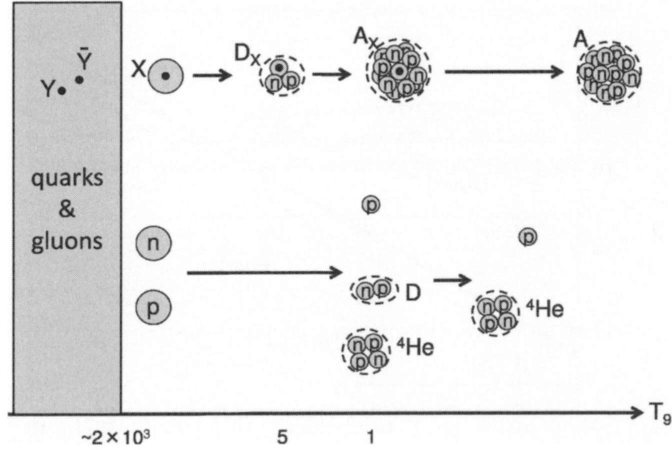

Figure 1. Nucleosynthesis in the presence of the long-lived strongly interacting X particle.

2. Model

Some particle models include long-lived heavy colored particles, denoted by Y (Kang *et al.* 2008). If the hypothetical colored particle Y exists, they would have experienced production and annihilation in the early universe. When the temperature of the universe decreases to $T < 180$ MeV, Y gets confined in exotic hadrons, which we call X (see Fig. 1). Two X particles can form bound states and the bound states decay into lower energy states gradually. Y particles in X particles could annihilate eventually when two X particles approach. Taking account of this process, Kang *et al.* (2008) estimated the relic abundance of X to be $n_X \sim 10^{-8} n_b$, where n_X and n_b are number densities of X and baryon, respectively. If the strongly interacting X particle had existed in BBN epoch, it would have affected the nucleosynthesis. We study effects of X on BBN. The X particle is assumed to be of spin 0, charge 0 and mass much larger than nuclear mass of $O(1\ \mathrm{GeV})$. The strength of interaction between an X^0 particle and nuclei is supposed to be similar to that between a nucleon and the nuclei as a rough approximation.

Binding energies between nuclides and an X^0 are calculated. The adopted nuclear potential between a nucleon and an X is well reproducing the binding energy of $n+p$ system when used for the $n+p$ system. The Woods-Saxon potential is adopted for systems of nuclides and an X. Two-body Shrödinger equations are solved by a variational calculation, and binding energies are derived. The binding energies are of $O(10\ \mathrm{MeV})$, and this strong binding leads to bound state formations early in the BBN epoch.

Thermonuclear reaction rates and β-decay rates of X-bound nuclei (i.e., X-nuclei or A_X) are, then, calculated. Reaction Q-values are derived taking account of binding energies of X-nuclei. The rates associated with X-nuclei are estimated using available cross sections for normal nuclei by correcting for Q-values and mass numbers of reactants.

We have put radiative X capture reactions by nuclides, nuclear reactions and β-decay of X-nuclides and all inverse reactions into the SBBN network code (Kawano 1992) and solved a set of rate equations. See Kusakabe *et al.* (2009b) for details on this study.

3. Result

Figure 1 shows a schematic view of nucleosynthesis as a function of $T_9 \equiv T/(10^9\ \mathrm{K})$ with temperature T. The relic abundance of X would be rather small ($Y_X = n_X/n_b \sim 10^{-8}$) although it depends on the mass and interaction strength of the exotic colored

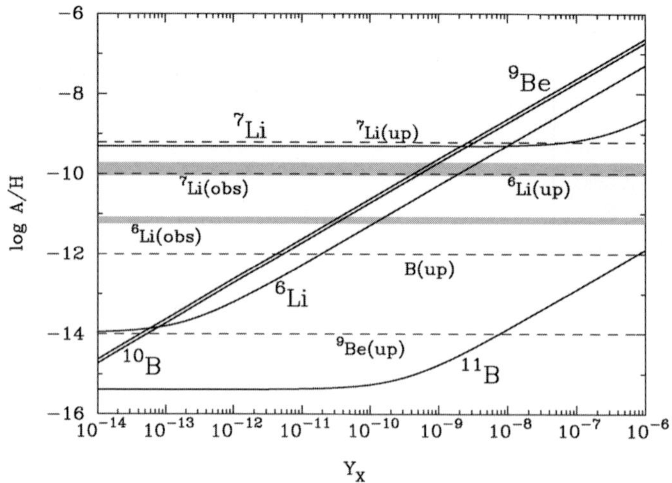

Figure 2. Calculated abundances of ^{6,7}Li, ^{9}Be and ^{10,11}B (solid lines) as a function of the X abundance Y_X. The grey bands correspond to the observed mean values of ^{7}Li abundances and ^{6}Li abundances of MPSs with detection of the isotope. The dashed lines indicate the adopted upper limits on the primordial abundances (see text).

Y particle. The main story of BBN, therefore, does not change even if the strongly interacting X particle exists. Effects on abundant light elements, i.e., H and He are small, while those on minor elements, i.e., Li, Be and B are significant. At high temperature of $T_9 > 5$, the X particles exist in the free state. When the temperature decreases to $T_9 \sim 5$, X particles radiatively capture neutrons and protons to form bound states. n_X and H_X thus formed are processed to ^{2}H$_X$ and ^{3}He$_X$. At $T_9 \sim 5$, the normal D abundance temporarily increases during the ^{4}He production. An efficient nuclear process of X-nuclei then occurs through consecutive nonradiative strong D-capture reactions, i.e., (d,p) and (d,n). Finally the X particles decay and heavy nuclides are left. From the fact that all X particles form X-nuclei before their decay, it can be seen that strongly interacting exotic particles have a great impact on BBN (Kusakabe *et al.* 2009b). One reason is that binding energies between an X and nuclides are large, and cross sections of captures of nucleons by X are large, leading to early formation of bound states between nucleons and an X. Another reason is that ^{5}He, ^{5}Li and ^{8}Be which are unstable to particle decays can be stabilized against the decays when they are bound to X particles, so that heavy X-nuclides can form through those stabilized states.

Figure 2 shows calculated abundances of ^{6,7}Li, ^{9}Be and ^{10,11}B (solid lines) as a function of the initial X abundance Y_X in the case of very long lifetime, τ_X, compared to the BBN time scale. Resulting yields through the X-catalyzed nucleosynthesis are all proportional to Y_X. The prediction of primordial abundances in this scenario is as follows:

$$^6\text{Li/H} \sim {}^9\text{Be/H} \sim {}^{10}\text{B/H} \sim 10^{-9}\left(Y_X/10^{-8}\right). \tag{3.1}$$

$$^7\text{Li/H} \sim 10^{-11}\left(Y_X/10^{-8}\right). \tag{3.2}$$

$$^{11}\text{B/H} \sim 10^{-14}\left(Y_X/10^{-8}\right). \tag{3.3}$$

The grey bands correspond to the observed mean values of ^{7}Li abundances (Ryan *et al.* 2000) and ^{6}Li abundances of MPSs with detection of the isotope (Asplund *et al.* 2006). The dashed lines show the adopted upper limits on the primordial abundances from observations: ^{7}Li/H$< 5 \times 1.23 \times 10^{-10}$ allowing for large depletion factors of up to five, ^{6}Li/H$< 5 \times 10^{-10}$ allowing for depletion factors up to ~ 10, ^{9}Be/H$< 10^{-14}$ (Ito *et al.*

Table 1. Possibility to solve the Li problems and expected signatures in elemental abundances.

Model	^{6}Li Problem solved ?	^{7}Li Problem solved ?	Possible Signatures on Other Nuclides
radiative decay	**YES**	NO	NO
leptonic X^-	**YES**	**YES**	**NO**
strongly interacting X^0	**NO**	**NO**	^{9}Be and/or ^{10}B
early cosmic ray	**YES**	NO	^{9}Be and ^{10,11}B

2009) and B/H$< 10^{-12}$ (Garcia Lopez *et al.* 1998). Y_X values leading to overproductions of light nuclides in abundances above the observational upper limits should be ruled out. The upper limits on the primordial Be abundance is found to give the most stringent constraint on X abundance, i.e., $Y_X < 10^{-13}$ in the case of long lifetime.

When the lifetime τ_X is not much longer than the BBN time scale, the decay of X should be considered in the numerical calculation. A constraint on the initial abundance Y_X and the lifetime τ_X is derived (Kusakabe *et al.* 2009b). A solution to the ^{6}Li or ^{7}Li problems was not found in this paradigm. As seen in Fig. 2, the light element abundances monotonically increase by the existence of X. A reduction of ^{7}Li abundance is thus NOT possible. The ^{6}Li production is realized with overproduction of ^{9}Be and ^{10}B, which is excluded from observations of Be and B. A ^{6}Li production is then NOT possible.

Constraints on the Y_X and τ_X are derived from limits on Li, Be, B abundances depending on lifetime ranges. Two important predictions of this scenario are found. First, ^{9}Be and B can be produced in amounts more than SBBN predictions. Future observations of Be and B abundances in MPSs might show plateaus suggesting a primordial origin. Second, ^{10}B tends to be produced much more than ^{11}B, i.e., a resulting isotopic ratio ^{10}B/^{11}B is very high (see Fig. 2). This preferential production of ^{10}B has never been predicted in other models of Be production like the Galactic or cosmological CR nucleosynthesis (^{10}B/^{11}B$\sim$ 0.4: e.g., Ramaty *et al.* 1997, Kusakabe 2008) or the supernova neutrino process (^{10}B/^{11}B$\ll$ 1: e.g., Woosley & Weaver 1995, Yoshida *et al.* 2005).

A constraint on τ_X is derived from this result and estimations of initial abundance of X. If $Y_X \sim 10^{-8}$ (Kang *et al.* 2008) is adopted, its lifetime should be $\tau_X < 200$ s.

4. Summary

The BBN scenario in the presence of long-lived strongly interacting relic particle X is described. Particles which interact with nuclei as a nucleon do capture nucleons to form X-nuclei in an early stage of BBN. The X-nuclei capture deuterons and increase their mass numbers. This nuclear catalysis of X activates new reaction paths to heavy nuclei. Constraints on the lifetime and abundance of X are derived. The lifetime should be less than 200 s.

Table 1 summarizes possibilities for four models to be solutions to the ^{6}Li or ^{7}Li problems and expected signatures in abundances of other light nuclides. In light of all available observational data, nonstandard BBN processes (first two models) possibly operate to produce ^{6}Li and/or reduce ^{7}Li. The existence of long-lived relic particles should be ascertained with future astronomical observations and collider experiments. The cosmological CR nucleosynthesis might also have produced light elements in an early epoch of the structure formation. Observations of primordial abundances of Li, Be and B are very important to constrain the possible processes in the early universe.

Acknowledgements

This work was supported by the Mitsubishi Foundation, the MEXT KAKENHI (20105004, 20244035), the JSPS Core-to-Core Program EFES, and the Grant-in-Aid for JSPS Fellows (21.6817). Work at the University of Notre Dame was supported by the U.S. Department of Energy under Nuclear Theory Grant DE-FG02-95-ER40934.

References

Aoki, W. *et al.* 2009, *ApJ*, 698, 1803
Asplund, M. *et al.* 2006, *ApJ*, 644, 229
Bird, C., Koopmans, K., & Pospelov, M. 2008, *Phys. Rev. D*, 78, 083010
Boesgaard, A. M. *et al.* 1999, *AJ*, 117, 1549
Boesgaard, A. M. & Novicki, M. C. 2006, *ApJ*, 641, 1122
Bonifacio, P. *et al.* 2007, *A&A*, 462, 851
Cayrel, R. *et al.* 2007, *A&A*, 473, L37
Cunha, K., Smith, V. V., Boesgaard, A. M., & Lambert, D. L. 2000, *ApJ*, 530, 939
Cyburt, R. H., Fields, B. D., & Olive, K. A. 2008, *J. Cosmol. Astropart. Phys.*, 11, 12
Duncan, D. K. *et al.* 1997, *ApJ*, 488, 338
Dunkley, J. *et al.* 2009, *ApJS*, 180, 306
Evoli, C., Salvadori, S., & Ferrara, A. 2008, *MNRAS*, 390, L14
Garcia Lopez, R. J. *et al.* 1998, *ApJ*, 500, 241
García Pérez, A. E. *et al.* 2009, *A&A*, 504, 213
Ito, H., Aoki, W., Honda, S., & Beers, T. C. 2009, *ApJ* (Letters), 698, L37
Kamimura, M., Kino, Y., & Hiyama, E. 2009, *Progress of Theoretical Physics*, 121, 1059
Kang, J., Luty, M. A., & Nasri, S. 2008, *J. High Energy Phys.*, 9, 86
Kawano, L. 1992, *NASA STI/Recon Technical Report N*, 92, 25163
Korn, A. J. *et al.* 2006, *Nature*, 442, 657; 2007, *ApJ*, 671, 402
Kusakabe, M., Kajino, T., & Mathews, G. J. 2006, *Phys. Rev. D*, 74, 023526
Kusakabe, M. *et al.* 2007, *Phys. Rev. D*, 76, 121302
Kusakabe, M. *et al.* 2008, *ApJ*, 680, 846
Kusakabe, M. 2008, *ApJ*, 681, 18
Kusakabe, M. *et al.* 2009a, *Phys. Rev. D*, 79, 123513
Kusakabe, M., Kajino, T., Yoshida, T., & Mathews, G. J. 2009b, *Phys. Rev. D*, 80, 103501
Lind, K. *et al.* 2009, *A&A*, 503, 545
Meléndez, J. & Ramírez, I. 2004, *ApJ* (Letters), 615, L33
Pospelov, M. 2007, *Phys. Rev. Lett.*, 98, 231301
Prantzos, N. 2006, *A&A*, 448, 665
Primas, F., Duncan, D. K., Peterson, R. C., & Thorburn, J. A. 1999, *A&A*, 343, 545
Primas, F., Molaro, P., Bonifacio, P., & Hill, V. 2000, *A&A*, 362, 666
Primas, F., Asplund, M., Nissen, P. E., & Hill, V. 2000, *A&A*, 364, L42
Ramaty, R., Kozlovsky, B., Lingenfelter, R. E., & Reeves, H. 1997, *ApJ*, 488, 730
Rich, J. A. & Boesgaard, A. M. 2009, *ApJ*, 701, 1519
Richard, O., Michaud, G., & Richer, J. 2005, *ApJ*, 619, 538
Rollinde, E., Vangioni, E., & Olive, K. 2005, *ApJ*, 627, 666; 2006, *ApJ*, 651, 658
Rollinde, E. *et al.* 2008, *ApJ*, 673, 676
Ryan, S. G. *et al.* 2000, *ApJ* (Letters), 530, L57
Shi, J. R. *et al.* 2007, *A&A*, 465, 587
Smiljanic, R. *et al.* 2009, *A&A*, 499, 103
Spite, F. & Spite, M. 1982, *A&A*, 115, 357
Steffen, M. *et al.* 2009, arXiv:0910.5917
Tan, K. F., Shi, J. R., & Zhao, G. 2009, *MNRAS*, 392, 205
Woosley, S. E. & Weaver, T. A. 1995, *ApJS*, 101, 181
Yoshida, T., Kajino, T., & Hartmann, D. H. 2005, *Phys. Rev. Lett.*, 94, 231101

Light Elements in the Universe
Proceedings IAU Symposium No. 268, 2009
C. Charbonnel, M. Tosi, F. Primas & C. Chiappini, eds.

© International Astronomical Union 2010
doi:10.1017/S1743921310003844

Primordial nucleosynthesis in higher dimensional cosmology

S. Chatterjee

Relativity and Cosmology Research Centre, Jadavpur University, Kolkata - 700032, India, and also at IGNOU, New Alipore College, Kolkata 700053 email: chat_ sujit1@yahoo.com

Abstract. We investigate nucleosynthesis and element formation in the early universe in the framework of higher dimensional cosmology. We find that temperature decays less rapidly in higher dimensional cosmology, which we believe may have nontrivial consequences *vis-a-vis* primordial physics.

Keywords. Cosmology: theory, cosmological parameters, early universe

1. The element formation

The work is primarily motivated by the consideration that both light element formation and higher dimensional cosmology are particularly relevant in the context of early universe.

From Einstein's field equations generalised to $(n+2)$dim. (Chatterjee & Bhui 1990) we get the following equations as

$$\frac{n(n+1)}{2}\frac{\dot{R}^2+k}{R^2} = \rho \tag{1.1}$$

$$-\frac{n\ddot{R}}{R} - \frac{n(n-1)}{2}\frac{\dot{R}^2+k}{R^2} = p \tag{1.2}$$

where ρ and p are the homogeneous mass density and pressure respectively. In the early universe one takes the radiation dominated case as our equation of state such that for a (n+2) dim. universe, $p = \frac{\rho}{n-1}$. One of the phenomenal successes favoring big bang cosmology is the almost correct prediction of the primeval nucleosynthesis, particularly the observed abundances of the light nuclei in the current universe. Now it can be shown via the field equations(1) and (2) that for the radiation dominated case we get

$$\rho = \frac{2n(n+1)}{(n+2)^2}\left(\frac{1}{t^2}\right) \tag{1.3}$$

Assuming absence of any dissipative mechanisms (for example, viscosity, friction etc.) and also that the laws of thermodynamics are valid in the early universe also with such huge temperature one gets from elementary thermodynamical considerations (Alvarez & Gavela 1983) that

$$\rho = \sigma T_{rad}^{n+2} \tag{1.4}$$

where σ is the higher dimensional Steffan's constant, hence it follows that

$$T_{rad} = \frac{2n(n+1)}{(n+2)^2\sigma}t^{-\frac{2}{n+2}} \tag{1.5}$$

where T_{rad} is the temperature of radiation and 't' is the age of the universe. For the usual

40   S. Chatterjee

4D spacetime it reduces to the well known relation $T_{kelvin} = 1.52 \times 10^{10} t^{-1/2}$ s. Equation (5) is the key equation for our attempt to investigate the effect of extra dimensions on the process of nucleosynthesis in early universe. We here study the situation when elementary particles have already materialized allowing us to take the low temperature approximation, $T << m_\mu c^2 = T_\mu$ for their distribution function. Here m_μ is the mass of a particular species. We here try to study the equilibrium condition for neutrinos with other species. As the reactions involving the neutrinos fall within the category of weak interactions and $T < T_\mu$ the cross section of a typical reaction is of the order of

$$A = f^2 h^{-4} (kT) \tag{1.6}$$

where f is the weak coupling constant. For simplicity it is further assumed that the constant A does not depend on the number of spatial dimensions. Moreover the number densities of the participating particles (say, muons) are, for (n+1) spatial dimensions, of the order of $(T/ch)^{n+1}$ and for reactions involving muons at low temperatures an exponential damping factor of $exp(-T/T_\mu)$ should also be considered. In the cosmological context one should also consider the rate of expansion of the background, which from equations (1) and (2) give

$$H^2 = \frac{\dot{R}^2}{R^2} \sim t^{-2} = T^{n+2} \tag{1.7}$$

Thus the ratio of the reaction rate to the expansion rate now becomes

$$\frac{Q}{H} \sim \left(\frac{T}{10^{10} K}\right)^{\frac{n+4}{2}} exp\left(\frac{10^{12} K}{T}\right) \tag{1.8}$$

One recovers the familiar 4D form when $n = 2$. As the temperature falls below the critical level of $10^{10} K$, the exponential decays rapidly. One can at this stage call attention to a significant quantitative difference from the 4D case. From equations (5) and (8) it is tempting to suggest that as the temperature falls less rapidly in higher dimensional cosmology than the analogous 4D situation it takes relatively more time for the elementary particles to cool below the threshold temperature. More importantly the quotient $\frac{Q}{H}$ is more sensitive to temperature fluctuations in multidimensional universe. Thus Q is larger than H for $T > 10^{12}$ depending on the number of extra dimensions and in these temperature ranges the neutrinos would be in thermal equilibrium with the rest of the other species. As the temperature falls further below $10^{10} K$ both the terms in the rhs of equation(8) drop rapidly, which means that the reactions involving neutrinos run at a slower rate than compared to the expansion of the universe. This triggers the so called decoupling of the neutrinos from the rest of the other constituents of matter and as pointed earlier in the higher dimensional universe it takes relatively more time for this decoupling phase of the neutrinos to occur. However, the theoretical and observational consequences of this supposed time lag for the initiation of the decoupling era need to be worked out in more detail before any definite inferences could be drawn.

2. Acknowledgments

The financial support of UGC, New Delhi in the form of a MRP award is acknowledged.

References

Chatterjee, S. & Bhui, B. 1990 *MNRAS*, 247, 57
Alvarez, E. & Gavela, M. 1983 *Phys. Rev. Lett.* 51, 931

Session II

Abundances of D, ^{3}He and ^{4}He: Observations

Paolo Ventura

Gary Ferland

Light Elements in the Universe
Proceedings IAU Symposium No. 268, 2009
C. Charbonnel, M. Tosi, F. Primas & C. Chiappini, eds.
© International Astronomical Union 2010
doi:10.1017/S1743921310003856

Measurements of deuterium in the Milky Way

Kenneth Sembach[1]

[1]Space Telescope Science Institute
3700 San Martin Dr., Baltimore, Maryland, USA
email: sembach@stsci.edu

Abstract. In this article I review measurements of deuterium in the Milky Way.

Keywords. ISM: abundances, Galaxy: abundances, Galaxy: solar neighborhood

1. Introduction

The abundance of deuterium in galactic environments depends upon both the primordial abundance created during Big Bang nucleosynthesis and the subsequent chemical evolution of the gas as a function of time. The primordial abundance is inferred from measurements of the angular power spectrum of the cosmic microwave background with the *Wilkinson Microwave Anisotropy Probe* (Spergel *et al.* 2007; Dunkley *et al.* 2009) and from ground-based measurements of D I absorption in a small number of quasar absorption systems at redshifts $z > 2$ (see Kirkman *et al.* 2003; O'Meara *et al.* 2006; Pettini *et al.* 2008; and references therein). A review of the extragalactic measurements and theory is given by Steigman (2007), who adopts $(D/H)_p = (2.68^{+0.27}_{-0.25}) \times 10^{-5}$.

The deuterium nucleus has a binding energy of only ~ 2.2 MeV, and thus is easily destroyed in the following step of the proton-proton chain reaction occurring in the interiors of stars:

$$^{2}_{1}D + p \rightarrow {}^{3}_{2}He + \gamma + 5.49 \text{ MeV}$$

This astration of deuterium leads to a net decrease in its abundance with time. Galactic chemical evolution models typically predict astration factors $f_D \sim 1.4 - 1.8$, leading to expected present-day values of $D/H \sim (1.4 - 1.9) \times 10^{-5}$ (see Tosi *et al.* 1998; Steigman, Romano, & Tosi 2007; and references therein). These models incorporate estimates of the production of heavier elements (C, N, O, Fe, etc.), as well as varying degrees of interstellar mixing and infall of unprocessed material from outside the Galaxy. Most models assume the interstellar material is chemically homogeneous, a reasonable assumption for regions of 100-200 parsecs in size but perhaps less certain for larger volumes (see the discussion in Moos *et al.* 2002). In conjunction with such models, measurements of deuterium in the Milky Way can provide valuable insights into the efficiency of mixing and the role of infalling gas in the evolving interstellar medium. They can also provide some guidance on the possible explanations for the large scatter in D/H values observed at higher redshifts.

There are two primary means by which the deuterium abundance in the neutral interstellar medium can be estimated. The most common is measurement of the D I Lyman series absorption in the ultraviolet spectra of stars, which began with pioneering observations by the *Copernicus* satellite in the 1970s (Rogerson & York 1973; York & Rogerson 1976; Vidal-Madjar *et al.* 1977), followed by extensive work with the *Hubble Space Telescope (HST)* and the *Far Ultraviolet Spectroscopic Explorer (FUSE)* (see Linsky *et al.* 2006 for a relatively recent review). Observations with these latter facilities have been particularly important in making precise measurements for nearby gas, increasing the number of sight lines examined, and extending the path length over which the absorption is probed. D I Lyman series observations were a key science driver for the design and operation of the *FUSE* mission (Moos *et al.* 2002). Although limited in number, important observations have also been made with experiments on the Shuttle-based *ASTRO-SPAS* platform; these include high resolution measurements with the Princeton *IMAPS* experiment (Jenkins *et al.* 1999; Sonneborn *et al.* 2000) and echelle spectroscopy with the *ORFEUS* telescope (Bluhm *et al.* 1999).

The other primary means for estimating the deuterium abundance is measuring the emission of the D I hyperfine transition at 91.6 cm (327 MHz). Work to detect this radio line began with Weinreb (1962) and has continued sporadically to the present day (Anantharamaiah & Radhakrishnan 1979; Blitz & Heiles 1987; Lubowich, Anantharamaiah, & Pasachoff 1989; Heiles, McCullough, & Glassgold 1993; Chengalur, Braun, & Burton 1997), albeit with limited success until recently (Rogers *et al.* 2005, 2007). The main limitation here is one of sensitivity, and as these latter two studies have shown this can be overcome with appropriate instrumentation tailored to detect the weak D I signal. Sadly, Tom Bania reported at this conference that the antenna array used for these observations at Haystack Observatory has been moth-balled and is unlikely to be resurrected anytime soon.

A secondary means for estimating the deuterium abundance is radio emission observations of deuterated molecules in the molecular component of the interstellar medium. However, even simple deuterium chemistry networks in molecular clouds become complicated quickly (Watson 1976; Dalgarno & Lepp 1984). Chemical fractionation makes it very difficult to determine D/H from molecular observations in most dark cloud environments because of the large number of molecular species and transitions that must be observed to account for each element (see, for example, Parise *et al.* 2009). This is a topic worthy of review, but too extensive to do so here.

Work on determining D/H in the Milky Way has slowed in the past few years since *FUSE* is no longer operating, but the debate over what the results mean continues. Figure 1 shows the value of D/H measured for sight lines in the local interstellar medium, the solar neighborhood, and the Galactic disk. These regions are delineated with dashed vertical lines. As can be seen in this figure, there is considerable variation in the D/H ratio from sight line to sight line, particularly at higher column densities. The measurements shown are estimates of the D/H ratio in the gas and do not account for any deuterium that may be locked into dust grains, a topic to which I will return briefly later. In the sections that follow, I summarize the measurements of atomic deuterium in the Milky Way that have been made to date and discuss some of their implications. I refer the reader to other articles in this proceedings by Linsky, Hébrard, Pradanović, and others for specific details on some of the topics discussed below.

2. The Local Bubble $[d \lesssim 100 \; pc]$

The local interstellar medium within about 100 pc of the Sun is often referred to as the Local Bubble. Within this cavity the properties of the interstellar medium are relatively homogeneous. The boundary of the Local Bubble appears to occur at hydrogen column densities of log N(H) $\sim 19.2 - 19.3$ (Sfeir *et al.* 1999). Within the Local Bubble D/H $\approx (1.5 - 1.6) \times 10^{-5}$ (Wood *et al.* 2004; Linsky *et al.* 2006), with some variation seen. Most of the Local Bubble measurements shown in Figure 1 are for sight lines to nearby, hot white dwarf stars with strong UV continua (e.g., Hébrard *et al.* 2002; Kruk *et al.* 2002; Lehner *et al.* 2002) or for later-type stars for which it is possible to measure

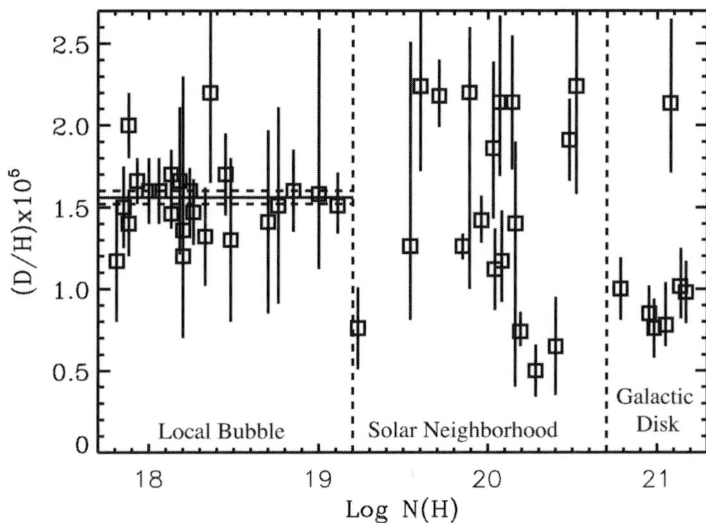

Figure 1. D/H as a function of hydrogen column density for sight lines observed with *Copernicus*, *IMAPS*, *HST*, and *FUSE* (adapted from Linsky *et al.* 2006 and Oliveira & Hébrard 2006). The vertical dashed lines denote the approximate column densities [N(H) = N(H I)+2N(H$_2$)] distinguishing the Local Bubble and Solar Neighborhood from more distant gas.

the H I and D I absorption against chromospheric emission lines (see, e.g., Linsky *et al.* 1995).

Vidal-Madjar *et al.* (1998) first suggested the possible detection of spatial variations in the D/H ratio within the Local Bubble. For many years debate raged over whether such variations were real or simply due to measurement error. Today, it seems reasonable to conclude that there are sight line to sight line variations of D/H in the Local Bubble, but for the most part such variations are small, especially when compared to those of more extended sight lines (see Figure 1). At this conference, Hébrard raised the interesting point that both the D/O and O/H ratios for Local Bubble sight lines tend to show less scatter than D/H, suggesting that there may be some systematic effects that could give rise to the (albeit small) variations seen in D/H.

3. The solar neighborhood [$100 \lesssim d(pc) \lesssim 500$]

There are now approximately 20 estimates of D/H for sight lines extending beyond the Local Bubble out to a few hundred parsecs. Most of the background sources are either white dwarf stars near the edge of the Local Bubble (e.g., Sonneborn *et al.* 2002), sub dwarfs (Friedman *et al.* 2002), or more distant OB stars (e.g., Jenkins *et al.* 1999; Sonneborn *et al.* 2000). These sight lines show a much stronger variation in D/H than the nearby stars, with roughly half showing larger values and half showing lower values than the Local Bubble average. There are no obvious correlations of D/H with Galactic longitude or latitude, nor with any known large interstellar structures. It is interesting that the mean value of D/H = 1.5×10^{-5} is indistinguishable from that of the Local Bubble.

On these scales, one would expect some variations around the Local Bubble value as a result of different gas mixing histories and deuterium astration. Several of these sight lines pass into known star-forming regions for which the chemical history of the gas is likely to be different than for the intervening interstellar medium along the sight lines. However, the magnitude of the variations seen, a factor of 2.5–3 for some sight line combinations, has led several authors to suggest that they might be due to deuteration of polycyclic aromatic hydrocarbons (PAHs $\rightarrow$ PADs; see Draine 2006; Linsky *et al.* 2006). Since astration and differential depletion into grains both serve to lower the gas-phase abundance of deuterium, this suggestion raises the obvious question: Why are some values of D/H so much higher than the Local Bubble values?

4. The Galactic disk [$d \gtrsim 500 \ pc$]

Deuterium absorption has been measured in the Galactic disk beyond the solar neighborhood along only seven sight lines for which the hydrogen column density exceeds log N = 20.7. Of these, only four (HD 41161, HD 53975, HD 90087, HD 191877) extend beyond 1 kpc (Hoopes *et al.* 2003; Hébrard *et al.* 2005; Oliveira & Hébrard 2006), and all lie within 3 kpc of the Sun. Several hundred extended sight lines have been observed with *FUSE* for a variety of purposes (see, e.g., Bowen *et al.* 2008), but few have proven useful for studies of D/H. The complex velocity structure arising from the superposition of gas clouds makes it difficult to disentangle the D I absorption from the very strong H I Lyman series absorption. Contamination of the D I lines by molecular hydrogen absorption is also problematic in many cases. We are fortunate to have seven cases for which N(D I) can be measured, but it is important to note that both the D I and H I measures are averages integrated over each sight line. The modest spectral resolution of *FUSE* and the great breadth of the H I absorption do not allow component by component analyses.

The D/H ratios for 6 of the 7 sight lines in this region of Figure 1 lie well below the Local Bubble average and are quite consistent with the average of the low D/H subset for the solar neighborhood. The single high D/H value (HD 41161) is similarly offset from the Local Bubble average and is consistent with the average of the high D/H subset for the solar neighborhood. This suggests that whatever processes serve to introduce variations in the solar neighborhood sample likely operate at larger distances as well. To first order, the variation in absorption beyond ~ 1 kpc does not seem dramatically larger (or smaller) than in the $100 \lesssim d(pc) \lesssim 500$ gas. Unfortunately, it isn't possible with the present data to separate out the foreground solar neighborhood contributions from the more distant cloud contributions for these sight lines.

A separate estimate for the D/H ratio in the Galactic plane has been made by Rogers *et al.* (2005, 2007), who report measurements of the 92 cm line emission in the Galactic plane at longitudes $l = 171°$, $183°$, and $195°$. Using an array of 24 small radio telescopes equipped with low-noise amplifiers to conduct these observations, they found values of D/H $= (2.4 \pm 0.3, 1.9 \pm 0.2,$ and $1.8 \pm 0.5) \times 10^{-5}$ in the three directions, with a final estimate of $(2.1 \pm 0.7) \times 10^{-5}$ (3σ) after accounting for uncertainties in the spin temperature of the gas. This value is very similar to that for the average of the high solar neighborhood and Galactic disk values inferred from the absorption-line measurements.

Rodgers *et al.* estimate that the deuterium emission is spread over approximately 5 kpc, which would make this the longest integrated path for which deuterium has been measured in the Galaxy. It may also well be the best average value for any region beyond the Local Bubble since the beam width of $14°$ samples a much larger volume of gas than the absorption-line observations. The fact that the value of D/H found is high does not rule out the possibility that some of the gas has low D/H, but it does suggest that the average value for the Galactic disk and low halo in the Galactic anti-center is higher than the local value and has a relatively low astration factor, $f_D \lesssim 1.3$. Similar measures for other directions, particularly those for extended directions observed by *FUSE* would be highly desirable but do not appear to be feasible at this time.

5. The Galactic halo $[|z| \gtrsim 200 \ pc]$

High latitude sight lines with measurable D I absorption have been observed, but most of these are very nearby and do not extend far enough to probe the warm neutral medium of the Galactic halo, which has a different history than warm gas in the Galactic disk. In particular, the neutral Galactic halo gas has higher gas-phase abundances of refractory elements, due in large part to grain destruction caused by the passage of shocks that transport material from the disk into the halo (Sembach & Savage 1996).

Savage *et al.* (2007) report the detection of D I in the Galactic halo gas toward the high latitude QSO HE 0226-4110 observed with *FUSE*. Absorption is present in several lines of the Lyman series. They find D/H $= (2.1 \pm^{0.8}_{0.6}) \times 10^{-5}$ for the warm neutral medium of the Galactic halo, after correcting for absorption by foreground gas in the Local Bubble. The quoted value for the halo clouds is a weighted average over all clouds more distant than the Local Bubble. In these clouds, the gas-phase oxygen abundance is approximately solar and the gas-phase iron abundance is 1/10 solar; these differential depletion results are typical of halo clouds and are less severe than the depletions seen for the average warm neutral medium in the Galactic disk (Savage & Sembach 1996).

The value of D/H toward HE 0226-4110 is consistent with the higher values of D/H found in the Galactic disk, again suggesting a low astration factor that pushes up against the limits of predictions of galactic chemical evolution models. The past dynamical history of the gas does not help alleviate this problem since the transport time for disk gas into the low halo ($|z| \lesssim 3$ kpc) is on the order of a few hundred Myr. The high value of D/H also leaves little, if any, room for depletion of deuterium into dust grains, which is consistent with the expectation that the grain mantles in such clouds should be highly processed. An extremely low abundance of molecular hydrogen along the sight line also implies that there is little molecular material in which to incorporate deuterium.

6. The high velocity cloud Complex C $[d = 10 \pm 2 \ kpc]$

Sembach *et al.* (2004) report the detection of D I absorption in high velocity cloud Complex C, a large parcel of gas in the northern Galactic sky at a distance of 10 ± 2 kpc and an altitude of ~ 8 kpc (Thom *et al.* 2008). Complex C has a metallicity of 0.1-0.25 solar, exhibits no evidence of dust, and has a low nitrogen abundance thereby implying that it is chemically young. It is composed of gas falling onto the Milky Way and is unlikely to be Galactic disk or halo gas ejected to large distances. The mass of Complex C is on the order of 10^7 M_o.

The velocity separation between Complex C and the Galaxy in the *FUSE* spectrum of QSO PG 1259+593 allows for an analysis of D/H in the high velocity gas. Sembach *et al.* (2004) find a value of D/H $= (2.2 \pm 0.7) \times 10^{-5}$, similar to that obtained for the warm halo clouds toward HE 0226-4110. Complex C contains metals and has therefore undergone some chemical evolution. Therefore, the similarity of its D/H value to some of the values found for the disk and halo of the Milky Way is somewhat perplexing since the oxygen abundances in Complex C and these other regions differ by roughly a factor of 5-10.

7. The outer planets

Several estimates of D/H have been made for Jupiter and Saturn, with the most reliable being those for Jupiter. There have been three different techniques used to measure D/H on Jupiter: weak optical H_2 and HD absorption lines (Trauger *et al.* 1973), in-situ *Galileo*

Table 1. D/H Summary

Site	D/H [ppm]	Possible Importance of Dust Depletion	No. of Meas.	Reference or Compilation
Primordial gas	$26.8 \pm^{2.7}_{2.5}$	Not important	...	Steigman 2007
IGM	28.5 ± 7.6 (SD)	Not important	7	Pettini *et al.* 2008
Complex C	22 ± 7 (1σ)	Not important	1	Sembach *et al.* 2004
Galactic halo	$22 \pm^{8}_{6}$ (1σ)	Not important	1	Savage *et al.* 2007
Galactic disk (92cm)	21.0 ± 2.3 (1σ)	Probably not important	3	Rodgers *et al.* 2007
Galactic disk (high)	$21.4 \pm^{5.1}_{4.3}$ (1σ)	Probably not important	1	Oliveira & Hébrard 2006
Galactic disk (low)	9.0 ± 1.2 (SD)	Perhaps important	6	Oliveira & Hébrard (2006)
Solar neighborhood (high)	21.1 ± 1.5 (SD)	Probably not important	8	Oliveira *et al.* 2006 Oliveira & Hébrard 2006
Solar neighborhood (low)	10.3 ± 3.3 (SD)	Perhaps important	10	Oliveira *et al.* 2006 Oliveira & Hébrard 2006
Local Bubble	15.8 ± 2.1 (SD)	Perhaps important	22	Linsky *et al.* 2006
Jupiter	23.0 ± 2.6 (1σ)	Unknown	3	see Section 7

Note: This table is an update of a similar table that first appeared in Savage *et al.* (2007). "SD" indicates that the simple standard deviation of the mean was adopted in computing the error.

mass spectrometer measurements (Mahaffy *et al.* 1998), and *Infrared Space Observatory* observations of H_2 and HD emission (Lellouch *et al.* 2001). Following Savage *et al.* (2007), I adopt here an error-weighted average of these values: D/H $= (2.30 \pm 0.26) \times 10^{-5}$.

8. D/H summary

Table 1 and Figure 2 summarize the results for D/H in various Galactic environments, together with the values available for QSO absorption-line systems at high redshift. Data for each of these points comes from the references cited in Table 1 and references therein. For regions for which multiple measurements exist, the values shown are straight averages of the measurements, and the error quoted is the standard deviation about the mean value. Otherwise, the 1σ error on the mean value is quoted from the original source.

9. Variations in D/H and the role of dust

Variations in D/H and D/O have been discussed extensively elsewhere (see Linsky *et al.* 2006; Oliveira & Hébrard 2006; Oliveira *et al.* 2006), so I will not delve into the details here. Suffice it to say that variations in D/H, O/H, and D/O exist to differing degrees for the sight lines measured, and that one possible explanation for some of these variations is extreme deuteration of hydrocarbon molecule chains (Draine 2006). It is tempting to jump on the "dust bandwagon" to explain all of these variations, but it seems equally plausible that differing levels of astration and incomplete mixing play important roles as well. It is probably fair to say that what we presently (don't) know about these sight lines leaves room for all of these possibilities. Indeed, in the one region we expect to have a reasonably uniform chemical history and be well mixed, the Local Bubble, the variations in D/H and D/O are small.

Table 1 contains a subjective assessment of whether or not depletion of deuterium onto dust is likely to be important for the regions represented. Some regions with high D/H values, such as the QSO absorbers and Complex C, should contain very little dust and are unlikely to harbor PAHs or PADs. But what about the regions with lower D/H

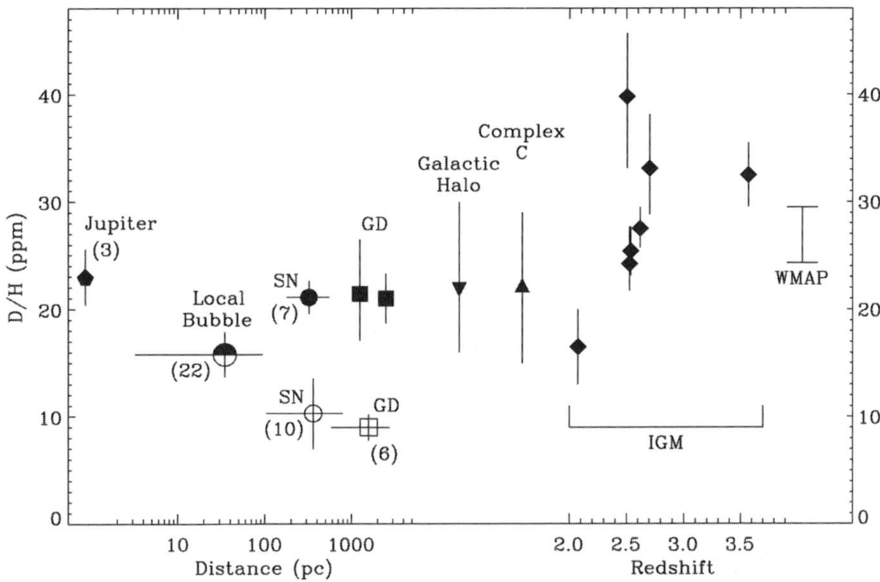

Figure 2. D/H measurements for a variety of environments. Values for the data points shown are given in Table 1, with symbol coding corresponding to the types of regions considered ("SN" = solar neighborhood. "GD" = Galactic disk). Numbers in parentheses indicate the number of measurements included for each data point if greater than one. Filled symbols indicate those measurements for which depletion of deuterium into dust is unlikely to be important.

values? Oliveira *et al.* (2006) find an anti-correlation between D/H and n(H), the sight line averaged density of H, as one typically sees for elements incorporated into dust (see Jenkins, Savage, & Spitzer 1986), but this result depends strongly on the measurements for a few sight lines.

It has been proposed that a positive correlation of D/H with the gas-phase abundances of highly refractory elements would indicate that deuterium is subject to inclusion in dust grains, as the gas-phase abundance of the refractories is highly dependent on the presence and processing of dust. Higher observed abundances of elements such as Si, Fe, and Ti would imply less dust and should be associated with higher values of D/H. Similarly, the more refractory elements are depleted from the gas phase, the better the chance that deuterium is also depleted. Linsky *et al.* (2006) checked this hypothesis using measures of Fe II and Si II along 38 of the Galactic disk sight lines with D/H measures (see their Figures 3 and 4) and found a positive correlation, as expected, with considerable scatter. The correlation between $(Si/H)_{gas}$ and D/H is strongly dependent on a few data points, but the correlation with $(Fe/H)_{gas}$ appears to be more secure.

Follow-up observations of the abundance of Ti from the ground have provided preliminary evidence for a correlation with D/H. Early results for 7 sight lines by Prochaska, Tripp, & Howk (2005) showed a strong correlation between $(Ti/H)_{gas}$ and D/H, providing further support to the idea that the scatter in D/H is linked to differential depletion. More extensive studies (Ellison, Prochaska, & Lopez 2007; Lallement, Hébrard, & Welsh 2008) confirm a correlation, but indicate that it is weaker than originally found and that the gradient observed for Ti/D is less than for Fe/D, contrary to expectations. It is possible that the results are subject to sight line selection effects (see below).

10. Some things to consider

It is tempting to take the existing deuterium results and generalize them to the Galaxy as a whole. However, that would surely be a dangerous thing to do, as the existing samples are subject to a variety of selection effects. The majority of deuterium measurements are obtained through observations of D I Ly-series absorption at far-ultraviolet wavelengths. Some possible caveats and concerns to keep in mind when considering the broader applicability of these results include:

(a) *Volume probed*: The volume of the Galaxy probed by the absorption-line measurements is very small, even with *FUSE*. Most of the sight lines are confined to within a few hundred parsecs of the Sun, a region which is not necessarily typical of the rest of the Galaxy. Only a few sight lines extend outside the local spiral arm, and only 1 or two sight lines extend as far as the nearest (Sagittarius) spiral arm.

(b) *Sight line conditions*: The types of regions probed are *not* conducive to the formation of molecules or dust grains. The sight lines observed have average densities typical of very diffuse clouds and were chosen for their simple velocity structure. The molecular content of the sight lines is orders of magnitude smaller than the atomic content. As a result, there is a strong selection against dark cloud environments where grain growth and molecular chemistry occurs.

(c) *Biased line strength*: There is probably a bias to report results for which the D I absorption is strong enough to detect, but not so strong as to be overwhelmed by H I absorption in close spectral proximity. This may limit the range of D/H probed, and sight line to sight line variations larger than those reported may exist. I am unaware of any systematic search through the existing archives for sight lines for which D I absorption should have, and could have, been seen but isn't.

(d) *Line saturation and unresolved velocity structure*: At the velocity resolution of *FUSE*, about 20 km s^{-1}, it is not possible to distinguish one absorbing region from another along most sight lines. Therefore, the values of N(D I) and N(O I) are necessarily averages for the sight line, weighted by the gas content of the individual clouds encountered. Similarly, only the total sight line column density of H I can be measured from the Ly-series lines. Narrow velocity structure resulting in line saturation could depress the values of N(D I) and N(O I) below their true values, even for cases for which multiple transitions can be observed. This effect can be quantified but depends strongly on the actual velocity structure encountered.

(e) *Not all grains are the same*: PADs and "normal" silicate or carbonaceous grains may have very different histories. Comparisons of D/H or D/O to the gas-phase abundances of refractory elements to infer the depletion of D are most meaningful if the deuterium and refractories are incorporated into the same grains. The relatively small variation in refractory element abundance with changes in D/H is somewhat surprising. Take Ti as an example. For diffuse clouds like those observed toward ζ Oph, this element has 999 of 1000 atoms locked into dust grains, with only 0.1% of the atoms in the gas phase (see Savage & Sembach 1996). To change the gas-phase depletion of Ti by two requires only 1 more of those 1000 atoms to be liberated into the gas, whereas for deuterium which has a depletion no more than a factor of a few, roughly half of the D atoms would have to be liberated to change the gas-phase abundance of D by a factor of two. Why isn't there more variation in the abundances of the refractory elements along these sight lines if dust content and the processing of dust are responsible for the variations in D/H? I suspect it is because the grains are already highly processed, and I remain somewhat skeptical that the observed variations are due mainly to differential depletion.

(*f*) *Local sources of deuterium*: All methods of deuterium production require either extreme environments or special conditions, with Big Bang nucleosynthesis being the only viable source for cosmological quantities (see Epstein, Lattimer, & Schramm 1976; Jedamzik 2002). However, localized sources could potentially account for some variability of D/H within the Galaxy. Possible sites of production include stellar flares (Mullan & Linsky 1999; Prodanović & Fields 2003) or supernova shock waves (see above references). Unfortunately, independent confirmation of deuterium production in such sources is notoriously difficult to come by, so for now it is difficult to quantify their importance.

(*g*) *Unknown unknowns*: There may be unknown systematic measurement effects leading to the apparent bifurcation of D/H into high and low branches beyond the Local Bubble. The absence of many points near the mean value for the Local Bubble, which is also the mean for all of the solar neighborhood points, is surely telling us something important about either the nature of the absorption or our ability to measure it.

11. Concluding remarks

Over the years, the focus of D/H determinations has evolved from trying to determine the primordial abundance of deuterium to trying to understand chemical evolution and the detailed physics of the sight lines along which deuterium can be measured. Despite their limitations, observations of deuterium in the Galaxy have proven to be extremely interesting and thought provoking. The lively discussion at this conference was direct proof of that!

The prospects for obtaining further D I absorption-line measurements within the Milky Way in the near future are not very bright. However, with the installation of the Cosmic Origins Spectrograph in *HST* in May 2009, there is again a chance that it will be possible to measure D/H in the low-redshift intergalactic medium along one or more of the many QSO sight lines that will be observed in the coming years. The key to doing this will be to find a strong enough Lyman-limit system at $z \lesssim 0.5$ in which to measure the D I Lyα absorption. The COS Team is also investigating whether it might be possible to use COS to observe at wavelengths below 1100 Å. Initial results using the low resolution grating (G140L) demonstrate that the *HST* optics transmit light at these wavelengths (McCandliss *et al.* 2010). Further tests using the G130M medium resolution grating, which could potentially reopen the possibility of measuring deuterium absorption in the Galaxy, may be undertaken.

I thank the organizers of this conference for the invitation to participate in this conference and to share my thoughts in this article. I acknowledge travel support from grant COS GTO grant NNX08AC14G.

References

Anantharamaiah, K. R. & Radhakrishnan, V. 1979, *A&A*, 79, L9
Blitz, L. & Heiles, C. 1987, *ApJ*, 313, L95
Bluhm, H., Marggraf, O., de Boer, K. S., Richter, P., & Heber, U. 1999, *A&A*, 352, 287
Bowen, D. V. *et al.* 2008, *ApJS*, 176, 59
Chengalur, J. N., Braun, R., & Burton, W. B. 1997, *A&A*, 318, L35
Dalgarno, A. & Lepp, S. 1984, *ApJ*, 287, L47
Draine, B. T. 2006, in *Astrophysics in the Far Ultraviolet: Five Years of Discovery with FUSE*, ASP Conf. Ser. 48, eds., G. Sonneborn, H. W. Moos, & B.-G. Andersson (San Francisco: ASP), 58
Dunkley, J. *et al.* 2009, *ApJS*, 180, 360
Ellison, S. L., Prochaska, J. X., & Lopez, S. 2007, *MNRAS*, 380, 1245

Epstein, R. I., Lattimer, J. M., & Schramm, D. N. 1976, *Nature*, 263, 198

Friedman, S. D. *et al.* 2002, *ApJS*, 140, 37

Hébrard *et al.* 2002, *ApJS*, 140, 103

Hébrard *et al.* 2005, *ApJ*, 635, 1136

Heiles, C., McCullough, P. R., & Glassgold, A. E. 1993, *ApJS*, 89, 271

Hoopes, C. G., Sembach, K. R., Hébrard, G., Moos, H. W., & Knauth, D. C. 2003, *ApJ*, 586, 1094

Jedamzik, K. 2002, *Planetary & Space Sci.*, 50, 1239

Jenkins, E. B., Savage, B. D., & Spitzer 1986, *ApJ*, 301, 355

Jenkins, E. B., Tripp, T. M., Woźniak, P. R., Sofia, U. J., & Sonneborn, G. 1999, *ApJ*, 520, 182

Kirkman, D., Tytler, D., Suzuki, N., O'Meara, J., & Lubin, D. 2003, *ApJS*, 149, 1

Kruk, J. W. *et al.* 2002, *ApJS*, 140, 19

Lallement, R., Hébrard, G., & Welsh, B. Y. 2008, *A&A*, 481, 381

Lehner, N. *et al.* 2002 *ApJS*, 140, 81

Linsky, J. L. *et al.* 1995, *ApJ*, 451, 335

Linsky, J. L. *et al.* 2006, *ApJ*, 647, 1106

Lubowich, D. A., Anantharamaiah, K. R., & Pasachoff, J. M. 1989, *ApJ*, 345, 770

Lellouch, E. *et al.* 2001, *A&A*, 370, 610

Mahaffy, P. R. *et al.* 1998, *Space Sci. Rev.*, 84, 251

McCandliss, S. R. *et al.* 2010, *ApJ*, 709, L183

Moos, H. W. *et al.* 2002, *ApJS*, 140, 3

Mullan, D. J. & Linsky, J. L. 1999, *ApJ*, 511, 502

Oliveira, C. & Hébrard, G. 2006, *ApJ*, 653, 345

O'Meara, J. *et al.* 2006, *ApJ*, 649, L61

Parise, B., Leurini, S., Schilke, P., Roueff, E., Thorwirth, S., & Lis, D. C. 2009, *A&A*, 508, 737

Pettini, M., Zych, B. J., Murphy, M. T., Lewis, A., & Steidel, C. C. 2008, *MNRAS*, 391, 1499

Prochaska, J. X., Tripp, T. M., & Howk, J. C. 2005, *ApJ*, 620, L39

Prodanović, T. & Fields, B. D. 2003, *ApJ*, 597, 48

Rogers, A. E. E. *et al.* 2005, *ApJ*, 630, L41

Rogers, A. E. E., Dudevoir, K. A., & Bania, T. M. 2007, *AJ*, 133, 1625

Rogerson, J. B. & York, D. G. 1973, *ApJ*, 186, L95

Savage, B. D., Lehner, N., Fox, A., Wakker, B., & Sembach, K. R. 2007, *ApJ*, 659, 1222

Savage, B. D. & Sembach, K. R. 1996, *ARA&A*, 34, 279

Sembach, K. R. & Savage, B. D. 1996, *ApJ*, 457, 211

Sembach, K. R. *et al.* 2004, *ApJS*, 150, 387

Sfeir, D. M., Lallement, R., Crifo, F., & Welsh, B. Y. 1999, *A&A*, 346, 785

Sonneborn, G. *et al.* 2000, *ApJ*, 545, 277

Sonneborn, G. *et al.* 2002, *ApJS*, 140, 51

Spergel, D. N. *et al.* 2007, *ApJS*, 170, 377

Steigman, G. 2007, *Ann. Rev. Nucl. Part. Sci.*, 57, 463

Steigman, G., Romano, D., & Tosi, M. 2007, *MNRAS*, 378, 576

Thom, C. *et al.* 2008, *ApJ*, 684, 364

Tosi, M., Steigman, G., Matteucci, F., & Chiappini, C. 1998, *ApJ*, 498, 226

Trauger, J. T., Roessler, F. L., Carelton, N. P., & Traub, W. A. 1973, *ApJ*, 184, L137

Vidal-Madjar, A., Laurent, C., Bonnet, R. M., & York, D. G. 1977, *ApJ*, 211, 91

Vidal-Madjar, A. *et al.* 1998, *A&A*, 338, 694

Watson, W. D. 1976, *Rev. Mod. Phys.*, 48, 513

Weinreb, S. 1962, *Nature*, 195, 367

Wood, B. E. *et al.* 2004, *ApJ*, 609, 838

York, D. G. & Rogerson, J. B. 1976, *ApJ*, 203, 378

Light Elements in the Universe
Proceedings IAU Symposium No. 268, 2009
C. Charbonnel, M. Tosi, F. Primas & C. Chiappini, eds.
© International Astronomical Union 2010
doi:10.1017/S1743921310003868

The total deuterium abundance in the local Galactic disk: decisions and implications

Jeffrey L. Linsky[1]

[1]JILA, University of Colorado and NIST,
Campus Box 440, Boulder, CO 80309-0440, USA
email: jlinsky@jilau1.colorado.edu

Abstract. Analyses of FUSE spacecraft spectra have provided measurements of D/H in the gas phase of the interstellar medium for many lines of sight extending to several kpc from the Sun. These measurements, together with the earlier Copernicus, HST, and IMAPS data, show a wide range of D/H values that have challenged both observers and chemical evolution modellers. I believe that the best explanation for the diverse D/H measurements is that deuterium can be sequestered on to carbonaceous grains and PAH molecules and thereby removed from the interstellar gas. Grain destruction can raise the gas phase D/H value to approximately the total D/H value. Supernovae and stellar winds, however, can decrease the total D/H value along lines of sight on time scales less than mixing time scales. I will summarize the theoretical and observational arguments for this model and estimate the most likely range for the total D/H in the local Galactic disk. This range in total D/H presents a constraint on realistic Galactic chemical evolution models or the primordial value of D/H or both.

Keywords. ISM: abundances; cosmology: cosmological parameters; ultraviolet: ISM

---

1. Introduction

It has long been recognized that accurate measurements of the primordial abundance of deuterium provides the best constraint on the properties of the early universe (100–1000 seconds after the Big Bang) and, in particular, the ratio of baryons to photons (η_{10}). Deuterium plays this critical role because it is an ideal relic; since its initial creation there has been no known process for creation of a significant amount of deuterium, but only destruction by nuclear reactions in stars. Analysis of quasar absorption lines (QAL) in seven lines of sight led Pettini *et al.* (2008) to infer $(D/H)_{QAL} = 28.2^{+2.0}_{-1.8}$ parts per million (ppm). Analysis of the cosmic microwave background signal leads to a measurement of η_{10} and thus inference of the primordial D/H ratio. In the most recent analysis of the 5-year WMAP data set, Dunkley *et al.* (2009) obtain the parameter $100\,\Omega_{b,0} = 2.273 \pm 0.062$, which corresponds to $\eta_{10} = 6.225 \pm 0.0170$ and $(D/H)_{WMAP} = 25.2 \pm 1.1$. While the two values for the primordial or near-primordial D/H ratio, $(D/H)_{QAL}$ and $(D/H)_{WMAP}$, are consistent, they are sufficiently different and likely to be subject to different systematic errors that they should not be averaged to obtain a more "robust" value of $(D/H)_{prim}$.

Since the first measurements of the interstellar deuterium Lyman lines by the Copernicus satellite, there have been many attempts to measure accurate deuterium abundances in the interstellar medium of our Galaxy. These have included measurements of the deuterium and hydrogen Lyman-α line with IUE and HST, as well as measurements of the higher Lyman lines with the Interstellar Medium Absorption Profile Spectrograph (IMAPS). More recently, the Far Ultraviolet Spectrograph Explorer (FUSE) has provided many new deuterium measurements along lines of sight extending to nearly 3pc.

Until recently, the factor of four range in measured D/H values for different lines of sight (Fig. 1) extending to stars beyond the Local Bubble posed a severe problem as no chemical evolution model of the Galaxy could explain this wide range. A solution to the problem proposed by Jura (1982) and developed by Draine (2003) and Draine (2006) is that deuterium is more tightly bound to carbonacious grains and PAH molecules than hydrogen because the C-D bond is slightly larger than the C-H bond. This is important because interstellar grains are cold ($T \sim 20$ K) and can thus sequester significant amounts of deuterium from the gas phase. Linsky *et al.* (2006) provided support for this explanation with three lines of evidence: (1) very high D/H ratios measured in interstellar carbonaceous dust grains embedded in interplanetary dust particles (a proof of concept), (2) correlation of high D/H values in interstellar gas with large depletions of refractory metals (Fig. 2), implying that deuterium and metals both deplete onto and evaporate from dust grains although the story is more complex, and (3) D/H abundances increase with interstellar gas temperature as inferred from excitation of H_2. Thus the total D/H ratio in the interstellar medium $(D/H)_{total}$ can be significantly larger than the gas phase measurements $(D/H)_{gas}$. Since even in recently shocked gas, some of the grains containing deuterium may still be present, one must conclude that $(D/H)_{total} \geqslant (D/H)_{gas}$. This statement is strengthened by recognizing that lines of sight likely include both recently shocked and unshocked gas containing grains with deuterium.

Many authors have now accepted the deuterium-depleted dust hypothesis as the explanation for the measured low values of $(D/H)_{gas}$. In this talk I would like to address the high values of $(D/H)_{gas}$ including recent measurements, infer a sensible value for

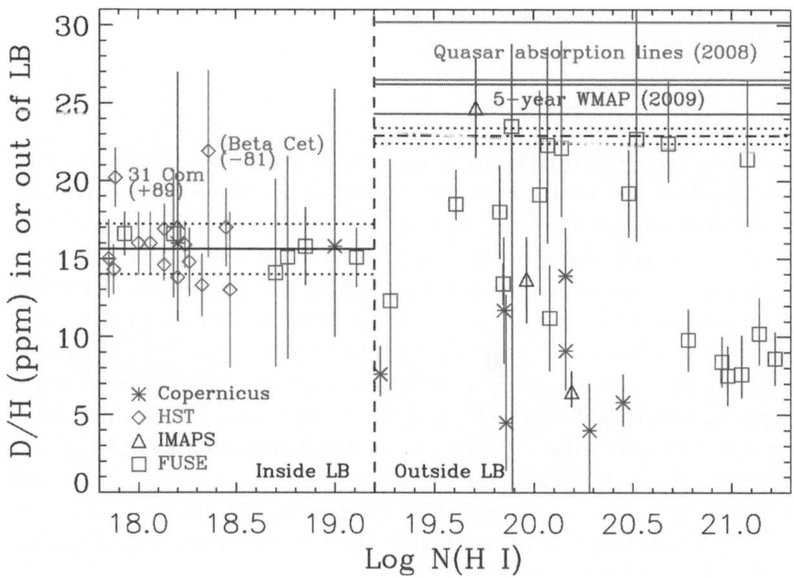

Figure 1. Measurements of the $(D/H)_{gas}$ for the 24 lines of sight located entirely inside of the Local Bubble (to the left of the vertical dashed line) and the gas located outside of the Local Bubble, $(D/H)_{gas-LB}$, (to the right of the vertical dashed line). The error bars are 1σ, and the horizontal axis is the hydrogen column density to the star at the end of the line of sight. The symbols are coded by the spacecraft that obtained the Lyman line spectra. The range of $(D/H)_{prim}$ values obtained from quasar absorption lines (Pettini *et al.* 2008) and from WMAP (Dunkley *et al.* 2009) are shown in the upper right. Just below is the range of the total deuterium abundance in the local Galactic disk, $(D/H)_{high} = 22.9 \pm 0.5$ ppm.

(D/H)$_{\rm total}$ in the Galactic disk, and relate this high value of (D/H)$_{\rm total}$ to the measurements of primordial D/H.

2. Analysis of the high values of (D/H)$_{\rm gas}$

Linsky *et al.* (2006) assembled all of the then-available published values of (D/H)$_{\rm gas}$ obtained with Copernicus, IUE, HST, IMAPS, and FUSE for lines of sight toward 47 stars. Since that paper was published, (D/H)$_{\rm gas}$ measurements for five additional lines of sight observed by FUSE have became available (Table 1) and are included in the following analysis. I consider only the direct measurements of N(HI) and N(DI) for which the hydrogen and deuterium components of the molecules H$_2$ and HD are added when measured. There are alternative methods for evaluating D/H, in particular from the product (D/O)x(O/H). As described by Hébrard (2010), this method minimizes errors in N(HI) due to the high opacity of the Lyman lines. However, this method introduces additional uncertainties associated with measurements of the oxygen lines, including depletion of oxygen on interstellar dust grains.

Linsky *et al.* (2006) separated the (D/H)$_{\rm gas}$ measurements into two groups depending upon whether the target star is located inside of or beyond the Local Bubble. The Local Bubble has been described as a superbubble of 10^6 K gas produced by recent supernovae in the Scorpio-Centaurus Association of young stars, although Welsh & Shelton (2009) argue that most of the volume of the gas is highly ionized but far cooler than 10^6 K. The N(HI) and N(DI) measurements toward stars inside of the Local Bubble refer to warm gas (4,000–10,000 K) located in clouds near the Sun (Redfield & Linsky 2008). The Local Bubble is mostly surrounded by cold gas detected in NaI beginning at about log N(HI) = 19.2, but there is no cold gas near the Galactic poles (Fig. 3) allowing the Local Bubble gas to interact with the halo. Most lines of sight to the 24 stars located inside of the Local Bubble show (D/H)$_{\rm gas}$ ratios consistent with the mean value of

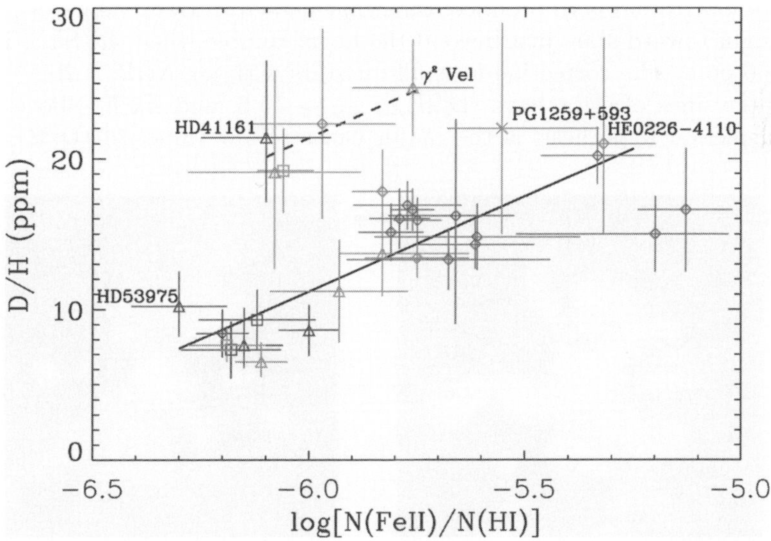

Figure 2. Plot of (D/H)$_{\rm gas}$ vs. depletion of iron for the lines of sight observed by HST, IMAPS, and FUSE. The coding symbols are the same as in Fig. 1. The solid line is the best fit to all of the data, and the dashed line is a displacement of this line upward by 8.5 ppm to be consistent with the five data points that lie well above the mean fit line. The five high points may be lines of sight where weak shocks have removed deuterium, but not iron, from the grains.

Table 1. Recent measurements of $(D/H)_{\mathrm{gas}}^1$.

Target	l	b	$d(pc)$	$\log \mathrm{N(HI)}^2$	$\log \mathrm{N(DI)}^3$	$(D/H)_{\mathrm{gas}}$	$(D/H)_{\mathrm{gas-LB}}$
REJ 1738+665	97	+32	243	19.83 ± 0.05	15.08 ± 0.04	$17.8^{+2.8}_{-2.5}$	18 ± 3
HD 41161	165	+13	1253	21.08 ± 0.08	16.41 ± 0.05	$21.4^{+5.1}_{-4.3}$	$21.4^{+5.1}_{-4.3}$
HD 53975	226	-02	1318	21.14 ± 0.06	16.15 ± 0.07	$10.2^{+2.3}_{-2.0}$	$10.2^{+2.3}_{-2.0}$
HD 93521[4]	183	+62		19.61 ± 0.055	$14.85^{+0.05}_{-0.02}$	$17.4^{+2.0}_{-0.8}$	$18.5^{+2.2}_{-1.0}$
LSE 234	329	-21	460 ± 120	$20.68^{+0.025}_{-0.05}$	16.02 ± 0.045	$22.0^{+4}_{-2.5}$	$22.4^{+4}_{-2.5}$

Notes:
[1] All errors are $\pm 1\sigma$. References: Dupuis *et al.* (2009) for REJ 1738+665, Oliveira & Hébrard (2006) for HD 41161 and HD 53975, Kruk *et al.* (2006) for HD 93521, and Lecavelier des Etangs *et al.* (2006) for LSE 234.
[2] N(HI) includes $2\mathrm{N(H_2)}$ and N(HD).
[3] N(DI) includes N(HD).
[4] HD 93521 is located in the Galactic halo at z = 1.5 kpc.

15.6 ± 0.4 ppm, but the two highest points toward 31 Com and β Cet are for stars located at high Galactic latitudes ($+89°$ and $-81°$). I include the line of sight to 31 Com in the Local Bubble measurements as the N(HI) and N(DI) gas in this line of sight has a velocity consistent with the North Galactic Pole cloud of warm gas located within 8.5 pc of the Sun (Redfield & Linsky 2008). I also include the line of sight to β Cet in the Local Bubble measurements as the velocities of the H(I) and D(I) gas are consistent with the Microscopium and Ceti clouds located within 5.1 and 15.5 pc of the Sun.

Since the inner boundary of the Local Bubble lies at $\log \mathrm{N(HI)} \approx 19.2$ (Sfeir *et al.* 1999, Lallement *et al.* 2003) in nearly all directions, it is sensible to subtract the Local Bubble foreground from the column densities toward stars that lie beyond the Local Bubble to infer $(D/H)_{\mathrm{gas}}$ ratios for the ISM not modified by deuterium-depleted gas from the Local Bubble supernovae. Assuming $(D/H)_{\mathrm{gas}} = 15.6 \pm 0.4$ ppm, the Local Bubble foreground is $\log \mathrm{N(HI)_{LB}} = 19.20$ and $\log \mathrm{N(DI)_{LB}} = 14.39$. The ratios $(D/H)_{\mathrm{gas-LB}}$ in Table 1 and in Table 3 of Linsky *et al.* (2006) include this correction. The effect of subtracting the Local Bubble foreground is to produce somewhat higher or lower values of $(D/H)_{\mathrm{gas-LB}}$ for lines of sight toward stars just beyond the Local Bubble when $(D/H)_{\mathrm{gas}}$ is greater or less than 15.6 ppm. The correction is small for stars with $\log \mathrm{N(HI)} > 20.5$.

In Fig. 1, ten lines of sight have $(D/H)_{\mathrm{gas-LB}} > 15.6$, and six lie above 20.0. If the deuterium-depletion hypothesis is the major cause of low values of $(D/H)_{\mathrm{gas-LB}}$, then

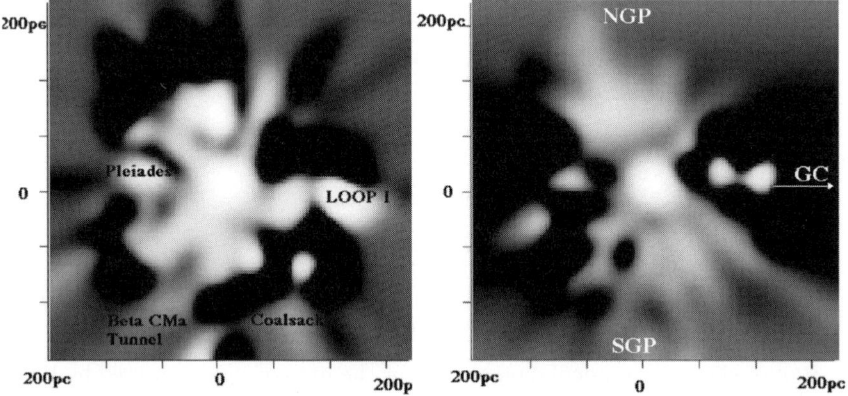

Figure 3. Two views of the Local Bubble from Welsh (2009) and Lallement *et al.* (2003). Dark indicates cold gas seen in the NaI, and white indicates no cold gas. Left: view of the Galactic plane from the North Galactic pole. Right: view from Galactic longitude $270°$ and latitude $0°$.

the highest credible value or values of $(D/H)_{gas-LB}$ should be the best approximation to a lower limit to $(D/H)_{total}$. Given that the data have errors, one can only estimate the highest credible value of $(D/H)_{gas-LB}$ statistically. At this symposium, Steigman (2010), has proposed another statistical method for estimating $(D/H)_{total}$. In Fig. 4, I plot the value of $(D/H)_{high}$ estimated as the mean of the $(D/H)_{gas-LB}$ values starting with the highest value and sequentially including lower values and weighting each data point by the inverse errors. Also plotted is σ, the error of the mean for each number of lines of sight included. The value of σ goes through a minimum at seven lines of sight **and** increases thereafter. For these seven lines of sight, $(D/H)_{high} = 22.9 \pm 0.5$ ppm. I therefore propose that the total D/H ratio (gas plus dust) in the Galactic disk near the Sun is $(D/H)_{total} \geqslant 22.9 \pm 0.5$ ppm.

3. Decisions and implications

Comparison of our estimate of the total D/H ratio in the nearby Galactic disk outside of the Local Bubble with estimates of the primordial D/H ratio places constraints on the evolution of deuterium over the lifetime of the Galaxy. Models of the chemical evolution of the Galaxy (e.g., Romano *et al.* 2006) include both the destruction of deuterium by nuclear reactions in stars and estimates of the infall of gas from captured galaxies and the intergalactic medium containing near primordial D/H. For different assumptions, these models compute the deuterium astration factor, $f_D = (D/H)_{prim}/(D/H)_{high}$. In Table 2, I list f_D computed from the QAL and WMAP values for $(D/H)_{prim}$. Also, given are the corresponding fractions of deuterium atoms today in the nearby Galactic disk that have not been processed through stars.

The values of f_D listed in the table computed from the QAL value of $(D/H)_{prim}$ and especially the WMAP value challenge recent Galactic chemical evolution models for the galactocentric distance of the Sun. The models of Chiappini *et al.* (2002), for example, predict that $f_D \approx 1.5$ for a wide range of assumptions. The more recent models of Romano *et al.* (2006) can accommodate values of f_D as low as ~ 1.3. While there may

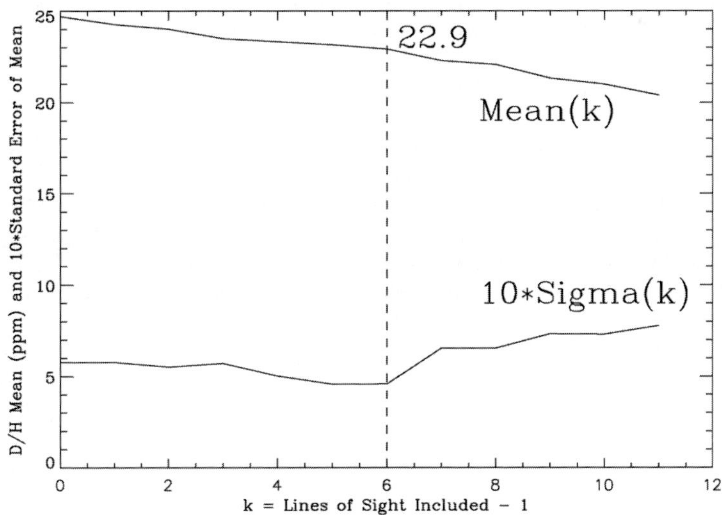

Figure 4. Mean values of $(D/H)_{high}$ obtained as a weighted average of $(D/H)_{gas-LB}$ values starting with the highest and including next lower values in sequence. The error in the mean, σ, increases above seven lines of sight, implying that the best value of the total D/H ratio in the local Galactic disk is $(D/H)_{high} \geqslant 22.9 \pm 0.5$ ppm.

 J. L. Linsky

Table 2. Is Concordance Possible?

Method	D/H (ppm)	$100\Omega_{b,0}h^2$	η_{10}	f_{D}	Fraction
ISM (high 7)	$\geqslant 22.9 \pm 0.5$	$\leqslant 2.411 \pm 0.33$	$\leqslant 6.604 \pm 0.090$		
QAL (7 LOS)	$28.2^{+2.0}_{-1.8}$	2.13 ± 0.09	5.84 ± 0.27	$\leqslant 1.23 \pm 0.09$	$\geqslant 81 \pm 7$ %
WMAP (5 years)	25.2 ± 1.1	2.273 ± 0.062	6.225 ± 0.170	$\leqslant 1.10 \pm 0.07$	$\geqslant 91 \pm 6$ %

not be an inconsistency with the QAL data, the WMAP value for f_{D} severely challenges even the Romano *et al.* (2006) models. At this point, it is not clear to me which are the major causes for the inconsistency, but important unknowns are the accretion rate of near primordial gas and the time scale for mixing of this gas with the ISM. Prodanović & Fields (2008) argue that an accretion rate of $\sim 1 M_\odot$ yr^{-1} could explain the high percentage of unprocessed deuterium. This topic is becoming increasingly interesting.

References

Chiappini, C., Renda, A., & Matteucci, F. 2002, *A&A*, 395, 789

Draine, B. T. 2003, *ARAA*, 41, 241

Draine, B. T. 2006, in A. McWilliam & M. Rauch (eds.), *Origin and Evolution of the Elements*, (Cambridge Univ. Press), p. 317

Dunkley, J. *et al.* 2009, *ApJS*, 180, 306

Dupuis, J., Oliveira, C. M., Hébrard, G. H., Moos, H. W., & Sonnentrucker, P. 2009, *ApJ*, 690, 1045

Hébrard, G. 2010, in C. Charbonnel, M. Tosi, F. Primas, & C. Chiappini (eds), *Light Elements in the Universe* (Cambridge Univ. Press), this volume

Jura, M. 1982, in Y. Kondo, J. M. Meade, & R. D. Chapman (eds), *Advances in UV Astronomy: 4 Years of IUE Research*, (NASA CP 2238), p. 54

Kruk, J. W., Oliveira, C., Sembach, K. R., & Savage, B. D. 2006, in G. Sonneborn, H. W. Moos, & B.-G. Andersson (eds), *Astrophysics in the Far Ultraviolet, Five Years of Discovery with FUSE, ASP-CS*, 348, p. 85

Lallement, R., Welsh, B., Vergeley, J. L., Crifo, F., & Sfeir, D. 2003, *A&A*, 411, 447

Lecavelier des Etangs, A., Hébrard, G., & Williger, G. M. in G. Sonneborn, H. W. Moos, & B.-G. Andersson (eds), *Astrophysics in the Far Ultraviolet, Five Years of Discovery with FUSE, ASP-CS*, 348, p. 88

Linsky, J. L. *et al.* 2006, *ApJ*, 647, 1106

Oliveira, C. M. & Hébrard, G. 2006, *ApJ*, 653, 345

Pettini, M., Zych, B. J., Murphy, M. T., Lewis, A., & Steidel, C. C. 2008, *MNRAS*, 391, 1499

Redfield, S. & Linsky, J. L. 2008, *ApJ*, 673, 283

Prodanović, T. & Fields, B. D. 2008, *J. Cosmology & Astroparticle Physics*, 09, 003

Romano, D., Tosi, M., Chiappini, C., & Matteucci, F. 2006, *MNRAS*, 369, 295

Sfeir, D. M., Lallement, R., Crifo, F., & Welsh, B. Y. 1999, *A&A*, 346, 785

Steigman, G. 2010, in C. Charbonnel, M. Tosi, F. Primas, & C. Chiappini (eds), *Light Elements in the Universe* (Cambridge Univ. Press), this volume

Welsh, B. Y. & Shelton, R. L. 2009, *AP&SS*, 323, 1

Light Elements in the Universe
Proceedings IAU Symposium No. 268, 2009
C. Charbonnel, M. Tosi, F. Primas & C. Chiappini, eds.
© International Astronomical Union 2010
doi:10.1017/S174392131000387X

What the D/O ratio tells us about the interstellar abundance of deuterium?

Guillaume Hébrard

Institut d'Astrophysique de Paris, UMR7095 CNRS, Université Pierre & Marie Curie,
98bis boulevard Arago, 75014 Paris, France
email: hebrard@iap.fr

Abstract. The ionization balances for HI, OI and DI being locked together by charge exchange, the deuterium-to-oxygen ratio is considered to be a good proxy for the deuterium-to-hydrogen ratio, in particular within the interstellar medium. As the DI and OI column densities are of similar orders of magnitude for a given sight line, comparisons of the two values are generally less subject to systematic errors than comparisons of DI and HI. Moreover, D/O is additionally sensitive to astration, because as stars destroy deuterium, they should produce oxygen. D/O measurements are now available for tens of lines of sight in the interstellar medium, most of them from *FUSE* observations. The D/H and D/O ratios show different pictures, D/H being clearly more dispersed than D/O. The low, homogeneous D/O ratio measured on distant lines of sight suggests a deuterium abundance representative of the present epoch that is about two times lower than this measured within the local interstellar medium.

Keywords. ISM: abundances; Galaxy: abundances; cosmology: cosmological parameters; ultraviolet: ISM

---

1. Introduction

In the Big Bang standard model, deuterium is produced in significant amounts only during the Big Bang nucleosynthesis. Among the elements created in BBN, the abundance of deuterium is the most sensitive to the baryonic density of the Universe. As deuterium is destroyed in stellar interiors, its abundance D/H is expected to continuously decrease, from its primordial value $(D/H)_{prim}$ to the value characteristic of the present epoch, $(D/H)_{PE}$. This $(D/H)_{PE}$ ratio is measured within the interstellar medium and should be characteristic of cosmic material after ~ 14 Gyrs of Galactic evolution.

$(D/H)_{PE}$ is a key ratio. It serves as a reference baseline for $(D/H)_{prim}$, providing a lower limit for $(D/H)_{prim}$. It also yields important constraints for chemical evolution models of the Galaxy. The depletion factor due to astration, $f_{ev} = (D/H)_{prim} / (D/H)_{PE}$, depends on the star formation and infall rates, and possibly other processes such as early Galactic wind. Models predict values around $f_{ev} \simeq 1.5$ (e.g. Chiappini *et al.* 2002). In addition, $(D/H)_{PE}$ can be studied more thoroughly than the other deuterium abundances, because it is measured in the interstellar medium. Several tens of sight lines suitable for high quality deuterium abundance measurements are available, and potentially, small sample size should not be an issue.

The determination of the canonical value of $(D/H)_{PE}$ has been the subject of considerable debate over the years. Numerous measurements have been performed through Lyman absorption-line observations in the far-ultraviolet spectral range. By observing hydrogen and deuterium directly in their atomic form, they provide accurate column density determinations that are not dependent on ionization or chemical fractionation. Among them, the ratio $D/H = 1.6 \times 10^{-5}$ measured toward Capella (Linsky *et al.* 1995) has often been used as a benchmark of $(D/H)_{ISM}$. Although it has been performed in a particular cloud, namely the Local Interstellar Cloud (Lallement & Bertin 1992), it has been taken as the canonical $(D/H)_{PE}$ value by numerous theoretical studies of the chemical evolution of the Galaxy. However, several lines of sight observed with *Copernicus* revealed values outside this range (see, e.g., Laurent *et al.* 1979; York 1983). It was uncertain whether the dispersion was the signature of spatial variations in $(D/H)_{ISM}$, or due to unknown systematic errors. The *FUSE* mission has brought significant progress on these issues. Tens of targets observed with *FUSE* have allowed D/H and D/O ratios to be measured in the the interstellar medium.

2. D/O as a proxy for D/H

One of the challenges of D/H measurements is to evaluate for the same line of sight the HI and DI column densities, $N(HI)$ and $N(DI)$, which differ by about five orders of magnitude. Such a large difference is a potential source of systematic errors. For example, all lines from the same species may lie on the non-linear part of the curve of growth, or HI column densities may be detectable in clouds for which DI are below the detection limit (see, e.g., Linsky & Wood 1996, Lemoine *et al.* 2002, or Vidal-Madjar & Ferlet 2002).

Many of the difficulties associated with obtaining accurate D/H ratios may be avoided by measuring the deuterium-to-oxygen ratio, D/O (Timmes *et al.* 1997, Hébrard & Moos 2003). First of all, the D/O ratio is of order of a few percent rather than $\sim 10^{-5}$ as for D/H. Many OI and DI transitions with different oscillator strengths are present in the *FUSE* bandpass, allowing measurement of $N(OI)$ and $N(DI)$ over a wide range of values. OI is believed to be a good tracer of HI in the nearby Galactic disk (Meyer et al. 2001; André *et al.* 2003; Cartledge *et al.* 2004). The neutral forms of oxygen and hydrogen likely dominate over their ions for many sight lines in the diffuse interstellar medium. In any case, because both species have nearly the same ionization potential, their ionization balances are strongly coupled to each other by charge exchange reactions (Jenkins *et al.* 2000). Thus, no corrections from ionization models are required, and we can use $N(DI)/N(OI)$ for D/O. Finally, D/O is particularly sensitive to stellar activity, because of both deuterium destruction (deuterium is burned in stellar interiors at temperatures as low as 6×10^{6} K) and oxygen production (oxygen is mainly produced by type II supernovae). Hence, spatial variations of the deuterium abundance due to different astration rates at different locations would translate in even higher D/O spatial variations.

3. Deuterium within the Local Bubble

The first published *FUSE* $(D/H)_{ISM}$ results focussed on the local interstellar medium (e.g. Moos *et al.* 2002; Friedman *et al.* 2002; Hébrard *et al.* 2002; Lemoine *et al.* 2002; Sonneborn *et al.* 2002). These early studies have shown that $(D/H)_{ISM}$ likely presents a single value in the Local Bubble. Oliveira *et al.* (2003) reported $(D/H)_{LB} = (1.52\pm0.07)\times 10^{-5}$ from *FUSE* data only. Wood *et al.* (2004) included previous LB measurements, and reported $(D/H)_{LB} = (1.56 \pm 0.04) \times 10^{-5}$. The Local Bubble is a $\sim 100\,\mathrm{pc}$-size low-density cavity which includes the LIC in which the Solar System is embedded, and other interstellar clouds (e.g. Snowden *et al.* 1998). The spectral resolution of *FUSE* is too low to distinguish these different individual clouds. However, as the $(D/H)_{ISM}$ integrated along the studied sight lines show no variations, it seems secure to conclude that the different clouds probed within the LB present an homogeneous $(D/H)_{ISM}$ ratio.

The homogeneity of the deuterium abundance in the different interstellar clouds within the LB was also shown from *FUSE* D/O measurements (Hébrard & Moos 2003). D/O was found to be constant within the Local Bubble, with the averaged value $(D/O)_{LB} =$

$(3.84\pm0.16)\times10^{-2}$ (see Figure 1, left panel). This homogeneity of $(D/O)_{LB}$ argues against variations of D/H in the LB. Indeed, the only possible way for a stable D/O together with a varying D/H would be for D/H and O/H to be correlated. Moreover, they would have to precisely vary in such a way that D/O remains constant. That seems unlikely for two different reasons: (i) O/H appears to be uniform in the interstellar medium over paths of several hundred parsecs (Meyer *et al.* 2001; André *et al.* 2003; Cartledge *et al.* 2004), and (ii) astration processes should lead to an anti-correlation of DI and OI abundances.

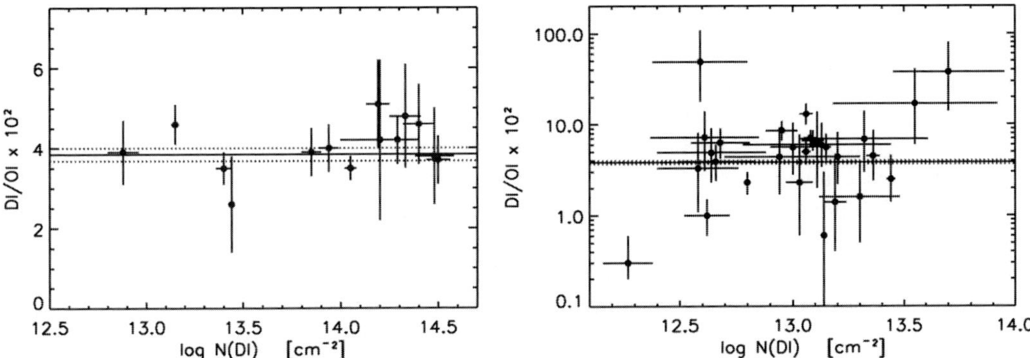

Figure 1. D/O in the Local Bubble from *FUSE* (left) and HST (right) measurements.

Using the interstellar O/H ratio from Meyer (2001), the $(D/O)_{LB}$ ratio translates into $(D/H)_{LB} = (1.32\pm0.08)\times10^{-5}$. According to the error bars, this result is $2.5\,\sigma$ lower than the direct $(D/H)_{LB}$ measurement reported above, which is significant. The explantion is unlikely to be due to the O/H used to translate D/O into D/H, as Oliveira *et al.* (2005) found a $(O/H)_{LB}$ ratio in good agreement with the O/H ratio measured in more distant interstellar medium by Meyer (2001). Whatever is the explanation, the homogeneity of the local D/O measurements implies that the spatial variations of D/H in the Local Bubble must be extremely small, if any (Hébrard & Moos 2003).

It seems now that the homogeneity of the deuterium abundance within the Local Bubble has reached consensus. One can note however the $N(DI)$ and $N(OI)$ measurements reported within 100 pc by Redfield & Linsky (2004), using HST data. The corresponding D/O ratios are plotted in Figure 1 (right panel). A wide dispersion appears of about 2 orders of magnitude. Apparently, the *FUSE* and HST studies produce opposite conclusions. It is unlikely that a malicious systematic effect disturbs the *FUSE* measurements in a way that erases actual variations. More probably, the HST measurements are perturbed in a random way by uncontroled systematics. The probable cause is the $N(OI)$ measurements, performed with the $\lambda1302\text{Å}$ saturated transition in the HST spectra, whereas $N(OI)$ is measured from unsaturated lines in the *FUSE* bandpass. Hébrard *et al.* (2005) and Friedman *et al.* (2006) have shown examples of systematic effects on column densities due to saturated lines.

4. Deuterium in the distant interstellar medium

Whatever the true $(D/H)_{LB}$ is, 1.32×10^{-5} or 1.56×10^{-5}, it now appears clear that this ratio should not be considered as a canonical value for $(D/H)_{ISM}$. $(D/H)_{LB}$ is not representative of the cosmic material at the present epoch. Indeed, in addition to the early *Copernicus* results reported above, *FUSE* has provided extra clues for a signifi- cant dispersion in $(D/H)_{ISM}$ beyond the Local Bubble (Friedman *et al.* 2002; Hoopes

62 G. Hébrard

et al. 2003; Wood *et al.* 2004; Williger *et al.* 2005; Hébrard *et al.* 2005; Friedman *et al.* 2006; Oliveira & Hébrard 2006; Oliveira *et al.* 2006; Dupuis *et al.* 2009). It is difficult to decide which one of the different values is representative of the present epoch, and even if a canonical value for $(D/H)_{ISM}$ does exist. There is certainly no reason to preferentially adopt $(D/H)_{LB}$.

If a $(D/H)_{ISM}$ canonical value exists, the best way to determine it might be to measure deuterium along the long lines of sight with high column densities. More material is probed, and thus localized anomalies are likely to be averaged out. Following that approach, Hébrard & Moos (2003) reported a trend mainly based on D/O: the deuterium abundance is lower for the most distant lines of sight and the highest column densities. This trend was reinforced since then by the additional *FUSE* results, as shown on the left panel of Figure 2. This figure plots the lines of sight from Hébrard & Moos (2003) and all the subsequent *FUSE* measurement published after and quoted above. It does not include the HST measurements presented in Redfield & Linsky (2004). On this plot, the local D/O is homogeneous; the distant D/O is also homogeneous, but with a value about two times lower. There are just two lines of sight with high D/O (Oliveira *et al.* 2006), but the OI column density is here possibly affected by saturation, as shown in the cases discussed in Hébrard *et al.* (2005) and Friedmand *et al.* (2006).

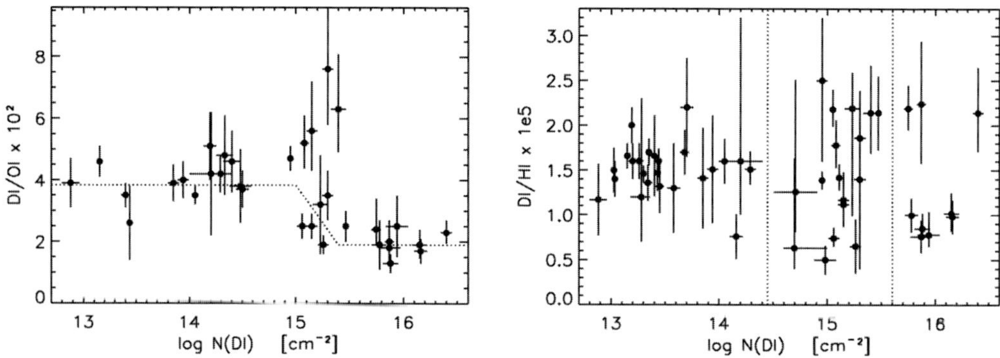

Figure 2. D/O and D/H as a function of DI column density.

This trend could suggest that the local deuterium abundance might be abnormally high, whereas the actual value of $(D/H)_{PE}$ may be significantly below 1×10^{-5} (Hébrard & Moos 2003). On the other hand, following an idea proposed by Jura (1982) and updated by Draine (2004; 2005), Wood *et al.* (2004) argued that deuterium can be preferentially depleted onto dust grains. This would imply total $(D/H)_{ISM}$ ratios larger than those measured in the gas phase, and a $(D/H)_{PE}$ value significantly larger than 2×10^{-5}. Thus, presently, two possibilities are proposed for the present-epoch deuterium abundance.

5. Two scenarios for $(D/H)_{PE}$

If deuterium is significantly depleted onto dust grains, $(D/H)_{PE}$ can only be measured in interstellar clouds in which all deuterium has been released in the gas phase by recent shocks. Assuming this is the case for the few high $(D/H)_{ISM}$ values measured to date, Linsky & Wood (2004) claimed $(D/H)_{PE} = (2.3 \pm 0.4) \times 10^{-5}$. Following this idea, this value was updated to $(D/H)_{PE} \geqslant (2.3 \pm 0.2) \times 10^{-5}$ by Linsky *et al.* (2006), and to a similar vaue $(D/H)_{PE} \geqslant (2.0 \pm 0.1) \times 10^{-5}$ using a Bayesian approach by Prodanovic *et al.* (2009). According to this scenario, lower D/H ratios are found for the sight lines which probe primarily interstellar clouds that have not been recently shocked. That may explain the three D/H regimes apparently seen by Linsky *et al.* (2006). Locally, the D/H is homogeneous but lower than $(D/H)_{PE}$ because the Local Bubble has been shocked recently, but not recently enough, so that part of the deuterium has been incorporated in the solid phase. At intermediate column densities, D/H shows variations, depending on the nature of the clouds, which may or may not have been shocked. Thirdly, at the largest column densities, deuterium is on average significantly depleted in the interstellar clouds probed, except for a few recently shocked clouds.

However, this third regime presenting an homogeneous, low D/H was not confirmed by subsequent *FUSE* measurements. This is shown in right panel of Figure 2, which includes all the measurements available now. The two vertical lines show the limits of the three regimes, as presented in Linsky *et al.* (2006). Rather than a 3-regime picture, this plot shows an homogeneous D/H ratio locally, and an increasing dispersion together with the increasing column density.

Moreover, the D/O measurements are difficult to fit with the depletion scenario. Indeed, D/H and D/O measurements are presently available for 9 distant targets (i.e. clearly outside the Local Bubble): HD 195965 and HD 191877 (Hoopes *et al.* 2003), LSS 1274 (Hébrard & Moos 2003; Wood *et al.* 2004), JL 9 (Wood *et al.* 2004), PG 0038+199 (Williger *et al.* 2005), HD 90087 (Hébrard *et al.* 2005), LSE 44 (Friedman *et al.* 2006), HD 41161 and HD 53975 (Oliveira & Hébrard 2006). The χ^2 are 35.8 for D/H and 5.9 for D/O, for 8 degrees of freedom. Thus, D/O measurements show less dispersion than D/H. Here again, it is difficult to guess a malicious effect able to erase actual variations in the deuterium abundance. In fact, the D/O measurements suggest a different scenario than the "depletion" hypothesis. In Figure 2 (left panel), D/O is homogeneous, with values near 3.8×10^{-2} for $\log N(DI) \leqslant 15$, i.e. in the Local Bubble, and is homogeneous with about two times lower value for $\log N(DI) \geqslant 15.5$. The intermediate region ($15 \leqslant \log N(DI) \leqslant 15.5$) appears more as a transition between these two different plateaus than a particular regime in itself. The lines of sight showing high D/H show low D/O (Williger *et al.* 2005; Friedman *et al.* 2006; Oliveira & Hébrard 2006).

The values of D/O for sight lines to distant targets are homogeneous, around $(2.0 \pm 0.5) \times 10^{-2}$. Because these lines of sight probe large amounts of material, this ratio may be characteristic of the present epoch. If so, it would imply $(D/H)_{PE} = (7 \pm 2) \times 10^{-6}$. Then, the reason for a high local D/H value is uncertain; the possibility of unknown deuterium enrichment process(es) or local infall of less evolved material must be considered. One can note that Cartledge *et al.* (2004) invoked local infall to possibly explain their O/H measurements.

6. Conclusions

The "depletion argument" and the "D/O argument" lead to two different values for $(D/H)_{PE}$: $\sim 2.3 \times 10^{-5}$ and $\sim 7 \times 10^{-6}$, respectively. As shown in Figure 3, the high $(D/H)_{PE}$ would imply a low deuterium destruction through astration, even possibly no destruction at all. Conversely, the low $(D/H)_{PE}$ would imply larger depletion factors f_{ev}, especially in the latest stages of the Universe evolution.

It appears clearly that the local value 1.5×10^{-5} should no longer be considered as a canonical value for $(D/H)_{PE}$. Up to now, neither the "depletion" nor the "D/O" hypothesis (or both of them) can be firmly ruled out. Their is no hope of extra deuterium measurements in the distant interstellar medium, as *FUSE* now stopped operations. In addition, it won't be possible to probe deuterium on column densities larger than those detected with *FUSE*, as the denser ones are at the limit where unsaturated deuterium

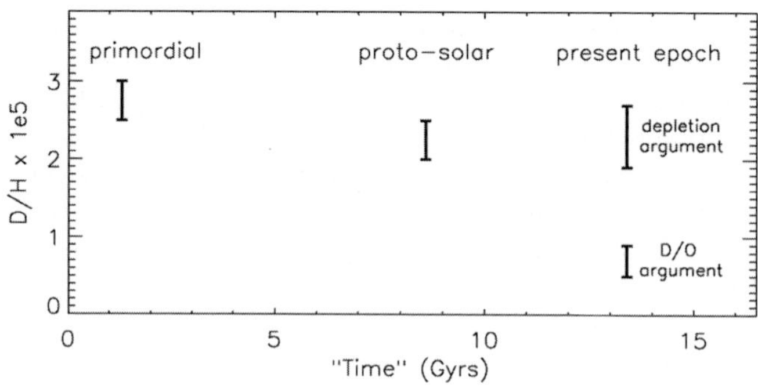

Figure 3. Deuterium abundance as a function of time after BBN.

lines are detectable. Additional deuterium measurements are thus unlikely to provide new constraints on this issue for a while.

References

André, M. K., Oliveira, C. M., Howk, J. C., *et al.* 2003, ApJ, 591, 1000

Cartledge, S., Lauroesch, J. , Meyer, D., & Sofia, U. 2004, ApJ, 613, 1037

Chiappini, C., Renda, A., & Matteucci, F. 2002, A&A, 395, 789

Dupuis, J., Hébrard, G., Oliveira, C., Moos, H. W., & Sonnentrucker, P., 2009, ApJ, 690, 1045

Friedman, S. D., Howk, J. C., Chayer, P., *et al.* 2002, ApJS, 140, 37

Friedman, S. D., Hébrard, G., Tripp, T. M., *et al.* 2006, ApJ, 638, 847

Hébrard, G., Lemoine, M., Vidal-Madjar, A., *et al.* 2002, ApJS, 140, 103

Hébrard, G. & Moos, H. W. 2003, ApJ, 599, 297

Hébrard, G., Tripp, T. M., Chayer, P., *et al.* 2005, ApJ, 635, 1136

Hoopes, C. G., Sembach, K. R., Hébrard, G., *et al.* 2003, ApJ, 586, 1094

Lallement, R. & Bertin, P. 1992, A&A, 266, 479

Laurent, C., Vidal-Madjar, A., & York, D. G. 1979, ApJ, 229, 923

Lemoine, M., Vidal-Madjar, A., Hébrard, G., *et al.* 2002, ApJS, 140, 67

Linsky, J. L., Diplas, A., Wood, B. E., *et al.* 1995, ApJ, 451, 335

Linsky, J. L. & Wood, B. E. 1996, ApJ, 463, 254

Linsky, J. L., Draine, B. T., Moos, H. W., *et al.*, 2006, ApJ, 647, 1106

Meyer, D. M. 2001, XVIIth IAP Colloquium, Paris, Edited by R. Ferlet *et al.*, p. 135

Moos, H. W., Sembach, K. R., Vidal-Madjar, A., *et al.* 2002, ApJS, 140, 3

Oliveira, C. M., Hébrard, G., Howk, J. C., *et al.* 2003, ApJ, 587, 235

Oliveira, C. M. & Hébrard, G., 2006, ApJ, 653, 345

Oliveira, C. M., Moos, H. W., Chayer, P., & Kruk, J. W. 2006, ApJ, 642, 283

Prodanovic, T., Steigman, G., & Fields, B. D. 2009, arXiv:0910.4961

Redfield, S. & Linsky, J. L. 2004, ApJ, 602, 776

Snowden, S. L., Egger, R., Finkbeiner, D. P., *et al.* 1998, ApJ, 493, 715

Sonneborn, G., André, M., Oliveira, C., *et al.* 2002, ApJS, 140, 51

Timmes, F. X., Truran, J. W., Lauroesch, J. T., & York, D. G. 1997, ApJ, 476, 464

Vidal-Madjar, A. & Ferlet, R. 2002, ApJ, 571, L169

Williger, G., Oliveira, C., Hébrard, G., Dupuis, J., Dreizler, S., & Moos, H. W. 2005, ApJ 625, 210

Wood, B. E., Linsky, J. L., Hébrard, G., *et al.* 2004, ApJ, 609, 838

York, D. G. 1983, ApJ, 264, 172

Light Elements in the Universe
Proceedings IAU Symposium No. 268, 2009
C. Charbonnel, M. Tosi, F. Primas & C. Chiappini, eds.

© International Astronomical Union 2010
doi:10.1017/S1743921310003881

(Un)true deuterium abundance in the Galactic disk

Tijana Prodanović[1], Gary Steigman[2] and Brian D. Fields[3]

[1]Department of Physics, University of Novi Sad,
Trg Dositeja Obradovića 4, 21000 Novi Sad, Serbia
email: prodanvc@df.uns.ac.rs

[2]Department of Physics and Astronomy, Ohio State University,
191 W. Woodruff Ave., Columbus OH 43210-1117, USA
email: steigman@mps.ohio-state.edu

[3]Department of Astronomy, University of Illinois,
1002 W. Green St., Urbana IL 61801, USA
email: bdfields@illinois.edu

Abstract. Deuterium has a special place in cosmology, nuclear astrophysics, and galactic chemical evolution, because of its unique property that it is only created in the big bang nucleosynthesis while all other processes result in its net destruction. For this reason, among other things, deuterium abundance measurements in the interstellar medium (ISM) allow us to determine the fraction of interstellar gas that has been cycled through stars, and set constraints and learn about different Galactic chemical evolution (GCE) models. However, recent indications that deuterium might be preferentially depleted onto dust grains complicate our understanding about the meaning of measured ISM deuterium abundances. For this reason, recent estimates by Linsky *et al.* (2006) have yielded a lower bound to the "true", undepleted, ISM deuterium abundance that is very close to the primordial abundance, indicating a small deuterium astration factor contrary to the demands of many GCE models. To avoid any prejudice about deuterium dust depletion along different lines of sight that are used to determine the "true" D abundance, we propose a model-independent, statistical Bayesian method to address this issue and determine in a model-independent manner the undepleted ISM D abundance. We find the best estimate for the *gas-phase* ISM deuterium abundance to be $(D/H)_{ISM} \geqslant (2.0 \pm 0.1) \times 10^{-5}$. Presented are the results of Prodanović *et al.* (2009).

Keywords. ISM: abundances, ISM: dust, Galaxy: abundances, Galaxy: evolution

1. Introduction

Deuterium is only created in significant amounts during the big bang nucleosynthesis (BBN) while all other processes destroy it (Epstein *et al.* 1976, Prodanović & Fields 2003, Steigman 2007). Consequently, the deuterium abundance should monotonically decrease after the Big Bang, where the main destruction channel is the stellar processing. Because of this unique property deuterium can be used as a cosmic baryometer but also to discriminate between different Galactic chemical evolution (GCE) models (see eg. Steigman *et al.* 2007, Prodanović & Fields 2008, Vangioni-Flam *et al.* 1994). Specifically, given that deuterium is mostly destroyed in stars, its abundance ($y_D \equiv 10^5 (D/H)$) can be used to probe the fraction of the interstellar gas that has never been processed through stars (Steigman & Tosi 1992, 1995), which is often quantified with the astration factor $f_D \equiv y_{DP}/y_{D,ISM}$, the ratio of the primordial D abundance (the D to H ratio by number) to the ISM D abundance.

The primordial D abundance, y_{DP}, predicted by the BBN model (Cyburt *et al.* 2003, Steigman 2007) based on the cosmic microwave background observations (Spergel *et al.*

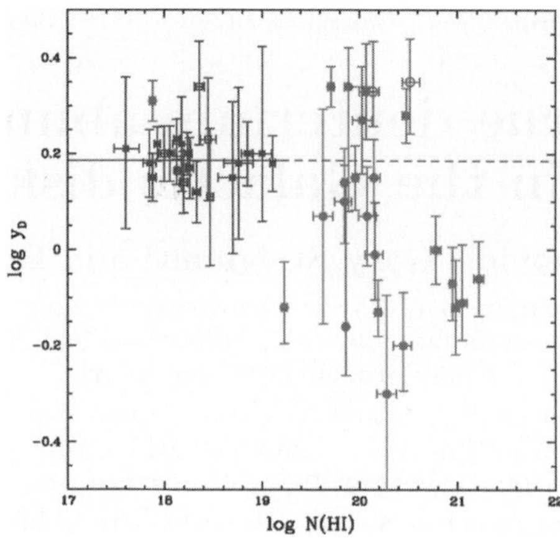

Figure 1. The logs of the deuterium abundances versus the logs of the H I column densities [cm^{-2}] for the 46 LOS from Table 2 of Linsky *et al.*(2006). The filled symbols are for the 38 LOS which have iron abundance data, while the open symbols are for the 8 LOS which lack iron abundances. The squares (blue) are for the LOS within the Local Bubble (LB) and the circles (red) are for the non-LB (nLB) LOS. The solid line is at the mean D abundance for the LB LOS (log $y_{D,LB} = 0.19$); the dashed line is its extension to the nLB LOS.

2007), is in excellent agreement with D abundance observations at high-redshift systems (Pettini *et al.* 2008). However, observations of the *gas-phase* deuterium abundance in the interstellar medium (ISM) have revealed large variations over different lines of sight (LOS) by factors of ~ 4 ($0.5 \lesssim y_{D,ISM} \equiv 10^5 (D/H)_{ISM} \lesssim 2.2$; Jenkins *et al.* 1999, Hébrard *et al.* 2002, Sonneborn *et al.* 2000, Hoopes *et al.* 2003). The range of these variations is best seen on Figure 1, where we have plotted the logs of the gas-phase D/H ($\log y_D \equiv 5 + \log N(D I) - \log N(H I)$) as a function of the logs of the H I column densities toward 46 different LOS inside and outside of the Local Bubble (LB) where the data has been taken from Linsky *et al.*(2006). Blue squares represent 21 LB LOS, while red circles represent 25 LOS from outside of the LB. Filled symbols represent those LOS for which iron abundance has also been measured. As an explanation for the observed scatter, it has been proposed that the gas-phase deuterium is depleted onto dust grains preferential to hydrogen (Jura 1982, Draine 2006). This hypothesis is in agreement with findings that the gas phase D/H correlates positively with refractory elements (Prochaska *et al.* 2005, Linsky *et al.* 2006, Steigman *et al.* 2007).

Arguing that deuterium is severely depleted onto dust grains, Linsky *et al.*(2006) have used the five *highest* D/H ratios to determine the *lower bound* to the true, gas-phase ISM deuterium abundance $y_{D,ISM} \geqslant 2.31 \pm 0.24$, which is at the level of $\sim 80\%$ of the primordial deuterium abundance $y_{DP} = 2.82^{+0.20}_{-0.19}$ (Pettini *et al.* 2008). Adopting this new estimate of the ISM D abundance, and the primordial abundance of Pettini *et al.*(2008), one finds the upper bound to the astration factor $f_D \leqslant 1.22 \pm 0.15$, which is inconsistent with many GCE models that require $1.4 \lesssim f_D \lesssim 1.8$ (Steigman *et al.* 2007). In order to account for such high ISM deuterium abundance and a low astration factor, most GCE models would require large infall rates of close to pristine material (Romano *et al.* 2006, Prodanović & Fields 2008).

Because the observed variations and a possibly severe dust-depletion complicate determination of the true, gas-phase, deuterium abundance in the ISM, we have proposed a Bayesian statistical approach (§2) in analyzing the deuterium data which yields the maximum likelihood estimate of the true, undepleted, *gas-phase* ISM deuterium abundance $y_{D,max}$(Prodanović *et al.* 2009). The underlying assumption of this analysis is that the observed spread in abundance measurements is the result of incompletely homogenized dust depletion. Our results are presented in §3, and conclusions in §4.

2. Bayesian analysis

To limit ourself to fewest assumptions possible, we have analyzed the FUSE ISM gas-phase deuterium abundance measurements (46 LOS from Table 2. of Linsky *et al.* 2006) by using a statistical Bayesian approach, which was first developed by Hogan *et al.*(1997) for the purpose of determining the primordial helium abundance that best fits the observations. The main assumption of this approach is that, in our case, depletion of gas-phase deuterium abundance is present at some level, but where there is no further assumption about the nature, level or spatial distribution of this depletion. In this sense, the Bayesian statistical approach is model independent. Thus, the available data are analyzed in an unbiased, statistical manner, which as a result determines the two parameters – the un-depleted (or a lower bound to it), gas-phase ISM D abundance $y_{\mathrm{D,ISM}} \gtrsim y_{\mathrm{D,max}}$, and the depletion parameter $w \equiv y_{\mathrm{D,max}} - y_{\mathrm{D,min}}$.

The values of the two parameters, $y_{\mathrm{D,max}}$ and w, that best fit the set of measurements $y_{\mathrm{D},i}$, with errors δ_i, are found by determining the maximum value of the likelihood function

$$\mathscr{L}(y_{\mathrm{D,max}}; w) \;=\; \prod_i P(y_{\mathrm{D},i}|y_{\mathrm{D,max}}; w) \tag{2.1}$$

$$=\; \prod_i \int dy_{\mathrm{D},i,\mathrm{T}}\, P(y_{\mathrm{D},i}|y_{\mathrm{D},i,\mathrm{T}})P(y_{\mathrm{D},i,\mathrm{T}}|y_{\mathrm{D,max}}; w) \tag{2.2}$$

The probability distribution $P(y_{\mathrm{D},i}|y_{\mathrm{D},i,\mathrm{T}})$ represents the probability of measuring the abundance $y_{\mathrm{D},i}$ along the ith LOS , given the true, error-free (but possibly depleted!) abundance $y_{\mathrm{D},i,\mathrm{T}}$ along that LOS. This distribution reflects the fact that real observed data have errors, and thus we assume that the shape of this distribution is a Gaussian of width σ_i. We treat the data with asymmetrical errors by assuming that σ_{i+} corresponds to $y_{\mathrm{D},i} > y_{\mathrm{D},i,\mathrm{T}}$ and σ_{i-} to $y_{\mathrm{D},i} < y_{\mathrm{D},i,\mathrm{T}}$. The probability distribution $P(y_{\mathrm{D},i,\mathrm{T}}|y_{\mathrm{D,max}}; w)$ is the probability of finding the true, depleted, LOS deuterium abundance $y_{\mathrm{D},i,\mathrm{T}}$, given the values of the maximum and minimum gas phase ISM deuterium abundances, $y_{\mathrm{D,max}}$ and $y_{\mathrm{D,min}} = y_{\mathrm{D,max}} - w$ respectively. If we assume that depletion is due to dust, then $P(y_{\mathrm{D},i,\mathrm{T}}|y_{\mathrm{D,max}}; w)$ represents the distribution of dust depletion along different LOS w_i. Given that we know very little about this distribution, and to limit ourselves to the fewest assumptions possible, we adopt three simplest forms of this depletion distribution: a top-hat (all levels of depletion are equally probable), a positive-bias (favors low depletion) and a negative-bias (favors large depletion) function. Functional forms of the three depletion distributions for $y_{\mathrm{D,min}} \leqslant y_{\mathrm{D},i,\mathrm{T}} \leqslant y_{\mathrm{D,max}}$ are given in Eq. (2.3), while outside of this range $P(y_{\mathrm{D},i,\mathrm{T}}|y_{\mathrm{D,max}}; w) = 0$. All depletion distributions are normalized to unity when integrated over $y_{\mathrm{D},i,\mathrm{T}}$.

$$P_t(y_{\mathrm{D},i,\mathrm{T}}|y_{\mathrm{D,max}}; w) = \begin{cases} 1/w, & \text{Top-hat} \\ 2(y_{\mathrm{D},i,\mathrm{T}} - y_{\mathrm{D,min}})/w^2, & \text{Positive-bias} \\ 2(y_{\mathrm{D,max}} - y_{\mathrm{D},i,\mathrm{T}})/w^2, & \text{Negative-bias} \end{cases} \tag{2.3}$$

With a choice of depletion distribution and a set of deuterium abundance measurements with errors $y_{\mathrm{D},i}, \delta_i$, from Equation (2.2) one can numerically calculate the likelihood function $\mathscr{L}(y_{\mathrm{D,max}}; w)$ for a choice of parameter $(y_{\mathrm{D,max}}; w)$ values. Likelihood is then calculated for a range of parameter values, where the combination of values that results in the maximum likelihood determines the most likely value of undepleted, gas-phase ISM deuterium abundance $y_{\mathrm{D,max}}$ and the corresponding depletion parameter w, that best fit the observed abundances (with their errors) with a spread that results from depletion of deuterium onto dust (Prodanović *et al.* 2009).

3. Results

Because the spread between the observed abundances is very different for only LB and only non-LB LOS (Fig. 1), we have first analyzed the two subsets of data separately. The analyzed data have been taken from Table 2 of Linsky *et al.*(2006), and there are 21 LB LOS and 25 non-LB LOS (about LB and non-LB LOS assignment see Linsky *et al.* 2006, Prodanović *et al.* 2009).

Measured LB deuterium abundances show very little scatter, and thus it is not surprising that all three shapes of depletion distribution given in Eq. (2.3) yield almost the same maximum likelihood udepleted, gas-phase deuterium abundance $y_{\mathrm{D,max}} \approx 1.5$ which is consistent with no depletion $w \approx 0$. The likelihood contours for the top-hat case and for the LB data are shown on the top panel of Fig. 2. This deuterium abundance that best fits the LB observations results in the astration factor $f_{\mathrm{D}} \leqslant 1.8$ (Prodanović *et al.* 2009) which is consistent with a range of successful GCE models presented in Steigman *et al.*(2007). In terms of the GCE models presented in Prodanović & Fields (2008) where the infall rate of the pristine material is taken to be proportional to the star formation rate with the proportionality constant α, our new ISM deuterium abundance for this case, is consistent with low infall rates $\alpha \lesssim 0.1$ and return fractions of $R \lesssim 0.5$ (fraction of the stellar mass that is returned to the ISM), which can now accommodate even the more recent initial mass functions.

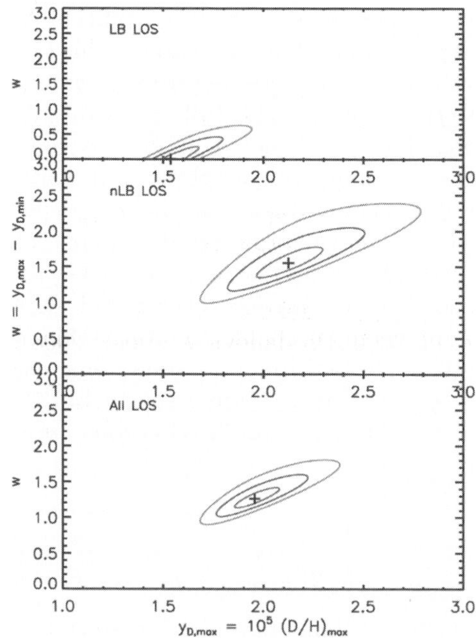

Figure 2. Likelihood contours (68%, 95%, 99%) in the $w - y_{\mathrm{D,max}}$ plane for the 21 LB LOS (top panel), the 25 nLB LOS (middle panel), and all 46 LOS (bottom panel), using the top-hat prior.

On the other hand, observations of the gas-phase D abundances along different LOS outside of the LB, show significant scatter which can be seen from Fig. 1. The maximum likelihood values for the non-LB data set and the top-hat depletion distribution are $y_{D,max} = 2.1$ and $w = 1.6$. This result is presented on the middle panel of Figure 2 (Prodanović *et al.* 2009). The resulting astration factor $f_{D,LB} \leqslant 1.3$ is within the errors consistent with some GCE models discussed in Steigman *et al.*(2007). Within our new ISM D abundance in this case, a wide range of infall rates required by the GCE models of Prodanović & Fields (2008) is now allowed with the infall parameter $0 \lesssim \alpha \lesssim 1$ where the corresponding return fractions can be found in the range are $0.2 \lesssim R \lesssim 0.4$, which can again accommodate even the modern initial mass functions.

Finally, when the full data set with all 46 LOS deuterium observations is considered with a choice of a top-hat depletion distribution, the maximum likelihood value of the undepleted, gas-phase deuterium abundance that best fits the observed data is found to be $y_{D,max} = 2.0$ with a non-zero depletion of $w = 1.3$. This result is presented on the bottom panel of Figure 2 (Prodanović *et al.* 2009). The corresponding astration factor is found to be $f_D \leqslant 1.4$ which is consistent with some but not all GCE models adopted Steigman *et al.*(2007). Similarly to the non-LB case, a wide range of the GCE models presented in Prodanović & Fields (2008) is allowed with our new estimate of the undepleted, gas-phase D abundance.

The two subsets of data, only LB and only non-LB, as well as the complete set of all 46 LOS measurements, have all been analyzed with all three depletion probability distributions from Eq. (2.3). In all cases, it was found that the top-hat distribution results in the largest maximum likelihood value, i.e. that it fits the spread observed in the data the best (Prodanović *et al.* 2009).

4. Conclusion

Observations of deuterium in the ISM have revealed large variations (Jenkins *et al.* 1999, Hébrard *et al.* 2002, Sonneborn *et al.* 2000, Hoopes *et al.* 2003) which complicates the determination of the unique, ISM deuterium abundance (see Figure 1). It was proposed that the observed variations originate from inhomogeneous depletion of deuterium onto dust grains preferential to hydrogen (Jura 1982, Draine 2006). In the light of this hypothesis, Linsky *et al.*(2006) have proposed that most of the deuterium in the ISM has been depleted and that the true, undepleted D abundance is $y_{D,ISM} \geqslant 2.31 \pm 0.24$ which is at the $\sim 80\%$ level of the primordial D abundance (Pettini *et al.* 2008). If the ISM D abundance is indeed so high this in turn creates tension with otherwise successful GCE modes (see eg. Romano *et al.* 2006, Prodanović & Fields 2008).

Because of the importance of knowing the undepleted, gas-phase ISM D abundance, we have employed a model-independent Bayesian statistical analysis first used by Hogan *et al.*(1997) to determine the primordial helium abundance. By determining the maximum likelihood value of the likelihood function for the assumed depletion distribution function, this approach yields the values of two parameters, undepleted, gas-phase deuterium abundance $y_{D,max}$ and the depletion parameter w, that best fit the observed data and scatter which is assumed to be due to different, in our case dust, depletion levels. We have investigated three different depletion distribution functions: a top-hat that equally favors both high and low depletion levels, a positive-bias that favors low-depletion and a negative-bias function that favors large depletion. Of the three depletion distributions the uniform top-hat distribution results in the largest maximum likelihood and thus fits the data best.

We have applied the Bayesian statistical analysis to the data set of deuterium abundance measurements along 46 different LOS, where the data has been adopted from Linsky *et al.*(2006). Figure 2 shows likelihood contours for the top-hat depletion distribution and for Bayesian analysis applied to 3 data subsets: LB only, non-LB only and the full data set. We found that the LB deuterium abundance measurements are consistent with having zero depletion $w = 0$ and D abundance of $y_{\mathrm{D,LB}} = 1.5(1 \pm 0.03)$. Unlike the LB, the non-LB data require significant depletion. Because it is still unclear whether the LB D is undepleted while some or all non-LB LOS have been enriched by unmixed infall, or if LB is uniformly depleted, we find that the best estimate of the undepleted, gas-phase ISM deuterium abundance is that which follows from analyzing the full 46 LOS data set, with a top-hat depletion distribution function, and is $y_{\mathrm{D,ISM}} \geqslant y_{\mathrm{D,max}} = 2.0 \pm 0.1 = 2.0(1 \pm 0.05)$ (Prodanović *et al.* 2009). With the primordial deuterium abundance adopted from Pettini *et al.*(2008), we find that our new ISM D abundance corresponds to the astration factor $f_{\mathrm{D}} \leqslant 1.4 \pm 0.1$ that is consistent with some successful GCE models (Steigman *et al.* 2007, Prodanović & Fields 2009), which releases some tension with GCE models that is created when a high ISM deuterium abundance, such as that of Linsky *et al.*(2006), is required.

Acknowledgements

The work of TP is supported in part by the Provincial Secretariat for Science and Technological Development, and by the Ministry of Science of the Republic of Serbia under project number 141002B.

References

Cyburt, R. H., Fields, B. D., & Olive, K. A. 2003, *Phys. Lett. B*, 567, 227

Draine, B. T. 2006, *Astrophysics in the Far Ultraviolet: Five Years of Discovery with FUSE*, 348, 58

Epstein, R. I., Lattimer, J. M., & Schramm, D. N. 1976, *Nature*, 263, 198

Hébrard, G., *et al.* 2002, *ApJS*, 140, 103

Hogan, C. J., Olive, K. A., & Scully, S. T. 1997, *ApJL*, 489, L119

Hoopes, C. G., Sembach, K. R., Hébrard, G., Moos, H. W., & Knauth D. C. 2003, *ApJ*, 586, 1094

Jenkins, E. B., Tripp, T. M., Woźniak, P. R., Sofia, U. J., & Sonneborn, G. 1999, *ApJ*, 520, 182

Jura, M. 1982, *Advances in Ultraviolet Astronomy*, Kondo, Y., Mead, J., Chapman, R. D., eds., NASA, Washington, p. 54

Linsky, J. L., *et al.* 2006, *ApJ*, 647, 1106

Pettini, M., Zych, B. J., Murphy, M. T., Lewis, A., & Steidel, C. C. 2008, *MNRAS*, 391, 1499

Prochaska, J. X., Tripp, T. M., & Howk, J. C. 2005, *ApJ*, 620, L39

Prodanović, T. & Fields, B. D. 2003, *ApJ*, 597, 48

Prodanović, T. & Fields, B. D. 2008, *JCAP*, 9, 3

Prodanović, T., Steigman, G., & Fields, B. D. 2009, *submitted to MNRAS*, arXiv:0910.4961

Romano, D., Tosi, M., Chiappini, C., & Matteucci, F. 2006, *MNRAS*, 369, 295

Sonneborn, G., Tripp, T. M., Ferlet, R., Jenkins, E. B., Sofia, U. J., Vidal-Madjar, A., & Woźniak, P. R. 2000, *ApJ*, 545, 277

Spergel, D. N., *et al.* 2007, *ApJS*, 170, 377

Steigman, G. & Tosi, M. 1992, *ApJ*, 401, 150

Steigman, G. & Tosi, M. 1995, *ApJ*, 453, 173

Steigman, G., Romano, D., & Tosi, M. 2007, *MNRAS*, 378, 576

Steigman, G. 2007, *Ann. Rev. Nucl. Part. Sci*, 57, 463

Vangioni-Flam, E., Olive, K. A., & Prantzos, N. 1994, *ApJ*, 427, 618

Light Elements in the Universe
Proceedings IAU Symposium No. 268, 2009
C. Charbonnel, M. Tosi, F. Primas & C. Chiappini, eds.

© International Astronomical Union 2010
doi:10.1017/S1743921310003893

Abundances of hydrogen and helium isotopes in the Protosolar Cloud

Johannes Geiss[1] and George Gloeckler[2]

[1]International Space Science Institute, Hallerstrasse 6, CH-3012 Bern, Switzerland,
email: geiss@issibern.ch

[2]Department of Atmospheric, Oceanic and Space Sciences, University of Michigan
2455 Hayward St., Ann Arbor, MI 48109, USA
email: gglo@umich.edu

Abstract. For our understanding of the origin and evolution of baryonic matter in the Universe, the Protosolar Cloud (PSC) is of unique importance in two ways: 1) Up to now, many of the naturally occurring nuclides have only been detected in the solar system. 2) Since the time of solar system formation, the Sun and planets have been virtually isolated from the galactic nuclear evolution, and thus the PSC is a galactic sample with a degree of evolution intermediate between the Big Bang and the present.

The abundances of the isotopes of hydrogen and helium in the Protosolar Cloud are primarily derived from composition measurements in the solar wind, the Jovian atmosphere and "planetary noble gases" in meteorites, and also from observations of density profiles inside the Sun. After applying the changes in isotopic and elemental composition resulting from processes in the solar wind, the Sun and Jupiter, PSC abundances of the four lightest stable nuclides are given.

Keywords. Sun: solar wind, corona, abundances

1. Introduction

Our concepts of nucleosynthesis were originally derived from solar system abundances, and even to this day, our knowledge of the production of most stable and long-lived nuclei is based on elemental and isotopic composition measurements in solar system material. Without isotopic abundances in meteorites and terrestrial samples, could we clearly distinguish the s- and r-processes, recognize the p-nuclei or establish a nuclear chronology for the Galaxy? For most nuclear species, Galactic evolution models are really models for the evolution of the matter that made up the PSC some 4.6 Gyr ago. Extensions to the present for most nuclei would be mere extrapolations, if we had not reliable Galactic *and* solar system data for several elements and a few isotopic ratios. Among these are, most importantly, the isotopes of hydrogen and helium. Composition measurements in the solar wind, the Jovian atmosphere and to a lesser extend - the "planetary gas" in meteorites provide the main evidence from which the protosolar abundances of these nuclides are derived (Table 1).

2. Protosolar (D+^{3}He)/H derived from ^{3}He/^{4}He in the solar wind

In the early Sun, deuterium was converted into ^{3}He by the reaction

$$D + H = {}^3He + \gamma. \tag{2.1}$$

The short D-burning phase, involving the whole Sun, preceded the H-burning epoch (Ezer & Cameron, 1965; Mazzitelli & Moretti, 1980). In the material of the Outer Convective Zone (OCZ) of the present Sun, ^{3}He has not been further processed, as can be surmised

Table 1. Measurements primarily used in this paper for deriving Solar and Protosolar Abundances.

SOLAR WIND
^{3}He/^{4}He measurements with SWICS/Ulysses, SWC/Apollo, old solar wind in lunar material
$\rightarrow$ Extrapolation to ^{3}He/^{4}He in the OCZ (<u>O</u>uter <u>C</u>onvective <u>Z</u>one of the Sun)
$\rightarrow$ Protosolar (D + ^{3}He)/H $\rightarrow$ Protosolar D/H

METEORITES
^{3}He/^{4}He measurement in the "Planetary Gas" Component of Meteorites
$\rightarrow$ Protosolar ^{3}He/^{4}He

SUN
Density Profile from SOHO Seismology
$\rightarrow$ Protosolar He/H

JUPITER
^{3}He/^{4}He, He/H measurement with Galileo Entry-Probe
D/H measurement with ISO
$\rightarrow$ Protosolar D/H, ^{3}He/^{4}He, (D + ^{3}He)/H

---

from the abundance there of beryllium (Geiss & Reeves, 1972). At any temperature ^{9}Be is destroyed much faster by thermonuclear reactions than ^{3}He (Figure 1). Thus, ^{3}He in the OCZ represents not only protosolar ^{3}He, but the sum of protosolar D and ^{3}He (see Geiss, Eberhardt & Signer, 1966).

Systematic investigations of ^{3}He/^{4}He in the solar wind have been carried out using various space experiments: SWC/Apollo, ICI/ISEE-3, SWICS/Ulysses, SWICS/ACE and recently Genesis. The results have revealed significant variations in solar wind composition. Theoretical studies have shown that the variations are caused by separation

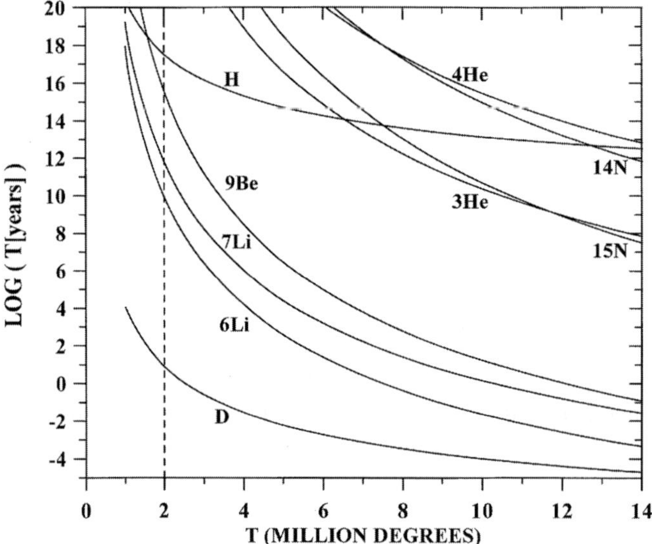

Figure 1. Life-time of light nuclides as a function of temperature. Given are the life-times due to thermonuclear reactions with density and composition normalized to the present conditions at the bottom of the Outer Convective Zone (OCZ) of the Sun. The dashed line marks the present temperature at the bottom of the OCZ. Note: D-burning proceeds well before H-burning commences due to the temperature difference at which the two hydrogen isotopes are fusing (Ezer & Cameron,1965; Mazzitelli & Moretti, 1980). ^{9}Be is destroyed much faster than ^{3}He at all temperatures that could have occurred during the life of the Sun. Thus the presence of Beryllium (^{9}Be) in the OCZ at protosolar abundance precludes the loss there of ^{3}He by nuclear reactions (Geiss & Reeves, 1972).

Table 2. Protosolar $(D+{}^3He)/{}^4He$ from ${}^3He/{}^4He$ in the solar wind.

Solar Wind (SWICS/Ulysses)	${}^3He/{}^4He$	Figure 1
Solar Wind (SWC/Apollo)	${}^3He/{}^4He$	Figure 1
Sun, Outer Convective Zone	${}^3He/{}^4He$	$(3.83 \pm 0.25) \times 10^{-4}$
Protosolar Cloud	$\mathbf{(D +{}^3He)/{}^4He}$	$\mathbf{(3.68 \pm 0.25) \times 10^{-4}}$

processes in the corona (Geiss, Hirt & Leutwyler, 1970; Bürgi & Geiss, 1986; Isenberg & Hollweg, 1983) as well as ion-neutral separation processes in or near the chromosphere (Geiss, 1982; von Steiger & Geiss, 1989). Several physical processes were identified that cause or control charging, separation and transport of major and minor ion species from the chromosphere to the corona and into the supersonic solar wind. Their varying influence produces the observed variations in space and time. Theory alone cannot exactly predict the changes in composition caused by different solar and solar wind conditions, but theory can predict which pairs of ion species should causally correlate and could be used to obtain OCZ abundances from solar wind data by extrapolation.

So far empirical extrapolations from abundances in the slow solar wind and the high-speed streams coming out of the polar coronal holes give the best results. We follow here the method introduced by Gloeckler & Geiss (2000). Figures 2a and 2b show, respectively, the correlations of ${}^3He/{}^4He$ with Si/O and with H/He measured by SWICS/Ulysses in several periods each of in-ecliptic solar wind and high speed streams. Extrapolating by linear regression to Si/O observed in the photosphere and to H/He = 11.9 determined for the OCZ (Pérez Hernández & Christensen-Dalsgaard, 1994) gives $({}^3He/{}^4He)_{OCZ} = (3.78 \pm 0.18) \times 10^{-4}$ and $({}^3He/{}^4He)_{OCZ} = (3.88 \pm 0.14) \times 10^{-4}$ respectively (Figure 2). The 1-σ errors include statistical uncertainties as well as the spread due to solar wind variability. The systematic instrumental uncertainties are estimated to be 0.2×10^{-4} (Gloeckler & Geiss, 2000). The two extrapolation methods give essentially the same result, leading to an average of $({}^3He/{}^4He)_{OCZ} = (3.83 \pm 0.25) \times 10^{-4}$ (Table 2), which is in remarkable agreement with earlier values of $({}^3He/{}^4He)_{OCZ}$ obtained by other methods (see Geiss & Gloeckler, 1998).

During the in-ecliptic periods, steady low-speed solar wind prevailed, free of CME events. Winds related to CMEs would compromise the correlation between ${}^3He/{}^4He$ and H/He, because they combine low H/He ratios with relatively high ${}^3He/{}^4He$ ratios, i.e. opposite to the correlation in Figure 2b.

We have plotted in Figure 2 also the average ${}^3He/{}^4He$ ratio of the SWC/Apollo experiments that collected the solar wind during five periods in 1969-1972 (Geiss et al., 2004, see also Geiss *et al.* 1970). These were slow wind periods (320-510 km/s), and there are no indications of a CME influence. The H/He and Si/O ratios for the SWC exposure periods are estimates. Solar wind composition data were rudimentary during the Apollo epoch (1969-1972). Thus the corresponding estimates of H/He and Si/O are crude by necessity.

Recently Heber *et al.* (2009) published precise isotopic abundances of noble gases including the ${}^3He/{}^4He$ ratio obtained from the DOS Genesis target. The Genesis value is 9 % higher then the SWC/Apollo average i.e. well within the known variability of in-ecliptic ${}^3He/{}^4He$ ratios. The collection period of the DOS target lasted 2.3 years and included slow solar wind, high speed streams and some CME events. Therefore, their results are not directly applicable to our extrapolation method.

The ${}^3He/{}^4He$ value derived from the extrapolations shown in Figure 2 refers to the present-day OCZ. Two processes, however, could have changed this ratio during the life of the Sun.

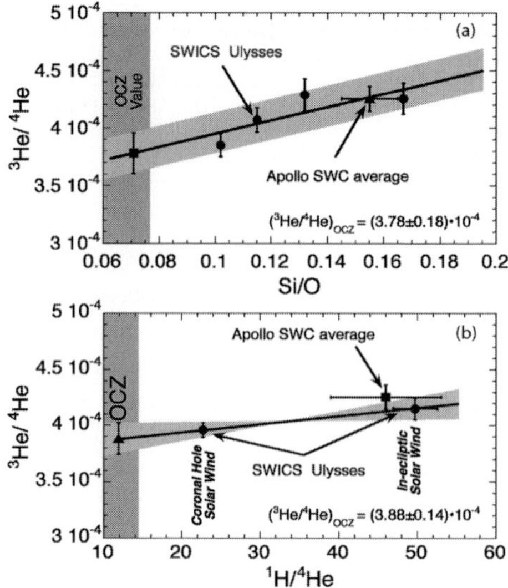

Figure 2. ^{3}He/^{4}He versus Si/O (2a) and ^{1}H/^{4}He (2b)) measured in the Solar Wind with SWICS/Ulysses and extrapolated to the corresponding values in the Outer Convective Zone (OCZ) of the Sun (adapted from Gloeckler & Geiss, 2000). The OCZ value for H/He was given by Pérez Hernández & Christensen-Dalsgaard (1994), the Si/O value by Grevesse & Sauval (1998). Combining these methods of extrapolation gives ^{3}He/^{4}He $= (3.83 \pm 0.25) \times 10^{-4}$ for the OCZ (Table 2). The ^{3}He/^{4}He ratio measured in 1969-1972 with Apollo SWC is identical to the "in-ecliptic" ^{3}He/^{4}He ratio measured with SWICS/Ulysses during quiet slow solar wind conditions.

Solar models show that He/H in the OCZ is 16% lower than it was in the PSC (e.g. Bahcall, Pinsonneault & Wasserburg, 1995). The difference is interpreted as being due to settling of helium out of the OCZ into deeper layers of the Sun. ^{3}He settles more slowly than ^{4}He, resulting in an increase in the present-day $(^3\mathrm{He}/^4\mathrm{He})_\mathrm{OCZ}$ ratio of 2 to 3% (Gautier & Morel, 1997).

The second possible change of $(^3\mathrm{He}/^4\mathrm{He})_\mathrm{OCZ}$ over solar history is due to solar mixing. During the lifetime of the Sun, the pp-reaction produces additional ^{3}He outside the solar core at intermediate depth in the Sun (Figure 3). H-burning at low temperature is controlled by the weak interaction reaction

$$^1\mathrm{H} + {}^1\mathrm{H} \to {}^2\mathrm{H} + e^+ + \nu. \tag{2.2}$$

Deuterium (D or ^{2}H) is immediately converted into ^{3}He (Eq. 2.1). Further fusion of ^{3}He becomes effective only at substantially higher temperature (see Figure 1).

A significant increase of ^{3}He in the OCZ could have been caused by mixing of pp-produced ^{3}He into the OCZ. Theoretical studies show this increase to be small, dependent on solar models and solar rotation (Vauclair, 1998; Turck-Chièze et al., 2001). Recent changes in solar abundances imply changes in opacity. Whether this has a significant effect on solar mixing is under investigation.

Addition of pp-produced ^{3}He to the OCZ has also been investigated by comparing solar wind helium trapped in very old and more recent lunar surface material. Using this method, Wieler & Heber (2003) concluded, that the increase of ^{3}He/^{4}He in the OCZ by solar mixing is at most 5%. Combining the two effects, settling of helium out of the OCZ

Table 3. Protosolar ^{3}He/^{4}He from ^{3}He/^{4}He in Jupiter and the Meteoritic "Planetary Gas".

	^{3}**He**/^{4}**He** $\times 10^{-4}$
Planetary Component in Meteorites[1,2]	$\sim 1.5 \pm 0.2$
Q-Phase of Carbonaceous Chrondite Isna[3]	1.23 ± 0.02
Jupiter Entry-Probe[4]	1.66 ± 0.06
Protosolar Cloud[5]	$\mathbf{1.66^{+0.06}_{-0.10}}$

Notes: [1]Eberhardt (1974); [2]Frick & Moniot (1977); [3]Busemann, Baur & Wieler (2000); [4] Mahaffy *et al.* (1998); [5]see text.

and solar mixing, we thus adopt a correction of $-(4 \pm 2)\%$ for $(^3\text{He}/^4\text{He})_{\text{OCZ}}$ and obtain $[(\text{D} +^3\text{He})/^4\text{He}]_{\text{PSC}} = (3.68 \pm 0.25) \times 10^{-4}$ (Table 2).

3. Protosolar ^{3}He/^{4}He

Before data of the Galileo Entry Probe were available, the ^{3}He/^{4}He ratio in the "planetary gas" of meteorites (e.g. Eberhardt, 1974; Frick and Moniot, 1977) served as a proxy for the isotopic abundance of helium in the PSC. More recently, it was realized that this planetary component is a mixture of components differing somewhat in composition. In Table 3 we have included the ^{3}He/^{4}He ratio in the "Q-phase" of the carbonaceous chondrite Isna (e.g., Busemann *et al.*, 2000). The situation changed when the ^{3}He/^{4}He ratio in the Jovian atmosphere was measured by the Galileo Probe Mass Spectrometer (GPMS), (see Niemann *et al.*, 1996). The ratio of $(1.66 \pm 0.06) \times 10^{-4}$ obtained by Mahaffy *et al.* (1998) given in Table 3 is presently considered to be the best value for estimating the protosolar value.

In the atmosphere of Jupiter helium is depleted by $\sim 18\%$ relative to the protosolar cloud (von Zahn, Hunten & Lehmacher, 1998). The degree, of helium depletion in the OCZ of the Sun is similar, but the causes could be quite different: diffusive separation in the case of the Sun, and probably descent of helium-rich droplets in the case of Jupiter (Stevenson & Salpeter, 1976; von Zahn, Hunten & Lehmacher, 1998). If the droplet hypothesis is correct, the Jovian process should fractionate isotopes much less than the solar process, and ^{3}He/^{4}He fractionation in the Jovian atmosphere would be less than the $\sim 2\%$ fractionation that has occurred in the OCZ of the Sun (Gautier & Morel, 1997; Vauclair, 1998). Thus, we adopt here the Jovian ^{3}He/^{4}He ratio obtained by Mahaffy *et al.* (1998) as the protosolar value, giving it, however, an asymmetric error.

The ^{3}He/^{4}He ratios in the "planetary gas" of meteorites are typically lower by (10-20)% than the Jovian ratio (Table 3). If ^{3}He/^{4}He in the Jovian atmosphere is identical to the OCZ value, a depletion of ^{3}He/^{4}He by (10-20)% in the planetary component in meteorites is certainly not high, considering the large mass difference of the two helium isotopes (cf. Geiss & Gloeckler, 2003).

4. Protosolar deuterium abundance

The deuterium abundance in the PSC has been derived in two ways: One method is based on a direct deuterium abundance measurement in Jupiter, either by remote spectroscopy (e.g. Beer & Taylor, 1973; Drossart *et al.*, 1982; Encrenaz *et al.*, 1996; Lellouch *et al.*, 2001) or by in-situ mass spectrometry (Niemann *et al.*, 1996; Mahaffy *et al.*, 1998). The values obtained by infrared spectroscopy have been given lower errors than those obtained by the Galileo Probe Mass Spectrometer (cf. Table 4). Lellouch

Table 4. Protosolar D/H from D/H in the Jovian Atmosphere.

	D/H $\times 10^{-5}$
Galileo Probe 1	2.6 ± 0.7
ISO 2	
H2	2.4 ± 0.4
CH4	2.2 ± 0.7
H2, CH4 Combined	2.25 ± 0.35
Corrected for D-Excess in Ice-Phase	2.1 ± 0.4
Protosolar D/H from Jovian D/H Data	$\mathbf{2.1 \pm 0.4}$

et al. (2001) measured D/H $= (2.40 \pm 0.4) \times 10^{-5}$ in molecular hydrogen and D/H $= (2.2 \pm 0.7) \times 10^{-5}$ in methane. Recently, Bézard (2002) obtained a D/H ratio in the methane below $< 2 \times 10^{-5}$. We adopt here the results of Lellouch *et al.* (2001) who combined their two determinations and obtained D/H $= (2.25 \pm 0.35) \times 10^{-5}$ as representative for Jupiter.

For the other method, the relation

$$(D/H)_{\mathrm{Protosolar}} = \tag{4.1}$$
$$\{[(D + {}^3He)/{}^4He]_{\mathrm{Protosolar}} - ({}^3He/{}^4He)_{\mathrm{Protosolar}}\} \times (He/H)_{\mathrm{Protosolar}}$$

is used. With protosolar $(D + {}^3He)/{}^4He = (3.68 \pm 0.25) \times 10^{-5}$, protosolar ${}^3He/{}^4He = (1.66^{+0.06}_{-0.10}) \times 10^{-4}$ (cf. Table 3) and protosolar H/He $= 10.28$ (Bahcall, Pinsonneault & Wasserburg (1995)) we obtain $(D/H)_{\mathrm{Protosolar}} = (1.97 \pm 0.3) \times 10^{-5}$ (Table 5).

Jupiters D/H ratio is probably enriched above the protosolar value by the admixture of deuterium-rich ices to the nebular gas during the planets formation (cf. Owen, 2003). According to Guillot (1999) the enrichment is about $(5\text{-}10)\%$. Lellouch *et al.* (2001) have corrected their ISO value for this effect and obtain $(D/H)_{\mathrm{Protosolar}} = (2.1 \pm 0.4) \times 10^{-5}$. The agreement between the D/H ratio determined using the solar wind value and that based on Jupiter measurements is excellent. We adopt $(2.0 \pm 0.3) \times 10^{-5}$ as the protosolar D/H ratio (Table 5).

5. Remarks and Conclusions

The protosolar abundances of hydrogen and helium isotopes are summarized in Table 5. We give "recommended abundances" whenever there are two independent values for the same abundance ratio. Our results are not very different from those in our earlier publications (Gloeckler & Geiss, 2000; Geiss & Gloeckler, 2003).

Deriving OCZ abundances from solar wind measurements could be improved, if data from experiments with different capabilities can be sensibly combined. Using Figure 2b, we present here an example. The SWC/Apollo ${}^3He/{}^4He$ ratio is 2.9% above the SWICS/Ulysses correlation line. There are two extremes we could consider: (1) Assuming that the 2.9% difference is due to a systematic error of the ${}^3He/{}^4He$ ratios determined with SWICS Ulysses and the difference of 2.9% was the same for all solar wind conditions covered in Figure 2b, the Apollo data would extrapolate to $({}^3He/{}^4He)_{\mathrm{OCZ}} = 3.99 \times 10^{-4}$. This value lies well within the systematic error limit assumed for the $({}^3He/{}^4He)_{\mathrm{OCZ}}$ ratio in Table 2. (2) As Figure 2a suggests, we assume that there are no systematic deviations between the ${}^3He/{}^4He$ values measured by the two instruments. In this case we correlate the SWC/Apollo slow wind data point with the high speed data point measured by SWICS/Ulysses and obtain $({}^3He/{}^4He)_{\mathrm{OCZ}} = 3.83 \times 10^{-4}$, about 1.3% below the value given in Table 2. These examples demonstrate the potential of combining the composition data of two experiments.

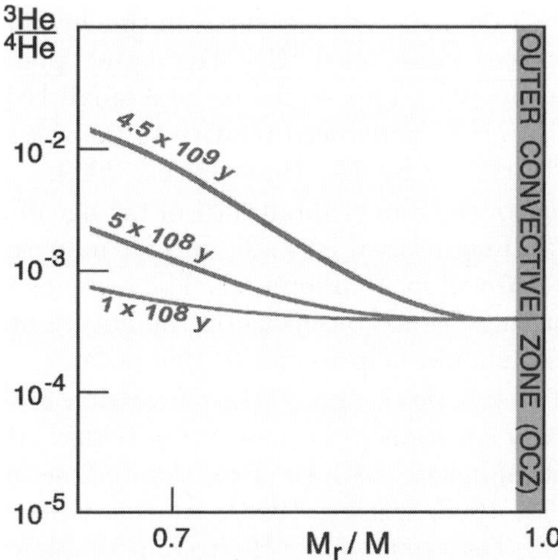

Figure 3. The secular buildup of ^{3}He at intermediate depth in the Sun as a function of M_r, (where M_r is the mass radius), calculated for a non-mixing Sun (Bochsler & Geiss, 1973) The buildup does not get close to the OCZ, but thorough experimental and theoretical investigations are needed to exclude contamination by ^{3}He produced at intermediate depth of the Sun.

Table 5. Protosolar Abundances.

From Solar Wind (SW) Data	$[(D+^3He)/H]_{Protosolar}$	$(3.58 \pm 0.25) \times 10^{-5}$
From Jupiter Data	$[(D+^3He)/H]_{Protosolar}$	$(3.72 \pm 0.42) \times 10^{-5}$
Recommended	$\mathbf{[(D+^3He)/H]_{Protosolar}}$	$\mathbf{(3.65 \pm 0.30) \times 10^{-5}}$
From Jupiter Data	$\mathbf{[^3He/H]_{Protosolar}}$	$(1.66^{+0.06}_{-0.10}) \times 10^{-4}$
From Jupiter D/H	$[D/H]_{Protosolar}$	$(2.1 \pm 0.4) \times 10^{-5}$
From $(^3He/^4He)_{SW}$ and $(^3He/^4He)_{Jup}$	$[D/H]_{Protosolar}$	$(1.97 \pm 0.3) \times 10^{-5}$
Recommended	$\mathbf{[D/H]_{Protosolar}}$	$\mathbf{(2.0 \pm 0.3) \times 10^{-5}}$

Presently, the uncertainties in the protosolar abundances are not so much due to measurement errors, but they result mainly from uncertainties in the transformation of solar wind or Jupiter data to protosolar values.

The systematic difference in H/He between OCZ and high-speed streams is caused mainly by ion-atom separation in or near the chromosphere (Geiss, 1982). Therefore, models applicable to corona ion fractionation cannot be applied to extend the $(^3He/^4He)$ versus H/He correlation from the high-speed wind data to the OCZ abundance. Theories for ion-atom separation depending on photo-ionization time predict only very minor fractionation for $^3He/^4He$ (von Steiger & Geiss, 1989). Since the difference between $^3He/^4He$ in the High-Speed Streams and the OCZ is only 2% (see Figure 2), the error introduced by the extrapolation ought to be small.

The contamination of the OCZ with pp-produced ^{3}He from below is one of the more serious uncertainties, but what could be done about this problem? The ^{9}Be abundance gives us a solid argument against loss of ^{3}He in the OCZ due to nuclear reactions. There is no comparable argument against a raise of $^3He/^4He$ in the OCZ by admixture of pp-produced ^{3}He. It is the other way around: a raise in time of ^{3}He in the OCZ or the solar wind is the most sensitive indicator for admixture of material to the OCZ from below, as was recognized long ago (Schatzman, Maeder, Angrand & Glowinski, 1981). Only by further theoretical studies and analyses of solar wind particles trapped in ancient dust and breccias could we further reduce the uncertainty of the protosolar $(D+^3He)/H$ ratio.

If we accept ^{3}He/^{4}He $= 1.66 \times 10^{-4}$ as measured in the Jovian atmosphere by Mahaffy *et al.* (1998) as the protosolar value, then ^{3}He/^{4}He in the "planetary gas" of meteorites would have been reduced by 10%. This seems to be a small reduction, considering that in the "planetary gas" ^{20}Ne/^{22}Ne is reduced relative to the solar abundance by (25-30)% (Geiss *et al.* 2004) and ^{36}Ar/^{38}Ar by 2% (Heber *et al.*, 2009).

Considering that the age of the Sun is about 35% of the age of the Galaxy, the chemical evolution during this 35% time interval is small, and for many nuclear species, evolution in time disappears in the noise of local differences. The levelling of evolution is attributed to infall (Tosi, 1998), but there is no consensus on the nature of this infall. Studying the evolution of dwarf galaxies should help to clarify this point.

The principal effect of stellar processing is the conversion of D into ^{3}He with the sum (D $+$ ^{3}He) remaining nearly constant (e.g. Geiss & Gloeckler, 2003). This was recognized when ^{3}He and D were determined in both the Protosolar Cloud and the Local Interstellar Cloud (Gloeckler & Geiss, 1996; Linsky, 1998). At the same time, theoretical studies showed (Charbonnel, 1998; Tosi, 1998) that ^{3}He from incomplete hydrogen burning could not have a large effect on the chemical evolution in the Galaxy. Since then, progress in understanding late Galactic evolution has slowed. Determining abundances of a larger number of nuclides in one and the same Galactic sample could correct this impasse better than further collecting scattered composition data. This could be accomplished with a mission similar to Ulysses but with the primary objective of measuring the composition of the Local Interstellar Cloud by determining abundances of elements and isotopes in the gas and dust that is entering the heliosphere. A study to identify the nuclear species in the Local Interstellar Cloud that could be determined in this way would be most worthwhile. Advanced solar wind composition measurement would be a second objective of such a mission. Moreover, comparing the differing paths of the physical and the chemical evolution of our Galaxy and dwarf galaxies would lead to identifying the nature and origin of the infall into the Milky Way.

Acknowledgements

The authors thank Rudolf von Steiger and Rudolf Treumann for discussions. J. G. is grateful to the ISSI directors and staff for their support. The work of G. G. was supported in part by NASA's Ulysses Heliophysics Investigation NNX09AH726 and the ACE Data Analysis Contract, 44A-1080828.

References

Bahcall, J. N., Pinsonneault, M. H., & Wasserburg, G. J. 1995, *Rev. Mod. Phys.*, 67, 781

Beer, R. & Taylor, F. W. 1973, *ApJ*, 179, 309

Bézard, B. 2002, in: *Highlights in Astronomy* (Astron. Soc. Pacific: San Francisco) Vol. 12, pp. 607

Bochsler, P. & Geiss, J. 1973, *Solar Phys.*, 32, 3

Bürgi, A. & Geiss, J. 1986, *Solar Phys.*, 103, 347

Busemann, H., Baur, H., & Wieler, R. 2000, *Meteorites & Planetary Sci.*, 35, 949

Charbonnel, C. 1998, in: *Space Science Series of ISSI*, SSSI-4, and *Space Sci. Rev.*, 84, 199

Drossart, P., Encrenaz, T., Combes, M., Kunde, V., & Hanel, R. 1982, *Icarus*, 49, 416-426

Eberhardt, P. 1974, *Earth Planet. Sci. Lett.*, 24, 182

Encrenaz, Th., de Graauw, Th., Schaeidt, S., Lellouch, E., Feuchtgruber, H., Beintema, D. A., Bézard, B., Drossart, P., Griffin, M., Heras, A., Kessler, M., Leech, K., Morris, P., Roelfsema, P. R., Roos-Serote, M., Salama, A., Vandenbussche, B., Valentijn, E. A., Davis, G. R., & Naylor, D. A. 1996, *A&A*, 315, L397

Ezer, D. & Cameron, A. G. W. 1965, in *Solar Wind Three* (University of California Press: Los Angeles), pp. 58

Frick, U. & Moniot, R. K. 1977, in *Eighth Lunar Science Conference* (Pergamon Press: New York), Vol. 1, pp. 229

Gautier, D. & Morel, P. 1997, *A&A*, 323, L9

Geiss, J. 1982, *Space Sci. Rev.*, 33, 201

Geiss, J. & Gloeckler, G. 1998, in *Space Science Series of ISSI*, SSSI-4, and *Space Sci. Rev.*, 84, 239

Geiss, J. & Gloeckler, G. 2003, in *Space Science Series of ISSI*, SSSI-16, and *Space Sci. Rev.*, 106, 3

Geiss, J. & Reeves, H. 1972, *A&A*, 18, 126

Geiss, J., Eberhardt, P., & Signer, P. 1966, *Proposal to NASA for Apollo experiment S-080*, (NASA archive, Record Number 14759)

Geiss, J., Eberhardt, P., Bühler, F., Meister, J., & Signer, P. 1970, *J. Geophys. Res.*, 75, 5972

Geiss, J., Hirt, P., & Leutwyler, H. 1970, *Solar Phys.*, 12, 458

Geiss, J., Bühler, F., Cerutti, H., Eberhardt, P., Filleux, Ch., Meister, J., & Signer, P. 2004, *Space Sci. Rev.*, 110, 307

Gloeckler, G. & Geiss, J. 1996, *Nature*, 381, 210

Gloeckler, G. & Geiss, J. 2000, in *The Light Elements and their Evolution*, Proc. IAU-Symp. Vol. 198, (IAU: Natal, Brazil) pp. 224

Grevesse, N. & Sauval, A. J. 1998, in *Space Science Series of ISSI*, SSSI-5, and *Space Sci. Rev.*, 85, 161

Guillot, T. 1999, *Planet. Space Sci.*, 47, 1183

Heber, V. S. Wieler, R., Baur, H., Olinger, C., Friedmann, T. A., & Burnett, D. S. 2009, *Geochimica et Cosmochimica Acta*, 73, 7414

Isenberg, P. A. & Hollweg, J. V. 1983, *J. Geophys. Res.*, 88, 3923

Lellouch, E., Bézard, B., Fouchet, T., Feuchtgruber, H., Encrenaz, T., & de Graauw, T. 2001, *A&A*, 370, 610

Linsky, J. L. 1998, in *Space Science Series of ISSI*, SSSI-4, and *Space Sci. Rev.*, 84, 285

Mahaffy, P. R., Donahue, T. M., Atreya, S. K., Owen, T. C., & Niemann, H. B. 1998, in *Space Science Series of ISSI*, SSSI-4, and *Space Sci. Rev.*, 84, 251

Mazzitelli, I. & Moretti, M. 1980, *ApJ*, 235, 955

Niemann, H. B., Atreya, S. K., Carignan, G. R., Donahue, T. M., Haberman, J. A., Harpold, D. N., Hartle, R. E., Hunten, D. M., Kasprzak, W. T., Mahaffy, P. R., Owen, T. C., Spencer, N. W., & Way, S. H. 1998, *Science*, 272, 846

Owen, T. 2003, in *Space Science Series of ISSI*, SSSI-28, and *Space Sci. Rev.*, 138, 301

Pérez Hernández, F. & Christensen-Dalsgaard, J. 1994, *MNRAS*, 269, 475

Schatzman, E., Maeder, A., Angrand, F., & Glowinski, R. 1981, *A&A*, 96, 1

von Steiger, R. & Geiss, J. 1989, *A&A*, 225, 222

Stevenson, D. J. & Salpeter, E. E. 1976, in *Jupiter* (University of Arizona Press: Tucson), pp. 85

Tosi, M. 1998, in *Space Science Series of ISSI*, SSSI-4, and *Space Sci. Rev.*, 84, 207

Turck-Chièze, S., Nghiem, P., Couvidat, S., & Turcotte, S. 2001, *Solar Phys.*, 200, 323

Vauclair, S. 1998, in *Space Science Series of ISSI*, SSSI-4, and *Space Sci. Rev.*, 84, 256

Wieler, R. & Heber, V. S. 2003, in *Space Science Series of ISSI*, SSSI-16, and *Space Sci. Rev.*, 106, 197

von Zahn, U., Hunten, D. M., & Lehmacher, G. 1998, *J. Geophys. Res.*, 103, 22815

Johannes Geiss, Corinne Charbonnel, Hubert Reeves

Manuel Peimbert & Tom Bania

Light Elements in the Universe
Proceedings IAU Symposium No. 268, 2009
C. Charbonnel, M. Tosi, F. Primas & C. Chiappini, eds.

© International Astronomical Union 2010
doi:10.1017/S174392131000390X

Measurements of ^{3}He in Galactic H II regions and planetary nebulae

T. M. Bania[1], R. T. Rood[2] and D. S. Balser[3]

[1]Department of Astronomy, Boston University
725 Commonwealth Ave., Boston, MA 02215, USA
email: bania@bu.edu

[2]Astronomy Department, University of Virginia
P.O. Box 400325, Charlottesville, VA 22904, USA
email: rtr@virginia.edu

[3]National Radio Astronomy Observatory,
520 Edgemont Rd., Charlottesville, VA 22903
email: dbalser@nrao.edu

Abstract. The cosmic abundance of the ^{3}He isotope has important implications for many fields of astrophysics. We are using the 8.665 GHz hyperfine transition of ^{3}He$^+$ to determine the ^{3}He/H abundance in Milky Way H II regions and planetary nebulae. Here we review the 30 year history of our ^{3}He program, report on its current status, and describe our future plans.

Keywords. Cosmology: cosmological parameters; The Galaxy: abundances, evolution; ISM: abundances, evolution, H II regions; Stars: AGB, post-AGB

1. The ^{3}He Problem

The ^{3}He abundance in Milky Way sources impacts stellar evolution, chemical evolution, and cosmology. The abundance of ^{3}He is derived from measurements of the hyperfine transition of ^{3}He$^+$ which has a rest wavelength of 3.46 cm (8.665 GHz). As with all the light elements, the present interstellar ^{3}He abundance results from a combination of Big Bang Nucleosynthesis (BBN) and stellar nucleosynthesis (Wilson & Rood 1994). We are measuring the ^{3}He abundance in Milky Way H II regions and planetary nebulae (PNe). H II regions are examples of zero-age objects that are young relative to the age of the Galaxy. Their ^{3}He abundances chronicle the results of billions of years of Galactic chemical evolution (GCE). PNe ^{3}He abundances arise from material that has been ejected from low-mass ($M \leqslant 2\,M_\odot$) and intermediate-mass ($M \sim 2$–$5\,M_\odot$) stars. Because the Milky Way ISM is optically thin at centimeter wavelengths, our source sample probes a larger volume of the Galactic disk than does any other light element tracer of GCE. The sample is currently comprised of 60 H II regions and 12 PNe.

The present ^{3}He abundance is an important GCE diagnostic. H II regions sample the result of the chemical evolution of the Milky Way since its formation. The ^{3}He/H abundance ratio is expected to grow with time and to be higher in those parts of the Galaxy where there has been substantial stellar processing. If the ^{3}He yields predicted by standard models of stellar nucleosynthesis (SSN) had been correct then we should have detected extremely strong ^{3}He$^+$ emission in Galactic H II regions (RST: Rood, Steigman, & Tinsley 1976). Our ^{3}He$^+$ experiment would then have taken about two weeks. After 30 years of effort (Rood, Wilson, & Steigman 1979; Bania *et al.* 2007) our observations still yield ^{3}He abundances for H II regions that are inconsistent with these expectations, leading to "The ^{3}He Problem" (Galli *et al.* 1995).

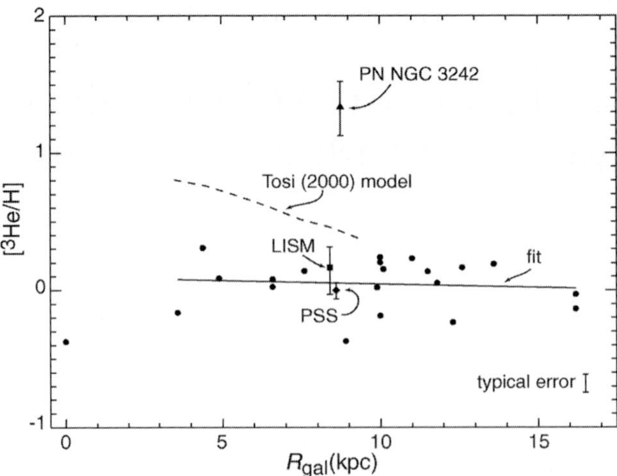

Figure 1. "The ^{3}He Problem" ^{3}He/H abundances as a function of Galactic radius. The [^{3}He/H] abundances by number for the BRB H II region sample are given with respect to the solar ratio. Shown also are the abundances for the planetary nebula NGC 3242 (triangle), the local interstellar medium (LISM—square), and protosolar material (PSS—diamond). There is no gradient in the ^{3}He/H abundance with Galactic position.

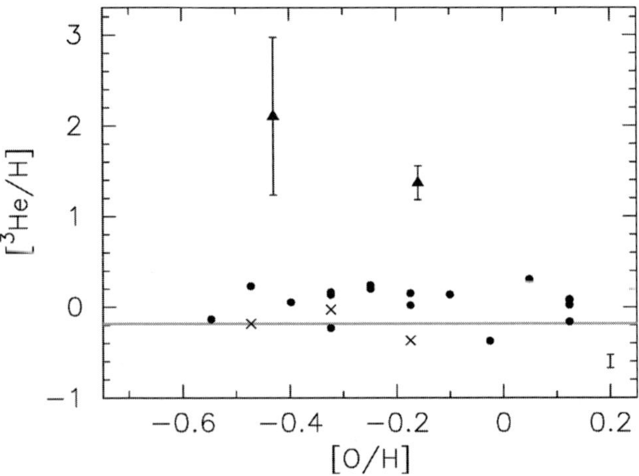

Figure 2. "The ^{3}He Plateau" [^{3}He/H] abundances as a function of source metallicity for the "simple" H II region sample. The gray line is the WMAP result. The ~ 0.15 dex typical error is shown in the right hand corner. The triangles denote abundances for the PNe J 320 (left) and NGC 3242. There is no trend in the ^{3}He/H abundance with source metallicity.

2. H II region abundances

Until its decommissioning in 1999 we used the NRAO 140-Foot telescope to observe ^{3}He$^+$ in a sample of 60 Galactic H II regions (Rood, Bania, & Wilson 1984; Bania, Rood, & Wilson 1987; Balser *et al.* 1994; Rood *et al.* 1995; Bania *et al.* 1997; Balser *et al.* 1999a; Bania, Rood, & Balser 2002; Bania *et al.* 2007). The $\sim 200''$ resolution (FWHM "beam") of the 140-Foot at the ^{3}He$^+$ frequency is comparable to the angular size of typical Galactic H II regions, making this the instrument of choice at the time.

Besides detecting ^{3}He$^+$ emission from an H II region and accurately determining the line intensity and line width, in order to derive the ^{3}He$^+$/H$^+$ abundance ratio we also need to measure the intensity of the continuum emission produced by thermal free-free

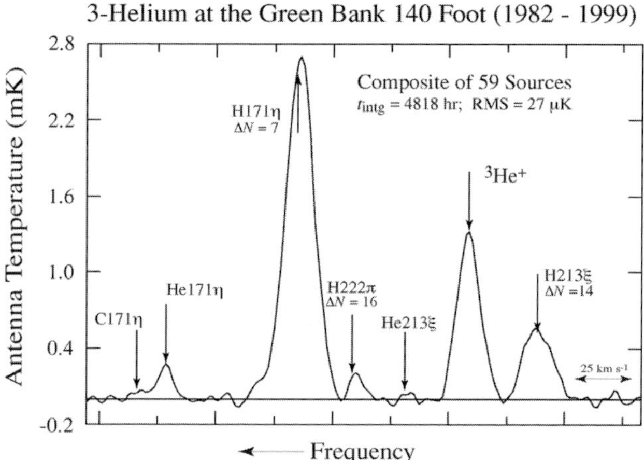

Figure 3. NRAO 140-Foot H II region composite ^{3}He spectrum. This 200 day integration is the most sensitive cm-wavelength spectrum ever taken. Besides the ^{3}He$^+$ emission, a variety of recombination lines from H, ^{4}He, and C are seen. Many of these arise from transitions resulting from large changes in principle quantum number, e.g. H 171 η, Δ N = 7, and H 213 ξ, Δ N = 14.

brehmstrahlung in the nebular plasma. A nebula's ^{3}He$^+$/H$^+$ abundance thus depends on its ^{3}He$^+$ line to thermal continuum intensity ratio. For typical Galactic H II regions the observed line-to-continuum ratio is $\sim 10^{-4}$ to $\sim 10^{-5}$. The 140-Foot's conventional on-axis optics are, moreover, a very poor design for the ^{3}He experiment. Typically, the ^{3}He emission lines have intensities of a few mK and line widths of $\sim 1\,$MHz and this weak, wide line emission must be measured in the presence of the far stronger continuum emission. For conventional blocked aperture optics such as those of the 140-Foot, this strong continuum emission is reflected and scattered by the secondary mirror, its support legs, the receiver cabin, etc., causing direct and multipath standing waves that lead to frequency structure in the observed spectra. This instrumental spectral baseline frequency structure is extremely complex and impossible to model a priori because these standing waves are not pure tones and they vary on short time scales as a source is tracked. Some of the baseline frequency structure can be removed by purposefully defocussing by $\pm \lambda/8$, but much structure remains and needs to be modeled empirically. Baseline modeling errors contribute to the uncertainty in the ^{3}He$^+$ line parameters that we can measure. For H II regions, such errors are too small to compromise our ^{3}He$^+$ detections.

Deriving an accurate ^{3}He/H abundance ratio also requires modeling the nebular density and ionization structure. In doing so we identified a special class of "simple" H II regions for which accurate ^{3}He/H abundances can be determined (Balser *et al.* 1999a). This source sample provides a strong constraint on GCE models. Standard evolution models predict that: (1) the protosolar ^{3}He/H value should be less than that found in the present ISM; (2) the ^{3}He/H abundance should grow with source metallicity; and (3) there should be a ^{3}He/H abundance gradient in the Galactic disk with the highest abundances occurring in the highly-processed inner Galaxy. None of these predictions is confirmed by observations.

Specifically, measurements of ^{3}He/H in protosolar material (Geiss 1993), the local solar neighborhood (Gloecker & Geiss 1996), and Galactic H II regions (Rood *et al.* 1995) all indicate a value of ^{3}He/H $\sim 2 \times 10^{-5}$ by number. Thus the H II regions show no evidence for stellar ^{3}He enrichment during the last 4.5 Gyr. There is no significant ^{3}He abundance gradient across the Milky Way (Fig. 1). And, finally, there is no trend of ^{3}He abundance with source metallicity (Fig. 2). To be compatible with this result GCE models require that $\sim 90\%$ of solar analog stars are non-producers of ^{3}He (Tosi 2000). Finally, BRB argued that the "The ^{3}He Plateau" ^{3}He/H abundance is an upper limit for the cosmological production of ^{3}He. They derived $\eta_{10} = 5.4\ (+2.2/-1.2)$, a year before and in complete concordance with the WMAP value (Spergel $et\ al.$ 2003; Romano $et\ al.$ 2003).

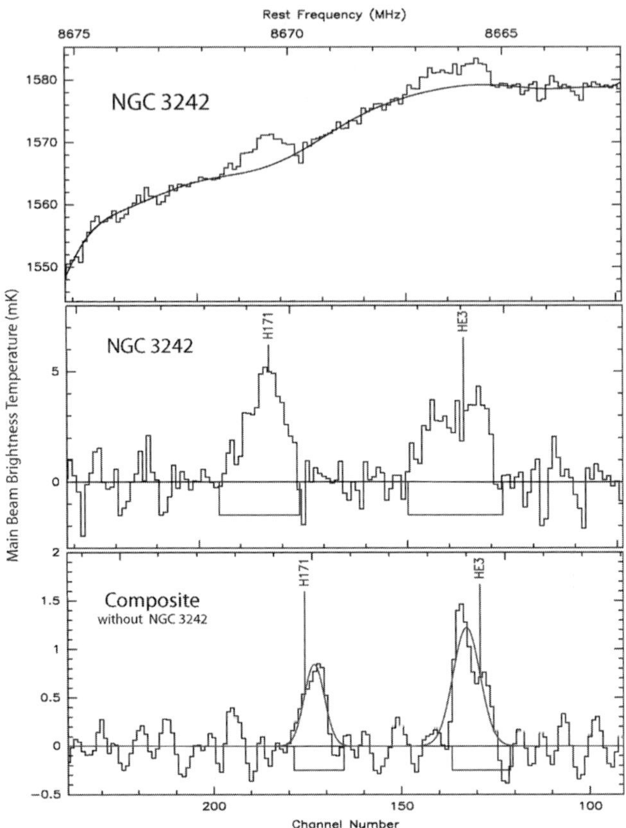

Figure 4. MPIfR 100-m ^{3}He$^+$ spectra for Galactic PNe. **TOP:** The NGC 3242 average spectrum (106 hr integration) with the baseline model superimposed. The H 171 η and ^{3}He$^+$ features are clearly seen. **MIDDLE:** The NGC 3242 spectrum after the baseline model is subtracted. **BOTTOM:** Composite ^{3}He$^+$ average spectrum (443 hr integration) for 6 PNe (NGC 6543, NGC 6720, NGC 7009, NGC 7662, & IC 289).

3. Planetary nebula abundances

Standard stellar evolution theory predicts not only that common solar-type stars produce ^{3}He but also that the mass lost from winds generated at advanced stages of their evolution and the final planetary nebulae should be substantially enriched in ^{3}He. Planetary nebula ^{3}He abundances are therefore important tests of stellar evolution theory since these low-mass, evolved objects are expected to be significant sources of ^{3}He. It is thus crucial to see if stars actually do produce ^{3}He. Detecting ^{3}He$^+$ in the nebulae surrounding PNe is an even more challenging experimental problem than that for H II regions. Finding ^{3}He in PNe challenges the sensitivity limits of all existing radio telescopes. PNe nebulae contain $\sim 1\ M_\odot$ of gas whereas H II region plasmas have masses ~ 100's to ~ 1000's $M_\odot$. The detection of ^{3}He in *any* PN, however, will require ^{3}He/H $\sim 10^{-4}$, which is the abundance predicted by standard stellar models.

We are observing a sample of 12 PNe that is purposefully biased to maximize the likelihood of finding ^{3}He. Between 1987 and 1997 we used the 100-m telescope of the Max Planck Institut für Radioastronomie (MPIfR) to observe ^{3}He$^+$ in a sample of 6 PNe (Rood, Bania, & Wilson 1992; Balser *et al.* 1997). The $\sim 85''$ 100-m beam is a good match to the angular extent of typical Galactic PNe. We reported a ^{3}He$^+$ detection for NGC 3242 and found significant ^{3}He$^+$ emission in a composite spectrum resulting from the average of 6 PNe (Fig. 4). Even though the continuum emission from PNe is significantly weaker than that from H II regions, the MPIfR optics, which have 22% geometric blockage, produce extremely complex instrumental baseline frequency structure as can be seen in Fig. 4 (top). It was important to verify our ^{3}He$^+$ detection with another telescope. Despite the fact that PNe with their small angular sizes were not good NRAO 140-Foot targets we were able to verify our NGC 3242 result at the $\sim 4\,\sigma$ level (Fig. 5). The MPIfR and NRAO NGC 3242 line shapes were consistent and they suggested that much of the emission ^{3}He$^+$ emission comes from a large, diffuse halo. Subsequently we detected ^{3}He$^+$ emission at the $\sim 4\,\sigma$ level from the PN J 320 with the NRAO Very Large Array (VLA) (Balser *et al.* 2006). It seems that at least *some* (i.e., > 2) PNe produce significant amounts of ^{3}He that survives the PN stage and enriches the ISM.

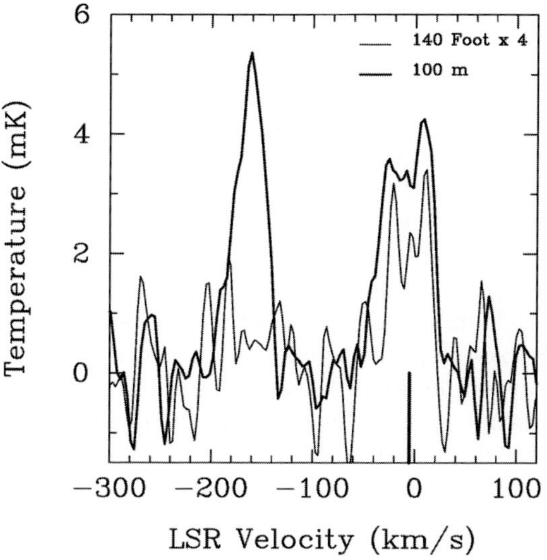

Figure 5. NRAO 140-Foot ^{3}He$^+$ spectrum for NGC 3242 compared with the MPIfR result. The ^{3}He$^+$ emission is flagged at $-5.3\,\mathrm{km\,sec}^{-1}$. This 270 hr integration is one of the longest cm-wavelength spectra ever made toward a single source.

4. Current status of the ^{3}He experiment

The ^{3}He$^+$ experiment challenges the sensitivity limits of all exisiting radio telescope spectrometers. We are now studying ^{3}He with the NRAO Green Bank Telescope (GBT) and NAIC's Arecibo Observatory. We are soon to begin using the NRAO Expanded VLA (EVLA) and ATNF's Parkes 64-m telescope. The EVLA will be a factor of 10 better than the VLA in all relevant experimental parameters. It will be a powerful new tool for studying ^{3}He in PNe. With Parkes it is feasible to detect ^{3}He$^+$ in the 30 Doradus

H II region in the Large Magellanic Cloud. Doing so would extend the metallicity range of the ^{3}He plateau ^{3}He/H abundances.

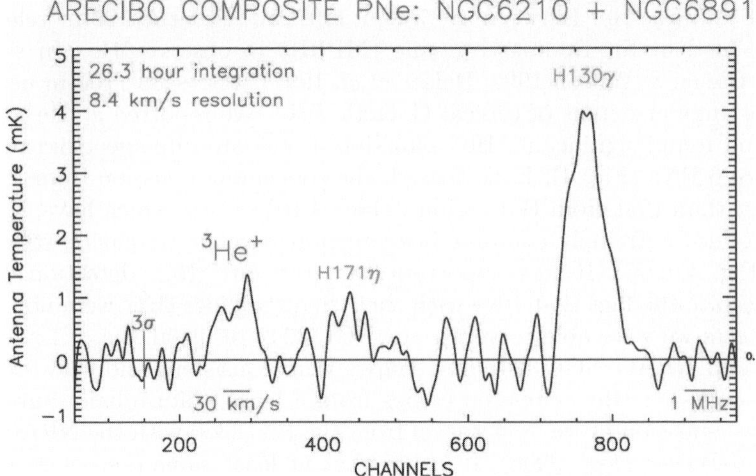

Figure 6. Arecibo ^{3}He$^+$ spectrum for a composite average of 2 Galactic PNe, NGC 6210 & NGC 6891, aligned to a common LSR velocity. A baseline model was removed and the spectrum smoothed to 8.4 km sec^{-1} resolution. The H130γ recombination and ^{3}He$^+$ spin-flip transitions are clearly seen.

We are using the 305-m Arecibo telescope to study ^{3}He$^+$ in PNe. Fig. 6 shows a composite ^{3}He$^+$ spectrum for two PNe. We choose targets to match the properties of the particular telescope we are using. For the most part our Arecibo target PNe were chosen to have angular sizes well matched to its $\sim 45''$ beam. Even though not on our initial Arecibo target list, our VLA results suggested that J 320 was also a plausible target. Thus, to confirm our VLA detection we began observing J 320 in January 2009. At Arecibo we are observing at the high frequency limit set by the surface accuracy of the primary mirror. The primary is fixed in place, which makes Arecibo a semi-transit instrument, so as a source is tracked across the sky different areas of the primary mirror are illuminated. Because of this the telescope gain varies significantly as a source is tracked, which makes intensity calibration difficult. The semi-transit nature of Arecibo also means that progress will be slow. We need significant integration times for our sources, ~ 50–100 hrs, but a full week of observing, 7 sidereal passes of a source, yields 14 hrs of source integration time.

We were part of the 100-m GBT project from its inception through commissioning and are now using it to study ^{3}He$^+$ in H II regions and PNe. Our ^{3}He$^+$ program thus far has focussed on the S 209 H II region and the NGC 3242 PN. Located in the outer Galactic disk at $R_{\rm gal} \sim 16.9$ kpc, S 209 provides an important constraint on BBN and GCE. Confirming the NGC 3242 ^{3}He abundance has important consequences for SSN and GCE. We are also observing the W 3 H II region and the PNe NGC 6543, NGC 6826, and NGC 7009.

The GBT's clear aperture, unblocked optics are unique; it has the potential to be much more sensitive than any cm-wavelength spectrometer ever built. This increased sensitivity results from the 100-m aperture, the unblocked optics, which allow for unprecedented dynamic range, and the unique capabilities of the 3 cm receiver and autocorrelation spectrometer (ACS). For the ^{3}He$^+$ experiment we simultaneously measure 8 different frequency bands at two orthogonal polarizations (RR and LL). Each band is 50 MHz

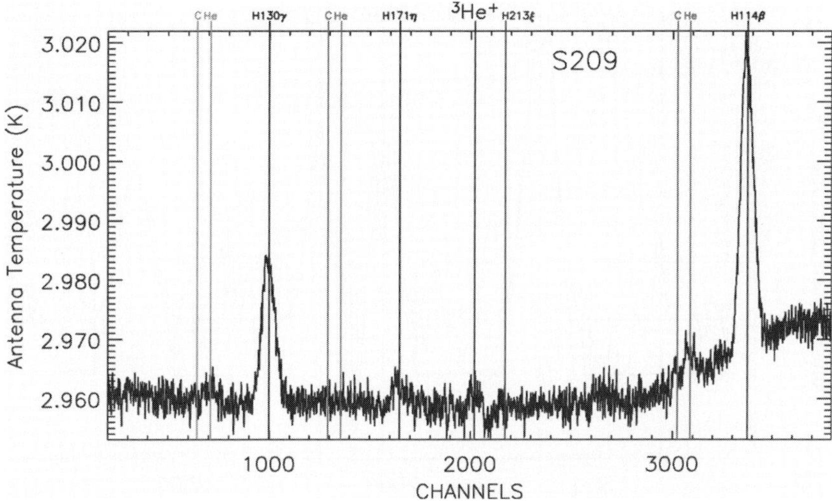

Figure 7. First epoch GBT $^3\mathrm{He}^+$ spectrum for H II region S 209. This calibrated, but otherwise unprocessed spectrum, shows that the GBT's unblocked optics have eliminated instrumental standing waves. There is a clear $^3\mathrm{He}^+$ signal in this 14.5 hr spectrum. Numerous recombination line transitions of H, and $^4\mathrm{He}$ can be seen, including a clear detection of the H 213 ξ line.

wide and is sampled by 4096 channels. We tune two bands to $^3\mathrm{He}^+$ in order to assess the GBT's electronics by sampling the identical signal through two independent paths. We thus measure a source's emission over 350 MHz of spectrum (7×50 MHz). The different bands are tuned to a host of radio recombination lines (RRLs) of H, $^4\mathrm{He}$, and C. Our targets' fully ionized plasmas will emit all possible RRL transitions, but at a vast range of intensities. We analyze more than 50 RRL transitions, spanning principal quantum number, N, from 90 through 224 and order, ΔN, from 1 through 16.

As we have done for all the $^3\mathrm{He}^+$ experiment telescopes, we use this plethora of RRL measurments to assess the performance of the GBT spectrometer. For example, the RRL line intensities should not vary with time, the $^4\mathrm{He}$/H intensity ratio should not be a function of principle quantum number, and the lines should only be seen in emission. RRL theory predicts line intensities for all these transitions, which can then be used to assess spectrometer performance. For example, in the $^3\mathrm{He}^+$ band, LTE models for diffuse H II regions predict that the H 213 ξ intensity should be $\sim 25\%$ the H 171 η intensity (see, e.g., Fig. 3, Fig. 4, & Fig. 5). For any given nebula the entire ensemble of RRLs should give an astrophysically self-consistent set of measured line parameters. Any deviations or inconsistencies are caused by instrumental effects which, experience shows, inevitably appear at some point as one attains ever increasing sensitivity levels. At the sensitivity required by the $^3\mathrm{He}$ experiment, characterizing the spectrometer's performance is a on-going, evolving process. We are just now in a position, for example, to begin exploring possible diurnal and seasonal effects on the GBT's instrumental baseline frequency structure.

The current status of the GBT $^3\mathrm{He}^+$ experiment for S 209 and NGC 3242 is shown in Fig. 8 (25.8 receiver-hr integration) and Fig. 9 (91.6 receiver-hr integration), respectively. Our GBT $^3\mathrm{He}^+$ result for S 209 is completely consistent with our 140-Foot measurement, confirming the $^3\mathrm{He}$/H abundance of "The $^3\mathrm{He}$ Plateau."

Extensive analysis of the RRLs from S 209 shows that at our current sensitivity level all spectral features with $\gtrsim 1$ mK intensities are reliably detected. Below this intensity, however, the observed line parameters become uncertain and we begin to see evidence for

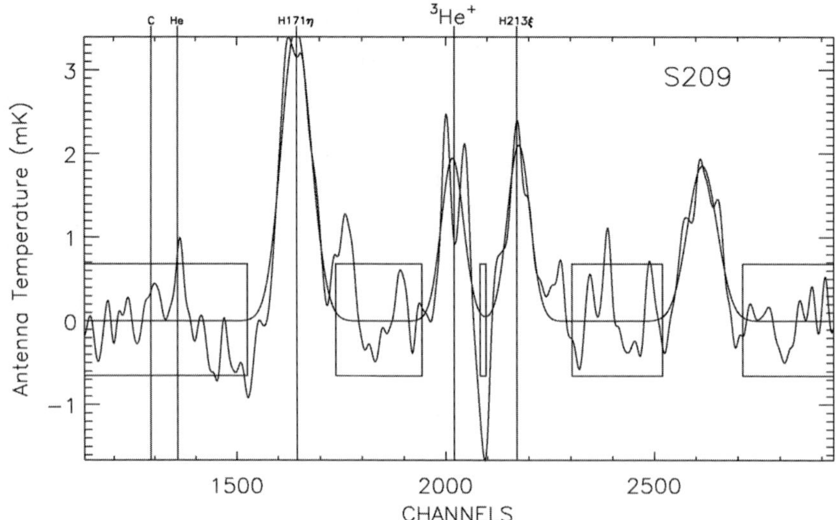

Figure 8. GBT ^{3}He$^+$ average spectrum for H II region S 209 as of November 2009.

instrumental baseline frequency structure (BFS). In the entire 350 MHz spectral range that we measure, we find only a few cases of instrumental BFS that mimic the weak, ~ 1 mK, wide, ~ 1 MHz, lines characteristic of ^{3}He$^+$ emission. Alas, we are very, very unlucky in that the two strongest such instrumental features not only occur within the ^{3}He$^+$ band, but also lie atop the H 213 ξ and H 203 μ RRLs! These instrumental BFS components make the observed H 213 ξ and H 203 μ intensities much too strong (Fig. 8 & Fig. 9). We currently estimate that instrumental BFS enhances these transition intensities by $\sim 20\%$ up to a factor of a few, increasing with source continuum intensity.

Our GBT ^{3}He$^+$ measurements for the PN NGC 3242 (Fig. 9) do not confirm our MPIfR result: the GBT intensities for ^{3}He$^+$ and H 171 η are both lower by a factor of ~ 4. Interestingly, our 140-Foot observations, Fig. 3, also suggested the MPIfR intensities were too high. At the time we attributed the differences between the 100 m and 140-Foot to be due to the different beam sizes: the observed spectra sampled very different volumes surrounding the PN. Density structure, in particular a large, low density, expanding halo could plausibly account for the intensity differences. This can no longer be the case since the GBT and MPIfR beams sizes are essentially identical.

Nonetheless, NGC 3242 *does* have a large, low density, expanding halo which produces a large, ~ 50 km sec^{-1} wide ^{3}He$^+$ line. As can be seen in Fig. 9 this wide ^{3}He$^+$ line blends with the H 213 ξ emission. Moreover, the proximity of H 203 μ, together with the instrumental BFS associated with these RRL transitions makes the ^{3}He$^+$ line parameters uncertain.

We explored a range of baseline models for the NGC 3242 ^{3}He$^+$ emission. Currently, at best, the GBT data are consistent with a ^{3}He$^+$ abundance ratio of $\sim 10^{-4}$ by number, with an uncertainty of $\sim 50\%$. We also studied a variety of composite ^{3}He$^+$ spectra for our GBT PNe sources. The best such cases are consistent with the composite PNe spectra obtained at the MPIfR and Arecibo. Again, these composites are consistent with a ^{3}He$^+$ abundance ratio of $\sim 10^{-4}$ by number.

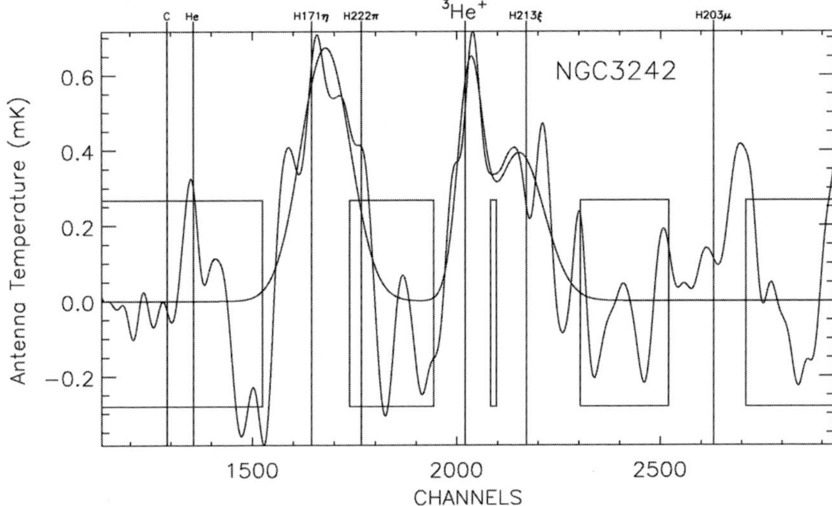

Figure 9. GBT ^{3}He$^+$ average spectrum for NGC 3242 as of November 2009.

5. Solving "The ^{3}Helium Problem"

Rood, Bania, & Wilson (RBW: 1984) suggested that the ^{3}He problem could be related to striking chemical anomalies in red giant stars. Much accumulated observational evidence shows that low-mass RGB stars undergo an extra-mixing event. This extra-mixing adds to the standard first dredge-up to modify the surface abundances. Efforts by many groups during the past 25 years have confirmed that RBW were on the right track. Significantly, all the relevant data indicate that the extra-mixing occurs in $\sim$ 90 to 95% of the low-mass stars. Thermohaline instability and rotation-induced mixing now seem to be able to account for the drastically reduced ^{3}He yields produced by low-mass red giants (Charbonnel & Zahn 2007a,b; Charbonnel & Lagarde 2010; Eggleton, Dearborn, & Lattanzio 2006).

6. Summary

The NRAO GBT result for S 209 is consistent with *all* previous Galactic H II region ^{3}He/H abundance determinations. Thus "The 3-Helium Plateau" ^{3}He/H abundance, important for primordial nucleosynthesis and Galactic chemical evolution, stands. The ^{3}He/H abundance by number, which is not seen to be a function of either source metallicity or Galactic position, is $\sim 2 \times 10^{-5}$.

Attempting to detect ^{3}He in PNe challenges the capabilities of all existing radio spectrometer systems. We reported ^{3}He detections in NGC 3242 with the MPIfR 100 m and NRAO 140-Foot telescopes. Our NRAO GBT observations are, however, inconsistent with the MPIfR 100 m result, implying a ^{3}He/H abundance that is $\sim$ 25% of that we previously reported. We find ^{3}He in J 320 at the 4 σ level with the NRAO VLA. Confirmation observations are underway using the NAIC Arecibo telescope. Composite PNe ^{3}He$^+$ spectra, made from MPIfR, Arecibo, and GBT observations, consistently show ^{3}He$^+$ emission at the $\sim$ 1 mK level. As a class our sources imply a ^{3}He/H abundance by number of $\sim 10^{-4}$, which suggests that some PNe produce ^{3}He.

Resolving "The ^{3}He Problem" requires that the vast majority of low-mass stars fail to enrich the ISM with ^{3}He due to extra-mixing in the RGB stage. GCE models can account for "The ^{3}He Plateau" only if $\gtrsim$ 90% of solar analog stars are non-producers of

^{3}He. Thermohaline instability and rotation-induced mixing are able to account for the drastically reduced ^{3}He yields produced by these stars. Our PNe target sample is purposefully biased to contain objects whose progenitor stars underwent no extra-mixing, which has been suppressed by some yet to be determined mechanism (but see the conjecture by Charbonnel & Zahn 2007b; Charbonnel & Lagarde 2010). These rare PNe seem to be returning ^{3}He to the ISM.

Acknowledgements

We thank the international light element abundances community for their collegiality and support over the years.

References

Balser, D. S., Bania, T. M., Brockway, C. J., Rood, R. T., & Wilson, T. L. 1994, *ApJ*, 430, 667

Balser, D. S., Bania, T. M., Rood, R. T., & Wilson, T. L. 1997, *ApJ*, 483, 320

Balser, D. S., Bania, T. M., Rood, & Wilson, T. W. 1999, *ApJ*, 510, 759

Balser, D. S., Rood, R. T., & Bania, T. M. 1999, *ApJ*, 522, L73 [N3242]

Balser, D. S., Goss, W. M., Bania, T. M., & Rood, R. T. 2005, *ApJ*, 640, 360 [J320]

Bania, T. M., Balser, D. S., Rood, R. T., Wilson, T. L., & Wilson, T. J. 1997, *ApJS*, 415, 54

Bania, T. M., Balser, D. S., Rood, R. T., Wilson, T. L., & LaRocque, J. M. 2007, *ApJ*, 664, 915

Bania, T. M., Rood, R. T., & Wilson, T. L. 1987, *ApJ*, 323, 30

Bania, T. M., Rood, R. T., & Balser, D. S. 2002, *Nature*, 415, 54

Charbonnel, C. & Zahn, J. P. 2007a, *A&A*, 467, L15

Charbonnel, C. & Zahn, J. P. 2007b, *A&A*, 476, L29

Charbonnel, C. & Lagarde, N. 2010, in: C. Charbonnel, M. Tosi, F. Primas, & C. Chiappini (eds.), *Light Elements in the Universe*. Proc. IAU Symposium No. 268, (Cambridge: CUP), p. XX

Eggleton, P. P., Dearborn, D. S. P., & Lattanzio, J. C. 2006, *Science*, 314, 1580

Galli, D., Palla, F., Ferrini, F., & Penco, U. 1995, *ApJ*, 443, 536

Galli, D., Stanghellini, L., Tosi, M., & Palla, F. 1997, *ApJ*, 456, 478

Geiss, J. 1993, in: N. Prantzos, E. Vangioni-Flam, & M. Casse (eds.), *Origin and Evolution of Elements*. (Cambridge: CUP), p. 89

Gloecker, G. & Geiss, J. 1996, *Nature*, 381, 210

Palla, F., Galli, D., Marconi, A., Stanghellini, L., & Tosi, M. 2002, *ApJ*, 568, L57

Romano, D., Tosi, M., Matteucci, F., & Chiappini, C. 2003, *MNRAS*, 346, 295

Rood, R. T., Bania, T. M., & Wilson, T. L. 1984, *ApJ*, 280, 629

Rood, R. T., Bania, T. M., & Wilson, T. L. 1992, *Nature*, 355, 618

Rood, R. T., Bania, T. M., Wilson, T. L., & Balser, D. S. 1995, in: P. Crane (ed.), *ESO/EIPC Workshop on the Light Elements*. (Heidelberg: Springer), p. 201

Rood, R. T., Steigman, G., & Tinsley, B. M. 1976, *ApJ*, 207, L57 [RST]

Rood, R. T., Wilson, T. L., & Steigman, G. 1979, *ApJ*, 227, L97

Spergel, D. N. *et al.* 2003 *ApJS*, 148, 175

Tosi, M. 2000, in: L. da Silva, M. Spite, & J. R. de Medeiros (eds.), *The Light Elements and Their Evolution*, Proc. IAU Symposium No. 198, (San Francisco: ASP), p. 525

Wilson, T. L. & Rood, R. T. 1994, *ARAA*, 32, 191

Light Elements in the Universe
Proceedings IAU Symposium No. 268, 2009
C. Charbonnel, M. Tosi, F. Primas, & C. Chiappini, eds.

© International Astronomical Union 2010
doi:10.1017/S1743921310003911

Measurements of ^{4}He in metal-poor extragalactic H II regions: the primordial Helium abundance and the $\Delta Y/\Delta O$ ratio

Manuel Peimbert[1], A. Peimbert[1], L. Carigi[1], and V. Luridiana[2]

[1]Instituto de Astronomía, Universidad Nacional Autónoma de México, Apdo. postal 70-264, México D.F. 04510, Mexico
email: peimbert@astroscu.unam.mx

[2]Instituto de Astrofísica de Canarias, c/ Vía Láctea s/n, 38205 La Laguna, Spain

Abstract. We present a review on the determination of the primordial helium abundance Y_p, based on the study of hydrogen and helium recombination lines in extragalactic H II regions. We also discuss the observational determinations of the increase of helium to the increase of oxygen by mass $\Delta Y/\Delta O$, and compare them with predictions based on models of galactic chemical evolution.

Keywords. ISM: abundances – H II regions – galaxies: abundances, evolution, irregular – Galaxy: disk – early universe

1. Overview

The determinations of the helium abundance by mass, Y, from metal-poor extragalactic H II regions provide the best method to obtain the primordial helium abundance. During their evolution galaxies produce a certain amount of helium and oxygen per unit mass that we will call ΔY and ΔO. The galaxies less affected by chemical evolution are those that present a large fraction of their baryonic mass in gaseous form and a small fraction of their baryonic mass in stellar form. These galaxies when experiencing bursts of star formation present bright metal poor H II regions that have been used to determine their chemical composition.

From a set of Y and O values and assuming a linear relationship it is possible to obtain Y_p and $\Delta Y/\Delta O$ from the following equation:

$$Y_p = Y - O\frac{\Delta Y}{\Delta O}. \tag{1.1}$$

The determinations of Y_p and $\Delta Y/\Delta O$ are important for at least the following reasons: (a) Y_p is one of the pillars of Big Bang cosmology and an accurate determination of Y_p permits to test the Standard Big Bang Nucleosynthesis (SBBN), (b) the models of stellar evolution require an accurate initial Y value; this is given by Y_p plus the additional ΔY produced by galactic chemical evolution, which can be estimated based on the observationally determined $\Delta Y/\Delta O$ ratio, (c) the combination of Y_p and $\Delta Y/\Delta O$ is needed to test models of galactic chemical evolution.

Recent reviews on primordial nucleosynthesis have been presented by Steigman (2007), Olive (2008), Weinberg (2008), and Pagel (2009). A review on the primordial helium abundance has been presented by Peimbert (2008) and a historical note on the primordial helium abundance has been presented by Peimbert & Torres-Peimbert (1999).

2. Recent Y_p determinations

The best Y_p determinations are those by Izotov, Thuan, & Stasińska (2007) that amount to 0.2516±0.0011 and Peimbert, Luridiana, & Peimbert (2007a) that amounts to 0.2477 ± 0.0029. The procedures used by both groups to determine Y_p are very different and it is not easy to make a detailed comparison of all the steps carried out by each of them. We consider that the error presented by Izotov *et al.* is a lower limit to the total error because it does not include estimates of some systematic errors. The difference between the central Y_p values derived by both groups is mainly due to the treatment of the temperature structure of the H II regions. Izotov *et al.* adopt temperature variations that are smaller than those derived by Peimbert *et al.*

To be more specific we can define the temperature structure of the H II regions by means of an average temperature, T_0, and a mean square temperature fluctuation, t^2 (Peimbert 1967). The value of t^2 derived by Izotov *et al.* (2007) for their sample is about 0.01; while Peimbert *et al.* (2007a) obtain a t^2 of about 0.026. From the observations adopted by Peimbert *et al.* and assuming $t^2 = 0.000$ we obtain $Y_p = 0.2523 \pm 0.0027$, and for $t^2 = 0.01$ we obtain $Y_p = 0.2505$.

The small value of t^2 derived by Izotov *et al.* (2007) is due to the parameter space used in their Monte Carlo computation where they permitted $T(\text{He I})$ to vary from 0.95 to 1.0 times the $T(4363/5007)$ value, which yields a t^2 of about 0.01. By allowing their $T(\text{He II})$ to vary from 0.80 to 1.0 times the $T(4363/5007)$ value their t^2 result would have become higher. This can be seen from their Table 5 where 71 of the 93 spectra corresponded to the lowest $T(\text{He I})$ allowed by the permitted parameter space of the Monte Carlo computation. A higher t^2 value for the Izotov *et al.* (2007) sample produces a lower Y_p, reducing the difference with the Peimbert *et al.* (2007a) Y_p value.

Two other systematic problems with the Izotov *et al.* (2007) determination related with the temperature structure are: (a) that to compute the $N(\text{O}^{++})$ abundance they adopted the $T(4363/5007)$ value which is equivalent to adopt $t^2 = 0.00$, and (b) to compute the once ionized oxygen and nitrogen abundances, $N(\text{O}^+)$ and $N(\text{N}^+)$, they adopted the $T(4363/5007)$ value, but according to photoionization models and observations of O-poor objects, the temperature in the O^+ regions, $T(\text{O}^+)$, is considerably smaller than in the O^{++} regions, typically by about 2000 K for objects with $T(\text{O}^{++}) = 16000$ K, and reaching 4000 K for the metal poorest H II regions (*e. g.* Peimbert, Peimbert, & Luridiana 2002; Stasińska 1990).

2.1. *Recombination coefficients of the helium I lines*

There are two recent line emissivity estimates to derive the He abundance from recombination lines: one due to Benjamin, Skillman, & Smits (1999, 2002), and another by Bauman *et al.* (2005) and Porter *et al.* (2007, 2009). The difference in the Y values derived from both sets of data amount to about 0.0040, the emissivities by the first group yielding values smaller than those of the second group. According to Porter, Ferland, & MacAdam (2007), the error introduced in their emissivities by interpolating in temperature the equations provided by them is smaller than 0.03%, which translates into an error in Y_p considerably smaller than 0.0001. Moreover according to Porter *et al.* (2009) the expected error in their line emissivities amounts to about 0.0010 in the Y derived values. In this review we are adopting the Bauman *et al.* and Porter *et al.* emissivities in the presented Y_p values.

2.2. *Beyond case B*

There are at least four processes that modify the level populations of H and He atoms relative to case B and consequently the Y_p determination: the optical depth of the He I

Table 1. Cosmological predictions based on SBBN and observations for $\tau_n = 885.7 \pm 0.8$ s

Method	Y_p	D_p	η_{10}	$\Omega_b h^2$
Y_p	0.2477 ± 0.0029^a	$2.93 + 2.53 - 1.06^b$	5.625 ± 1.81^b	0.02054 ± 0.00661^b
D_p	0.2479 ± 0.0007^b	2.82 ± 0.28^a	5.764 ± 0.360^b	0.02104 ± 0.00132^b
$WMAP$	0.2487 ± 0.0006^b	2.49 ± 0.13^b	6.226 ± 0.170^b	0.02273 ± 0.00062^a

[a] Observed value. [b] Predicted value. References: τ_n Arzumanov *et al.* (2000); Y_p Peimbert *et al.* (2007a); D_p O'Meara *etal.*(2006); $WMAP$ Dunkley *et al.* (2009).

Table 2. Cosmological predictions based on SBBN and observations for $\tau_n = 878.5 \pm 0.8$ s

Method	Y_p	D_p	η_{10}	$\Omega_b h^2$
Y_p	0.2477 ± 0.0029^a	$2.22 + 1.46 - 0.71^b$	6.688 ± 1.81^b	0.02442 ± 0.00661^b
D_p	0.2462 ± 0.0007^b	2.82 ± 0.28^a	5.764 ± 0.360^b	0.02104 ± 0.00132^b
$WMAP$	0.2470 ± 0.0006^b	2.49 ± 0.13^b	6.226 ± 0.170^b	0.02273 ± 0.00062^a

[a] Observed value. [b] Predicted value. References: same as in Table 1 with the exception of τ_n that comes from Serebrov *et al.* (2005, 2008).

lines, the collisional excitation from the He 2^3S metastable level, the collisional excitations from the H ground level, and the fluorescent excitation of the H I and the He I lines (case D of Luridiana *et al.* 2009). The last two processes are the least studied of the four.

Luridiana *et al.* (2009) have introduced case D into the study of gaseous nebulae. Case D increases slightly the emissivities of the H I and He I lines affecting the accuracy of the Y_p determination. There are no published estimates of the importance of this effect but it depends on the spectra of the ionizing stars, particularly in the region of the H I and He I lines, the spatial distribution of the ionizing stars, the gaseous electron density distribution, the fraction of ionizing photons that escape the nebula, the radial velocity of the gas relative to that of the stars, the nebular turbulence, and the region of the nebula observed. Therefore it requires tailor-made models for each observed object. Peimbert *et al.* (2007a) presented a list of thirteen sources of error in the Y_p determination, each of them with systematic and statistical components but dominated by one or the other. The systematic error produced by not having considered Case D should be added to that list.

3. Comparison of the directly determined Y_p with the Y_p values computed under the assumption of the SBBN and the observations of D_p and WMAP

To compare the Y_p value with the primordial deuterium abundance D_p (usually expressed as $10^5(D/H)_p$) and with the WMAP results, we will use the framework of the SBBN. The ratio of baryons to photons multiplied by 10^{10}, η_{10}, is given by (Steigman 2006, 2007):

$$\eta_{10} = (273.9 \pm 0.3)\Omega_b h^2, \qquad (3.1)$$

where Ω_b is the baryon closure parameter, and h is the Hubble parameter. In the range $4 < \eta_{10} < 8$ (corresponding to $0.2448 < Y_p < 0.2512$), Y_p is related to η_{10} by (Steigman 2006, 2007):

$$Y_p = 0.2483 \pm 0.0005 + 0.0016(\eta_{10} - 6). \qquad (3.2)$$

In the same η_{10} range, the primordial deuterium abundance is given by (Steigman 2006, 2007):

$$10^5 (D/H)_p = D_p = 46.5(1 \pm 0.03)(\eta_{10})^{-1.6}. \tag{3.3}$$

From the Y_p value by Peimbert *et al.* (2007a), the D_p value by O'Meara *et al.* (2006), the $\Omega_b h^2$ value by Dunkley *et al.* (2009), and the previous equations we have produced Table 1. From this table, it follows that within the errors Y_p, D_p, and the WMAP observations are in very good agreement with the predicted SBBN values.

Equations (2), (3), and (4) were derived under the assumption of a neutron lifetime, τ_n, of 885.7 ± 0.8 s (Arzumanov *et al.* 2000). A recent result by Serebrov *et al.* (2005, 2008) yielded a τ_n of $878.5 \pm 0.7 \pm 0.3$ s. This result would lead to a SBBN Y_p value of 0.2470 for the WMAP $\Omega_b h^2$ determined value (Steigman 2007, Mathews *et al.* 2005). Mathews *et al.* obtain for $\tau_n = 881.9 \pm 1.6$ s, mainly the average of the results by Arzumanov *et al.* (2000) and Serebrov *et al.* (2005, 2008), a Y_p value 0.0009 smaller than for $\tau_n = 885.7 \pm 0.8$.

The 9σ difference between both τ_n determinations probably indicates that at least one of them includes systematic errors that have not yet been sorted out.

From the Y_p by Peimbert *et al.* (2007a), the D_p by O'Meara *et al.* (2006), and SBBN it is found that for $\tau_n = 881.9 \pm 1.6$ the number of effective neutrino families, N_{eff}, is equal to 3.12 ± 0.23. Based on the production of the Z particle by electron-positron collisions in the laboratory and taking into account the partial heating of neutrinos produced by electron-positron annihilations during SBBN Mangano *et al.* (2002) find that $N_{eff} = 3.04$. The N_{eff} value derived from Y_p, D_p, and SBBN is in excellent agreement with the value derived by Mangano *et al.* (2002).

The Y_p derived from WMAP is based on the very strong assumption of SBBN. It is also possible to derive Y_p from the study of the microwave radiation without assuming SBBN: from the cosmic microwave background radiation (*i.e.* WMAP + ACBAR + CBI + BOOMERANG), Ichikawa, Sekiguchi and Takahashi (2008a,b) obtain that $Y_p < 0.44$, when they also include the information obtained from BAO + SN + HST (baryon acoustic oscillations in the distribution of galaxies, the distance measurements from type Ia supernovae, and the HST value for H_0) the constrain improves to $Y_p = 0.25^{+0.10}_{-0.07}$. Including the expected data from the Planck satellite they predict a reduction on the Y_p error of about a factor of four to seven, an error still about four to six times higher than the one estimated from the best Y_p determinations based on metal poor H II region observations.

4. The $\Delta Y / \Delta O$ ratio

To determine the Y_p value from a set of metal poor H II regions it is necessary to estimate the fraction of helium, present in the interstellar medium of the galaxy where each H II region is located, produced by galactic chemical evolution. From observations of metal poor extragalactic H II regions it has been found that the Y versus O observations can be fitted with a straight line given by $\Delta Y / \Delta O$ and equation (1) has been used often to derive Y_p. A straight line is predicted by chemical evolution models of metal poor galaxies with the same initial mass function, a given set of stellar yields, and different star formation rates. To obtain different $\Delta Y / \Delta O$ from the models it is necessary to change the initial mass function (for example the maximum mass allowed or the slope at high masses), or the adopted yields.

The observational Y_p, ΔY, and ΔO are affected by different amounts in the presence of temperature variations, while Y_p diminishes by 0.0046 due to temperature variations in the sample of Peimbert *et al.* (2007a), ΔY is slightly affected and ΔO is strongly affected by temperature variations (*e. g.* Carigi & Peimbert 2008, Table 3). In addition a fraction of O is embedded in dust grains, fraction that needs to be estimated to compare with models of galactic chemical evolution.

The importance of $\Delta Y/\Delta O$ is two fold: it permits us to obtain a more accurate Y_p value, and permits us to test for the presence of large temperature variations in gaseous nebulae when comparing nebular values with stellar ones.

4.1. *The gaseous O/H determination*

There are two methods to derive the gaseous O/H ratio, from the $I(4363)/I(5007)$ ratio together with the $I(3727)/I(\mathrm{H}\beta)$ and the $I(5007)/I(\mathrm{H}\beta)$ line ratios, the so called $T(4363)$ method, and from the intensity ratio of O II recombination lines to H I recombination lines that has been called the O II_{RL} method by Peimbert *et al.* (2007b). The O II_{RL} method usually provides higher O/H ratios by 0.15 to about 0.3 dex, this difference is due to temperature variations inside the observed volume and is smaller for metal poor H II regions and higher for metal rich H II regions. The $T(4363)$ method in the presence of temperature variations produces a systematic effect that lowers the O/H abundances relative to the real ones. On the other hand the O II_{RL} method is independent of the temperature structure.

4.2. *The total O/H determination*

Another factor that has to be taken into account to obtain the total O/H ratio is the fraction of O atoms trapped in dust grains. Esteban *et al.* (1998), based on the depletion of Fe, Mg and Si in the Orion nebula, estimated that the fraction of oxygen atoms trapped in dust grains amounts to 0.08 dex. Mesa-Delgado *et al.* (2009) have estimated that the fraction of O atoms trapped in dust grains in the Orion nebula amounts to 0.12 ± 0.03 dex; this result is based on three different methods: a) by comparing the O abundances of the B stars of the Orion association with the O abundance of the Orion nebula, b) from the depletion of Fe, Mg and Si in the nebula and assuming that these atoms are combined with molecules containing O, and c) by comparing the shock and nebular abundances of the material associated with the Herbig-Haro object 202.

Rodríguez & Rubin (2005) have estimated the fraction of Fe atoms in galactic and extragalactic H II regions in the gaseous phase, the depletions derived for the different objects define a trend of increasing depletion at higher metallicities; while the Galactic H II regions show less than than 5% of their Fe atoms in the gas phase, the extragalactic ones (LMC 30 Doradus, SMC N88A, and SBS 0335-052) have somewhat lower depletions. Izotov *et al.* (2006) find a slight increase of Ne/O with increasing metallicity, which they interpret as due to a moderate depletion of O onto grains in the most metal-rich galaxies, they conclude that this O/Ne depletion corresponds to $\sim 20\%$ of oxygen locked in the dust grains in the highest-metallicity H II regions of their sample, while no significant depletion would be present in the H II regions with lower metallicity. Peimbert & Peimbert (2010) based on the Fe/O abundances of Galactic and extragalactic H II regions estimate that for objects in the $8.3 < 12 + \log \mathrm{O/H} < 8.9$ range the fraction of O atoms trapped by dust grains amounts to 0.12 ± 0.03 dex, for objects in the $7.7 < 12 + \log \mathrm{O/H} < 8.3$ range amounts to 0.09 ± 0.03 dex, and for objects in the $7.3 < 12 + \log \mathrm{O/H} < 7.7$ range amounts to 0.06 ± 0.03 dex.

4.3. *Extrapolation of the Y determinations to the value of Y_p, or the O $(\Delta Y/\Delta O)$ correction*

From chemical evolution models of different galaxies it is found that $\Delta Y/\Delta O$ depends on the initial mass function (IMF), the star formation rate, the age, and the O value of the galaxy in question. Peimbert *et al.* (2007b) have found that $\Delta Y/\Delta O$ is well fitted by a constant value for objects with the same IMF, the same age, and an O abundance smaller than 4×10^{-3}. This result is consistent with the custom of using a constant value for $\Delta Y/\Delta O$ to fit observational data.

To obtain an accurate Y_p value, a reliable determination of $\Delta Y/\Delta O$ for O-poor objects is needed. The $\Delta Y/\Delta O$ value derived by Peimbert, Peimbert, & Ruiz (2000) from observational results and models of chemical evolution of galaxies amounts to 3.5 ± 0.9. More recent results are those by Peimbert (2003) who finds 2.93 ± 0.85 from observations of 30 Dor and NGC 346, and by Izotov & Thuan (2004) who, from the observations of 82 H II regions, find $\Delta Y/\Delta O = 4.3 \pm 0.7$. Peimbert *et al.* (2007a) have recomputed this value by taking into account two systematic effects not considered by Izotov & Thuan: the fraction of oxygen trapped in dust grains, and the increase in the O abundances due to the presence of temperature variations. From these considerations they obtained for the Izotov & Thuan sample that $\Delta Y/\Delta O = 3.2 \pm 0.7$.

On the other hand Peimbert *et al.* (2007b) from chemical evolution models with different histories of galactic inflows and outflows for objects with $O < 4 \times 10^{-3}$ find that $2.4 < \Delta Y/\Delta O < 4.0$. From the theoretical and observational results Peimbert *et al.* (2007a) adopted a value of $\Delta Y/\Delta O = 3.3 \pm 0.7$, that they used with the Y and O determinations from each object to obtain the Y_p value.

4.4. *Comparison of the oxygen abundances of the ISM of the solar vicinity with those of the Sun and F and G stars of the solar vicinity*

In addition to the evidence presented in section 2 in favor of large t^2 values, and consequently in favor of the O II_{RL} method, there is another independent test that can be used to discriminate between the $T(4363)$ method and the O II_{RL} method that consists in the comparison of stellar and H II region abundances of the solar vicinity.

Esteban *et al.* (2005) determined that the gaseous O/H value derived from H II regions of the solar vicinity amounts to $12 + \log (O/H) = 8.69$, and including the fraction of O atoms tied up in dust grains it is obtained that $12 + \log (O/H) = 8.81 \pm 0.04$ for the O/H value of the ISM of the solar vicinity. Alternatively from the protosolar value by Asplund *et al.* (2009), that amounts to $12 + \log(O/H) = 8.71$, and taking into account the increase of the O/H ratio due to galactic chemical evolution since the Sun was formed, that according to the chemical evolution model of the Galaxy by Carigi *et al.* (2005) amounts to 0.13 dex, we obtain an O/H value of 8.84 ± 0.04 dex, in excellent agreement with the value based on the O II_{RL} method. In this comparison we are assuming that the solar abundances are representative of the abundances of the solar vicinity ISM when the Sun was formed.

There are two other estimates of the O/H value in the ISM that can be made from observations of F and G stars of the solar vicinity. According to Allende-Prieto *et al.* (2004) the Sun appears deficient by roughly 0.1 dex in O, Si, Ca, Sc, Ti, Y, Ce, Nd, and Eu, compared with its immediate neighbors with similar iron abundances; by assuming that the O abundances of the solar immediate neighbors are more representative of the local ISM than the solar one, and by adding this 0.1 dex difference to the solar value by Asplund *et al.* (2009) we obtain that the present value of the ISM has to be higher than $12 + \log O/H = 8.81$. A similar result is obtained from the data by Bensby & Feltzing (2005) who obtain for the six most O-rich thin-disk F and G dwarfs of the solar vicinity an average $[O/H] = 0.16$; by assuming their value as representative of the present day ISM of the solar vicinity we find $12 + \log O/H = 8.87$. Both results are in good agreement with the O/H value derived from the O II_{RL} method.

4.5. *Comparison of $\Delta Y/\Delta Z$ values of the Galactic H* II *region M17 with K dwarfs of the solar vicinity*

The best Galactic H II region to determine the He/H ratio is M17 because it contains a very small fraction of neutral helium and the error introduced by correcting for its presence is very small. Carigi & Peimbert (2008) obtained from observations for M17 (and adopting a Y_p value) a value of $\Delta Y/\Delta Z = 1.97 \pm 0.41$ for $t^2 = 0.036 \pm 0.013$, where Y and Z are the helium and heavy elements by unit mass; by correcting this value considering that the fraction of O trapped in dust amounts to 0.12 dex instead of 0.08 dex (Mesa-Delgado *et al.* 2009) we obtain $\Delta Y/\Delta Z = 1.77 \pm 0.37$. This $\Delta Y/\Delta Z$ value is in agreement with two independent $\Delta Y/\Delta Z$ determinations derived from K dwarf stars of the solar vicinity that amount to 2.1 $\pm$ 0.4 (Jiménez *et al.* 2003) and to 2.1 $\pm$ 0.9 (Casagrande *et al.* 2007).

On the other hand the value $\Delta Y/\Delta Z = 3.6 \pm 0.68$ derived from collisionally excited lines of M17 under the assumption of $t^2 = 0.00$ by Carigi and Peimbert (2008) (value corrected for a fraction of O trapped by dust grains of 0.12 dex) is not in agreement with the values derived from K dwarf stars of the solar neighborhood.

4.6. *Comparison of Y_p and $\Delta Y/\Delta O$ with models of chemical evolution for the disk of the Galaxy*

To compare Galactic chemical evolution models of Y and O with observations we need to use the best determinations available of the abundances of these elements. We consider that the two most accurate Galactic Y and O determinations are the presolar values (Asplund *et al.* 2009) and the M17 H II region values (Carigi & Peimbert 2008).

Carigi *et al.* (2005) presented chemical evolution models for the disk of the Galaxy that fit the slope and the absolute value of the O/H gradient, these models are also successful

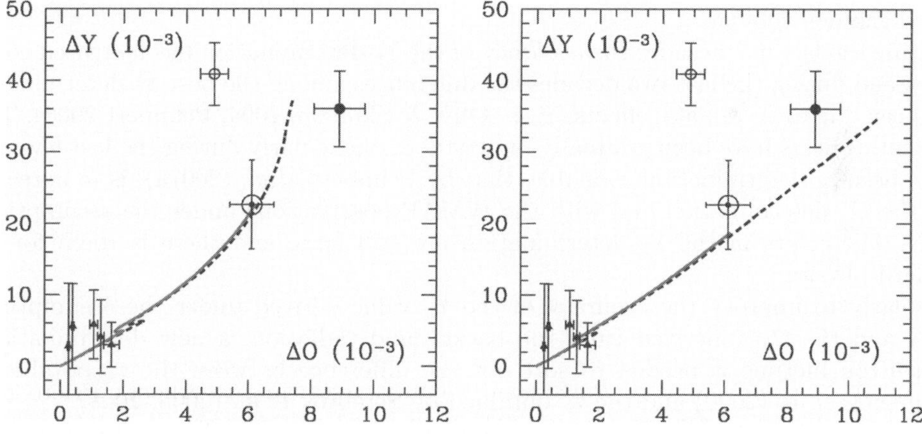

Figure 1. In the two panels the origin corresponds to $Y_p = 0.2477$, the triangles represent the Y and O values derived by Peimbert *et al.* (2007a) for five metal poor extragalactic H II regions, the open and filled circles represent the M17 values for $t^2 = 0.00$, and $t^2 = 0.036$ respectively from Carigi & Peimbert (2008), the large open circle with a dot in the center corresponds to the presolar values by Asplund *et al.* (2009). The left panel presents two chemical evolution models by Carigi & Peimbert (2010), the solid line is a model for the Galactic disk with a time span from the formation of the Galaxy to the formation of the Sun at a galactocentric distance of 8 kpc, while the dashed line corresponds to a model for the Galactic disk with a time span from the formation of the Galaxy to the present at a galactocentric distance of 6.75 kpc, the distance of M17 to the galactic center; the stellar yields for the two panels are different, see subsection 4.6 for further details.

in reproducing the C/O gradient derived from H II regions and the C/O versus O/H evolution history of the solar vicinity obtained from stellar observations. In Figure 1 we present chemical evolution models by Carigi & Peimbert (2010) for the Y and O abundances of the Galactic disk for two sets of stellar yields where they have added the presolar values by Asplund *et al.* (2009) for comparison. These models are based on the same assumptions than those adopted by Carigi and Peimbert (2008). The only difference between the models in the left panel and those in the right one, is that for massive stars with $Z > 0.004$ the ones in the left use the yields by Maeder (1992) with high mass loss, while the ones on the right use the yields by Hirschi, Meynet, & Maeder (2005) with low mass loss. There are at least three conclusions that we can extract from the figure: (a) the fit for the presolar values is very good for the two sets of yields, (b) the fit for M17 for $t^2 = 0.036 \pm 0.013$ is good, but it can be improved by assuming a set of yields intermediate between the two sets used; a similar result in favor of an intermediate set of yields was obtained by Cescutti *et al.* (2009) based on C/O observations for stars in the bulge of the Galaxy, (c) the models do not fit M17 for $t^2 = 0.00$, this result is in agreement with those by Esteban *et al.* (2005, 2009) and Peimbert *et al.* (2007a) who find large t^2 values for Galactic and extragalactic H II regions. Moreover the O and Y abundances for the presolar material and for M17 are derived from independent methods, and they are fitted by the same chemical evolution model. This fit provides us with a consistency check on the gaseous nebulae helium and oxygen abundance determinations based on large t^2 values.

5. Conclusions

During the last 50 years the determination of Y_p has been very important for the study of cosmology, stellar evolution, and the chemical evolution of galaxies. To determine Y_p it is necessary to determine accurate atomic parameters and the physical conditions inside ionized gaseous nebulae.

During the last five decades the accuracy of the Y_p determination has increased considerably, and during the last two decades the differences among the best Y_p determinations have been due to systematic effects (*e. g.* Olive & Skillman 2004, Peimbert 2008). These systematic effects have been gradually understood, particularly during the last few years.

The best Y_p determination available, that by Peimbert *et al.* (2007a), is in agreement with the D_p determination and with the WMAP observations under the assumption of SBBN. The errors in the Y_p determination are still large and there is room for non-standard physics.

Similarly to improve the accuracy of the Y_p value derived under the assumption of SBBN and the $\Omega_b h^2$ derived from the background radiation, a new determination of the neutron lifetime is needed to sort out the difference between the τ_n obtained by Arzumanov *et al.* (2000) and the τ_n obtained by Serebrov *et al.* (2005, 2008).

To improve the accuracy of the Y_p determination based on metal poor H II regions, the following steps should be taken in the near future: (a) to obtain new observations of high spectral resolution of metal poor H II regions, those with $0.0005 < Z < 0.001$ to reduce the effect of the collisional excitation of the Balmer lines, which for the present day Y_p determinations is one of the two main sources of error; (b) to determine the temperature of the H II regions based on a large number of He I lines observed with high accuracy, and to try to avoid the use of $T(4363/5007)$ temperatures that weigh preferentially the regions of higher temperature than the average one, effect that artificially increases the Y_p determinations; (c) additional efforts should be made to understand the mechanisms that produce temperature variations in giant H II regions and once they are understood they should be incorporated into photoionization models; (d) the He I recombination coefficients should be computed again with an accuracy higher than that of the last two determinations; (e) the computation of tailor-made photoionization models for extragalactic H II regions including the effect of case D on the intensity of the H I and He I lines should be carried out.

The observed $\Delta Y/\Delta O$ ratios provide us with strong constrains for models of galactic chemical evolution. The gaseous O abundances have to be corrected by the fraction of oxygen embedded in dust grains and by the effect of temperature variations, these two corrections together amount to about 0.2 to 0.4 dex. Chemical evolution models for the Galactic disk are able to reproduce the observed Y and O presolar values and the Y and O values derived for M17 based on H, He and O recombination lines, but not the M17 Y and O values derived from $T(4363/5007)$ and O collisionally excited lines under the assumption of $t^2 = 0.00$. This result provides a consistency check in favor of the presence of large temperature variations in H II regions and in favor of the Y_p determinations based on the $T(\text{He I})$ and t^2 values derived from He I recombination lines.

Acknowledgements

We wish to thank Gary Ferland, Evan Skillman and Gary Steigman for several fruitful discussions. We are grateful to the SOC for an outstanding meeting and to the LOC for their warm hospitality. We would also like to acknowledge partial support received from CONACyT grant 46904.

References

Allende-Prieto, C., Barklem, P. S., Lambert, D. L., & Kunha, K. 2004, *A&A*, 420, 183

Arzumanov, S., Bondarenko, L. Chernyavsky, S., *et al.* 2000, *Physics Letters B*, 483, 15

Asplund, M., Grevesse, N., Sauval, A. J., & Scott, P. 2009, *ARAA*, 47, 481

Bauman, R. P., Porter, R. L., Ferland, G. J., & MacAdam, K. B. 2005, *ApJ*, 628, 541

Benjamin, R. A., Skillman, E. D., & Smits, D. P. 1999, *ApJ*, 514, 307

Benjamin, R. A., Skillman, E. D., & Smits, D. P. 2002, *ApJ*, 569, 288

Bensby, T. & Feltzing, S. 2006, *MNRAS*, 367, 1181

Carigi, L. & Peimbert, M. 2008, *Rev. Mexicana AyA*, 44, 311

Carigi, L. & Peimbert, M. 2010, in preparation

Carigi, L., Peimbert, M., Esteban, C., & García-Rojas, J. 2005, *ApJ*, 623, 213

Casagrande, L., Flynn, C., Portinari, L., Girardi, L. & Jiménez, R. 2007, *MNRAS*, 382, 1516

Cescutti, G., Matteucci, F., McWilliam, A., & Chiappini, C. 2009, *A&A*, 505, 605

Dunkley, J., Komatsu, E., Nolta, M. R., *et al.* 2009, *ApJS*, 180, 306

Esteban, C., Bresolin, F., Peimbert, M., *et al.* 2009, *ApJ*, 700, 654

Esteban, C., García-Rojas, J., Peimbert, M., *et al.* 2005, *ApJ*, 618, L95

Esteban, C., Peimbert, M., Torres-Peimbert, S., & Escalante, V. 1998, *MNRAS*, 295, 401

Hirschi, R., Meynet, G., & Maeder, A. 2005, *A&A*, 433, 1013

Ichikawa, K., Sekiguchi, T., & Takahashi, T. 2008a, *Phys. Rev. D*, 78, 043509

Ichikawa, K., Sekiguchi, T., & Takahashi, T. 2008b, *Phys. Rev. D*, 78, 083526

Izotov, Y. I., Stasińska, G., Meynet, G., Guseva, N. G., & Thuan, T. X. 2006, *A&A*, 448, 955

Izotov, Y. I. & Thuan, T. X. 2004, *ApJ*, 602, 200

Izotov, Y. I., Thuan, T. X., & Stasińska, G. 2007, *ApJ*, 662, 15

Jiménez, R., Flynn, C., MacDonald, J., & Gibson, B. K. 2003, *Science*, 299, 1552

Luridiana, V., Simón-Díaz, S., Cerviño, M., *et al.* 2009, *ApJ*, 691, 1712

Maeder, A. 1992, *A&A*, 264, 105

Mangano, G., Miele, G., Pastor, S., & Peloso, M. 2002, *Physics Letters B*, 534, 8

Mathews, G. J., Kajino, T., & Shima, T. 2005, *Phys. Rev. D*, 71, 021302

Mesa-Delgado, A., Esteban, C., García-Rojas, J., *et al.* 2009, *MNRAS*, 395, 855

Olive, K. A. 2008, in: *Chemical Evolution Across Space & Time: From the Big Bang to Prebiotic Chemistry* L. Zaikowski & J. M. Friedrich, (eds.). ACS Symposium Series Vol. 981, 16

Olive, K. A. & Skillman, E. D. 2004, *ApJ*, 617, 290

O'Meara, J. M., Burles, S., Prochaska, J. X., & Prochter, G. E. 2006, *ApJ*, 649, L61

Pagel, B. E. J. 2009, *Nucleosynthesis and Chemical Evolution of Galaxies, Second Edition*, Cambridege University Press

Peimbert, A. 2003, *ApJ*, 584, 735

Peimbert, A. & Peimbert, M. 2010, in preparation

Peimbert, A., Peimbert, M., & Luridiana, V. 2002, *ApJ*, 565, 668

Peimbert, M. 1967, *ApJ*, 150, 825

Peimbert, M. 2008, *Current Science*, 95, 1165

Peimbert, M., Luridiana, V., & Peimbert, A. 2007a, *ApJ*, 666, 636

Peimbert, M., Luridiana, V., Peimbert, A., & Carigi, L. 2007b, in: *From Stars to Galaxies: Building the Pieces to Build Up the Universe* A. Vallenari, R. Tantalo, L. Portinari, & A. Moretti, (eds.), ASP Conference Series Vol. 374, 81

Peimbert, M., Peimbert, A., & Ruiz, M. T. 2000, *ApJ*, 541, 688

Peimbert, M. & Torres-Peimbert, S. 1999, *ApJ* (Centennial Issue), 525C, 1143

Porter, R. L., Bauman, R. P., Ferland, G. J., & MacAdam, K. B. 2005, *ApJ*, 622, L73

Porter, R. L., Ferland, G. J., & MacAdam, K. B. 2007, *ApJ*, 657, 327-337

Porter, R. L., Ferland, G. J., MacAdam, K. B., & Storey, P. J. 2009, *MNRAS*, 393, L36

Rodríguez, M. & Rubin, R. H. 2005, *ApJ*, 626, 900

Serebrov, A. P., Varlamov, V. E., Kharitonov, A. G., *et al.* 2005, *Physics Letters B*, 605, 72

Serebrov, A. P., Varlamov, V. E., Kharitonov, A. G., *et al.* 2008, *Phys. Rev. C*, 78, 035505

Stasińska, G. 1990, *A&AS*, 83, 501

Steigman, G. 2006, *Int. J. Mod. Phys. E*, 15, 1

Steigman, G. 2007, *Annu. Rev. Nucl. Part. Sci.*, 57, 463

Weinberg, S. 2008, *Cosmology*, Oxford University Press

Light Elements in the Universe
Proceedings IAU Symposium No. 268, 2009
C. Charbonnel, M. Tosi, F. Primas & C. Chiappini, eds.

© International Astronomical Union 2010
doi:10.1017/S1743921310003923

^{4}He abundances: Optical versus radio recombination line measurements

Dana S. Balser[1], Robert T. Rood[2] and T. M. Bania[3]

[1] National Radio Astronomy Observatory, 520 Edgemont Road, Charlottesville,
VA 22903-2475 USA
email: `dbalser@nrao.edu`

[2] Astronomy Department, University of Virginia, P.O.Box 3818, Charlottesville VA
22903-0818, USA
email: `rtr@virginia.edu`

[3] Institute for Astrophysical Research, 725 Commonwealth Avenue, Boston University,
Boston MA 02215, USA
email: `bania@bu.edu`

Abstract. Accurate measurements of the ^{4}He/H abundance ratio are important in constraining Big Bang nucleosynthesis, models of stellar and Galactic evolution, and H II region physics. We discuss observations of radio recombination lines using the Green Bank Telescope toward a small sample of H II regions and planetary nebulae. We report ^{4}He/H abundance ratio differences as high as $15 - 20\%$ between optical and ratio data that are difficult to reconcile. Using the H II regions S206 and M17 we determine ^{4}He production in the Galaxy to be $dY/dZ = 1.71 \pm 0.86$.

Keywords. Galaxy: abundances, ISM: planetary nebulae, H II regions, radio lines: ISM

1. Introduction

Observations of optical recombination lines (ORLs) are the primary diagnostics used to determine ^{4}He abundances in ionized nebulae such as H II regions and planetary nebulae (PNe). At optical wavelengths there are also many bright collisionally excited lines that can be used to probe the physical conditions in these objects and to derive their metallicity. Peimbert *et al.* (2010) discuss the main uncertainties in calculating accurate ^{4}He/H abundance ratios from ORLs. Much of the focus has been on deriving ^{4}He abundances in metal poor objects, such as blue compact galaxies, to deduce the primordial ^{4}He/H abundance (e.g., see Izotov 2010; Skillman 2010). Because there is no way to directly measure neutral helium within the H II region it is difficult to determine accurate ^{4}He/H abundance ratios for most Galactic H II regions due to the relatively soft ionizing radiation field for most objects. One exception is M17 which contains several O3 type stars and is estimated to contain little neutral helium (e.g., Carigi & Peimbert 2008).

Observations of recombination lines at other wavelengths (e.g., radio and infrared) give important checks to ORLs since deriving ^{4}He/H abundance ratios using these transitions will have different systematic uncertainties. Moreover, recombination lines at radio and infrared wavelengths are not obscured by dust and may be used to probe ^{4}He abundances throughout the Galactic disk.

Radio recombination lines (RRLs) are relatively weak when compared to ORLs and emission measures of $\int n_e^2 d\ell \sim 10^3$ cm^{-6} pc are required to detect these lines with current instrumentation compared with ~ 1cm^{-6} pc at optical wavelengths. Recent improvements in radio spectrometers, however, now allow the simultaneous observation of multiple RRLs. Adjacent RRLs have line intensities and widths that are different by only a few percent. Therefore, averaging adjacent RRLs can be used to increase the signal-to-noise ratio without significantly increasing the error in the line parameters.

Besides sensitivity, the main problem in making accurate ^{4}He measurements using RRLs has been poor spectral baselines. For single-dish telescopes reflections from the super-structure cause standing waves that are not easy to model and that do not integrate down with time. (Although the spectral baselines are much better for radio interferometers, until recently the spectral resolution and bandwidth were not sufficient to produce accurate line parameters in most cases.) The off-axis, clear aperture design of the Green Bank Telescope (GBT) has significantly improved the spectral baselines, and when combined with a flexible spectrometer provides a significant improvement over previous telescopes when measuring weak, wide spectral lines (e.g., Balser 2006; Bania *et al.* 2010).

One advantage to using RRLs over ORLs is that the high n states of hydrogen and helium should be altered by radiative and collisional effects in the same way so that the ratio of the line areas is equal to the ionic abundance ratio. At high n, where the levels are controlled by collisions and are in LTE, this should certainly be the case. H II region models indicate that near $n \sim 90$ (9 GHz) the RRLs are approximately in LTE (Shaver 1980; Balser *et al.* 1999). Early RRL observations of Galactic H II regions, however, produced ^{4}He$^+$/H$^+$ abundance ratios that varied with n (Lockman & Brown 1982). Peimbert *et al.* (1992) studied α, β, and γ RRL transitions near 9 GHz with the same spatial resolution and derived ^{4}He$^+$/H$^+$ abundance ratios consistent within the errors. These data had better sensitivity and improved spectral baselines over previous RRL measurements (Balser *et al.* 1994).

Ionization and density structure must be understood when using either optical or radio recombination lines to calculate ^{4}He/H abundance ratios. Since there is no transition of neutral helium other diagnostics must be used to determine if any neutral helium resides within the H II region. Therefore, corrections for ionization structure often contribute significantly to the uncertainty in calculating the ^{4}He/H abundance ratio (e.g., Baldwin *et al.* 1991). This is why H II regions with very hard radiation fields are often chosen for ^{4}He studies (e.g., blue compact galaxies). Nevertheless, a reverse ionization correction can result in these objects wherein neutral hydrogen exists within the He II region (Ballantyne *et al.* 2000; Gruenwald *et al.* 2002; Sauer & Jedamzik 2002). Moreover, when combined with density structures the ionization correction factor will change (Viegas *et al.* 2000). Also, density structure alone can alter the radiative transfer and thus effect ^{4}He analyses (Mathis & Wood 2005).

2. Observations

We made GBT RRL observations toward four PNe (NGC 3242, NGC 6826, NGC 6543, and NGC 7009) and two Galactic H II regions (M17 and S206). All observations were made at X-band ($8 - 10$ GHz) with a spatial resolution (beam) of 80 arcsec and a spectral resolution of 0.4 km s^{-1}. The PNe observations were part of the ^{3}He experiment wherein the 91α and 92α H and He RRLs were observed simultaneously with the ^{3}He$^+$ hyperfine transition (see Bania *et al.* 2010). These PNe consist of multiple shells with large, extended halos. The RRL emission arises from the shell structure which is unresolved by the GBT beam. Because the hot central star produces a hard ionizing radiation field there should be little or no neutral helium in these nebulae.

The H II region observations were part of a Galactic H II survey measuring H, He and C RRL emission. Seven adjacent alpha RRLs were observed simultaneously ($87\alpha - 93\alpha$). M17 and S206 are expected to have little neutral helium and therefore are good candidates

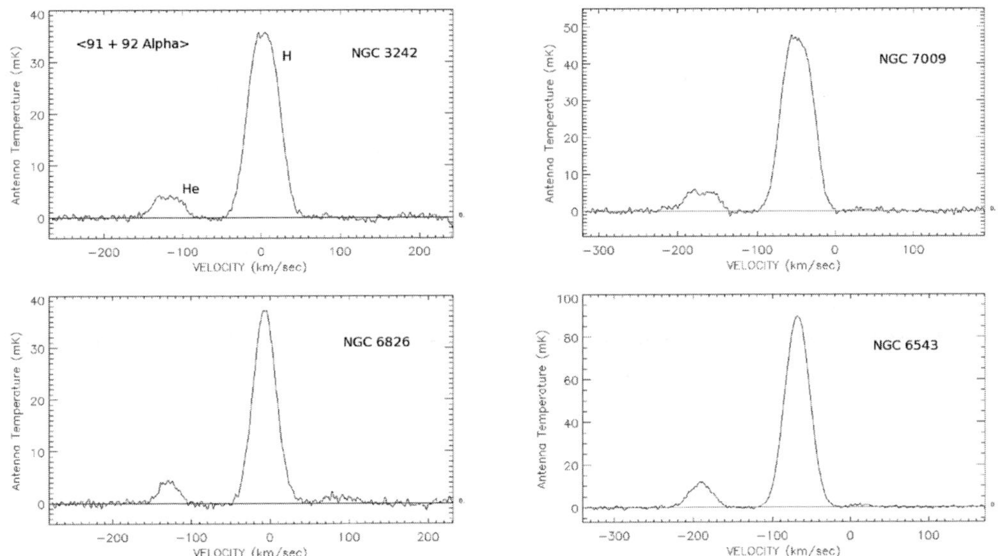

Figure 1. Planetary nebulae GBT RRL spectra. Plotted is the antenna temperature in mK versus the LSR velocity in $\mathrm{km\,s^{-1}}$. The 91α and 92α RRLs have been averaged and the spectra smoothed to a velocity resolution of $2.0\ \mathrm{km\,s^{-1}}$. The continuum emission has been subtracted.

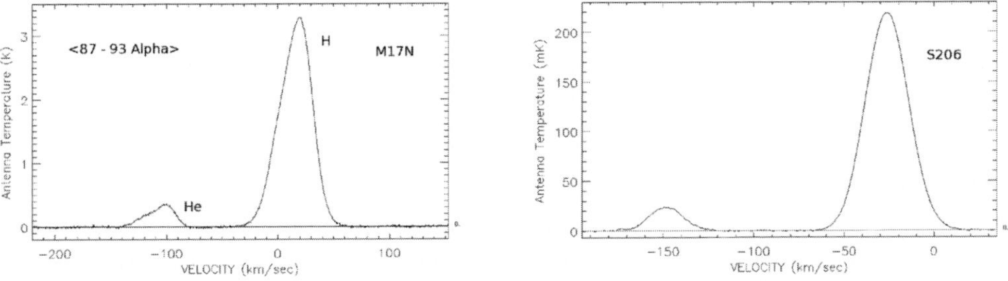

Figure 2. H II region GBT RRL spectra. Plotted is the antenna temperature in mK versus the LSR velocity in $\mathrm{km\,s^{-1}}$. RRLs spanning 87α to 93α have been averaged. The continuum emission has been subtracted.

for measuring ^{4}He/H (e.g., Deharveng *et al.* 2000; Carigi & Peimbert 2008). Both M17 and S206 are extended relative to the GBT beam and thus we are only probing part of these nebulae. Results for S206 have already been published (Balser 2006). The M17 data consist of observations toward a northern (M17N) and a southern (M17S) component. Since M17S is optically obscured, we discuss M17N here to make comparisons with ORLs. Our M17N J2000 position (RA $= 18{:}17{:}46.5$, Dec $= -16{:}10{:}33.1$) overlaps with some optical positions.

The PNe spectra are summarized in Figure 1. The H and ^{4}He line profiles for NGC 3242 and NGC 7009 are square-shaped, consistent with an optically thin, unresolved, expanding nebulae. When the expansion velocity is small compared with the thermal and turbulent velocity dispersion then the line profiles will become Gaussian shaped. This appears to be the case for NGC 6826 and NGC 6543 since the line profiles for these objects are well modeled by a Gaussian function.

The H II region spectra are summarized in Figure 2. The line profiles are well fit by Gaussian components. M17N is best fit by two components and S206 by one component.

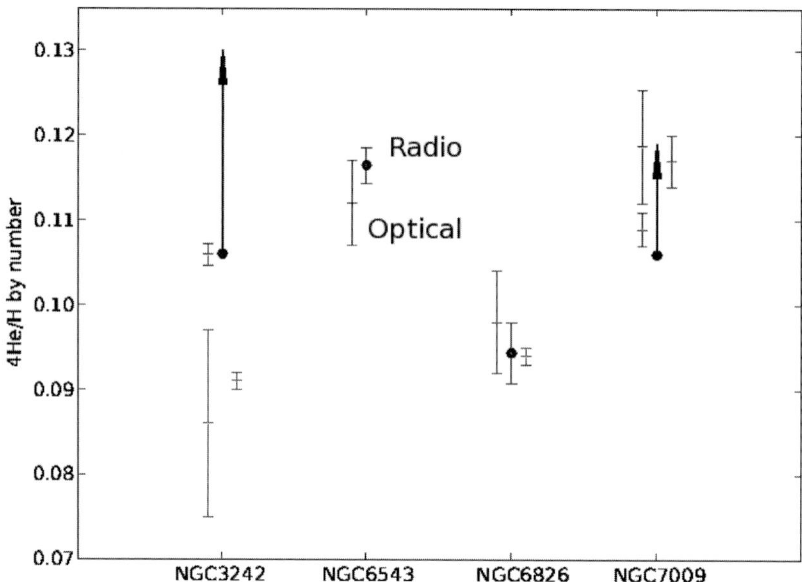

Figure 3. ^{4}He/H abundance ratios by number for the planetary nebulae sample. Filled circles are GBT results, while the plus sign symbols are from ORLs in the literature (Barker 1983, 1985, 1988; Perinotto *et al.* 2004; Krabbe & Copetti 2006).

In both nebulae carbon RRL emission is detected. The signal-to-noise ratio is excellent and the spectral baselines well behaved.

3. Results and discussion

The ^{4}He abundances for our PNe sample are shown in Figure 3. The GBT ^{4}He/H abundance ratios are calculated assuming ^{4}He/H $=$ ^{4}He$^+$/H$^+$. For NGC 3242 and NGC 7009 ^{4}He^{++} has been detected and thus our abundance estimates are only limits as indicated by the arrows in Figure 3. The tip of the arrow is an estimate of ^{4}He/H using the ^{4}He^{++}/^{4}He$^+$ ratios from Cahn *et al.* (1992) and assuming a uniform distribution of ^{4}He$^+$ and ^{4}He^{++}. The optical and radio data are in excellent agreement except for NGC 3242. This PN has significant amounts of ^{4}He^{++}, however, that may not be distributed uniformly. Therefore, the estimate of ^{4}He/H for the GBT data shown in Figure 3 may be smaller than indicated.

Figure 4 shows the derived abundance by mass, Y, plotted as a function of nebular metallicity, Z. Both Galactic and extragalactic H II region ^{4}He abundances are plotted as symbols together with an estimate of ^{4}He processing, dY/dZ, shown by the solid lines. For the GBT H II region data we again assume ^{4}He/H $=$ ^{4}He$^+$/H$^+$. The optical data shown for M17 are based on two different positions: M17-14 and M17-123 (an average of M17-1, M17-2, and M17-3) (Peimbert *et al.* 1992). The high ^{4}He optical value shown in Figure 4 is from M17-123 and is expected to be a better estimate of ^{4}He/H since this region has a higher degree of ionization and therefore less neutral helium (Carigi & Peimbert 2008). Although the radio ^{4}He abundance is the same within the errors as the optical value for position M17-14, the GBT position overlaps with position M17-3 and not with M17-14. Yet the GBT ^{4}He/H abundance ratio is significantly smaller than the optical value for position M17-3. This discrepancy may result because the large GBT beam averages over regions with different properties.

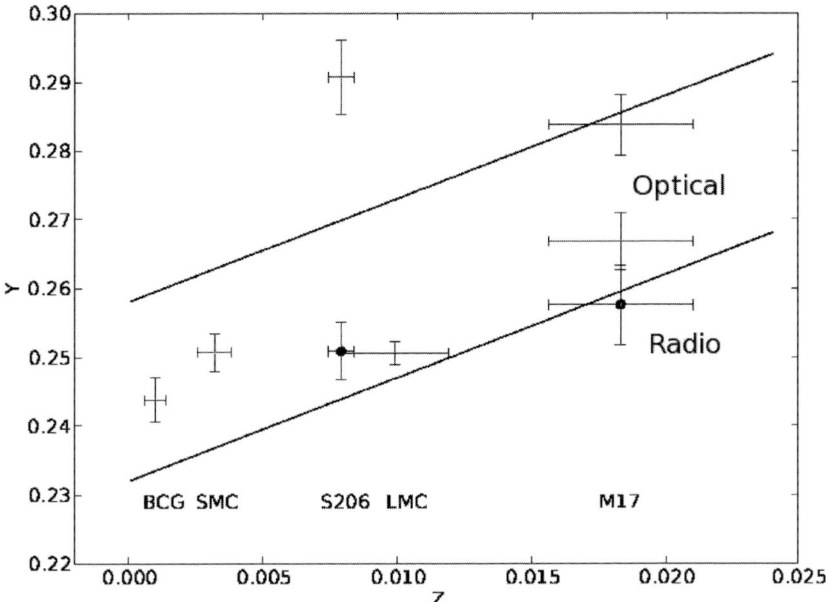

Figure 4. Evolution of ^{4}He/H. The helium abundance by mass, Y, is plotted against the metallicity, Z. GBT results are shown as filled circles for M17 and S206 while ORL results are plus signs for M17 (Carigi & Peimbert 2008; García-Rojas *et al.* 2007), the Small Magellanic Cloud (Peimbert *et al.* 2007), the Large Magellanic Cloud (Peimbert 2003), and a sample of blue compact galaxies (Izotov & Thuan 2004). The solid lines assume the range of primordial ^{4}He/H abundance ratios from Olive & Skillman (2004) and $dY/dZ = 1.5$ (Chiappini *et al.* 2002).

The ^{4}He abundances for S206 are puzzling. Deharveng *et al.* (2000) measure no variation in ^{4}He/H across S206 and values of O^{++}/O that range from 0.7 to 0.85; they conclude that there is no neutral helium in the nebula. S206 is ionized primarily by an O5 type star—the nebula structure is simple compared with M17 and contains less dust. Therefore, S206 should be an excellent candidate for measuring ^{4}He/H.

But the optical value for ^{4}He/H is 20% higher than the radio value. This discrepancy is difficult to reconcile and the optical result is not consistent with our understanding of Galactic chemical evolution. We expect there to be a net production of ^{4}He in stars and therefore $dY/dZ > 0$. This is illustrated by the two solid lines in Figure 4 where we have adopted the ^{4}He Galactic production of $dY/dZ = 1.5$ from Chiappini *et al.* (2002) and the conservative range for the ^{4}He primordial abundance from Olive & Skillman (2004). For the Milky Way we derive $dY/dZ = 1.71 \pm 0.86$ by using only the radio ^{4}He/H value for S206, both the optical and radio ^{4}He/H values for M17, and the primordial ^{4}He/H abundance ratio from Steigman (2007).

References

Baldwin, J. A., Ferland, G. J., Martin, P. G., Corbin, M. R., Cota, S. A., Peterson, B. M., & Slettebak, A. 1991, *ApJ*, 374, 580

Ballantyne, D. R., Ferland, G. J., & Martin, P. G. 2000, *ApJ*, 536, 773

Balser, D. S. 2006, *AJ*, 132, 2326

Balser, D. S., Bania, T. M., Brockway, C. J., Rood, R. T., & Wilson, T. L. 1994, *ApJ*, 430, 667

Balser, D. S., Bania, T. M., Rood, R. T., & Wilson, T. L. 1999, *ApJ*, 510, 759

Bania, T. M., Rood, R. T., & Balser, D. S. 2010, in: *Light Elements in the Universe* C. Charbonnel, M. Tosi, F. Primas, & C. Chiappini (eds.), Proc. IAU Symposium No. 268, (Cambridge: CUP), p. XX

Barker, T. 1983, *ApJ*, 267, 630

Barker, T. 1985, *ApJ*, 294, 193

Barker, T. 1988, *ApJ*, 326, 164

Cahn, J. H., Kaler, J. B., & Stanghellini, L. 1992, *A&AS*, 94, 399

Carigi, L. & Peimbert, M. 2008, *Rev. Mexicana AyA*, 44, 311

Chiappini, C., Renda, A., & Matteucci, F. 2002, *A&A*, 395, 789

Deharveng, L., Peña, M., Caplan, J., & Costero, R. 2000, *MNRAS*, 311, 329

García-Rojas, J., Esteban, C., Peimbert, A., Rodríguez, M., Peimbert, M., & Ruiz, M. T. 2007, *Rev. Mexicana AyA*, 43, 3

Gruenwald, R., Steigman, G., & Viegas, S. M. 2002, *ApJ*, 567, 931

Izotov, Y. I. 2010, in: *Light Elements in the Universe* C. Charbonnel, M. Tosi, F. Primas, & C. Chiappini (eds.), Proc. IAU Symposium No. 268, (Cambridge: CUP), p. XX

Izotov, Y. I. & Thuan, T. X. 2004, *ApJ*, 602, 200

Krabbe, A. C. & Copetti, M. V. F. 2006, *A&A*, 450, 159

Lockman, F. J. & Brown, R. L. 1982, *ApJ*, 259, 595

Mathis, J. S. & Wood, K. 2005, *MNRAS*, 360, 227

Peimbert, A. 2003, *ApJ*, 584, 735

Peimbert, M., Luridiana, V., & Peimbert, A. 2007, *ApJ*, 666, 636

Peimbert, P., Peimbert, A., Carigi, L., & Luridiana, V. 2010, in: *Light Elements in the Universe* C. Charbonnel, M. Tosi, F. Primas, & C. Chiappini (eds.), Proc. IAU Symposium No. 268, (Cambridge: CUP), p. XX volume

Peimbert, M., Rodríguez, L. F., Bania, T. M., Rood, R. T., & Wilson, T. L. 1992a, *ApJ*, 395, 484

Peimbert, M., Torres-Peimbert, S., & Ruiz, M. T. 1992b, *Rev. Mexicana AyA*, 24, 155

Perinotto, M., Morbidelli, L., & Scatarzi, A. 2004, *MNRAS*, 349, 793

Olive, K. A. & Skillman, E. D. 2004, *ApJ*, 617, 29

Sauer, D. & Jedamizik, K. 2002, *A&A*, 381, 361

Shaver, P. A. 1980, *A&A*, 91, 279

Skillman, E. 2010, in: *Light Elements in the Universe* C. Charbonnel, M. Tosi, F. Primas, & C. Chiappini (eds.), Proc. IAU Symposium No. 268, (Cambridge: CUP), p. XX

Steigman, G. 2007, *Ann. Rev. Nucl. Part. Sci.*, 57, 463

Viegas, S. M., Gruenwald, R., & Steigman, G. 2002, *ApJ*, 531, 813

Light Elements in the Universe
Proceedings IAU Symposium No. 268, 2009
C. Charbonnel, M. Tosi, F. Primas & C. Chiappini, eds.

© International Astronomical Union 2010
doi:10.1017/S1743921310003935

The primordial abundance of ^{4}He from a large sample of low-metallicity H II regions

Yuri I. Izotov

Main Astronomical Observatory,
Ukrainian National Academy of Sciences,
Zabolotnoho 27, Kyiv 03680, Ukraine
email: `izotov@mao.kiev.ua`

Abstract. We determine the primordial helium mass fraction Y_p using 1700 spectra of low-metallicity extragalactic H II regions. This sample is selected from the Data Release 7 of the Sloan Digital Sky Survey, from European Southern Observatory archival data and from our own observations. We have considered known systematic effects which may affect the ^{4}He abundance determination. They include collisional and fluorescent enhancements of He I recombination lines, underlying He I and hydrogen stellar absorption lines, collisional excitation of hydrogen lines, temperature and ionization structure of the H II region. Monte Carlo methods are used to solve simultaneously the above systematic effects. We find a primordial helium mass fraction $Y_p = 0.2512 \pm 0.0006$(stat.) ± 0.0020 (syst.). This value is higher than the value given by Standard Big Bang Nucleosynthesis (SBBN) theory. If confirmed, it would imply slight deviations from SBBN.

Keywords. galaxies: abundances; galaxies: irregular; galaxies: ISM; ISM: H II regions; ISM: abundances

1. Introduction

The determination of the primordial ^{4}He (hereafter He) abundance and some other light elements (D, ^{3}He, ^{7}Li) plays an important role in testing cosmological models. In the standard theory of big bang nucleosynthesis (SBBN), given the number of light neutrino species, the abundances of these light elements depend only on one cosmological parameter, the baryon-to-photon number ratio η.

While a single good baryometer like D is sufficient to derive the baryonic mass density from BBN, accurate measurements of the primordial abundance of He are required to check the consistency of SBBN. The primordial abundance of He, can in principle be derived accurately from observations of the helium and hydrogen emission lines from low-metallicity H II regions. Several groups have used this technique to derive the primordial He mass fraction Y_p, with somewhat different results. In the most recent study Izotov *et al.* (2007) based on a large sample of 93 H II regions derived $Y_p = 0.2472 \pm 0.0012$ and $Y_p = 0.2516 \pm 0.0011$, using Benjamin *et al.* (1999,2002) and Porter *et al.* (2005) He I emissivities, respectively. On the other hand, Peimbert *et al.* (2007) obtained $Y_p = 0.2477 \pm 0.0029$ based on the sample of 5 H II regions, Porter *et al.* (2005) He I emissivities, and adopting non-zero temperature fluctuations. Fukugita & Kawasaki (2006) derived $Y_p = 0.250 \pm 0.004$ for a sample of 31 H II regions by Izotov & Thuan (2004), adopting Benjamin *et al.* (1999,2002) He I emissivities.

Although He is not a sensitive baryometer (Y_p depends only logarithmically on the baryon density), its primordial abundance depends much more sensitively on the expansion rate of the Universe and is very sensitive to any small deviation from SBBN. However, to detect small deviations from SBBN and make cosmological inferences, Y_p

has to be determined to a level of accuracy of less than one percent. In particular, many known systematic effects need to be taken into account to transform the observed He I line intensities into a He abundance. These effects are: (1) reddening, (2) underlying stellar absorption in the He I lines, (3) collisional excitation of the He I lines which make their intensities deviate from their recombination values, (4) fluorescence of the He I lines which also make their intensities deviate from their recombination values, (5) collisional excitation of the hydrogen lines (hydrogen enters because the helium abundance is calculated relative to that of hydrogen), (6) possible departures from case B in the emissivities of H and He I lines, (7) the temperature structure of the H II region and (8) its ionization structure. All these corrections are at a level of a few percent except for effect (3) that can be much higher, exceeding 10% in the case of the He I $\lambda 5876$ emission line in hot and dense H II regions. All these effects were analyzed and taken into account by Izotov et $al.$ (2007) for a sample of 93 H II regions from Izotov & Thuan (2004) (hereafter HeBCD sample).

In this contribution we are aiming to significantly increase the sample of H II regions for the analysis of the systematic effects and determine the primordial He abundance with more accuracy. For this we select spectra of bright H II regions from the Data Release 7 (DR7) of the Sloan Digital Sky Survey (hereafter SDSS sample) (Abazajian et $al.$ 2009). In addition, we use available spectroscopic data from the ESO archive. The use of these additional data greatly increases the sample of objects suitable for the primordial He determination by a factor up to ~ 20 compared to the HeBCD sample of H II regions.

In the present determination of the primordial He abundance, we take into account all eight effects discussed above. The method has been detailed in Izotov et $al.$ (2007) and is more concisely discussed in this paper.

2. The sample

We determine Y_p for the sample which includes three different subsamples, HeBCD, VLT, and SDSS. The HeBCD subsample, composed of 93 spectra has been discussed by Izotov et $al.$ (2007). The SDSS subsample was selected from the SDSS DR7 and consists of 1534 spectra of H II regions with the Hβ line flux greater than 4×10^{-15} erg s^{-1} cm^{-2}. Finally, we selected 73 spectra of H II regions from the ESO archive which were obtained with the spectrographs UVES, FORS1 and FORS2 mounted on the VLT. The majority of objects in the HeBCD, VLT and SDSS subsamples are low-metallicity H II regions in dwarf emission-line galaxies.

3. The method

3.1. *Linear regressions*

As in our previous work (see Izotov et $al.$ 2007 and references therein), we determine the primordial He mass fraction Y_p by fitting the data points in the $Y - $ O/H plane with linear regression line of the form (Peimbert & Torres-Peimbert 1974,1976)

$$Y = Y_p + \frac{\mathrm{d}Y}{\mathrm{d(O/H)}}(\mathrm{O/H}), \qquad (3.1)$$

where

$$Y = \frac{4y(1 - Z)}{1 + 4y} \qquad (3.2)$$

is the He mass fraction, Z is the heavy element mass fraction, $y = (y^+ + y^{2+}) \times ICF(\mathrm{He^+ + He^{2+}})$ is the He abundance, $y^+ \equiv \mathrm{He^+/H^+}$ and $y^{2+} \equiv \mathrm{He^{2+}/H^+}$ are respectively the abundances of singly and doubly ionized He, and $ICF(\mathrm{He}^+ + \mathrm{He}^{2+})$ is the ionization correction factor for He.

We also take into account depletion of oxygen on dust grains. Izotov *et al.* (2006) demonstrated that the Ne/O abundance ratio for low-metallicity BCDs is not constant but is increased with increasing oxygen abundance. This effect is small, with $\Delta\log\mathrm{Ne/O}$ = 0.1 when the oxygen abundance is changed from $12+\log\mathrm{O/H}$ = 7.0 to 8.6, and it is attributed to oxygen depletion. We correct oxygen abundance for such depletion, using regression log Ne/O versus oxygen abundance found by Izotov *et al.* (2006) and assuming that depletion is absent in galaxies with $12+\log\mathrm{O/H}$ = 7.0.

To derive the parameters of the linear regressions, we use the maximum-likelihood method (Press *et al.* 1992) which takes into account the errors in Y, and O/H for each object.

The derived y^+ abundances depend on the He I emissivities, the fraction $\Delta I(\mathrm{H}\alpha)/I(\mathrm{H}\alpha)$ of the Hα emission line flux due to collisional excitation, the electron number density $N_e(\mathrm{He}^+)$, the electron temperature $T_e(\mathrm{He}^+)$, the equivalent widths $\mathrm{EW}_{abs}(\lambda3889)$, $\mathrm{EW}_{abs}(\lambda4471)$, $\mathrm{EW}_{abs}(\lambda5876)$, $\mathrm{EW}_{abs}(\lambda6678)$ and $\mathrm{EW}_{abs}(\lambda7065)$ of He I stellar absorption lines, and the optical depth $\tau(\lambda3889)$ of the He I $\lambda3889$ emission line. To determine the best weighted mean value of y^+_{wm} , we use the Monte Carlo procedure described in Izotov & Thuan (2004) and Izotov *et al.* (2007), randomly varying each of the above parameters within a specified range, excluding He I which are not varied.

In those cases when the nebular He II $\lambda4686$ emission line was detected, we have added the abundance of doubly ionized helium $y^{2+} \equiv \mathrm{He}^{2+}/\mathrm{H}^+$ to y^+.

3.2. *Set of parameters for the He abundance determination*

For the determination of He abundance we adopt Porter *et al.* (2005) He I emissivities and take into account the following systematic effects: 1) reddening; 2) the temperature structure of the H II region, i.e. the temperature difference between $T_e(\mathrm{He}^+)$ and $T_e(\mathrm{O}$ III); 4) underlying stellar He I absorption; 4) collisional and fluorescent excitation of He I lines; 5) collisional excitation of hydrogen lines; and 6) the ionization structure of the H II region.

The set of parameters, which vary in the range of physically reasonable values and is the same as that considered by Izotov *et al.* (2007), is defined in the following way:

1. We adopted the reddening law by Whitford (1958). Izotov *et al.* (2007) have shown that He abundances are not sensitive to the adopted reddening law. In particular, the use of the Cardelli *et al.* (1989) reddening curve is resulted in He and other element abundances, which are similar to those obtained with the Whitford (1958) reddening law. The extinction coefficient $C(\mathrm{H}\beta)$ is derived from the observed hydrogen Balmer decrement after subtraction of the contribution due to the collisional excitation from the the Hα and Hβ line fluxes. Finally, all emission lines are corrected for reddening adopting the derived $C(\mathrm{H}\beta)$.

2. The electron temperature of the He$^+$ zone is varied in the range $T_e(\mathrm{He}^+)$ = (0.95 – 1.0)$\times T_e(\mathrm{O}$ III). This assumption is adopted following Guseva *et al.* (2006,2007) who derived the electron temperature in the H$^+$ zone from the Balmer and Paschen discontinuities in spectra of more than 100 H II regions and showed that $T_e(\mathrm{H}^+)$ differs from $T_e(\mathrm{O}$ III) by not more than 5%.

3. Oxygen abundances are calculated adopting an electron temperature equal to $T_e(\mathrm{O}$ III) and $T_e(\mathrm{He}^+)$.

4. $N_e(\mathrm{He}^+)$ and $\tau(\lambda3889)$ vary respectively in the ranges $10 - 450$ cm^{-3} and $0 - 5$, typical for extragalactic H II regions.

5. The fraction of Hα emission due to collisional excitation is varied in the range 0% – 5%, in accordance with Stasińska & Izotov (2001). The fraction of Hβ emission due to the collisional excitation is assumed to be 1/3 that of Hα emission.

6. The equivalent width of the He I λ4471 absorption line is fixed to EW$_{abs}$(λ4471) = 0.4Å. The equivalent widths of the other absorption lines are fixed according to the ratios EW$_{abs}$(λ3889) / EW$_{abs}$(λ4471) = 1.0, EW$_{abs}$(λ5876) / EW$_{abs}$(λ4471) = 0.3, EW$_{abs}$(λ6678) / EW$_{abs}$(λ4471) = 0.1 and EW$_{abs}$(λ7065) / EW$_{abs}$(λ4471) = 0.1. This set of the EWs is justified by Izotov $et\ al.$ (2007) as the most likely one.

7. He I emissivities from Porter $et\ al.$ (2005) are adopted.

8. The He ionization correction factor ICF(He$^+$+He^{++}) is adopted from Izotov $et\ al.$ (2007).

4. The primordial He mass fraction Y_p

Our sample constitues the largest sample for the primordial He determination and thus greatly reduces the uncertainties caused by the statistical errors in He abundances. This allows us to study systematic effects on firm statistical grounds.

Linear regressions Y – O/H for the total sample of 1700 spectra with the basic set of parameters are shown in Figs. 1a and 1b. The difference between these two regressions is that the oxygen abundance in Fig. 1a is derived adopting the temperature of the O^{++} zone equal to T_e(He$^+$), while the temperature T_e(O III) derived from the [O III] λ4363/(λ4959+λ5007) line flux ratio is used in Fig. 1b to obtain the oxygen abundance. The primordial value obtained from the regression in Fig. 1a of Y_p = 0.2512 $\pm$ 0.0006 is very close to the value of Y_p = 0.2516 $\pm$ 0.0011 obtained for HeBCD sample by Izotov $et\ al.$ (2007) for the same set of parameters. It is seen from this comparison that the use of the large sample reduces statistical errors in Y_p by a factor of two. The value of Y_p derived here is by 1.3% greater than the SBBN value obtained from the WMAP data.

The variation of the parameter ranges, similar to that done by Izotov $et\ al.$ (2007), shows that the primordial He abundance is always larger than the standard one and varies in the range $\sim$ 0.249 – 0.252. Finally, we adopt as the best the regression in Fig. 1a with Y_p= 0.2512 $\pm$ 0.0006. This regression is obtained with the basic set of parameters and the temperature T_e(He$^+$) for the oxygen abundance determination. Adding a systematic error caused by the uncertainties in He I emissivities (Porter $et\ al.$ 2009) we obtain Y_p= 0.2512 $\pm$ 0.0006(stat.) $\pm$ 0.0020(syst.).

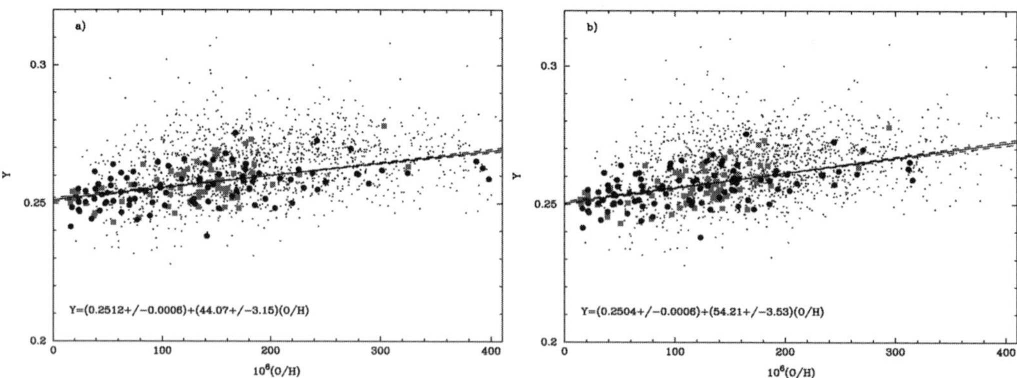

Figure 1. Linear regressions helium mass fraction Y vs. oxygen abundance O/H. The HeBCD H II regions are shown by filled circles, the archival VLT data by grey squares and the SDSS objects by dots. The oxygen abundance O/H is derived adopting (a) T_e(He$^+$) and (b) T_e(O III).

5. Cosmological implications

We now investigate whether our derived values of Y_p are consistent with the predictions of SBBN and whether the baryonic mass density corresponding to Y_p agrees with the one derived from measurements of the CMB. We follow here the analysis by Izotov *et al.* (2007) with our new value $Y_p = 0.2512 \pm 0.0006$. With this value, and with an equivalent number of light neutrino species equal to 3, the SBBN model gives $\eta_{10} = 10^{10}\eta = 8.7 \pm 0.5$, where the error bars denote 1σ errors. This value corresponds to a baryonic mass fraction $\Omega_b h^2 = 0.032 \pm 0.002$ and is significantly higher than $\Omega_b h^2 = 0.02273 \pm 0.00062$ derived from measurements of the fluctuations of the microwave background radiation by WMAP (Dunkley *et al.* 2009). Thus, the deviations from the SBBN are likely present. However, if the systematic error of 0.0020 in Y_p is taken into account because of uncertainties in the He I emissivities, then our value within 2σ is consistent with the SBBN value $Y_p = 0.2482^{+0.0003}_{-0.0004} \pm 0.0006$ (syst.) inferred by Spergel *et al.* (2007).

Deviations from the standard rate of Hubble expansion in the early Universe can be caused by an extra contribution to the total energy density (for example by additional flavors of active or sterile neutrinos) which can conveniently be parameterized by an equivalent number of neutrino flavors N_ν. Combining $\Omega_b h^2 = 0.002273 \pm 0.00062$ obtained by WMAP (Dunkley *et al.* 2009) with $Y_p = 0.2512 \pm 0.0006$, we obtain $N_\nu \sim 3.25$ (Steigman 2005,2007).

6. Summary

We present the determination of the primordial helium mass fraction Y_p by linear regressions of a large sample of low-metallicity extragalactic H II regions.

In the determination of Y_p, we have considered several known systematic effects. We have used Monte Carlo methods to take into account the effects of collisional and fluorescent enhancements of He I recombination lines, of collisional excitation of hydrogen emission lines, of underlying stellar He I absorption, of the difference between the temperature $T_e(He^+)$ in the He^+ zone and the temperature $T_e(O\ \text{III})$ derived from the [O III]$\lambda4363/(\lambda4959+\lambda5007)$ flux ratio, and of the ionization correction factor $ICF(He^+ + He^{2+})$.

We have obtained the following results:

1. For the determination of the primordial He abundance we construct a sample of 1700 spectra of low-metallicity extragalactic H II regions which includes the HeBCD sample Izotov *et al.* (2007), archival VLT spectra and spectra from the Data Release 7 (DR7) of the Sloan Digital Sky Survey (SDSS). This sample is the largest data sets in existence for the determination of Y_p.

2. We obtain $Y_p = 0.2512 \pm 0.0006$(stat.) ± 0.0020(syst.), corresponding to $\Omega_b h^2 = 0.032 \pm 0.002$(stat.), significantly larger than the $\Omega_b h^2$ values derived from the deuterium abundance and microwave background radiation fluctuation measurements. If we take the higher value of Y_p at its face value, then this would imply the existence of small deviations from SBBN. In order to bring the high value of Y_p into agreement with the deuterium and WMAP measurements, we would need an equivalent number of neutrino flavors equal to 3.25 instead of the canonical 3.

References

Abazajian, K., *et al.* 2009, *ApJS*, 182, 543
Benjamin, R. A., Skillman, E. D., & Smits, D. P. 1999, *ApJ*, 514, 307

Benjamin, R. A., Skillman, E. D., & Smits, D. P. 2002, *ApJ*, 569, 288
Cardelli, J. A., Clayton, G. C., & Mathis, J. S. 1989, *ApJ*, 345, 245
Dunkley, J., *et al.* 2009, *ApJS*, 180, 306
Fukugita, M. & Kawasaki, M. 2006, *ApJ*, 646, 691
Guseva, N. G., Izotov, Y. I., & Thuan, T. X. 2006, *ApJ*, 644, 890
Guseva, N. G., Izotov, Y. I., Papaderos, P., & Fricke, K. J. 2007, *A&A*, 464, 885
Izotov, Y. I. & Thuan, T. X. 2004, *ApJ*, 602, 200 (IT04)
Izotov, Y. I., Thuan, T. X., & Lipovetsky, V. A. 1994, *ApJ*, 435, 647
Izotov, Y. I., Thuan, T. X., & Lipovetsky, V. A. 1997, *ApJS*, 108, 1
Izotov, Y. I., Stasińska, G., Meynet, G., Guseva, N. G., & Thuan, T. X. 2006, *A&A*, 448, 955
Izotov, Y. I., Thuan, T. X., & Stasińska, G. 2007, *ApJ*, 662, 15
Peimbert, M. & Torres-Peimbert, S. 1974, *ApJ*, 193, 327
Peimbert, M. & Torres-Peimbert, S. 1976, *ApJ*, 203, 581
Peimbert, M., Luridiana, V., & Peimbert, A. 2007, *ApJ*, 666, 636
Porter, R. L., Bauman, R. P., Ferland, G. J., & MacAdam, K. B. 2005, *ApJ*, 622, L73
Porter, R. L., Ferland, G. J., MacAdam, K. B., & Storey, P. J. 2009, *MNRAS*, 393, L36
Press, W. H., Teukolsky, S. A., Vetterling, W. T.,& Flannery, B. P. 1992, *Numerical Recipes in C, The Art of Scientific Computing /Second Edition/*, Cambridge University Press
Spergel, D. N., Bean, R., Doré, O., *et al.* 2007, *ApJS*, 170, 377
Stasińska, G. & Izotov, Y. I. 2001, *A&A*, 378, 817
Steigman, G. 2005, *Physica Scripta*, T121, 142
Steigman, G. 2007, *Annual Review of Nuclear and Particle Science*, 57, 463
Whitford, A. E. 1958, *AJ*, 63, 201

Light Elements in the Universe
Proceedings IAU Symposium No. 268, 2009
C. Charbonnel, M. Tosi, F. Primas & C. Chiappini, eds.
© International Astronomical Union 2010
doi:10.1017/S1743921310003947

Uncertainties in nebular helium abundances

Evan D. Skillman[1]

[1]Astronomy Department, University of Minnesota,
116 Church St. SE, Minneapolis, MN, USA
email: skillman@astro.umn.edu

Abstract. Efforts to determine the primordial helium abundance via observations of metal poor HII regions have been limited by significant uncertainties. Because of a degeneracy between the solutions for density and temperature, the precision of the helium abundance determinations is limited. Spectra from the literature are used to show the effects of new atomic data and to demonstrate the challenges of determining precise He abundances. Several suggestions are made for meeting these challenges.

Keywords. nucleosynthesis, abundances, cosmological parameters, early universe, ISM: HII regions

1. Introduction: In defense of Olive & Skillman (2004)

Fig. 1 shows the historical evolution of the observational determinations of the primordial helium abundance compared to the present day value inferred from WMAP observations and BBN calculations. One well known effect shown in the plot is the small error bars relative to the differences between determinations and relative to the currently favored value. In 2004, Keith Olive and I (Olive & Skillman 2004, hereafter OS04) attempted to account for most of the important statistical and systematic uncertainties, and, in a reanalysis of data from the literature (Izotov & Thuan 1998; Peimbert, Peimbert, & Ruiz 2000), found significantly larger uncertainties for helium abundance determinations for individual objects and a resulting larger uncertainty in the primordial helium abundance. In the plot, this result appears as a giant step backwards as the uncertainty in the primordial helium abundance is comparable to uncertainties associated with determinations from the 1980's. Part of the large uncertainty is due to concentrating on a small sample of "high quality" spectra of the lowest metallicity objects. This was to avoid two statistical effects which we felt were unsound. Including a large number of objects will reduce the error in the primordial helium abundance, even if some of those objects have relatively lower quality spectra. Additionally, including objects at higher metallicity will reduce the error on the primordial helium abundance because the extrapolation to zero metallicity will have a smaller error. However, this increases the dependence on the assumption of a linear relationship between the helium abundance and the oxygen (or nitrogen) abundance. By avoiding these two effects, we felt that the derived uncertainty in the primordial helium abundance better characterized the true uncertainty. Notably, a significant factor in the increased uncertainty of OS04 was simply larger uncertainties on the helium abundances of individual objects due to including neglected terms and running Monte Carlo analyses to account for degeneracies in minimization solutions.

An important result of the Monte Carlo analysis was the discovery of degeneracies in the sense that the physical conditions in the HII regions were not always independently constrained (Olive & Skillman 2001; Olive & Skillman 2004). In particular, the temperature and density exhibited a trade-off whereby increasing the temperature while decreasing the density would leave the χ^2 relatively unchanged. The main reason for this

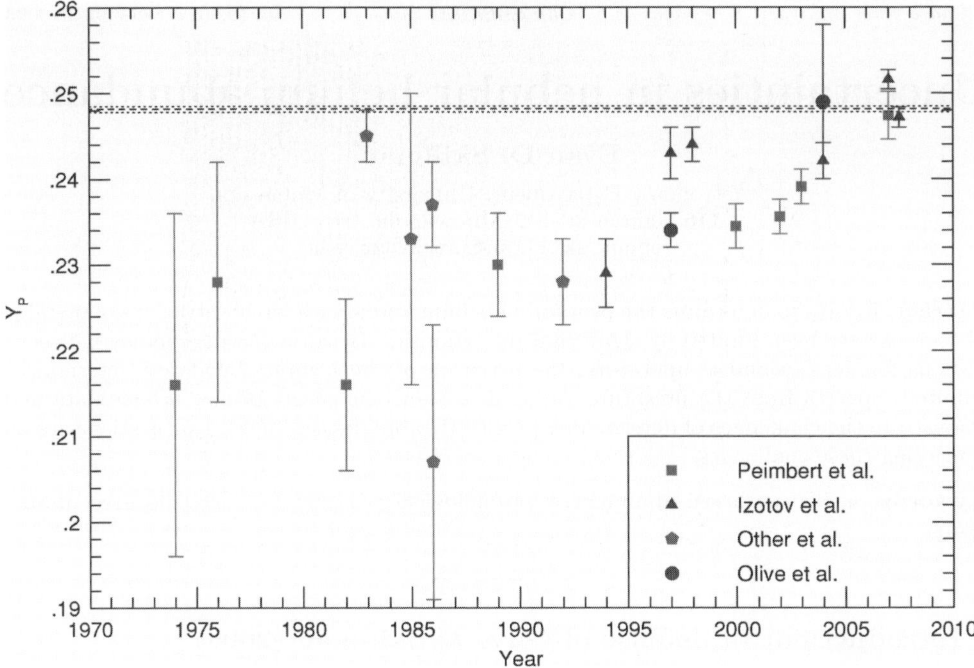

Figure 1. Historical evolution of the observational determinations of the primordial helium abundance. The solid horizontal line with dashes shows the value inferred from WMAP observations and BBN calculations.

is that four of the HeI emission lines used to derive the helium abundance show similar exponential dependencies on temperature and density.

We demonstrate the importance of the degeneracy between temperature and density in Figure 2. Here, we have generated a synthetic spectrum using typical physical conditions and a helium abundance. Values of the helium abundance and a corresponding χ^2 were then derived from this spectrum keeping all parameters fixed, but varying the density and temperature. The impact of trading the density against the temperature can be clearly seen in Figure 2 which illustrates a well constrained quotient of density and temperature but a $\triangle\chi^2 = 2.3$ boundary spanning a temperature range of 5500 K, a density range of $275\ \mathrm{cm}^{-3}$, and an abundance variation of 10%. The shallow, extended minimum results in a large range of best fit parameters for the density and temperature upon the Monte Carlo perturbation, and the abundance gains a much larger uncertainty (see Figure 3). Only by solving for these two parameters simultaneously can this degeneracy be discovered.

2. New advances in helium abundance analysis

With University of Minnesota graduate student Erik Aver, Keith Olive and I have investigated improvements to the code that we used in 2004. This was, in part, in response to recent improvements to the atomic data (Porter, Ferland, & MacAdam 2007) which have resulted in significant changes to the derived helium abundances of HII regions (Izotov *et al.* 2007; Peimbert, Luridiana, & Peimbert 2007). Additionally, we have investigated combining the minimizations, previously based on the hydrogen lines and the helium lines separately, into a single minimization, which is inherently self-consistent. The single minimization allows us to investigate the correction for the collisional emission of a small population of neutral hydrogen (e.g., Davidson & Kinman 1985; Skillman &

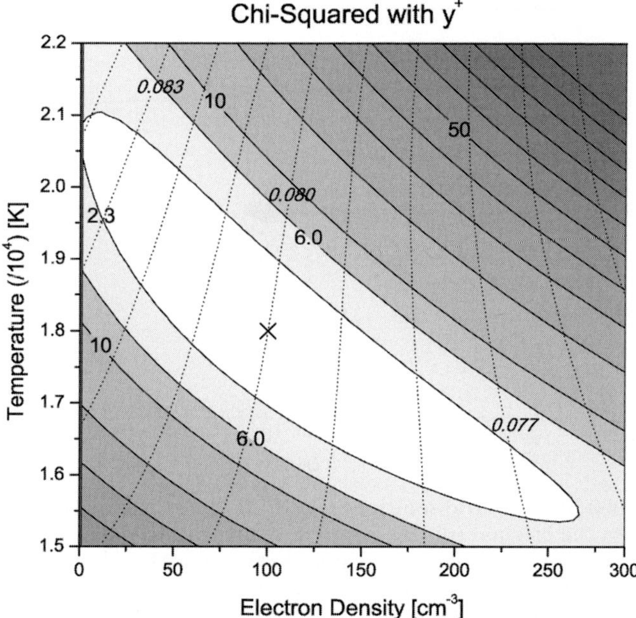

Figure 2. A plot of the derived helium abundance as a function of the temperature and density for a synthetically generated spectrum. Impressed on this diagram is a contour plot of χ^2 versus density and temperature for the same spectrum. The extension of the $\chi^2 = 2.3$ contour with a strong negative correlation highlights the degeneracy between density and temperature. That the χ^2 and abundance contours are nearly perpendicular demonstrates the impact of the degeneracy on the abundance determination ($\pm 5\%$). From Aver, Olive, & Skillman (2010).

Kennicutt 1993; Stasińska & Izotov 2001; Luridiana *et al.* 2003). Finally, the wavelength dependence in underlying stellar absorption can be included (initial previous work assumed identical values of equivalent width of underlying absorption for all H I or He I emission lines, regardless of wavelength) as has been investigated previously by Izotov *et al.* (2007). Detailing the incorporation of these four effects and investigation of their impact is the primary aim of a forthcoming paper (Aver, Olive, & Skillman 2010), and here I will provide a summary of the main results.

The updated emissivities, including the collisional corrections, in general raise the calculated abundance of helium (in agreement with Izotov *et al.* 2007; Peimbert, Luridiana, & Peimbert 2007). However, because the changes in different lines are qualitatively different, the update does not categorically raise the abundance. For $\lambda 4471$, 5876, and 6678, the Hβ to He emissivity is increased rather uniformly (2-3%) across the relevant temperature range, and all three lines exhibit a gradual increase with temperature. The behavior of $\lambda 4026$ is similar except that the new emissivity is shifted to lower values. $\lambda 3889$ is more complicated in that the emissivity is reduced at low temperatures but is enhanced at $T \geqslant 16{,}000$ K. $\lambda 7065$ is similar though the enhancement begins at 13,000 K. Broadly though, both $\lambda 3889$ and $\lambda 7065$ as well as $\lambda 4471$ track the BSS emissivities more closely (within $\sim 2\%$) than the other three lines. The three strongest lines with the smallest relative errors will produce a larger abundance, thus raising the average abundance. However, the changes to $\lambda\lambda 3889$, 7065, and 4471 affect the derived physical conditions in the nebula, and, in some cases, drive the derived abundance to lower values, opposite to the trend for the three strongest lines.

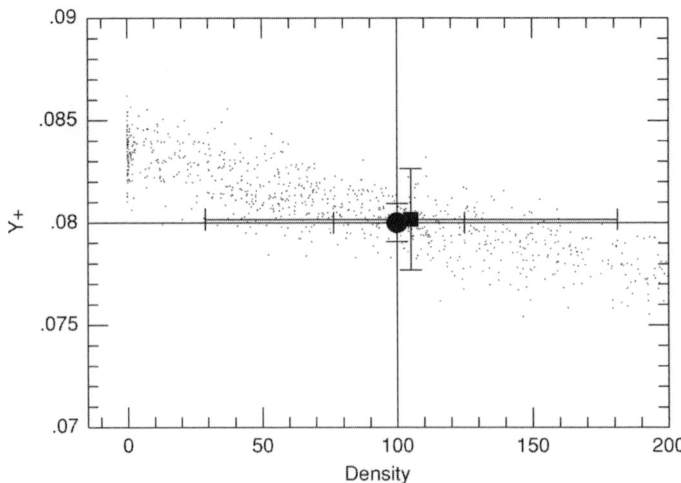

Figure 3. Plot of 1000 Monte Carlo solutions based upon a synthetic spectrum. The solid circle with error bars marks the original (direct) solution; while the solid square with error bars marks the average of the Monte Carlo solutions. Upon performing the Monte Carlo, the large range in density solutions (and correspondingly temperature solutions), gives a much larger density uncertainty, resulting in a marked increase in the abundance uncertainty (from 1% to 3%). From Olive, & Skillman (2001).

The changes to the atomic data for helium emissivities were much larger than ever accounted for in previous work. In most cases these uncertainties were regarded as insignificant when compared with the other uncertainties. Benjamin, Skillman, & Smits (1999) estimated a minimum uncertainty of 1.5% and this value is comparable to the changes due to the use of the new emissivities. This change alone validates the main conclusion of Olive & Skillman (2004) that the uncertainties in the derived helium abundances of individual objects were underestimated.

The combining of both the helium and hydrogen lines into a single minimization program has only small impact *as long as collisional excitation of neutral hydrogen is not included.* This is because the relative fluxes of the hydrogen emission from recombination has very low dependencies on density and temperature, so the hydrogen lines do not play into the density/temperature degeneracy. However, once collisional excitation of neutral hydrogen is included, then the strengths of the hydrogen lines gain a new temperature dependence, and it is crucial to include them in the minimization. A three way degeneracy between electron density, temperature, and the neutral hydrogen fraction emerges. These degeneracies are shown in Figure 4 which shows the Monte Carlo results for NGC 346. Plotted are the individual results for density, temperature, and neutral hydrogen fraction for 1000 MC realizations. As one can see, the solutions map out a wide range for these parameter values. The neutral hydrogen fraction can be increased to compensate for a decrease in the collisional to recombination rate which is itself only temperature dependent. Furthermore, the neutral hydrogen collisional emission correction can have a surprisingly significant effect in raising the abundance. Since the collisional correction for $H\beta$ multiplies each helium line abundance, it multiplicatively raises the abundance but does not change the value of the corresponding χ^2. As a result, the helium abundances of all the objects that we studied increased.

The uncertainty in the choice of correction for underlying absorption is chiefly mitigated by selecting only spectra with very high emission line equivalent widths. The main problems regarding underlying absorption arise when corrections are not included at all

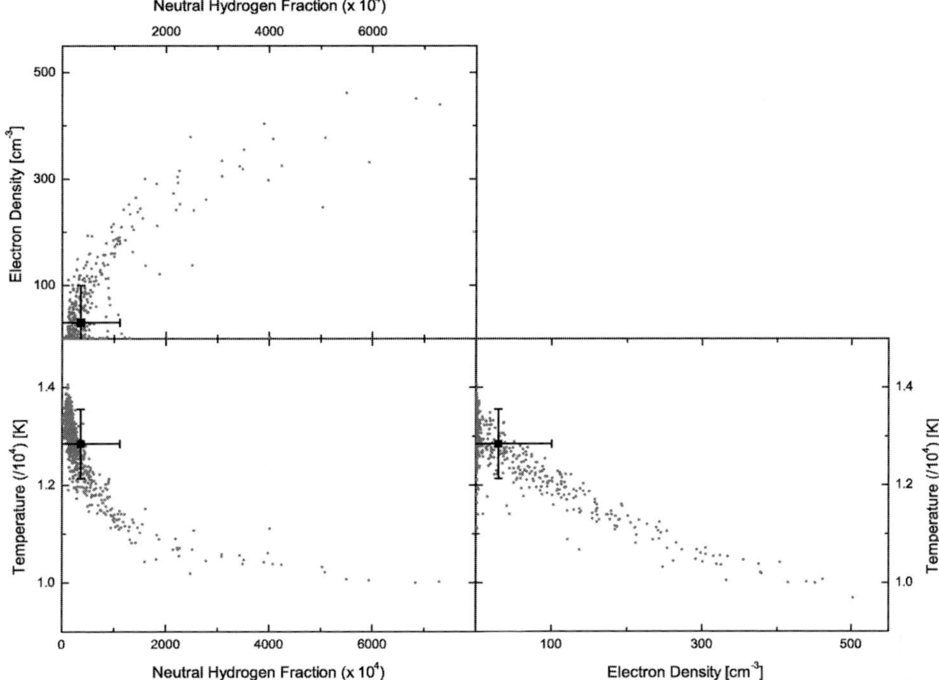

Figure 4. Plots of Monte Carlo results (1000 points) of density, temperature, and neutral hydrogen fraction for NGC 346. The strong correlation between the three parameters demonstrates the difficulty in solving for them independently. From Aver, Olive, & Skillman (2010).

(cf. Skillman, Terlevich, & Terlevich 1998). Even in the case of NGC 346, where a filament with no obvious exciting stars was observed (Peimbert, Peimbert, & Ruiz 2000), Olive & Skillman (2004) found evidence for significant underlying absorption. This was confirmed by Peimbert, Luridiana, & Peimbert (2007). Once again, this underlines the importance of obtaining high quality spectra of the most promising targets.

Extending the previous work to allow for self-consistent analysis reflecting the common temperature and reddening parameters between the hydrogen and helium lines does not diminish the importance of the systematic uncertainties or the case for the larger errors determined via Monte Carlo. In fact, the addition of the neutral hydrogen collisional emission, introduced with a new physical parameter, the neutral hydrogen fraction, highlights that abundance errors may be even larger than previously thought. In addition to the need for observations to constrain systematic uncertainty as much as possible, we have found that relatively small systematic errors in the observed fluxes, even of the weakest lines, can have a significant impact on the derived value of y^+. Clearly the highest quality spectra are critical for a meaningful analysis.

Ultimately, the sensitivity of the determination of the primordial helium abundance demands that analysis be self-consistent and comprehensive; unfortunately efforts in providing such an analysis yield calculated uncertainties that limit the utility of that determination. Nevertheless, fully understanding the terrain of potential obstacles is necessary to direct further measurement efforts so as to provide a reliable calculation of the primordial helium abundance. Apart from improving the quality of the spectra, the neutral hydrogen collisional correction for the weaker hydrogen lines can be easily improved from extended principal quantum number collisional strengths. Furthermore the use of an additional hydrogen line would benefit the neutral hydrogen fraction and hydrogen absorption since those two parameters are determined exclusively through the three hydrogen lines. The next Balmer line, H7 (λ3971) is blended with Ne III (λ3967), and H8, blended with HeI λ3889, would not be independent for analysis. This leaves the very weak H9 and H10 as the best candidates. The use of other emission lines may also be useful in lifting the error increasing degeneracy between density and temperature. However, this requires accurate metal line fluxes whose ratio is sensitive to the electron density or temperature in the applicable range. Most candidates are impaired by the weakness of their emission lines and ionization populations that are not characteristic of the majority of the HII region.

References

Benjamin, R. A., Skillman, E. D., & Smits, D. P. 1999, *ApJ*, 514, 307

Davidson, K. & Kinman, T. D. 1985, *ApJS*, 58, 321

Izotov, Y. I. & Thuan, T. X. 1998, *ApJ*, 500, 188

Izotov, Y. I., Thuan, T. X., & Stasinska, G. 2007, *ApJ*, 662, 15 (ITS07)

Luridiana, V., Peimbert, A., Peimbert, M., & Cerviño, M. 2003, *ApJ*, 592, 846

Olive, K. A. & Skillman, E. D. 2001, *New Astronomy*, 6, 119

Olive, K. A. & Skillman, E. D. 2004, *ApJ*, 617, 29

Peimbert, M., Luridiana, V., & Peimbert, A. 2007, *ApJ*, 666, 636 (PLP07)

Peimbert, M., Peimbert, A., & Ruiz, M. T. 2000, *ApJ*, 541, 688

Porter, R. L., Ferland, G. J., & MacAdam, K. B. 2007, *ApJ*, 657, 327 (PFM)

Skillman, E. D. & Kennicutt, R. C. 1993, *ApJ*, 411, 655

Skillman, E. D., Terlevich, E., & Terlevich, R. 1998, *Spa. Sci. Rev.*, 84, 105

Stasińska, G. & Izotov, Y. I. 2001, *A&A*, 378, 817

Light Elements in the Universe
Proceedings IAU Symposium No. 268, 2009
C. Charbonnel, M. Tosi, F. Primas & C. Chiappini, eds.

© International Astronomical Union 2010
doi:10.1017/S1743921310003959

The quite complex "Simple Stellar Populations" of globular clusters

Angela Bragaglia[1]

[1]INAF-Osservatorio Astronomico di Bologna,
via Ranzani 1, 40127 Bologna, Italy
email: `angela.bragaglia@oabo.inaf.it`

Abstract. There is compelling observational evidence that globular clusters (GCs) are quite complex objects. A growing body of photometric results indicate that the evolutionary sequences are not simply isochrones in the observational plane -as believed until a few years ago- from the main sequence, to the subgiant, giant, and horizontal branches. The strongest indication of complexity comes however from the chemistry, from internal dispersion in iron abundance in a few cases, and in light elements (C, N, O, Na, Mg, Al, etc.) in *all* GCs. This universality means that the complexity is *intrinsic* to the GCs and is most probably related to their formation mechanisms. The extent of the variations in light elements abundances is dependent on the GC mass, but mass is not the only modulating factor; metallicity, age, and possibly orbit can play a role. Finally, one of the many consequences of this new way of looking at GCs is that their stars may show different He contents.

Keywords. Stars: abundances, Population II – Galaxy: globular clusters

1. It's not so simple

Globular Clusters (GCs) have long been considered the best approximations of a *Simple Stellar Population* (Renzini & Buzzoni 1986), and this may still be valid for some purposes. However, they are not truly simple. We know that GCs contain a fraction of binaries (e.g., Meylan & Heggie 1987). But now we also know that their stars are not strictly coeval; for old clusters, like the Galactic globulars, the age differences are so small to be hardly detectable, but the same is not true for the Magellanic Clouds ones. And we also know that the initial chemical composition of the stars we presently observe was not the same.

There is a growing body of observational evidence of the complexity of GCs, both from photometry (with multiple, or split, or wide sequences) and from spectroscopy (with large differences in light elements and even, in a few cases, with spreads in metallicity).

Of course, ω Cen is the first example that comes to mind, even if for its characteristics (or better, because of its characteristics) it has often been labelled as the nucleus of an ancient dwarf spheroidal galaxy. However, ω Cen is only the tip of the iceberg and there are many other interesting cases, like M54, which lies in the nucleus of the Sagittarius dwarf galaxy (Ibata *et al.* 1994, Bellazzini *et al.* 2008a), and which resembles ω Cen in several aspects. But also more "normal", lower mass clusters show peculiarities, like NGC 2808 which, among many other oddities, presents three well separated main sequences (Bedin *et al.* 2004, Sollima *et al.* 2007). I will come back to these three objects later.

Among other evident examples we find for instance M22 (NGC 6656) which had long been suspected to have a dispersion in metallicity (from photometry, e.g. Hesser *et al.* 1977 or spectroscopy, e.g. Brown & Wallerstein 1992, but see also Ivans *et al.* 2004)

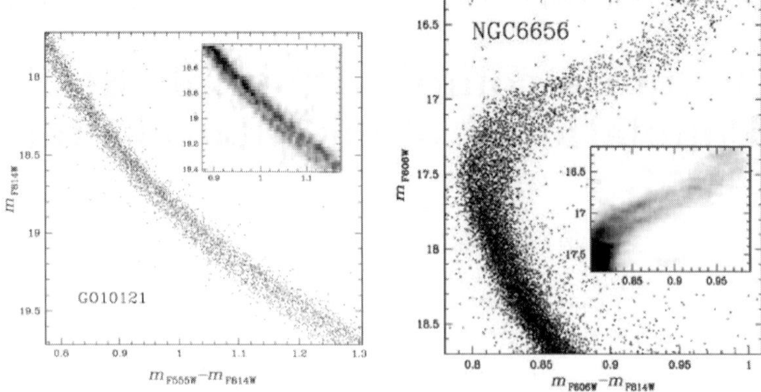

Figure 1. Examples of split main sequences or subgiant branches recently found in Galactic GCs using the high precision photometry of ACS@HST: NGC 6752 (Milone *et al.* 2009), and M22 (courtesy of A. Milone).

and that also displays a split in the SGB and RGB (see Fig. 1), or NGC 1851 with its split SGB (Milone *et al.* 2008) and RGB (Han *et al.* 2009), and its anomalous chemical abundances (e.g. Yong *et al.* 2009).

Interestingly, thanks to the exquisite precision of the ACS camera on HST (whose photometry can reach the depth and precision to detect even small colour and magnitude differences), more clusters were found to show wide or even split evolutionary sequences, from the main sequence and up to the SGB and RGB. The newest entries are 47 Tuc, where Anderson *et al.* (2009) find both a split SGB and a wide faint main sequence, and NGC 6752, unsuspected until now, which shows a split main sequence (Milone *et al.* 2009); for both, see Fig. 1. It seems that more and more GCs are beginning to unveil their complex photometric sequences; for a more extended discussion, see the earlier reviews by Piotto (2008, 2009).

Even if the evidence from photometry is the easiest to see, even for non-specialists, the strongest proof that GCs harbour at least two stellar generations comes from spectroscopy. Not only we may see stars with the same evolutionary status but with very different chemical composition in GCs, at least for some light elements, but also this situation is not limited to a few "freaks", as ω Cen or M22† were considered for a long time. The chemical signatures we used to call "anomalies" are widespread and show up in *all* clusters studied. Bimodality -and anticorrelation- in CN and CH strengths, or anticorrelations between other light elements, like Na and O, or Mg and Al, have long been observed in evolved stars, where they could be explained, although with some difficulties, considering extra-mixing episodes (see e.g. the reviews by Smith 1987 and Kraft 1994 where the extra-mixing *vs* the primordial-enrichment hypotheses are discussed).

However, these same so-called chemical anomalies were later found also in non evolved, main sequence stars (e.g., Cannon *et al.* 1998 found a bimodality in CN, CH in 47 Tuc, and other authors in many other GCs, see the review by Gratton *et al.* 2004).

Once established that the Na-O anticorrelation was present also in unevolved stars (Gratton *et al.* 2001 found it for the first time in NGC 6752 and NGC 6397, followed

---

† Recently, the claim of a dispersion in metallicity in M22 has gained substance, and two papers appeared. One is based on a sample of 17 stars studied with high resolution spectra (Marino *et al.* 2009), and indicates differences both in metallicity and heavy elements. The second paper, by Da Costa *et al.* (2009), presents a larger sample, but metallicity is derived from the Calcium triplet. Both show a bimodality in metallicity.

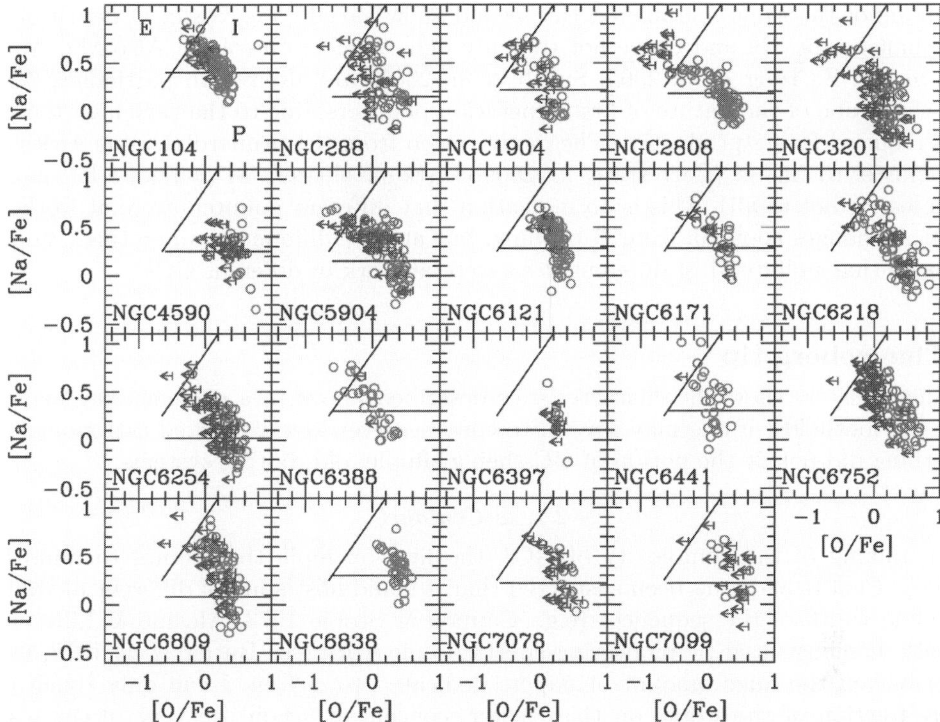

Figure 2. Na-O anticorrelations in 19 GCs observed with FLAMES@VLT, with separation among Primordial, Intermediate, and Extreme populations (see Carretta *et al.* 2009a for details).

by Ramirez & Cohen 2002 in M71, and by Carretta *et al.* 2004 in 47 Tuc), it became clear that another explanation was required. These chemical signatures are the result of H-burning at high temperature (ON, NeNa, MgAl cycles: see Denisenkov & Denisenkova 1989, Langer et al. 1993); the resulting chemical patterns cannot be produced in low mass, main sequence stars like those we are presently seeing in GCs. So the typical chemical signature of high N, Na, Al, and low C, O, Mg *must have originated in a previous generation of more massive stars, that polluted the gas from which part of the GC stars formed later.*

Perhaps the most famous pair of elements showing such variations, anti-correlated with each other, are O and Na. This Na-O anticorrelation was first found among cluster giants, mainly by the Lick-Texas group, by Kraft, Sneden, and many collaborators. They studied two-three tens of targets per cluster, observing stars one by one; for a review of their work, see Kraft (1994) and Gratton *et al.* (2004). The availability of efficient spectrographs at 8-10m telescopes, and their multi-object capabilities have permitted to extend this kind of study both to faint, unevolved stars and to much larger samples. About three years ago, there were about 200 giants studied in literature, and about 50 unevolved stars, as shown in Carretta *et al.* (2006) in their presentation of the Na-O anticorrelation in about 100 stars in NGC 2808, studied with FLAMES@VLT.

The work by Carretta and collaborators (e.g. Carretta *et al.* 2009a) has further demonstrated the universality of the Na-O anticorrelation; all GCs studied show it, as evident in the 19 GCs displayed in Fig. 2. However, this anticorrelation is not the same in all clusters; the shape and the extension vary from cluster-to-cluster. On the basis of the Na and O abundances, Carretta and collaborators separated the cluster stars in first-generation and second-generation ones. The first are the ones with Na and O similar to field stars of the same metallicity, the second show varying degrees of O depletion and Na enhancement. Na and O are not the only light elements involved. Also Mg, Al (and even Si and F, Yong *et al.* 2005, Smith *et al.* 2005) are altered. In particular, Al is a powerful probe of the nature of first-generation polluters, due to the very high temperatures required for its production. The modification from the primordial value varies a lot from cluster to cluster (Carretta *et al.* 2009b: in some clusters Al changes a lot, in other much less or not at all). This is an indication that different polluters were at work. The chemical changes come all from H-burning, but at very different temperatures, and this indicates that polluters† of different mass were at work in different GCs.

2. The iceberg tip

A few clusters show the characteristics described above in a extreme way and have been the footholds, if one may say so, to convince even the distracted astronomer that something did not fit the notion of GCs being simple, old, boring systems.

2.1. ω Centauri

When talking of the complexity of GCs, the first example that comes to mind is of course ω Cen. It had long been suspected that it could host stars of different metallicity, given the width of the sequences (e.g., Cannon & Stobie 1973, Alcaino & Liller 1987) and also demonstrated by pioneering spectroscopic work (see Butler *et al.* 1978, Cohen 1981). Given the huge amount of papers dedicated to ω Cen, I can only touch upon a tiny fraction of the works on this cluster, concentrating on the very recent results, which show not only dispersion in colour or metallicity, but actual discrete separation in different sub-samples of the total cluster population. Initially the clear indication of multiple populations came from the RGB (Lee *et al.* 1999, Pancino *et al.* 2000, Ferraro *et al.* 2004), see Fig. 3. Several metallicities, and perhaps different ages were deduced for the different RGB sequences, in particular for the very metal-rich "anomalous RGB", see for instance the high resolution spectroscopic study of 40 giants by Norris & Da Costa (1995), the work on Ca abundances of about 500 giants observed at low spectroscopic resolution by Norris et al (1996), the Strömgren photometry by Hilker & Richtler (2000), the high resolution analysis of 6 stars on the "anomalous" RGB by Pancino *et al.* (2002), the near-IR spectroscopy of about 20 giants by Origlia *et al.* (2003), the study of main sequence stars by Stanford *et al.* (2007), the large survey at intermediate spectroscopic resolution by Johnson *et al.* (2009).

However, the differences go deeper, and with data obtained with HST, it was possible to detect also separate main sequences (Bedin *et al.* 2004, see Fig. 3), to which I will come back later.

2.2. *NGC 2808*

Even more "normal", less massive clusters show striking features. NGC 2808, among many other marked peculiarities, presents three well separated main sequences (Piotto *et al.* 2005). It also has a very complex Horizontal Branch (HB), with three main groups of stars, that cannot be explained under standard assumptions. Both these features seem to require He enhancement in part of its stars (see next Section), which would also naturally agree very well with the observed Na-O anticorrelation (the third most extended one,

---

† While no definitive consensus has been reached on the actual nature of the polluters, the most promising candidates are fast rotating massive stars (e.g. Decressin *et al.* 2007), and asymptotic giant branch stars (e.g., Ventura *et al.* 2001). Since they are discussed in other contributions, I will not say more on the subject.

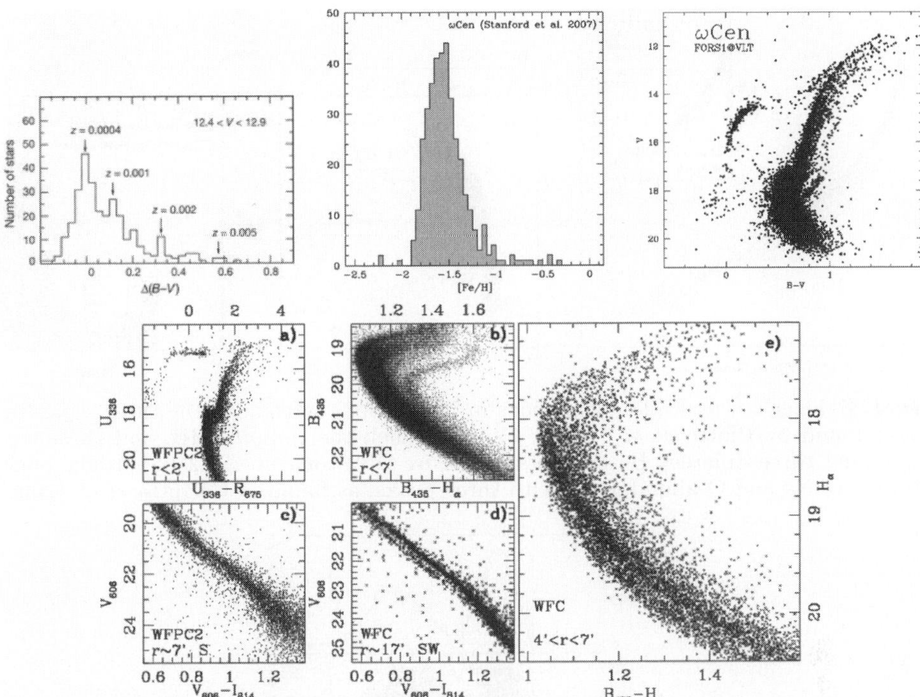

Figure 3. Evidence of multiple populations in ω Cen: Upper left panel: distribution in colour of RGB stars from Lee *et al.* (1999). Central upper panel: distribution in [Fe/H] of main sequence stars, adapted from Stanford *et al.* (2007). Upper right panel: distinct RGBs, from Ferraro *et al.* (2004). Lower panel: collection of CMDs showing the various populations, both for evolved and main sequence stars, from Bedin *et al.* (2004).

after ω Cen and M54), since Na-rich and O-poor stars should also be He-rich (the main outcome of H-burning is, of course, He). Fig. 4 shows the three main sequences, a plausible interpretation of its HB (D'Antona *et al.* 2005), and the Na-O anticorrelation (Carretta *et al.* 2006).

2.3. *M54*

M54 is the second most massive cluster of the Galaxy, and lies at the center of the disrupting Sgr dwarf galaxy. It has been suspected to have a dispersion in metallicity since the observations by Sarajedini & Layden (1995), whose CMD shows a wide RGB, compatible with a dispersion of 0.16 dex in [Fe/H], see Fig. 5. This has been recently confirmed by low resolution spectroscopy of a very large sample of M54 and Sgr stars (Bellazzini *et al.* 2008a, see Fig. 5). The very recent results obtained by Carretta *et al.* (2010) using FLAMES spectra of about 80 RGB stars further confirm this: M54 has a dispersion in metallicity of the order of about 0.2 dex, well above the errors (see Fig. 5). Furthermore, it has a very extendend Na-O anticorrelation, more extended for the metal-rich than for the metal-poor stars. Carretta *et al.* (2010) also noticed that the same happens in ω Cen. M54 deserves more study, but it's clear that it resembles ω Cen; maybe, as it has been suggested (Bellazzini *et al.* 2008a, Carretta *et al.* 2010), we see it now as ω Cen was a long time ago, before the dwarf galaxy around it dispersed.

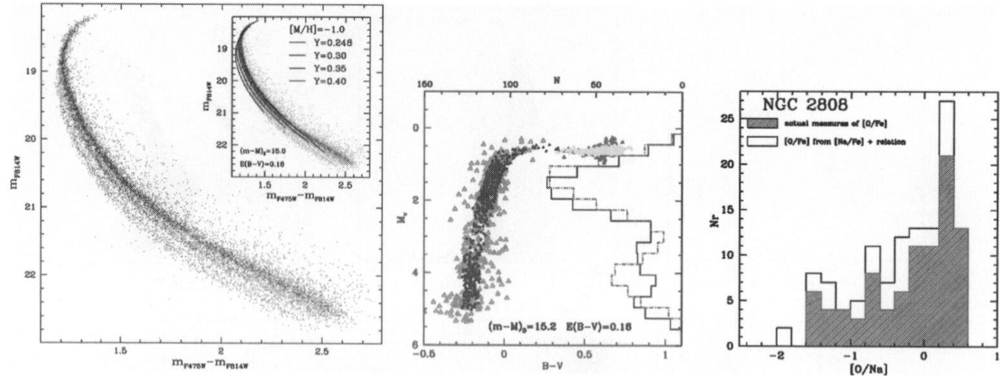

Figure 4. Evidence of multiple populations in NGC 2808. Left panel: the three separate main sequences found by Piotto *et al.* (2007). Central panel: the complex HB, and the interpretation assuming three different He contents, made by D'Antona *et al.* (2005). Right panel: the distribution of Na and O abundances, with three peaks, as found in Carretta *et al.* (2006).

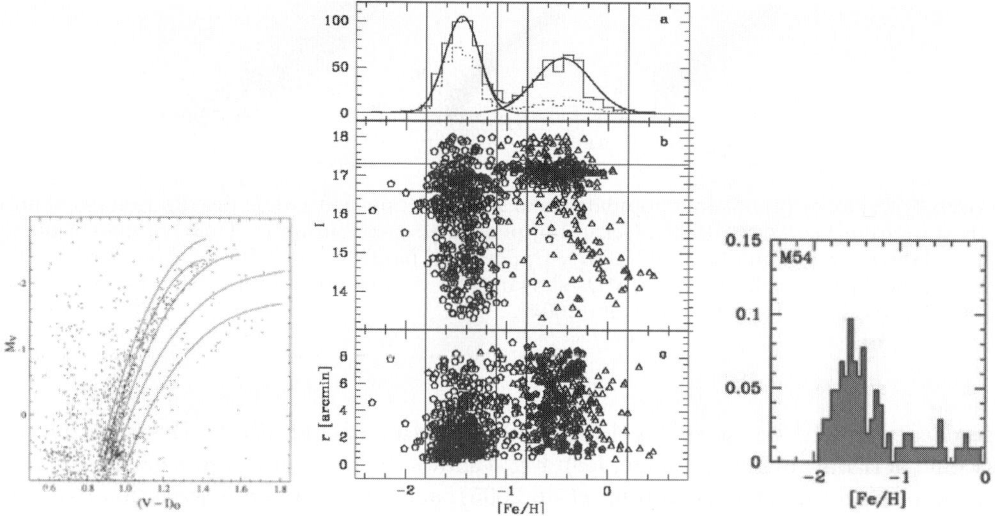

Figure 5. Dispersion in metallicity in M54. Left panel: Sarajedini & Layden (1995) found a colour dispersion in the RGB. Central panel: Separation of M54 from the Sgr population in Bellazzini *et al.* (2008a), based on Calcium triplet observations of several hundreds of stars; notice that both M54 and Sgr show a noticeable dispersion in [Fe/H]. Right panel: results obtained by Carretta *et al.* (2010) from 76 star in M54 (peaked at [Fe/H]~ -1.6) and 25 stars in Sgr (extending in [Fe/H] from about -1 to solar) observed with FLAMES@VLT.

3. Are there different He abundances in GCs ?

The main problem for establishing if He is variable in GCs it that it is difficult to see the effect of small variations in the CMDs, apart from the HB, which is a sort of "amplifier". HBs suffer from the notorious "second-parameter" problem: metallicity is of course the first parameter explaining their structure, age is most probably the second, but their combination cannot explain all HBs, in particular at the bluer, hotter extremes.

A solution is to bring also He into the problem, since an higher He content means brighter and bluer HBs. As shown in Fig. 4, D'Antona *et al.* (2005) were able to reproduce the observed distribution of HB stars in NGC 2808 assuming three different He contents (from a "primordial" value of Y $= 0.25$ for the red HB, to Y $= 0.40$ for the extreme blue

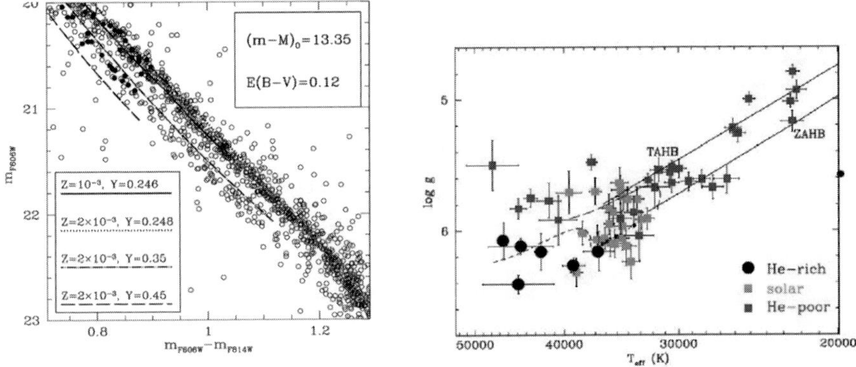

Figure 6. He abundances in ω Cen. Left panel: the two main sequences in Piotto *et al.* (2005), with star observed with FLAMES indicated by filled red and blue circles. The isochrones shown have been computed for $Z = 1 \times 10^{-3}$, $Y = 0.25$ and $Z = 2 \times 10^{-3}$, $Y = 0.35$. Right panel (adapted from Moehler *et al.* (2007)): blue and extreme HB stars for which Moehler *et al.* obtained low resolution spectra and for which determined temperatures, gravities, and He abundances; part of the extreme HB stars are He-enhanced.

HB). A similar exercise has been done by Busso *et al.* (2007) and D'Antona & Caloi (2008) for NGC 6388, producing similar results. This is a strong indication that GC stars are not so homogeneous in He as we thought: in the same cluster we may have stars with "normal" He and stars with different levels of He-enhancement, even if perhaps this doesn't happen in all clusters, or at least not at this high level.

It could be very interesting to directly measure He abundances in GC stars; however, this is possible only on the HB, and with many limitations. There is a small range in temperature (9500-11500 K) where He lines can be observed (even if they are tiny, at a few percent of the continuum level) and He abundances are not altered by dilution and mixing. Recently, Villanova *et al.* (2009) have obtained high resolution, high S/N of a few HB stars in NGC 6752, but their results are not decisive: they could measure He only in four stars, all of them Na-poor, O-rich (hence expected to be He-"normal", that is what they found), while they could not do so for the only Na-rich, O-poor star (expected to be He-enhanced). On the other hand, He has been found to be enhanced in some very blue HB stars in NGC 2808 and ω Cen (see Moehler *et al.* 2007 and Fig. 3), even if the cause of He-enhancement could also be mixing following a late He-flash (Castellani & Castellani 2003) for these extreme HB stars.

In the present volume there is also a discussion (see Bragaglia *et al.* 2010 for a lengthier presentation) on how to determine differences in He abundances from RGB stars using colours, metallicities, and magnitude of the RGB bump.

Finally, maybe the strongest case for He-enhancement of part of GC stars comes from two massive objects. The first one is again ω Cen, with its two separate main sequences (see Fig. 3). Piotto *et al.* (2005) obtained spectra of moderate resolution for 17 stars on the blue and 17 stars on the red sequences; surprisingly, the blue sequence turned out to be more metal-rich by about 0.3 dex. This could be explained only if we assume that the blue sequence is also much more He-rich than the red one ($Y \sim 0.35$-0.38 *vs* 0.25) as seen from isochrone fit. Of course we have to remember that the different metallicities are actually measured, while the He-enhancement is only inferred.

The split main sequence is even more spectacular in NGC 2808 (Fig. 4), and again the three sequences can be well fit assuming the same age but three different He levels (i.e., $Y \sim 0.25$, 0.30, and 0.38, see Piotto *et al.* 2007).

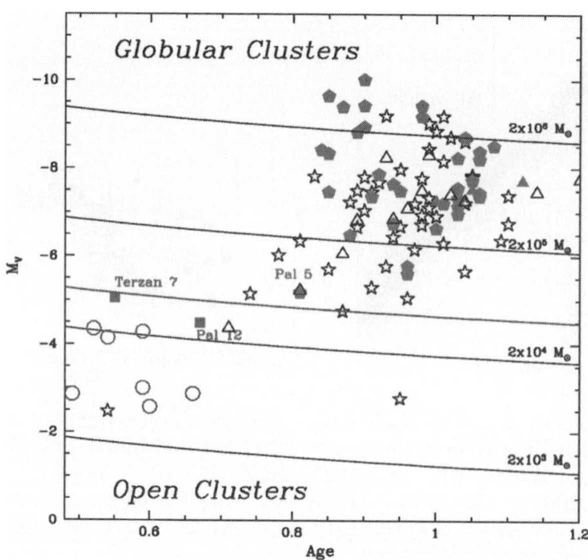

Figure 7. Relative Age parameter vs absolute magnitude M_V for globular and old open clusters. Red filled pentagons and triangles are GCs where Na-O anticorrelation has been observed, in the Milky Way or the LMC respectively; green squares are clusters which do not show (yet?) evidence for O-Na anticorrelation (Terzan 7 and Pal 12, both in the Sgr dwarf galaxy). Open stars and triangles mark clusters for which not enough data is available. Finally, open circles are old open clusters (data from Lata *et al.* 2002). Superimposed are lines of constant mass (light solid lines, see Bellazzini *et al.* 2008b). The heavy blue solid line (at a mass of 4×10^4 $M_\odot$) is the proposed separation between globular and open clusters. This figure is taken from Carretta *et al.*, submitted

4. Summary and perspectives

We have seen that there is photometric evidence of multiple populations in many GCs. This generally happens among the most massive ones in our Galaxy, but not exclusively (see the cases of NGC 6752, M4). Mass is an important factor. It has been shown that many properties correlate with cluster mass, for instance, the maximum temperature reached on the HB (Recio-Blanco *et al.* 2006)

However, mass is not all the story. We have seen that all GCs display the Na-O anticorrelation, but if we quantify its extension (using e.g., the Interquartile range, see Carretta 2006) and plot it against total cluster mass, or integrated magnitude, we see that, preferentially, only high-mass clusters have an extended anticorrelation. This is, however, only a necessary condition; a notable counterexample is the massive GC 47 Tuc, which has a short anticorrelation. Some other factor, maybe metallicity, or age, or cluster orbit have to be involved.

Finally, I recall that the Na-O (and similar) anticorrelations seem to represent an intrinsic property of GCs: each time Na and O have been measured, they anti-correlate, while this does not happen in open clusters (see Fig. 7) or for field stars. So maybe we have an operative definition of the separation between globular and open clusters: GCs are those aggregates massive enough to sustain self-pollution, hence able to host at least two stellar generations and to develop a Na-O anticorrelation. This has of course to be related to the mechanism of cluster formation.

The financial support from the IAU and the Swiss National Science Foundation is gratefully acknowledged. Many thanks to Eugenio Carretta, Raffaele Gratton, Valentina D'Orazi, Sara Lucatello, and Antonino Milone for their help and suggestions.

References

Alcaino, G. & Liller, W. 1981, *AJ*, 94, 1585

Anderson, J., Piotto, G., King, I. R., Bedin, L. R., & Guhathakurta, P. 2009, *ApJ*, 697, L58

Bedin, L. R., Piotto, G., Anderson, J., Cassisi, S., King, I. R., Momany, Y., & Carraro, G. 2004, *ApJ*, 605, L125

Bellazzini, M., *et al.* 2008a, *AJ*, 136. 1147

Bellazzini, *et al.* 2008b, *Memorie della Societa Astronomica Italiana*, 79, 663

Bragaglia, A., Carretta, E., Gratton, R., D'Orazi, V., Cassisi, S., & Lucatello, S. 2010, *A&A*, subm.

Brown, J. A. & Wallerstein, G. 1992, *AJ*, 104, 1818

Busso, G., *et al.* 2007, *A&A*, 474, 105

Butler, D., Dickens, R. J., & Epps, E. 1978, *ApJ*, 225, 148

Cannon, R. D. & Stobie, R. S. 1987, *MNRAS*, 162, 207

Carretta, E. 2006, *AJ*, 131, 1766

Carretta, E., Carretta, E., Gratton, R. G., Bragaglia, A., Bonifacio, P., & Pasquini, L. 2004, *A&A*, 416, 925

Carretta, E., Bragaglia, A., Gratton, R. G., Leone, F., Recio-Blanco, A., & Lucatello, S. 2006, *A&A*, 450, 523

Carretta, E., *et al.* 2009a, *A&A*, 505, 117

Carretta, E., Bragaglia, A., Gratton, R. G., & Lucatello, S. 2009b, *A&A*, 505, 139

Carretta, E., *et al.* 2010, *ApJ*, subm.

Castellani, M. & Castellani, V. 1993, *ApJ*, 407, 649

Cohen, J. G. 1981, *ApJ*, 247, 869

D'Antona, F. & Caloi, V. 2008, *MNRAS*, 390, 693

D'Antona, F., Bellazzini, M., Caloi, V., Pecci, F. F., Galleti, S., & Rood, R. T. 2005, *ApJ*, 631, 868

Da Costa, G. S., Held, E. V., Saviane, I., & Gullieuszik, M. 2009, *ApJ*, 705, 1481

Denisenkov, P. A. & Denisenkova, S. N. 1989, *A.Tsir.*, 1538, 11

Decressin, T., Meynet, G., Charbonnel C. Prantzos, N., & Ekstrom, S. 2007, *A&A*, 464, 1029

Ferraro, F. R., Sollima, A., Pancino, E., Bellazzini, M., Straniero, O., Origlia, L., & Cool, A. M. 2004, *ApJ*, 603, L81

Gratton, R., Sneden, C., & Carretta, E. 2004, *ARAA*, 42, 385

Gratton, R., *et al.* 2001, *A&A*, 369, 87

Gratton, R. G., Carretta, E., Bragaglia, A., Lucatello, S., & D'Orazi, V. 2010, *A&A*, subm.

Han, S.-I., Lee, Y.-W., Joo, S.-J., Sohn, S. T., Yoon, S.-J., Kim, H.-S., & Lee, J.-W. 2009, *ApJ*, 707, L190

Hesser, J. E., Hartwick, F. D. A., & McClure,R. D. 1977, *ApJS*, 33, 471

Hilker, M. & Richtler, T. 2000, *A&A*, 326, 895

Ibata, R. A., Gilmore, G., & Irwin, M. J. 1994, *Nature*, 370, 194

Ivans, I. I., Sneden, C., Wallerstein, G., Kraft, R. P., Norris, J. E., Fulbright, J. P., & Gonzalez, G. 2004, *Memorie della Societa Astronomica Italiana*, 75, 286

Johnson, C. J., Pilachowski, C. A., Rich, M. R., & Fulbright, J. P. 2009, *ApJ*, 698, 2048

Kraft, R. 1994, *PASP*, 106, 553

Langer, G. E., Hoffman, R., & Sneden, C. 1993, *PASP*, 105, 301

Lata, S., Pandey, A. K., Sagar, R., & Mohan, V. 2002, *A&A*, 388, 158

Lee, Y.-W., Joo, J.-M., Sohn, Y.-J., Rey, S.-C., Lee, H.-C., & Walker, A. R. 1999, *Nature*, 402, 55

Marino, A. F., Milone, A. P., Piotto, G., Villanova, S., Bedin, L. R., Bellini, A., & Renzini, A. 2009, *A&A*, 505, 1099

Meylan, G. & Heggie, D. C. 1987, *A&ARv*, 8, 1

Milone, A. P., *et al.* 2008, *ApJ*, 673, 241

Milone, A. P., *et al.* 2009, *ApJ*, in press (arXiv:0912.2055)

Moehler, S., Dreizler, S., Lanz, T., Bono, G., Sweigart, A. V., Calamida, A., Monelli, M., & Nonino, M. 2007, *A&A*, 475, L5

Norris, J. E., Freeman, K. C., & Mighell, K. J. 1996, *ApJ*, 462, 241

Norris, J. E. & Da Costa, G. S. 1995, *ApJ*, 447, 680

Origlia, L., Ferraro, F. R., Bellazzini, M., & Pancino, E. 2003, *ApJ*, 591, 916

Pancino, E., Ferraro, F. R., Bellazzini, M., Piotto, G., & Zoccali, M. 2000, *ApJ*, 534, L83

Pancino, E., Pasquini, L., Hill, V., Ferraro, F., & Bellazzini, M. 2002, *ApJ* (Letters), 568, L101

Piotto, G. 2008, *IAU Symposium*, 246, 141

Piotto, G. 2009, *IAU Symposium*, 258, 233

Piotto, G., *et al.* 2005, *ApJ*, 621, 777

Piotto, G., *et al.* 2007, *ApJ*, 61, L53

Ramirez, S. V. & Cohen, J. G. 2002, *AJ*, 123, 3277

Recio-Blanco, A., Aparicio, A., Piotto, G., de Angeli, F., & Djorgovski, S. G. 2006, *A&A*, 452, 875

Renzini, A. & Buzzoni, A. 1986, *Spectral Evolution of Galaxies*, 122, 195

Sarajedini, A. & Layden, A. C. 1995, *AJ*, 109, 1112

Smith, G. H. 1987, *PASP*, 99, 67

Smith, V. V., Cunha, K., Ivans, I. I., Lattanzio, J. C., Campbell, S., & Hinkle, K. H. 2005, *ApJ*, 633, 392

Sollima, A., Ferraro, F. R., Bellazzini, M., Origlia, L., Straniero, O., & Pancino, E. 2007, *ApJ*, 654, 915

Stanford, L. M., Da Costa, G. S., Norris, J. E., & Cannon, R. D. 2007, *ApJ*, 667. 911

Ventura, P., D'Antona, F., Mazzitelli, I., & Gratton, R. 2001, *ApJ*, 550, L65

Villanova, S., Piotto, G., & Gratton, R. G. 2009, *A&A*, 499, 755

Yong, D., Grundahl, F., Nissen, P. E., Jensen, H. R., & Lambert, D. L. 2005, *A&A*, 438, 875

Yong, D., Grundahl, F., D'Antona, F., Karakas, A. I., Lattanzio, J. C., & Norris, J. E. 2009, *ApJ*, 695, L62

Light Elements in the Universe
Proceedings IAU Symposium No. 268, 2009
C. Charbonnel, M. Tosi, F. Primas & C. Chiappini, eds.

© International Astronomical Union 2010
doi:10.1017/S1743921310003960

Revisiting the helium abundance in globular clusters with multiple main sequences

Luca Casagrande[1], Laura Portinari[2] and Chris Flynn[2]

[1] Max Planck Institue for Astrophysics, Karl Schwartzschild Straße 1, Garching, Germany
email: `luca@mpa-garching.mpg.de`

[2] Tuorla Observatory, Department of Physics and Astronomy, University of Turku, Finland
email: `lporti@utu.fi,cflynn@utu.fi`

Abstract. For nearby K dwarfs, the broadening of the observed Main Sequence at low metallicities is much narrower than expected from isochrones with the standard helium–to–metal enrichment ratio $\Delta Y/\Delta Z \sim 2$. A much higher value, of order 10, is formally needed to reproduce the observed broadening, but it returns helium abundances in awkward contrast with Big Bang Nucleosynthesis. This steep enrichment ratio resembles, on a milder scale, the very high $\Delta Y/\Delta Z$ estimated from the multiple Main Sequences observed in some metal-poor Globular Clusters. We argue that a revision of low Main Sequence stellar models, suggested from nearby stars, could help to reduce the overwhelmingly high $\Delta Y/\Delta Z$ deduced so far for those clusters. Under the most favourable assumptions, the estimated helium content for the enriched populations may decrease from $Y \simeq 0.4$ to as low as $Y \simeq 0.3$, with intermediate values being plausible.

Keywords. Stars: fundamental parameters, late-type – subdwarfs – globular clusters: individual (ωCen, NGC 2808, 47 Tuc)

1. Introduction

Helium (Y) is the second most abundant element in the Universe, having been produced by Big Bang Nucleosynthesis (BBN) with a universal mass fraction $Y_P = 0.24 - 0.25$ and topped up by successive stellar generations which also synthesize metals (Z). In spite of its large abundance, it is an elusive element. It can be measured directly, from spectroscopic lines, only for $T_{\rm eff} \gtrsim 8000$ K: young, massive stars or blue Horizontal Branch stars though, apart from a small temperature range, their surface abundance may not trace the original one, due to helium sedimentation and metal levitation (e.g. Villanova, Piotto & Gratton 2009). Only indirect methods can be used for stars of low mass and cooler temperatures, which constitute the bulk of the Galaxy's stellar population. One such method relies on the broadening of the low Main Sequence (MS), where evolutionary effects do not need to be taken into account. At a fixed metallicity Z, an increase of Y makes a given mass on the isochrone hotter and brighter, so that the net effect of varying $\Delta Y/\Delta Z$ is to affect the broadening of the lower MS (e.g. Faulkner 1967; Fernandes *et al.* 1996).

Previous studies based on Hipparcos parallaxes concluded $\Delta Y/\Delta Z = 2 - 3$ (Pagel & Portinari 1998; Jimenez *et al.* 2003). More recently, Casagrande *et al.* (2006, 2007) further improved the analysis by compiling a much larger sample of K dwarfs with bolometric magnitudes and effective temperatures derived via InfraRed Flux Method (IRFM). The implementation adopted for the IRFM resorted to homogeneous multi–band photometry to determine in a fully self-consistent, (almost) model independent way $T_{\rm eff}$ and $M_{\rm Bol}$. The study of the broadening of the lower MS could thus be performed in the theoretical

HR diagram where the effects of $\Delta Y/\Delta Z$ are more prominent, also bypassing uncertainties related to synthetic colour-temperature transformations†.

Around solar metallicities this analysis yielded $\Delta Y/\Delta Z \sim 2$ but at lower Z, Casagrande et al. (2007) found that the observed MS is narrower than expected from stellar models computed under this standard $\Delta Y/\Delta Z$ (i.e. the stars are cooler than the isochrones), rather implying a much steeper value of ~ 10. This change in slope is at odds with Galactic chemical evolution models, which predict a ratio substantially constant with Z (Carigi & Peimbert 2008; Casagrande 2008); a constant $\Delta Y/\Delta Z \sim 2$ is also found from HII regions (e.g. Peimbert et al. 2007). More importantly, $\Delta Y/\Delta Z \sim 10$ at low metallicities implies a helium content for nearby subdwarfs much lower than Y_P, in awkward contrast with the cosmological floor set by standard BBN.

This result points toward inadequacies in MS stellar models of low metallicity (see also Lebreton et al. 1999), an issue which is worth investigating since it may lead to reconsider the problem of the helium enrichment in Globular Clusters (GCs). In fact, at magnitudes comparable to those of the local stars studied in Casagrande et al. (2007, $M_V \sim 5.5-7.5$), multiple MSs have been discovered in some Globular Clusters and interpreted as evidence for huge helium enhancement in a sub-population (Piotto 2009 and references therein). The required $\Delta Y/\Delta Z \gtrsim 100$ is extremely difficult to explain for stellar nucleosynthesis and chemical evolution models (Yi 2009 and references therein). While it is likely that a significant helium enhancement is present in those stellar populations, our purpose is to show that the revision of low metallicity MS stellar models, needed to cure the problem of the high $\Delta Y/\Delta Z$ and sub-primordial Y deduced in nearby K dwarfs, may significantly reduce current estimates of $\Delta Y/\Delta Z$ in Globular Clusters with multiple MS, easing their theoretical interpretation. Full details on the work presented here are given in Portinari, Casagrande & Flynn (2010).

2. Homology relations

Homology relations, holding for largely radiative structures (Cox & Giuli 1968) such as MS stars of $M \lesssim 1\ M_\odot$, express the distance, in the theoretical HR diagram of a Zero Age Main Sequence (ZAMS) of composition (Y, Z) with respect to another reference ZAMS of composition (Y_r, Z_r):

$$\Delta \log T_{\rm eff} = \quad -0.0935 \log \left[1 - \tfrac{\delta}{X_r}(Z - Z_r)\right]$$
$$-0.196 \log \left[1 - \tfrac{5\delta+1}{(3+5X_r-Z_r)}(Z - Z_r)\right]$$
$$-0.051 \log \left[1 - \tfrac{\delta}{(1+X_r)}(Z - Z_r)\right]$$
$$-0.051 \log \left(\tfrac{100Z+1}{100Z_r+1}\right) \tag{2.1}$$

where $\delta = 1 + \frac{\Delta Y}{\Delta Z}$ and $X_r = 1 - Y_r - Z_r$. Notice that the same relation holds also for a generic combination of (Y, Z) replacing $\delta\,(Z - Z_r) \to (Y - Y_r) + (Z - Z_r)$.

---

† It is worth recalling that the adopted $T_{\rm eff}$ scale has been recently validated by interferometric angular diameters and HST spectrophotometry (Casagrande et al. 2010). Without entering into further details, suffice here to say that this scale is hotter than other IRFM renditions, and hotter than or close to various spectroscopic scales. Therefore, any other $T_{\rm eff}$ scale will just worsen the comparison with the isochrones discussed in this paper, with real stars even cooler than the models.

2.1. *Application to Globular Clusters*

In recent years, thanks to deep photometric observations, multiple, distinct MSs have been detected in ωCen and NGC 2808 (Piotto 2009 and references therein). Recent observations also seem to support a MS broadening in 47 Tuc (Anderson *et al.* 2009). Since helium crucially affects the morphology of the HR diagram, it is very tempting to interpret multiple stellar populations in terms of different helium contents (Catelan, Valcarce & Sweigart 2009 and references therein).

For ωCen, detailed isochronc analysis indicates that a helium enrichment of $\Delta Y \simeq 0.15$ is needed to reproduce the observed split between the blue (bMS, [Fe/H] $= -1.3$) and red (rMS, [Fe/H] $= -1.6$, $Y_{\rm rMS} = 0.246$) main sequence. Homology relations (Eq. 2.1) indeed yield a similar result: using the colour–temperature–metallicity scale of Casagrande *et al.* (2006) to translate the dereddened maximum observed colour split between the bMS and rMS returns $\Delta \log T_{\rm eff} = 0.0185$ implying $\Delta Y = 0.144$ and $Y_{\rm bMS} = 0.39$ (Fig. 2). These values are in excellent agreement with the results obtained by Norris (2004), Piotto *et al.* (2005), Lee *et al.* (2005) and Sollima *et al.* (2007) by means of isochrone analysis.

Another striking example of multiple populations is NGC 2808, where three MSs are found at virtually the same metallicity [Fe/H] $= -1.1 \pm 0.03$ dex (Carretta *et al.* 2006) and helium abundances of $Y = 0.248$ (the value assumed for the red MS, which includes the bulk of the population), $Y = 0.30$ (middle MS) and $Y = 0.37$ (blue MS; D'Antona *et al.* 2005; Lee *et al.* 2005; Piotto *et al.* 2007). In this case $\Delta Z \to 0$, and one should just focus on ΔY. Translating the dereddened maximum split of the bMS and rMS with respect to the mMS and using the $T_{\rm eff}$ scale of Casagrande *et al.* (2006) returns $\Delta \log T_{\rm eff} = \pm 0.009$. For an assumed $Y_{\rm rMS} = 0.248$, homology relations return $Y_{\rm mMS} = 0.31$ and $Y_{\rm bMS} = 0.37$ again in excellent agreement with isochrone analysis (Piotto *et al.* 2007).

Finally, if the observed broadening of the MS in 47 Tuc is interpreted in terms of helium variation, homology relations reproduce the measured $\Delta \log T_{\rm eff} = 0.003$ with $\Delta Y = 0.023$, again close to the conclusion based on isochrones (Anderson *et al.* 2009).

2.2. *Application to isochrones*

We have shown that theoretical homology relations, applied to the multiple (or broadened) MSs of Globular Clusters, provide a similar interpretation of the data in terms of ΔY and $\Delta Y/\Delta Z$ as detailed isochrone analysis. We now compare directly homology relations to current stellar models by fitting isochrones with a homology–like relation.

An interpolation formula for isochrones directly inspired by Eq. (2.1) involves at least 4 independent parameters, but for fitting purposes, a simplified form of Eq. (2.1) is more suitable. We verified that the result is still a very good approximation of the rigorous homology relations, especially for metallicities $Z \leqslant Z_\odot$ and even for extremely high $\Delta Y/\Delta Z > 100$, which are of interest for Globular Clusters. Therefore we seek a three–parameters fitting formula for isochrones (and, later below, for real stars) of the kind:

$$\Delta \log T_{\rm eff} = \ -P_1 \log \left[1 - \frac{\delta}{(0.6 + X_r)}(Z - Z_r) \right]$$
$$-P_2 \log \left(\frac{P_3 Z + 1}{P_3 Z_r + 1} \right) \tag{2.2}$$

where P_i are 3 free fitting parameters. It is our experience that this formula provides an adequate fit to isochrones, as good as (and more robust than) other homology-like fitting formulæ with 4 or more parameters. This simplified formulation condenses together the first three (helium dependent) terms of the original relation and has a clearcut separation between the helium– and metallicity–dependent part, which proves to be handy for the empirical calibration discussed in Section 3.

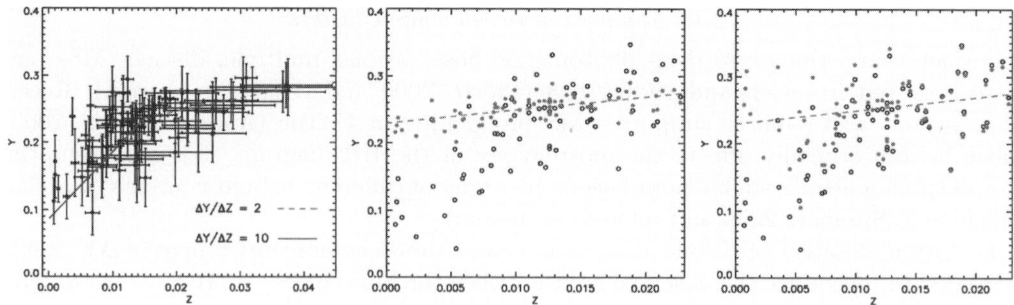

Figure 1. *Left panel:* metal versus helium mass fraction for nearby field dwarfs obtained from Casagrande *et al.* (2007). Lines with two values of $\Delta Y/\Delta Z$ (break at $Z = 0.015$) are plotted to guide the eye. *Central panel:* zooming on metallicities around and below solar. Open circles are derived applying *numerical homology relations* and look remarkably similar to those in the left panel. Full circles are obtained using *empirical homology relations* with $P_1 = 1.5$. *Right panel:* same as central panel, but using *empirical homology relations* with $P_3 \sim 150$.

We consider both the Padova isochrones computed specifically for the analysis in Casagrande *et al.* (2007), and the more recent release by Bertelli *et al.* (2008) with varying $\Delta Y/\Delta Z$ and $0.0001 \leqslant Z \leqslant 0.04$, which fully brackets the metallicity range relevant for the present study. In Casagrande *et al.* (2007) we had checked that other sets of isochrones (Yonsei–Yale, Teramo, McDonald) were very similar in the lower MS.

We find that the behaviour of isochrones, at least in the range $M_{\mathrm{Bol}} = 5.4 - 7.0$ (where the effect of $\Delta Y/\Delta Z$ is expected to be maximal) is well described by homology–like relations with the following set of parameters: $P_1 = 0.50 \pm 0.03$, $P_2 = 0.064 \pm 0.005$ and $P_3 = 670 \pm 200$. When using these parameters we will speak of *numerical homology relations*. In Casagrande *et al.* (2007) the metal and helium mass fraction of the stars were computed iteratively, with a star–by–star isochrone fitting: despite the less sophisticated approach provided by homology relations, with the aforementioned parameters nearly identical results are obtained, including the change in slope above and below $Z \sim 0.015$ (Fig. 1).

3. Empirical homology relations and Globular Clusters

In the previous section we have shown that theoretical homology relations agree very well with the isochrone–based interpretation of the MS splits observed in ωCen and NGC 2808, and adequately reproduce the behaviour of isochrones in general, as a function of $\Delta Y/\Delta Z$. However, isochrones are known to fail the interpretation of the HR diagram of nearby low–Z MS stars: while for $Z \gtrsim 0.015$ the broadening of the main sequence is well reproduced with the standard $\Delta Y/\Delta Z \sim 2$, with current isochrones a much higher $\Delta Y/\Delta Z \sim 10$ is needed to fit lower metallicities. While in principle $\Delta Y/\Delta Z$ may not necessarily be constant, taken at face value, this result implies a helium content in local metal poor stars as low as $Y = 0.1$. Such a striking contrast with BBN is to be ascribed to inadequacies of low metallicity MS stellar models.

While awaiting for a solution to this problem by improved stellar physics, we now define *empirical homology relations*, calibrated to reproduce the broadening of the local MS, and extrapolate the consequences for the interpretation of Globular Clusters. Here we adopt a very pragmatic approach: we *assume* $\Delta Y/\Delta Z \sim 2$ also for $Z < Z_\odot$, and empirically calibrate homology relations to fulfill this constraint for nearby stars. Our assumption is very reasonable, since HII region measurements and chemical evolution models support a constant $\Delta Y/\Delta Z \sim 2$ at all Z, as does the following simple argument: taking the solar

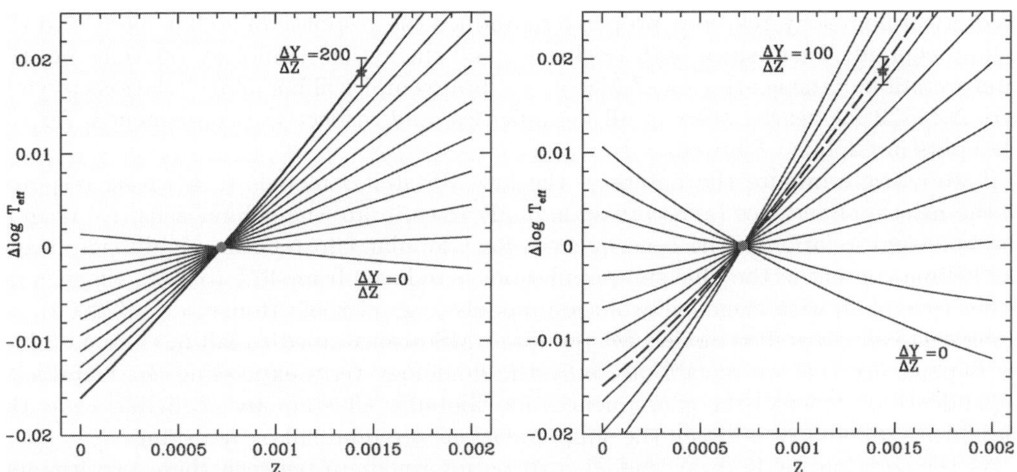

Figure 2. *Left panel:* $\Delta Y/\Delta Z$ obtained applying homology relations to the measured $\Delta \log T_{\mathrm{eff}}$ between the rMS (dot) and bMS (star) in ωCen. *Right panel:* same as left, but using *empirical homology relations* with maximized effect on helium (i.e. $P_1 = 1.5$)

bulk abundances (Asplund *et al.* 2009) and the primordial Y_P, one derives $\Delta Y/\Delta Z \sim 2$ for $Z < Z_\odot$.

To define *empirical homology relations*, one can act on either terms of Eq. (2.2). The first one includes the effects of the helium content, while the second term (stemming essentially from the metallicity dependence of opacity) is independent of $\Delta Y/\Delta Z$. The second term is therefore irrelevant for the multiple MSs in GCs, where metallicity differences are small or vanishing. Therefore, if we calibrate our empirical homology relations onto nearby stars acting only on the first term (via the parameter P_1), we maximize the change in the role of helium, and the consequences for the interpretation of GC's.

Keeping P_2 and P_3 fixed and optimizing P_1 so that the inferred helium content of nearby K dwarfs with $Z < Z_\odot$ follows $\Delta Y/\Delta Z = 2$ (Fig. 1, central panel) yields $P_1 = 1.5$ and the helium abundances required to describe the multiple MSs decrease to $Y_{\mathrm{bMS}} = 0.30$ (ωCen, Fig. 2), $Y_{\mathrm{mMS}} = 0.265$ and $Y_{\mathrm{bMS}} = 0.283$ (NGC 2808) and $\Delta Y = 0.006$ (47 Tuc).

Acting instead on the second term of the homology relations has no impact on the interpretation of the multiple MS of Globular Clusters. But for nearby K dwarfs and subdwarfs, the metallicity range is significant and exploring the effects of this second term is worthwhile. The parameters P_2 and P_3 are degenerate, in the sense that a given change in the second term can be obtained by modifying either of the two parameters. Therefore, we choose to discuss the role of the second term by optimizing P_3. Keeping P_1 and P_2 fixed we find that $P_3 \sim 150$ is the optimal value to obtain $\Delta Y/\Delta Z = 2$ for nearby low–Z stars. However, while we do obtain $\Delta Y/\Delta Z = 2$ on average, this solution has considerable scatter and many stars remain with uncomfortably low helium abundances (Fig. 1, right panel).

4. Conclusions

The purpose of the present exercise has been to draw attention on the possible connection between the HR diagram of Globular Clusters with multiple MSs and the broadening of the low MS defined by nearby stars. We estimate the possible extent of the required revision of low MS stellar models on the base of homology relations. First we show that theoretical homology relations properly reproduce the response of stellar models to the helium content. Then, since both isochrones and theoretical homology relations fail the interpretation of the nearby low–Z MS, we calibrate empirical homology relations to yield $\Delta Y/\Delta Z \sim 2$ for nearby stars of all metallicities, and inspect the consequences for the MS splits in Globular Clusters.

If we entirely impute the failure of the low MS stellar models to a wrong response to the helium abundance (and correspondingly re-calibrate the helium–sensitive term of the homology relations), the consequences for Globular Clusters are highly significant: the helium content of the blue sub-populations is reduced from 40% to 30%, which is far easier to explain with chemical evolution models (e.g. Yi 2009; Romano *et al.* 2009).

Alternatively, if stellar models for the lower MS are assumed to fail in their metallicity dependence (i.e. we re-calibrate only the homology term expressing the metallicity dependence of opacity) the consequences for Globular Clusters are negligible — as the metallicity differences between the subpopulations are minimal or vanishing.

As the solution for K dwarf models can be intermediate between these two extreme assumptions, we suggest that the helium rich populations in Globular Clusters are likely to have a helium content Y in the range 0.3–0.4; but altogether, there is room to decrease their estimated helium content from the extreme $Y = 0.4$ that is the commonly quoted value. Improvements in the physics of low MS stellar models are thus potentially important also for the riddle of the helium self-enrichment in Globular Clusters.

References

Anderson, J., Piotto, G., King, I. R., Bedin, L. R., & Guhathakurta P. 2009, *ApJ*, 697, L58

Asplund, M., Grevesse, N., Sauval, A. J., & Scott, P. 2009, ARA&A, 47, 481

Bertelli, G., Girardi, L., Marigo, P., & Nasi, E. 2008, *A&A*, 484, 815

Carigi, L. & Peimbert, M. 2008, *RMxAA*, 44, 341

Carretta, E., Bragaglia, A., Gratton, R. G. *et al.* 2006, *A&A*, 450, 523

Casagrande, L. 2008, *PhD Thesis*, Annales Universitatis Turkuensis, Scr. A1 t., 382

Casagrande, L., Portinari, L., & Flynn, C. 2006, *MNRAS*, 373, 13

Casagrande, L., Flynn, C., Portinari, L., Girardi, L., & Jimenez, J. 2007, *MNRAS*, 382, 1516

Casagrande, L., Ramírez, I., Meléndez, J., Bessell, M., & Asplund, M. 2010, *A&A*, in press

Catelan, M., Valcarce, A. A. R., & Sweigart, A. V. 2009, *IAU Symp.* 266, (arXiv:0910.1367)

Cox, J. P. & Giuli, R. T. 1968, *Principles of Stellar Structure*, Gordon & Breach, New York

D'Antona, F., Bellazzini, M., Caloi, V., Pecci, F. F., Galleti, S., & Rood, R. T. 2005, *ApJ*, 631, 868

Faulkner, J. 1967, ApJ 147, 617

Fernandes, J., Lebreton, Y., & Baglin, A. 1996, *A&A*, 311, 127

Jimenez, R., Flynn, C., McDonald, J., & Gibson, B. 2003, *Science*, 299, 1552

Lebreton, Y., Perrin, M.-N., Cayrel, R., Baglin, A., & Fernandes, J. 1999, *A&A*, 350, 587

Lee, Y.-W., Joo, S.-J., Han, S.-I. *et al.* 2005, *ApJ*, 621, L57

Norris, J. E. 2004, *ApJ*, 612, 25

Pagel, B. E. J. & Portinari, L. 1998, *MNRAS*, 298, 747

Peimbert, M., Luridiana, V., & Peimbert, A. 2007, *ApJ*, 666, 636

Piotto, G. 2009, in The Ages of Stars *IAU Symp.* 258, 233

Piotto, G., Villanova, S., Bedin, L. R., Gratton, R., Cassisi, S. *et al.* 2005, *ApJ*, 621, 777

Piotto, G., Bedin, L. R., Anderson, J., King, I. R., Cassisi, S. *et al.* 2007, *ApJ*, 661, L53

Portinari, L., Casagrande, L., & Flynn, C. 2010, submitted

Romano, D., Tosi, M., & Cignoni, M. *et al.* 2009, *MNRAS*, in press (arXiv:0910.1299)

Sollima, A., Ferraro, F. R., Bellazzini, M. *et al.* 2007, *ApJ*, 654, 915

Villanova, S., Piotto, G., & Gratton, R. G. 2009, *A&A*, 499, 755

Yi, S.-K. 2009, in The Ages of Stars *IAU Symp.* 258, 253

Light Elements in the Universe
Proceedings IAU Symposium No. 268, 2009
C. Charbonnel, M. Tosi, F. Primas & C. Chiappini, eds.

© International Astronomical Union 2010
doi:10.1017/S1743921310003972

Helium-rich stars in globular clusters: constraints for self-enrichment by massive stars

Thibaut Decressin[1], G. Meynet[2] and C. Charbonnel[2,3]

[1] Argelander Institute for Astronomy (AIfA), Auf dem Hügel 71, D-53121 Bonn, Germany
email: **decressin@astro.uni-bonn.de**

[2] Geneva Observatory, University of Geneva, 51 ch. des Maillettes, 1290 Versoix, Switzerland
email: **georges.meynet@unige.ch**

[3] Laboratoire d'Astrophysique de Toulouse-Tarbes, CNRS UMR 5572, Université de Toulouse,
14 Av. E. Belin, 31400 Toulouse, France
email: **corinne.charbonnel@unige.ch**

Abstract. Globular clusters exhibit peculiar chemical patterns where Fe and heavy elements are constant inside a given cluster while light elements (Li to Al) show strong star-to-star variations. This pattern can be explained by self-pollution of the intracluster gas by the slow winds of fast rotating massive stars. Besides, several main sequences have been observed in several globular clusters which can be understood only with different stellar populations with distinct He content. Here we explore how these He abundances can constrain the self-enrichment in globular clusters.

Keywords. stars: abundances, evolution, horizontal-branch, mass loss – globular clusters: general

1. Chemical enrichment of globular clusters

Globular clusters are self-gravitating aggregates of tens of thousands to millions of stars that have survived over a Hubble time. Many observations show that these objects are composed of (at least) two distinct stellar populations. The first evidence rests on the chemical analysis that reveals large star-to-star abundance variations in light elements in all individual clusters studied so far, while the iron abundance stays constant (for a review see Gratton *et al.* 2004). These variations include the well-documented anticorrelations between C-N, O-Na, Mg-Al, Li-Na and F-Na (Kraft 1994; Carretta *et al.* 2007; Gratton *et al.* 2007; Carretta *et al.* 2006; Pasquini *et al.* 2007; Bonifacio *et al.* 2007; Lind *et al.* 2009). This global chemical pattern requires H-burning at high temperature around 75×10^6 K (Arnould *et al.* 1999; Prantzos *et al.* 2007). As the observed chemical pattern is present in low-mass stars both on the red giant branch (RGB) and at the turn-off that do not reach such high internal temperatures, the abundance anomalies must have been inherited at the time of formation of these stars.

Further indications for multiple populations in individual GCs come from deep photometric studies that have revealed multiple giant branches or main sequences. In ω Cen a blue main sequence has been discovered (Bedin *et al.* 2004) that is presumably related to a high content in He (Piotto *et al.* 2005; Villanova *et al.* 2007). A triple main sequence has been discovered in NGC 2808 (Piotto *et al.* 2007). The additional blue sequences are explainable by a higher He content of the corresponding stars which shifts the effective temperatures towards hotter values. He-rich stars have also been proposed to explain the morphology of extended horizontal branch (hereafter HB) seen in many globular clusters (see e.g., Caloi & D'Antona 2005). Whereas no direct observational link between abundance anomalies and He-rich sequences has been found, this link is easily understood theoretically as abundance anomalies are the main result of H burning to He.

These observed properties lead to the conclusion that globular clusters born from giant gas clouds first form a generation of stars with the same abundance pattern as field stars. Then a polluting source enriches the intracluster-medium with H-burning material out of which a chemically-different second stellar generation forms. This scheme can explain at the same time the abundance anomalies in light elements and He-enrichment.

Two main candidates that reach the right temperature for H-burning have been proposed to be at the origin of the abundance anomalies (Prantzos & Charbonnel 2006): (a) intermediate mass stars evolving on the thermal pulses along the asymptotic giant branch (hereafter TP-AGB), and (b) main sequence massive stars. After being first proposed by Cottrell & Da Costa (1981) the AGB scenario has been extensively studied (see e.g., Ventura & D'Antona 2008b,a, 2009, see Ventura, this volume) and has been seriously challenged by rotating AGB models that predict unobserved CNO enrichment in low-metallicity globular clusters (Decressin et al. 2009).

On the other hand, as being suggested by Brown & Wallerstein (1993) and Wallerstein et al. (1987), massive stars can also pollute the interstellar medium (ISM) of a forming cluster (see Smith 2006; Prantzos & Charbonnel 2006). In particular Decressin et al. (2007b) show that fast rotating massive stars (with a mass higher than ~ 25 $M_\odot$) are good candidates for the self-enrichment of globular clusters. In this wind of fast rotating massive stars (WFRMS) scenario, rotationally-induced mixing transports H-burning products (and hence matter with correct abundance signatures) from the convective core to the stellar surface, and, providing initial rotation is high enough, the stars reach the break-up on the main sequence evolution. As a result a mechanical wind is launched from the equator that generates a disk around the star similar to that of Be stars (e.g. Townsend et al. 2004). Later, when He-burning products are brought to the surface, the star has already lost a high fraction of its initial mass and angular momentum, so that it no longer rotates at the break-up velocity. Matter is then ejected through a classical fast isotropic radiative wind. From the matter ejected in the disk, a second generation of stars may be created with chemical pattern in agreement with observations.

2. He-rich stars evolution

As explained in Decressin et al. (2007a), matter stored in the discs around massive stars is heavily enriched with He. If the second generation stars form locally around each massive star from a mixture of matter stored in the slow winds and the ISM, we expect the following distribution: a main peak is present at normal He-abundance (Y = 0.245, in mass fraction) and it extends up to 0.4. However a long tail toward higher Y-values is also present with around 12% of the stars with initial He value between 0.4 and 0.72. Here we want to explore in more details the consequence of the super He-rich stars for the properties of globular clusters.

To assert these implications we have computed a grid of low-mass stellar models from 0.2 to 0.9 $M_\odot$ at a metallicity of $Z = 0.0005$ (similar to the metal-poor globular cluster NGC 6752) for initial He mass fraction between 0.245 and 0.72 with the stellar evolution code STAREVOL V2.92 (see Siess et al. 2000; Siess 2006, for more details). These models have been computed without any kind of mixing except for an instantaneous mixing in convection zones. All models have been computed from the pre-main sequence to the end of the central He-burning phase.

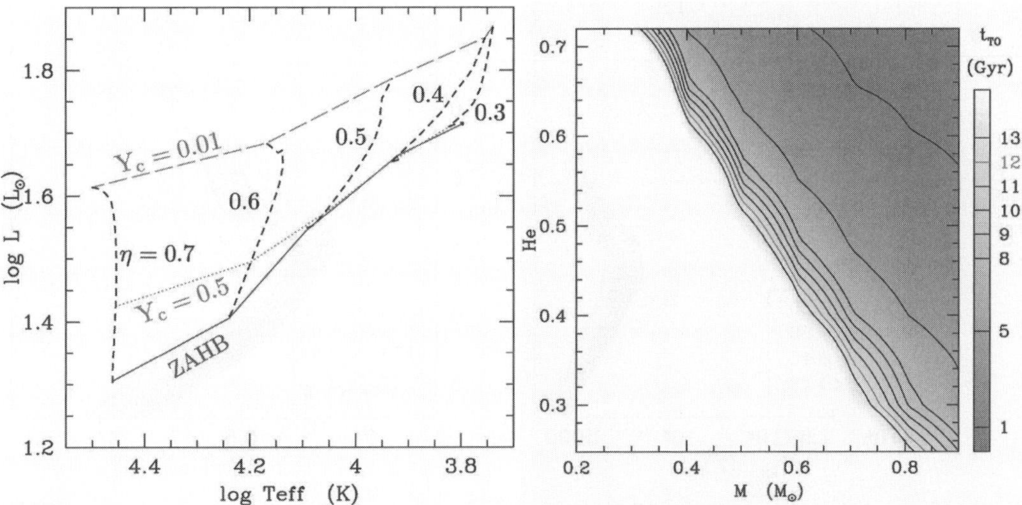

Figure 1. *Left:* Location of central He-burning stars with initial mass of 0.83 $M_\odot$ reaching the TO after 13 Gyr. with various mass loss parameters on the RGB (η_R = 0.3 to 0.7). Full, dotted, and long-dashed lines indicate respectively places where the stars arrive on the HB and where the central He abundance is 0.5, and 0.01. *Right:* Age at the turn-off for low-mass stars as a function of their initial mass (0.2–0.9 $M_\odot$) and initial He value (0.245-0.72, mass fraction). White area indicates stars still on the main sequence after 15 Gyr of evolution.

The mass-loss rate is the only physical parameter that can change the evolution of standard stellar models. This is of peculiar importance in globular clusters as both mass-loss and He-enrichment can produce blue HB. Here we adopt the mass-loss rate following Reimers (1975) prescription (without metallicity dependence). In Fig. 1 (left panel) we present the location of the stars on the HB used to calibrate the mass-loss. Models with He-normal composition (Y_{ini} = 0.245) are computed with different values, from 0.3 to 0.7, for the η parameter in Reimers (1975) prescription. A high value for η (i.e., a high mass-loss rate on the RGB) gives bluer stars on the HB. Below η = 0.5 the stars stay near the ZAHB in the first part of central He-burning phase. As time passes the luminosity of the H-burning shell (hereafter HBS) decreases while the luminosity of the He-burning core increases, leaving almost no variation of surface luminosity. On the other hand stars with high mass loss have a weak HBS on the ZAHB and the increase of surface luminosity reflects the one in the He-core.

From the observational point of view only few studies have been able to assess the He abundance for HB stars. In the GC M 3, Catelan *et al.* (2009) show that for stars with effective temperature below 10 000 K, only normal He abundance agreed with the surface gravity. Besides in NGC 6752, Villanova *et al.* (2009) find through spectroscopic measurements that stars with effective temperature in the range 9 000–8 500 K have a He composition compatible with normal He abundance. From Fig. 1 (left panel) we see that a value of η around 0.4 implies HB stars with standard He abundance and effective temperature below 10 000 K. If we assume the same mass-loss rate for He-rich stars, they will have a smaller mass on the HB and hence will only have hotter temperature. Therefore, higher η values than 0.4 will produce too hot HB stars without stars in the red part of the HB. On the other hand, lower η values will require high He abundance HB stars with effective temperature below 10 000 K in contradiction with the observations. In our computation we use η = 0.4 for the parameter of Reimers (1975) law for all models independently of the initial He content.

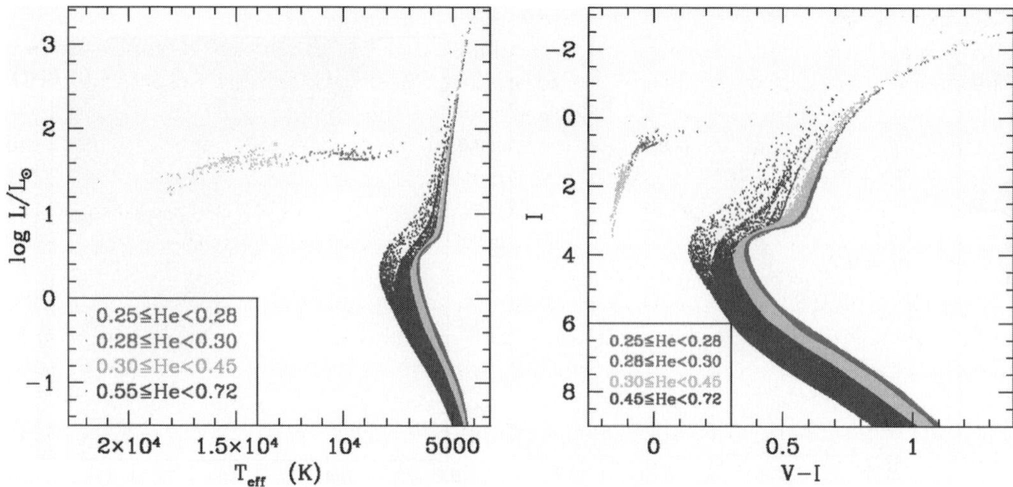

Figure 2. Synthetic diagram of a 13 Gyr old globular cluster in HR diagram (right) and CMD (left). Magenta, red, orange and blue colours (from right to left) indicate the initial abundance of He. He-rich stars are present on the blue part of the MS, RGB and HB, while super-He rich stars are only seen on the MS and RGB.

For a given stellar mass, He-rich stars evolve faster on the main-sequence due to their lower initial H-content and to their higher luminosity. Figure 1 (right panel) illustrates this point showing the turn-off age as a function of the initial mass and He mass fraction of stars. After 12 Gyr, stars of 0.85 $M_\odot$ with standard helium ($Y = 0.245$) as well as He-rich stars of 0.4 $M_\odot$ ($Y = 0.6$) are leaving the main sequence. Thus the distribution of the initial He abundance will be reflected by a distribution of stellar mass for stars at the same evolutionary stage.

3. Effects on colour-magnitude diagram of GCs

Figure 2 shows synthetic colour-magnitude diagrams (in the $T_{\rm eff}$ vs. $\log L$ and I vs. V-I planes) of a 13 Gyr old globular cluster with the initial spread in He following the distribution obtained by Decressin *et al.* (2007a) for a cluster enriched by fast rotating massive stars. The age is chosen to be of 13 Gyr, in agreement with that determined for NGC 6752 (13.8 ± 1.1 Gyr, Gratton *et al.* 2003). This CMD has been computed with a modified version of the program used by Meynet (1993) to investigate supergiant populations. We use the colours transformation from VandenBerg & Clem (2003) which considered only He-normal abundance. This adds another uncertainty to the position of He-rich stars in right panel of Fig. 2.

The spread in He converts into a spread in mass at the turn-off. The luminosity increase of He-rich stars is mainly compensated by their shorter lifetime so that the turn-off luminosity is almost constant. Besides due to their differences in opacity and to their compactness they are also hotter. Thus the He-rich main-sequence and RGB stars are shifted to the left side of the CMD.

While being present on blue side of MS and RGB, super He-rich stars are not laying in the blue part of the HB. This is due to the combined effects of mass-loss during the RGB phase with a smaller initial mass which produce a He-core too light to be able to start He-burning. Thus stars with He content higher than ~ 0.45–0.5 will end as He-WD instead of blue HB stars and CO-WD.

Recently Milone *et al.* (2009) analyse deep photometry of NGC 6752, revealing a broadening of the main sequence which could be attributed neither to binaries nor to photometric errors. If we compare our theoretical CMD with the one observed by those authors we note some discrepancies. In particular the theoretical width of the main sequence at the turn-off is too large compared to that observed for this cluster. Let us stress however that we use calibration relations for normal He content while stars of different He content are present. This might artificially distort the MS. Interestingly the extended blue loop of NGC 6752 can be reproduced by our theoretical CMD with different He-value ranging from 0.245 to 0.45.

A possibility to avoid super He-rich stars is to consider stronger dilution factors. For instance, if we assume a global dilution to produce a mean value we will have $Y \sim 0.37$, which is quite consistent with the blue main sequence in ω Cen and the triple main sequences detected in NGC 2808. However this situation is not completely satisfactory as it leads to a reduction of the amplitude of star-to-star variations of light elements that may then disagree with the observations.

Another way to solve the discrepancy comes from the stellar evolution models of fast rotating massive stars. As already explained, during the main sequence the star reaches the break-up velocity and we assume that the matter is ejected through a slow equatorial wind. However the matter with a very high He-abundance is released only at the end of the main-sequence and the beginning of the central-He burning phase. During these phases, the stars is still at the break-up velocity but this time the radiation pressure contribute significantly to it. This is the so-called $\Omega\Lambda$ limit (Maeder & Meynet 2000). In previous studies we have assumed that the matter released has still a small velocity and will be stored also in the disc to maximise the amount of H-burning processed matter released by fast rotating massive stars. However if the radiation starts to dominate we can instead expect faster winds that will escape the cluster potential well so that this super He-rich matter is lost for the self-enrichment of globular clusters. In this case it would be possible to produce both large abundance variations in light elements along with a moderate increase in He abundance.

Acknowledgements

We acknowledge support from the Swiss National Science Foundation (FNS) and from the "Programme National de Physique Stellaire" of CNRS/INSU, France.

References

Arnould, M., Goriely, S., & Jorissen, A. 1999, *A&A*, 347, 572
Bedin, L. R., Piotto, G., Anderson, J. *et al.* 2004, *ApJ (Letters)*, 605, L125
Bonifacio, P., Pasquini, L., Molaro, P. *et al.* 2007, *A&A*, 470, 153
Brown, J. A. & Wallerstein, G. 1993, *AJ*, 106, 133
Caloi, V. & D'Antona, F. 2005, *A&A*, 435, 987
Carretta, E., Bragaglia, A., Gratton, R. G. *et al.* 2006, *A&A*, 450, 523
Carretta, E., Bragaglia, A., Gratton, R. G., Lucatello, S., & Momany, Y. 2007, *A&A*, 464, 927
Catelan, M., Grundahl, F., Sweigart, A. V., Valcarce, A. A. R., & Cortés, C. 2009, *ApJ (Letters)*, 695, L97
Cottrell, P. L. & Da Costa, G. S. 1981, *ApJ (Letters)*, 245, L79
Decressin, T., Charbonnel, C., & Meynet, G. 2007a, *A&A*, 475, 859
Decressin, T., Charbonnel, C., Siess, L. *et al.* 2009, *A&A*, 505, 727
Decressin, T., Meynet, G., Charbonnel, C., Prantzos, N., & Ekström, S. 2007b, *A&A*, 464, 1029
Gratton, R., Sneden, C., & Carretta, E. 2004, *ARAA*, 42, 385
Gratton, R. G., Bragaglia, A., Carretta, E. *et al.* 2003, *A&A*, 408, 529

Gratton, R. G., Lucatello, S., Bragaglia, A. *et al.* 2007, *A&A*, 464, 953
Kraft, R. P. 1994, *PASP*, 106, 553
Lind, K., Primas, F., Charbonnel, C., Grundahl, F., & Asplund, M. 2009, *A&A*, 503, 545
Maeder, A. & Meynet, G. 2000, *A&A*, 361, 159
Meynet, G. 1993, in The Feedback of Chemical Evolution on the Stellar Content of Galaxies, ed. D. Alloin & G. Stasińska, 40–+
Milone, A. P., Piotto, G., King, I. R. *et al.* 2009, ArXiv e-prints
Pasquini, L., Bonifacio, P., Randich, S. *et al.* 2007, *A&A*, 464, 601
Piotto, G., Bedin, L. R., Anderson, J. *et al.* 2007, *ApJ* (Letters), 661, L53
Piotto, G., Villanova, S., Bedin, L. R. *et al.* 2005, *ApJ*, 621, 777
Prantzos, N. & Charbonnel, C. 2006, *A&A*, 458, 135
Prantzos, N., Charbonnel, C., & Iliadis, C. 2007, *A&A*, 470, 179
Reimers, D. 1975, Circumstellar envelopes and mass loss of red giant stars (Problems in stellar atmospheres and envelopes.), 229–256
Siess, L. 2006, *A&A*, 448, 717
Siess, L., Dufour, E., & Forestini, M. 2000, *A&A*, 358, 593
Smith, G. H. 2006, *PASP*, 118, 1225
Townsend, R. H. D., Owocki, S. P., & Howarth, I. D. 2004, *MNRAS*, 350, 189
VandenBerg, D. A. & Clem, J. L. 2003, *AJ*, 126, 778
Ventura, P. & D'Antona, F. 2008a, *MNRAS*, 385, 2034
Ventura, P. & D'Antona, F. 2008b, *A&A*, 479, 805
Ventura, P. & D'Antona, F. 2009, *A&A*, 499, 835
Villanova, S., Piotto, G., & Gratton, R. G. 2009, *A&A*, 499, 755
Villanova, S., Piotto, G., King, I. R. *et al.* 2007, *ApJ*, 663, 296
Wallerstein, G., Leep, E. M., & Oke, J. B. 1987, *AJ*, 93, 1137

Light Elements in the Universe
Proceedings IAU Symposium No. 268, 2009
C. Charbonnel, M. Tosi, F. Primas, C. Chiappini, eds.

© International Astronomical Union 2010
doi:10.1017/S1743921310003984

What helium and lithium can tell us about CEMP stars?

Georges Meynet[1], Raphael Hirschi[2,3], Sylvia Ekstrom[1], André Maeder[1], Cyril Georgy[1], Patrick Eggenberger[1], and Cristina Chiappini[1]

[1]Geneva Observatory, Geneva University,
CH–1290 Sauverny, Switzerland
email: georges.meynet@unige.ch

[2]Astrophysics group, Keele University,
Lennard-Jones Lab., Keele, ST5 5BG, UK

[3]IPMU, University of Tokyo,
Kashiwa, Chiba 277-8582, Japan

Abstract. We show that the peculiar surface abundance patterns of Carbon Enhanced Metal Poor (CEMP) stars has been inherited from material having been processed by H- and He-burning phases in a previous generation of stars (hereafter called the "Source Stars"). In this previous generation, some mixing must have occurred between the He- and the H-burning regions in order to explain the high observed abundances of nitrogen. In addition, it is necessary to postulate that a very small fraction of the carbon-oxygen core has been expelled (either by winds or by the supernova explosion). Therefore only the outermost layers should have been released by the Source Stars. Some of the CEMP stars may be He-rich if the matter from the Source Star is not too much diluted with the InterStellar Medium (ISM). Those stars formed from nearly pure ejecta would also be Li-poor.

Keywords. stars: AGB, early-type, evolution – supernovae: general – Galaxy: halo – nucleosynthesis

1. Introduction

Observations have revealed in these last years a galactic halo much more complex than previously thought. The existence of two halos, an inner one and an outer one (Carollo *et al.* 2008) showed that, both from the point of view of kinematics and chemical composition, the galactic halo cannot be viewed as a single entity. At the smaller scale of the globular clusters, there are now ample evidences for the succession of at least two and maybe more stellar generations in clusters, each leaving a very peculiar imprint on the chemical composition of stars (see the talks by Bragaglia and Decressin in the present volume and references therein). In that respect it is striking to note that stellar populations in clusters and in the field have each their own "anomalous" counterpart. In globular clusters the "anomalous" component is made up of stars essentially made from material processed by only H-burning reactions. In the field, the "anomalous" component is made up of stars presenting the nucleosynthetic signature of both H- and He-burning processes. In this paper, we focus our attention on the "anomalous" population observed in the field of the halo. In that population, we find the most iron poor star known today, HE 1327–2326, with a [Fe/H] as low as –5.96 (Frebel *et al.* 2008). In this star, one counts 72 000 atoms of carbon for one atom of iron, while in the Sun, there is only 9 atoms of carbon per atom of iron! Hence the name of Carbon Enhanced Metal Poor Stars

(CEMPs). The CEMP stars represent about 1 in 5 stars at metallicity below [Fe/H] < –2.5 (Lucatello *et al.* 2006). In HE 1327–2326, the numbers of nitrogen and oxygen atoms with respect to the number of iron atoms are respectively 20 000 times and 2500 times greater than in the Sun. Thus this star could have also been named a N(itrogen)EMP or an Ox(ygen)EMP star! This star is either at the end of the Main-Sequence phase or just after the Main-Sequence (a subgiant), thus these abundances cannot have been produced in the star itself but must have been mainly inherited from the interstellar cloud from which it formed about 13 billion years ago† or be acquired through a mass transfer episode in a close binary system. Whatever the scenario chosen, the main question is what was the nature of the "Source Stars" which either enriched the natal cloud in such a peculiar way or transferred its envelope to its less massive companion conferring it its status of CEMP star? Was it a massive star? An intermediate mass star? Can we deduce some of its properties? This is the point we want to discuss in the present paper.

2. Constraints from nucleosynthesis

There are a few deductions that can be done without reference to any peculiar stellar models. These are the following:

• The CEMP stars present a great scatter in the abundances of the CNO elements (see the hatched zones in the left panel of Fig. 1). This is an indication that the chemical characteristics of the material from which these stars inherited their peculiar composition were different from case to case. This can be understood if the peculiar abundances reflect the chemical abundance of one or at a most a very small number of nucleosynthetic events.

• The very high overabundances with respect to iron of carbon, nitrogen and oxygen imply that the material responsible for these enhancements was mainly processed by the nuclear reactions involving nuclear reaction chains typical of H- (enhancement of N) and He-burnings (increases of C and O). In both nuclear phases, iron is not synthesized. Thus either iron has been produced in stars belonging to generations having occurred before the "Source Stars", or it has been produced, in part or in totality, by the "Source Stars" themselves in advanced nuclear phases. In that case, the "Source Stars" should produce much smaller quantities than those predicted by type II supernova explosions which are expected to produce [O/Fe] around 0.50 or less (see for instance table 1 in Tominaga *et al.* 2007), while the Frebel star, for instance, show [O/Fe] of the order of 3.4!

• The mixing of material having been processed by H-burning and of material processed by He-burning is however not a sufficient condition to reproduce the observed surface abundances of CEMP stars. To see what would be obtained by simple addition of material having been processed by these two nuclear phases *without allowing any mixing between them in the "Source Star"*, we can look at the dotted, short and long dashed curves plotted in the left panel of Fig. 1. These curves show the chemical composition of the mixed outer envelope of a non-rotating 7 $M_\odot$ at $Z = 10^{-5}$ at the early-AGB phase. Only the CNO elements are shown because these models were computed without the Ne-Na, Mg-Al chains. When only the envelope above 1.1 $M_\odot$ is ejected, we see matter processed only by H-burning mixed with some unaltered material in the outer envelope. In case the initial metallicity of the model would have been chosen equal to $Z = 10^{-6}$ (maximum metallicity of stars which could have participated to the production of the material from which HE 1327–2326 could form), the whole curve would have been shifted downward by 1 dex. We see that material enriched in H-burning material cannot account

---

† Microscopic diffusion may have altered the surface abundances however (Korn *et al.* 2009), but this process cannot be responsible for the huge excesses in CNO elements.

for the observed abundances in CEMP stars. When matter above 0.9 $M_\odot$ is ejected, some material processed by the He-burning reactions is added into the mixture. The abundance of nitrogen is not much changed, while those of carbon and oxygen increase a lot despite the fact that only 0.2 $M_\odot$ has been added with respect to previous case! The trend is reinforced in case a still lower mass cut is considered. Considering a lower initial metal content, would in those cases, shift downward the nitrogen abundances without much changing the positions of the carbon and oxygen abundances. We see thus that such models have no chance to reproduce the observed patterns. Similar conclusions can be reached looking at other initial mass models. In order to achieve high abundances for all the three main CNO isotopes, nitrogen must also be produced from helium as carbon and oxygen. This is possible if some carbon and oxygen produced in the helium core diffuse in the H-burning shell and are transformed there in nitrogen. *Thus the simultaneous large increases of the three CNO elements are an indication that some mixing occurred between the He- and the H-burning regions.*

• Let us suppose that such a mixing occurred. The increase of nitrogen in the H-burning shell cannot be as high as the one of carbon and oxygen in the He-burning core. Diffusion is indeed not as efficient as convection and will never allow to homogenize carbon and oxygen between the He- and the H-burning regions. To see what happens when such a mixing occurs let us consider the left panel of Fig. 1. The upper continuous and long-short dashed curves show the abundances in the mixed outer envelope of a rotating 7 $M_\odot$ stellar model with an initial value of Ω/Ω_c equal to ~ 0.8 where Ω is the surface angular velocity and Ω_c the critical angular velocity at the surface†. The model is in the early AGB phase. In such a model, during the whole core He-burning phase, carbon and oxygen diffuse from the He-core into the H-burning shell. We see that the envelope above the CO core (the CO core mass is about 1.3 $M_\odot$) is strongly enriched in CNO elements. Interestingly it would also be strongly enriched in fluorine, neon (actually ^{22}Ne), in ^{23}Na, in magnesium (mainly ^{26}Mg), and ^{27}Al. Mixing one part of this envelope material from the "Source Star" with 100 parts of ISM would shift the curve downward by 2 dex providing a very good fit to the mean values of the observed patterns shown in that figure. Since all these elements are formed either directly or indirectly from transformation of helium, the overabundances shown here are not so much dependant on the metallicity and would occur in a similar way in more metal poor stars. Also similar conclusions can be obtained from different initial mass models (Meynet *et al.* 2006; Hirschi 2007). However, depending on the initial mass of the model considered for the "Source Star", the degree of dilution required to fit the observed values can be different.

3. The spinstar scenario

In the "spinstar scenario" (Meynet *et al.* 2006; Hirschi 2007; Meynet *et al.* 2009), the required mixing is induced by the instabilities triggered by axial rotation of stars. Rotation can indeed induce mixing in radiative zones and thus allow for instance carbon and oxygen to diffuse from the He core into the H-burning zone. This produces important amounts of nitrogen (primary nitrogen since it is produced from carbon and oxygen synthesized by the star itself) and of primary ^{13}C. It is interesting to note that observations of the surface abundances of "normal" metal poor halo field stars (i.e. non CEMP stars) indicate that actually important amounts of primary nitrogen need to be

---

† The critical velocity is such that when this velocity is reached, the centrifugal acceleration at the equator compensates for the gravity there.

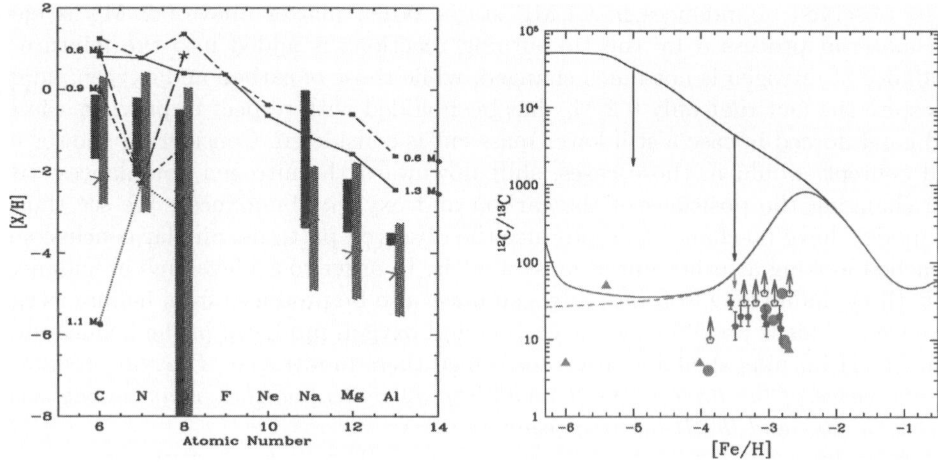

Figure 1. *Left panel :*The continuous and long-short dashed curves with filled squares show the composition obtained by mixing the outer envelope above the langrangian mass coordinate 1.3 $M_\odot$ and 0.6 $M_\odot$ at the E-AGB phase of our $Z = 10^{-5}$ rotating 7 $M_\odot$ stellar model. The dotted, short and long-dashed lines correspond to the composition of the envelope above the langrangian mass coordinate 1.1, 0.9 and 0.6 $M_\odot$ respectively at the E-AGB phase of our non-rotating 7 $M_\odot$ stellar model. The CO core in the rotating and non rotating models have masses around 1.3 and 0.9 $M_\odot$ respectively. The star symbols show the observed values for the most iron poor star known today (Frebel *et al.* 2008). The vertical hatched zones show the range of observed values for CEMP-no stars (i.e. those with no evidence of s-process element enhancements, see references in Masseron *et al.* 2009). The dark left hatched zone corresponds to non-evolved stars, the right light grey zone to giants. *Right panel :*Predicted evolution of the ^{12}C/^{13}C ratio according to chemical evolution models (CEM) for the halo computed with different stellar models for metallicities below $Z = 10^{-5}$. The upper solid line corresponds to CEM models without fast rotators, the dashed and lower solid lines correspond to CEM models with fast rotators. The data are the unmixed stars of Spite *et al.* (2006). The open symbols represent lower limits. The descending arrows indicate the final ^{12}C/^{13}C ratio obtained after the first dredge-up in giants starting from the initial values given by the CEM models (see details in Chiappini *et al.* 2008). The filled triangles (lower limits) and circles correspond to CEMP stars showing no s-process element enhancements and to the Frebel star (most iron poor star at [Fe/H] $= -5.96$).

produced in very metal poor massive stars. Chiappini *et al.* (2006, 2008) showed that rotating massive star models can reproduce the amount of primary nitrogen required by observation provided they began their life with an initial angular momentum content of the same magnitude as the one in solar metallicity massive stars showing an averaged surface rotation rate during the MS of 200 km s^{-1}. In this model, the "normal" halo field stars are formed from a well mixed reservoir enriched by stars of different initial masses and of different initial metallicities. A consequence of this model is that low ratios of the ^{12}C/^{13}C should be observed at the surface of non-evolved metal poor halo field stars (see the right panel of Fig. 1).

If now, instead of considering stars formed from material taken from a well mixed reservoir, as above, we consider stars formed from material ejected by one single source, which kind of composition can we expect? This is shown in the left panel of Fig. 1 and was already discussed in the previous section. A good fit is obtained with the CEMP abundance patterns. Thus we see that, in the frame of this model, the main difference between the "normal" and the "anomalous" population is not the nature of the "Source Stars" which in both cases are rotating stars, but the fact that the normal populations are born from a well mixed reservoir, while the CEMP-stars are formed from a much more localized and specifically enriched reservoir.

An interesting feature of the present scenario is that rotational mixing can not only help in producing interesting abundance patterns, it can also be responsible for the loss through winds of the outermost layers. Indeed, due to rotational mixing, the radiative envelope is enriched in CNO elements, its global opacity is increased and this may trigger line driven stellar winds, both in the case of massive stars in the supergiant phase and in the case of intermediate mass stars along the AGB. Such an opacity increase only occurs when the stars have evolved beyond the Main-Sequence. Only at that time, can the surface be CNO enriched with values well above the initial metallicity. This will only occur at very low metallicity, let's say at metallicities below about 10^{-5} . Why? For two reasons, first when the metallicity decreases rotational mixing is more efficient (Maeder & Meynet 2001), second the mass losses, both due to line radiation winds and to mechanical winds due to the reaching of the critical velocity are very weak at these low metallicities (see Meynet *et al.* 2009), preventing the H-burning shell to disappear when the He-burning core is in activity. Thus only at low metallicities can we expect to have winds enriched in both H- and He-burning products. At higher metallicities $(10^{-5} < Z < 10^{-3})$, fast rotating models lose important amounts of matter through mechanical equatorial winds during the Main-Sequence (Decressin *et al.* 2007, see also the contribution of Decressin in the present volume). This mechanical wind is enriched only in H-burning products. This may contribute to explain the origin of the different anomalous populations observed in the field and in the clusters of the halo: the clusters being more metal rich would be enriched by mechanical winds rich in H-burning products, while field halo stars may be enriched by line driven winds rich in both H- and He-burning products.

A prediction of the present scenario is that a very low $^{12}C/^{13}C$ ratio can be obtained in CEMP stars. In the present models both isotopes are produced as primary elements, the ratio is thus not much sensitive to the dilution factor (unless we consider so large dilution factors that it would also erase any strong overabundances in the CNO elements!). Very low ratios are obtained in case the "Source Star" is a massive star having lost its outer envelope (below about 10). Higher ratios are obtained in the envelope of early AGB models (of the order of 100, see Table 2 in Meynet *et al.* 2009). Thus this ratio may be useful both for providing clues supporting the present "spinstar" scenario and also for disentangling between massive and intermediate mass stars (early AGB) as the possible "Source Stars". In the right panel of Fig. 1, we have plotted the $^{12}C/^{13}C$ ratios observed at the surface of the CEMP-no stars. We see that even with some CEMP stars being dwarf stars (as for instance the filled circle at [Fe/H] equal to -4.015), the CEMP stars are in general below the $^{12}C/^{13}C$ ratio observed in giants (which should have lower $^{12}C/^{13}C$ ratios than dwarf stars since they went through the dredge-up episode). Moreover these CEMP stars are also below the predictions of the chemical evolution model which describes the evolution of the composition of stars made up from the well mixed reservoir. These two points support the view of a different origin of the CEMP star with respect to the normal halo stars based only from the $^{12}C/^{13}C$ ratio, second it shows that very low $^{12}C/^{13}C$ ratios are observed in some dwarf CEMP star supporting, in view of the large ^{12}C overabundances, a primary origin for ^{13}C as predicted by the rotating models.

4. Constraints from Li and He

Let us now come to the question asked in the title, namely how measures of the Li and He abundances in CEMP stars can be used to constrain the above scenario? The material of the "Source Stars" are Li-poor (actually Li is completely destroyed) and He-rich (see Table 2 in Meynet *et al.* 2009). Thus CEMP stars made of pure "Source Star" material would be Li-poor and He-rich. Note that the same line of reasoning implies that stars presenting enhancements of Na and depletion of O in globular clusters should also be He-rich. Both in globular clusters and in the field, the actual level of He overabundance depends on the degree of dilution with interstellar medium (see the contribution by Decressin in the present volume and references therein, see also Maeder & Meynet 2006).

Some Li observations exist for CEMP stars. Many of the CEMP-no stars show very low Li or even only upper limits compatible with no Li at all. Of course these very low Li abundances can be due to depletion having occurred in the CEMP star itself and thus may not be the signature of a formation process from Li-depleted material. A high helium abundance in the whole star (and not only at the surface) would be a much stronger constraint pointing towards a small dilution factor. The finding of such a star would indeed be extremely interesting since in that case one would have an object providing a nearly direct insight into the chemical composition of the ejecta of the first stars. Of course the difficulty here resides in how to measure helium abundance in a cool star. Although this is a real challenge, the situation may not be completely desperate. A high helium abundance would imply a different initial mass associated to a given observed position in the HR diagram than the one obtained assuming a normal (here a cosmological helium) abundance (see the contribution by Decressin in the present volume). The discovery of eclipsing binaries whose components would be CEMP stars would be a way to associate observationnally determined masses to given positions in the HR diagram and thus to test the He-rich hypothesis. Asteroseismology would also be an interesting way to address that question. Finally, helium lines may be observable in cool stars, whose strength might be related to the He abundance (Moehler *et al.* 2000). While these possibilities appear quite exciting and will probably be explored in the future, it is however important to mention that the absence of He-rich stars among the CEMP stars would not be an argument against the present spinstar scenario. High helium abundances imply small dilution factors (a factor of a few), while normal helium abundances imply larger dilution factors (greater than an order of magnitude as in the case shown in the left panel of Fig. 1), but in both cases spinstars are required to explain the high CNO elements! It is also interesting to mention that the observed amounts of Li, Be and B produced by spallation reactions at very low metallicities give also some support to the present "spinstar" scenario (see the review by Prantzos in this volume).

References

Carollo, D., Beers, T. C., Lee, Y. S. *et al.* 2008, Nature, 451, 216
Chiappini, C., Ekström, S., Meynet, G. *et al.* 2008, A&A, 479, L9
Chiappini, C., Hirschi, R., Meynet, G. *et al.* 2006, A&A, 449, L27
Decressin, T., Meynet, G., Charbonnel, C., Prantzos, N., & Ekström, S. 2007, A&A, 464, 1029
Frebel, A., Collet, R., Eriksson, K., Christlieb, N., & Aoki, W. 2008, ApJ, 684, 588
Hirschi, R. 2007, A&A, 461, 571
Korn, A. J., Richard, O., Mashonkina, L. *et al.* 2009, ApJ, 698, 410
Lucatello, S., Beers, T. C., Christlieb, N. *et al.* 2006, ApJ, 652, L37
Maeder, A. & Meynet, G. 2001, A&A, 373, 555
Maeder, A. & Meynet, G. 2006, A&A, 448, L37
Masseron, T., Johnson, J. A., Plez, B. *et al.* 2009, ArXiv e-prints 0901.4737
Meynet, G., Ekström, S., & Maeder, A. 2006, A&A, 447, 623
Meynet, G., Hirschi, R., Ekstrom, S. *et al.* 2009, ArXiv e-prints 0910.3856
Moehler, S., Sweigart, A. V., Landsman, W. B., & Heber, U. 2000, A&A, 360, 120
Spite, M., Cayrel, R., Hill, V. *et al.* 2006, A&A, 455, 291
Tominaga, N., Umeda, H., & Nomoto, K. 2007, ApJ, 660, 516

Light Elements in the Universe
Proceedings IAU Symposium No. 268, 2009
C. Charbonnel, M. Tosi, F. Primas & C. Chiappini, eds.

© International Astronomical Union 2010
doi:10.1017/S1743921310003996

The Helium contribution from massive AGBs

Paolo Ventura

INAF - Observatory of Rome, 00040, Monte Porzio Catone (RM), Italy
email: ventura@oa-roma.inaf.it

Abstract. The helium produced by AGB and super-AGB stars is a key quantity to understand whether these objects may have been the main polluters of the interstellar medium within globular clusters, and originate a second generation of stars with a chemistry showing the imprinting of their ejecta. Helium is the most important element for this topic, as any difference in the original helium between the two populations would determine clearly distinguishable features both in the morphology of the Horizontal Branches and in the Main Sequences. We present the helium yields from massive AGB stars, and show that the results are rather robust, being approximately independent of the various uncertainties that affect the description of the evolution of these stars. The implications for the self-enrichment scenario are discussed and commented.

Keywords. Stars: abundances, AGB and post-AGB

1. Introduction

The traditional paradigma that stars in globular clusters (GC) are a classic example of a single stellar population, with the same age and chemistry, has been challenged in the last decades by spectroscopic evidences, showing that star-to-star differences exist in their surface chemistry, involving all the "light elements", i.e. all those species lighter than aluminium (Kraft 1994). The original idea (Denissenkov & Weiss 2001) that these differences were due to some "in-situ" mechanism, associated to non-canonical extra-mixing from the bottom of the convective envelope, was disregarded by the discovery that the same patterns were also present in main sequence stars (Gratton *et al.* 2001), for which no active advanced nucleosynthesis could be invoked.

This opened the way to studies focused on the understanding of whether these chemical anomalies, that were surely present in the gas from which these stars formed, could be generated by the winds of intermediate-mass stars during their Asymptotic Giant Branch (AGB) phase (Ventura *et al.* 2001): in fact, the most massive of these objects are known to experience Hot Bottom Burning (HBB), i.e. the base of their external mantle becomes hot enough to ignite an advanced nucleosynthesis, whose products would be almost instantaneously transported to the surface, due to the rapidity of convective motions (Blöcker & Schönberner 1991). The formation of new stellar generations in GC from the gas expelled by rapidly evolving stars belonging to the original population is commonly referred to as the "self-enrichment" scenario.

One predictions of these investigations is that massive AGB stars are helium producers, thus the gas from which the new generations of stars form is expected to be helium-rich (Ventura *et al.* 2001).

Based on this expectation, the stellar evolution group in Rome suggested in a series of papers that differences in the original helium content was the natural explanation of the so called "second parameter" in the interpretation of the different morphologies of the Horizontal Branches (HB) of GC, the bluest and less luminous clumps being populated by stars helium enhanced (Caloi & D'Antona 2005, Caloi & D'Antona 2007, D'Antona & Caloi 2004, D'Antona & Caloi 2008).

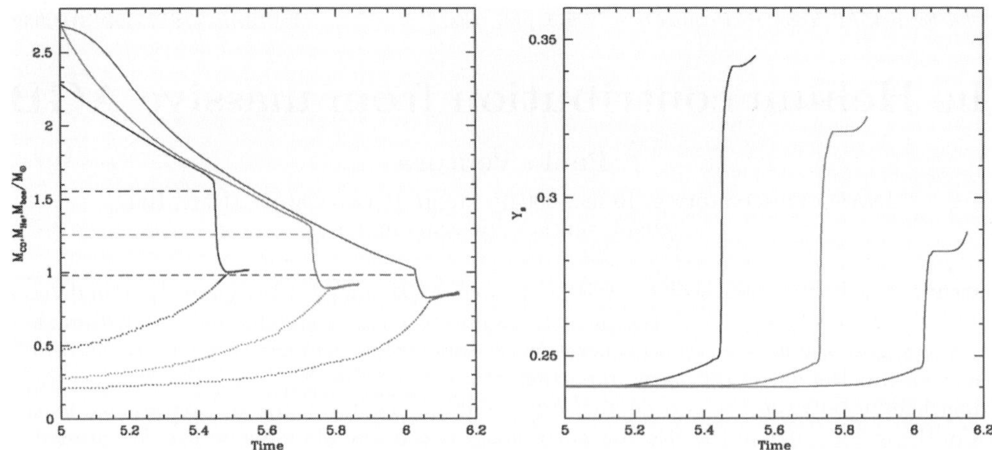

Figure 1. Left: Variation with time (in logarithmic scale, and counted from the helium exhaustion in the core) of the bottom of the convective envelope (solid), the H/He interface (dashed), and the external border of the CO core (dotted) in models with initial masses 4 (right), 5 (middle), and $6M_\odot$ (left). Right: The helium increase following the second dredge-up for the models shown in the right panel.

This idea received a robust confirmation by the most recent photometric analysis, showing the presence of multiple main sequences in many massive GC, that can be understood only by assuming the existence of a stellar population enriched in helium (Piotto *et al.* 2007).

In this contribution we present the helium yields from massive AGBs, i.e. stars with mass in the range $3 - 6.5M_\odot$, that never reach temperatures sufficiently high to burn carbon in the core. We discuss the major sources of uncertainty, and the physical processes most relevant for this particular issue. We compare these findings with the helium content of the ejecta of super-AGB stars, and find a trend monotonically increasing with the core mass.

We eventually discuss how these results may be used in the interpretation of the extendend blue tails in the HB of many GC and the presence of blue main sequences, in the context of the self-enrichment scenario for the formation and evolution of GC.

2. The enrichment of the surface helium in intermediate-mass stars

In stars of intermediate mass, helium burning is activated in the core in conditions of non-degeneracy; the relic of this phase is a core, made up of carbon and oxygen, evolving to conditions of progressively more enhanced degeneracy, and a helium burning buffer just above it. The ignition of this shell favours the expansion of the outer layers, with the consequent temporary extinction of the CNO burning region. During this phase the external convection sinks inwards, penetrating to layers previously touched by CNO burning, in what is commonly called "the second dredge-up" (SDU).

The left panel of Fig. 1 refers to models with masses 4, 5, and $6M_\odot$; times are counted from the exhaustion of central helium. The chemistry is typical of GC with intermediate metallicity, i.e. $Z = 0.001$, $Y = 0.24$. The evolution of the bottom the convective envelope (solid), the border of the H-rich region (dashed), and the external border of the CO core (dotted) are shown. The H/He interface is initially unchanged with time, because the CNO shell is not active. The base of the external mantle eventually reaches such interface, and penetrates inwards, down to the He-burning shell. This episode is accompanied by

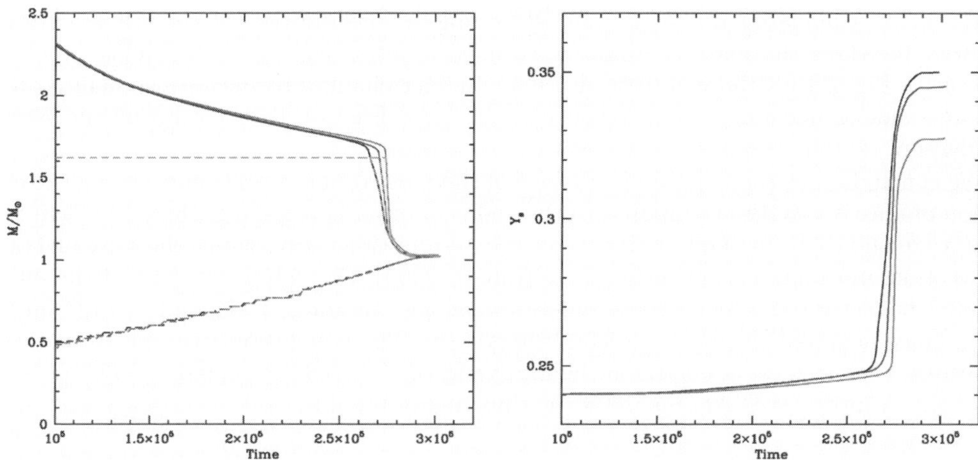

Figure 2. Left: Inwards penetration of the external envelope in a 5$M_\odot$ models calculated with different extent of the overshooting region, i.e. with a choice of the free parameter $\zeta = 0.01$ (right), 0.02 (middle), 0.05 (right). Right: The increase in the surface helium in the same models presented in the left panel.

a sudden increase in the surface helium content (see the right panel of Fig. 1), which is higher in the $6M_\odot$ model, due to the largest width of the He-rich buffer.

We will see that the increase in the surface helium achieved during this phase will be crucial to determine the overall helium content of the ejecta of these stars. We therefore discuss how these results may depend on the way with which the convective/radiative interface is treated.

Modelling the AGB phase demands a diffusive approach, in which nuclear burning and mixing of chemical are coupled self-consistently (Cloutman & Eoll 1976). The reason for this is that a non-negligible percentage of the overall nuclear energy is generated within the convective envelope, due to a partial overlapping of the external mantle with the CNO burning shell (Mazzitelli *et al.* 1999). Within the diffusive context, overshooting from any convective border, fixed via the Schwartzschild criterion, is modelled by an exponential decay of convective velocities within the stable region, with an e-folding distance that is commonly parametrized as ζH_p: ζ is therefore the free parameter entering the description, and providing the extent of the extra-mixing region.

The ζ adopted in this investigation is $\zeta = 0.02$, in agreement with a calibration based on the observed main sequences of open clusters, given in Ventura *et al.* (1998). To study the sensitivity of the helium enhancement on the details of the overshooting description, we calculated three models with initial mass $5M_\odot$ with $\zeta = 0.01, 0.02$ and 0.05. The results of these simulations are shown in Fig. 2. We see in the left panel, showing the same quantities reported in the left panel of Fig. 1, that a greater extent of the extra-mixed region favours a faster inwards penetration of the convective envelope, though the innermost point reached, corresponding to the location of the He-burning shell, is approximately unchanged. The final surface helium reached at the end of this phase is thus only mildly dependent on ζ, showing an increase of $\delta Y \sim 0.01$ for ζ exceeding by more than 100% the value expected on the basis of empirical calibrations.

The surface helium reached after the second dredge-up by intermediate-mass stars will be the starting point for the following phase, during which these stars experience a series of thermal pulses, provoked by the periodic ignition of a helium-rich buffer in conditions of thermal instability. Fig. 3 shows the evolution of the surface helium during the whole

AGB phase, calculated up to the almost complete consumption of the whole external mantle. We show the mass as abscissa, to have a better feeling of the yields expected. We see in Fig. 3 that there is little space for further changes in the surface helium, that remains almost unchanged with respect to the mass fraction reached during the second dredge-up.

We conclude that the helium yield is one of the most robust results concerning the evolution of massive AGBs. Compared to the yields of other species, e.g. the CNO elements, that were shown to depend critically on the assumptions concerning the treatments of convection and mass loss (Ventura & D'Antona 2005), the helium content of the ejecta is only modestly touched by these indeterminations, the final Y depending only on the abundance achieved during the second dregde-up. The conclusion from previous investigations, that suggest a maximum helium yield of $Y \sim 0.35$ expected from these stars (see, e.g., Ventura & D'Antona 2008), is thus highly reliable.

3. The self-enrichment scenario

We showed that massive AGBs are expected to be efficient helium producers, providing ejecta with an average helium content up to $Y \sim 0.35$. These results confirm the speculation by Pumo et al. (2008), that the helium yields of the most massive AGBs are extremely close to the corresponding yields from the less massive super-AGBs, i.e. those stars that undergo an off-centre, degenerate, carbon ignition (Siess 2006). This fixes an increasing trend of Y with mass (or, more precisely, with core mass), the largest helium predicted being $Y \sim 0.40$ (see Fig. 1 in Pumo et al. 2008).

We note that this is the helium invoked by D'Antona & Caloi (2004) to explain the presence of a detached clump of low-luminosity, blue stars, populating the HB of NGC 2808, and it is also the same quantity required to interpret the bluest main sequence observed in the same cluster (Piotto et al. 2007).

These results open the way to new tests of the self-enrichment scenario framework. It will be possible to explain any peculiar morphology of the HBs of GC or any intrinsic width of the observed main sequences as the overlapping of an original stellar component, with the helium coming from the Big Bang, and an additional generation, born from gas that was contaminated by the winds of AGB and super-AGB stars. The degree of dilution of this gas with a pristine component is the relevant quantity to determine the helium content of the second generation, hence the features expected on the HB and the spread of the MS. The total mass and the degree of concentration of the cluster is expected to determine the extent of such a dilution process (D'Ercole et al. 2008).

4. Conclusions

We presented a detailed investigation of the helium yields expected from intermediate-mass stars during their evolutionary history. We find that it is mainly during the second dredge-up that most of the helium enrichment is produced, with little increase in the following phase during which the stars experience a series of thermal pulses.

The helium yields show up to be scarcely sensitive to the uncertainties associated to convective overshooting, since the innermost point reached by the penetration of the convective envelope during the second dredge-up, i.e. the location of the helium-burning-shell, is almost independent of the overshooting assumed. Even the impact of mass loss is negligible, because most of the mass is lost after the major phase of helium enhancement.

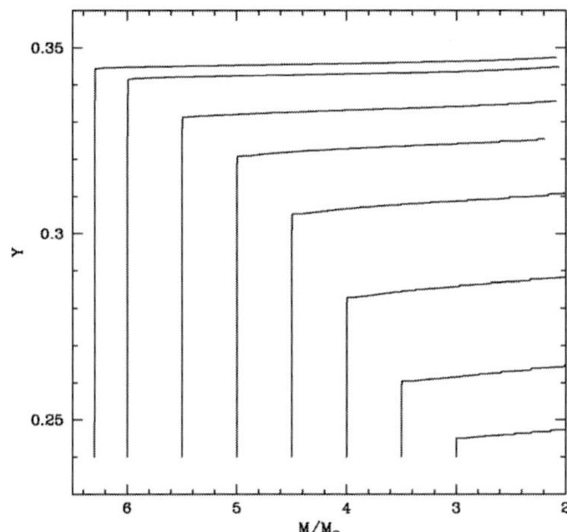

Figure 3. Evolution of the surface helium content within intermediate-mass models. Mass was chosen as abscissa, to have an idea of the chemical content of the ejecta. We note a sudden increase in the surface helium, associated to the second dredge-up, and little change in the following phase of thermal pulses.

These results indicate that helium is produced during the evolution of stars of intermediate mass, the highest abundances, $Y \sim 0.35$, being reached by the highest masses not achieving carbon-burning, i.e. $\sim 6M_\odot$, this limit depending on the overshooting from the core during the H-burning phase. A higher helium enhancement is found in slightly higher mass models, the super-AGB stars, that undergo carbon ignition in a degenerate layer out of the centre.

The combination of AGB + super-AGB models produce yields with a helium content gradually increasing with mass, with a maximum of $Y \sim 0.4$. This is the value that was invoked to interpret the existence of extended blue tails in the HBs of some GC, and, more recently, the presence of a blue main sequence in NGC 2808, that is supposed to be populated by stars with a high helium content.

This result indirectly supports the self-enrichment scenario hypothesis, i.e. that in GC we observe the overlapping of more that an individual generation of stars, because new stars were born from the ashes left by the evolution of stars of intermediate mass.

References

Blöcker, T. & Schönberner, D. 1991, *A&A* (Letters), 244, 43

Caloi, V. & D'Antona, F. 2005, *A&A*, 435, 987

Caloi, V. & D'Antona, F. 2007, *A&A*, 463, 949

Cloutman, L. & Eoll, J. G. 1976, *ApJ*, 206, 548

D'Antona, F. & Caloi, V. 2004, *ApJ*, 611, 871

D'Antona, F. & Caloi, V. 2008, *MNRAS*, 390, 693

Denissenkov, P. & Weiss, A. 2001, *ApJ* (Letters), 559, 115

D'Ercole, A., Vesperini, E., D'Antona, F., McMillan, S. L. W., & Recchi, S. 2008, *MNRAS*, 391, 825

Gratton, R., Bonifacio, P., Bragaglia, A., Carretta, E., Castellani, V. *et al.* 2001, *A&A*, 369, 87

Kraft, R. P. 1994, *PASP*, 106, 553

Mazzitelli, I., D'Antona, F., & Ventura, P. 1999, *A&A*, 348, 846

Piotto, G., Bedin, L.R., Anderson, J., King, I. R., Cassisi, S. *et al.* 2007, *ApJ* (Letters), 661, 53

Pumo, M. L. P., D'Antona, F., & Ventura, P. 2008, *ApJ* (Letters), 672, 25
Siess, L. 2006, *A&A*, 448, 717
Ventura, P., D'Antona, F., Mazzitelli, I., & Gratton, R. 2001, *ApJ* (Letters), 550, 65
Ventura, P. & D'Antona, F. 2005, *A&A*, 279, 288
Ventura, P. & D'Antona, F. 2008, *A&A*, 479, 805
Ventura, P., Zeppieri, A., Mazzitelli, I., & D'Antona, F. 1998, *A&A*, 334, 953

Light Elements in the Universe
Proceedings IAU Symposium No. 268, 2009
C. Charbonnel, M. Tosi, F. Primas & C. Chiappini, eds.

© International Astronomical Union 2010
doi:10.1017/S174392131000400X

Discussion A: On the abundance of deuterium in the local interstellar medium and in high-redshift systems

Monica Tosi

INAF - Osservatorio Astronomico di Bologna
Via Ranzani 1, I-40127, Bologna, Italy
email: monica.tosi@oabo.inaf.it

Abstract. The first discussion session held at the IAU Symposium 268 focussed on the deuterium content in the local interstellar medium (LISM) and in high-redshift systems. There were two key questions proposed to the audience: 1) what should be taken as representative abundance of D in the LISM, and 2) how can we explain the dispersion of the D abundance measured in high-redshift, very low metallicity environments? While on the latter point people seem to agree that observational and data analyses uncertainties are the most likely explanation, on the former question no consensus was reached. The historical and observational background at the basis of these questions and the discussion are schematically reported here.

Keywords. ISM: abundances; Galaxy: evolution; cosmology: observations; quasars: absorption lines

1. Introduction

Four discussion sessions were held at the IAU Symposium 268 on the most debated issues related to the light elements. The first of these special sessions was devoted to our current understanding of the deuterium abundance in the local interstellar medium (LISM) and in high-redshift systems. Two key questions were selected by the SOC as the current hottest topics on deuterium and opened to discussion: 1) what should be taken as representative abundance of D in the LISM, and 2) how can we explain the dispersion of the D abundance measured in high-redshift, very low metallicity environments. In the recent literature there have been quite interesting debates on both instances and the lively discussion which took place at the meeting reflected the deep involvement of the community.

The very circumstance that these questions need to be asked is the positive result of the efforts and achievements of scientists exploiting modern, high-performance instruments, such as ground-based high-resolution spectrographs at 10 m class telescopes and those on the HST and FUSE satellites. On the high-redshift side, the data currently available on D have been summarized in the recent papers by O'Meara *et al.* (2006) and Pettini *et al.* (2008). On the Galactic side, Geiss, Hébrard, Linsky and Sembach have comprehensively described here (this volume) the available measurements. From their presentations, it is apparent that the increasing number of accurate measures has led to a much larger coverage of different environments. As summarized by Savage *et al.* (2007), we now have D measured in the solar system, in the LISM, in a couple of fields in the Galactic disk and halo, in a high-velocity cloud (Complex C) and in several Damped Lyman-α systems (DLAs). Yet, we have reached neither a clear understanding of the deuterium distribution in different environments, nor a consensus on its most representative values.

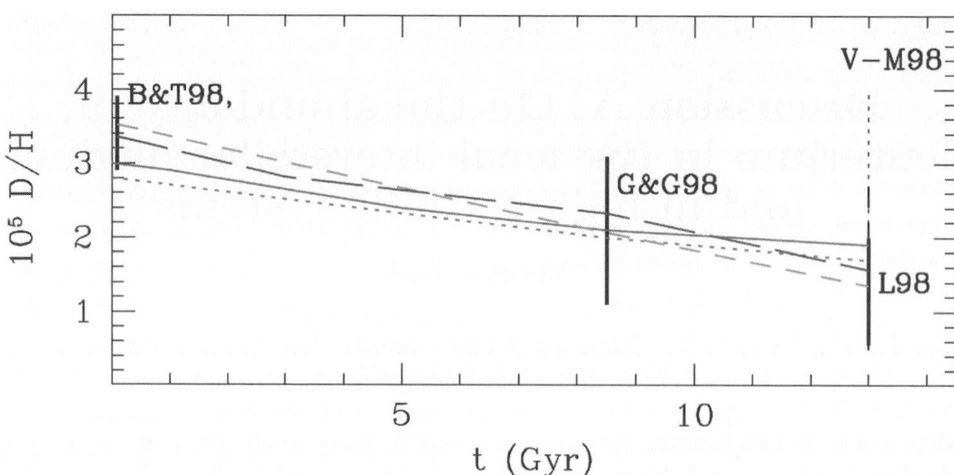

Figure 1. What we knew of the deuterium abundances by number ten years ago (see Tosi 2000). The coloured curves show the abundance variation with time as predicted by chemical evolution models for the solar neighbourhood. These curves refer to models by different groups, all able to reproduce the majority of the Galaxy observational constraints (see for references Tosi 2000). The vertical bars show at 2σ the abundances estimated by Burles & Tytler (1998) for high redshift absorbers, by Geiss & Gloeckler (1998) for the pre-solar cloud and by Linsky (1998) (solid bar) and Vidal-Madjar *et al.* (1998) (dashed and dotted bars) for the LISM.

Ten years ago the situation looked much simpler and stable. At the IAU Symposium 198 on the Light Elements, held in Natal (Brasil) in 1999, the main players in the D measurement game showed D/H values with rather small dispersions, almost undistinguishable from those presented two years earlier at the meeting on the same topic organized by the International Space Science Institute (ISSI) in Bern (Switzerland). These values are schematically plotted in Fig. 1 (showed at the IAU Symp.198) and display a steady and moderate decrease from the close-to-primordial D/H of high-z absorbers (Burles & Tytler 1998), to the proto-solar cloud (PSC) value inferred from solar-system data (Geiss & Gloeckler 1998), to the abundance (Linsky 1998) in the Local Interstellar Cloud (LIC). Assuming the high-z values as typical of proto-galactic clouds 13 Gyr ago, the PSC value as typical of the LISM at the time of the Sun formation 4.5 Gyr ago, and the LIC value as typical of the LISM at the present epoch, the plotted trend traces the evolution of D in the solar neighbourhood during the Galaxy lifetime.

Ten years ago, Tytler and collaborators had just demonstrated that the cases of high-z absorbers where D/H ratios almost an order of magnitude higher had been claimed to exist (e.g. Songaila *et al.* 1994) were actually misinterpretations of interloopers or of the continuum level in the observed spectra. Chemical evolution models able to reproduce the vast majority of the Galaxy observed properties were nicely consistent with these D/H data (coloured lines in Fig. 1), including the value 9 times lower than the local ones measured by Lubowich *et al.* (2000) in the Galactic center, $(D/H)_{GC} = 1.7 \pm 0.3$ ppm. People felt reassured and satisfied.

There were actually voices of warning (Vidal-Madjar *et al.* 1998), arguing that D/H in the LISM appeared to vary significantly from one line-of-sight (LOS) to the other, but the establishment tended to disclaim those arguments, although admitting that the constant values were actually confined within the quite small region (100 pc radius) of the Local Bubble (LB). The dashed and dotted vertical bars in Fig. 1 show respectively the likely range of LISM values and the less likely one including the maximum, possibly wrong, D/H = 4 ppm ever estimated (see Vidal-Madjar *et al.* 1998 for references).

2. What do we currently know of the local D abundance?

In ancient Greece mythology Cassandra's profecies were never believed either by the Trojans or by the Achaeans, but she was always right. So were the skeptics about the homogeneity of LISM deuterium. If we look at the distribution of D/H with column density N(HI) (which can be considered a proxy for distance), as resulting now from years of analyses of both old and new data (see Fig. 2, taken from Linsky *et al.* 2006), it is apparent that D/H varies significantly. Only the data within the LB are tight to the D/H value which was attributed to the LISM ten years ago. Regions with log N(HI) $\geqslant$ 20.7 seem to have much lower D/H (lower D or higher H ?) and regions with intermediate column densities show an impressive spread, which is now convincingly explained in terms of spatially varying D depletion onto dust grains (Jura 1982, Draine 2004, Linsky *et al.* 2006). So, what value, if any, could actually be taken as the current "average" LISM deuterium? Is it the upper undepleted value 23 ± 2.4 ppm of the intermediate region, or a lower value (say 20 ± 1 ppm) allowing for observational errors, or the very low value 9.8 ± 1.9 ppm of the highest column density regions? Different authors (Linsky *et al.* 2006, Prodanovich *et al.* 2009, Hébrard *et al.* 2005, respectively) have suggested each of these possibilities and the question is still open.

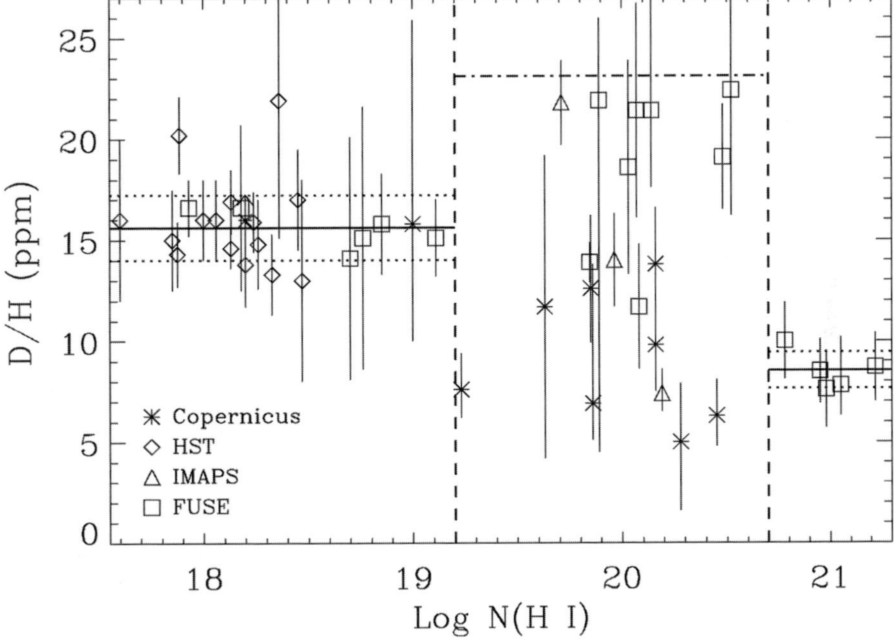

Figure 2. From Linsky *et al.* (2006): deuterium abundance vs hydrogen column density measured along several LOS in the LISM.

Each of these possibilities also has side effects on our understanding of the chemical evolution of the Galaxy. If the LISM D/H is as high as 23 ± 2.4 ppm, at face value it is higher than in the PSC (21 ± 5 ppm, Geiss & Gloeckler 1998) and implies either a way to enhance D in the last 4.5 Gyr (quite unlikely) or the need to consider the Sun not representative of the local medium at the time of its formation (a fairly recurrent theme, never settled with satisfactory arguments). Moreover, if the primordial D/H is $(D/H)_P = 26.1 \pm 3$ ppm as implied by the first modelling of the WMAP data (Spergel *et al.* 2003) the total deuterium astration factor from the Big Bang to a present LISM D/H $= 23$ ppm would be 1.13, quite low even for chemical evolution models allowing for the continuous accretion of large amounts of primordial extragalactic gas, but still consistent with the Galaxy properties (Steigman *et al.* 2007). Once we consider that the infalling gas is probably not primordial and more likely with a D content similar to that of high velocity clouds (Sembach *et al.* 2004 estimated a D/H $= 22 \pm 7$ ppm in Complex C), it becomes clear that no viable model can account for such values. If $(D/H)_P = 28.2$ as recently suggested by Pettini *et al.* (2008), the astration factor to D/H $= 23$ ppm would be 1.22, still challenging, but not impossible to achieve. Models consistent with the various Galactic constraints predict astration factors larger than 1.3 (e.g. Romano *et al.* 2006, Steigman *et al.* 2007) and can reproduce a LISM D/H of 19–20 ppm or lower with standard assumptions on metal poor infall. Most likely one should consider spatial variations of depleting dust grains and of the accreted metal-poor gas to account for the empirical inhomogeneity of the D abundances.

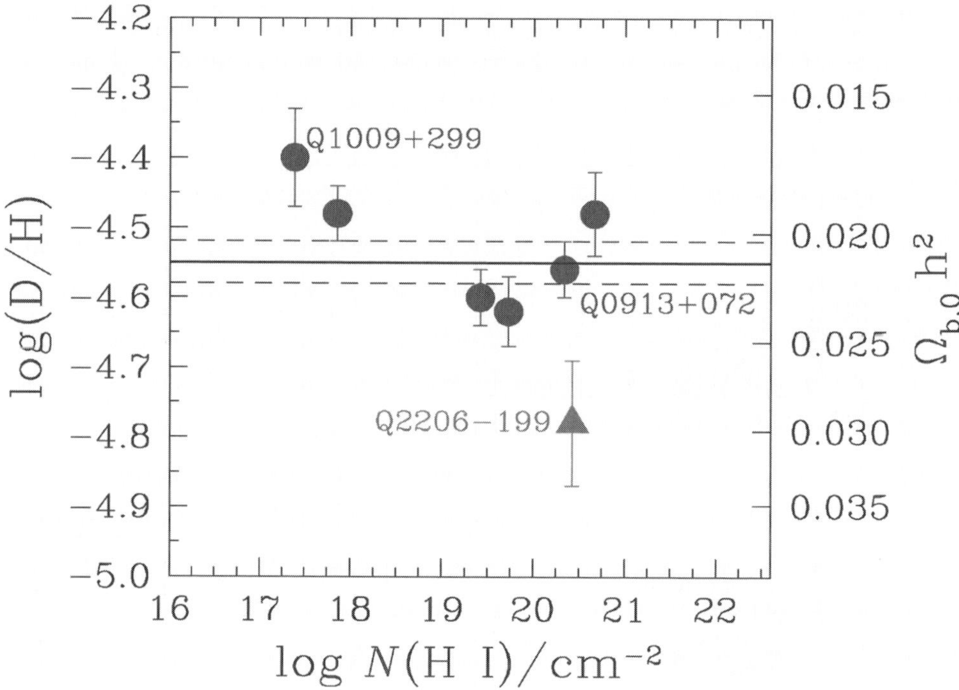

Figure 3. From Pettini *et al.* (2008): The deuterium abundance measured in high-redshift DLAs. The horizontal lines show their mean value (solid) $\pm 1\sigma$ (dashed). On this scale the primordial D inferred from WMAP is -4.59.

3. What do we currently understand of high-redshift deuterium?

Outside our Galaxy deuterium is measured in the absorption lines of gas systems falling on the LOS of QSOs with redshift lower than 4. Fig. 3 (taken from Pettini *et al.* 2008) illustrates these data and the corresponding mean deuterium value, 28.2 ppm. All measurements are consistent at 1σ with this mean, except two values, which happen however to be the most uncertain ones, one because of the lower S/N of its spectra (the red triangle in Fig. 3) and the other because D is detected only in the Ly-α line.

Should this mean value of 28.2 ppm be taken as the best estimate of the primordial D/H? At face value, it is slightly larger than – although consistent with – the values $(D/H)_P = 26.1\pm3$ ppm, or $(D/H)_P = 25.7\pm1.5$ ppm estimated from WMAP after the first year and the third year data release (Spergel *et al.* 2003 and Spergel *et al.* 2007, respectively). If real, this difference would be unexplainable, because DLAs have very low but non zero metals. Since metals are produced by stars, and stars always destroy D, DLAs are supposed to have D somewhat lower than primordial. If the DLA measurements can be considered sufficiently robust, Pettini's heuristical and backward approach of using the observed high-z D/H as a prior in the analysis of WMAP data appears very reasonable. Can we follow his arguments and conclude that $(D/H)_P$ is more likely 28.2 ppm?

If we consider both the deuterium and oxygen abundances of these systems, we find that DLAs with higher O may also have higher D, contrary to basic nucleosynthesis principles. Is this also due to observational issues or should we worry? And, more generally, are the apparent differences in D/H from one absorption system to the other real? If so, what is the physical meaning of the high-z values different from the average value?

4. Discussion

First of all, during the discussion session the participants, sollicited by Ken Sembach and me, agreed that one cannot assume the LISM D value as representative of the value for the Galaxy as a whole, or even for the local spiral arm. It is not clear whether we can actually identify a mean LISM value, but it is evident that such value could not be representative of other Galactic regions with different conditions and evolution.

The quest for the actual D abundances here and now (LISM), and there and then (high-z absorbers) was then debated at length.

Jeff Linsky noticed that in many papers authors implicitly assume that a measurement of D/H in the gas phase is a measure of the total D/H ratio, and urged authors to make a clear distinction between the gas phase D/H and the total D/H that include D in dust grains. Following up on this point, Donatella Romano asked if, on the other hand, there could be undetected molecular hydrogen along some FUSE LOS. If so, could the contribution from this unaccounted H_2 help to bring the highest observed values of D/H in the local ISM in agreement with the predictions on D evolution from standard chemical evolution models? Linsky answered that all H_2 was included in the estimate of log N(H) and that, in any case, it provides only a negligible fraction of the total H column density.

Gary Steigman described in some detail his and Tijana Prodanovic' attempts on using both D and Fe to try to find the "true" (i.e., gas plus dust) ISM D abundance. Based on $\log(y_D)$ vs. $\log(y_{Fe})$ plots [where $y_D = 10^5(D/H)$ and $y_{Fe} = 10^6(Fe/H)$], he argued that the LB D abundance shows no evidence for any variation, in contrast to the LB Fe abundance, suggesting that D may not be depleted in there (i.e., no correlation between D and Fe). Most of the non-LB LOS have lower Fe abundances than in the LB, and for most of them there is an apparent correlation between D and Fe. However, there are 6 remaining non-LB LOS which have D abundances in excess of the LB value, and he suggested that these high D and (mostly) low Fe abundances might have resulted from incompletely mixed infall, so that the "true" ISM D abundance might be the LB value $y_{D,LB} = 1.5$, corresponding to a D astration factor of 1.8. An alternative interpretation of the data is that the "true" ISM D abundance is the value ($y_{D,ISM} = 2.0$) found by Prodanovic *et al.* (2009) from Bayesian analysis. The potential problem with this choice is that the LB D abundance is uniform but depleted compared to the ISM value by a factor of $2.0/1.5 = 1.3$. In response to this speculation that the high D abundances might be due to infall, Guillaume Hébrard remarked however that the oxygen abundance along these LOS is "normal", and not low as would be expected for metal-poor gas.

Sembach asked what future observations of D or other species might help resolve the issue of deuterium depletion onto dust, but no encouraging answer was provided by the audience, either in terms of currently usable instrumentation or of tracing D through other elements. He himself did not see many future prospects for FUV DI absorption measures for gas within the Milky Way for two reasons. First, there is no obvious follow-on to FUSE in NASA's strategic plan (HST/COS could make some breakthroughs in the low-reshift IGM arena, however, with very interesting output). Second, within the Milky Way these types of measurements are confusion limited. Thus, it may be that nature simply doesn't provide a simple enough velocity structure along the vast majority of extended sight lines to be able to infer D/H or D/O much beyond a kiloparsec from the Sun. Higher resolution ($R \geqslant 50,000$) observations might help in understanding this limitation better and may help push the distance envelope a factor of two or three.

Tom Bania replied reporting on the first solid detection of the D hyperfine transition at 327 MHz in the Milky Way ISM (Rogers *et al.* 2005, Rogers *et al.* 2007). Unfortunately, the array is now dismissed and there are no plans to either repeat or extend these measurements. In contrast to all other D/H abundance determinations these measurements are not made toward specific targets but rather probe specific Galactic directions. The three fields are all in the Galactic anticenter direction and latitude 0 deg. The average D/H abundance derived for these three fields is $<D/H> = 21 \pm 7$ ppm, where the error is $\pm 3\sigma$ and contains an estimate for the uncertainty in the HI excitation temperature. In one direction, D/H = 24 ppm is found at a SNR of 8.2 with a total integration time of 17.5 years. No one is going to supersede these 92 cm measurements any time soon. Bania and coworkers estimate that D is distributed over $\sim$5 kpc in these Galactic anti-center directions, with no significant difference between the three. This makes this path length as long as any of the measured optical paths for D LOS in the Milky Way. For this $\sim$5 kpc path the HI column density is $3 \times 10^{21} \mathrm{cm}^{-2}$. Sembach commented that, while the FUSE results apply over a very limited range of distances, the measures of the DI 92 cm emission might have more applicability in defining an "average" Galactic value, since those types of measurements can potentially sample large volumes of the Galaxy.

With reference to the two questions raised before, (1) how much molecular hydrogen is likely to be found in these LOS, and (2) is this a gas phase measurement only, Bania argued that each of these LOS probes DI emission from an enormous volume of the ISM, unlike the pencil beams absorbing the continuum radiation from a background target object. At a distance of 5 kpc the beam extends to ± 610 pc from the Galactic plane. This is $\sim$15 times the 40 pc scale-height of the molecular gas disk of the Milky Way. Thus most of the volume probed by the DI measurements should be devoid of molecular gas. There should be, on average, very little molecular hydrogen contamination on the corresponding D/H abundance derivation. The possibility that significant amounts of D has frozen out onto grains cannot be discounted, but Bania believes that their LOS may actually be the cleanest available in this regard, because most of their probed lies far above the dense, cold material in the Galactic plane. Most of their observed volume contains warm neutral medium (WNM) gas at temperatures between 1,000 and 4,000 K. Grains mixed into this environment are unlikely to be cold enough to freeze out D. Hence, the fraction of observed volume wherein D can freeze out onto grains is too small to significanly skew their D/H abundance determination.

Given the uncertainties in accurately deriving D/H, Guillaume Hébrard argued that a comparison of D with O may be more reliable. He pointed out that the D/O ratio appears much more homogeneous than D/H. He thus suggested that the distant low D/O and D/H ratios are more likely representative of the local current values, whereas the high D/H for which no high D/O are measured could be due to systematics. Linsky agreed that it is important to estimate D/H by different techniques, but believes that estimating D/H through another element like oxygen (D/H = D/O×O/H) introduces a whole new set of poorly understood problems in addition to the problems in understanding D/H. One needs to know the ionization equilibrium, depletion, and perhaps also the distribution of the element along the LOS which will be different than for D and H. Also, chemical evolution, depletion, and mixing could be different for this element than for H and D.

Chris Howk mentioned that for Galactic measurements, there is almost certainly some bias due to the manner in which sight lines were chosen for analysis from the FUSE database. Given the work required for each individual sight line, there is a tendency to choose sight lines with visible deuterium absorption. This means the DI is separated from the HI absorption, but it potentially also biases the results against low D sight lines. How large a bias this represents is unknown. Sembach confirmed that the sight lines sampled by FUSE for both D/H and D/O measurements have strong selection effects. These are kinematically simple sight lines that probe the warm diffuse ISM in which there is little molecular hydrogen. Dust grains in these types of regions show no evidence of icy mantles. These types of regions tend to have mantles that are highly processed by shocks. These regions are not representative of the types of regions where grains mantles are built. Rather, they are regions grains inhabit after they leave darker cloud environments. He wondered if anyone has done (or will do) a systematic assessment of the FUSE archive to determine if there are any sight lines for which DI absorption could have been detected but wasn't. In other words, are there any "D-free" sight lines that might be telling us something interesting? Nobody seemed able to answer this question.

Sembach also pointed out that a correlation between the inferred depletion of deuterium and heavier elements such as Fe and Ti is not surprising, but one should be puzzled by how little variation there is about the trend. Given that Fe or Ti are depleted by factors of 100-1000, compared to at most a factor of 2 for D, even a small change in the mantle properties should produce a large change in the gas-phase abundance of Fe or Ti. In other words, one could double the gas-phase abundance of Ti or Fe easily by releasing only 1 part in 100 or 1000 of the Fe or Ti back into the gas, whereas a doubling of the D abundance would require essentially all of the D to be released back into the gas phase. The observed data seem at odds with there being a simple distribution of D and the heavier elements within dust grains if this is indeed the explanation for the variations in the D/H ratio. These arguments made him skeptical, as they suggest there could still be considerable unknown (systematic) uncertainties associated with the values of the HI column densities along some of these sight lines.

Finally, Sembach recalled that, although difficult to test observationally, some of the LOS to LOS variation in D/H could be due to local sources of the type discussed during the conference. These sources are insufficient to produce cosmological quantities of D, but he saw no particular reason they couldn't add to the variation in deuterium abundances seen in different directions. The importance of such sources remains indeed to be determined.

To summarize, it is apparent that there is no consensus on which is the actual deuterium abundance of the LISM, or even on whether a representative value can exist. Clearly further studies are necessary to better understand this issue, but adequate instruments are currently missing. As pointed out by Linsky, we need a new spectrograph for the 912-1200 A spectral range, with higher sensitivity and spectral resolution than FUSE. This would allow a deeper insight on the local abundance distribution as well as D/H measurements beyond 1 kpc providing information on the dependence of D/H with radial position in the disk and halo of the Galaxy and in neighboring galaxies like the Magellenic Clouds. Radio arrays able to measure the ground-state spin-flip transition of D at 327 MHz (92 cm) in several Galactic longitudes, to add to those described by Rogers *et al.* (2007) could be useful too.

Concerning high-z absorbers, Joanna Dunkley argued that, when comparing results for the deuterium abundance from WMAP and the Pettini *et al.* (2008) measurements, it is worth noting that there is a less than 2σ difference between the two. Since they are statistically consistent, at this stage it may not be worth investing much time in choosing one or the other. However, to try to compare the two measurements, one should consider how robust the WMAP result is. WMAP has very low systematics, leading to a precise measurement of the angular power spectrum. Inferring the baryon density from the power spectrum also relies on linear physics that we understand. We can numerically compute the expected spectrum for a given cosmological model to high precision, so the estimated density should not include additional uncertainties. There is a small dependence of the estimated baryon density on the cosmological model assumed; for example, the estimate moves by about 1σ if we extend the standard ΛCDM cosmological model to marginalize over a running spectral index, or a varying equation of state of dark energy (Komatsu *et al.* 2009). However, there is currently no evidence that these extended models are favored over ΛCDM. The final step, i.e. inferring the D abundance from the baryon density, assumes the standard Big Bang model. While this is consistent with the CMB observations, there could be extensions to this model to be invoked to explain the apparent discrepancy in the lithium abundance (discussed in Jedamzik's talk, this volume), that could modify this inference.

Finally, Paolo Molaro remarked that after almost two decades of 10m telescope efforts we remain with only 8 measurements. Moreover, these measurements show a dispersion which exceeds the reported errors, thus suggesting either the presence of a scatter in the D/H or an underestimate of the errors. He recalled two aspects to consider. First, the measurements are obtained towards absorbers with neutral hydrogen column densities differing by more than two orders of magnitude, which normally are referred to different classes of objects. Given that the major source of error is the estimate of the hydrogen column density, he would regard the determination obtained towards damped wings of the hydrogen lines in the DLAs somewhat more reliable than in the other systems. A second aspect is that these systems show low metallicities but within a quite large range $-3 \leqslant [\text{Si/H}] \leqslant -1$. At these metallicities he believes that no significant astration can have occurred and there is no hint of correlation between D/H and metallicities. However, considering that the measurements in the Galaxy show a correlation with the Fe abundances, suggesting a depletion of D into dust grains, it would be important to check if any correlation exists between the D/H abundance and the gas depletion factor. Unfortunately, a robust estimate of depletion can be obtained only from Zn abundances, which are not available for the observed systems: certainly a check to be done in more detail in future observations.

Acknowledgements

The Swiss National Science Foundation and IAU are gratefully acknowledged for financial support. Partial support comes also from the Italian PRIN-MIUR-2007. I'm grateful to Max Pettini for useful and updated information, and to Tom Bania, Joanna Dunkley, Guillaume Hébrard, Chris Howk, Jeff Linsky, Paolo Molaro, Donatella Romano, Ken

Sembach, and Gary Steigman for their active participation to the discussion and for their contribution to this report. Fruitful and pleasant meetings and discussions on this topic held at ISSI in Bern are also acknowledged.

References

Burles, S. & Tytler, D. 1998, *in Primordial Nuclei and their Galactic evolution*, N. Prantzos, M. Tosi, R. von Steiger eds., *Space Science Reviews* 84, 65

Draine, B. T. 2004, *in Origin and evolution of the elements*, A. McWilliams, M. Rauch eds, CUP, p.317

Geiss, J. & Gloeckler, G. 1998, *in Primordial Nuclei and their Galactic evolution*, N. Prantzos, M. Tosi, R. von Steiger eds., *Space Science Reviews* 84, 239

Hébrard, G., Tripp, T. M., Chayer, P., Friedman, S. D., Dupuis, J., Sonnetrucker, P., Williger, G. M., & Moos, H. W. 2005, *ApJ*, 65, 1136

Jura, M. 1982, *in Advances in UV astronomy*, Y. Kondo, J. Mead, R. D. Chapman eds., NASA, p. 54

Komatsu, E., Dunkley, J. *et al.* 2009, *ApJS*, 180, 330

Linsky, J. L. 1998, *in Primordial Nuclei and their Galactic evolution*, N. Prantzos, M. Tosi, R. von Steiger eds., *Space Science Reviews* 84, 285

Linsky, J. L., Draine, B. T., Moos, H. W., Jenkins, E. B. *et al.* 2006, *ApJ* 647, 1106

Lubowich, D. A., Pasachoff, J. M., Balonek, T. J., Millar, T. J., Tremonti, C., Roberts, H., & Galloway, R. P. 2000, *Nature*, 405, 1025

O'Meara, J. M., Burles, S., Prochaska, J. X., Prochter, G. E., Bernstein, R. A., & Burgess, K. M. 2006 *ApJ*, 649, L61

Pettini, M., Zych, B. J., Murphy, M. T., Lewis, A., & Steidel, C. C. 2008, *MNRAS*, 391, 1499

Prodanovic, T., Steigman, G., & Fields, B. D. 2009, arXiv:0910.4961

Rogers, A. E. E., Dudevoir, K. A., Carter, J. C., Fanous, B. J., Kratzenberg, E., & Bania, T. M. 2005, *ApJ*, 630, L41

Rogers, A. E. E., Dudevoir, K. A., & Bania, T. M. 2007, *AJ*, 133, 1625

Romano, D., Tosi, M., Chiappini, C., & Matteucci, F. 2006, *MNRAS*, 369, 295

Savage, B. D., Lehner, N., Fox, A., Wakker, B., & Sembach, K. 2007, *ApJ*, 659, 1222

Sembach, K. *et al.* 2004, *ApJS*, 150, 387

Songaila, A., Cowie, L. L., Hogan, C. J., & Rugers, M. 1994, *Nature*, 368, 599

Spergel, D. N. *et al.* 2003, *ApJS*, 148, 175

Spergel, D. N. *et al.* 2007, *ApJS*, 170, 377

Steigman, G., Romano, D., & Tosi, M. 2007, *MNRAS*, 378, 576

Tosi, M. 2000, *in The light elements and their evolution*, IAU Symp.198, *ASP Conf.Ser.*, L. DaSilva, M. Spite, J. R. do Medeiros eds., p. 525

Vidal-Madjar, A., Ferlet, R., & Lemoine, M. 1998, *in Primordial Nuclei and their Galactic evolution*, N. Prantzos, M. Tosi, R. von Steiger eds., *Space Science Reviews* 84, 297

Monica Tosi & Angela Bragaglia

Georges Meynet & Urs Frischknecht

Light Elements in the Universe
Proceedings IAU Symposium No. 268, 2009
C. Charbonnel, M. Tosi, F. Primas, & C. Chiappini, eds.

© International Astronomical Union 2010
doi:10.1017/S1743921310004011

What is ^{4}He from H II regions? What needs to be done to better understand the systematic errors?

Gary J. Ferland[1], Yuri Izotov[2], Antonio Peimbert[3], Manuel Peimbert[3], Ryan L. Porter[4], Evan Skillman[5] and Gary Steigman[6]

[1] Physics, University of Kentucky, Lexington KY 40506, USA,
email: gary@pa.uky.edu

[2] Main Astronomical Observatory, 27 Acad. Zabolotnoho St, UA Kyiv 03680 Ukraine
email: izotov@mao.kiev.ua

[3] Instituto de Astronomia, UNAM Apartado Postal 70-264 Mexico 04510, D.F., Mexico,
email: antonio@astroscu.unam.mx peimbert@astroscu.unam.mx

[4] Astronomy, University of Michigan, Ann Arbor, MI USA
email: ryanlporter@gmail.com

[5] Dept. of Astronomy, U of Minnesota, 116 Church St. SE, Minneapolis, MN 55455
email: skillman@astro.umn.edu

[6] Depts. of Physics and Astronomy, The Ohio State University, Columbus, OH 43210. USA
email: steigman.1@osu.edu

Abstract. Here we summarize the discussion that took place after the presentation of papers discussing measurement of abundances from emission lines in H II regions. A cosmological test using ^{4}He is harder than those using the other light elements due to the small dynamic range over which He/H varies for cosmologically interesting parameters. A precision of much better than 1% is needed. This precision challenges both observations and theory. Several of the major uncertainties were identified and methods of improving the accuracy were described.

Keywords. ISM: abundances, HII regions, Galaxy: abundances

1. Introduction

Ferland began the discussion with Don Osterbrock's reminder of how hard this problem is. Don's 1961 PASP paper (Osterbrock & Rogerson 1961) was the first to suggest that much of the current helium originated in the creation of the universe. He recounts the events leading to that paper in *The Helium Content of the Universe* (Osterbrock 2009). He concludes with the statement

> I have heard lectures, and seen cosmological papers, in which values of X and Y derived from nebular spectrophotometry are quoted and used to three significant figures. But all the available calculations of the H I and He I emission-line intensities that I know are based on simplified models. . . . The "mean" values may not represent these conditions to high accuracy . . .

From my own personal knowledge, Don thought that there was roughly a 10% error on He/H determinations from emission lines. I also fear that I was the one who gave the talk Don mentions in his review.

We need to identify the error sources that affect He/H determinations in H II regions, quantify their size, and establish priorities for reducing them to the level needed for precision cosmology. There are five broad sources of error – observing methods, correcting for stellar absorption beneath the emission lines, converting emission-line intensities into ionic abundances, converting ionic abundances into total abundances, and, finally, correcting for stellar contributions to the current helium content. These are enumerated in Davidson & Kinman (1985) and Peimbert *et al.* (2003). Unfortunately many of the errors are systematic rather than statistical and so are hard to quantify.

The discussion was organized into small presentations by active researchers. Each discussed what they saw as the major problems and the way to make progress. The session was then opened to the audience. The following summarizes this discussion.

2. Comments by Gary Ferland

Ryan Porter and I have worked on aspects of the physics of helium emission-line formation on the microscopic level. This includes the atomic physics of the line formation and the physical properties of the ionized gas where He I and H I lines form.

Porter *et al.* (2005) revisited the formation of He I lines under the Case B approximation (Osterbrock & Ferland 2006, hereafter AGN3) conditions. We found emissivities that differ by, in some cases, several percent from the previous high-accuracy calculation by Benjamin *et al.* (1999). These were mainly due to the treatment of non-hydrogenic small-l levels where improved atomic data or methods had become available since the publication of Benjamin *et al.* The critical atomic rates, which have significant remaining uncertainties, are the photoionization cross sections (used to get the radiative recombination coefficient from the Milne relation, AGN3) and l-changing collisions for small-l levels.

Porter *et al.* (2009) did an analysis of the error sources that enter into the calculation of the theoretical emissivities. We found that the photoionization cross sections and l-changing collisions for small-l levels introduce a significant uncertainty. Paradoxically, because of how the lines form, the stronger lines, like $\lambda 5876$, can have the greater uncertainty. An abundance analysis which relies on only a handful of the stronger lines will have a significant error, as much as several percent, introduced by these uncertainties. It must be stressed that this will be a systematic error that will introduce spurious conclusions if the uncertainties in the atomic physics are not taken into account.

The best way to quantify the uncertainties is to carefully examine a bright H II region where many He I lines can be measured. We did this in Porter *et al.* (2007), using one of the deepest published spectra of the Orion Nebula ever obtained. We found that the observational errors, as indicated by line ratios that were set by the atomic physics, were much larger than was given in the original paper. The intensities of 22 of the highest quality He I lines are consistent with an optimized model of the Orion environment and an observational error of 3.8% percent. The strongest He I line, $\lambda 5876$, a line with one of the larger theoretical uncertainties in the emissivity, was matched to within 6.6%. The conclusion was that the uncertainties were substantially larger than originally estimated.

What to make of this? Don's guess of a 10% uncertainty in He/H may be realistic. My opinion is that we should understand Orion first. This is the brightest and simplest H II region we know of, close to the Earth and predominantly ionized by a single star. The giant H II region 30 Dor should come next. It is the nearest and brightest object that is a close kin to the galaxies used to derive the primordial helium abundance. It would be good if independent data sets could be obtained and compared to form an impartial estimate of the observational errors. Finally high-precision photoionization cross sections and collisional rates are needed to improve the accuracy of the theoretical emissivities.

3. Comments by Yuri Izotov

Underlying stellar absorption is one of the most important systematic effects that should be taken into account for the primordial ^{4}He determination. Recently, Gonzalez Delgado *et al.* (2005) produced evolutionary stellar population synthesis models at high resolution for single stellar populations. These data allows one to take into account properly the underlying hydrogen and helium line absorption in spectra of H II regions. Another important source of systematic errors are collisional and fluorescent enhancements of He I emission lines. Recently, Porter *et al.* (2005) calculated new He I emissivities in the case B approximation using improved radiative and collisional data. However, for the practical use, simple fits of the emissivities including the contribution of the collisional and fluorescent excitation would be very useful.

I think that one or a few bright H II regions are not enough to derive the primordial ^{4}He abundance. The sample should be large enough to reduce the observational uncertainties. In particular, the spectrum of the bright H II region NGC 346 in the Small Magellanic Cloud mentioned by G.Steigman was obtained with a high signal-to-noise ratio. However, this object is extended and serious problems arise with the subtraction of the night sky foreground. Therefore, it is unlikely to reach very good accuracy in the determination of Y for this object. Bright compact sources are more preferable.

Evan Skillman presented a non-parametric method for corrections of the systematics using HeI lines themselves. Potentially, this method could be one of the best for estimation of the systematic errors. However, I am not very happy to see large error bars in Y values, the large spread of points and the absence of a correlation between helium mass fraction and oxygen abundance. Perhaps, some other constraints should be considered to reduce the large error in Y_p. In particular, the electron temperature and electron number density could be derived from the Balmer jump. Furthermore, larger samples could be considered in order to reduce statistical errors.

4. Comments by Antonio & Manuel Peimbert

By considering that the temperature structure of a gaseous nebula can be represented by the average temperature, T_0, and the mean square temperature variation, t^2, it is possible to a second approximation to determine observationally the temperature structure of a gaseous nebula.

For the 5 metal poor extragalactic H II regions, used by Peimbert, Luridiana, & Peimbert (2007) to derive Y_p, they obtained an average $< t^2 >$ of 0.026. With this value they derive $Y_p = 0.2477$, while by adopting $t^2 = 0.00$ they obtain $Y_p = 0.2523$, a difference of 0.0046. This difference is critical for deriving an accurate Y_p value.

What is the evidence in favor of $< t^2 > = 0.026$? The typical t^2 values derived from photoionization models like CLOUDY are in the 0.003 to 0.010 range. Many astronomers, mainly based on photoionization models and the simplicity of the equations for abundance determinations for constant temperature, adopt $t^2 = 0.000$ to derive chemical abundances. We consider that the observations of high quality that provide t^2 values higher than 0.020 are reliable and that the presence of temperature variations has to be considered in the abundance determinations.

To determine T_0 and t^2 you need to combine line intensities that have a different dependence on the electron temperature. There are at least seven methods that have been used in the literature to determine T_0 and t^2. The four most commonly used are: (a) the combination of [O III] collisionally excited lines with O II recombination lines, (b) the combination of [C III] collisionally excited lines with C II recombination lines,

(c) the combination of collisionally excited lines of [O II] and [O III] with the ratio of the Balmer continuum to a Balmer recombination line, and (d) the combination of about 10 He I lines with [O II] and [O III] collisionally excited lines to obtain t^2, N(He I), $\tau(3889)$, and Y simultaneously.

There are four well-observed objects where the four methods mentioned above have been used simultaneously, two H II regions and two planetary nebulae: the Orion nebula, 30 Doradus, NGC 5315, and NGC 6543; the derived $< t^2 >$, values for each object are 0.022 ± 0.002, 0.033 ± 0.005, 0.051 ± 0.004, and 0.028 ± 0.005, respectively. For any given object each of the four methods gives a t^2 in agreement with the average value. This result gives a very strong support to each of the methods, and to the existence of large temperature variations in each of these objects.

The presentation by Peimbert, Peimbert, Carigi, and Luridiana, in these proceedings, includes some observational evidence supporting the large t^2 value (0.036) derived for the galactic H II region M17.

5. Comments by Evan Skillman

My main comment was that the current work on He abundances in extragalactic H II regions indicates that the mean temperature (as indicated by the fluxes of the He I emission lines) is significantly lower than the temperature derived from the [O III] lines. This result is significant beyond the determination of the primordial He abundance, as a lower temperature implies higher abundances of the heavy elements in these regions. Since the uncertainty in the temperature in these H II regions is a dominant source of uncertainty, due to the degeneracy between temperature and density in the minimization, it is very important to sort this out.

Izotov *et al.* (2007) justify the assumption of low temperature fluctuations (and thus the suitability of the [O III] temperature) based on agreement between [O III] temperatures and temperatures derived from the Balmer decrement in a sample of blue compact dwarfs. However, deriving electron temperatures from the Balmer decrement is intrinsically difficult and the history of this technique in the literature is not consistent. I consider this question to be the major uncertainty in the determination of helium abundances in individual nebular objects.

6. Comments by Gary Steigman

In my opinion the dominant sources of uncertainty in the helium abundance determinations are systematic, not statistical. Neither the statistical error nor the uncertain extrapolation to zero metallicity limit the accuracy of our current estimate of the primordial helium abundance and the cosmological constraints that follow from it. To illustrate the significance of a concerted observational (and theoretical) effort to limit the systematic errors, I pointed out the value of a few (one!) good helium abundance determinations.

In my invited review talk I noted that if the value of the baryon density parameter inferred from the inferred primordial deuterium abundance is combined with that derived from the WMAP CMB observations, the estimate of the effective number of neutrinos is $N_{\text{eff}} = 4.0 \pm 0.7$, which is within 1.4σ of the standard model value of 3. In contrast, if we simply adopt as an **upper limit** to the primordial helium abundance the Peimbert, Luridiana, Peimbert (2007) estimate of the helium abundance (and its uncertainty) for the SMC H II region NGC 346, $Y_{\text{SMC}} = 0.2507 \pm 0.0042$ (where their statistical and systematic errors have been added linearly), and combine this with the observationally inferred primordial deuterium abundance, we would find an **upper bound** to the effective number of neutrinos, $N_{\text{eff max}} \leqslant 3.2 \pm 0.3$. The key point of this example is to illustrate that if the PLP error estimate is realistic, the **uncertainty** in N_{eff} may be reduced by more than a factor of two. A few more carefully observed and analysed H II regions like NGC 346 could go a long way in pinning down the value of the primordial helium abundance and tightening the constraints that follow from it.

7. Conclusions

The discussion illuminated several sources of uncertainty in the precision analysis of emission lines emitted by H II regions. A variety of approaches were suggested as ways to improve the accuracy of the derived abundances. It would be productive if all were attempted. Gary Steigman's comment that even a very few high-precision measurements would be cosmologically significant established a realizable goal. Clearly much work, both observational and theoretical, remains to be done.

References

Benjamin, Robert A., Skillman, Evan D., Smits, & Derck, P. 1999, *ApJ*, 514, 307

Davidson, K. & Kinman, T. D. 1985, *ApJS*, 58, 321

Gonzalez Delgado, R. M., Cervio, M., Martins, L. P., Leitherer, C., & Hauschildt, P. H. 2005, *MNRAS*, 357, 945

Izotov, Y., Thuan, T., & Stasinska, G. 2007, *ApJ*, 662, 15

Osterbrock, D. E. & Ferland, G. J. 2006 *The Astrophysics of Gaseous Nebulae and Active Galactic Nuclei*, 2nd Edition (Mill Valley; University Science Press) (ANG3)

Osterbrock, D. E. & Rogerson, J. B. 1961, *PASP*, 73, 129

Osterbrock, D. E., 2009, in *Finding the Big Bang*, P. J. E. Peebles, L. A. Page & R. B. Partridge eds (Cambridge; Cambridge University Press), p. 86

Peimbert, Manuel, Luridiana, Valentina & Peimbert, Antonio 2007, *ApJ*, 666, 636

Peimbert, Manuel, Peimbert, Antonio, Luridiana, Valentina, & Ruiz, Maria Teresa 2003, *ASPC*, 297, 81

Porter, R. L., Bauman, R. P., Ferland, G. J., & MacAdam, K. B. 2005, *ApJ*, 622, L73

Porter, R. L., Ferland, G. J., & MacAdam, K. B. 2007, *ApJ*, 657, 327

Porter, R. L., Ferland, G. J., MacAdam, K. B., & Storey, P. J. 2009, *MNRAS*, 393L, 36

Roberto Costa

Piercarlo Bonifacio

Light Elements in the Universe
Proceedings IAU Symposium No. 268, 2009
C. Charbonnel, M. Tosi, F. Primas, & C. Chiappini, eds.

© International Astronomical Union 2010
doi:10.1017/S1743921310004023

He-rich and He-poor populations in RGB stars. Results on a sample of 19 globular clusters

Angela Bragaglia[1], Valentina D'Orazi[2], Raffaele Gratton[2], Eugenio Carretta[1], Santi Cassisi[3] and Sara Lucatello[2]

[1]INAF-Osservatorio Astronomico di Bologna,
via Ranzani 1, 40127 Bologna, Italy
email: `angela.bragaglia@oabo.inaf.it`, `eugenio.carretta@oabo.inaf.it`

[2]INAF-Osservatorio Astronomico di Padova,
vicolo Osservatorio 5, 35122 Padova, Italy
email: `valentina.dorazi@oapd.inaf.it`, `raffaele.gratton@oapd.inaf.it`,
`sara.lucatello@oapd.inaf.it`

[3]INAF-Osservatorio Astronomico di Collurania,
via M. Maggini, I-64100, Teramo, Italy
emal: `cassisi@oa-teramo.inaf.it`

Abstract. We use about 1400 red giant branch stars observed in 19 Galactic Globular Clusters (GCs) to compare colours, metallicities, and RGB bump luminosities of stars assigned to first and second generations. We find subtle differences which we attribute to the different He content. In general these differences are visible only when we consider the extreme second generation stars, with the exception of NGC 2808. When using various indicators, the implied helium enhancements are similar, but the absolute calibration is still uncertain.

Keywords. Stars: abundances, Population II – Galaxy: globular clusters

---

1. He differences from globular clusters red giant stars

We exploit the unprecedented database of about 1400 stars on the Red Giant Branch (RGB) observed with FLAMES@VLT in 19 Galactic Globular Clusters (GCs) in the course of our project on the Na-O anticorrelation. For a description of our survey and for results see Carretta *et al.* (2009a), Carretta *et al.* (2009b); for a description of the influence of He, see Gratton *et al.* (2010) and Bragaglia *et al.* (2010).

Stars with different He are expected to have different temperatures (i.e., different colours), slightly different [Fe/H] values, and different magnitudes of the RGB bump. All these differences are small, but our study has the necessary precision, statistics, and homogeneity to detect them.

We derive the differences in colours, metallicities and RGB bump magnitudes in the Primordial (P), Intermediate (I) and Extreme (E) population (Carretta *et al.* 2009a), which are associated with different levels of pollution and correspond to different He enhancements, as is expected in a scenario of intra-cluster pollution by material processed through hot H-burning (e.g. CNO cycle) in a previous generation of stars. We find that P stars show significant differences in colour and metallicity only compared to E stars (see Fig. 1, left panels). If we interpret this as an effect of He, these differences imply a $\Delta Y \sim 0.05$-0.10. NGC 2808 is an exception, since differences are visible both between P and I stars and between P and E stars (indicating a $\Delta Y \gtrsim 0.10$-0.12).

A. Bragaglia *et al.*

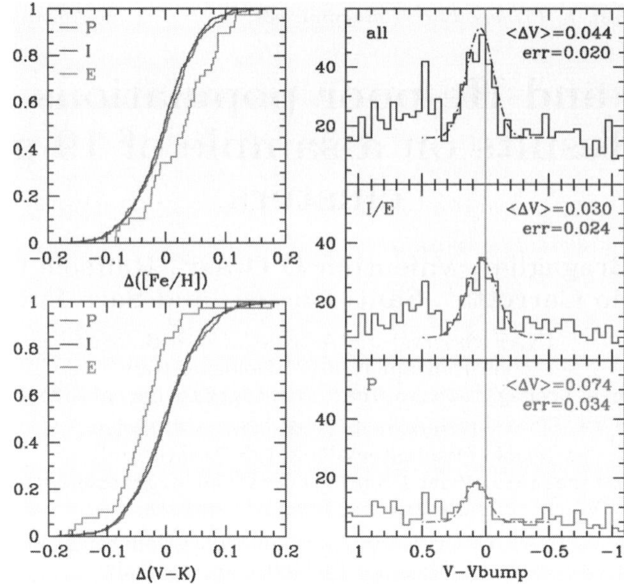

Figure 1. Left panels: Cumulative distributions of the differences in metallicity (upper panel) and in colour (lower panel) for the three populations: the P and I ones are similar, while the P and E ones are significantly different. Right panels: Histograms of the difference in magnitude between individual V values and V_{bump} (for each cluster), zoomed near the RGB bump. We show for al stars (upper panel), second-generation I and E, and first-generation P stars (middle and lower panel, respectively), with gaussians fitting each bump. In each panel the peak of the bump and the associated error are indicated.

We find that the level of the RGB bump defined by first-generation P stars is fainter than the one defined by second-generation I and E stars (see Fig. 1, right panels). The interpretation in terms of different He is complex. We explore two cases with different mixtures of heavy elements: one with "normal" α-enhancement (Pietrinferni *et al.* 2006) and one that takes into account the observed anticorrelations in the chemical abundances observed in GCs (Pietrinferni *et al.* 2009). We obtain results in fair agreement with those obtained with the other methods.

We conclude that *i*) RGB stars in GCs show indication of different He levels, *ii*) that different methods produce similar results, but that *iii*) the absolute calibration of the derived ΔY is still uncertain.

AB is grateful to the IAU and the Swiss National Science Foundation for financial support. SL acknowledges the DFG cluster of excellence "Origin and Structure of the Universe" for support.

References

Bragaglia, A., Carretta, E., Gratton, R., D'Orazi, V., Cassisi, S., & Lucatello, S. 2010, *A&A*, subm.

Carretta, E., Bragaglia, A., Gratton R. G., Leone, F., Recio-Blanco, A., & Lucatello, S. 2006, *A&A*, 450, 523

Carretta, E., *et al.* 2009a, *A&A*, 505, 117

Gratton, R. G., Carretta, E., Bragaglia, A., Lucatello, S., & D'Orazi, V. 2010, *A&A*, subm.

Pietrinferni, A., Cassisi, S., Salaris, M., & Castelli, F. 2006, *ApJ*, 642, 797

Pietrinferni, A., Cassisi, S., Salaris, M., Percival, S., & Ferguson, J. W. 2009, *ApJ*, 697, 275

Light Elements in the Universe
Proceedings IAU Symposium No. 268, 2009
C. Charbonnel, M. Tosi, F. Primas, & C. Chiappini, eds.

© International Astronomical Union 2010
doi:10.1017/S1743921310004035

Helium abundances in inner Galaxy planetary nebulae

Oscar Cavichia[1,2], Roberto D. D. Costa[1], and Walter J. Maciel[1]

[1]IAG, University of São Paulo, 05508-900, São Paulo-SP, Brazil
[2]email: cavichia@astro.iag.usp.br

Abstract. New helium abundances of planetary nebulae located towards the bulge of the Galaxy were derived, based on observations made at OPD (Brazil). We present accurate helium abundances for 56 PNe located towards the galactic bulge. The data show good agreement with other results in the literature, in the sense that the distribution of the abundances is similar to previous works. Furthermore, the radial helium gradient is extended towards the galactic center. The results show that no trend can be identified when comparing the internal gradient (R $\leqslant$ 4 kpc) to the whole galactic disk.

Keywords. ISM: planetary nebulae: general; Galaxy: abundances, evolution

1. Introduction

Planetary nebulae (PNe) are an important tool to study the chemical abundances in the Galaxy. As an offspring of intermediate mass stars, their helium abundances suffer from contamination from the evolution of the progenitor star. Even when the contamination of the progenitor star is considered, the existence of a He/H gradient in the galactic disk is very uncertain (see Maciel 2000). The derived He/H gradients from PNe abundances are negligible, and can be written as $d(He/H)/dR = 0.0000 \pm 0.0004$ (Maciel 2000).

Since bulge and disk may have formed in different ways such as represented by the disk inside-out formation model (Chiappini *et al.* 2001), or the model of multiple infalls onto the bulge (Costa *et al.* 2008), we would expect that these differences should appear in the abundance distributions of these structures. This way, in the present work we have extended the radial He/H gradient of the disk (R $\geqslant$ 4 kpc) towards the galactic center, in order to compare the He/H gradient of both regions. This was done by using a new set of helium abundances of PNe located towards the galactic bulge derived by our group.

2. Method

Spectrophotometry observations in the optical domain were made at OPD observatory (Brazil) for a sample of 56 planetary nebulae located in the direction of the galactic bulge. The data were reduced following standard reduction procedures with the IRAF software (see Escudero *et al.* 2004). Details of the observation and data reduction procedures are described by Cavichia *et al.* (2009). The disk sample consists of 73 PNe compiled from our group (see for example Costa *et al.* 2004 and references therein).

In order to study the He/H gradient, PNe distances were taken from Stanghellini *et al.* (2008). The heliocentric distances were converted to galactocentric distances adopting $R_\odot = 8$ kpc. Additionally, new distances were derived for 46 PNe using the statistical distance scale proposed by Stanghellini *et al.* (2008), and based on the 5 GHz radio flux and optical angular diameter. In a first approach, no correction was considered for the contamination of the progenitor star evolution on the original helium abundance.

3. Results and discussion

The He/H radial gradient is shown in Figure 1. For R $\leqslant$ 4 kpc we addopted two different binnings in order to minimize the effect of the bin selection on the radial gradient.

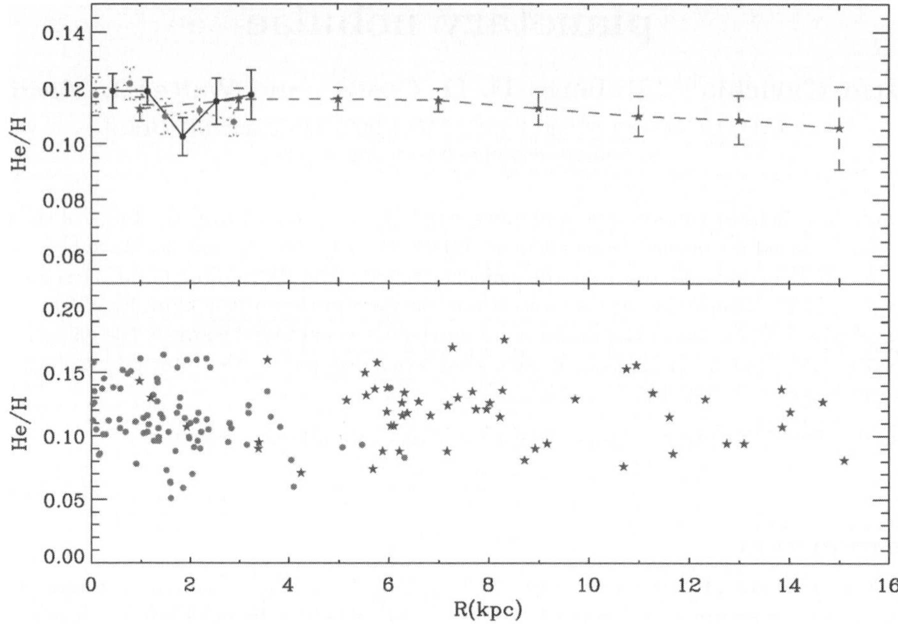

Figure 1. Top panel: the helium abundance as a function of the galactocentric distance for the bulge/inner disk sample (continuous and dotted lines represent the two different binnings) and for the disk sample (dashed line). For each bin the mean and the standard error of the mean (error bars) are given. Bottom panel: the radial distribution of helium abundances without considering mean abundances. Circles represent bulge/inner disk data and stars represent disk.

It can be seen in the figure that the He/H gradient for R $\leqslant$ 4 kpc is negligible, as it is in the outer part of the disk. As a conclusion, more reliable abundances and distances, especially in the inner region of the Galaxy, are necessary in order to investigate the radial dependence of the helium abundance.

References

Cavichia, O., Costa, R. D. D., & Maciel, W. J. 2009, *Rev. Mexicana AyA*, (Submitted)
Chiappini, C., Matteucci, F., & Romano, D. 2001, *ApJ*, 554, 1044
Costa, R. D. D., Uchida, M. M. M., & Maciel, W. J. 2004, *A&A*, 423, 199
Costa, R. D. D., Maciel, W. J., & Escudero, A. V. 2008, *Baltic Astronomy*, 17, 321
Escudero, A. V., Costa, R. D. D., & Maciel, W. J. 2004, *A&A*, 414, 211
Maciel, W. J. 2000, in: L. da Silva, R. de Medeiros, & M. Spite (eds.), *The Light Elements and their Evolution*, Proc. IAU Symposium No. 198, p. 204
Stanghellini, L., Shaw, R. A., & Villaver, E. 2008, *ApJ*, 689, 194

Light Elements in the Universe
Proceedings IAU Symposium No. 268, 2009
C. Charbonnel, M. Tosi, F. Primas, & C. Chiappini, eds.

© International Astronomical Union 2010
doi:10.1017/S1743921310004047

Primordial helium abundance of the SMC: a view from intermediate mass stars

Roberto D. D. Costa[1], Walter J. Maciel[1] and Thais E. P. Idiart[1]

[1]IAG, University of São Paulo,
Rua do Matão 1226, 05508-090, São Paulo/SP, Brazil
email: `roberto@astro.iag.usp.br`, `maciel@astro.iag.usp.br`, `thais@astro.iag.usp.br`

Abstract. Helium abundances were derived for a sample of planetary nebulae of the Small Magellanic Cloud. These abundances were corrected from the chemical evolution of the galaxy as well as from stellar nucleosynthesis, and the primordial helium abundance then was estimated for the sample. Results indicate that the resulting average value for the sample is consistent with the expected values for primordial helium from SBBN and other values derived from HII regions in different stellar systems, even varying the enrichment ratio within its uncertainty range.

Keywords. ISM: planetary nebulae; stars: abundances; galaxies: Magellanic Clouds

1. Introduction

We derived the chemical composition for a sample of planetary nebulae of the Small Magellanic Cloud. Helium abundances were estimated as accurately as possible, including correction of collisional effects using collision-to-recombination correction factors. Based on luminosities and effective temperatures derived for the central star of each nebula, the masses, and then ages of the progenitor stars were estimated using isochrones and mass-age relationships. Using these results, helium abundance for each nebula was corrected for the contamination from the evolution of their progenitor star as well as from the chemical evolution of the interstellar medium, by using yields for intermediate mass stars and a helium-to-heavy elements abundance ratio. Therefore we determined the helium enrichment in the SMC and used these results to estimate the pregalactic helium abundance of this galaxy.

2. Results and discussion

The observational sample consisted in 49 PNe of the SMC. Observations were made at ESO/La Silla and OPD/Brazil. The observation and analysis procedures are described by Idiart *et al.* (2007). Primordial helium was estimated for each nebula using the expression

$$Y_P = Y - \frac{\Delta Y}{\Delta Z}Z - \Delta Y_S$$

where Y was derived from the helium abundance, Z was estimated from the chemical abundances, ΔY_S are the helium yields from van den Hoek & Groenewegen (1997) and $\Delta Y/\Delta Z = 2.0 \pm 0.6$ was taken from Peimbert *et al.* (2007).

The results are in Table 1, that shows the average values for primordial helium derived from the expression above, using the described PNe sample and excluding objects with neutral helium, without available yield and type-I PNe. The primordial abundances were calculated for $\Delta Y/\Delta Z = 2.0 \pm 0.6$. The error is mainly due to uncertainties in the mass determination of the progenitor stars, that were derived from their effective temperatures and luminosities using isochrones and mass-age relationships.

Table 1. Average values for the primordial helium abundance.

$\Delta Y/\Delta Z$	$\overline{Y_P}$	Error
2.0	0.234	0.009
2.6	0.234	0.009
1.4	0.241	0.008

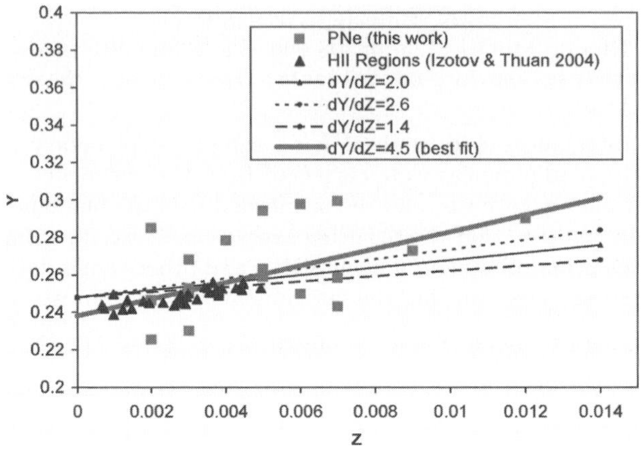

Figure 1. Y vs Z relation derived from our data, compared to a sample of HII regions in blue compact galaxies.

Figure 1 shows the Y vs Z relation derived from our data, compared to a sample of HII regions in blue compact galaxies, that are usually adopted to derive primordial helium abundances, taken from Izotov & Thuan (2004). It can be seen that, despite their larger dispersion, the data from PNe fit well the abundance distribution of HII regions. The thin lines correspond to $\Delta Y/\Delta Z = 2.0 \pm 0.6$, using the theoretical value of $Y_P = 0.248$ (Steigman 2007) as the $Z = 0$ value. The thick line corresponds to the best fit for both samples combined, which resulted in $\Delta Y/\Delta Z = 4.5$. It is important to note that this steeper value is consistent with that reported by Izotov & Thuan (2004): $\Delta Y/\Delta Z = 3.7 \pm 1.2$. The figure also shows that, despite their different Z ranges, enrichment ratios derived from HII regions and PNe are compatible.

From this result it is possible to see that He/H abundances derived from planetary nebulae can be used to study helium abundances throughout a stellar system like the SMC, and the resulting Y_P values calculated as described above are consistent with the expected values for primordial helium from SBBN and other values derived from HII regions in different stellar systems, even varying the enrichment ratio within its uncertainty range. Moreover, our results indicate that despite their metallicities are distributed in different Z ranges, enrichment ratios of HII regions and PNe are compatible, although a larger sample of PNe is required to improve these results.

References

Idiart, T. P., Maciel, W. J., & Costa, R. D. D. 2007, *A&A*, 472, 101
Izotov, Y. I., Thuan, T. X. 2004, *ApJ*, 602, 200
Peimbert, M., Luridiana, V., Peimbert, A., & Carigi, L. 2007, *ASPC*, 374, 81
Steigman, G. 2007, *Ann.Rev.Nuc.Part.Sci.* 57, 463
van den Hoek, L. B., & Groenewegen, M. A. T. 1997, *A&AS*, 123, 305

Light Elements in the Universe
Proceedings IAU Symposium No. 268, 2009
C. Charbonnel, M. Tosi, F. Primas, & C. Chiappini, eds.

© International Astronomical Union 2010
doi:10.1017/S1743921310004059

The helium spread among the stars of 47Tuc

Marcella Di Criscienzo

INAF-Osservatorio Astronomico di Roma,
Monte Porzio Catone, Roma 00100, Italy
email: dicrisci@oa-roma.inaf.it

Abstract. We show that the current data of the HB branch of 47 Tuc show a particular feature that cannot be explained if a single population with an unique mechanism of mass loss is considered. We find that a spread in helium abundance among the stars is necessary, of ~ 0.02. We indicate that the same variation in helium is present among the sub giant branch stars and suggest that is responsible of the spread in luminosity of the bright sub giant branch, while only a small part of the second generation is characterized by C+N+O increase and gives the faint sub giant branch.

Keywords. globular clusters: individual (47 Tuc); stars: horizontal-branch, evolution, abundances

1. Introduction

Spectroscopic data show chemical anomalies in most of 47 Tucanae stars consisting in bimodal CN band strengths (Briley 1997) and Na-O anticorrelations(Carretta *et al.* 2009).

These results indicate that the abundance spread in 47 Tuc is not due to an evolutionary effect, rather to the existence of an original stellar population (first generation, FG) and of a second generation (SG). Stars in the SG formed from material processed through the hot CNO cycle in the progenitors, belonging to the FG, but not enriched in the heavy elements expected in supernova ejecta, according to todays' new paradigm for the formation of galactic GCs (D'Ercole *et al.* 2008 and reference therein).

The recent photometric data by Anderson *et al.* (2009) seem to confirm the existence of multiple populations in 47 Tuc. In particular, they found a splitted sub giant branch (SGB) with at least two distinct components: a brighter one with a small and real spread in magnitude (~ 0.06 mag) and a second one containing about 10% of star a little (~ 0.05 mag) fainter.

Our idea is that CN-strong stars all belong to the SG, and were formed from gas showing the signature of CNO processing and a helium enrichment, but that only a small percentage of SG stars are increased in the overall CNO content.

2. The results

Following the suggestion of D'Antona & Caloi (2008) we decided to investigate this hypothesis using the morphology of the HB of 47 Tuc, where a population with a larger helium abundance should emerge. In Figure 1 we show a zoom of the CMD of 47 Tuc at the HB level (filled circle) based on images of 5 sec in F606W and F814W from ACS (GO-10775, PI Sarajedini).

In particular we note that at $m_{F606W}-m_{F814W} \sim 0.675$ mag there is a particular feature in the data: a step in luminosity between the blue and brighter and red and dimmer part

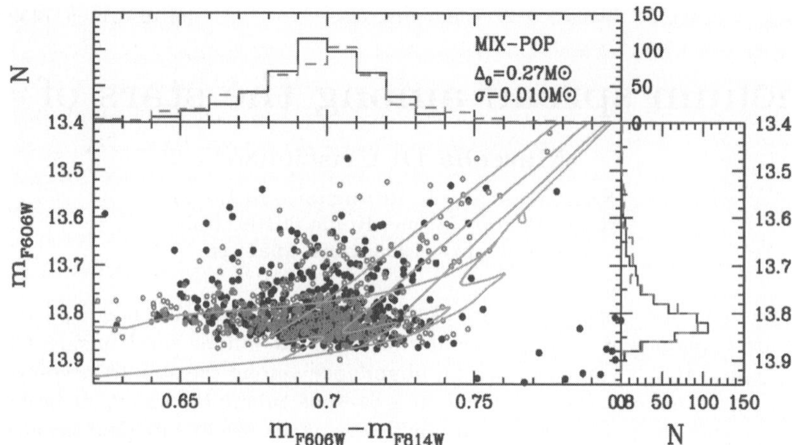

Figure 1. Comparison between observed (filled circle) and synthetic HB stars (open circle-magenta in the electronic version) calculate as described in the text considering an helium spread among stars of $\Delta Y \sim 0.02$. The dashed histograms refer to synthetic populations while solid ones are those relative observations.

of HB. Our simulations show that a single mechanism mass loss can't reproduce this feature, while if we consider two different populations with two different helium contents it is possible to do a good job. The solid lines in the Figure are ZAHBs calculated from our models and reported in the observational plane using a reddening E(606-814)=0.038 mag. An apparent distance modulus $(m–M)_{F606W}=13.09$ mag was chosen in order to fit a ZAHB of pristine Y(=0.25) to the lower envelope. ZAHB for Y=0.28 is also reported together with HB evolutionary tracks of stars with masses M=0.63, 0.66, 0.70 and 0.80 solar masses.

Open circles are synthetic stars obtained using a gaussian mass loss around the central value $\Delta M_0=0.27 M_\odot$ and $\sigma=0.010 M_\odot$. The synthetic population is built under the hypothesis that 70 % of stars have a N(Y) constant between Y=0.25 and Y=YUP=0.27. Other details of the model are reported in Di Criscienzo *et al.* (2010, in preparation). The two histograms for both magnitudes and colours show the comparison between the data and our simulations. The most interesting result is that the same variation in helium N(Y) can also explain the spread of the bright SGB observed by Anderson *et al.* 2009 while the faint SGB is made of stars with higher C+N+O (for details see Di Criscienzo *et al.* 2010).

We interpret these results as the confirmation that SG in 47 Tuc consist of about 70 % of stars, as suggested by spectroscopic studies.

References

Anderson, J.; Piotto, G.; King, I. R.; Bedin, L. R.; & Guhathakurta, 2009, 2009, *ApJ*, 697, L58
Briley, M., 1997, *AJ*, 114, 1051
Carretta, E.; Bragaglia, A.; Gratton, R. G. *et al.*, 2009, *A&A*, 505, 117
D'Antona, F. & Caloi, V. 2008, *MNRAS*, 390, 693
D'Ercole, A., Vesperini, E., D'Antona, F., McMillan, S. L. W., & Recchi, S. 2008, *MNRAS*, 391, 825

Light Elements in the Universe
Proceedings IAU Symposium No. 268, 2009
C. Charbonnel, M. Tosi, F. Primas, & C. Chiappini, eds.

© International Astronomical Union 2010
doi:10.1017/S1743921310004060

Lithium and proton-capture elements in globular clusters: the case of 47 Tucanae

Valentina D'Orazi[1], Sara Lucatello[1,2], Raffaele Gratton[1], Angela Bragaglia[3], and Eugenio Carretta[3]

[1]INAF-Osservatorio Astronomico di Padova, vicolo dell'Osservatorio 5, I-35122, Padova
email: `valentina.dorazi@oapd.inaf.it`, `sara.lucatello@oapd.inaf.it`,
`raffaele.gratton@oapd.inaf.it`

[2]Excellence Cluster Universe, Boltzmannstr. 2, D-85748, Garching, Germany
[3]INAF-Osservatorio Astronomico di Bologna, via Ranzani 1, I-40127, Bologna
email: `angela.bragaglia@oabo.inaf.it`; `eugenio.carretta@oabo.inaf.it`

Abstract. The determination of lithium (Li) abundances in Globular Clusters (GCs), along with proton-capture elements (Na, O, Mg, Al), offer a key tool to address the pollution scenario and its mechanisms, the dilution process acting within each star and the first phases in the lifetime of GCs. We present our results on Na, O and Li abundance determination in a large sample of dwarf stars in the GC 47 Tucanae (NGC 104). While we found a clear Na-O anti-correlation, in perfect agreement with giant members by Carretta *et al.* (2009a,b), Li abundance appears neither positively correlated with oxygen, nor anti-correlated with sodium. Our finding unveils an intrinsic scatter in Li content, independent of intra-cluster pollution by a first generation of more massive, faster evolving stars.

Keywords. Stars: abundances – Galaxy: globular clusters: individual: 47 Tuc

1. Introduction

The traditional assumption of GCs as simple stellar populations has been proven too simplistic. In the last years a lot of observational studies revealed the presence of anti-correlations between C, O, Mg and N, Na, Al, respectively, pointing out to self-enrichment processes within GCs, and to the existence of multiple stellar generations. In this context, Li abundances can provide very powerful tracers of cluster formation and evolution, because this element is easily destroyed in the stellar interiors. Regardless of the nature of the polluters (AGB stars -Ventura & D'Antona 2009- or fast rotating massive stars -Decressin *et al.* 2007), the Li content in polluting matter must be very close to zero (although in some cases AGB stars can also produce Li). The determination of Li variations in GCs allows us to uniquely constrain the nature and the extent of the dilution process between pristine and polluted material. Specifically two main issues can be addressed: (1) Do entirely polluted (Li$\simeq$0) stars exist? (2) Is the minimum detectable Li the same in all GCs or, otherwise, can it significantly vary from cluster-to-cluster? The latter point has also strong implications on the GC formation and early evolution. Here we present our results on the metal-rich ([Fe/H]=-0.76, Carretta *et al.* 2009c) GC 47 Tuc.

2. Sample and analysis

We retrieved from the ESO Archive (Programme 081.D-0287, PI P. Shen) FLAMES-GIRAFFE spectra of 109 unevolved, turnoff (TO) member stars of 47 Tuc. Gratings HR15n, HR18 and HR20a were employed to cover the Li I (6707 Å), O I (7771, 7773,

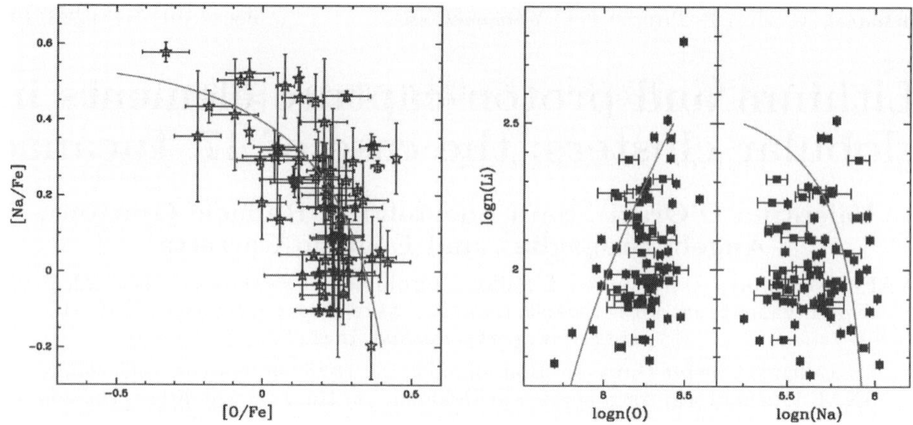

Figure 1. Na-O anticorrelation (left panel) and Li-Na-O distributions in dwarf stars of 47 Tuc (right panels).

7774 Å), and Na I (8183, 8194 Å) features, respectively. The analysis was carried out using the ROSA code (developed by R. Gratton) and Kurucz model atmospheres; for Na and O, we derived abundances from the equivalent widths, applying corrections due to NLTE effects; LTE Li abundances were instead obtained through spectral synthesis.

3. Results and discussion

As shown in Fig. 1, the unevolved cluster stars reveal a very clear Na-O anti-correlation, confirming what has been obtained for giant members (Carretta *et al.* 2009a). The Na-O anti-correlation is well described by a simple dilution model (red solid line in Fig. 1), like the one proposed by Prantzos & Charbonnel (2006). On the other hand, Li abundances show a wide scatter, in both the Li-O and Li-Na planes, with values often well below those expected from a simple dilution model. Furthermore, the cluster behaviour seems different from that of NGC 6397 (Lind *et al.* 2009), a GC much more metal-poor than 47 Tuc ([Fe/H]=-1.99, Carretta *et al.* 2009c) and for which Lind *et al.* detected a significant Li-Na anti-correlation. Note that turnoff stars in 47 Tuc are cooler ($T_{eff} \sim$5700-5800 K) than those in NGC 6397 ($T_{eff} \sim$6100-6300 K), being well below the limit usually considered for the Spite plateau. Finally, the large scatter observed in 47 Tuc seems reminiscent of the wide Li variations observed in the old open cluster M 67 (Randich *et al.* 2000) and, in general, in old thick disk stars with $T_{eff} \approx$5800K (Ryan *et al.* 2001): this is the first time that this phenomenon is observed in GC stars.

References

Carretta, E., Bragaglia, A., Gratton, R., *et al.* 2009a, *A&A*, 505, 117
Carretta, E., Bragaglia, A., Gratton, R., & Lucatello, S. 2009b, *A&A*, 505, 139
Carretta, E., Bragaglia, A., Gratton, R., D'Orazi, V., & Lucatello, S. 2009c, *A&A*, 508, 695
Decressin, T., Charbonnel C., Prantzos, N. & Ekstrom, S. 2007, *A&A*, 464, 1029
Lind, K., Primas, F., Charbonnel, C., Grundahl, F., & Asplund, M. 2009, *A&A*, 503, 545
Prantzos, N. & Charbonnel, C. 2006, *A&A*, 458, 135
Randich, S., Pasquini, L., & Pallavicini, R. 2000, *A&A*, 356, 25
Ryan, S., Kajino, T., & Beers, T. 2001, *ApJ*, 549, 55
Ventura, P. & D'Antona, F. 2009, *A&A*, 499, 835

Light Elements in the Universe
Proceedings IAU Symposium No. 268, 2009
C. Charbonnel, M. Tosi, F. Primas, & C. Chiappini, eds.

© International Astronomical Union 2010
doi:10.1017/S1743921310004072

The Galactic deuterium gradient

Donald Lubowich[1] and Jay M. Pasachoff[2]

[1] Dept. of Physics and Astronomy, Hofstra University,
Hempstead, NY 11549
email: donald.lubowich@hofstra.edu

[2] Dept. of Astronomy, Williams College
Williamstown, MA 01267
email: jay.m.pasachoff@williams.edu

Abstract. The Galactic deuterium abundance gradient has been determined from observations of DCN in Galactic molecular clouds. This is the only way to observe D throughout the Galaxy because the molecular clouds are not limited to the 2 kpc region around the Sun observed with FUSE and from DI. We used an astrochemistry model and the DCN/HCN ratios to estimate the underlying D/H ratios in 16 molecular clouds including five in the Galactic Center. The resulting positive Galactic D gradient and reduced Galactic Center D/H ratio imply that there are no significant Galactic sources of D, there is continuous infall of low-metallicity gas into the Galaxy, and that deuterium is cosmological.

Keywords. Galaxy: abundances; ISM: molecules, abundances; radio lines: galaxies

1. Introduction

Deuterium has been extensively studied because it is not produced via stellar nucleosynthesis and is thought to be primarily produced in the big-bang so its abundance will decrease with time and metallicity unless there are any additional sources of D. The abundance of D depends on the temperature and baryonic density during the epoch of nucleosynthesis (first 3 min). Thus any Galactic source of D would undermine its use to estimate the baryonic density of the universe and place constraints on big-bang nucleosynthesis models. In homogeneous inflationary or other flat models, the D/H ratio gives the amount of dark matter and an upper limit to the number of ν families.

"Independently of the model for chemical evolution, it is predicted that the D abundance (D/H) increases with Z if D is produced in stars, but decreases as Z increases if it is primordial: these correlations hold as a function of time and for gradients of abundances at the present time. A firm observation of the correlation of D/H with Z (e.g., a gradient in the Galaxy) would therefore tell whether D is mostly made during galactic evolution before most metals were synthesized; this test could not distinguish between production in the Big-Bang and production in the first generation of stars." (Audouze & Tinsley 1976). If D is produced via any stellar or Galactic process, then the D/H ratio would be a maximum in the GC. Conversely, if there are no Galactic sources of D, then astration would reduce the D abundance in the GC to 3×10^{-12} (Audouze *et al.* 1976).

2. Observations and results

We observed DCN and $HC^{15}N$ in 16 Galactic molecular clouds (including five Galactic Center clouds). $HC^{15}N$ was observed because the HCN line is saturated. Thus DCN/HCN $= (DCN/HC^{15}N)(^{15}N/^{14}N)$. As in Lubowich *et al.* (2000) a model with 5300 chemical reactions was used to determine the underlying D/H ratios. Our results (a positive gradient

Table 1. D/H ratios in Galactic Molecular Clouds

Source	R (kpc)	D/H (ppm)
CND	0.002	< 1
Sgr A 20 km/s	0.01	2.4 0.7
Sgr A 50 km/s	0.01	2.3 0.6
GC Arc G0.13-0.13	0.035	2.5 0.5
Sgr B2	0.100	2.9 0.9
W33	4.2	9.82.4
G34.26	5.3	4.00.8
M17	5.9	2.81.1
DR21(OH)	8.1	7.10.7
Ori KL	8.2	8.72.2
NGC 2024	8.2	9.22.0
S140	8.3	143.5
NGC 2264	8.8	7.01.4
NGC 7538	9.4	173.9
W3(OH)	9.8	205.1
Edge 2 cloud	20	< 32

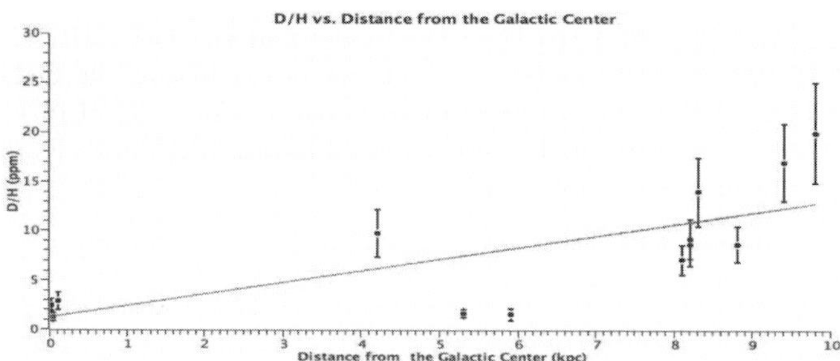

Figure 1. The Galactic Deuterium Abundance Gradient.

in the D/H ratio) are shown in Table 1 and Fig. 1 (compare to Pasachoff & Vidal-Madjar 1989). Our detection of DCN in GC molecular clouds shows D within 10-100 pc of the GC. The Sgr A clouds are 10^7 years old and contain reprocessed gas from recent star formation. Since fractionation enhances deuterated molecules, these results confirm our previous GC observations that there are no significant Galactic sources of D and that D is the result of infall of primordial gas. The low D/H ratios of 4 ppm and 2.8 ppm at 5.3 and 5.9 kpc may reflect astration in spiral arms. However, the D/H = 9.8 ppm in W33 at 4.2 kpc is not explained. At 10 kpc our results are in agreement with ISM FUSE (Linsky *et al.* 2006, 23 ppm), anticenter DI (Rogers, Dudevoir, & Bania 2007, 21 ppm, and low-metallicity halo-cloud (Sembach *et al.* 2004, 22 ppm,) D/H ratios. Better models of grain depletion, infall, astration, and astrochemistry are needed to explain the similar Galactic, quasar, and WMAP D/H ratios (26 ppm, Table 1, Linsky *et al.* 2006) while allowing large variation in the abundances of other elements. These observations are strong evidence that D is cosmological with no other significant sources of D.

References

Audouze, J., Lequeux, J., Reeves, H., & Vigroux, L. 1976, *ApJ* (Letters), 208, L51
Audouze, J. & Tinsley, B. M. 1976, *ARAA*, 14, 43
Linsky, J. L. *et al.* 2006, *ApJ* 647, 1106
Lubowich, D. A., Pasachoff, J. M. *et al.* 2000, *Nature*, 405, 1025
Pasachoff, J. M. & Vidal-Madjar, A. 1989 *ComAp*, 14, 61
Rogers, A. E. E., Dudevoir, K. A., & Bania, T. M. 2007, *AJ* 133, 1625
Sembach, K. R. *et al.* 2004, *ApJS* 150, 387

Light Elements in the Universe
Proceedings IAU Symposium No. 268, 2009
C. Charbonnel, M. Tosi, F. Primas, & C. Chiappini, eds.

© International Astronomical Union 2010
doi:10.1017/S1743921310004084

Helium abundances in planetary nebulae: Nucleosynthesis and chemical evolution

W. J. Maciel, R. D. D. Costa and T. E. P. Idiart

University of São Paulo, Astronomy Department,
Rua do Matão 1226, Cidade Universitária, São Paulo SP, CEP 05509-0900, Brazil
email: maciel@astro.iag.usp.br, roberto@astro.iag.usp.br, thais@astro.iag.usp.br

Abstract. We have obtained a large sample of PN with accurately determined helium abundances, as well as abundances of several heavy elements. The nebulae are located in the solar neighbourhood, in the galactic bulge, disk and anticentre, and in the Magellanic Clouds. The abundances are analyzed both in terms of the nucleosynthesis of intermediate mass stars and the chemical evolution of the host galaxies. In particular, correlations between the He/H ratio and the abundances of N and O are used as constraints of the nucleosynthetic processes occurring in the progenitor stars.

Keywords. ISM: planetary nebulae: general; Galaxy: abundances

1. Introduction

Helium abundances can be accurately measured in photoionized nebulae, comprising planetary nebulae (PN) and HII regions. The abundances measured in PN include the original helium content previous to the formation of the progenitor stars and the contamination during the nuclear processes in these objects. As a consequence, PN can be used to study the nucleosynthetic processes in intermediate mass stars and the chemical evolution of the host systems. In this work, we analyze the helium abundances of a large sample of PN in different systems, namely, the disk and bulge of the Milky Way (MW), and both the Small (SMC) and Large (LMC) Magellanic Clouds.

2. Average abundances and abundance distributions

We have analyzed several samples of PN with elemental abundances obtained in a homogeneous way. We have considered our own data, the IAG sample (see Maciel *et al.* 2009 and references therein, and Cavichia *et al.*, in preparation), plus some recent data from the literature (Stasińska *et al.* 1998, SRM, Leisy & Dennefeld 2006, LD). Maciel *et al.* (2009) made a detailed comparison of these results concerning the heavy elements, and concluded that the merged sample maintains the homogeneity of the individual samples, in view of the similar methods employed. Average helium abundances are in the range $0.093 \leqslant$ He/H $\leqslant 0.119$. These averages are similar in all systems, within the uncertainties, but the SMC abundances are slightly lower than either the LMC and MW, being similar to the HII region value, as in the Orion nebula or 30 Dor. The MW disk seems slightly richer in He than the bulge, which may be due to a lack of intermediate mass stars near the upper mass limit in the latter. Concerning the He/H distributions, there is a general agreement among the different samples. The peaks of the distributions are in the range He/H $= 0.080$ to 0.100 (SMC) and He/H $= 0.080$ to 0.120 (LMC), while the MW data show a peak between He/H $= 0.120$ and 0.140 for the disk and He/H $= 0.080$ to 0.120 for the bulge.

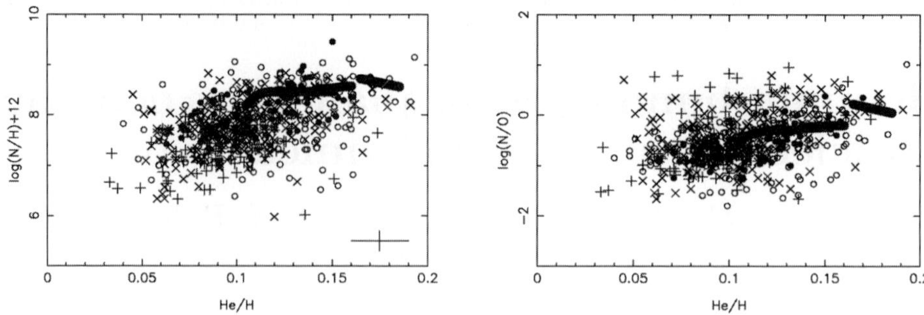

Figure 1. He and N abundances in PN in the Milky Way disk (dots), Bulge (empty circles), SMC (crosses) and LMC (x signs). Thick lines are models for TP-AGB stars with masses in the range 1.1 to 5 $M_\odot$ by Marigo *et al.* (2003).

3. Abundance correlations: Observations and theory

Figure 1 shows the N/H vs He/H and N/O vs He/H plots for the merged sample. A few objects with He/H < 0.03 are not included, as some contribution from neutral helium is probably unaccounted for. It is apparent that all systems have a similar behaviour, with similar results for the N/O ratio, although the metallicity range may be different in these systems. For example, the MW shows a larger range in nitrogen abundances, as expected, since our galaxy is more metal rich than the Magellanic Clouds. However, the evolution of the abundances is consistent with a similar trend for all systems, since the slopes are similar. In Fig. 1 we have included the predictions of theoretical models by Marigo *et al.* (2003), as shown by the thick lines. These are synthetic evolutionary models for the thermally-pulsing Asymptotic Giant Branch stars (TP-AGB) with masses in the range 1.1 to 5 $M_\odot$. In these objects, up to three dredge-up episodes occur, apart from hot-bottom burning (HBB) for the most massive objects. All of these processes affect the He/H ratio, and in fact most objects present some enhancement compared to the solar values. According to Marigo *et al.* (2003), progenitors having 0.9 to 4 $M_\odot$ and solar composition can explain the "normal" abundances, He/H $\leqslant$ 0.15, while for those objects with higher enhancements (He/H > 0.15) larger masses are needed, in the range 4 to 5 $M_\odot$, plus an efficient HBB. The latter are represented in Fig. 1 by the slanted black thick lines at the upper right corner of the figures. It can be seen that the agreement is very good, especially for the N/O ratio. For the N/H ratio the plot shows a better agreement for the Milky Way, as expected, since this galaxy is somewhat more metal rich than the Magellanic Clouds, so that we would expect the theoretical predictions to be located in the upper part of the plot, as is indeed the case. We found a continuous transition between the "normal" and He-richer nebulae, which is probably due to the fact that we have a much larger sample compared to Marigo *et al.* (2003). These results suggest that the nucleosynthetic processes occurring in these systems are similar, even though the global metallicity may be different and the chemical evolution may be affected by different star formation rates.

Acknowledgements. This work was partially supported by FAPESP and CNPq.

References

Leisy, P. & Dennefeld, M. 2006, *A&A*, 456, 451
Maciel, W. J., Costa, R. D. D., & Idiart, T. E. P. 2009, *Rev. Mexicana AyA*, 45, 127
Marigo, P., Bernard-Salas, J., Pottasch, S. R., *et al.*, 2003, *A&A*, 409, 619
Stasińska, G., Richer, M. G., & McCall, M. 1998, *A&A*, 336, 667

Light Elements in the Universe
Proceedings IAU Symposium No. 268, 2009
C. Charbonnel, M. Tosi, F. Primas, & C. Chiappini, eds.
© International Astronomical Union 2010
doi:10.1017/S1743921310004096

Chemical composition of stellar populations in ω Centauri

A. F. Marino[1,2]**, G. Piotto**[1]**, R. Gratton**[3]**, A. P. Milone**[1]**, M. Zoccali**[2]**, L. R. Bedin**[4]**, S. Villanova**[5] **and A. Bellini**[1,4]

[1]Dept. of Astronomy, University of Padova, Vicolo dell'osservatorio 3, 35122, Padova, Italy
email: `anna.marino@unipd.it`
[2]P. Univ. Católica de Chile, Dept. de Astronomía y Astrofísica, Casilla 306, Santiago 22, Chile
[3]INAF-Osservatorio Astronomico di Padova, Vicolo dell'Osservatorio 5, 35122 Padova, Italy
[4]Space Telescope Science Institute, 3700 San Martin Drive, Baltimore, MD 21218, USA
[5] Dept. de Astronomia, Universidad de Concepcion, Casilla 160-C, Concepcion, Chile

Abstract. We derive abundances of Fe, Na, O, α and s-elements from GIRAFFE@VLT spectra for more than 200 red giant stars in the Milky Way satellite ω Centauri. Our preliminary results are that: (i) we confirm that ω Centauri exhibits large star-to-star metallicity variation (~ 1.4 dex); (ii) the metallicity distribution reveals the presence of at least five stellar populations with different [Fe/H]; (iii) a distinct Na-O anticorrelation is clearly observed for the metal-poor and metal-intermediate stellar populations while apparently the anticorrelation disappears for the most metal rich populations. Interestingly the Na level grows with iron.

Keywords. galaxy: stellar content, Globular clusters: individual (Omega Centauri)

1. Introduction

Omega Centauri (ω Cen) is among the most studied and enigmatic Milky Way satellites. It has always been considered a Globular Cluster (GC), but a number of peculiarities, like the mass, the kinematics, and the complexity of its numerous populations identified by both spectroscopic and photometric investigations (Cannon & Stobie 1973, Lee *et al.* 1999, Pancino *et al.* 2000, Bedin *et al.* 2004, Piotto *et al.* 2005, Villanova *et al.* 2007), suggest that it may be the remnant of a larger stellar system.

Recently, it has been shown that some peculiar features observed in ω Cen are shared with other GCs: A bimodality in s-elements is present in NGC 1851 (Yong *et al.* 2008) and M22 (Marino *et al.* 2009), and intrinsic variations in [Fe/H] were detected in M22 (Marino *et al.* 2009), and M54 (Sarajedini & Layden 1995; Bellazzini *et al.* 2008; Carretta *et al.* 2010). However ω Cen still remains a unique GC in the Milky Way as concerns the complexity of its stellar populations.

Here we present preliminary results of our project aimed to characterize the evolutionary connections of the sub-populations in ω Cen, by studying their chemical content.

2. Observations and data reduction

We analyzed FLAMES/GIRAFFE HR09 and HR13 spectra for a sample of more than 200 red giant stars. Iron abundances are obtained from an equivalent width analysis by using the Local Thermodynamical Equilibrium program MOOG (Sneden 1973), while the other elements are measured by comparing observed spectra with synthetic ones. More details on the abundance measurements can be found in Marino *et al.* (2008, 2009).

3. Results

We obtained that the [Fe/H] ranges from ~ -2.1 to ~ -0.7 dex, with at least five distinct stellar groups in the iron distribution as shown in Fig. 1 (left panel). In the top right panels of Fig. 1 we represent Na and O abundances for the five sub-populations, selected on the basis of different iron content and position on the color magnitude diagram (CMD). In the lower panels the position on the B-$(B-R)$ CMD from Bellini *et al.* (2009) for the different selected groups is shown, with the corresponding NaO anticorrelation in each upper panel. We superimposed to each Na-O plane, a fiducial traced by hand on the NaO anticorrelation for stars with $-1.80 < [Fe/H] < -1.34$. We note that the NaO anticorrelation is well defined for stars belonging to the metal intermediate populations $(-1.80 < [Fe/H] < -1.34)$, and some hints of a probably less extended one are present for the more metal poor stars with $[Fe/H] < -1.80$. Apparently, the anticorrelation disappears for stars with $[Fe/H] > -1.00$. Note that the Na level grows with Fe.

From the analysis of the s-elements La and Ba we derive that the s element abundance grows with increasing iron, and interestingly enough, in each group defined as in Fig. 1, apparently the s element abundance increases also with Na.

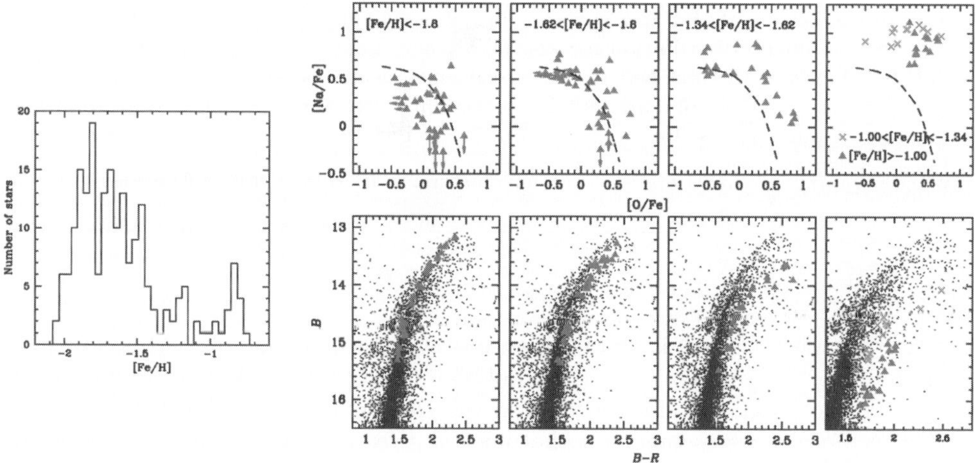

Figure 1. *Left panel*: Distribution of the iron content for our analyzed sample. *Right panels*: Na-O anticorrelation for the different groups of stars selected on the basis of the iron content and the position on the CMD represented in lower panels.

References

Bedin, L. R., *et al.*, 2004, *ApJ*, 605, L125
Bellazzini, M., *et al.*, 2008, *AJ*, 136, 1147
Bellini, A., *et al.*, 2009, *A&A*, 493, 959
Cannon, R. D. & Stobie R. S., 1973, *MNRAS*, 162, 207
Carretta, E., *et al.*, 2010, *ApJL* in press
Marino, A. F., *et al.*, 2008, *A&A*, 490, 625
Marino, A. F., *et al.*, 2009, *A&A*, 505, 1099
Lee, Y.-W., *et al.* 1999, *Nature*, 402, 55
Pancino, E., *et al.*, 2000, *ApJ*, 534, L83
Piotto, G., *et al.*, 2005, *ApJ*, 621, 777
Sarajedini, A. & Layden, A. C., 1995, *AJ*, 109, 1086
Sneden, C. A., 1973, Ph.D. Thesis
Villanova, S. *et al.*, 2007, *ApJ*, 663, 296

Light Elements in the Universe
Proceedings IAU Symposium No. 268, 2009
C. Charbonnel, M. Tosi, F. Primas, & C. Chiappini, eds.

© International Astronomical Union 2010
doi:10.1017/S1743921310004102

On the total O/H abundance ratio in Galactic and extragalactic H II regions

Antonio Peimbert[1] **and Manuel Peimbert**[1]

[1]Instituto de Astronomía, Universidad Nacional Autónoma de México, Apdo. postal 70-264, México D.F. 04510, Mexico
email: antonio@astroscu.unam.mx

Abstract. To determine the primordial helium abundance and to study the chemical evolution of galaxies it is necessary to derive the total O/H ratio in H II regions. To determine the total O/H ratio in H II regions it is necessary to add to the gas-phase component the dust-phase component of O atoms. Based on the Fe/O ratio and other considerations we estimate the dust-phase fraction as a function of O/H.

Keywords. ISM: abundances – H II regions – galaxies: abundances, evolution, irregular – Galaxy: disk – early universe

1. Introduction

In this note we present a preliminary discussion on the fraction of O trapped in dust grains, elsewhere we present a full discussion of this problem (Peimbert & Peimbert 2010a). Mesa-Delgado *et al.* (2009) based on three different methods have estimated that the fraction of O atoms trapped in dust in the Orion nebula amounts to 0.12 ± 0.03 dex.

2. The Fe/O ratio

In Figure 1 we present the Fe/O versus O/H ratio compiled from many sources in the literature. Part of the scatter in Figure 1 could be due to errors in the determinations of the gas-phase Fe/O ratios and part to the different star formation histories of the different galaxies. The closer in time to a recent burst of star formation the lower the total Fe/O ratio in the ISM. In the solar vicinity at present all the O abundance is due to core collapse supernovae, while about 40% of the Fe is due to core collapse supernovae and the other 60% to Type Ia supernovae (e.g. Pagel 2009). There is a time delay in the Fe formation relative to the O formation, and consequently the Fe/O ratio depends on: the star formation rate, the initial stellar mass function, and the gas flows from and into the intergalactic medium. There are two well established Fe/O ratios from observations: the one when the Sun was formed, called the protosolar ratio that amounts to -1.19 dex (Asplund *et al.* 2009), and the present value in the solar vicinity based on observations of B stars that amounts to -1.32 dex (Przybilla, Nieva, & Butler 2008). From the previous discussion we will adopt for the ISM of the objects in Figure 1 a total value of Fe/O = -1.3 dex.

From Figure 1, and assuming that the total Fe/O ratio amounts to -1.3 dex, we obtain that for the Galactic and extragalactic H II regions with abundances in the $8.35 < 12 + \log O/H < 8.85$ range the average fraction of Fe in the gaseous-phase is about 4% . For the extragalactic H II regions with abundances in the $7.75 < 12 + \log O/H < 8.35$ range the fraction of Fe in the gaseous-phase is about 20%. For the extragalactic H II regions

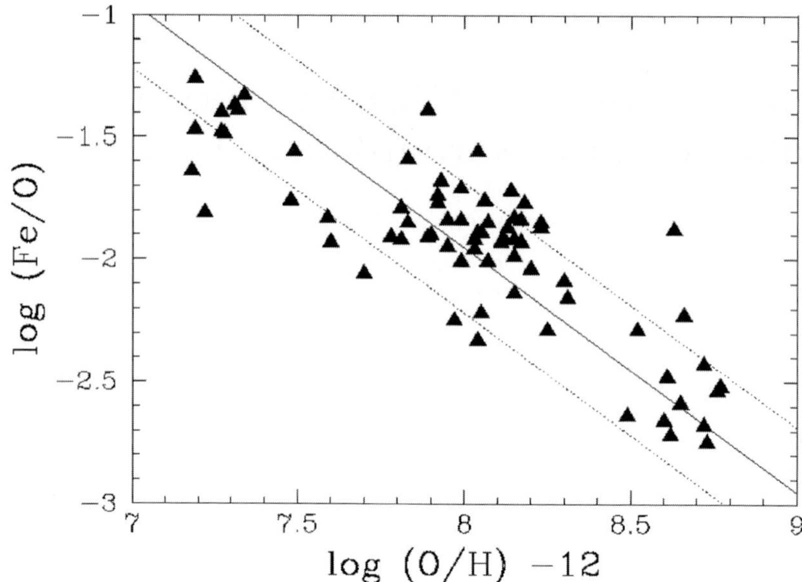

Figure 1. Log Fe/O versus 12 + log O/H gas-phase abundance ratios. The data comes from the literature. The straight line corresponds to 12 + log Fe/H = 6.05 ± 0.27.

in the $7.15 < 12 + \log O/H < 7.75$ range the fraction of Fe in the gaseous phase is about 40%.

3. Conclusions

We find that the gaseous $12 + \log Fe/H$ ratio is typically in the 6.05 ± 0.27 range for H II regions with $12 + \log O/H$ in the 7.35 to 8.85 range. The almost constancy of the gas-phase Fe/H ratio reflects the efficiency of the processes of dust formation and dust destruction. It probably implies that there is a minimum threshold for dust formation given by a gas-phase $12 + \log Fe/H$ ratio of about 5.7.

We estimate that the dust-phase fraction of O in Galactic and extragalactic H II regions is in the 0.08 ± 0.03 to 0.12 ± 0.03 dex range. We also consider that the H II region abundances derived from the $T(4363)$ method underestimate the O/H ratio by about 0.2 to 0.4 dex (Peimbert & Peimbert 2010b, and references therein). These two effects taken together imply that to obtain the gas-phase plus the dust-phase O/H abundance ratio it is necessary to increase the O/H gas values derived from the $T(4363)$ method by about 0.25 to 0.50 dex.

References

Asplund, M., Grevesse, N., Sauval, A. J., & Scott, P. 2009, *ARAA* 47, 481

Mesa-Delgado, A., Esteban, C., García-Rojas, J., *et al.* 2009, *MNRAS* 395, 855

Pagel, B. E. J. 2009, *Nucleosynthesis and Chemical Evolution of Galaxies, Second Edition,* Cambridge University Press

Peimbert, A. & Peimbert, M. 2010a, in preparation

Peimbert, M. & Peimbert, A. 2010b, *Rev. Mexicana AyA*, arXiv: 0912.3781, in press

Przybilla, N., Nieva, M. F., & Butler, K. 2008, *ApJ* 688, L103

Light Elements in the Universe
Proceedings IAU Symposium No. 268, 2009
C. Charbonnel, M. Tosi, F. Primas, & C. Chiappini, eds.

© International Astronomical Union 2010
doi:10.1017/S1743921310004114

On the origin of the helium-rich population in the peculiar globular cluster Omega Centauri

Donatella Romano[1,2], M. Tosi[2], M. Cignoni[1,2], F. Matteucci[3], E. Pancino[2] and M. Bellazzini[2]

[1]Dept. of Astronomy, Bologna University,
Via Ranzani 1, I-40127, Bologna, Italy
email: `donatella.romano@oabo.inaf.it`

[2]INAF-Bologna Observatory,
Via Ranzani 1, I-40127, Bologna, Italy

[3]Dept. of Physics, Trieste University,
Via Tiepolo 11, I-34143, Trieste, Italy

Abstract. In this contribution we discuss the origin of the extreme helium-rich stars which inhabit the blue main sequence (bMS) of the Galactic globular cluster Omega Centauri. In a scenario where the cluster is the surviving remnant of a dwarf galaxy ingested by the Milky Way many Gyr ago, the peculiar chemical composition of the bMS stars can be naturally explained by considering the effects of strong differential galactic winds, which develop owing to multiple supernova explosions in a shallow potential well.

Keywords. globular clusters: individual (Omega Centauri) – galaxies: dwarf, evolution – stars: abundances, chemically peculiar

1. A dwarf galaxy progenitor suffering strong galactic winds

It has been suggested – and it is now widely accepted – that the kinematical, dynamical and chemical properties of the most massive globular cluster of the Milky Way, Omega Centauri (ω Cen, NGC 5139), can all be understood if it is the surviving remnant of a larger system, captured and partially disrupted by the Milky Way many Gyrs ago (see, e.g., Romano *et al.* 2007, and references therein). However, the origin of the extreme He-rich stars hosted on its blue main sequence (bMS) still awaits a satisfactory explanation. Here we propose a possible solution, in the framework of a chemical evolution model which reproduces other major observed properties of the cluster, namely, its stellar metallicity distribution function (MDF), age-metallicity relation (AMR), trends of several abundance ratios with metallicity and Na-O anticorrelation (Fig. 1; see Romano *et al.* 2007, 2010).

In order to reproduce all the relevant observations, the parent galaxy must experience both infall of gas of primordial chemical composition and outflow of processed matter. The key assumption that allows the formation of extreme He-rich stars is that, while supernova (SN) ejecta are efficiently lost through the galactic outflow, elements restored to the interstellar medium (ISM) by gentle winds from both asymptotic giant branch (AGB) and fast rotating massive stars (FRMSs) are mostly retained in the cluster potential well (see Romano *et al.* 2010 and references therein for details).

He and Na are dispersed in the ISM through slow stellar winds by both AGBs and FRMSs. O, instead, as well as other heavy elements, is expelled by massive stars through fast polar winds. Hence, in the framework of our model, He and Na are preferentially retained inside the cluster potential well, while O is mainly vented out through the galactic outflow. This naturally produces the Na-O anticorrelation and He-rich bMS population, as observed (Fig. 1, right-hand panels). The coexistence of populations with 'normal' and 'high' He content at [Fe/H] $\geqslant$ -1.4 is expected if He is removed with different intensities from different regions of the proto-cluster.

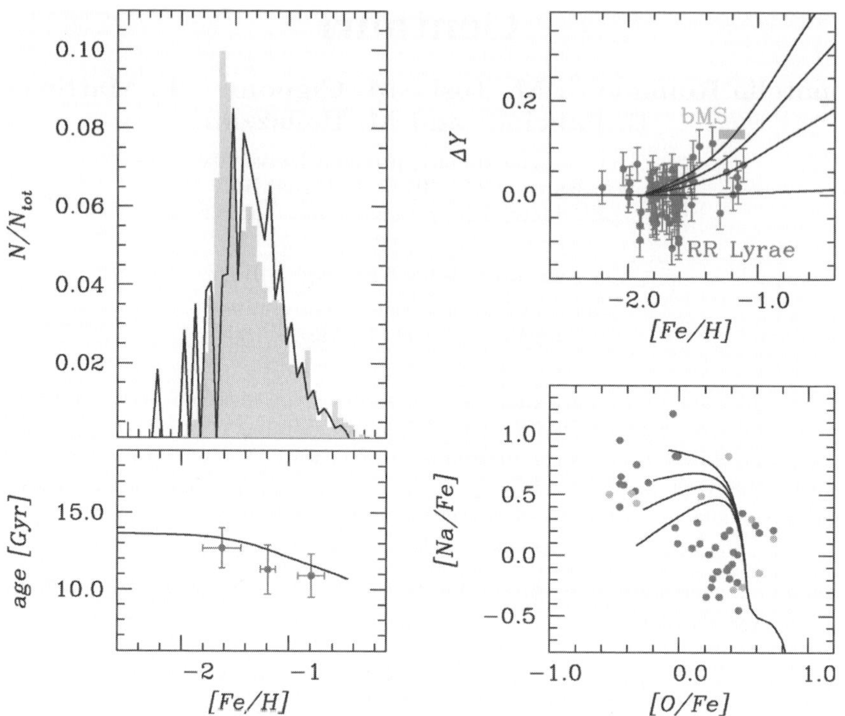

Figure 1. Predicted (thick solid lines) MDF (top left-hand panel), AMR (bottom left-hand panel), relative He enrichment (top right-hand panel) and Na-O anticorrelation (bottom right–hand panel) for ω Cen stars compared to observations from Sollima *et al.* (2005; shaded histogram, top left-hand panel), Hilker *et al.* (2004; dots with error bars, bottom left-hand panel), Sollima *et al.* (2006; dots, top right-hand panel), Norris (2004; box, top right-hand panel), Norris & Da Costa (1995) and Smith *et al.* (2000) (dots, bottom right-hand panel). The upper (lower) curves in the right-hand panels correspond to models computed with lower (higher) efficiencies of He and Na entrainment in the galactic outflow (see Romano *et al.* 2010).

References

Hilker, M., Kayser, A., Richtler, T., & Willemsen, P. 2004, *A&A*, 422, L9

Norris, J. E. 2004, *ApJ*, 612, L25

Norris, J. E. & Da Costa, G. S. 1995, *ApJ*, 447, 680

Romano, D., Matteucci, F., Tosi, M., Pancino, E., Bellazzini, M., Ferraro, F.R., Limongi, M., & Sollima, A. 2007, *MNRAS*, 376, 405

Romano, D., Tosi, M., Cignoni, M., Matteucci, F., Pancino, E., & Bellazzini, M. 2010, *MNRAS*, in press (arXiv:0910.1299)

Smith, V. V., Suntzeff, N. B., Cunha, K., Gallino, R., Busso, M., Lambert, D. L., & Straniero, O. 2000, *AJ*, 119, 1239

Sollima, A., Ferraro, F. R., Pancino, E., & Bellazzini, M. 2005, *MNRAS*, 357, 265

Sollima, A., Borissova, J., Catelan, M., Smith, H. A., Minniti, D., Cacciari, C., & Ferraro, F. R. 2006, *ApJ*, 640, L43

Session III

Abundances of Li, Be and B:
Observations

Livia Schnyder, Elisa Delgado Mena, Jonay Gonzalez Hernandez

Martin Asplund & Karin Lind

Light Elements in the Universe
Proceedings IAU Symposium No. 268, 2009
C. Charbonnel, M. Tosi, F. Primas & C. Chiappini, eds.
© International Astronomical Union 2010
doi:10.1017/S1743921310004126

The light elements in the light of 3D and non-LTE effects

Martin Asplund[1] and Karin Lind[2]

[1] Max-Planck-Institut für Astrophysik, Postfach 1317, D-85741 Garching, Germany
email: `asplund@mpa-garching.mpg.de`

[2] European Southern Observatory, Karl-Schwarzschild-Strasse 2, D-85748 Garching, Germany
email: `klind@eso.org`

Abstract. In this review we discuss possible systematic errors inherent in classical 1D LTE abundance analyses of late-type stars for the light elements (here: H, He, Li, Be and B). The advent of realistic 3D hydrodynamical model atmospheres and the availability of non-LTE line formation codes place the stellar analyses on a much firmer footing and indeed drastically modify the astrophysical interpretations in many cases, especially at low metallicities. For the $T_{\rm eff}$-sensitive hydrogen lines both stellar granulation and non-LTE are likely important but the combination of the two has not yet been fully explored. A fortuitous near-cancellation of significant but opposite 3D and non-LTE effects leaves the derived ^{7}Li abundances largely unaffected but new atomic collisional data should be taken into account. We also discuss the impact on 3D non-LTE line formation on the estimated lithium isotopic abundances in halo stars in light of recent claims that convective line asymmetries can mimic the presence of ^{6}Li. While Be only have relatively minor non-LTE abundance corrections, B is sensitive even if the latest calculations imply smaller non-LTE effects than previously thought.

Keywords. convection, line: formation, radiative transfer, Sun: abundances, Sun: atmosphere, stars: abundances, stars: atmospheres, stars: Population II, Galaxy: abundances

1. Stellar model atmospheres and spectral line formation

Stellar chemical compositions are not observed: to decipher the spectral fingerprints in terms of abundances requires realistic models for the stellar atmospheres and the line formation processes. Traditionally abundance analyses of late-type stars have been carried out relying on 1D, time-independent, hydrostatic model atmospheres, which treat convection with the rudimentary mixing length theory. All of these are dubious approximations as even a casual glance at the solar atmosphere will immediately reveal. The solar atmosphere, as for other late-type stars, is dominated by granulation, which is the observational manifestation of convection: an evolving pattern of broad, warm upflows in the midst of narrow cool downdrafts. Because of the great temperature sensitivity of the opacity, the temperature drops precipitously as the ascending gas nears the optical surface before it overturns and is accelerated downwards. The temperature contrast is therefore very pronounced in the photospheric layers, amounting to $> 1000\,$K at the optical surface for the Sun (even when averaged over surfaces of equal optical depths these rms-differences amount to $\approx 400\,$K). In addition to ignoring such atmospheric inhomogeneities, 1D model atmospheres can not be expected to have the correct mean temperature stratification because of the simplified convection treatment and the neglect of convective overshoot. Even small temperature differences can propagate to very large changes in the emergent stellar spectrum because of the non-linearities in the radiative transfer and the extreme opacity variations.

Over the past decade or so, 3D, time-dependent, hydrodynamical model atmospheres for a range of stellar parameters have started to be developed and applied to stellar abundance work (e.g. Asplund 2005, and references therein). Such 3D models solve the standard hydrodynamical conservation equations coupled with a simultaneous solution of the 3D radiative transfer equation and therefore self-consistently predict the convective and radiative energy transport (see e.g. Nordlund $et\ al.$ 2009, for further details). To make the 3D modelling computationally tractable, the radiative transfer is not solved for the many thousands of wavelength points as routinely done in 1D model atmosphere codes. Instead the frequencies are sorted in opacity and more recently also in wavelength space for $\approx 4-20$ bins for which the radiative transfer is solved. The resulting total radiative heating/cooling as a function of atmospheric depth is surprisingly well reproduced. The 3D models are based on similarly realistic microphysics (equation-of-state and continuous/line opacities) as employed in standard 1D model atmospheres. Currently there are mainly four different codes being used to develop 3D models – STAGGER (e.g. Nordlund $et\ al.$ 2009), CO5BOLD (e.g. Ludwig $et\ al.$ 2009b), MURAM (e.g. Vögler $et\ al.$ 2005) and ANTARES (e.g. Muthsam $et\ al.$ 2009) – although essentially all abundance related work to date has been performed within the first two collaborations.

Being more sophisticated in the modelling does not automatically translate to being more realistic. Over the past few years, substantial effort has therefore been dedicated to verify the suitability of the 3D models for quantitive stellar spectroscopy using an arsenal of observational diagnostics. Some of the striking successes are that the 3D models accurately predict the detailed solar granulation properties (e.g. Nordlund $et\ al.$ 2009) as well as spectral line profiles, including their asymmetries and shifts (e.g. Asplund $et\ al.$ 2000), which strongly suggests that the 3D modelling captures the essence of the real atmospheric structure and macroscopic gas motions. Another crucial test is the continuum center-to-limb variation, which is an excellent probe of the mean temperature stratification in the solar atmosphere. As shown in Fig. 1, the latest generation of 3D models computed with the STAGGER code reproduces the observations extremely well (Pereira $et\ al.$ 2010); the CO5BOLD solar model does a similarly good job (H.-G. Ludwig, private communication). This is a remarkable achievement. The 3D solar model not only outperforms all tested theoretical 1D model atmospheres like the Kurucz, MARCS and PHOENIX flavours in this respect, it also does noticeably better than the Holweger & Müller (1974) semi-empirical model, which was constructed largely to fulfill this observational constraint. The 3D solar model also performs very well when confronted with other tests, such as the spectral energy distribution, H lines (Pereira $et\ al.$ 2010) and spatially resolved line profiles (Pereira $et\ al.$ 2009a,b). In all aspects the most recent 3D solar models are clearly highly realistic and can therefore safely be trusted for abundance analysis purposes (Asplund $et\ al.$ 2009).

The success in the solar case gives some credence to the 3D modelling being similarly realistic also for other stars for which far fewer indisputable observational tests are available. 3D models now exist for a sizable portion of the HR-diagram covering essentially spectral types A, F, G, K, M and even later (e.g. Collet $et\ al.$ 2007; Ludwig $et\ al.$ 2009b). Both dwarfs, subgiants, giants and even some supergiants have been simulated and for a wide range of metallicities. Qualitatively the granulation behaves similarly to the Sun for most of these stars but typically the convection becomes more vigorous towards higher $T_{\rm eff}$ and lower $\log g$. Perhaps the most marked difference compared with 1D models are apparent at low [Fe/H], where 3D models have much cooler temperatures in the optically thin regime (Asplund $et\ al.$ 1999). This is a natural consequence of the dominance of adiabatic heating over radiative heating in absence of a heavy line-blanketing at low metallicities. The low temperatures are bound to affect especially minority species, low

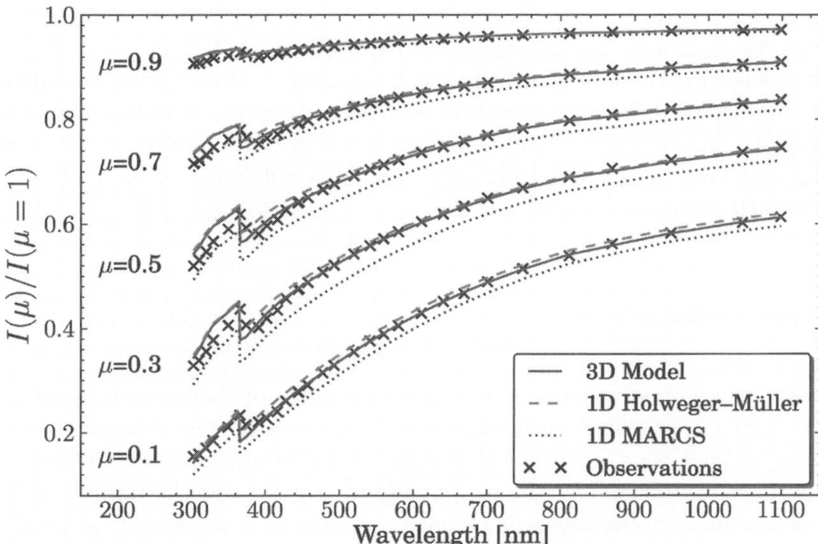

Figure 1. Comparison of the observed continuum center-to-limb variation as a function of wavelength (Neckel & Labs 1994) against the predictions for different solar model atmospheres. The 3D solar model employed by Asplund *et al.* (2009) outperforms the 1D theoretical MARCS model and even the 1D semi-empirical Holweger & Müller (1974) model atmosphere.

excitation lines and molecules, with 3D LTE abundance corrections often amounting to $> +0.5$ dex (e.g. Asplund 2005). One should be aware though that the steep temperature gradients may significantly enhance the non-LTE effects compared to the 1D case.

With a stellar model atmosphere it is possible to compute the emergent spectrum, which normally for late-type stars is done within the LTE approximation and in 1D. It should not come as a surprise that this approach often leads to severe systematic errors. In recent years, efforts towards rectifying these shortcomings have mainly focused on 1D non-LTE investigations and 3D LTE calculations, but the combination of 3D and non-LTE is still largely unexplored (e.g. Asplund 2005, and references therein). When solving the non-LTE problem one must solve the rate equations for all relevant levels involving both radiative and collisional transitions coupled with a simultaneous solution of the radiative transfer equation for all necessary frequencies. Needless to say, it rapidly becomes far more cumbersome than in LTE with a great deal of additional atomic data that must be considered. While the radiative processes like transition probabilities and photo-ionization cross-sections are now in reasonably good shape – especially for the non-complicated cases like the light elements – the main uncertainty in non-LTE calculations often stems from the poorly known collisional excitation and ionization due to electrons and H as well as charge transfer reactions. Especially the inelastic H collisions are a major problem in stellar abundance analyses. Mostly simple but likely erroneous classical recipes like the Drawin (1968) formula are applied, often with an additional ad-hoc scaling factor. For a few elements, Li being a notable case in point, detailed quantum mechanical or laboratory measurements exist (e.g. Belyaev & Barklem 2003), which reveal that the Drawin (1968) formula typically overestimates the real cross-sections by several orders of magnitude. An alternative approach is to attempt to calibrate the unknown collisional rates using various observations, such as obtaining consistent results from different lines or using center-to-limb variations. Not surprisingly, such empirical procedures suggest a range of scaling factors, from $\leqslant 0.1$ for Mn (Bergemann & Gehren 2008), to ≈ 1 for O

(Pereira *et al.* 2009a) or even > 1 for Fe (Korn *et al.* 2003). Whether these are realistic values, or merely artificial results due to too simplified modelling (e.g. 1D) remains to be seen. It is also worth remembering that a common scaling factor for all transitions within the same species, let alone between different elements, is highly unlikely. In spite of the remaining uncertainties introduced by our still poor handle on all atomic data, the alternative of relying on the LTE approximation will most often lead to even more severe systematic errors.

2. The light elements

2.1. *Hydrogen*

As the most abundant element and the main provider of continuous opacity, it is not possible to derive the stellar H abundance spectroscopically. The H lines are still of great use in stellar physics as a probe of the atmospheric conditions. In late-type stars the wings of the Balmer lines reflect the effective temperature and the temperature gradient near the continuum forming layers with only a relatively small gravity dependence. The great temperature sensitivity arises from the high excitation potential of the lower level ($\chi_{\mathrm{exc}} = 10.2\,\mathrm{eV}$) of the Balmer lines. Fortunately, both the Stark broadening and self-broadening are now well established (e.g. Barklem *et al.* 2000; Allard *et al.* 2008); the new broadening results in significantly lower T_{eff}, especially at low [Fe/H].

The H lines are however sensitive to the convection treatment, which in 1D normally is estimated by the mixing length theory (MLT) and its inherent four free parameters. Before comparing the best-fitting α_{MLT} and T_{eff} between different studies it is therefore important to bear in mind the various flavours of MLT existing in different 1D model atmospheres. Given the rudimentary nature of MLT to describe the correct convective energy transport and overshoot near/in the optical surface, 1D-based T_{eff}-scales from H lines will always be uncertain in an absolute sense even if highly accurate relative values can be obtained (e.g. Asplund *et al.* 2006). As explained above, 3D hydrodynamical models do provide an attractive alternative through their self-consistent computation of the convection without the need to invoke free parameters. Ludwig *et al.* (2009a) have investigated the 3D LTE formation of Hα, Hβ and Hγ lines for six different 3D models, including the Sun and metal-poor dwarfs/subgiants and compared with 1D models computed with identical microphysics but with different MLT implementations; see also Sbordone *et al.* (2010) for further discussions. Ludwig *et al.* (2009a) found a very complex dependence on the 3D corrections to the 1D-based estimates of T_{eff} with differences amounting to $\pm300\,\mathrm{K}$ depending on the line in consideration and the T_{eff}, [Fe/H] and α_{MLT} of the 1D model. An extra complication arises from the different line shapes in 3D and 1D, which can not be directly translated to a T_{eff} difference without specifying exactly which wavelength regions have been used for the comparison. The prospect for using pre-tabulated 3D corrections to 1D-based T_{eff} estimates is therefore not encouraging. It would be advantageous to directly compare observations with 3D predictions, which should be possible in the near future with the advent of grids of 3D models.

Until recently it has always been argued that the wings of the H Balmer wings are formed in LTE. Barklem (2007) has investigated whether this is indeed correct and found that it is not necessarily so. In particular he found that electron collisions are not sufficient to establish LTE. In terms of T_{eff}, the non-LTE calculations would imply $\geqslant 100\,\mathrm{K}$ higher values than in LTE. This however depends crucially on the still uncertain inelastic H collisions – improved atomic physics calculations are clearly needed to resolve this issue. It would also be necessary to perform 3D non-LTE H line computations, especially at low metallicity. Such calculations have recently been performed in the solar case with encouraging results (Pereira *et al.* 2010).

2.2. *Helium*

While in hot stars the very highly excited transitions of He I enable a determination of the He abundance (see Kaufer, these proceedings), in late-type stars the corresponding atomic levels are not sufficiently populated at typical photospheric temperatures. Instead the He I lines (e.g. 1083.0 nm) have a chromospheric origin and can neither probe the He abundance nor the photospheric temperature structure.

2.3. *Lithium*

Stellar Li abundance are usually determined only from the resonance line at 670.7 nm but in exceptional cases also from the weaker subordinate line at 610.4 nm. Contrary to most elements, the simplicity of the Li atom has enabled quantum mechanical calculations of all atomic data necessary for non-LTE investigations, especially of cross-sections for collisional excitation and charge transfer with neutral hydrogen (Barklem *et al.* 2003). The widely used study of Carlsson *et al.* (1994) mapped out the non-LTE effects in late-type stars and their detailed variation with metallicity, effective temperature, surface gravity, and lithium line strength. Recently, Lind *et al.* (2009a) revisited the 1D non-LTE line formation with improved collisional data and treatment of line-blocking (see Fig. 2). The investigations have shown that two competing effects govern the population levels of Li. Over-ionization from the first excited level of Li I is prominent at low effective temperatures due to a strong $J_\nu - B_\nu$ excess bluewards of the photo-ionization threshold at 350 nm. When the Li line is weak, the balance between UV over-ionization from the first excited state and over-recombination to higher excited levels determines the population levels of Li I. The resulting non-LTE corrections are typically minor, smaller than ± 0.1 dex, but increasing up to $+0.5$ dex in red giants. However, when the line is close to saturation, the loss of line photons establishes an efficient recombination and de-excitation ladder, which increases the population of low-excited levels of neutral Li. Furthermore, the scattering-dominated resonance line source function drops far below B_ν. The non-LTE corrections then change sign and reach -0.5 dex in extreme cases. Lind *et al.* (2009a) have made available convenient routines to interpolate the non-LTE corrections for a wide range of stellar parameters and Li abundances.

In LTE the Li I level populations depend on the local conditions, especially the gas temperature. The much cooler mean temperature stratifications in 3D models than corresponding 1D models at low metallicity translate to a 3D LTE abundance correction of ≈ -0.3 dex for the Li I 670.8 nm resonance line in metal-poor dwarfs and subgiants (e.g. Asplund *et al.* 1999). In addition, the presence of atmospheric inhomogeneities typically also increase the line strength. One should be aware however that the steeper temperature gradients in the 3D models are likely to boost the over-ionization. Spatially resolved observations of the solar surface show how the line actually is weaker in the bright, granular upflows, despite the drastic drop in temperature that they undergo. This can be understood since the intense UV radiation field in the granules produces pronounced over-ionization (Kiselman 1997).

Asplund *et al.* (2003) performed full 3D non-LTE computations for the Sun and two metal-poor stars using a 21-level Li model atom. They concluded that the line strengthening due to the cooler temperatures in the upper atmospheric layers on the one hand and line weakening from increased over-ionization on the other hand largely cancel each other. Barklem *et al.* (2003) updated these results in light of new collisional data, especially charge transfer reactions and reached similar conclusions. The 3D non-LTE calculations

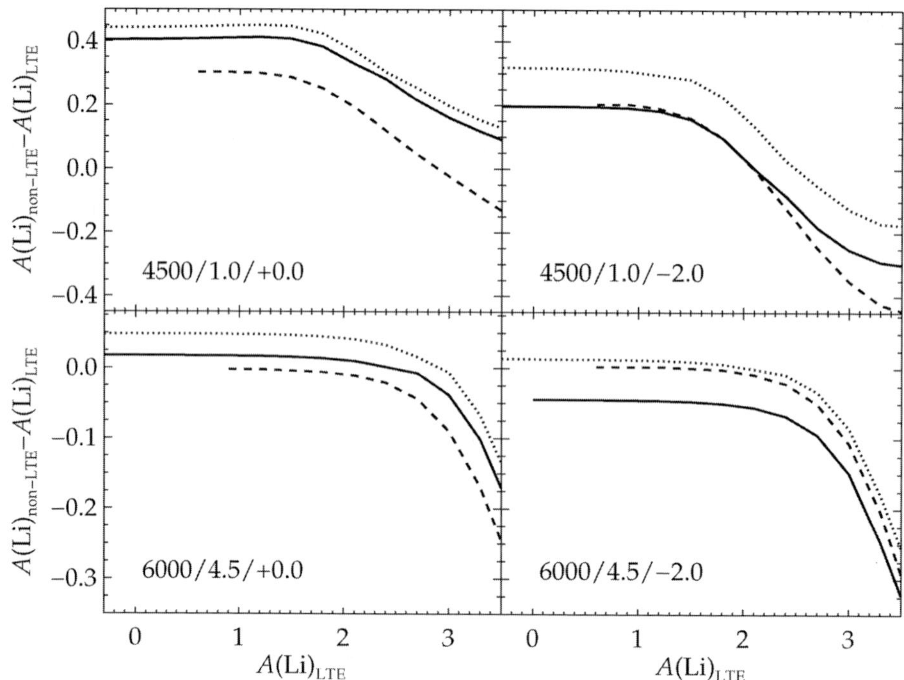

Figure 2. The 1D non-LTE abundance corrections for a selection of stellar parameters (labelled with $T_{\rm eff}$ / $\log g$/[Fe/H]) from the study of Lind *et al.* (2009a). The solid line denotes the case with all transitions considered while the dotted line correspond to the case when the charge transfer reactions of Barklem *et al.* (2003) are ignored. Also shown as dashed lines are the non-LTE corrections from Carlsson *et al.* (1994).

thus agree with the 1D non-LTE results to within 0.05–0.1 dex. In fact, even better agreement between the two is seen in the recent study of extremely metal-poor stars by Sbordone *et al.* (2010) with difference of < 0.03 dex, at least compared with the particular family of 1D model atmospheres they used. Sbordone *et al.* (2010) used a smaller atom (8 levels) but computed for more 3D models. They also provided a very handy functional relation between the Li abundance and the line strength, which allows direct computation of the 3D non-LTE based abundance without invoking any particular 1D model atmosphere. The Spite & Spite (1982) Li-plateau as inferred from 1D analyses is thus unlikely to be seriously in danger although the exact slope with for example [Fe/H] may be affected (e.g. Asplund *et al.* 2006; Sbordone *et al.* 2010), which deserves to be studied closer with improved 3D non-LTE calculations.

2.4. $^6Li/^7Li$

The minor isotope ^{6}Li can be detected through the isotopic shift in the Li I 670.8 nm line. The distortion of the line profile is very small and therefore necessitates extremely high-quality spectra ($S/N > 400$, $\lambda/\Delta\lambda > 100,000$ and minimal fringing). At solar metallicity, possible blending lines must also be carefully evaluated (Israelian *et al.* 2003), a problem which however disappears at low [Fe/H]. The first positive detection in a halo star was claimed by Smith *et al.* (1993) for HD 84937. Asplund *et al.* (2006) boosted the number of $> 2\sigma$ detections to 10 (including HD 84937) using high-quality VLT/UVES spectra analysed both in 1D and 3D LTE. The derived ^{6}Li abundance at [Fe/H]< -2.5 can not be easily accommodated with standard Galactic cosmic ray production, which has spurred a number of studies of more speculative production channels, including non-standard Big Bang nucleosynthesis with supersymmetric particles (e.g. Jedamzik & Pospelov 2009).

Several potential problems linger over the published ^{6}Li/^{7}Li analyses. García Pérez *et al.* (2009) have suggested that residual fringing in the observed spectra can cause large uncertainties. Their Subaru/HDS spectra are, however, much more inflicted by fringing than the VLT/UVES spectra of Asplund *et al.* (2006), who furthermore carefully assessed their possible impact on the results. More worrisome is the restriction to 1D or 3D LTE line formation. The atmospheric convective motions typically result in C-shaped line asymmetries (e.g. Asplund *et al.* 2000), which can thus mimic the presence of ^{6}Li. Indeed, Cayrel *et al.* (2007) have argued based on 3D non-LTE calculations that the Asplund *et al.* (2006) results must be reevaluated. Steffen *et al.* (2010) have considered this in more detail and concluded that the derived ^{6}Li/^{7}Li ratios in a 1D analysis are overestimated by typically 0.015, which if true could explain the tendency for most stars with non-significant detections in the Asplund *et al.* (2006) sample to cluster around ^{6}Li/^{7}Li$\approx$ 0.01. Furthermore, Steffen *et al.* (2010) argue that by taking this convective effect into account, only 4 out of 10 stars would remain as $> 2\sigma$ detections.

We argue that the Steffen *et al.* (2010) results should not be over-interpreted, since there are subtle but crucial differences in the analyses. Steffen *et al.* (2010) rely only on the Li line itself to determine all parameters: ^{7}Li and ^{6}Li abundances, wavelength shift and intrinsic line broadening (rotation in 3D but in 1D also macro-/microturbelence) while Asplund *et al.* (2006) rely on a range of Fe, Ca etc lines to first determine the line broadening which is then fixed for the Li profile fitting. Since the other lines are also asymmetric, this is accounted for in the profile fitting through a slightly larger required broadening, which then compensates for the use of symmetric Li lines in 1D. Asplund *et al.* (2006) also did a 3D LTE analysis for all lines and found very similar results as in 1D but with much improved profile fits. We have also performed 3D non-LTE line calculations for Li similar to Steffen *et al.* (2010) and can in fact exactly reproduce their results and their claimed 0.015 effect in ^{6}Li/^{7}Li when only using the Li line and not any other lines. We argue though that this does not make full use of the information encoded in the spectra due to the degeneracy between the four fitting parameters. Indeed, their procedure is akin to determining the D abundance in QSO absorption systems only from the Lyα line without first resolving the intrinsic velocity structure of the H I clouds from other metallic lines. Similarly problematic is to rely on 3D non-LTE for Li but 3D LTE for the other lines, in view of the likely pronounced non-LTE effects. The only way forward to convincingly demonstrate whether or not ^{6}Li is present in detectable amounts in the atmospheres of halo turn-off stars is to perform full 3D non-LTE calculations for all lines considered. We are currently working on this and will report the results elsewhere.

In summary, it is not yet possible to say that ^{6}Li has definitely been detected but it is definitely too early to say that ^{6}Li has not been detected. We also note that even Steffen *et al.* (2010) agree that ^{6}Li is present in HD 84937 and a few other stars. Since ^{6}Li is always destroyed and more so than ^{7}Li, one would still end up with a very significant cosmological ^{6}Li problem when invoking stellar depletion to solve the cosmological ^{7}Li problem (e.g. Asplund *et al.* 2006; Korn *et al.* 2006; Lind *et al.* 2009b).

2.5. Beryllium

In late-type stars the Be abundance can in practice only be derived from the resonance doublet of Be II at 313 nm. In contrast to Li and B, Be is normally present in significant amounts in both neutral and singly ionized form due to its high ionization potential, which makes Be less prone to over-ionization. The Be II line formation proceeds under non-LTE conditions but because the two dominant non-LTE effects – over-ionization and over-excitation – largely compensate each other the resulting non-LTE abundance corrections are typically minor (Garcia Lopez *et al.* 1995). Both the lower and upper level of the doublet are over-populated relative to the LTE prediction. Be I tends to be somewhat over-ionized due to a UV radiation excess ($J_\nu/B_\nu > 1$), which simultaneously produce an over-excitation of the upper level of the Be II 313 nm transitions. García Pérez *et al.* (2010) have investigated the non-LTE line formation of the Be II 313 nm doublet across a wide range of stellar parameters. They found that at solar metallicity the two effects almost perfectly balance each other with the resulting non-LTE abundance corrections amounting to $|\Delta \log \epsilon_{\mathrm{Be}}| \leqslant 0.05$ dex. At low metallicity the non-LTE corrections are positive ($\Delta \log \epsilon_{\mathrm{Be}} \approx 0.1$ dex) for $T_{\mathrm{eff}} \geqslant 6000$ K and negative ($\Delta \log \epsilon_{\mathrm{Be}} \approx -0.1$ dex) for lower T_{eff}. While the non-LTE effects will not qualitatively change the conclusions for the metallicity evolution of Be (Primas and Boesgaard, these proceedings), accounting for the new non-LTE calculations will tend to make the slope somewhat shallower.

The 3D line formation of the Be II lines have so far only been investigated for the Sun (Asplund *et al.* 2009). The Be II lines are not very sensitive to 3D effects at least within the LTE approximation. The 3D non-LTE case has not yet been investigated but given the relatively modest non-LTE effects in 1D one would not expect a great difference in 3D, although this prediction should be verified with detailed calculations.

2.6. *Boron*

As for Be, both over-ionization and over-excitation are at play in the formation of the B I resonance lines at 249.7 nm and 209.0 nm but because B I is the minority ionization stage for $T_{\mathrm{eff}} \geqslant 6000$ K the effects on the resulting line profiles are much more dramatic (Kiselman & Carlsson 1996). The over-ionization is largely driven by photo-ionization from the excited B I levels since $J_\nu/B_\nu > 1$ at the relevant wavelengths. It is also assisted by pumping in the B I resonance lines, which produces a sufficient over-population of these upper levels. The same pumping results in a substantial increase in the line source function $S_\nu/B_\nu \approx J_\nu/B_\nu > 1$ for the 249.7 and 209.0 nm lines. Thus, in contrast to the case of Be, for the B I resonance lines the two non-LTE effects work in tandem to decrease the non-LTE line strengths substantially compared with the LTE prediction.

Kiselman & Carlsson (1996) computed non-LTE abundance corrections for the B I lines for a grid of MARCS 1D model atmospheres. The non-LTE effects grow in size towards higher T_{eff} and lower [Fe/H]. Indeed, they found non-LTE abundance corrections as large as $> +0.5$ for typical turn-off halo stars. As both non-LTE processes feed on the UV radiation field, it is important to include background line opacities both for the calculations of the photo-ionization rates and the resonance lines, which Kiselman & Carlsson (1996) did in a somewhat approximate manner. Recently, Tan *et al.* (2010) confirmed the importance of over-ionization and pumping but they obtained markedly less severe non-LTE effects, which can be traced to a more complete treatment of the line-blocking. Their estimated non-LTE abundance corrections amount to $\approx +0.3$ dex at the lowest [Fe/H], when not including the highly uncertain inelastic H collisions according to Drawin (1968); the corrections would be $\approx +0.1$ dex when this formula would be applied without any further scaling factor. As shown in Fig. 3, the new non-LTE results of Tan *et al.* (2010) make the resulting evolution of B as a function of metallicity somewhat steeper with a slope of ≈ 1.3. The reader is referred to Primas (these proceedings) for a discussion of the astrophysical implications of such a correlation.

With the exception of the Sun (Asplund *et al.* 2009), the 3D line formation of the B I lines has not been investigated, neither in LTE nor in non-LTE. In view of the prominent non-LTE effects and the general expectation that they would become even

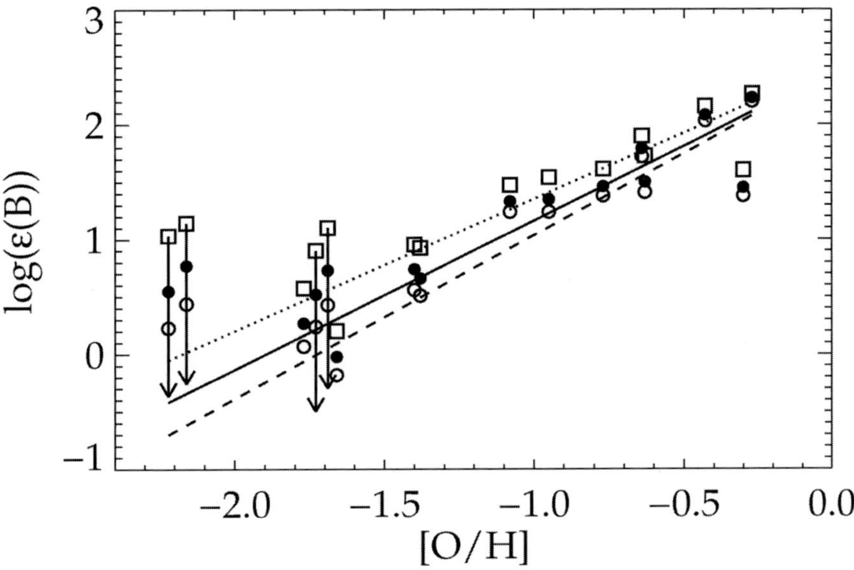

Figure 3. The derived stellar B abundances in halo stars (Tan *et al.* 2010) as a function of [O/H] for LTE (open circles with dashed line the least square fit) and non-LTE (filled circles and solid line); upper limits are marked with arrows. The O abundances have been corrected for non-LTE effects according to Fabbian *et al.* (2009). Also shown are the LTE-based B abundances of Tan *et al.* (2010) but corrected according to the non-LTE calculations of Kiselman & Carlsson (1996) (open squares and dotted line). Inelastic H collisions have not been included for the B non-LTE results but for O according to the Drawin (1968) formula.

more pronounced in 3D due to the atmospheric inhomogeneities, performing detailed 3D non-LTE calculations for B I should have a high priority.

3. Concluding remarks

Important progress has been made over the last decade in terms of improving the abundance analyses of the light elements in late-type stars by accounting for non-LTE as well as 3D effects. The non-LTE line formation of the resonance lines of Li I, Be II, and B I share common properties, but the final non-LTE abundance corrections are always element specific. Especially differences in ionization potential and the extent to which UV radiative transitions contribute to the excitation and ionization balance govern the LTE departures. There are however still several outstanding analysis problems.

Signs are that the new generation of 3D hydrodynamical model atmospheres are indeed highly realistic and thus trustworthy for abundance purposes as they successfully reproduce most, if not all, observational tests they have been exposed to. The key development for the future will be to make such 3D models generally available for a wide range of stellar parameters and actually to get stellar abundance practitioners to start using them routinely. The main obstacle here will likely be to overcome old habits rather than say lack of computational power, since it is now possible to perform a 3D-based solar/stellar abundance analysis for thousands of lines, at least in LTE (Asplund *et al.* 2009). On the non-LTE side, the challenges are two-fold: improving the necessary atomic data, especially for collisions, and to couple the non-LTE line formation with 3D model atmospheres, which is still very much unchartered territory. This is now certainly feasible but will require additional (wo)manpower. Only then can we with some degree of confidence argue that our analysis efforts match the investments in obtaining impressive observational data. As outlined above, such modelling efforts will no doubt lead to many reinterpretations of observations with profound implications for our understanding of stellar, galactic and cosmic evolution in general and the light elements in particular.

References

Allard, N. F., Kielkopf, J. F., Cayrel, R., & van't Veer-Menneret, C. 2008, *A&A*, 480, 581

Asplund, M. 2005, *ARAA*, 43, 481

Asplund, M., Carlsson, M., & Botnen, A. V. 2003, *A&A*, 399, L31

Asplund, M., Grevesse, N., Sauval, A. J., & Scott, P. 2009, *ARAA*, 47, 481

Asplund, M., Lambert, D. L., Nissen, P. E., Primas, F., & Smith, V. V. 2006, *ApJ*, 644, 229

Asplund, M., Nordlund, Å., Trampedach, R. *et al.* 2000, *A&A*, 359, 729

Asplund, M., Nordlund, Å., Trampedach, R., & Stein, R. F. 1999, *A&A*, 346, L17

Barklem, P. S. 2007, *A&A*, 466, 327

Barklem, P. S., Belyaev, A. K., & Asplund, M. 2003, *A&A*, 409, L1

Barklem, P. S., Piskunov, N., & O'Mara, B. J. 2000, *A&A*, 363, 1091

Belyaev, A. K. & Barklem, P. S. 2003, *Phys. Rev. A*, 68, 062703

Bergemann, M. & Gehren, T. 2008, *A&A*, 492, 823

Carlsson, M., Rutten, R. J., Bruls, J. H. M. J., & Shchukina, N. G. 1994, *A&A*, 288, 860

Cayrel, R., Steffen, M., Chand, H. *et al.* 2007, *A&A*, 473, L37

Collet, R., Asplund, M., & Trampedach, R. 2007, *A&A*, 469, 687

Drawin, H. 1968, Zeitschrift für Physik, 211, 404

Fabbian, D., Asplund, M., Barklem, P. S., Carlsson, M., & Kiselman, D. 2009, *A&A*, 500, 1221

Garcia Lopez, R. J., Severino, G., & Gomez, M. T. 1995, *A&A*, 297, 787

García Pérez, A., Asplund, M., & Kiselman, D. 2010, *A&A*, submitted

García Pérez, A. E., Aoki, W., Inoue, S. *et al.* 2009, *A&A*, 504, 213

Holweger, H. & Müller, E. A. 1974, *Solar Phys.*, 39, 19

Israelian, G., Santos, N. C., Mayor, M., & Rebolo, R. 2003, *A&A*, 405, 753

Jedamzik, K. & Pospelov, M. 2009, New Journal of Physics, 11, 105028

Kiselman, D. 1997, *ApJ* (Letters), 489, L107

Kiselman, D. & Carlsson, M. 1996, *A&A*, 311, 680

Korn, A. J., Grundahl, F., Richard, O. *et al.* 2006, *Nature*, 442, 657

Korn, A. J., Shi, J., & Gehren, T. 2003, *A&A*, 407, 691

Lind, K., Asplund, M., & Barklem, P. S. 2009a, *A&A*, 503, 541

Lind, K., Primas, F., Charbonnel, C., Grundahl, F., & Asplund, M. 2009b, *A&A*, 503, 545

Ludwig, H., Behara, N. T., Steffen, M., & Bonifacio, P. 2009a, *A&A*, 502, L1

Ludwig, H., Caffau, E., Steffen, M. *et al.* 2009b, *MemSAI*, 80, 711

Muthsam, H. J., Kupka, F., Loew-Baselli, B. *et al.* 2009, arXiv:0905.0177

Neckel, H. & Labs, D. 1994, *Solar Phys.*, 153, 91

Nordlund, Å., Stein, R. F., & Asplund, M. 2009, Living Reviews in Solar Physics, 6, 2

Pereira, T., Asplund, M., Trampedach, R., & Collet, R. 2010, *A&A*, in press

Pereira, T. M. D., Asplund, M., & Kiselman, D. 2009a, *A&A*, 508, 1403

Pereira, T. M. D., Kiselman, D., & Asplund, M. 2009b, *A&A*, 507, 417

Sbordone, L., Bonifacio, P., Caffau, E. *et al.* 2010, *A&A*, in press

Smith, V. V., Lambert, D. L., & Nissen, P. E. 1993, *ApJ*, 408, 262

Spite, F. & Spite, M. 1982, *A&A*, 115, 357

Steffen, M., Cayrel, R., Bonifacio, P., Ludwig, H., & Caffau, E. 2010, arXiv:1001.3274

Tan, K., Shi, J., & Zhao, G. 2010, *ApJ*, in press

Vögler, A., Shelyag, S., Schüssler, M. *et al.* 2005, *A&A*, 429, 335

Light Elements in the Universe
Proceedings IAU Symposium No. 268, 2009
C. Charbonnel, M. Tosi, F. Primas & C. Chiappini, eds.

© International Astronomical Union 2010
doi:10.1017/S1743921310004138

Li isotopes in metal-poor halo dwarfs: a more and more complicated story

Monique Spite and François Spite

GEPI, Observatoire de Paris, 92195 Meudon Cedex, CNRS UMR 8111
email: monique.spite@obspm.fr, francois.spite@obspm.fr

Abstract. The nuclei of the lithium isotopes are fragile, easily destroyed, so that, at variance with most of the other elements, they cannot be formed in stars through steady hydrostatic nucleosynthesis.

The ^{7}Li isotope is synthesized during primordial nucleosynthesis in the first minutes after the Big Bang and later by cosmic rays, by novae and in pulsations of AGB stars (possibly also by the ν process). ^{6}Li is mainly formed by cosmic rays. The oldest (most metal-deficient) warm galactic stars should retain the signature of these processes if, (as it had been often expected) lithium is not depleted in these stars. The existence of a "plateau" of the abundance of ^{7}Li (and of its slope) in the warm metal-poor stars is discussed. At very low metallicity ([Fe/H] < -2.7dex) the star to star scatter increases significantly towards low Li abundances. The highest value of the lithium abundance in the early stellar matter of the Galaxy ($\log\epsilon$(Li) $=$ A(^{7}Li) $=$ 2.2 dex†) is much lower than the the value ($\log\epsilon$(Li) $=$ 2.72) predicted by the standard Big Bang nucleosynthesis, according to the specifications found by the satellite WMAP. After gathering a homogeneous stellar sample, and analysing its behaviour, possible explanations of the disagreement between Big Bang and stellar abundances are discussed (including early astration and diffusion). On the other hand, possibilities of lower productions of ^{7}Li in the standard and/or non-standard Big Bang nucleosyntheses are briefly evoked.

A surprisingly high value (A(^{6}Li) $=$ 0.8 dex) of the abundance of the ^{6}Li isotope has been found in a few warm metal-poor stars. Such a high abundance of ^{6}Li independent of the mean metallicity in the early Galaxy cannot be easily explained. But are we really observing ^{6}Li ?

Keywords. Stars: abundances, Population II – nucleosynthesis – early universe

---

1. Introduction

1.1. *A first view*

In the eighties and the nineties, the abundance of ^{7}Li in the warm (5700 $<$ Teff $<$ 6500K) metal-poor dwarfs was found constant, independent of the temperature and of the metallicity, at least in the interval $-2.8 <$ [Fe/H] < -1.8 : Spite & Spite (1982a, and 1982b) and subsequent papers (e.g. Spite, Spite and Maillard 1984, Hobbs & Thorburn 1991, Molaro, Primas and Bonifacio 1995, Spite *et al.* 1996, Bonifacio and Molaro 1997, Smith *et al.* 1998). Later works, published up to 2005, are vividly reviewed in the section 2 of Charbonnel and Primas (2005) : the small differences between authors are discussed.

This "universal" behaviour of lithium suggested that ^{7}Li observed in the old metal poor dwarfs was synthesized during primordial nucleosynthesis (when the Universe was only a few minutes old) and that lithium, although a very fragile element, had survived unaltered in the atmosphere of warm metal-poor dwarfs. In the so called "standard" stellar models (i. e. without diffusion and mixing), the Li depletion is negligible (Deliyannis *et al.* 1990, Pinsonneault *et al.* 1992). However, already at that time, several theoreticians of stellar

---

† In the literature the lithium abundance, is noted indifferently by $\log\epsilon$(Li) or by A(Li); both notations are in logarithmic scale of number of atoms where $\log\epsilon$(H) $=$ A(H) $=$ 12

atmospheres thought that it was difficult to admit that the abundance of lithium in the atmosphere of the old very metal-poor dwarfs had remained unchanged during 13 billions years. Reciprocally, it was also difficult to explain a strictly uniform lithium depletion in the warm metal-poor dwarfs, whatever their temperature, mass and metallicity.

The primordial abundance of ^{7}Li had been then largely identified with the stellar abundance found on the plateau, the value being between $\log\epsilon(\text{Li}) = 2.0$ and $\log\epsilon(\text{Li}) = 2.3$ depending on the temperature scale adopted by the authors. Moreover, taking into account the errors of measurement and the uncertainties of the observed abundances, the simultaneous production of all the light elements (^{4}He, ^{3}He, D and Li) by the Big Bang, could be explained.

In spite of the obvious difficulty of disentangling the ^{6}Li and ^{7}Li profiles in stellar spectra, attempts were made at finding the abundance of ^{6}Li : it was possible only in a handful of metal-poor dwarfs. For the first time Smith, Lambert, & Nissen (1993), claimed a positive detection of ^{6}Li in a metal-poor star: HD84937 ([Fe/H] ≈ -2.1). Later Nissen *et al.* (1999) observed ^{6}Li in 2 metal-poor stars out of five; see also Cayrel *et al.* (1999). The existence of ^{6}Li (more fragile that ^{7}Li) in some warm metal-poor dwarfs reinforced the idea that ^{7}Li is not depleted in such stars.

1.2. *A second step*

Two astronomical satellites brought new points of view :
- in 1997, Hipparcos has provided accurate parallaxes for numerous stars, improving the determination of the parameters of the stellar atmospheres (and enabling to discriminate dwarfs and subgiants).
- in 2003, WMAP has determined (Bennett *et al.* 2003) the physical conditions of the Big Bang especially a precise value for the baryons to photons ratio : η, leading to precise values of the abundances of the light elements produced in the standard Big Bang, including lithium. This Li abundance was found much higher than the abundance measured in the warm metal-poor dwarfs and thus the interest for a precise re-determination of the lithium abundance was reactivated.

2. The lithium isotopes in the matter of the early Galaxy

2.1. *Formation of Lithium*

A peculiarity of lithium is that it is very fragile and destroyed at "low" temperatures. Unlike most of the other elements, when it is synthesized inside the stars by hydrostatic nucleosynthesis, it is immediately destroyed. Lithium is formed :
– during the Big Bang (primordial nucleosynthesis) (^{7}Li and negligible amount of ^{6}Li).
– by spallation (cosmic rays or in superbubbles: mainly ^{6}Li)
– possibly by the "ν" process in type II supernovae (^{7}Li)

Pulsations of AGB stars and novae explosions can also enrich the interstellar medium in ^{7}Li, but not in the early times of the Galaxy.

2.2. *Primordial abundance of ^{7}Li*

The ^{7}Li observed in the atmospheres of the very old galactic stars has been mainly formed by the primordial nucleosynthesis. Cosmic rays and ν process in massive supernovae could produce a small increase of $\epsilon(^7\text{Li})$ with metallicity (a slope).

In the standard Big-Bang, the production of lithium depends only on the baryons to photons ratio η. This ratio has been deduced, with precision, from the measurements of the cosmic microwave background radiation by the satellite WMAP : Bennett *et al.* (2003), Spergel *et al.* (2007), Cyburt *et al.* (2008) : $\eta = 6.23 \pm 0.17 \ 10^{-10}$.

Then, the primordial lithium abundance is $\log\epsilon(^7\mathrm{Li}) = 2.72 \pm 0.06$ for $\log\epsilon(\mathrm{H}) = 12$ (see also Komatsu *et al.* 2009, and Iocco *et al.* 2009, for the discussion of the uncertainties).

2.3. *Depletion of lithium in the atmosphere of the stars*

It is generally admitted that the abundances measured in the atmospheres of unevolved stars are representative of the abundances in the material from which the star has been originally formed.

However different phenomena are known to affect the superficial abundance of lithium during the life of the star.

Lithium is a very fragile element destroyed as soon as the temperature is higher than $2.5 \ 10^6 \mathrm{K}$ for $^7\mathrm{Li}$ and even $2.0 \ 10^6 \mathrm{K}$ for $^6\mathrm{Li}$.

If there is some mixing between the surface of the star and the hot deep layers, little by little lithium is destroyed and disappears from the atmosphere of the star. In giants, Li is strongly diluted after the first dredge-up. Even in dwarfs lithium is often depleted: in the atmosphere of the Sun after 4.5 Gyr, the lithium abundance is only $\log\epsilon(\mathrm{Li}) = 1.03$ (Caffau *et al.*, this volume) although in meteorites $\log\epsilon(\mathrm{Li})$ reaches ≈ 3.25 (a value representative of the abundance of lithium in the material that formed the Sun). However in the warm metal-poor dwarfs/turnoff stars, mixing is not as deep as in solar metallicity stars, and lithium could have been preserved.

Lithium is also supposed to slowly settle down into the stars by diffusion (gravitational settling), but the diffusion can be easily thwarted by turbulent mixing.

3. The behaviour of $^7\mathrm{Li}$ in the most metal poor stars in the Milky Way

The difference between the lithium abundance predicted by BBN + WMAP and the mean value observed in the atmosphere of the old galactic dwarfs, leads to a redetermination of the lithium abundance, with an effort towards better temperature determinations: the determination of lithium abundance is very sensitive to the choice of the temperature of the stellar atmosphere. Moreover, the precise behaviour of $\log\epsilon(\mathrm{Li})$ vs. temperature and metallicity and also the scatter around the mean relations may give a clue about the exact mechanism of lithium depletion.

Since 2005 several papers about the lithium abundance in very metal-poor stars were published. They all try to determine very carefully the temperatures of the stellar atmospheres, using high quality data: it is essential to discriminate between intrinsic scatter and determinations errors.

–Charbonnel & Primas (2005) have determined the temperatures from a very precise photometry (*uvby*) associated with the Alonso *et al.* (1996) calibration and taking into account very carefully the reddening of the stars. Meléndez *et al.* (2010 and this symposium) had a similar approach but they used a new implementation of the infrared flux method to determine the temperatures from multiband IR photometry.

–Boesgaard *et al.* (2005) and Hosford *et al.* (2009) based their analysis on excitation temperature (the abundances from the Fe I lines must be independent of the excitation potential of the line).

–All the other authors used the wings of the hydrogen lines (mainly Hα) to determine the temperatures: Asplund *et al.* (2006), Bonifacio *et al.* (2007), González Hernández *et al.* (2008), García Pérez *et al.* (2008, 2009), Aoki *et al.*(2009).

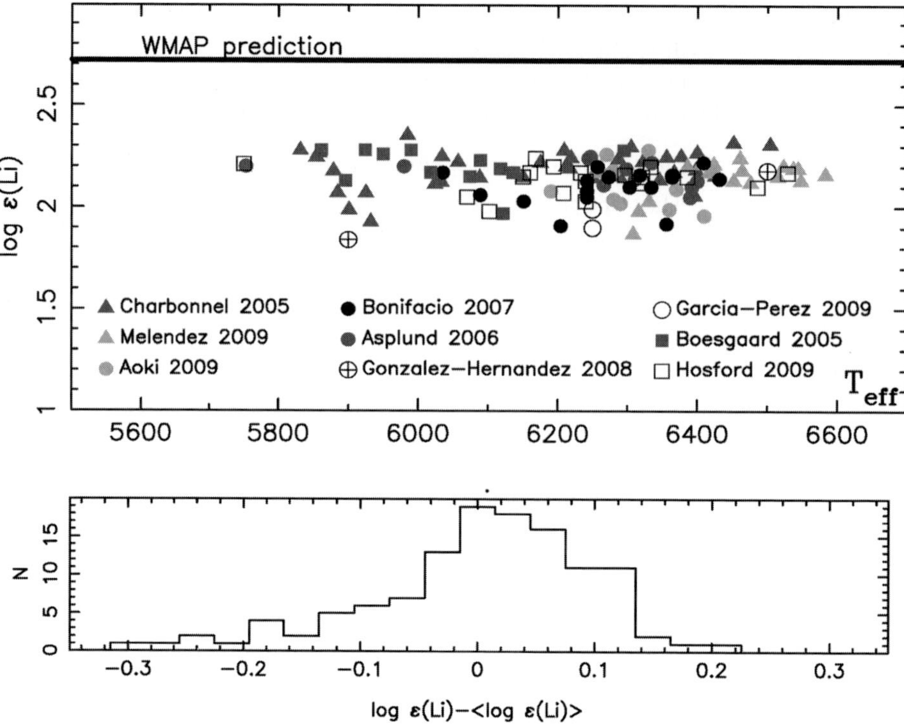

Figure 1. Lithium abundance versus Temperature for metal-poor stars ([Fe/H] < -2) and (below) histogram. Triangles indicate that the temperature has been determined from photometry, circles from the profile of the hydrogen lines and squares from the excitation temperature. In this interval of temperature and metallicity, the lithium abundance is independent of the temperature but the scatter is rather large. The mean value of the lithium abundance is $\log\epsilon(\mathrm{Li}) \approx 2.15$, more than 0.55 dex below the prediction of standard BB + WMAP (full black line). On the histogram showing the distribution of the distances of each point to the mean value of the lithium abundance it can be seen that the distribution is not gaussian.

In Fig. 1 we present $\log\epsilon(\mathrm{Li})$ versus T_{eff}, for the metal-poor dwarfs and turnoff stars with [Fe/H] < -2.0, measured in these different works. There is a rather good agreement between the different determinations, (even if some slight systematic differences appear between different authors). No significant trend of $\log\epsilon(\mathrm{Li})$ vs. T_{eff} is apparent. The mean value of the lithium abundance is $\log\epsilon(\mathrm{Li}) \approx 2.15$: about four times less than the abundance formed by the standard Big Bang nucleosynthesis.

This mean value of the lithium abundance measured in the old galactic stars is also very close to the value measured in ω Cen (see Bonifacio *et al.*, this symposium): $\log\epsilon(\mathrm{Li}) \approx 2.19 \pm 0.14$. The cluster ω Cen is now considered as the remnant of a captured dwarf galaxy. As a consequence $\log\epsilon(\mathrm{Li}) = 2.15$ could be a "universal" value of stellar Li abundance at low metallicity.

In Fig. 1 the scatter around the mean value of the lithium abundance is 0.10 dex. This scatter is not symmetric (see the histogram in Fig. 1) and is larger than the measurement errors. It is NOT mainly due to systematic differences between different authors, it exists even for the data of a given author.

What is the cause of this scatter of the lithium abundance when [Fe/H] < -2 ?

If we plot $\log\epsilon(\mathrm{Li})$ versus [Fe/H] (Fig. 2) a slight decrease of $\epsilon(\mathrm{Li})$ with the metallicity appears. As a consequence the scatter observed in Fig 1 should be partly due to a

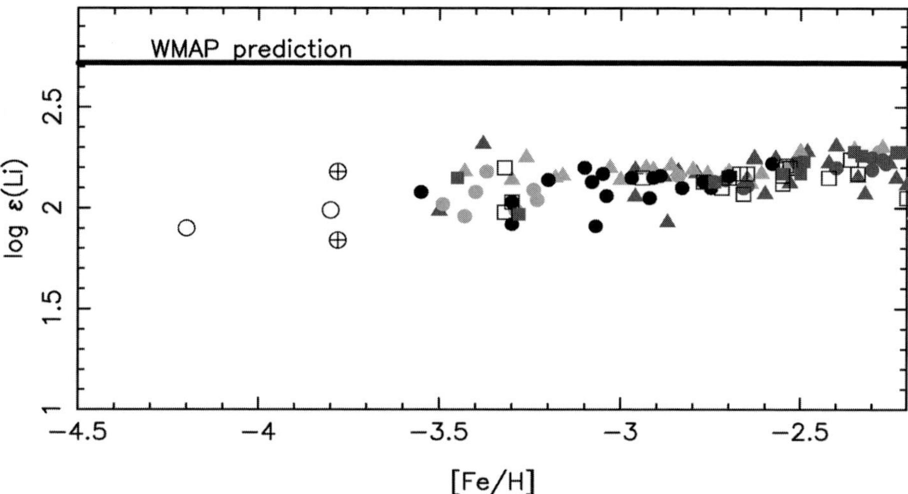

Figure 2. $\log\epsilon(\mathrm{Li})$ vs. [Fe/H], for turnoff stars ($\mathrm{T_{eff}} > 5800\mathrm{K}$). The symbols are the same as in Fig. 1. The slight decline of $\log\epsilon(\mathrm{Li})$ when the metallicity decreases appears for [Fe/H] < -3.

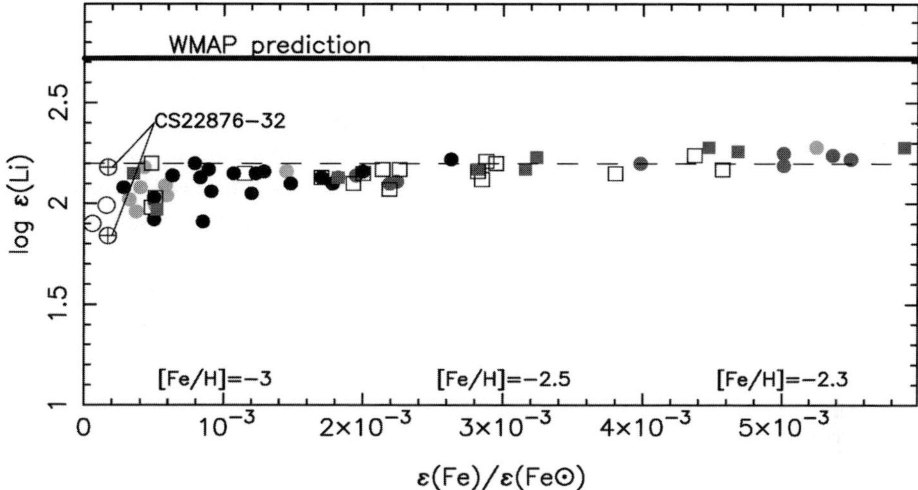

Figure 3. $\log\epsilon(\mathrm{Li})$ vs. the iron abundance $\epsilon(\mathrm{Fe}_*)/\epsilon(\mathrm{Fe}_\odot)$(linear scale) for turnoff stars ($\mathrm{T_{eff}} > 5800\mathrm{K}$). The symbols are the same as in Fig. 1. To improve the homogeneity, we have kept only the measurements of the lithium abundance based on a temperature determination independent of the reddening ($\mathrm{T_{Hydrogen}}$ or $\mathrm{T_{exc}}$). When $\epsilon(\mathrm{Fe}_*)/\epsilon(\mathrm{Fe}_\odot) > 0.001$ ([Fe/H] > -3), the scatter around the mean curve is very small (0.05 dex) but at a lower metallicity the scatter suddenly increases. The lithium abundances in the two components of the extremely metal-poor binary CS 22876-32 are different by a factor of two. Lithium would sometimes suffers an extra depletion in the atmospheres of the most metal-poor stars. Then, the pristine value of the lithium abundance would correspond to the higher envelope, $\log\epsilon(\mathrm{Li}) \approx 2.2$.

dependence of lithium abundance on metallicity. This decrease suggests that the extrapolated abundance of lithium could be even lower than 2.15 when extrapolating toward zero-metal stars (just after the Big Bang).

In Fig. 3 we have plotted $\log\epsilon(\mathrm{Li})$ versus $\epsilon(\mathrm{Fe})_*/\epsilon(\mathrm{Fe})_\odot$ (linear scale), and for a better homogeneity, we have kept only stars with a temperature determined spectroscopically (Hα profiles or excitation temperature) and thus independent of the interstellar reddening. In the interval $2 \times 10^{-3} < \epsilon(\mathrm{Fe})_*/\epsilon(\mathrm{Fe})_\odot < 6 \times 10^{-3}$ the lithium abundance increases steadily and the scatter around the mean curve is extremely small: 0.05 dex.

In the most metal-poor stars $(\epsilon(\mathrm{Fe})_*/\epsilon(\mathrm{Fe})_\odot < 2 \times 10^{-3}$, i.e. $[\mathrm{Fe/H}] < -2.7$, the mean value of the lithium abundance decreases suddenly to reach $\log\epsilon(\mathrm{Li}) \approx 2.0$ dex at $\epsilon(\mathrm{Fe}) = 0$. (zero-metal stars). Therefore $\log\epsilon(\mathrm{Li}) \approx 2.0$ could be the mean value of the lithium abundance in the early Galaxy. But in this region the scatter at a given metallicity increases strongly (see Sbordone et $al.$ 2010 and this symposium, for confirmation by additional data). This scatter could be interpreted as the result of a variable depletion, decreasing the lithium abundance from a natal value of about $\log\epsilon(\mathrm{Li}) = 2.2$ dex, this value would represent the pristine value of the lithium abundance in the early Galaxy. In both cases the pristine value of the lithium abundance is well below the cosmological value derived from the standard Big Bang (with the WMAP value of the η parameter).

In Fig 3, no stars are found above $\log\epsilon(\mathrm{Li}) = 2.2$ dex for $[\mathrm{Fe/H}] < -2.7$ and above $\log\epsilon(\mathrm{Li}) = 2.3$ dex for $-2.7 < [\mathrm{Fe/H}] < -2.2$ defining an upper envelope of the lithium abundance. (A handful of Li-rich halo stars have been found in the literature, but all are several times more metal-rich than the stars considered here.)

Several interpretations can be put forward for explaining the gap between the cosmological and the observed lithium and also the sudden scatter appearing at low metallicity.

3.1. *Astration*

It has been proposed that the "uniform" lithium abundance in halo dwarfs could be due to a general process affecting the whole Milky Way lowering the cosmological Li abundance down to the pristine abundance. Piau et $al.$ (2006) have proposed an early astration in very masssive stars Pop III stars (destroying Li). They could have destroyed 3/4 of the primordial lithium, 50% of the mass of the Galactic halo having been processed in these massive stars.

The scatter of the lithium abundance appearing in the most metal-poor stars could then be explained by a still incomplete mixing of the matter in the Galaxy, or by the fact that these most metal-poor stars would have been accreted from different faint dwarf galaxies or satellites (Frebel et $al.$ 2010) with different astration histories. There are objections to a strong initial astration in our Galaxy: the ejecta of the stars destroying Li would have induced (Prantzos 2007) a large early abundance of elements (e.g. C and O) that is not observed. But, in ω Cen, remnant of an accreted dwarf galaxy, the lithium abundance $(\log\epsilon(\mathrm{Li}) \approx 2.19$, Bonifacio et $al.$, this symposium), is very similar to the Galactic value for stars of the same metallicity, although they certainly had different astration histories. Probably, a very early global astration of the matter of the universe, by massive first stars, before the formation of the galaxies, would explain the plateau, but would encounter a similar objection about the early abundances of C and O in the Galaxy. However it could be possible that these massive Pop III stars rotate and thus eject only external layers with hydrogen and helium (Meynet , this symposium).

3.2. *Depletion in the stellar atmospheres*

The gap between the theoretical value of the lithium abundance predicted by the standard Big Bang and the observed value in the atmosphere of the stars, can be also the result of a depletion of lithium, even in the warm metal-poor dwarfs.

The stellar abundance now observed could be due to diffusion (Michaud et $al.$, 1984). Some calculations have been made (e.g. Richard et $al.$, 2005): a depletion such as figured in Fig, 3, could be matched by a diffusion carefully moderated by some turbulence. The very low scatter of the data around the mean (0.05 dex) in the interval $2 \times 10^{-3} <$

$\epsilon(\mathrm{Fe})_*/\epsilon(\mathrm{Fe})_\odot < 6 \times 10^{-3}$ brings severe constaints to the theoretical computations. Also the sudden downward scatter, at the left part of the Figure 3 needs to be explained.

The computation of lithium depletion by diffusion in a globular cluster (Korn *et al.*, 2007) carefully combined with turbulence, shows that the lithium depletion amounts to about 0.26 dex. If this correction is applied to the most metal-poor field stars the pristine (or natal) Li abundance of these stars should be (in the best case) about 2.46 dex, a value still significantly smaller than the cosmological value. As a consequence the depletion by diffusion moderated by turbulence cannot explain completely the gap. Moreover this depletion by diffusion/turbulence in globular clusters is questionned by González Hernández *et al.* (this symposium).

We have to remark (Fig. 2 and 3) that the two components of the extremely metal-poor ($[\mathrm{Fe/H}] \approx -3.8$) binary star CS 22176-32, analyzed by González Hernández *et al.* (2008) have a lithium abundance different by a factor of two, although their temperature is higher than 5800 K and that, as a consequence, no depletion of lithium is expected in these stars. This difference is above the measurement errors. The two stars were presumably born with the same lithium abundance and thus the only explanation of this large difference is that lithium has been depleted (at least) in CS 22176-32B.

"Extra-scatter" by "extra depletion" in the extremely metal-poor stars? This "extra-depletion" would appear only, and sometimes, at very low metallicity ($[\mathrm{Fe/H}] < -3$).

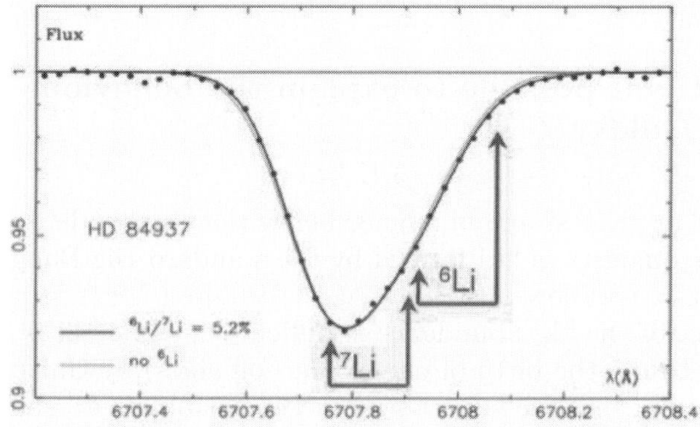

Figure 4. Profile of the doublets of ^{7}Li and ^{6}Li in a very metal poor turnoff star HD 84937, compared to a synthetic spectrum computed with no ^{6}Li and ^{6}Li/^{7}Li$=5\%$

4. Observation of ^{6}Li in the most metal poor stars in the Milky Way

Since ^{6}Li is not significantly formed by the standard Big Bang, if ^{6}Li is observed in very metal-poor stars it is supposed to be formed mainly by cosmic rays.

If ^{6}Li exists in these stars, its abundance is very small and it is thus very difficult to detect, in particular because the lines (doublets) of ^{6}Li and ^{7}Li are overlapping (see Fig. 4). Very high resolution and high S/N spectra are required as also very precise computations of the profiles.

Several groups have recently tried to measure ^{6}Li in metal-poor halo stars (see in particular Asplund *et al.* 2006, Asplund & Meléndez 2008, Cayrel *et al.* 2007, García Pérez *et al.* 2009).

Asplund *et al.* (2006, 2008) observed a sample of 27 turnoff stars with $-3.3 < [\mathrm{Fe/H}] < -1.2$. In twelve of them, they detected ^{6}Li. The abundance of ^{6}Li was constant $\log\epsilon(^6\mathrm{Li}) \approx$

0.8 dex and the ratio ^{6}Li/^{7}Li was between 4 and 10%. In all the 27 stars the abundance of lithium or its upper limit is compatible with a plateau with ^{6}Li/^{7}Li $\approx$ 5%.

This result remains fragile: Cayrel *et al.* (2007) have shown that, if a classical analysis of a given star would provide a ratio ^{6}Li /^{7}Li $= 4$% the result would be ^{6}Li /^{7}Li $= 0$% if the asymetry of the lines, due to convection, is taken into account (3D computations).

Recently, García Pérez *et al.* (2009) have insisted on the fact that the error of the ^{6}Li measurement is generally underestimated due to the uncertainty of the position of the continuum, the residual fringes etc. It is interesting to remark that Asplund & Meléndez (2008) found that the ^{6}Li /^{7}Li ratio in G64-37 is 11 % while García Pérez *et al.* (2009) found less than 1%.

4.1. *Have we really observed ^{6}Li in EMP stars ?*

In none of the stars, the ^{6}Li isotope is detected with a precision higher or equal to 3σ (Steffen *et al.*, this symposium). If the error is a little underestimated, and if there is in fact, no ^{6}Li in the old metal-poor stars, the noise would mimic an absorption of ^{6}Li in about half of the stars (detection) and an emission of ^{6}Li in the other half of the stars (no detection). This is not far from what is observed...

More precise computations (NLTE, 3D) about spectra with higher S/N ratios and a better defined continuum are necessary to firmly conclude about the detection and the abundance of ^{6}Li in the most metal-poor stars.

5. Conclusion : Is it possible to explain the behaviour of ^{6}Li and ^{7}Li in the early Galaxy ?

5.1. ^6Li

The absence of ^{6}Li, or at least an abundance below the limit of detection, would be easy to explain since the quantity of ^{6}Li formed by the standard Big Bang and by the cosmic rays is supposed to be very low.

The measurement of the ^{6}Li abundance is difficult. If some ^{6}Li is observed, it should have been formed (before the birth of our old turnoff stars) by Galactic Cosmic Rays or during the explosion of massive supernovae in superbubbles: we should then observe a clear increase of ^{6}Li with [Fe/H] like the increase of boron and beryllium (see Boesgaard 2004, Prantzos 2007). At the present time, a large abundance of ^{6}Li, more or less constant with [Fe/H], such as observed by Asplund *et al.* (2006, 2008): $\log\epsilon$(Li) $= 0.8$dex, cannot be understood in the frame of the standard theories.

5.2. ^7Li

Depending on the interpretation of Fig. 3, the pristine stellar abundance of ^{7}Li in our Galaxy is $\log\epsilon$(Li) $= 2.0$ or 2.2 dex: in both cases, it is far from the value (2.72 dex) of the ^{7}Li abundance produced by the standard Big Bang nucleosynthesis, according to the WMAP specifications. Some explanations can be proposed none of them completely satisfactory.

• **Early astration** - According to Piau *et al.* (2006), the primordial ^{7}Li, (such as computed by the Big Bang) would have been severely destroyed in the Galaxy before the formation of the old metal-poor stars by astration of the matter in massive Pop III stars, lowering the lithium abundance to the observed pristine abundance. But this theory cannot explain that the same pristine abundance of lithium is observed in the Milky Way and in ωCen (the remnant of an accreted dwarf galaxy) which certainly had different astration histories (see section 3.1). Moreover in this case a strong excess (not observed) of C and O in the matter of the early Galaxy would be, a priori, expected, However it has been remarked (Meynet , this symposium) that massive rotating Pop III stars would eject only external layers with hydrogen and helium (without excess of C and O).

• **Depletion / Diffusion** - A second hypothesis is that lithium has been depleted in the atmosphere of the observed turnoff stars during their long evolution, by atomic diffusion carefully partially compensated by turbulence (see section 3.2). The data gathered in Fig. 3, which show for $[\mathrm{Fe/H}] > -3$ a very small scatter (entirely explained by the determination errors) are a challenge to the computations of such an uniform depletion. The downward scatter of the lithium abundance in the extremely metal-poor stars remains to be explained (variable extra-depletion ?), as well as the absence (within our limits of T_{eff} and metallicity) of any (Li-rich) star in the "desert" between the high Big Bang lithium abundance and the plateau. Anyway, if the "depletion correction" computed by Korn *et al.* (2007) for the globular cluster NGC 6397 is applied to the field metal-poor dwarfs, the resulting value of the pristine lithium abundance (in the best case, about 2.46 dex) is again far from the cosmological predictions.

• **Gravity waves** - The behaviour of lithium should be evaluated taking into account the gravity waves (e.g. Talon & Charbonnel 2004, see also Talon *et al.*, this symposium): the results of this promising theory are eagerly awaited.

• **Uncertainties in the Big Bang theories** - The production of $^7\mathrm{Li}$ in the Big Bang could be lower in the standard nucleosynthesis if a resonance level (Cyburt & Pospelov, 2009) is taken ito account in the transformation of $^7\mathrm{Be}$ into $^9\mathrm{B}$.

Moreover, a number of theories have been proposed for various more or less *ad hoc* non-standard Big Bang nucleosyntheses (see e.g. Iocco *et al.* 2009).

As a conclusion, in addition to progresses in fundamental physics and nucleosynthesis, the use of NLTE, 3D model atmospheres for interpreting high quality spectra, analysing differential observations (comparison dwarfs-subgiants, binaries, delicate trends, small scatters and other such details) should shed some light on the always complex problem of the lithium abundance in the early Galaxy.

6. Acknowledgments

This text benefitted from conversations with P. Bonifacio, L. Sbordone and E. Caffau.

References

Aoki, W., Barklem, P. S., Beers, T. C., Christlieb, N., Inoue, S. *et al.* 2009, *ApJ*, 698, 1803
Alonso, A., Arribas, S., & Martinez-Roger, C. 2009, *A&A*, 313, 873
Asplund, M., Lambert, D. L., Nissen, P. E., Primas, F., & Smith, V. V. 2006, *ApJ*, 644, 229
Asplund, M. & Meléndez, J. 2008, AIP Conf. Proc. FIRST STARS III: First Stars II Conference, Vol. 990, p. 342
Bennett, C. L. *et al.* 2003, *ApJS*, 148, 1
Boesgaard, A. M. 2004, Carnegie Obs. Astrophys. Ser., Vol. 4: Origin and Evolution of the Elements, eds. A. McWilliam and M. Rauch (Cambridge: Cambridge Univ. Press)
Boesgaard, A. M., Stephens, A., & Deliyannis, C. P. 2005, *ApJ*, 63, 398
Bonifacio, P., Molaro, P., Sivarani, T., Spite, M., Spite, F. *et al.* 2007, *A&A*, 462, 851
Bonifacio, P. & Molaro, P. 1997, *MNRAS*, 285, 847
Cayrel, R., Spite, M., Spite, F., Vangioni-Flam, E., Cassé, M., & Audouze, J. 1999, *A&A*, 343, 923
Cayrel, R., Steffen, M., Chand, H., Bonifacio, P., Spite, M., Spite, F. *et al.* 2007, *A&A*, 473, L37
Charbonnel, C. & Primas, F. 2005, *A&A*, 442, 961

Cyburt, R. H., Fields, B. D., & Olive, K. A. 2008, *JCAP*, 11, 12

Cyburt, R. H. & Pospelov, M. 2009, arXiv(0906.4373)

Deliyannis, C. P., Demarque, P., & Kawaler, S. D. 1990 *ApJS*, 73, 21

Frebel, A., Simon, J. D., Geha, M., & Willman, B. 2010 *ApJ*, 708, 560

García Pérez, A. E., Aoki, W., Inoue, S., Ryan, S. G., Suzuki, T. K. *et al.* 2009, *A&A*, 504, 213

García Pérez, A. E., Christlieb, N., Ryan, S. G., Beers, T. C. *et al.* 2008, *Physica Scripta*, 133, 4036

González Hernández, J., Bonifacio, P., Ludwig, H.-G., Caffau, E., Spite, M., Spite, F. *et al.* 2008, *A&A*, 480, 233

Hobbs, L. M. & Thorburn, J. A. 1991, *ApJ*, 375, 116

Hosford, A., Ryan, S. G., García Pérez, A. E., Norris, J. E., & Olive, K. A. 2009, *A&A*, 493, 601

Iocco, F., Mangano, G., Miele, G., Pisanti, O., & Serpico, D. 2009, *Phys. Rep.*, 472, 1

Komatsu, E., Dunkley, J., Nolta, M. R. *et al.* 2009, *ApJS*, 180, 330

Korn, A. J., Grundahl, F., & Richard, O. 2007, *ApJ*, 671, 402 (Paper I)

Meléndez, J., Casagrande, L., Ramírez, I., & Asplund, M. 2010, Proceedings of the IAU symp. 265 "Chemical Abundances in the Universe: Connecting First Stars to Planets", K. Cunha, M. Spite & B. Barbuy, eds., Cambridge University Press, p. 71

Michaud, G., Fontaine, G., & Beaudet, G. 1984, *ApJ*, 282, 206

Molaro, P., Primas, F., & Bonifacio, P. 1995, *A&A*, 348, 211

Nissen Poul, E., Lambert, D. L., Primas, F., & Smith, V. V. 1999, *A&A*, 295, 47

Piau, L., Beers, T. C., Balsara, D. S., Sivarani, T., Truran, J. W., & Ferguson, J. W. 2006, *ApJ*, 653, 300

Pinsonneault, M. H., Deliyannis, C. P., & Demarque, P. 1992, *ApJS*, 78, 179

Prantzos, N. 2007, *Space Sci. Rev.*, 130, 27

Richard, O., Michaud, G., & Richer, J. 2005, *ApJ*, 619, 538

Sbordone, L., Bonifacio, P., Caffau, E., Ludwig, H.-G., Behara, N. *et al.* 2010, Proceedings IAU Symposium No. 265, "Chemical Abundances in the Universe: Connecting First Stars to Planets", K. Cunha, M. Spite & B. Barbuy, eds., Cambridge University Press, p. 75

Smith, V. V., Lambert, D. L., & Nissen Poul, E. 1998, *ApJ*, 408, 262

Smith, V. V., Lambert, D. L., & Nissen Poul, E. 1998, *ApJ*, 506, 405

Spergel, D. N., Bean, R., Doré, O. *et al.* 2007, *ApJS*, 170, 377

Spite, M. & Spite, F. 1982a, *Nature*, 297, 483

Spite, F. & Spite, M. 1982b, *A&A*, 115, 357

Spite, F., Spite, M. & Maillard, J. P. 1984, *A&A*, 141, 56

Spite, M., Francois, P., Nissen, P. E., & Spite, F. 1996, *A&A*, 307, 172

Steffen, M., Cayrel, R., Bonifacio, P., Ludwig, H.-G., & Caffau, E. 2010, Proceedings of the IAU symp. 265 "Chemical Abundances in the Universe: Connecting First Stars to Planets", K. Cunha, M. Spite & B. Barbuy, eds., Cambridge University Press, p. 23

Talon, S. & Charbonnel, C. 2004, *A&A*, 418, 1051

Light Elements in the Universe
Proceedings IAU Symposium No. 268, 2009
C. Charbonnel, M. Tosi, F. Primas & C. Chiappini, eds.
© International Astronomical Union 2010
doi:10.1017/S174392131000414X

Observational signatures for depletion in the Spite plateau: solving the cosmological Li discrepancy?

Jorge Meléndez[1], Luca Casagrande[2], Iván Ramírez[2], Martin Asplund[2] and William J. Schuster[3]

[1]Centro de Astrofísica, Universidade do Porto, Rua das Estrelas, 4150-762 Porto, Portugal
email: `jorge@astro.up.pt`

[2]Max-Planck-Institut für Astrophysik, Karl-Schwarzschild-Str. 1, Postfach 1317, D-85741 Garching, Germany

[3]Observatorio Astronómico Nacional, UNAM, Apartado Postal 877, Ensenada, BC, CP 22800, Mexico

Abstract. We present Li abundances for 73 stars in the metallicity range $-3.5 <$ [Fe/H] < -1.0 using improved IRFM temperatures (Casagrande *et al.* 2010) with precise *E(B-V)* values obtained mostly from interstellar NaI D lines, and high-quality equivalent widths ($\sigma_{EW} \sim 3\%$). At all metallicities we uncover a fine-structure in the Li abundances of Spite plateau stars, which we trace to Li depletion that depends on both metallicity and mass. Models including atomic diffusion and turbulent mixing seem to reproduce the observed Li depletion assuming a primordial Li abundance $A_{\mathrm{Li}} = 2.64$ dex (MARCS models) or 2.72 (Kurucz overshooting models), in good agreement with current predictions ($A_{\mathrm{Li}} = 2.72$) from standard BBN. We are currently expanding our sample to have a better coverage of different evolutionary stages at the high and low metallicity ends, in order to verify our findings.

Keywords. nucleosynthesis – cosmology: observations – stars: abundances, Population II

1. Introduction

One of the most important discoveries in the study of the chemical composition of stars was made in 1982 by M. and F. Spite, who found an essentially constant Li abundance in warm metal-poor stars (Spite & Spite 1982), a result interpreted as a relic of primordial nucleosynthesis. Due to its cosmological significance, there have been many studies devoted to Li in metal-poor field stars (e.g. Meléndez & Ramírez 2004; Boesgaard *et al.* 2005; Charbonnel & Primas 2005; Nissen *et al.* 2005; Asplund *et al.* 2006; Bonifacio *et al.* 2007; Shi *et al.* 2007; Hosford *et al.* 2009; Aoki *et al.* 2009), with observed Li abundances at the lowest [Fe/H] from as low as $A_{\mathrm{Li}} = 1.94$ to as high as $A_{\mathrm{Li}} = 2.37$.

Using the theory of big bang nucleosynthesis (BBN) and the baryon density obtained from WMAP data, a primordial Li abundance of $A_{\mathrm{Li}} = 2.72^{+0.05}_{-0.06}$ is predicted (Cyburt *et al.* 2008), which is a factor of 2–6 times higher than the Li abundance inferred from halo stars. There have been many theoretical studies on non-standard BBN trying to explain the cosmological Li discrepancy by exploring the frontiers of new physics (e.g. Coc *et al.* 2009; Jedamzik & Pospelov 2009; Kohri & Santoso 2009). Alternatively, the Li problem could be explained by a reduction of the original Li stellar abundance due to internal processes (i.e., by stellar depletion). In particular, stellar models including atomic diffusion and mixing can deplete a significant fraction of the initial Li content (e.g., Richard *et al.* 2005; Korn *et al.* 2006; Lind *et al.* 2009).

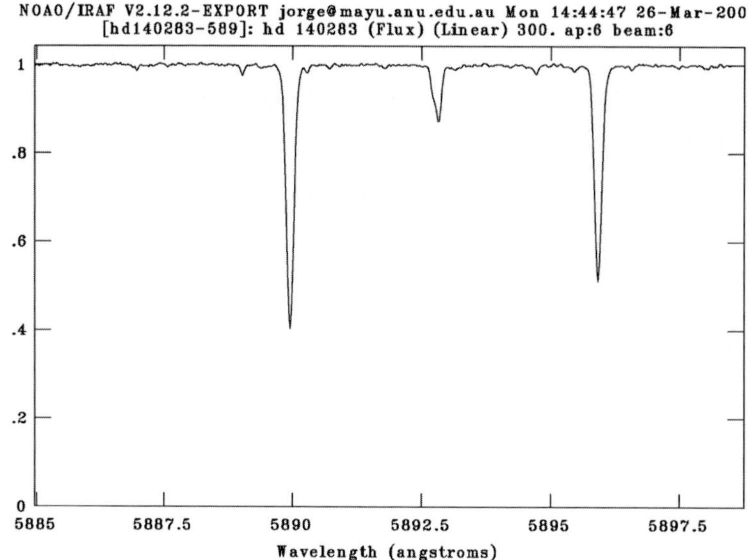

Figure 1. Keck HIRES spectrum around the NaD lines for the nearby (57 pc) metal-poor ([Fe/H] = −2.4) star HD 140283. The only stellar lines that can be easily seen are NaD (the two strongest features) and a blend of NiI/FeI around 589.3nm. The other features are mostly due to weak telluric water vapor lines. From the absence of IS NaD lines we infer E(B-V) = 0.000 for this star, although small amounts of reddening (at the level of 0.001 magnitudes) can not be excluded.

Due to the uncertainties in the Li abundances and to the limited samples available, only limited comparisons of models of Li depletion with stars in a broad range of mass and metallicities have been performed. We are performing such a study (Meléndez *et al.* 2010), achieving errors in Li abundance lower than 0.035 dex, for a large sample of metal-poor stars ($-3.5 <$ [Fe/H] < -1.0), for the first time with precisely determined T_{eff} (Casagrande *et al.* 2010) and masses in a relatively broad mass range (0.6–0.9 $M_\odot$).

Our new temperature scale (Casagrande *et al.* 2010) is highly accurate, since it has been calibrated using solar twins (Meléndez *et al.* 2009; Ramírez *et al.* 2009), and it has also been tested using stellar diameters and absolute flux spectra (Casagrande *et al.* 2010). Our temperatures are also very precise because for most stars the interstellar (IS) NaI D lines were used in the determination of E(B-V) (see e.g. Ramírez *et al.* 2006), implying thus in relatively low errors in our IRFM effective temperatures. An example of a nearby halo star (HD 140283) showing no detectable IS NaD lines (E(B-V) = 0.00) is presented in Fig. 1.

2. Li depletion in Spite plateau stars

Our work shows that Li is depleted in Spite plateau stars (Fig. 2). The spread of the Spite plateau at any metallicity is much larger than the error bar, as can be clearly seen in Fig. 2. Also, there is a correlation between Li and stellar mass at any probed metallicity (Fig. 3), showing thus that Li has been depleted in Spite plateau stars at any metallicity. In Fig. 3 we confront the stellar evolution predictions of Richard *et al.* (2005) with our inferred stellar masses and Li abundances. The models include the effects of atomic diffusion, radiative acceleration and gravitational settling but moderated by a parametrized turbulent mixing. The agreement is very good when adopting a turbulent model of T6.25 and an initial $A_{Li} = 2.64$. The stellar NLTE Li abundances used above

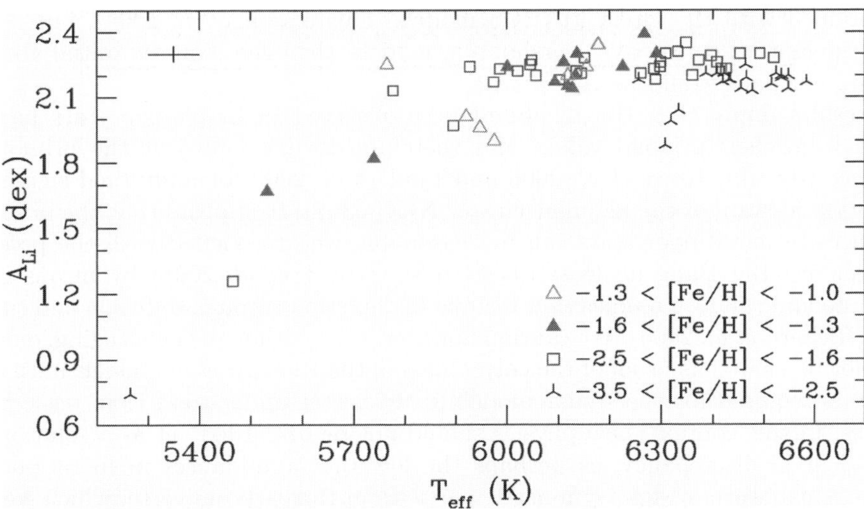

Figure 2. Li abundances vs. T_{eff} for our sample of metal-poor stars in different metallicity ranges. The spread at any given metallicity is much larger than the error bar. Figure taken from Meléndez *et al.* (2010).

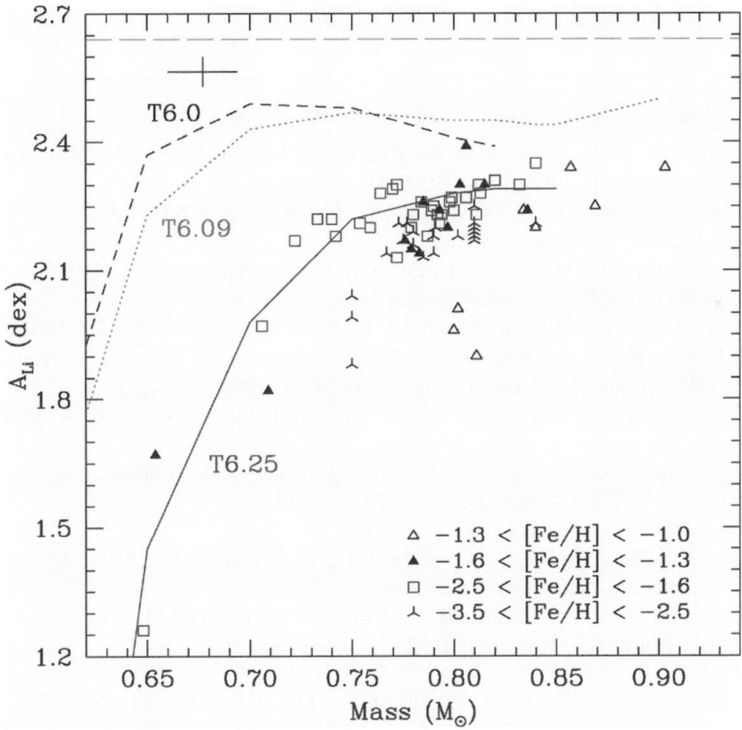

Figure 3. Li abundances as a function of stellar mass in different metallicity ranges. Models at [Fe/H] $= -2.3$ including diffusion and T6.0 (short dashed line), T6.09 (dotted line) and T6.25 (solid line) turbulence (Richard *et al.* 2005) are shown. The models have been rescaled to an initial A_{Li}=2.64 (long dashed line). Figure taken from Meléndez *et al.* (2010).

214   J. Meléndez *et al.*

were obtained with the latest MARCS models (Gustafsson *et al.* 2008), but if we use instead the Kurucz convective overshooting models, then the required initial abundance to explain our data would be $A_{\mathrm{Li}} = 2.72$.

Our results imply that the Li abundances observed in Li plateau stars have been depleted from their original values and therefore do not represent the primordial Li abundance (see also Korn *et al.* 2006 and Lind *et al.* 2009 for additional signatures of Li depletion in stars of the globular cluster NGC 6397). It appears that the observed Li abundances in metal-poor stars can be reasonably well reconciled with the predictions from standard Big Bang nucleosynthesis (e.g. Cyburt *et al.* 2008) by means of more realistic stellar evolution models that include Li depletion through diffusion and turbulent mixing (Richard *et al.* 2005). We caution however, that, although encouraging, our results should not be viewed as proof of the correctness of the Richard *et al.* models until the free parameters required for the stellar modeling are better understood from basic physical principles. In this context, new physics should not be discarded yet as a solution of the cosmological Li discrepancy, as perhaps the low Li-7 abundances in metal-poor stars might be a signature of supersymmetric particles in the early universe, which could also explain the Li-6 detections in metal-poor stars (e.g. Asplund *et al.* 2006; Asplund & Meléndez 2008).

We are expanding our sample to have a better coverage of different evolutionary stages at all metallicities. Our expanded sample (Meléndez *et al.* 2010) will allow us to verify if the Li plateau is indeed depleted at low and high metallicities.

References

Aoki, W. *et al.* 2009, *ApJ*, 698, 1803

Asplund, M., Lambert, D. L., Nissen, P. E., Primas, F., & Smith, V. V. 2006, *ApJ*, 644, 229

Asplund, M. & Meléndez, J. 2008, First Stars III, 990, 342

Boesgaard, A. M., Stephens, A., & Deliyannis, C. P. 2005, *ApJ*, 633, 398

Bonifacio, P. *et al.* 2007, *A&A*, 462, 851

Casagrande, L., Ramírez, I., Meléndez, J., Bessell, M., & Asplund, M. 2010, *A&A*, submitted

Charbonnel, C. & Primas, F. 2005, *A&A*, 442, 961

Coc, A., Olive, K. A., Uzan, J.-P., & Vangioni, E. 2009, *Phys. Rev. D.*, 79, 103512

Cyburt, R. H., Fields, B. D., & Olive, K. A. 2008, *J. Cosmology & Astro-Particle Phys.*, 11, 12

Gustafsson, B., Edvardsson, B., Eriksson, K. *et al.* 2008, *A&A*, 486, 951

Hosford, A., Ryan, S. G., García Pérez, A. E., Norris, J. E., & Olive, K. A. 2009, *A&A*, 493, 601

Jedamzik, K. & Pospelov, M. 2009, *New Journal of Physics*, 11, 105028

Kohri, K. & Santoso, Y. 2009, *Phys. Rev. D.*, 79, 043514

Korn, A. J. *et al.* 2006, *Nature*, 442, 657

Lind, K., Primas, F., Charbonnel, C., Grundahl, F., & Asplund, M. 2009, *A&A*, 503, 545

Meléndez, J. & Ramírez, I. 2004, *ApJ* (Letters), 615, L33

Meléndez, J., Asplund, M., Gustafsson, B., & Yong, D. 2009, *ApJ* (Letters), 704, L66

Meléndez, J., Casagrande, L., Ramírez, I., & Asplund, M. 2010, & W. J. Schuster *A&A*, submitted

Nissen, P. E., Akerman, C., Asplund, M., Fabbian, D., & Pettini, M. 2005, From Lithium to Uranium: Elemental Tracers of Early Cosmic Evolution, *IAU Symp.*, 228, 101

Ramírez, I., Allende Prieto, C., Redfield, S., & Lambert, D. L. 2006, *A&A*, 459, 613

Ramírez, I., Meléndez, J., & Asplund, M. 2009, *A&A*, 508, L17

Richard, O., Michaud, G., & Richer, J. 2005, *ApJ*, 619, 538

Shi, J. R., Gehren, T., Zhang, H. W., Zeng, J. L., & Zhao, G. 2007, *A&A*, 465, 587

Spite, F. & Spite, M. 1982, *A&A*, 115, 357

Light Elements in the Universe
Proceedings IAU Symposium No. 268, 2009
C. Charbonnel, M. Tosi, F. Primas & C. Chiappini, eds.
© International Astronomical Union 2010
doi:10.1017/S1743921310004151

Convection and ^{6}Li in the atmospheres of metal-poor halo stars

Matthias Steffen[1], R. Cayrel[2], P. Bonifacio[2], H.-G. Ludwig[3], and E. Caffau[2]

[1] Astrophysiklaisches Institut Potsdam, An der Sternwarte 16, D-14482 Potsdam, Germany
email: msteffen@aip.de

[2] GEPI – Observatoire de Paris, Paris, France

[3] ZAH-Landessternwarte, Königstuhl 12, D-69117 Heidelberg, Germany

Abstract. Based on 3D hydrodynamical model atmospheres computed with the CO^{5}BOLD code and 3D non-LTE (NLTE) line formation calculations, we study the effect of the convection-induced line asymmetry on the derived ^{6}Li abundance for a range in effective temperature, gravity, and metallicity covering the stars of the Asplund *et al.* (2006) sample. When the asymmetry effect is taken into account for this sample of stars, the resulting ^{6}Li/^{7}Li ratios are reduced by about 1.5% on average with respect to the isotopic ratios determined by Asplund *et al.* (2006). This purely theoretical correction diminishes the number of significant ^{6}Li detections from 9 to 4 (2σ criterion), or from 5 to 2 (3σ criterion). In view of this result the existence of a ^{6}Li plateau appears questionable. A careful reanalysis of individual objects by fitting the observed lithium 6707 Å doublet both with 3D NLTE and 1D LTE synthetic line profiles confirms that the inferred ^{6}Li abundance is systematically lower when using 3D NLTE instead of 1D LTE line fitting. Nevertheless, halo stars with unquestionable ^{6}Li detection do exist even if analyzed in 3D-NLTE, the most prominent example being HD 84937.

Keywords. stars: abundances, atmospheres – hydrodynamics – convection, radiative transfer – line: formation, profiles – stars: individual (G271-162, HD 74000, HD 84937)

1. Introduction

The spectroscopic signature of the presence of ^{6}Li in the atmospheres of metal-poor halo stars is a subtle extra depression in the red wing of the ^{7}Li doublet, which can only be detected in spectra of the highest quality. Based on high-resolution, high signal-to-noise VLT/UVES spectra of 24 bright metal-poor stars, Asplund *et al.* (2006) report the detection of ^{6}Li in nine of these objects. The average ^{6}Li/^{7}Li isotopic ratio in the nine stars in which ^{6}Li has been detected is about 4% and is very similar in each of these stars, defining a ^{6}Li plateau at approximately $\log n(^6\text{Li}) = 0.85$ (on the scale $\log n(\text{H}) = 12$). A convincing theoretical explanation of this new ^{6}Li plateau turned out to be problematic: the high abundances of ^{6}Li at the lowest metallicities cannot be explained by current models of galactic cosmic-ray production, even if the depletion of ^{6}Li during the pre-main-sequence phase is ignored (see reviews by e.g. Christlieb 2008, Cayrel *et al.* 2008, Prantzos 2010 [this volume] and references therein).

A possible solution of the so-called 'second Lithium problem' was proposed by Cayrel *et al.* (2007), who point out that the intrinsic line asymmetry caused by convection in the photospheres of metal-poor turn-off stars is almost indistinguishable from the asymmetry produced by a weak ^{6}Li blend on a presumed symmetric ^{7}Li profile. As a consequence, the derived ^{6}Li abundance should be significantly reduced when the intrinsic line asymmetry in properly taken into account. Using 3D NLTE line formation calculations based on 3D hydrodynamical model atmospheres computed with the CO^5BOLD code (Freytag *et al.* 2002, Wedemeyer *et al.* 2004, see also `http://www.astro.uu.se/~bf/co5bold_main.html`), we quantify the theoretical effect of the convection-induced line asymmetry on the resulting ^6Li abundance as a function of effective temperature, gravity, and metallicity, for a parameter range that covers the stars of the Asplund *et al.* (2006) sample.

A careful reanalysis of individual objects is under way, in which we consider two alternative approaches for fixing the residual line broadening, V_{BR}, the combined effect of macroturbulence (1D only) and instrumental broadening, for given microturbulence (1D only) and rotational velocity: (i) treating V_{BR} as a free parameter when fitting the Li feature, (ii) deriving V_{BR} from additional unblended spectral lines with similar properties as Li I 6707. We show that method (ii) is potentially dangerous, because the inferred broadening parameter shows considerable line-to-line variations, and the resulting ^6Li abundance depends rather sensitively on the adopted value of V_{BR}.

2. 3D hydrodynamical simulations and spectrum synthesis

The hydrodynamical atmospheres used in the present study are part of the CIFIST 3D model atmosphere grid (Ludwig *et al.* 2009). They have been obtained from realistic numerical simulations with the CO^5BOLD code which solves the time-dependent equations of compressible hydrodynamics in a constant gravity field together with the equations of non-local, frequency-dependent radiative transfer in a Cartesian box representative of a volume located at the stellar surface. The computational domain is periodic in x and y direction, has open top and bottom boundaries, and is resolved by typically $140 \times 140 \times 150$ grid cells. The vertical optical depth of the box varies from $\log \tau_{\mathrm{Ross}} \approx -8$ (top) to $\log \tau_{\mathrm{Ross}} \approx +7.5$ (bottom), and the radiative transfer is solved in 6 or 12 opacity bins. Further information about the models used in the present study is compiled in Table 1. Each of the models is represented by a number of snapshots, indicated in column (6), chosen from the full time sequence of the corresponding simulation.

These representative snapshots are processed by the non-LTE code NLTE3D that solves the statistical equilibrium equations for a 17 level lithium atom with 34 line transitions, fully taking into account the 3D thermal structure of the respective model atmosphere. The photo-ionizing radiation field is computed at 704 frequency points between $\lambda\,925$ and $32\,407$ Å, using the opacity distribution functions of Castelli & Kurucz (2004) to allow for metallicity-dependent line-blanketing, including the H I–H$^+$ and H I–H I quasi-molecular absorption near $\lambda\,1400$ and 1600 Å, respectively. Collisional ionization by neutral hydrogen via the charge transfer reaction $H(1s) + Li(n\ell) \leftrightarrow Li^+(1s^2) + H^-$ is treated according to Barklem *et al.* (2003). More details are given in Sbordone *et al.* (2009). Finally, 3D NLTE synthetic line profiles of the Li I $\lambda\,6707$ Å doublet are computed with the line formation code Linfor3D (`http://www.aip.de/~mst/linfor3D_main.html`), using the departure coefficients $b_i = n_i(\mathrm{NLTE})/n_i(\mathrm{LTE})$ provided by NLTE3D for each level i of the lithium model atom as a function of geometrical position within the 3D model atmospheres. As demonstrated in Fig. 1, 3D NLTE effects are very important for the metal-poor dwarfs considered here: they strongly reduce the height range of line formation such that the 3D NLTE equivalent width is smaller by roughly a factor 2 compared to 3D LTE. Ironically, the line strength predicted by standard 1D mixing-length models in LTE are close to the results obtained from elaborate 3D NLTE calculations. We note that the half-width of the 3D NLTE line profile, FWHM(NLTE) = 8.5 km/s, is larger by about 10%: FWHM(LTE) = 7.7 and 7.5 km/s, respectively, before and after reducing the Li abundance such that 3D LTE and 3D NLTE equivalent widths agree. This is because 3D LTE profile senses the higher photosphere where both thermal and hydrodynamical

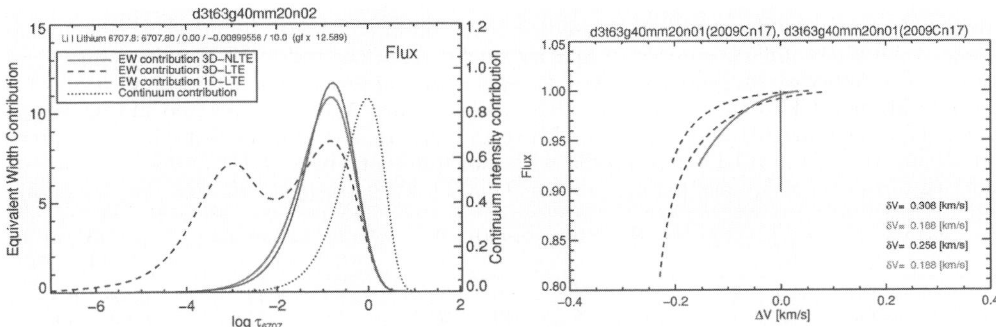

Figure 1. Comparison of 3D LTE (dashed), 3D non-LTE (thick solid), and 1D LTE (thin solid) equivalent width contribution functions (left) and line bisectors (right) of a single ^{7}Li component, computed for a typical metal-poor turn-off halo star ($T_{\rm eff} = 6215$ K, $\log g = 4.0$, [Fe/H] $= -2$). Non-LTE effects strongly reduce the height range of formation and equivalent width of the line (W: 35.5 $\rightarrow$ 15.6 mÅ), while differences between 1D LTE and 3D NLTE are much smaller. The line asymmetry is smaller in non-LTE: the velocity span of the bisector is δv(NLTE) $\approx$ 190 m/s, while δv(LTE) $\approx$ 310 and 260 m/s, respectively, before and after adjusting the equivalent width (by reducing the Li abundance) and the half-width (by Gaussian broadening) of the LTE line to match the values of the NLTE profile (compare the two dashed bisectors).

velocities are lower. However, the NLTE line profile is significantly less asymmetric than the LTE profile, even if the latter is broadened to the same half-width (Fig. 1, right panel).

3. ^{6}Li bias due to convective line asymmetry

As outlined above, the ^{6}Li abundance is systematically overestimated if one ignores the intrinsic asymmetry of the ^{7}Li line components. To quantify this bias theoretically, we rely on synthetic spectra. The idea is as follows: we represent the observation by the synthetic 3D NLTE line profile of the ^{7}Li line blend, computed with zero ^{6}Li content. Except for an optional rotational broadening, the only source of non-thermal line broadening is the 3D hydrodynamical velocity field, which also gives rise to a convective blue-shift and an intrinsic line asymmetry. Now this 3D ^{7}Li line blend is fitted by 'classical' 1D synthetic line profiles composed of intrinsically symmetric components of ^{6}Li and ^{7}Li. Four parameters are varied independently to find the best fit (minimum χ^2): in addition to the total ^{6}Li+^{7}Li abundance, A(Li), and the ^{6}Li/^{7}Li isotopic ratio, q(Li), which control line strength and line asymmetry, respectively, we also allow for a residual line broadening described by a Gaussian kernel with half-width $V_{\rm BR}$, and a global line shift, Δv. Note that the four fitting parameters are non-degenerate, since each one has a distinctly different effect on the line profile. The rotational line broadening is fixed to the value used in the 3D spectrum synthesis (we tried $v \sin i = 0$ and 2 km/s). The value q^*(Li) of the best fit is then identified with the correction that has to be *subtracted* from the ^{6}Li/^{7}Li isotopic ratio derived from the 1D analysis to correct for the bias introduced by neglecting the intrinsic line asymmetry: $q^{(3D)}$(Li) $= q^{(1D)}$(Li) - q^*(Li). The procedure properly accounts for radiative transfer in the lines, including saturation effects.

Two different sets of 1D profiles were used for this purpose: (a) NLTE line profiles based on the $\langle$3D$\rangle$ model, constructed by averaging the 3D model on surfaces of constant optical depth, and (b) LTE line profiles computed from a so-called LHD model, a 1D mixing-length model atmosphere that has the same stellar parameters and uses the same microphysics and radiative transfer scheme as the corresponding 3D model. For both kind of 1D models, the microturbulence was fixed at $\xi_{\rm mic} = 1.5$ km/s. The mixing length parameter adopted for the LHD models is $\alpha_{\rm MLT} = 0.5$.

Table 1. List of models used in the present study. Columns (2)-(6) give effective temperature, surface gravity, metallicity, number of opacity bins used in the radiation hydrodynamics simulation, and number of snapshots selected for spectrum synthesis. The equivalent width of the synthetic 3D non-LTE ^{7}Li doublet at $\lambda\,6707$ Å, assuming A(Li)=2.2 and no ^{6}Li, is given in column (7). Columns (8) and (9) show q_a^*(Li) and q_b^*(Li), the ^{6}Li/^{7}Li isotopic ratio inferred from fitting this 3D non-LTE line profile with two different kinds of 1D profiles, in each case assuming a rotational broadening of $v \sin i = 0.0$ / 2.0 km/s, respectively (see text for details).

Model	$T_{\rm eff}$[1] [K]	$\log g$	[Fe/H]	Nbin	Nsnap	W[2] [mÅ]	$q_a^* = $n(^{6}Li)/n(^{7}Li) $\langle$3D$\rangle$ NLTE [%]	$q_b^* = $n(^{6}Li)/n(^{7}Li) 1D LTE [%]
d3t59g40mm30n02	5846	4.0	-3.0	6	20	44.9	1.14 / 1.14	**0.88 / 0.88**
d3t59g45mm30n01	5924	4.5	-3.0	6	19	39.9	0.75 / 0.75	**0.63 / 0.64**
d3t63g40mm30n01	6269	4.0	-3.0	6	20	23.9	1.96 / 1.93	**1.86 / 1.83**
d3t63g40mm30n02	6242	4.0	-3.0	12	20	24.1	1.81 / 1.80	**1.63 / 1.62**
d3t63g45mm30n01	6272	4.5	-3.0	6	18	24.3	1.07 / 1.06	**1.02 / 1.00**
d3t65g40mm30n01	6408	4.0	-3.0	6	20	20.0	1.75 / 1.70	**1.70 / 1.66**
d3t65g45mm30n01	6556	4.5	-3.0	6	12	16.4	1.29 / 1.27	**1.25 / 1.22**
d3t59g35mm20n01	5861	3.5	-2.0	6	20	42.8	2.48 / 2.44	**2.02 / 2.01**
d3t59g40mm20n02	5856	4.0	-2.0	6	20	45.2	1.59 / 1.59	**0.96 / 0.98**
d3t59g45mm20n01	5923	4.5	-2.0	6	18	42.3	1.12 / 1.13	**0.45 / 0.46**
d3t63g35mm20n01	6287	3.5	-2.0	6	20	22.1	4.09 / 4.00	**4.04 / 3.97**
d3t63g40mm20n01	6278	4.0	-2.0	6	16	23.7	1.95 / 1.94	**1.79 / 1.78**
d3t63g40mm20n02	6215	4.0	-2.0	12	20	25.1	1.92 / 1.92	**1.66 / 1.67**
d3t63g45mm20n01	6323	4.5	-2.0	6	19	23.0	1.18 / 1.18	**0.97 / 0.97**
d3t65g40mm20n01	6534	4.0	-2.0	6	19	16.2	2.28 / 2.21	**2.22 / 2.16**
d3t65g45mm20n01	6533	4.5	-2.0	6	19	17.0	1.31 / 1.29	**1.21 / 1.19**
d3t59g40mm10n02	5850	4.0	-1.0	6	20	41.5	1.60 / 1.62	**1.45 / 1.47**
d3t59g45mm10n01	5923	4.5	-1.0	6	08	38.2	1.14 / 1.15	**0.83 / 0.85**
d3t63g40mm10n02	6261	4.0	-1.0	6	20	22.0	2.37 / 2.38	**2.33 / 2.33**
d3t63g40mm10n02	6236	4.0	-1.0	12	20	21.8	2.15 / 2.16	**2.05 / 2.06**
d3t63g45mm10n01	6238	4.5	-1.0	6	20	23.4	1.36 / 1.37	**1.23 / 1.24**
d3t65g40mm10n01	6503	4.0	-1.0	6	20	15.5	3.10 / 3.02	**3.14 / 3.06**
d3t65g45mm10n01	6456	4.5	-1.0	6	19	17.1	1.51 / 1.51	**1.44 / 1.43**

Notes: [1] averaged over selected snapshots; [2] averaged over selected 3D non-LTE spectra

We have determined q^*(Li) according to the method outlined above for our grid of 3D model atmospheres. The results are given in Table 1, both for fitting with the $\langle$3D$\rangle$ NLTE profiles (q_a^*(Li), col. (8)) and with the 1D LTE profiles of the LHD models (q_b^*(Li), col. (9)). At given metallicity, the corrections are largest for low gravity and high effective temperature. They increase towards higher metallicity. We also note that q^*(Li) is essentially insensitive to the choice of the rotational broadening.

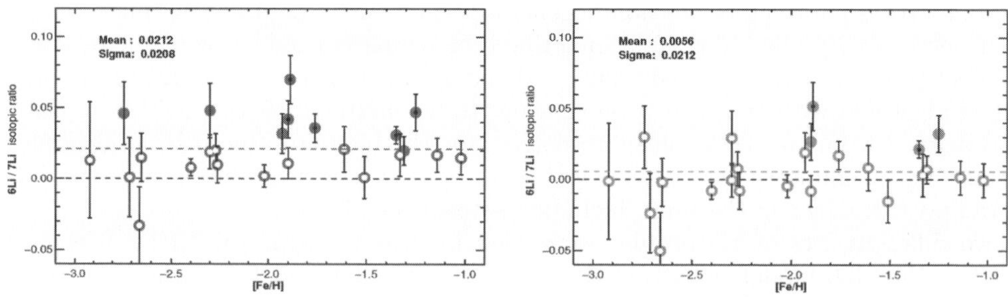

Figure 2. ^{6}Li/^{7}Li isotopic ratio, and $\pm 1\sigma$ error bars, as a function of metallicity as derived by Asplund *et al.* (2006) before (left) and after (right) subtraction of q_b^*(Li) (Table 1, col. (9)) to correct for the bias due to the intrinsic line asymmetry. Open circles denote non-detections, filled circles and filled circles with a black dot denote ^{6}Li detections above the 2σ and 3σ level, respectively.

The analysis of Asplund *et al.* (2006) utilizes 1D LTE profiles computed from MARCS model atmospheres. Hence, the correction $q_b^*(\mathrm{Li})$, computed with the 1D LTE profiles of the 1D LHD models, should be applied to their $^6\mathrm{Li}/^7\mathrm{Li}$ isotopic ratios. The resulting downward corrections are typically in the range $1\% < q_b^*(\mathrm{Li}) < 2\%$ for the stars of their sample (cf. Steffen *et al.* 2010, Fig. 2). After subtracting the individual $q_b^*(\mathrm{Li})$ for each of these stars, according to their T_{eff}, $\log g$, and [Fe/H], the mean $^6\mathrm{Li}/^7\mathrm{Li}$ isotopic ratio of the sample is reduced from 0.0212 to 0.0056, as illustrated in Fig. 2. Given the $^6\mathrm{Li}/^7\mathrm{Li}$ isotopic ratios and their 1σ error bars as determined by Asplund *et al.* (2006), the number of stars with a $^6\mathrm{Li}$ detection above the 2σ and 3σ level is 9 and 5, respectively, out of 24. After correction, the number of 2σ and 3σ detections is reduced to 4 and 2, respectively. One of the stars meeting the 3σ criterion after correction, HD 106038, survives only because of its particularly small error bar of ±0.006. The other one is G020-024, which shows the clearest evidence for the presence of $^6\mathrm{Li}$ ($q(\mathrm{Li}) = 0.052 \pm 0.017$), while HD 102200 remains the only clear 2σ detection ($q(\mathrm{Li}) = 0.033 \pm 0.013$). The spectra of these stars should be reanalyzed with 3D NLTE line profiles.

4. Analysis of observed spectra: 3D NLTE versus 1D LTE line fitting

As a further exercise, we have fitted the observed Li I $\lambda\,6707$ Å spectra of three halo turn-off stars (see Table 2), both with 3D NLTE and 1D LTE synthetic line profiles. Since the parameters of these three stars are very similar, the synthetic spectra were computed in all cases from hydrodynamical model d3t6300g40mm20n01 (see Table 1) and the corresponding 1D LHD model. The standard approach is to vary the four parameters $A(\mathrm{Li})$, $q(\mathrm{Li})$, V_{BR}, and Δv to find the optimum line profile fit, as described above (method I). The results are presented in Table 3. As expected, the 3D analysis yields lower $^6\mathrm{Li}/^7\mathrm{Li}$ isotopic ratios by about 1.7%. The numbers differ slightly from those given by Steffen *et al.* 2010 due to an upgrade of the line fitting procedure, but the conclusions remain unchanged: HD 74000 and G271$-$162 are considered non-detections, while HD 84937 remains a clear $^6\mathrm{Li}$ detection with $q^{(3\mathrm{D-NLTE})}(\mathrm{Li}) \approx 5\%$.

For completeness, Table 3 also shows the results of fitting the observed lines with 3D LTE synthetic profiles (not recommended). The $^6\mathrm{Li}$ abundances obtained in this way lie between the 3D NLTE and the 1D LTE results. It is not obvious why fitting with the more asymmetric 3D LTE profiles gives higher $^6\mathrm{Li}$ than fitting with the slightly more symmetric 3D NLTE profiles, but this is a robust result.

For HD 74000, we also tried an alternative approach (method II), where V_{BR} is determined from other spectral lines, and is then fixed during the fitting of the Li I $\lambda\,6707$ Å feature. For this purpose, we rely on a set of 6 clean Fe I lines with similar excitation potential between 2.4 and 2.6 eV: the 5 lines of Cayrel *et al.* (2007), Table 1, plus Fe I $\lambda\,6230.7$ Å, $E_i = 2.559$ eV. Their equivalent widths range from $W \approx 13$ to 30 mÅ, embracing the strength of the Li doublet. Fitting the 6 Fe I lines with 3D LTE synthetic line profiles, we obtain $V_{\mathrm{BR}}(\mathrm{Fe}) = 3.05\,(2.15)\pm0.16\,(0.23)$ km/s for $v \sin i = 0\,(2)$ km/s, in close agreement with $V_{\mathrm{BR}}(\mathrm{Li})$ inferred from the fitting of the Li I profile. This result may be taken as an indication that, in contrast to Li, the selected Fe I lines are not severely affected by departures from LTE. We note, however, a trend of $V_{\mathrm{BR}}(\mathrm{Fe})$

Table 2. Observed Li I $\lambda\,6707$ Å spectra of three metal-poor turn-off stars

Star	T_{eff} [K]	$\log g$	[Fe/H]	$R=\lambda/\Delta\lambda$	S/N	Instrument	Reference
HD 74000	6203	4.03	-2.05	120 000	600	ESO3.6 / HARPS	Cayrel *et al.* 2007
G271$-$162	6330	4.00	-2.25	110 000	600	VLT / UVES	Nissen *et al.* 2000
HD 84937	6300	4.00	-2.30	100 000	630	CFHT / GECKO	Cayrel *et al.* 1999

Table 3. Fitting the observed spectra of Table 2 with 3D and 1D synthetic line profiles. Columns (4)-(7) show the results for $v \sin i = 0.0\ /\ 2.0$ km/s.

Star	synthetic spectrum	fitting method	$A(\mathrm{Li})$ [1]	$n(^6\mathrm{Li})/n(^7\mathrm{Li})$ [%]	V_{BR} [2] [km/s]	Δv [km/s]
HD 74000	3D NLTE	I	2.25 / 2.25	**−1.1 / −1.1**	3.1 / 2.1	0.64 / 0.64
	(3D LTE)	I	1.85 / 1.85	**−0.4 / −0.5**	4.7 / 4.1	0.66 / 0.67
	1D LTE	I	2.23 / 2.23	**0.6 / 0.6**	5.9 / 5.4	0.43 / 0.43
	3D NLTE	II	2.25 / 2.25	**−1.1 / −1.3**	**3.1 / 2.2**	0.64 / 0.65
	(3D LTE)	II	1.84 / 1.84	**4.6 / 4.2**	**3.1 / 2.2**	0.46 / 0.47
	1D LTE	II	2.23 / 2.23	**1.7 / 1.7**	**5.6 / 5.1**	0.39 / 0.39
G271−162	3D NLTE	I	2.30 / 2.30	**0.7 / 0.6**	3.7 / 2.9	0.04 / 0.04
	(3D LTE)	I	1.89 / 1.89	**1.3 / 1.1**	5.2 / 4.6	0.06 / 0.07
	1D LTE	I	2.27 / 2.27	**2.3 / 2.4**	6.2 / 5.8	−0.17 / −0.17
HD 84937	3D NLTE	I	2.20 / 2.20	**5.2 / 5.2**	3.3 / 2.4	0.02 / 0.02
	(3D LTE)	I	1.80 / 1.80	**5.8 / 5.7**	4.8 / 4.3	0.05 / 0.05
	1D LTE	I	2.18 / 2.18	**6.9 / 6.9**	6.0 / 5.6	−0.19 / −0.19

Notes: [1] $\log\left[n(^6\mathrm{Li}) + n(^7\mathrm{Li})\right] - \log n(\mathrm{H}) + 12$; [2] Gaussian kernel, **bold**: fixed from Fe I lines.

increasing slightly with W, which remains to be understood. If instead the Fe I lines are fitted with 1D LTE synthetic profiles, we find $V_{\mathrm{BR}}(\mathrm{Fe}) = 5.61\ (5.13) \pm 0.07\ (0.09)$ km/s, systematically lower than $V_{\mathrm{BR}}(\mathrm{Li})$ by 0.3 km/s. Fixing $V_{\mathrm{BR}} = V_{\mathrm{BR}}(\mathrm{Fe})$, the best fit of the Li I doublet implies $q^{\mathrm{1D\ LTE}}(\mathrm{Li}) = 1.7\%$, which is 1.1% higher than with method I. Finally, Table 3 demonstrates that applying method II with 3D LTE fitting of Li I leads to a severe overestimation of the $^6\mathrm{Li}/^7\mathrm{Li}$ isotopic ratio: $q^{\mathrm{3D\ LTE}}(\mathrm{Li}) > 4\%$ compared to $q^{\mathrm{3D\ NLTE}}(\mathrm{Li}) = -1.1\%$!

5. Conclusions

The $^6\mathrm{Li}/^7\mathrm{Li}$ isotopic ratio derived from fitting of the Li I doublet with 3D NLTE synthetic line profiles is shown to be about 1% to 2% lower than what is obtained with 1D LTE profiles. Based on this result, we conclude that only 2 out of the 24 stars of the Asplund *et al.* (2006) sample would remain significant $^6\mathrm{Li}$ detections when subjected to a 3D non-LTE analysis, suggesting that the presence of $^6\mathrm{Li}$ in the atmospheres of galactic halo stars is rather the exception than the rule, and hence does not necessarily constitute a *cosmological* $^6\mathrm{Li}$ problem. If we adopt the approach by Asplund *et al.* (2006), relying on additional spectral lines to fix the residual line broadening, the difference between 3D NLTE and 1D LTE results increases even more, as far as we can judge from our case study HD 74000. Until the 3D NLTE effects are fully understood for all involved lines, we consider this method as potentially dangerous.

References

Asplund, M., Lambert, D. L., Nissen, P. E., Primas, F., & Smith, V. V. 2006, *ApJ*, 644, 229
Barklem, P. S., Belyaev, A. K., & Asplund, M. 2003, *A&A*, 409, L1
Castelli, F. & Kurucz, R. L. 2004, arXiv:astro-ph/0405087
Cayrel, R. *et al.* 1999, *A&A*, 343, 923
Cayrel, R. *et al.* 2007, *A&A*, 473, L37
Cayrel, R. *et al.* 2008, in Proceedings of *Nuclei in the Cosmos (NIC X)*
Christlieb, N. 2008, *Journal of Physics G: Nuclear Physics*, 35, 014001
Freytag, B., Steffen, M., & Dorch, B. 2002, *AN*, 323, 213
Ludwig, H.-G. *et al.* 2009, *MemSAI*, 80, 708
Nissen, P. E., Asplund, M., Hill, V., & D'Odorico, S. 2000, *A&A* 357, L49
Sbordone, L., Bonifacio, P., Caffau, E. *et al.* 2009, *A&A* (in press)
Steffen, M. *et al.* 2010, in: K. Cunha, M. Spite, and B. Barbuy (eds.), *Chemical Abundances in the Universe: Connecting First Stars to Planets*, Proc. IAU Symposium No. 265, in press
Wedemeyer, S., Freytag, B., Steffen, M., Ludwig, H.-G., & Holweger, H. 2004, *A&A*, 414, 1121

Light Elements in the Universe
Proceedings IAU Symposium No. 268, 2009
C. Charbonnel, M. Tosi, F. Primas & C. Chiappini, eds.

© International Astronomical Union 2010
doi:10.1017/S1743921310004163

Beryllium and boron in metal-poor stars

Francesca Primas[1]

[1]European Southern Observatory,
KarlSchwarzschild Strasse 2, D-85748, Garching b. München, Germany
email: fprimas@eso.org

Abstract. Knowledge of lithium, beryllium, and boron abundances in stars of the Galactic halo and disk plays a major role in our understanding of Big Bang nucleosynthesis, cosmic-ray physics, and stellar interiors. ^{9}Be and ^{10}B are believed to originate entirely from spallation reactions in the interstellar medium (ISM) between α-particles and protons and heavy nuclei like carbon, nitrogen, and oxygen (CNO), whereas ^{11}B may have an extra production channel via neutrino-spallation. Beryllium and boron are both observationally challenging, with their main resonant doublets falling respectively at 313 nm and at 250 nm. The advent of 8-10m class telescopes equipped with highly sensitive (in the near-UV/blue) spectrographs has opened up a new era of Be abundance studies. Here, I will review and discuss the most interesting results of recent observational campaigns in terms of formation and evolution of these two light elements.

Keywords. Galaxy: halo, abundances, evolution – stars: Population II, abundances, atmospheres, interiors

1. Introduction

After hydrogen (H), helium (He), and lithium (Li), beryllium (Be) and boron (B) complete the group of the so-called light elements, all covered in great detail during this Symposium. All these elements share a common feature concerning their origin: they can not be produced in stellar interiors, where in fact most of them get destroyed. Deuterium, He, and Li have a strong cosmological component responsible for their formation, hence they offer important insights on the primordial abundances of their isotopes in the early Universe. On the contrary, Be and B are formed via spallation reactions between CNO atoms and α+p particles.

For a long time, it was believed that these reactions take mostly place in the interstellar medium, where alphas and protons (carried by cosmic rays – CR – traveling across the Galaxy) would end up hitting target nuclei like CNO, spallating into atoms of beryllium and boron (cf Reeves *et al.* 1970, Meneguzzi *et al.* 1971). One of the earliest observational findings supporting this scenario was the beryllium-to-hydrogen ratios observed in young stars ($\sim 10^{-11}$) in the 70s (e.g. Boesgaard 1976, Boesgaard *et al.* 1977) which was found to match rather closely the theoretical prediction of Be formation when integrated along the entire history of the Galaxy. In other words, the abundances derived in a few Population I (Pop I) objects were approximately matching the product of the formation rate of Be† $\times$ the abundance ratio of these targets with respect to hydrogen ($\sim 10^{-3}$) in space $\times$ the age of the Galaxy.

Indeed, the observed CR flux, integrated over 10^{10} years of Galactic evolution, seems sufficient to produce the solar abundances of B, Be, and ^{6}Li, despite, e.g., the failure in reproducing the observed solar B isotopic ratio. CR spallation in the general ISM has

---

† described by Reeves *et al.* (1970) to be the product between the flux of high-energy protons (~ 16 cm^{-2} sec) $\times$ the cross sections for ^{9}Be formation by proton collision on the most abundant targets, ^{16}O and ^{12}C (~ 5 mb)

therefore been mostly accepted for 25 years as the main origin of these light elements. However, as I will show in the next section, the first observational studies carried out at 4m ground-based telescopes and with the Hubble Space Telescope (HST) showed some surprising results that triggered a revision of the above described scenario.

Last but not least, the light elements Li, Be, and B, when measured in the same object(s) can be successfully used as diagnostics in the study of stellar mixing. Because they burn at relatively low, but slightly different (and increasing) temperatures, they are expected to show different degrees of depletion and/or dilution at different times. Lithium, which is destroyed at the lowest temperature of all three elements ($T_B \simeq 2.5 \times 10^6$ K) should deplete first, followed by Be ($T_B \simeq 3.5 \times 10^6$ K) and B ($T_B \simeq 5 \times 10^6$ K). Unfortunately, there are large discrepancies in the size of the data samples that have been looked at, which clearly correlate with the observational difficulties in detecting a given transition: lithium, with its main resonant doublet falling in the optical, at 670 nm, is the most studied of the three, followed by beryllium (main resonant doublet very close to the atmospheric cut-off, at 313 nm), and boron (resonant doublet at 250 nm and single feature at 209 nm), which requires a telescope in space.

2. A bit of (observational) history

The first attempts to detect beryllium in Population II (Pop II) stars were made in the mid-late 80s (e.g. Molaro *et al.* 1984, Rebolo *et al.* 1988) with, respectively, the International Ultraviolet Explorer (IUE) and the 2.5m Isaac Newton Telescope. Unfortunately, these studies succeeded only in measuring beryllium in the Pop I stars of their targeted samples, but failed to detect it in the most metal-poor, Pop II objects, for which only upper limits were derived.

These attempts were then followed by (somewhat larger) observational campaigns in the early 90s (Ryan *et al.* 1992, Gilmore *et al.* 1992, Boesgaard & King 1993), which used 4m class telescopes, equipped with spectrographs with a higher sensitivity in the blue. Ryan *et al.* (1992) and Boesgaard & King (1993) found a quadratic dependence of Be abundances from metallicity and oxygen, thus confirming the CR spallation scenario as the main production channel for beryllium. On the contrary, the study by Gilmore *et al.* (1992) uncovered a different trend, a linear slope between Be and Fe,O which confirmed some of the very early results obtained with HST for stellar boron abundances (see below).

For boron, after the pioneering work by Boesgaard & Heacox (1973, 1978) who looked at early-type stars (similar to what was done in the very early works for beryllium), one has to wait until the launch of the Hubble Space Telescope for the first measurements of boron abundances in metal-poor stars (Duncan *et al.* 1992). The sample consisted of 3 metal-poor stars, which these investigators surprisingly found to follow a linear relation. According to the authors, this implied that the evolution of boron cannot depend on metallicity, hence the spallation reactions cannot possibly take place in the interstellar medium, whose metal content keeps increasing as the Galaxy evolves. Instead, they proposed a "reverse" spallation reaction scenario as the one dominating the early phases of the Galaxy, in which it is the CNO (especially oxygen) in the CRs (under the assumption that they remain constant) that hit protons and α-particles in the ISM. For the record, it is important to note that this reverse mechanism had already been mentioned by Reeves and collaborators, but thought to be negligible, hence it had basically been ignored.

Needless to say, both findings (Gilmore *et al.* 1992 for beryllium and Duncan *et al.* 1992 for boron) have had a very strong impact on this field of research. In the following sections, I will mostly focus on beryllium, since it is the element for which most progress

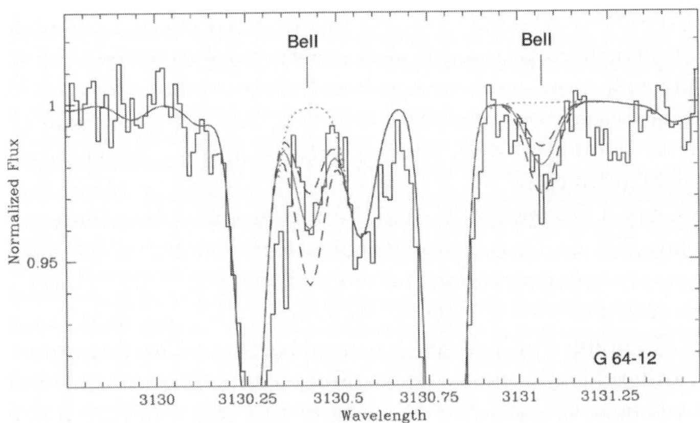

Figure 1. Best-fit spectrum synthesis (continuous curve) of the Be II doublet in one of the most metal-poor stars ever studied for beryllium, G 064-012 ([Fe/H]$\simeq -3.2$, histogram). Overplotted are also two spectrum syntheses computed with $+/-$ 0.15 dex with respect to the best-fit Be value (dashed curves) and one synthesis computed without any beryllium (dotted curve). (From Primas *et al.* 200b).

has been made in the last decade. I will come back to boron towards the end of this review.

3. The impact of 8-10m class telescopes and efficient spectrographs

The advent of high-resolution spectrographs very sensitive in the near-UV on 8-10m class telescopes marked a new observational era, especially for studies of beryllium abundances in metal-poor stars. The observational progress achieved at the Be wavelengths (313 nm) over a bit more than one decade is remarkable: one went from having to integrate 6,000 s on a $V_{mag} = 8$ star in order to obtain a spectrum with a resolving power of R$\simeq$11,000 and S/N $= 37$ (1988, Isaac Newton Telescope) to 3,000 s, R $\simeq 25,000$ and S/N $= 55$ (1993, Canada-France-Hawaii Telescope) and to 1,800 s, R $\simeq 45,000$ and S/N $= 120$ (2000, Very Large Telescope).

Boesgaard *et al.* (1999) presented the first systematic analysis of Be in a sample of 22 halo stars plus 5 disk stars observed with HIRES at Keck (Vogt *et al.* 1994) pushing the detection of Be in stars with metallicities approaching one thousandth less than solar. The major result from this study was a much more robust confirmation of the linear dependence between Be and Fe already found by Gilmore *et al.* (1992) based only on 6 objects. Furthermore, Boesgaard *et al.* (1999) confirmed a similar correlation also between Be and O.

Primas *et al.* (2000a,b) explored the capabilities of the high-resolution echelle spectrograph UVES (Dekker *et al.* 2000) at the VLT and recorded the first detection of Be in two stars with metallicities below one thousandth solar ([Fe/H] $= -3.15$, cf Figure 1).

Finally, Pasquini *et al.* (2004, 2007) detected for the first time with UVES at the VLT beryllium in turn-off stars of metal-poor globular clusters, like NGC 6397 and NGC 6752.

In more recent years, several Be studies have been conducted, which looked at different aspects of the formation of this light isotope, both in more metal-rich/solar-metallicity stars and metal-poor objects. Very recently, an upper limit has been derived on the Be content of a very metal-poor ([Fe/H] $= -3.7$), carbon-rich subgiant star (Ito *et al.* 2009). The three most recent analyses (Smiljanic *et al.* 2009, Rich & Boesgaard 2009, Tan *et al.* 2009) that have looked at stellar samples covering a wide range of metallicities, all addressed issues like the formation and evolution of Be along the Galactic history. All three studies have presented the trends of Be *vs* iron <u>and</u> *vs* oxygen (or versus an average ratio of the alpha elements) and similar results are found: the correlation coefficient between Be and Fe ranges between 0.97 and 1.2, and they all seem to find a steeper relation when Be is compared to O, which seems to imply a faster increase of Be in the Galaxy with respect to oxygen.

Based on our current understanding of the formation of beryllium, it is indeed more important to relate Be to O: oxygen atoms are the main target nuclei of spallation reactions, hence such a relation should be more straightforward to interpret. In practice, accurate measurements of stellar oxygen abundances remain hard to achieve, due to a variety of oxygen abundance indicators that need to be used in different types of stars and that suffer from different systematic uncertainties that have not always been carefully evaluated. For instance, UV OH lines, basically the only ones available in dwarf stars at very low metallicities, give a very different trend of [O/Fe] *vs* [Fe/H], which seems to be due to the usage of unsuitable model atmospheres and model atoms (ignoring NLTE and 3D effects). The oxygen triplet, on the other hand, possibly detectable in the most metal-poor stars for which a Be detection has been achieved, is highly sensitive to the assumed stellar effective temperature, which in this case becomes the largest source of error on the final determination of the oxygen abundance. The O forbidden line (at 630 nm), the cleanest of all indicators, becomes quickly undetectable in dwarfs as one reaches metallicities around one hundredth less than solar.

3.1. *Main recent findings and remaining open issues*

All most recent studies of beryllium abundances in Galactic stars seem to agree on a few main findings: *a)* Be closely follows Fe; *b)* the slope between Be and O seems to be slightly steeper than with respect to Fe; *c)* the dispersion around the trends is very small in the most metal-poor stars but it increases at higher metallicities. The metallicity range around the so-called halo-disk transition seems to be characterized by significant scatter.

As already pointed out above, the uncertainty affecting our current determination of stellar oxygen abundances remains one of the main issues in order to properly interpret the Be *vs* O trend, if the correlation is truly different from a linear one. This uncertainty affects also another important test that has been attempted in recent years: the utilization of Be as a cosmic clock. The idea is based on the fact that if Be is produced via a primary process, this process would then become a global one instead of a local one, like, e.g., the production of heavier nuclei in supernovae explosions. Beryllium abundances are then expected to show smaller dispersion compared to elements like O and/or Fe, making Be a possibly more reliable cosmo-chronometer than the commonly used ratios like [Fe/H] and/or [O/H]. Pasquini *et al.* (2005) were the first ones to test this idea, by investigating the evolution of the star formation rate in the early Galaxy using Be and O abundances. In order to do so, they compared O *vs* Be observed trends for stars belonging to two separate kinematical classes, the *accretion* and the *dissipative* components (roughly corresponding to the halo and thick disk populations), as identified by Gratton *et al.* (2003). They found out that these two samples seem to show different evolutions. Unfortunately, the sample size was rather limited and the sample splitting was not as clean as wished for such a comparison (for instance, they ended up with representatives of the *dissipative* component belonging to the halo, but with kinematical characteristics of the thick disk). This prevented these authors from drawing firm conclusions on the usefulness of Be as a cosmo-chronometer. Subsequent attempts also failed to fully prove the concept. Tan *et al.* (2009), following Pasquini *et al.* (2005) did not succeed in obtaining a clearer separation between the *dissipative* and the *accretion* components (cf their

Fig. 9). Smiljanic *et al.* (2009) presented [O/Fe] and [α/Fe] ratios as a function of Be for a large data sample, and tried to similarly distinguish between halo and thick disk stars. Indeed, they found a possible separation, but also here abundances were too scattered (especially the alpha abundances, since they were taken from the literature) to be able to draw firm conclusions.

4. Shedding new light on Be formation and evolution

Our observational campaign, aimed at investigating the formation and evolution of Be across the Galactic history, has been carried out at the VLT with the UVES spectrograph. The sample includes more than 50 stars, spanning a wide range in metallicities (from -3.3 to -0.5). The objects are mostly dwarf or subgiant stars, belonging to both the halo and disk populations. All observed spectra have a resolving power of at least 40,000 and S/N ratios are of the order of 100 or higher at the Be II transitions. At redder wavelengths (since we took advantage of the dichroic mode of UVES) both S/N ratios and resolving power are higher, respectively, above 2–300 and R $\simeq$ 55,000.

Stellar parameters of our sample were determined from Strömgren photometry (effective temperatures, also cross-checked both with V-K colour indices and with H-α fitting), from Hipparcos parallaxes (gravities), and from ionized Fe lines (metallicities). We used 1-D OSMARCS model atmospheres (Gustafsson *et al.* 2008) and we corrected our derived Be abundances for NLTE corrections (García Pérez, *priv. comm.*). We also derived lithium, oxygen, and other abundances, like Mg.

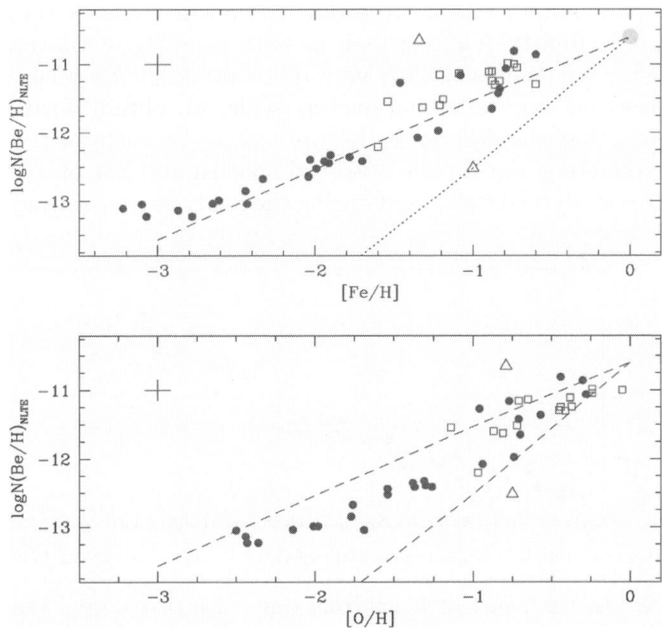

Figure 2. Our derived Be abundances *vs* metallicity (*top*) and [O/H] (*bottom*). Typical errorbars are shown in the upper left corner of each plot. Dashed and dotted line represent respectively a linear and quadratic relation. Symbols: filled circles represent halo stars, open squares disk objects. The two open triangles represent two particular disk objects, one depleted in its Be content and one instead with a very high content of Be (as well as many other elemental abundances).

Figure 2 presents our Be results at large. The Be *vs* Fe trend follows a linear slope, with very little scatter below a metallicity of [Fe/H] $= -1.5$, and a significantly increased dispersion around the halo-disk transition. If we display our Be abundances *vs* oxygen (which is a mix of abundances derived from the forbidden line at 630 nm and the triplet at 770 nm), the slope does not seem to change much but the trend may be more difficult to interpret, especially at higher metallicities (see caption for more details). Furthermore, we have added a few more Be detections below [Fe/H] = −3.0, which seem to fall a bit above the unitary slope trend identified by the rest of the sample. This feature, together with the need to correctly understand the larger dispersion at higher metallicities, are the main highlights of this study, thus they deserve to be discussed in more detail.

4.1. *The very metal-poor tail of the Be evolution*

The 4-5 most metal-poor stars of our sample seem to have slightly higher Be abundances with respect to the trend of unitary slope, i.e. the metal-poor tail of the Be *vs* Fe evolutionary trend seems to show some flattening (cf Fig. 2). What does this mean ? Certainly, it does not carry any significance for a possible primordial production of beryllium; rather, some of the theoretical scenarios proposed in the past for the production of Be atoms in the early Galaxy predicted the appearance of such a plateau, when the masses of the supernova progenitors were in the range of 40-60 M$_\odot$ (cf Vangioni-Flam *et al.* 1998).

However, before looking for exotic explanations for such feature, it is important to note that one other study that has succeeded in detecting Be in stars with a metallicity less than one thousandth solar does not agree with our finding. At least, not at first sight. Rich & Boesgaard (2009, but also Boesgaard, this volume), in fact, claim that their most metal-poor stars continue to follow the linear trend of the more metal-rich objects (cf their, e.g., Fig. 4). A closer look at both samples shows that most stars are actually in common, but analyzed with very different stellar parameters. If we correct their Be abundances on our stellar parameters scale, we obtain a satisfactory overlap, which demonstrates that the differences between these two sets of results rely only on different input parameters, not on the observed spectra and not on the different model atmospheres and/or analytical tools used during the analysis (reassuring!). Figure 3 shows the results of this test.

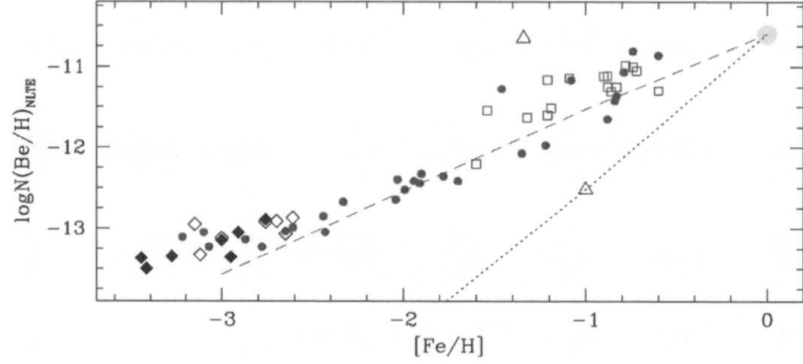

Figure 3. Same as the top panel of Fig. 2, this time with overplotted the most metal-poor data points from Rich & Boesgaard (2009): filled diamonds as in original paper, open diamonds corrected on our stellar parameters scale (see text for more details).

4.2. *Be at the halo-disk transition: real scatter ?*

The larger dispersion detected at higher metallicities is present in both plots of Be trends, *vs* Fe and O. The second plot is certainly more worrisome than the first one, since Be and O are expected to be well correlated. Figure 2 shows that all in a sudden, when a metallicity of [Fe/H] = −1.5 or [O/H] = −1.2 is reached, the tight correlation between, e.g., Be and O becomes looser: there are stars with the same content of oxygen, but with highly different contents of Be.

Since these metallicities map what we commonly call the halo-disk transition, the first attempt to shed some light into this region was to separate the objects kinematically, by associating them to the halo or to the disk component (cf caption of Fig. 2 for symbols coding). Unfortunately, not much was gained. This confirms earlier attempts made by Smiljanic *et al.* (2009) and Tan *et al.* (2009) who tried to associate stars to the *dissipative* vs *accretion* components and failed in finding a clear and convincing separation in their Be evolutionary trends.

Nissen & Schuster (2009) have determined very accurate abundances of several alpha elements in a large sample of halo-disk transition stars. The entire analysis has been carried out strictly differentially and this has allowed these authors to achieve very small error-bars on each data point and notably reduce spurious sources of scatter in the data. This work is an extension of Nissen & Schuster (1997), at a much higher accuracy. Their main finding (cf their Figs. 1 and 2) is a clear separation between halo-low and halo-high stars (where low/high refers to a low/high content of α-elements respectively) and with the halo-high overlapping the group of thick disk stars.

If we now use the accurate $[\alpha/\text{Fe}]$ ratios that Nissen & Schuster (2009) have derived for those stars for which we also have Be, we recover as well a very clear separation (in, e.g., a plot of $[\alpha/\text{Fe}]$ *vs* Be) between halo-low and halo-high, with the latter overlapping the thick disk stars. This splitting now helps us also to interpret the dispersion revealed by Fig. 2. The formerly very confusing halo-disk transition region now appears to be populated by objects that belong to distinct groups: the halo-low stars seem to continue the trend identified by the more metal-poor stars, possibly making the slope a bit steeper, whereas the halo-high stars seem to follow a shallower relation, and at similar oxygen contents have a much higher content of Be. The thick-disk partly overlaps with the halo-high but, with the exclusion of one object, seems to follow a slope more similar to the halo-lows.

Let's make one step further. If we now look at the plot of Be *vs* $[\alpha/\text{H}]$ (cf Fig. 4), the picture becomes even clearer and confirms what we just said. Halo-lows are a continuation of the more metal-poor stars trend, the thick-disk seems indeed to fall more closely the halo-low than the halo-high, though some overlap is present, especially at the highest metallicities; the halo-high show a flatter trend. As it can be seen in Fig. 4, there are some data points without any α-abundance-group identification (open circles), which are very important in order to confirm or not our preliminary findings. Still, this is an important step forward, in the interpretation of the dispersion now found by several recent studies.

5. Boron: fewer data and slower progress

For boron, the situation has not changed much during the last 10 years. Stellar boron abundances are inferred from the analysis of the B I resonant doublet at 250 nm, hence one needs to go into space. After a burst of studies that followed the launch of the Hubble Space Telescope and that lasted approximately a decade, very few abundance studies in metal-poor halo stars, if any, have been carried out during the last decade. More work has been carried out in stars at higher metallicities (cf Cunha, this volume) and in massive stars (cf Kaufer, this volume).

Therefore, when reviewing the formation and evolution of boron in the early Galaxy, one needs to go back to the second half of the 90s, basically to the works of Duncan *et al.* (1997), García López *et al.* (1998), and Primas *et al.* (1998, 1999). The B *vs*

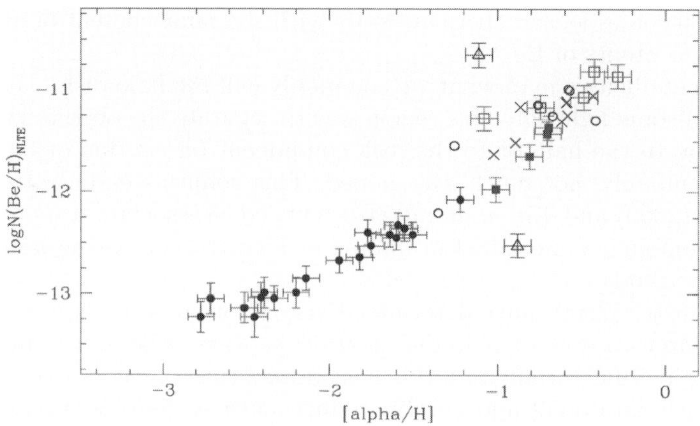

Figure 4. Be *vs* [α/H] trend, showing a much clearer separation (wrt, e.g., Fig. 2) among the three distinct α-element abundance groups, halo-low (filled squares), halo-high (open squares), and thick disk stars (crosses). Open circles represent disk stars which are not in the sample of Nissen & Schuster (2009); filled circles and open triangles as in Figs. 2 and 3.

Fe,O relations found by these authors confirmed the earlier results obtained by Duncan *et al.* (1992) from only three objects, i.e. a primary origin also for this spallative element.

One very recent progress possibly worth mentioning here is the recent re-calculation of the B I lines formation under NLTE conditions carried out by Tan et al (*priv. comm.*). These NLTE corrections seem to differ significantly from those applied in earlier studies (e.g. Kiselman 1994) and to reconcile the B trends (vs Fe and vs O) with the trends recently found also for Be.

6. Constraining stellar mixing with Li, Be, and B

Simultaneous knowledge of Li, Be, and B abundances in the same target(s) is an important diagnostic to investigate mixing mechanisms at work in the outermost layers of stellar photospheres. Because this elemental trio burns at different and slightly increasing temperatures, one expects that Li, which is saved in the outermost layers, will burn first, before Be starts to be affected by any depletion mechanism.

It is important to note that this type of investigations has so far assumed that a Spite-plateau lithium abundance (i.e. the canonical value found in metal-poor Galactic halo stars of A(Li) ≃ 2.2) represents an un-depleted abundance of the lithium stellar content of that given object. But after WMAP (Dunkley *et al.* 2009 and references therein), we now have a very precise prediction of the primordial lithium abundance in the Standard Big Bang framework (A(Li)= 2.72 ± 0.06, Cyburt *et al.* 2008), which instead hints at a rather large depletion of the lithium content that we measure today in the oldest stars of our Galaxy. This of course does not affect the basic depletion/dilution trend of the three light isotopes, but it may have significant implications on how we interpret the comparison of Li, Be, and B abundances in the same objects, especially as far as the amounts of depletion/dilution are concerned.

What still works well is of course a differential analysis among stars that share similar characteristics in, e.g., their stellar parameters. We know several pairs of stars that have an identical lithium content, that are indistinguishable as far as effective temperature, gravity and metallicity are concerned, but have very different Be and B abundances. Some of these have the same Be content, but differ in their B signatures (cf Fig. 5, dotted and thin lines – does this imply that there could be another production mechanism for

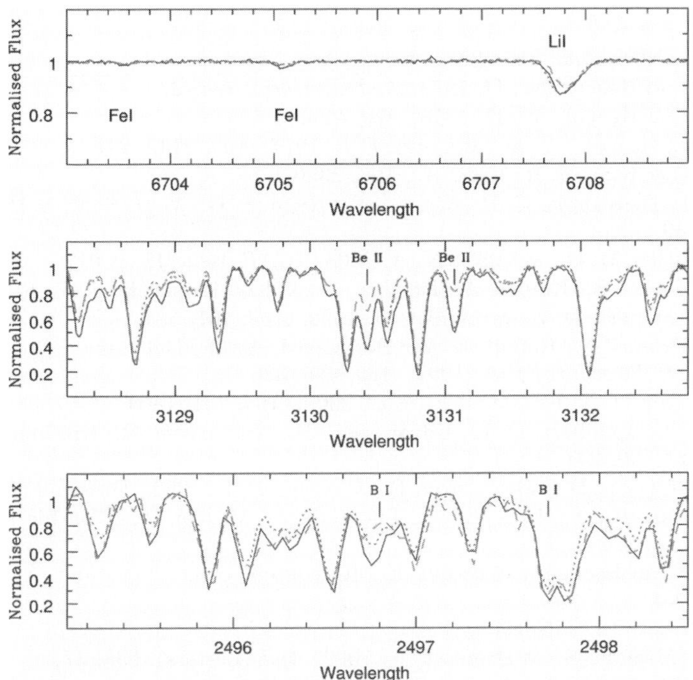

Figure 5. Observed spectra of three different stars overplotted respectively in the Li, Be, and B spectral regions (from *top* to *bottom*). The stars have very similar stellar parameters and Li contents, but show some differences in Be and/or B.

boron, like the often proposed ν-spallation?), others instead have different Be and B contents, for which it then becomes important to see how the degree of depletion of each element compares to the other (cf Fig. 5, dashed and thin line).

An example is given in Fig. 5, where observed spectra of three different objects are shown and overplotted. These three objects have very similar stellar parameters and Li contents, but one of them has much less beryllium and boron (dashed spectrum), and the other two share a similar content of beryllium but have a puzzling different B content!

7. Concluding remarks and acknowledgments

Remarkable observational progress has been made in the last decade, especially as far as Be studies are concerned. In this review, I have shown that the combination of higher quality data and highly accurate measurements of abundances (e.g., α-elements) is now starting to shed new light on the interpretation of some specific features of the Be evolutionary trend (e.g. the dispersion at the halo-disk transition). Progress in studies concerning B abundances has lagged behind, mostly due to the unavailability during the last 4-5 years of the Space Telescope Imaging Spectrograph on board of the Hubble Space Telescope (due to a mechanical/technical failure). But after the successful NASA Service Mission #4, STIS is again operational, therefore new abundance studies of boron in metal-poor stars will hopefully come soon.

Acknowledgments. The author warmly thanks Poul Erik Nissen, for providing the α-elements abundances discussed in S. 4, and the LOC for the financial support received through the Swiss National Research Foundation.

References

Boesgaard, A. M. 1976, *ApJ*, 210, 466

Boesgaard, A. M. & Heacox, W. D. 1973, *ApJ* (Letters), 185, 27

Boesgaard, A. M. & Heacox, W. D. 1978, *ApJ*, 226, 888

Boesgaard, A. M., Heacox, W. D., & Conti, P. S. 1977, *ApJ*, 214, 124

Boesgaard, A. M. & King, J. R. 1993, *AJ*, 106, 2309

Boesgaard, A. M., Deliyannis, C. P., King, J. R., Ryan, S. G., Vogt, S. S., & Beers, T. C. 1999, *AJ*, 117, 1549

Cyburt, R. H., Fields, B. D., & Olive, K. A. 2008, *JCAP*, Issue 11, p. 012

Dekker, H., D'Odorico, S., Kaufer, A., Delabre, B., & Kotzlowski, H. 2000, *SPIE*, 4008, 534

Duncan, D. K., Lambert, D. L., & Lemke, M. 1992, *ApJ*, 401, 584

Duncan, D. K., Primas, F., Rebull, L. M., Boesgaard, A. M., Deliyannis, C. P., Hobbs, L. M., King, J. R., & Thorburn, J. A. 1997, *ApJ*, 488, 338

Dunkley, J., Spergel, D. N. Komatsu, E. *et al.* 2009, *ApJ*, 701, 1804

García López, R. L., Lambert, D. L., Edvardsson, B., Gustafsson, B., Kiselman, D., & Rebolo, R. 1998, *ApJ*, 500, 241

Gilmore, G., Gustafsson, B., Edvardsson, B., & Nisen, P. E. 1992, *Nature*, 357, 379

Gratton, R. G., Carretta, E., Desidera, S., Lucatello, S., Mazzei, P., & Barbieri, M. 2003, *A&A*, 406, 131

Gustafsson, B., Edvardsson, B., Eriksson, K., JØrgensen, U. G., Nordlund, Å, & Plez, B. 2008, *A&A*, 486, 951

Kiselman, D. 1994, *A&A*, 286, 169

Ito, H., Aoki, W., Honda, S., & Beers, T. C. 2009, *ApJ* (Letters), 698, 37

Meneguzzi, M., Audouze, J., & Reeves, H. 1971 *A&A*, 15, 337

Molaro, P., Beckman, J. E., & Castelli, F. 1984, in Proceedings of *ESA 4th European IUE Conference*, p. 197

Nissen, P. E. & Schuster, W. E 1997, *A&A*, 326, 751

Nissen, P. E. & Schuster, W. E. 2009, in: J. Andersen, J. Bland-Hawthorn, & B. Nordström (eds.), *The Galaxy Disk in Cosmological Context* (Proceedings of the International Astronomical Union, IAU Symposium, Vol 254), p. 103

Pasquini, L., Bonifacio, P., Randich, S., Galli, D., & Gratton, R. G. 2004, *A&A*, 426, 651

Pasquini, L., Galli, D., Gratton, R. G., Bonifacio, P., Randich, S., & Valle, G. 2005, *A&A* (Letter), 436, 57

Pasquini, L., Bonifacio, P., Randich, S., Galli, D., Gratton, R. G., & Wolff, B. 2007, *A&A*, 464, 601

Primas, F., Duncan, D. K., & Thorburn, J. A. 1998, *ApJ* (Letters), 506, 51

Primas, F., Duncan, D. K., Peterson, R. C., & Thorburn, J. A. 1999, *A&A*, 343, 545

Primas, F., Molaro, P., Bonifacio, P., & Hill, V. 2000a, *A&A*, 362, 666

Primas, F., Asplund, M., Nissen, P. E., & Hill, V. 2000b, *A&A* (Letter), 364, 42

Rebolo, R., Molaro, P., Abia, C., & Beckman, J. E. 1988, *A&A*, 193, 193

Reeves, H., Fowler, W. A., & Hoyle, F. 1970, *Nature*, 226, 727

Rich, J. A. & Boesgaard, A. M. 2009, *ApJ*, 701, 1519

Ryan, S. G., Norris, J. E., Bessell, M. S., & Deliyannis, C. P. 1992, *ApJ*, 388, 184

Smiljanic, R., Pasquini, L., Bonifacio, P., Galli, D., Gratton, R. G., Randich, S., & Wolff, B. 2009, *A&A*, 499, 103

Tan, K. F., Shi, J. R., & Zhao, G. 2009, *MNRAS*, 392, 205

Vangioni-Flam, E., Ramaty, R., Olive, K. A., & Casse, M. 1998, *A&A*, 337, 714

Vogt, S. S., Allen, S. L., Bigelow, B. C. *et al.* 1994, *SPIE*, 2198, 362

Light Elements in the Universe
Proceedings IAU Symposium No. 268, 2009
C. Charbonnel, M. Tosi, F. Primas & C. Chiappini, eds.
© International Astronomical Union 2010
doi:10.1017/S1743921310004175

New Beryllium results in halo stars from Keck/HIRES spectra

Ann Merchant Boesgaard[1], Jeffrey A. Rich[1], Emily M. Levesque[1] and Brendan P. Bowler[1]

[1]University of Hawaii, Institute for Astronomy
2680 Woodlawn Drive, Honolulu, HI 96822, U.S.A.
email: boes@ifa.hawaii.edu, jrich@ifa.hawaii.edu,
emsque@ifa.hawaii.edu, bpbowler@ifa.hawaii.edu

Abstract. We have obtained high-resolution, high signal-to-noise Keck spectra to determine Be abundances in over 100 stars in the Galactic halo. The stellar metallicities range from [Fe/H] = -0.50 to -3.50. Using this large sample, we have examined the trends of Be with Fe and Be with O. We find a real dispersion in Be at a given [O/H] that indicates that Be may not be a good cosmochronometer. Our results indicate that the dominant production mechanism for Be changes as the Galaxy ages. In the early eras of the Galaxy, when massive stars become supernovae, Be is produced from the acceleration of energetic CNO atoms which bombard protons in the vicinity of supernovae. Later spallation reactions occur as high energy protons bombard CNO atoms in the interstellar gas. The change occurs near [Fe/H] = -2.2. We have found that Be is deficient in Li-deficient halo stars, which favors the blue straggler analog hypothesis.

Keywords. Stars: abundances, Population II, kinematics, late-type – Galaxy: halo

1. Introduction

The trio of the rare light elements provides excellent probes into many astrophysical issues: cosmology, the chemical evolution of the Galaxy, the origin of the light elements, element destruction in stars, stellar interiors and evolution, cosmic ray theory, hypernovae, etc. For many of these issues Be is the best probe. It has only one long-lived isotope, ^{9}Be; the non-LTE effects are negligible; there is only one source for production (spallation); it is less fragile than the Li isotopes; and it has the potential to be a good cosmochronometer.

For our Be studies we have been using the Keck 10-m telescope (with more than 50% more light-gathering power than 8-m telescopes) on the 4200-m Mauna Kea, which is above 40% of Earth's atmosphere. From 1993 to 2003 the CCD on HIRES (Vogt *et al.* 1994) had only 8% quantum efficiency near the Be II lines at 3130.4 and 3131.1 Å. However, after the upgrade of HIRES in 2004, the quantum efficiency on the new "blue chip" rose to 93%. (The upgraded version now has 3 CCDs with 15-μm pixels.) We have been able to take UV spectra of stars almost 3 magnitudes fainter. However, signal-to-noise ratios of more than 100 are needed to find Be abundances in the most metal-poor stars, where the Be II lines are very weak.

Due to the benefits of large aperture and high altitude, we have been able to achieve some interesting results with the original version of HIRES. In our first comprehensive study of Be in halo stars, we found a linear relationship between [Fe/H] and A(Be) (= log N(Be)/N(H) + 12.00) with a slope of $+0.96$ ±0.04 (Boesgaard *et al.* 1999). We discovered a Be dip like the Li dip in the Hyades F dwarfs (Boesgaard & King 2002), but unlike the

Li decline in the Hyades G dwarfs, there is no decline in Be in the G dwarfs (Boesgaard & King 2002). We followed that up with Be studies in F and G dwarfs in several other open clusters: Pleiades and α Per (Boesgaard *et al.* 2003a), Coma and UMa moving group (Boesgaard *et al.* 2003b), and Praesepe (Boesgaard *et al.* 2004a). In a large study of Li and Be in both field and cluster stars we found that the depletions of Li and Be are correlated in the stars in the Li-Be dip with a slope of $+0.38 \pm 0.03$ (Boesgaard *et al.* 2004b). This could result from rotationally induced mixing. By studying B in stars with large Be deficiencies, we were able to find a correlation between Be and B depletions (Boesgaard *et al.* 2005); the slope for that relationship is 0.22 ± 0.05.

2. Observations with the upgraded HIRES

The upgraded HIRES has a mosaic of 3 CCDs, each 2048×2048 pixels with a pixel size of 15 μm. Since September 2004, we have received 16 nights of Keck time with HIRES over a 41-month period to pursue research on Be. Only two of those nights were lost: one due to the closure of the road to the summit from a raging blizzard and one due to two deep undersea earthquakes measuring 6.7 and 6.0 on the Richter scale at just after 7 a.m. on the morning of our one-night run. They caused damage to both Keck telescopes and to the remote-observation rooms. On the other 14 nights, we obtained high signal-to-noise (S/N) spectra of over 100 stars for several different research goals. As Fig. 1 in Rich & Boesgaard (2009) shows, it is important to achieve high S/N in the most metal-poor stars due to the weakness of the Be II lines.

We determined the stellar parameters—T_{eff}, log g, [Fe/H], and microturbulence (ξ)—spectroscopically (see Rich & Boesgaard 2009). The values ranged from 5550 to 6400 K, 3.2 to 4.9 in log g, -0.5 to -3.5 in [Fe/H], and 0.9 to 1.5 km s^{-1} in ξ. The spectrum synthesis method was used to find abundances; an example is shown in Fig. 1, where both Be and O were varied to achieve the best fit. Oxygen abundances were found from three OH features. The two other OH features (3139, 3140 Å) can be seen in Fig. 3 of Rich & Boesgaard (2009). Each feature contains several OH transitions.

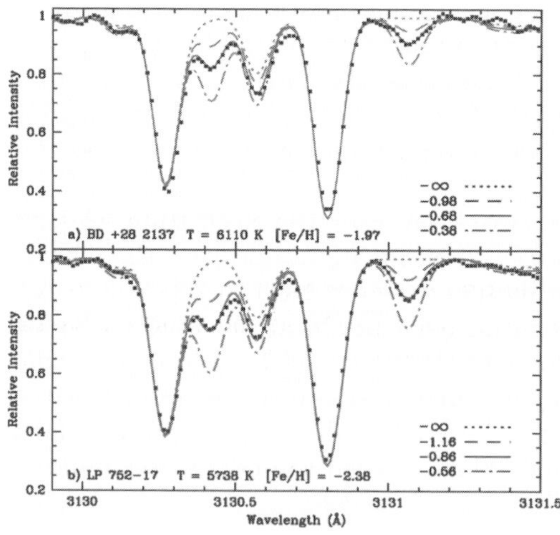

Figure 1. The spectrum synthesis fit for two of our stars: BD +28 2137 at [Fe/H] $= -1.97$ and LP 752-17 at [Fe/H] $= -2.38$. The dots are the data points and the solid line is the best fit. The dot-dash line has twice as much Be and 0.1 dex more O than the best fit. The dashed line has half the Be and 0.1 dex less O. The dotted line contain no Be and is another 0.1 dex lower in O.

3. Abundances and trends

Be trends with Fe and with O. We have been able to fit the trends of Be with Fe and Be with O with straight lines in the log-log plot. We prefer a two slope fit for both. These fits are shown in Fig. 2. Data from Smiljanic *et al.* (2009) are included in the figure with [Fe/H], but not in the figure with [O/H] as they did not determine O abundances.

A change in slope would be expected. In the oldest stars, Be would be formed mostly in the vicinity of SN II by the acceleration of CNO nuclei into protons, etc. Thus Be would be proportional to the instantaneous number of supernovae and therefore proportional to O with a slope of $\leqslant 1$. In the younger stars, the dominant formation mechanism would be traditional GCR spallation from high-energy cosmic rays bombarding CNO in the interstellar gas. The number of O atoms would depend on the cumulative number of supernovae, while the number of energetic cosmic rays is proportional to the instantaneous rate of SN II. The abundance of spallation products is $\int N dN = kN^2$ giving a slope of $\leqslant 2$. Due to effects such as mass outflow during star formation, the predicted slopes of 1 and 2 would be modified to lower values.

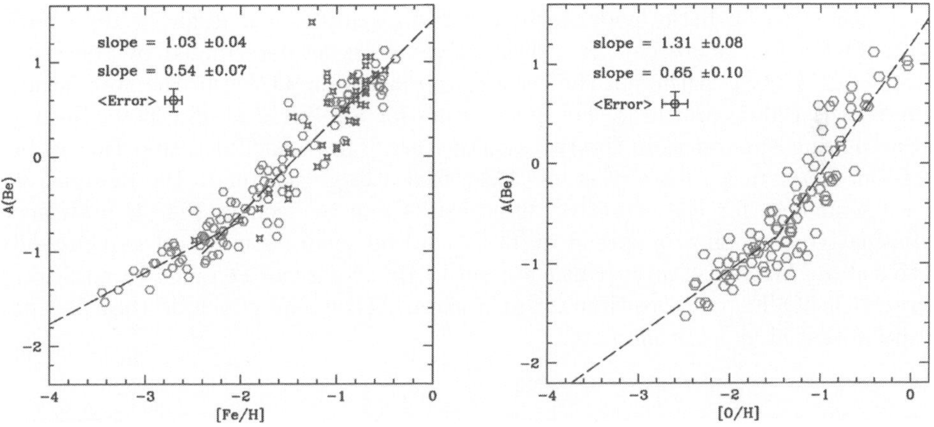

Figure 2. The two-slope fit of A(Be) with [Fe/H] and A(Be) with [O/H]. The open hexagons are from this work and the skeletal squares are from Smiljanic *et al.* (2009).

Beryllium plateau? We find no evidence for a Be plateau at low metallicities comparable to the Li plateau. There may be a steady increase in [Be/Fe] at the lowest values of [Fe/H], but this may only reflect the paucity of stars (only 5) observed for Be with [Fe/H] < −3.0.

Beryllium spread. We do find evidence for a real spread in A(Be) at a given [O/H] at the 4σ level and probably at a given [Fe/H] at the 3σ level. We have done statistical tests using a prediction interval. We start with the null hypothesis that there is no spread in Be abundances in a certain Fe or O interval. We derive slope and offset values from the points outside the metallicity interval under consideration. We then use those data to predict the range of Be values inside our metallicity interval for a given confidence level, and we compare that with the actual data. For example, if there are 25 points in our metallicity interval, and 7 lie outside the 95% confidence level, we can then use the binomial theorem to determine the probability of this occurring by chance. We can try different confidence levels (99%, 95%, 90%) and different size bins for [Fe/H] and [O/H] (±0.25, ±0.20, ±0.10). Fig. 3 shows an example of this. The probability that 7 of 25 stars fall outside the 95% confidence level by chance is 0.015%.

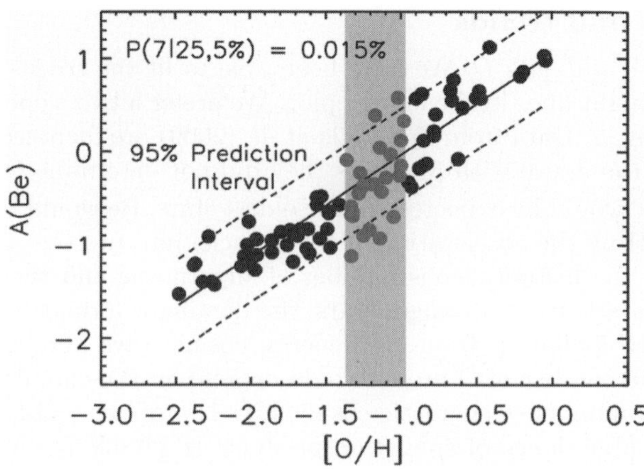

Figure 3. An example of the prediction-interval testing that we did to check for a spread in Be at a given value of [O/H] or [Fe/H].

There seems to be little evidence in our data sample of a different distribution of Be with [O/Fe] for "accretive" vs. "dissipative" stars as determined by the criteria of Gratton *et al.* (2003). Although the errors were large for [O/Fe] in the stars studied by Pasquini *et al.* (2005) and in [α/Fe] in the stars for Smiljanic *et al.* (2009), both groups found intriguing differences in the two sets of stars. They concluded that Be can be used as a chronometer in a subset of stars. They find a large scatter in the dissipative stars and two sequences for the accretive. Our results can be seen in Fig. 4; both accretive and dissipative stars show a spread in [O/H] and no separation of two sequences for the accretive stars. We found an intrinsic spread in Be at a given Fe or O. In addition there is a spread in [O/Fe] of more than 2σ at a given A(Be). We conclude than Be may not fulfill its potential as a chronometer.

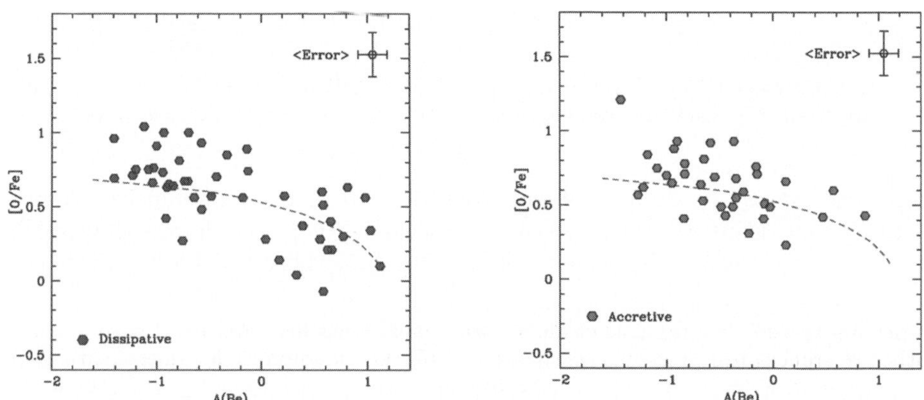

Figure 4. A plot of [O/Fe] vs. A(Be) in stars in the "accretive" and dissipative" populations (as defined by Gratton *et al.* 2003). We see no evidence for a difference in the distribution between the two populations.

4. Beryllium in ultra-Lithium-deficient stars

Although most metal-poor stars have A(Li) values between ∼1.9 to 2.5, there are a few that are deficient in Li and fall below this plateau. This can be seen in Fig. 5 (left).

These were dubbed "ultra-Li-deficient" stars by Ryan *et al.* (2001). There are at least two plausible explanations for the Li deficiencies. Ryan *et al.* (2001, 2002) suggest that they are blue-straggler analogs in which mass transfer or binary coalescence has occurred that destroyed the Li. Pinsonneault *et al.* (1999, 2002) hypothesize that they are a subset of (once) rapidly rotating stars which have depleted their Li as they spun down their rotation rate. These two ideas make different predictions about Be. In the blue straggler model, most or all of the Be would also be destroyed due to complete internal mixing. In the rotation model some or all of the Be would be preserved because Be is less fragile than Li.

Boesgaard & Novicki (2006) and Boesgaard (2007) determined Be abundances in seven of the nine Li-deficient stars that are indicated in the left panel of Fig. 5. They found that the Li-deficient stars are also Be-deficient as can be seen the in right panel of Fig. 5. Those Be deficiencies are larger than the predictions from the models of rotationally-induced mixing by Pinsonneault *et al.* (1992). This result thus favors the blue-straggler model which predicts large Be deficiencies. Newer models by Pinsonneault *et al.* (2002) predict *smaller* Li depletions, and probably smaller Be depletions as well.

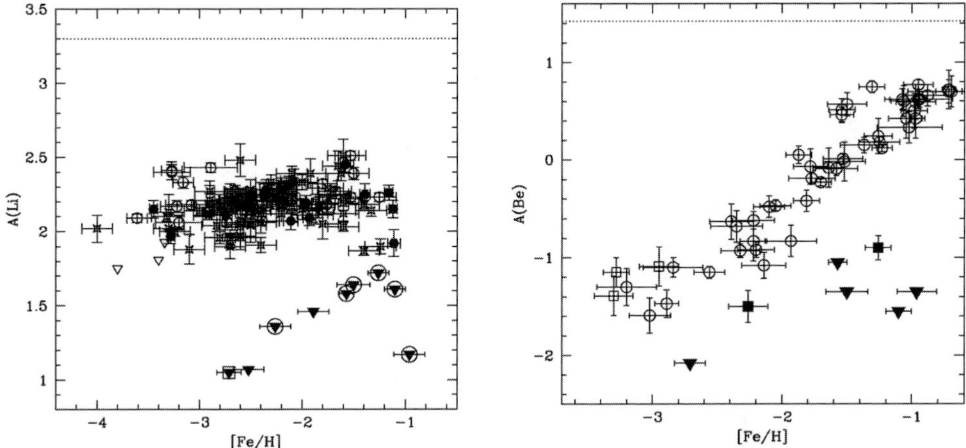

Figure 5. Left: The abundance of Li, A(Li) = log N(Li) − log N(H) +12.00, plotted against [Fe/H] for halo dwarfs. The upper limits for the Li-deficient stars are indicated by downward triangles; those observed for Be have circles around them (Boesgaard 2007) or a square (Boesgaard & Novicki 2006). See the caption on the similar figure in Figure 1 of Boesgaard (2007) for the detailed references. The horizontal dotted line is the meteoritic Li abundance, A(Li) = 3.30 (Grevesse & Sauval 1998). Right: The Be abundances/limits of the ultra-Li-deficient stars shown in the context of Be abundances in Li-normal stars as a function of [Fe/H]. The filled squares are the two ultra-Li-deficient stars with detected Be, while the filled triangles represent the upper limits on four ultra-Li-deficient stars. Open circles are from Boesgaard *et al.* (1999), and other references are given in Boesgaard (2007). Open squares (low-metallicity stars) are from Primas *et al.* (2000a, 2000b). Individual error bars on both A(Be) and [Fe/H] are shown. The horizontal dotted line is the meteoritic Be abundance, A(Be) = 1.42 (Grevesse & Sauval 1998).

5. Summary of Keck Be results

Our findings from our Keck Be data include the following:

- There are correlated depletions of Li with Be and of Be with B. These observations are well-matched by the predictions of rotationally induced mixing.

- There is a Be dip like the Li dip in the Hyades F stars and in other open clusters. However, for the G stars there is no decline in Be to mirror the decline in Li.

- We have found abundance trends between A(Be) and [Fe/H] and between A(Be) and [O/H] which can be well fit by one straight line or by two. The two-slope fits are preferred and expected if the dominant source of Be production changes during the course of Galactic evolution from core-collapse supernovae (bullets = CNO and targets = p, n) to GCR spallation (bullets = p, n and targets = CNO). There are also tight relationships between [Fe/H] and [O/H] and between [O/Fe] and [Fe/H].

- There appears to be a real spread in Be at a given [Fe/H] and a given [O/H]. We find no difference in the relationships between Be and [O/Fe] for the accretive sample and the dissipative sample. This would seem to indicate that Be cannot be relied on as a cosmochronometer.

- We have found Be abundances and upper limits in those halo stars where the upper limits on the Li abundances put them well-below the Li plateau. A plausible explanation is that these stars are analogs to blue stragglers and have destroyed most or all of their Li and Be though mass transfer or binary coalescence.

Acknowledgements

We gratefully acknowledge support from the National Science Foundation through grant AST-05-05899 to A. M. B.

References

Boesgaard, A. M. 2007, *ApJ*, 667, 1196

Boesgaard, A. M. & King, J. R. 2002, *ApJ*, 565, 587

Boesgaard, A. M. & Novicki, M. C. 2006, *ApJ*, 641, 1122

Boesgaard, A. M., Armengaud, E., & King, J. R. 2003a, *ApJ*, 582, 410

Boesgaard, A. M., Armengaud, E., & King, J. R. 2003b, *ApJ*, 583, 955

Boesgaard, A. M., Armengaud, E., & King, J. R. 2004a, *ApJ*, 605, 864

Boesgaard, A. M., Armengaud, E., King, J. R., Deliyannis, C. P., & Stephens, A. 2004b, *ApJ*, 613, 1202

Boesgaard, A. M., Deliyannis, C. P., King, J. R., Ryan, S. G., Vogt, S. S., & Beers, T. C. 1999, *AJ*, 117, 1549

Boesgaard, A. M., Stephens, A., & Deliyannis, C. P. 2005, *ApJ*, 633, 398

Gratton, R. G., Carretta, E., Claudi, R, Lucatello, S., & Barbien, M. 2003, *A&A*, 404, 187

Grevesse, N. & Sauval, A. J. 1998, *Space Sci. Rev.*, 85, 161

Pasquini, L., Galli, D., Gratton, R. G., Bonifacio, P., Randich, S., & Valle, G. 2005, *A&A* (Letters), 436, 57

Pinsonneault, M. H., Deliyannis, C. P., & Demarque, P. 1992, *ApJS*, 78, 179

Pinsonneault, M. H., Steigman, G., Walker, T. P., & Narayanan, V. K. 2002, *ApJ*, 574, 398

Pinsonneault, M. H., Walker, T. P., Steigman, G., & Narayanan, V. K. 1999, *ApJ*, 527, 180

Primas, F., Asplund, M., Nissen, P. E., & Hill, V. 2000a, *A&A* (Letters), 364, 42

Primas, F., Asplund, M, Bonifacio, P., & Hill, V. 2000b, *A&A*, 362, 666

Rich, J. A. & Boesgaard, A. M. 2009, *ApJ*, 701, 1519

Ryan, S. G., Beers, T. C., Kajino, T., & Rosolankova, K. 2001, *ApJ*, 547, 231

Ryan, S. G., Gregory, S. G., Kolb, U., Beers, T. C., & Kajino, T. 2002, *ApJ*, 571, 501

Smiljanic, R., Pasquini, L., Bonifacio, P., Galli, D., Gratton, R. G., Randich, S., & Wolff, B. 2009, *A&A*, 499, 103

Vogt, S. S. *et al.* 1994, *SPIE*, 2198, 362

Light Elements in the Universe
Proceedings IAU Symposium No. 268, 2009
C. Charbonnel, M. Tosi, F. Primas & C. Chiappini, eds.

© International Astronomical Union 2010
doi:10.1017/S1743921310004187

Boron abundances in diffuse interstellar clouds

Adam M. Ritchey[1], S. R. Federman[1], Y. Sheffer[1]† and D. L. Lambert[2]

[1]Department of Physics and Astronomy, University of Toledo,
2801 West Bancroft Street, Toledo, OH 43606, USA
email: `adam.ritchey@utoledo.edu`, `steven.federman@utoledo.edu`,
`ysheffe@utnet.utoledo.edu`

[2]W.J. McDonald Observatory, University of Texas at Austin,
1 University Station, Austin, TX 78712, USA
email: `dll@astro.as.utexas.edu`

Abstract. We present a comprehensive survey of B abundances in diffuse interstellar clouds from *HST*/STIS observations along 56 Galactic sight lines. Our sample is the result of a complete search of archival STIS data for the B II λ1362 resonance line, with each detection confirmed by the presence of absorption from other dominant ions at the same velocity. The data probe a range of astrophysical environments including both high-density regions of massive star formation as well as low-density paths through the Galactic halo, allowing us to clearly define the trend of B depletion onto interstellar grains as a function of gas density. Many extended sight lines exhibit complex absorption profiles that trace both local gas and gas associated with either the Sagittarius-Carina or Perseus spiral arm. Our analysis indicates a higher B/O ratio in the inner Sagittarius-Carina spiral arm than in the vicinity of the Sun, which may suggest that B production in the current epoch is dominated by a secondary process. The average gas-phase B abundance in the warm diffuse ISM [log ϵ(B) = 2.38 ± 0.10] is consistent with the abundances determined for a variety of Galactic disk stars, but is depleted by 60% relative to the solar system value. Our survey also reveals sight lines with enhanced B abundances that potentially trace recent production of ^{11}B either by cosmic-ray or neutrino-induced spallation. Such sight lines will be key to discerning the relative importance of the two production routes for ^{11}B synthesis.

Keywords. ISM: abundances, atoms – nuclear reactions, nucleosynthesis, abundances – ultraviolet: ISM

---

1. Introduction

Production of the two stable isotopes of boron results from spallation reactions between energetic particles and interstellar nuclei. The less abundant ^{10}B is mainly a product of Galactic cosmic-ray (GCR) spallation (e.g., Meneguzzi, Audouze & Reeves 1971; Ramaty *et al.* 1997), which typically involves relativistic protons and α-particles impinging on CNO nuclei in the interstellar medium (ISM) but can also occur as accelerated CNO nuclei are spalled from ambient interstellar H and He. While a significant amount of ^{11}B is produced through GCR spallation reactions, an additional source is required to raise the isotopic ratio ^{11}B/^{10}B from the value predicted by models of cosmic-ray spallation (2.4; Meneguzzi *et al.* 1971) to the value measured in carbonaceous chondrites (4.0;

† Present address: Department of Astronomy, University of Maryland, College Park, MD 20742, USA

Lodders 2003). The ν-process, or neutrino-induced spallation in Type II supernovae, has often been invoked to account for the discrepancy in the predicted versus measured boron isotopic ratios because, while the yields for ^{11}B are substantial, virtually no ^{10}B is produced (Woosley *et al.* 1990). Without the ν-process, enhanced synthesis of ^{11}B could be attributed to an increased flux of low-energy cosmic rays, which are unobservable from the Earth due to the modulating effect of the solar wind.

All of the processes that produce boron in significant quantities occur in, or are closely associated with, the interstellar medium. Even the ^{11}B synthesized in core collapse supernovae will quickly be injected into the surrounding interstellar gas. It therefore becomes a critical test of the theoretical ideas concerning boron nucleosynthesis to measure its interstellar abundance. The first detection of interstellar boron was reported by Meneguzzi & York (1980), who used the *Copernicus* satellite to measure B II λ1362 along the line of sight to κ Ori. They derived an interstellar abundance of log ϵ(B) $= 2.2 \pm 0.2$, which was consistent with the then-current stellar value of 2.3 ± 0.2 (Boesgaard & Heacox 1978), assumed to be the Galactic value. Federman *et al.* (1996a), using GHRS on board *HST*, presented the first measurement of ^{11}B/^{10}B outside the solar system. Their analysis, along with later work by Lambert *et al.* (1998), showed that the solar system ratio is not anomalous but probably representative of the local Galactic neighborhood. The survey by Howk, Sembach & Savage (2000) expanded the sample of interstellar boron abundances to the extended sight lines accessible to *HST*/STIS. These authors found clear evidence for boron depletion onto interstellar grains and derived a lower limit to the present-day total interstellar boron abundance of $\gtrsim 2.40 \pm 0.13$.

The discovery of newly synthesized lithium toward o Per, a line of sight that passes very near to the massive star-forming region IC 348 and has a low ^{7}Li/^{6}Li ratio consistent with predictions of GCR spallation (Knauth *et al.* 2000a; 2000b), prompted us to seek ^{11}B/^{10}B ratios along this and other nearby sight lines in the Per OB2 association with STIS. Ultimately, the acquired spectra toward 40 Per, o Per, ζ Per, and X Per lacked the signal-to-noise ratio required to yield meaningful results on ^{11}B/^{10}B but did provide accurate B column densities. Thus, we shifted our focus to determining elemental boron abundances for a much larger Galactic sample.

2. STIS archival survey

All archival STIS datasets employing either the E140H or E140M grating were examined in an effort to find unambiguous interstellar absorption from O I λ1355, Cu II λ1358, and Ga II λ1414. Subsequent searches for absorption from B II λ1362 at the same velocity resulted in a sample of 56 Galactic sight lines. All of the above species represent the dominant ionization stage of their particular element in neutral diffuse clouds and are thus expected to coexist. As in our previous work (Federman *et al.* 1996a; Lambert *et al.* 1998), the stronger O I, Cu II, and Ga II lines serve as templates of interstellar component structure when determining B II column densities. The STIS spectra acquired with the E140H grating are characterized by velocity resolutions in the range $\Delta v = 2.1-3.6$ km s^{-1}, while E140M spectra have resolutions of $6.5-7.9$ km s^{-1}. Thus, to help constrain the velocity structure along the lines of sight in our sample, we obtained high-resolution ($\Delta v = 1.6-1.8$ km s^{-1}) ground-based data on Ca II λ3933 and K I λ7698 for many directions either at McDonald Observatory or from the literature (e.g., Welty, Morton & Hobbs 1996; Welty & Hobbs 2001; Pan *et al.* 2004). We also incorporated into our analysis five sight lines with previous measurements of interstellar B from GHRS (Jura *et al.* 1996; Lambert *et al.* 1998; Howk *et al.* 2000) but did not rederive abundances in these directions.

The full boron sample probes a diverse array of astrophysical environments, including both high-density regions of massive star formation (e.g., Per OB2 and Cep OB2) as well as low-density paths through the Galactic halo (e.g., the sight lines to HD 156110 and HD 187311). Some directions sample quite local gas (e.g., in Sco OB2 at $d = 160$ pc), while others trace distant spiral arms (e.g., HD 104705, located beyond the Sagittarius-Carina arm at $d = 3.9$ kpc). These characteristics enable a detailed investigation of the effect of physical environment on the observed abundances of boron in diffuse clouds. More information on the boron sample can be found in Ritchey *et al.* (2010).

3. Abundance analysis

As a first step in the analysis, the absorption profiles of O I, Cu II, and Ga II were synthesized with the rms-minimizing code ISMOD (Y. Sheffer, unpublished). These fits yielded the column density, velocity, and b-value of each absorption component along the line of sight. The B II line was then fitted by holding the b-values, relative velocities, and relative component strengths constant and varying only the absolute velocity of the profile and the total column density. The results of the O I, Cu II, and Ga II fits were each applied separately to the B II line. Additionally, if Ca II and K I data were available, these lines were synthesized and the results were used as a high-resolution template for synthesizing the B II profile. Final B II column densities were determined by taking the mean of these fits, which in all cases were mutually self-consistent.

When a sight line exhibited multiple complexes of absorption components well separated in velocity, the various templates were constructed for each complex and fitted to that portion of the B II profile independently. This technique allows for the possibility that elemental abundance ratios may vary from complex to complex along the line of sight. Indeed, we find suggestive evidence for a higher B/O ratio in components tracing the inner Sagittarius-Carina spiral arm than in those sampling local gas in the same direction. Abundances of secondary elements increase relative to those of primary ones toward the Galactic center due to enhanced rates of star formation and stellar nucleosynthesis. If confirmed, an elevated B/O ratio toward the inner Galaxy would indicate the secondary nature of boron, which in turn would cast doubt on the efficiency of the ν-process, a primary production mechanism.

Under the assumption that all boron in diffuse clouds is singly ionized, the B II column densities derived above can be considered the total boron column densities in the gas phase. Elemental boron abundances can then be determined from knowledge of the total column densities of atomic and molecular hydrogen along the line of sight. The atomic hydrogen data come mainly from the archival study of Lyman-α absorption by Diplas & Savage (1994), while the majority of H_2 column densities were provided by Sheffer *et al.* (2008). In Figure 1, we plot gas-phase B abundances as a function of the average line-of-sight hydrogen density, defined as $\langle n_H \rangle = [N(\text{H I}) + 2N(H_2)]/d$, where d is the distance to the background star. Immediately apparent is the trend of decreasing gas-phase abundance with increasing line-of-sight density, a clear signature of the depletion of boron onto interstellar dust grains. Following Jenkins, Savage & Spitzer (1986), we calculated mean B abundances in the warm, low-density gas and in the cold, higher-density clouds based on an idealized model of the neutral interstellar medium (Spitzer 1985). The analysis shows that the depletion (relative to solar) increases from -0.40 dex in lines of sight with the lowest density to -1.20 dex for the directions sampling higher concentrations of cold clouds.

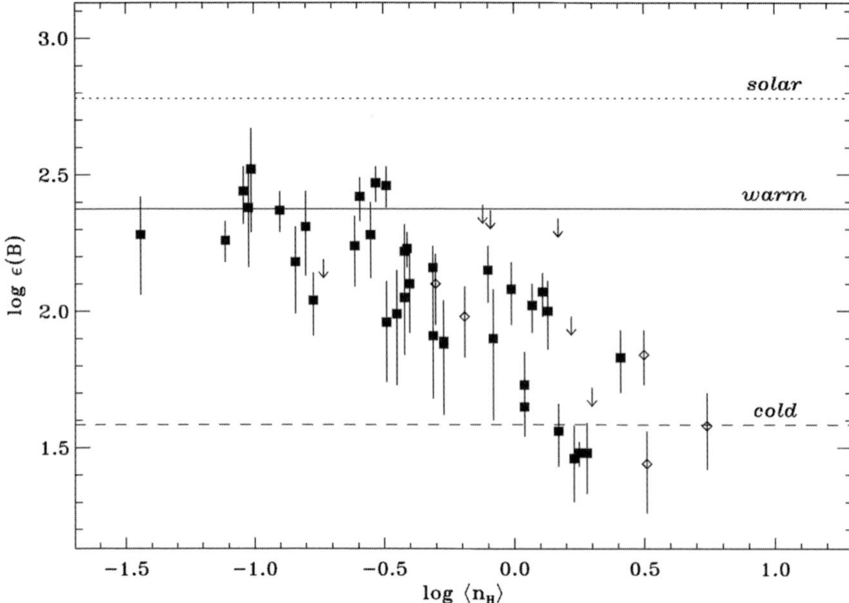

Figure 1. Gas-phase interstellar B abundances versus average line-of-sight hydrogen densities. Solid symbols (and upper limits) denote the STIS boron sample (this work). Open symbols represent GHRS measurements from the literature (see text). The dotted line indicates the solar system abundance (2.78±0.04; Lodders 2003), while the solid and dashed lines correspond to the mean abundances in warm gas (2.38±0.10) and cold clouds (1.58±0.16), respectively.

4. Discussion and conclusions

Our result for the mean B abundance in warm diffuse gas, $\log \epsilon(\mathrm{B}) = 2.38 \pm 0.10$, represents a lower limit to the total interstellar B abundance, since some depletion is expected even in the lowest-density phase of the diffuse ISM. This value agrees remarkably well with the lower limit derived by Howk *et al.* (2000) but is depleted by 60% relative to the solar system (meteoritic) value of 2.78 ± 0.04 (Lodders 2003). The solar abundance is traditionally used as a cosmic standard against which to measure interstellar depletion, although the abundances in F and G field dwarfs of solar metallicity or hot O and B stars are also appropriate references (e.g., Snow & Witt 1996; Sofia & Meyer 2001). Interestingly, our interstellar B abundance for warm gas is fairly consistent with the abundances in a variety of Galactic disk stars (see Table 1). At the very least, if one were to take the F and G dwarf sample of Cunha *et al.* (2000a) and the sample of B-type stars in Venn *et al.* (2002) to derive a cosmic B abundance of 2.5, then B would seem to be only lightly depleted along the lowest-density interstellar sight lines. Unless the various solar and stellar abundances can be reconciled, establishing the total interstellar abundance of boron will have to await deeper insight into the gas-grain interactions responsible for removing atoms in diffuse clouds from the gas phase.

While the absolute level of boron depletion will remain controversial without a definitive cosmic standard, the variation in relative depletion has clearly been demonstrated here (Figure 1). It then becomes possible to identify intrinsic variations in abundance superimposed on the general trend due to depletion. Enhanced boron abundances are expected, for example, in regions shaped by Type II supernovae, since these are the sources most likely responsible for the acceleration of Galactic cosmic rays and are also sites of the ν-process. Any recent production of ^{11}B in such a region, due either to cosmic-ray or neutrino-induced spallation, should manifest itself as a local enhancement in the total

Table 1. Stellar disk and interstellar boron abundances

	log ϵ(B)	Reference
ISM (warm)	2.38 ± 0.10	This Work
G stars (Orion)	2.35 ± 0.30	Cunha *et al.* (2000b)
FG stars	2.51 ± 0.20	Cunha *et al.* (2000a)
B stars	2.54 ± 0.17	Venn *et al.* (2002)
Solar (photospheric)	2.70 ± 0.17	Cunha & Smith (1999)
Solar (meteoritic)	2.78 ± 0.04	Lodders (2003)

boron abundance. Depending on the degree of localization, however, the observational signature may be difficult to discern.

Nevertheless, we did find evidence for enhanced boron abundances in a few directions from our large sample of interstellar sight lines. The abundance we derive of log ϵ(B) $= 2.47 \pm 0.06$ for the line of sight to HD 93222, a member of Collinder 228 in the Carina Nebula, is enhanced by 0.27 dex relative to sight lines with similar average densities and is significantly elevated compared to the values for three other sight lines in the Carina Nebula. The stars HD 93205, CPD$-$59 2603, and HDE 303308, all members of Trumpler 16, have interstellar B abundances of 2.28 ± 0.12, 2.23 ± 0.06, and 2.05 ± 0.14, respectively, and lie just $23'$ to the north of HD 93222. Walborn *et al.* (2007) discuss high-velocity expanding structures seen in interstellar absorption lines toward many of these cluster members in the context of a supernova remnant (SNR) in this direction. They note that the highest known interstellar velocities in the nebula occur in the spectrum of HD 93222.

For HD 43818, a member of the Gem OB1 association, we derive a B abundance of 2.07 ± 0.07. The line of sight to this star is characterized by a factor of 4 higher average density than those in Carina. Since higher depletion is therefore expected, the abundance in this direction represents an enhancement, which is found to be 0.26 dex over similarly dense sight lines. The proximity of this line of sight to IC 443, a young SNR known to be interacting with nearby molecular gas, suggests a possible nucleosynthetic origin for the enhancement. The presence of hadronic cosmic rays accelerated by the SNR and their interaction with ambient molecular material was revealed by VERITAS observations of very-high energy γ-ray emission (Acciari *et al.* 2009). HD 43818, however, lies considerably to the north of the γ-ray source and so may not be related. We are currently pursuing ^{7}Li/^{6}Li ratios and Li and Rb abundances toward stars closer to IC 443 with the Hobby-Eberly Telescope at McDonald Observatory in an effort to constrain the contribution from massive stars to the synthesis of these elements.

Finally, the line of sight to o Per, which is located just $8'$ to the north of the star-forming region IC 348, has a B abundance enhanced by 0.18 dex relative to the other three sight lines in Per OB2. While the enhancement is only modest (50%), it becomes significant in light of the fact that the other sight lines show very little scatter in log ϵ(B). For 40 Per, ζ Per, and X Per, we derive abundances of 1.48 ± 0.11, 1.46 ± 0.12, and 1.48 ± 0.04, while for o Per we find an abundance of 1.65 ± 0.09. Considering the low ^{7}Li/^{6}Li ratio in this direction (Knauth *et al.* 2000a; 2000b; 2003) and an enhanced cosmic-ray flux, which was inferred from measurements of interstellar OH (Federman, Weber & Lambert 1996b) and is consistent with an upper limit derived from observations of H$_3^+$ (Indriolo *et al.* 2007), evidence seems to be mounting of the effect of cosmic-ray spallation reactions on the interstellar abundances of Li and B near IC 348. Recently, Li data were obtained toward the fainter stars of IC 348, itself. Complementary data on interstellar B should now be acquired for these stars, perhaps with the Cosmic Origins Spectrograph, so that the abundance enhancements resulting from cosmic-ray interactions with interstellar clouds can be traced in more detail. In this way, a clearer picture of light element nucleosynthesis will emerge.

Acknowledgements

This research was funded by the Space Telescope Science Institute (STScI) through grant HST-AR-11247.01-A. The data were obtained from the Multimission Archive at STScI, operated by the Association of Universities for Research in Astronomy, Inc. under NASA contract NAS5-26555. A. M. R. would like to thank the Swiss National Science Foundation for a grant provided to cover attendance at the IAU Symposium 268.

References

Acciari, V. A., Aliu, E., Arlen, T., & VERITAS Collaboration. 2009, *ApJ*, 698, L133

Boesgaard, A. M. & Heacox, W. D. 1978, *ApJ*, 226, 888

Cunha, K. & Smith, V. V. 1999, *ApJ*, 512, 1006

Cunha, K., Smith, V. V., Boesgaard, A. M., & Lambert, D. L. 2000a, *ApJ*, 530, 939

Cunha, K., Smith, V. V., Parizot, E., & Lambert, D. L. 2000b, *ApJ*, 543, 850

Diplas, A. & Savage, B. D. 1994, *ApJS*, 93, 211

Federman, S. R., Lambert, D. L., Cardelli, J. A., & Sheffer, Y. 1996a, *Nature*, 381, 764

Federman, S. R., Weber, J., & Lambert, D. L. 1996b, *ApJ*, 463, 181

Howk, J. C., Sembach, K. R., & Savage, B. D. 2000, *ApJ*, 543, 278

Indriolo, N., Geballe, T. R., Oka, T., & McCall, B. J. 2007, *ApJ*, 671, 1736

Jenkins, E. B., Savage, B. D., & Spitzer, L. 1986, *ApJ*, 301, 355

Jura, M., Meyer, D. M., Hawkins, I., & Cardelli, J. A. 1996, *ApJ*, 456, 598

Knauth, D. C., Federman, S. R., Lambert, D. L., & Crane, P. 2000a, *Nature*, 405, 656

Knauth, D. C., Federman, S. R., Lambert, D. L., & Crane, P. 2000b, in: L. da Silva, M. Spite & J. R. de Medeiros (eds.), *The Light Elements and Their Evolution*, Proc. IAU Symposium No. 198 (San Francisco: ASP), p. 338

Knauth, D. C., Federman, S. R., & Lambert, D. L. 2003, *ApJ*, 586, 268

Lambert, D. L., Sheffer, Y., Federman, S. R., Cardelli, J. A., Sofia, U. J., & Knauth, D. C. 1998, *ApJ*, 494, 614

Lodders, K. 2003, *ApJ*, 591, 1220

Meneguzzi, M., Audouze, J., & Reeves, H. 1971, *A&A*, 15, 337

Meneguzzi, M. & York, D. G. 1980, *ApJ*, 235, L111

Pan, K., Federman, S. R., Cunha, K., Smith, V. V., & Welty, D. E. 2004, *ApJS*, 151, 313

Ramaty, R., Kozlovsky, B., Lingenfelter, R. E., & Reeves, H. 1997, *ApJ*, 488, 730

Ritchey, A. M., Federman, S. R., Sheffer, Y., & Lambert, D. L. 2010, *in preparation*

Sheffer, Y., Rogers, M., Federman, S. R., Abel, N. P., Gredel, R., Lambert, D. L., & Shaw, G. 2008, *ApJ*, 687, 1075

Snow, T. P. & Witt, A. N. 1996, *ApJ*, 468, L65

Sofia, U. J. & Meyer, D. M. 2001, *ApJ*, 554, L221

Spitzer, L. 1985, *ApJ*, 290, L21

Venn, K. A., Brooks, A. M., Lambert, D. L., Lemke, M., Langer, N., Lennon, D. J., & Keenan, F. P. 2002, *ApJ*, 565, 571

Walborn, N. R., Smith, N., Howarth, I. D., Kober, G. V., Gull, T. R., & Morse, J. A. 2007, *PASP*, 119, 156

Welty, D. E. & Hobbs, L. M. 2001, *ApJS*, 133, 345

Welty, D. E., Morton, D. C., & Hobbs, L. M. 1996, *ApJS*, 106, 533

Woosley, S. E., Hartmann, D. H., Hoffman, R. D., & Haxton, W. C. 1990, *ApJ*, 356, 272

Light Elements in the Universe
Proceedings IAU Symposium No. 268, 2009
C. Charbonnel, M. Tosi, F. Primas & C. Chiappini, eds.
© International Astronomical Union 2010
doi:10.1017/S1743921310004199

Boron abundances in the Galactic disk

Katia Cunha[1]

[1]NOAO
950 N. Cherry Ave, Tucson, Arizona
email: kcunha@noao.edu

Abstract. When compared to lithium and beryllium, the absence of boron lines in the optical results in a relatively small data set of boron abundances measured in Galactic stars to date. In this paper we discuss boron abundances published in the literature and focus on the evolution of boron in the Galaxy as measured from pristine boron abundances in cool stars as well as early-type stars in the Galactic disk. The trend of B with Fe obtained from cool F-G dwarfs in the disk is found to have a slope of 0.87 ± 0.08 (in a log-log plot). This slope is similar to the slope of B with Fe found for the metal poor halo stars and there seems to be a smooth connection between the halo and disk in the chemical evolution of boron. The disk trend of boron with oxygen has a steeper slope of 1.5. This slope suggests an intermediate behavior between primary and secondary production of boron with respect to oxygen. The slope derived for oxygen is consistent with the slope obtained for Fe provided that [O/Fe] increases as [Fe/H] decreases, as observed in the disk.

Keywords. stars: abundances – Galaxy: disk – ultraviolet: stars

1. Introduction

The light element boron is one of the few elements whose production is not dominated by nucleosynthesis in stars, nor by nucleosynthesis occurring in the Big Bang. In fact, it has been known now for almost 4 decades that Galactic Cosmic Rays are related to the formation of the light elements and, in particular, of boron (Reeves, Fowler & Hoyle 1970). The connection between boron production and cosmic rays spallation reactions, which involve C, N, O atoms, as well as protons and α particles, makes boron an interesting element whose abundance evolution in the Galaxy probes the history of cosmic rays in the galactic environment. In addition, boron is also proposed to be produced by neutrino nucleosynthesis occurring in core collapse of Supernovae Type II (Woosley *et al.* 1990).

Unveiling the underlying behavior of boron with metallicity (iron and oxygen abundances) is crucial in order to constrain models for boron production. One of the challenges in trying to pin down the behavior of boron with metallicity, however, comes first from the fact that boron is fragile and easily destroyed in stellar interiors (although sturdier than Li and Be) and, in addition, from the difficulty in obtaining boron observations. In this paper we discuss stellar boron abundance results mainly for disk stars which have been published in the literature. Unfortunately, no new boron observations were available in recent years due to the failure of STIS on board HST.

2. Boron transitions and abundance determinations

Boron abundance results are still sparse as boron abundances can only be measured from transitions which fall mainly in the ultraviolet. Boron abundance indicators in different temperature regimes are from three ionization stages: neutral boron in cool stars (B I at 2497.723Å); B II (at 1362Å) mostly in A-type stars and the B III resonance doublet (at 2060Å) in B-type stars.

244 K. Cunha

2.1. *Boron in cool stars and the Sun*

Boron was measured in the Sun by Kohl, Parkinson & Withbroe (1977). One of the pioneering studies of boron abundances in stars was by Boesgaard & Heacox (1978). A few studies appeared more than a decade later from observations obtained with the Hubble Space Telescope (Duncan, Lambert, & Lemke 1992; Duncan *et al.* 1997; Primas *et al.* 1999). The sample analyzed in Duncan *et al.* (1997) was mostly for halo stars. Their results were particularly important as they found that boron abundances scaled linearly (in log-log space) with the abundance of metals, in contrast with predictions from the standard models of cosmic ray production, which proposed a secondary behavior for boron with metallicity. These predictions from cosmic ray models had remained unchallenged for ~ 20 years.

Following studies focused on samples of stars with disk metallicities and, by the nature of their sample, these probed the behavior of boron in the most metal rich stars in the Galaxy (Boesgaard *et al.* 1998; Cunha & Smith 1999; Cunha *et al.* 2000; Boesgaard *et al.* 2004; Boesgaard *et al.* 2005). It is important to acknowledge, however, that the line list in the spectral region of the B I transition is a major challenge for the analysis of boron in solar-like stars as the spectral region to be synthesized is covered with strong blending lines for which, in many instances, there is no atomic data available. Note, however, that this does not represent a severe problem for the analysis of halo stars as the metal

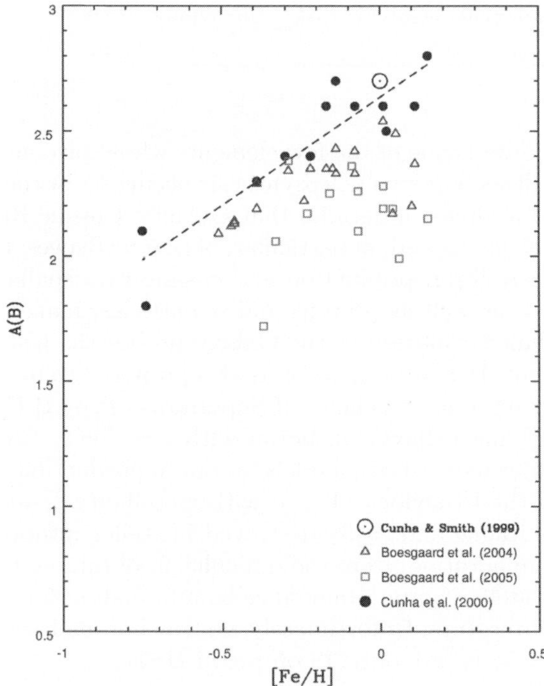

Figure 1. Boron results for the available samples of F-G disk stars. The solar abundance is also plotted. The targets in Boesgaard *et al.* (2004) and Cunha *et al.* (2000) are beryllium undepleted, while the stars analyzed in Boesgaard *et al.* (2005) are mixed showing a degree of Be depletion. For the Be undepleted stars it is clear that there is a systematic abundance offset between the studies of Cunha *et al.* (2000) and Boesgaard *et al.* (2004) which can be attributed to the differences in the line lists adopted in the analyses. The slope representing the trend of boron with iron for the disk is 0.87 ± 0.08 (dashed line); this defines the upper envelope (undepleted) of the distribution.

lines are vanishingly weak. For solar metallicity stars, different studies in the literature constructed and adopted different line lists which resulted in systematic differences in the derived boron abundances. In the following we briefly present some of the boron results in these studies.

Cunha *et al.* (2000) analyzed dwarf stars with [Fe/H] > -1.0 (T_{eff}'s between 5650 - 6700K) from HST archival data. One important aspect of that study in comparison with Boesgaard *et al.* (1998; 2004) is that the line list adopted in the calculation of model spectra in Cunha *et al.* (2000) was empirically adjusted in order to fit the Sun. In using the Sun as a benchmark in the study of solar-type stars, Cunha & Smith (1999) re-visited the analysis of boron in the solar photosphere. In particular, significant effort was put in that study into evaluating and updating the opacities which are important in the ultraviolet and which affect the derived boron abundances. This resulted in the revision upwards of the boron abundance in the solar photosphere, which was found to be in good agreement with the boron meteoritic value of A(B)= 2.79 $\pm$ 0.04 (see Lodders, Palme & Gail 2009). The agreement between the boron abundances in the solar photosphere and meteorites (implying an *absence* of boron depletion in the Sun) is an important result as it reconciles with the most recent assessment of the beryllium abundance in the solar photosphere by Asplund *et al.* (2009), indicating no beryllium depletion in the Sun. If Be is indeed not depleted in the solar photosphere, it follows that boron, which is less fragile than Be, cannot be depleted in the solar photosphere.

Boron abundances for disk dwarfs with effective temperatures close to solar are shown in Figure 1. Non-LTE corrections for the B I transition at 2497Å for this temperature range at solar metallicity are deemed to be small (Kiselman & Carlson 1996). In order to have all stars and the Sun on a consistent scale, all disk stars in Cunha *et al.* (2000) were analyzed homogeneously. The study by Boesgaard *et al.* (2004) used a different line list which was not fine tuned in order to fit the solar spectrum. It can be seen that the results from Cunha *et al.* (2000) and Boesgaard *et al.* (2004), all for targets with undepleted beryllium, have a small systematic offset. It is clear also that the targets analyzed Boesgaard *et al.* (2005) have significantly lower boron abundances, but this is expected as the target stars were selected in that study to be beryllium depleted in order to further study mixing.

2.2. *Boron in early-type stars*

Of the light element trio, boron is the only element whose abundance can be measured in early-type stars. One problem in using early-type stars to define the boron Galactic trend, however, is the varying amounts of boron depletion in OB-type stars, as depletion of boron is proportional to stellar mass, age and rotational velocity. In addition, unlike the case of observations of Li and Be in cool stars, there is not a sensitive monitor of depletion in early-type stars. Boron is burnt at temperatures which are lower than those at which the CN cycle takes place and is much more sensitive to mixing than nitrogen.

Some studies in the literature have derived boron abundances from HST observations obtained with the GHRS and STIS spectrographs in relatively small samples of early-type stars (Cunha *et al.* 1997; Venn *et al.* 2002; Mendel *et al.* 2006). Most of these boron abundances, however, were found to be somewhat mixed and therefore not representative of the chemical composition of the gas which formed these young stars. The larger sample analyzed by Proffitt & Quigley (2001) from IUE archival observations of the B III resonance line at 2066Å, although not having the same spectral quality as HST data, found some stars to be boron undepleted which helped define the disk trend.

3. Boron abundance trends in the disk

The evolution of boron with oxygen for metallicities covering the range spanned by the disk is shown in Figure 2. The blue filled circles represent the FG-dwarfs analyzed in Cunha *et al.* (2000) with the errorbars indicating the estimated abundance uncertainties. Boron results for early-type stars from Proffitt *et al.* (2001; filled red triangles) and Mendel *et al.* (2006; filled red squares) are also shown. Most of the B stars shown have roughly undepleted boron and on average follow the disk trend delineated by the cool stars. The general agreement between the results in cool and hot stars is pleasing given that the physical conditions in their stellar atmospheres are quite distinct. The behavior of boron with oxygen can be represented by a linear relation (in the log-log plot) with a slope ∼1.5, which can suggest an intermediate behavior between primary and secondary production for boron with respect to oxygen.

In Figure 3 we plot boron abundances versus [Fe/H] spanning the metallicity range from the halo to the disk. If we adopt the boron abundances for the Be undepleted stars from Cunha *et al.* (2000; filled circles) as representative of the disk value, there seems to be a smooth transition between the halo and disk which follows a slope of ∼ 0.9. This slope for the disk + halo is closer to a primary rather than a secondary behavior for boron production. In addition, it is important to note that the slope derived for oxygen (from Figure 2) is consistent with the one obtained for Fe provided that [O/Fe] increases as [Fe/H] decreases, as observed in the disk.

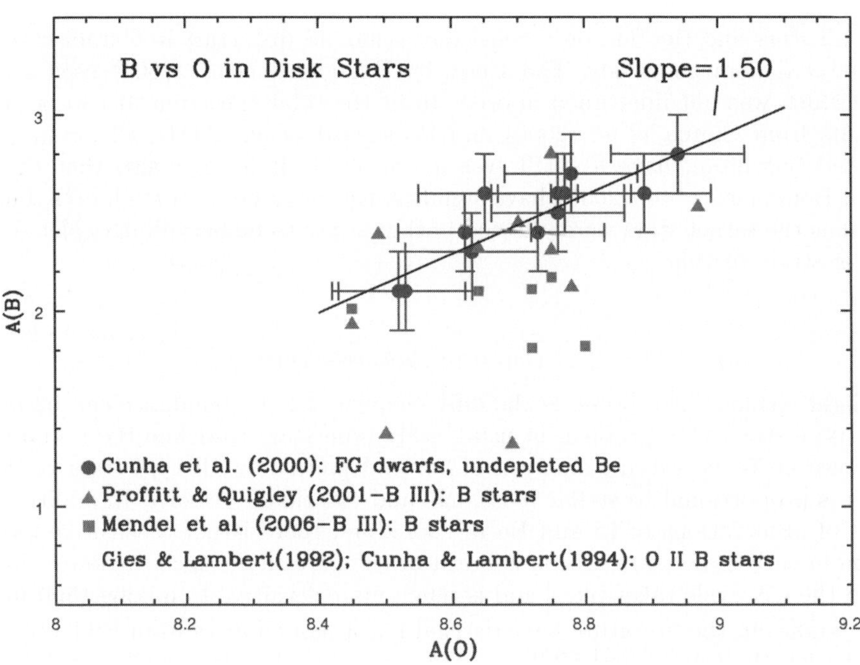

Figure 2. The evolution of boron and oxygen abundances in the Galactic disk. Boron results for cool disk FG-type dwarfs are taken from Cunha *et al.* (2000); these targets have undepleted Be abundances which indicate that their boron content is not mixed and representative of their natal clouds. The evolution of boron and oxygen from this dataset can be represented by linear relation with slope ∼1.5. Boron results for early-type stars from Proffitt & Quigley (2001) and Mendel *et al.* (2006) are also shown. The lower boron abundances in some of the B-type stars are due to internal mixing and astration.

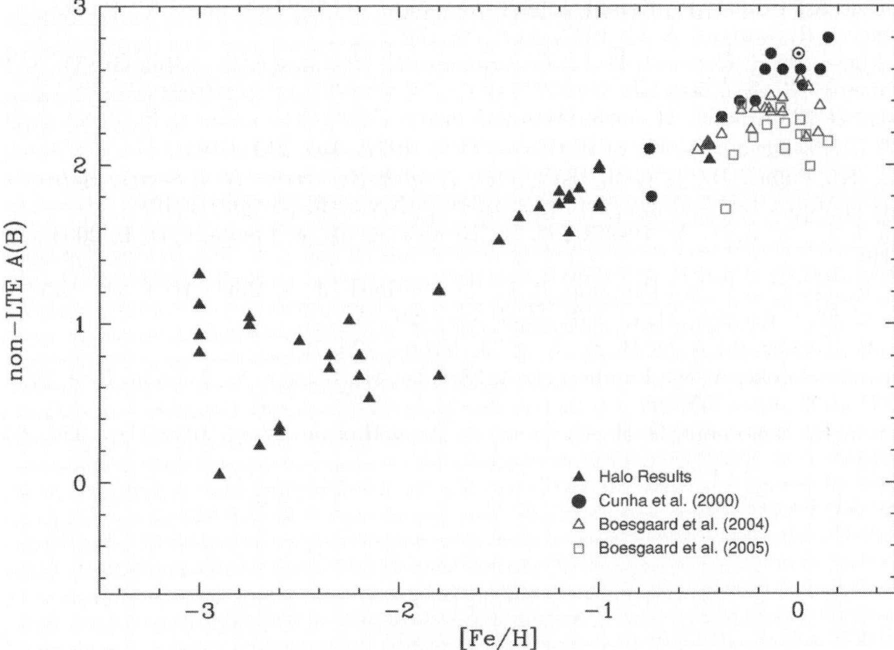

Figure 3. The behavior of boron with metallicity for the Galactic disk in comparison with the trend obtained for the more metal poor stars in the halo. If the abundance results in Cunha *et al.* (2000) are adopted as representative of the disk (filled circles), there seems to be a smooth transition between the disk and halo. The halo star abundances are taken from Duncan *et al.* (1997) and García López *et al.* (1998).

4. How do we move forward?

Probing the behavior of boron with metallicity is crucial for understanding boron production in the Galaxy and the relatively small number of stars analyzed to date could usefully be increased. It is good news that the STIS spectrograph has been recently fixed in a very successful NASA servicing mission (SM4) and that it can now be ready for more boron observations in the UV. In addition, the new UV Cosmic Origins Spectrograph (COS) was deployed on HST during SM4. We also need improvements in the abundance analysis and in particular improvements in the line lists to model the boron region. Full non-LTE treatment is needed, including non-LTE analysis of transitions of all elements contributing to the boron blend; and a complete hydrodynamic 3-D modelling of the stellar atmospheres is also something to look forward to in particular for the cool stars.

References

Asplund, M., Grevesse, N., Sauval, A. J., & Scott, P. 2009, *ARAA*, 47, 481
Boesgaard, A. M. & Heacox, W. D. 1978, *ApJ*, 226, 888
Boesgaard, A. M., Deliyannis, C. P., Stephens, A., & Lambert, D. L. 1998, *ApJ*, 492, 727
Boesgaard, A. M., McGrath, E. J., Lambert, D. L., & Cunha, K. 2004, *ApJ*, 606, 306
Boesgaard, A. M., Deliyannis, C. P., & Steinhauer, A. 2005, *ApJ*, 621, 991
Cunha, K., Lambert, D. L, Lemke, M., Gies, D. R., & Lewis, C. R. 1997, *ApJ*, 478, 211
Cunha, K. & Smith, V. V. 1999, *ApJ*, 512, 1006
Cunha, K., Smith, V. V., Boesgaard, A. M., & Lambert, D. L. 2000, *ApJ*, 530, 939
Duncan, D. K., Lambert, D. L., & Lemke, M. 1992, *ApJ*, 401, 584

Duncan, D. K., Primas, F., Rebull, L. M., Boesgaard, A. M., Deliyannis, C. P., Hobbs, L. M., King, J. R., & Ryan, S. G. 1997, *ApJ*, 488, 338

García López, R. J., Lambert, D. L., Edvardsson, B., Gustafsson, B., Kiselman, D., & Rebolo, Rafael 1998, *ApJ*, 500, 241

Kiselman, D. & Carlsson, M. 1996, *A&A* 311, 680

Kohl, J. L., Parkinson, W. H., & Withbroe, G. L. 1977, *ApJ*, 212, L101

Lodders, K., Palme, H., & Gail, H-P. 2009, *Landolt-Bornstein, New Series, Astronomy and Astrophysics*, Ed. Springer Verlag, in press (arXiv:astro-ph/0901.1149)

Mendel, J. T., Venn, K. A., Proffitt, C. R., Brooks, A. M., & Lambert, D. L. 2006, *ApJ*, 640, 1039

Primas, F., Duncan, D. K., Peterson, R. C., & Thorburn, J. A. 1999, *A&A*, 343, 545

Proffitt, C. R. & Quigley, M. F. 2001, *ApJ*, 548, 429

Reeves, H., Fowler, W. A., & Hoyle, F. 1970, *Nature*, 226, 727

Venn, K. A., Brooks, A. M., Lambert, D. L., Lemke, M., Langer, N., Lennon, D. J., & Keenan, F. P. 2002, *ApJ*, 565, 571

Woosley, S. E., Hartmann, D. H., Hoffman, R. D., & Haxton, W. C. 1990, *ApJ*, 356, 272

Light Elements in the Universe
Proceedings IAU Symposium No. 268, 2009
C. Charbonnel, M. Tosi, F. Primas & C. Chiappini, eds.
© International Astronomical Union 2010
doi:10.1017/S1743921310004205

Lithium in globular clusters

Andreas J. Korn

Dept. of Physics and Astronomy, Uppsala University,
Box 516, 75120 Uppsala, Sweden
email: `andreas.korn@fysast.uu.se`

Abstract. I review recent progress in mapping out and understanding the behaviour of stellar surface abundances of lithium as evidenced by spectroscopic studies of nearby globular clusters (GCs). It will become clear that these observations necessitate revisions to the canonical picture of stellar and globular-cluster evolution: stars evolve with additional non-convective mixing processes and GCs are not simple stellar populations. In spite of these complications, GCs are excellent test beds for chemical-abundance studies. Spectroscopic observations of GC stars of different evolutionary stages reveal systematic trends of surface abundances likely caused by atomic diffusion and mixing. Correcting for their combined effect on surface lithium, a stellar solution to the cosmological lithium discrepancy is likely, if not probable. However, a definitive answer can only be given once we know the effective temperatures of warm subdwarfs and subgiants to high accuracy and understand the processes which give rise to the mixing needed to moderate atomic diffusion.

Keywords. Stars: Population II, atmospheres, abundances – diffusion – line: formation – globular clusters: individual (M 92, NGC 6397, NGC 6752) – cosmology: early universe – techniques: spectroscopic

---

1. Introduction

Cosmology with stars. This was the implicit promise of the seminal discovery of a uniform lithium abundance among warm halo field stars in 1982 by Monique and Francois Spite (Spite & Spite 1982). A quarter century on, in the era of precision cosmology (PC), heralded by balloon experiments like MAXIMA and BOOMERANG and fully established through the WMAP-satellite all-sky measurements, we realize that the evolution of stellar-surface lithium abundances is more complicated and the inference of its primordial (or Big-Bang nucleosynthesis, BBN) value less direct. As early as 1984, Michaud, Fontaine & Beaudet (1984) cautioned that the Spite plateau of lithium was unlikely to represent the unaltered primordial abundance of lithium, as diffusive processes inside the stars would inevitably deplete the stellar surface layers of heavy elements over the course of the billion-year stellar lifetimes.

It is the intention of this review article to give an overview of the latter half of this time frame of Galactic lithium studies with potential cosmological implications. This is the period when the current generation of 8-10m telescopes became available giving routine access to much fainter Spite-plateau stars. Apart from singular heroic efforts with 4m-class telescopes (e.g. Pasquini & Molaro 1996), lithium in globular clusters (GCs) has been exclusively studied with efficient spectrographs on Keck, the VLT and their siblings. The TOP (turn-off point) stars in even the most nearby low-reddening GCs (e.g. NGC 6397) are no brighter than $V \approx 16.5^m$. In the very metal-poor GC M 92, the TOP stars have $V \approx 18.5^m$ which makes studies at high resolution ($R \approx 40\,000$) and high signal-to-noise ratio (S/N ≈ 100) a true challenge, even with 50+ square metres of light-collecting area and 90 % overall quantum efficiency.

For the sake of a historical assessment, I subdivide the roundabout 12 years to be covered here into two periods: the first six years (1998–2003, Sect. 2) are characterized by a single-star observational approach and small-number statistics, with in part contradicting results; the second six years (2004–2009, Sect. 3) have led to some far-reaching revisions based on multi-object spectroscopy coupled with better observations and better modelling. Some apparent contradictions are discussed. Work in progress is presented in Sect. 4 and concluding remarks are given in Sect. 5.

2. The HIRES and UVES years (1998-2003)

Scientists with access to Keck had a head start to the GC lithium business. Building on observations of Deliyannis, Boesgaard & King (1995), Boesgaard *et al.* (1998) analysed high-resolution, moderate-S/N spectra of a handful of subgiants in M 92 and concluded that there is a significant dispersion in lithium abundances among otherwise very similar stars. This dispersion was interpreted to be an indication for different levels of lithium depletion taking place in these stars, possibly correlated with the stars' angular-momentum history. This is not necessarily a far-fetched thought (cf. Pinsonneault, these proceedings). But to make sense of a marked dispersion in M 92 on the one hand and no dispersion (plus a few outliers) on the field-star Spite plateau would require to postulate very different angular-momentum conditions for these two groups of stars.

Bonifacio (2002) subsequently analysed the same M 92 data and concluded that there is no compelling evidence for dispersion beyond the level expected from observational uncertainties. Somewhat surprisingly, no one has to this day reinvestigated the chemical signatures of little evolved stars in M 92 (but see Sect. 4).

Across the Atlantic, a team of European researchers used large amounts of observing time at the VLT to systematically explore the nucleosynthesis within a variety of GCs (cf. Bragaglia, these proceedings, for an overview). One of the early targets was the smallish GC NGC 6397. Bonifacio *et al.* (2002) studied lithium in a dozen TOP stars in this GC and arrived at a common lithium abundance of $\log \varepsilon(\text{Li}) = 2.34 \pm 0.06$ on the customary logarithmic scale that equates $\log \varepsilon(\text{H})$ with 12. It is probably fair to say that in the early years of this century it looked as if the GC Spite plateau behaved just like that of the field stars.

This idea was overthrown by Pasquini *et al.* (2005) who observed a 0.45 dex tip-to-tip dispersion among nine TOP stars in NGC 6752 at [Fe/H] = −1.5. They went beyond Boesgaard and collaborators in showing that lithium in this cluster correlates with other elements: sodium, oxygen and nitrogen. Together with other inverse correlations traced down to the main-sequence TOP (O-Na, Mg-Al, Gratton *et al.* 2001), it thus became clear that a significant fraction of the GC stars we observe today suffer from intra-cluster pollution through a previous generation of (more massive) GC stars. This paradigm shift in GC research paved the way to research on multiple populations in GCs (Piotto *et al.* 2007, Lee *et al.* 2009).

3. The FLAMES years (2004-2009)

This period was characterized by several breakthroughs, theoretical and observational ones alike. It became clear that primordial lithium as predicted by standard WMAP-calibrated BBN and the Spite-plateau lithium abundance are irreconcilable at face value, the difference exceeding 0.3 dex (Coc *et al.* 2004). Building on 20 years of work within the Michaud school, Richard, Michaud & Richer (2005) tuned their stellar-structure models with additional mixing and proposed a stellar-physics solution: the surface lithium diffuses out of the convective envelope according to the physical laws for inhomogeneous gases in a gravitational field, but this gravitational settling is moderated by an unspecified parametrized mixing process below the convective envelope. It is this (turbulent) mixing that manages to keep the Spite plateau thin and flat, even though the surface abundance is reduced by as much as 0.4 dex. Since this postulated mixing lacks a physical interpretation, the models can be said to loose their predictive power. But it turns out that only a subset of models fulfills the observational constraints: in the language of mixing† this is the range from T6.0 to T6.28.

I started to look into the same issue from an observational point as early as 2003. With the vagaries of Paranal weather, it took two years to collect spectroscopic data of sufficient quality to take a fresh look at chemical abundances in NGC 6397. But what a data set it turned out to be! With the advances in instrumentation at the VLT, we not only got high-resolution data for 18 stars along the evolutionary sequence of NGC 6397 with UVES in fibre mode; we also got 100+ spectra of intermediate resolution ($R = 25{,}000$) using FLAMES-MEDUSA. This instrument has really propelled us into a new era of GC research!

In a series of papers, Korn *et al.* (2006, 2007) and Lind *et al.* (2008) showed that there are systematic trends of surface abundances with evolutionary stage. The TOP stars show systematically lower abundances (by up to 0.2 dex) than the RGB stars. This is indeed what one would expect from gravitational settling: it is most efficient when the convective envelope (in itself always fully mixed on a convective-turnover time scale) is thinnest and least massive. Once the stars reach the RGB, the convective envelope expands inward and mixes the settled elements back to the surface. With a notable exception.

The run of surface lithium from the TOP to the RGB is instead characterized by a steep surface dilution. Lithium that settles into layers whose temperature exceeds 2 million Kelvin will capture a proton and disintegrate into He-4 and He-3, this lithium is thus lost and cannot replenish the surface abundance. But before the surface dilution sets is, we can see that lithium did actually settle: in the middle of the subgiant branch (SGB), the lithium abundance is higher than among TOP stars. We can understand why by looking at the structure of our model star which apart from the convective envelope has a region below the convection zone that is rich in lithium. This region is created by the mixing we introduced into the model. When the convection zone expands inward, it first encounters this region and brings its lithium content to the surface, the surface lithium abundance increases a little. Indeed, this is seen as a ≈ 0.1 dex abundance difference in lithium between TOP and SGB stars.

Lind *et al.* (2009) took this work further by analysing several hundred NGC 6397 cluster members with FLAMES-MEDUSA, from the main sequence to beyond the RGB bump. The derived run of lithium as a function of evolution is a textbook result, clearly demonstrating the accuracy with which such work can nowadays be executed. The upturn in the middle of the SGB discussed above and the need for extra mixing around the bump (at $M_V = 0$ for NGC 6397) is seen very clearly. This extra mixing is now believed to be connected to a thermohaline instability that develops in connection with a mean-molecular-weight inversion (Charbonnel & Zahn 2007). We really need more work along these lines, ideally coupling different elements, to verify that we understand all the

---

† There are, in principle, two parameters one can tune: the strengths and the density dependence of mixing. Richard, Michaud & Richer (2005) chose to vary the strength and keep the run with density ρ constant, with an exponent of -3. The Tx.x notation then specifies the logarithm of the temperature T_0 at which the diffusion coefficient for mixing D_T is connected to the atomic-diffusion coefficient for He D_{He} according to $D_T = 400 D_{\mathrm{He}}(T_0)(\rho/\rho(T_0))^{-3}$.

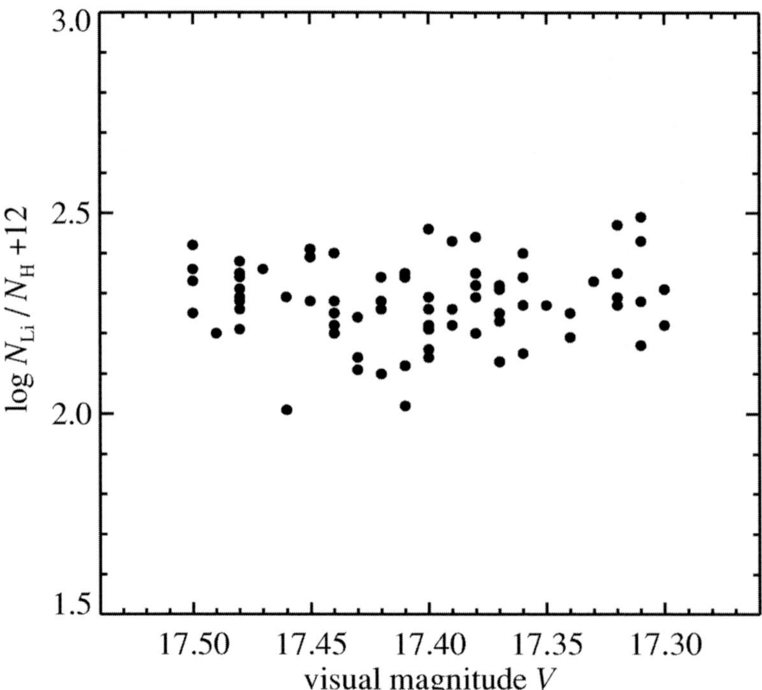

Figure 1. In the subgiant sample of González Hernández *et al.* (2009), there is little of a correlation between $\log \varepsilon$(Li) and V magnitude. If anything, there is a slight rise towards the middle of the SGB (brighter stars), like in the other works that studied NGC 6397.

variables at play here. For lithium production in more advanced stages of evolution, the reader is referred to Smith (these proceedings).

Finally, González Hernández *et al.* (2009) performed a vertical diffusion study comparing the lithium abundances between dwarfs and subgiants at a given $(B - V)$. Like Lind *et al.* (2009), they find a systematic difference between dwarfs and subgiants with the latter being 0.1 dex more lithium-rich. Working on a newly established effective-temperature scale (Balmer lines in 3D model atmospheres), they derive abundances as high as 2.4 dex for the subgiants. The also find trends with effective temperature that may or may not be significant (since this project was conceived as a vertical study, the $T_{\rm eff}$ range covered is not very large and does not include the TOP; cf. Lind, these proceedings). Without resorting to statistical tests, there seems to be little of a positive trend of $\log \varepsilon$(Li) with V in Fig. 1. It would be good to see González Hernández and collaborators perform their analysis on all spectroscopically observed stars in NGC 6397 and include other elements into the picture.

4. Ongoing analyses

New evidence in favour of and against atomic diffusion of lithium was presented at IAU 268. Meléndez showed how the field-star Spite plateau can be reconciled with WMAP-calibrated BBN predictions using a T6.25 diffusion-and-mixing model à la Richard. When plotted versus stellar mass, there is a specific morphology that is well-captured by the model predictions. As a cautionary note, it should be said that a single such model (one for [Fe/H] $= -2.3$) was used to explain lithium abundances over a large range of metallicities. This should be remedied. Bonifacio presented challenging observations of

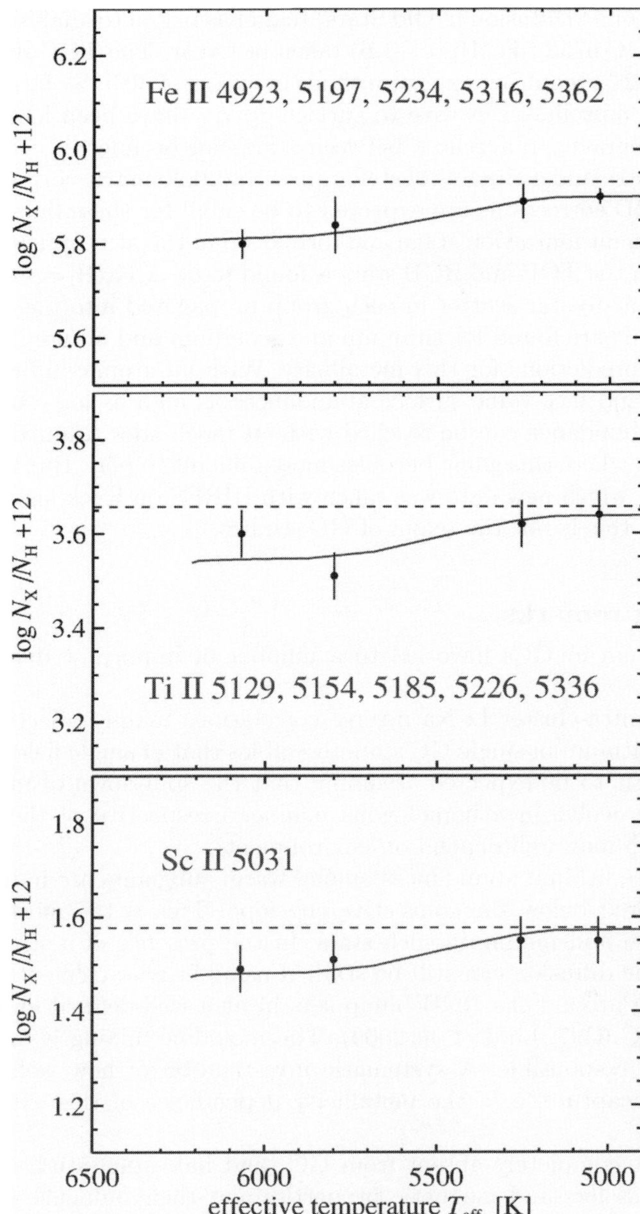

Figure 2. Measured abundance trends for stars in NGC 6752 ([Fe/H] $= -1.6$) versus predictions from a T6.2 Richard model. For iron, the star-to-star scatter in each group is given as an error bar, for titanium and scandium which are both measured on the group-averaged spectra, 0.05 dex is assumed.

Spite-plateau stars drawn from various populations in ω Cen. Identifying a common Spite plateau for the majority of these stars, he claimed that this must represent an unaltered stellar-birth value, as these different populations have different ages and should thus be affected differently by a temporal process like atomic diffusion. But as long as we do not know the age difference with certainty (and independent of the unknown helium-content difference), this argument rests on weak foundations.

The ADiOS (Atomic Diffusion in Old Stars) team has begun to analyse VLT/FLAMES-UVES data on NGC 6752 ([Fe/H] $= -1.6$) taken last year. The TOP stars in this cluster are faint ($V \approx 17.2^m$) requiring exposure times in excess of 30 h. So far, following a strict quality approach, only lines sensitive to surface gravity have been looked at: Fe II, Ti II and Sc II. Surface-gravity differences between stars can be inferred with high precision based on photometry and estimates of stellar mass and bolometric correction. In addition, both NLTE and 3D corrections are expected to be small for these lines, as they connect levels in the dominant ionization stage and form deep in the atmosphere. The abundance difference between the TOP and RGB stars is found to be Δ[Fe/H] $= -0.10 \pm 0.03$ where the error is the star-to-star scatter in each group propagated into the abundance difference. Similar trends are found for titanium and scandium and all are in good agreement with T6.2 model predictions for this metallicity. With an atomic-diffusion correction of 0.25 dex for Li-7 and face-value surface abundances as high as $\log \varepsilon(\text{Li}) = 2.5$, the primordial lithium abundance can be reached without much ado. Admittedly, it is towards lower metallicities where this game becomes more difficult to play. But beyond the metallicity of M 92 (for which new data was taken with HIRES on Keck last year (PI Cohen), analysis pending) this is not the realm of GC studies.

5. Concluding remarks

Studies of lithium in GCs have led to a number of important discoveries in recent years:

- Apart from intra-cluster Li-Na inverse correlations mainly affecting massive GCs, the behaviour of lithium in single GC stars resembles that of single field stars of the same mass range. This is to be expected assuming that the spin-down of metal-poor stars is fast, i.e. all stars evolve in a homologous manner irrespective of their initial angular momentum (which may well depend on environment).

- Higher surface lithium abundances among warm subgiants are indicative of dredge-up of lithium stored below the convective envelope. This is the most direct signature of atomic diffusion and mixing in such stars. In the presence of a strong Li-Na inverse correlation, atomic diffusion can still be studied using heavier elements.

- Extra mixing around the RGB bump sets in at a well-defined absolute magnitude ($M_V = 0$ for NGC 6397, Lind $et\ al.$ 2009). Thermohaline mixing is now believed to be the main process responsible. A systematic investigation of how well the best stellar-structure models capture, e.g., the metallicity dependence of this extra mixing is still lacking.

What is almost completely absent from GC (and halo-star) studies so far is insight into rotation, mass loss and magnetic properties and their influence on inferred abundances and stellar evolution. While these "complications" are generally believed to be less prominent in Population II stars, they may still have a non-negligible impact on the quantitative picture.

As is commonplace in modelling of any kind, there are some mismatches between the run of inferred GC surface abundances and those predicted by stellar-structure models with atomic diffusion and mixing. However, given the tests that the atomic-diffusion hypothesis has been subjected to (using different spectrographs, different model atmospheres, different effective-temperature scales, different line-formation theories, different abundance-analysis techniques, spectral lines with different sensitivities to T_{eff} and $\log g$), there can be little doubt that surface abundances of old, low-mass stars are an explicit function of time. Lithium happens to be one of the elements most affected by atomic diffusion. Surface depletions of 0.25-0.4 dex are predicted by the current generation of sophisticated stellar-structure models ("Richard models") without an accompanying effect on well-observed elements like calcium or iron exceeding ≈ 0.1 dex. It is difficult to unambiguously establish the reality of such shallow abundance trends (cf. the case of NGC 6752 with a T6.2 mixing model). But even evidence from lithium-only studies is accumulating (Meléndez, these proceedings).

If we convince ourselves that the well-observed TOP-part of the Spite plateau lies at $\log \varepsilon(\mathrm{Li}) = 2.25$, then the atomic-diffusion corrected surface abundance of these stars (around $\log \varepsilon = 2.5$ at [Fe/H]$=-2$) are in marginal (1.5σ) agreement with the currently favoured WMAP-calibrated primordial value of 2.72 ± 0.06 (Dunkley *et al.* 2009). There may, however, be biases both in the stellar-atmosphere, line-formation and stellar-structure modelling that could make the stellar abundances fall short. There may be remaining biases in the BBN predictions (cf. the relatively recent correction of the $^7\mathrm{Be}(d,p)2\alpha$ cross-section at BBN energies, Angulo *et al.* 2005), even though few dare to say so (that would be PC blasphemy!). As an example of the remaining uncertainties of the stellar-atmosphere side, let us recall that an effective-temperature scale hotter by $100\,\mathrm{K}$ (as favoured by recent 3D Balmer-profile analyses, González Hernández *et al.* 2009) would add a further 0.07 dex to the above-mentioned values†.

One may wish to add extra boundary conditions to the picture: the possibility of global Li-7 destruction by decaying super-symmetric particles early on in the Universe (indeed, during the very phase of BBN taking place on the natural time and energy scale for this to happen, cf. Jedamzik, these proceedings); the possibility of global Li-7 destruction by Population III stars (Piau *et al.* 2006); the observation of fragile Li-6 in some of the classical Spite-plateau halos stars which currently divides the hydro-modelling experts into believers and sceptics (Asplund *et al.* 2006, Cayrel *et al.* 2007; cf. the contributions by Asplund and Steffen, these proceedings); the possibility of global Li-6 production by the above-mentioned or a similar cosmological process (e.g. Jedamzik *et al.* 2006). All these make lithium a fascinating element to observe and model. But until solid evidence is provided for any of these effects, I prefer to settle for the physics we know (Principle of Parsimony): stellar structure and evolution as a remarkably precise theory. This does not free us of the need to develop this theory (and its atmospheric counterpart) to full radiation-hydrodynamic self-consistency.

References

Angulo, C., Casarejos, E., Couder, M., Demaret, P., Leleux, P., Vanderbist, F., Coc, A., Kiener, J., Tatischeff, V., Davinson, T., Murphy, A. S., Achouri, N. L., Orr, N. A., Cortina-Gil, D., Figuera, P., Fulton, B. R., Mukha, I., & Vangioni, E. 2005, *ApJ*, 630, 105

Asplund, M., Lambert, D. L., Nissen, P. E., Primas, F., & Smith, V. V. 2006, *ApJ*, 644, 229

Boesgaard, A. M., Deliyannis, C. P., Stephens, A., & King, J. R. 1998, *ApJ*, 493, 206

Bonifacio, P. 2002, *A&A*, 395, 515

Bonifacio, P., Pasquini, L., Spite, F., Bragaglia, A., Carretta, E., Castellani, V., Centurin, M., Chieffi, A., Claudi, R., Clementini, G., D'Antona, F., Desidera, S., Franois, P., Gratton, R. G., Grundahl, F., James, G., Lucatello, S., Sneden, C., & Straniero, O. 2002, *A&A*, 390, 91

Cayrel, R., Steffen, M., Chand, H., Bonifacio, P., Spite, M., Spite, F., Petitjean, P., Ludwig, H.-G., & Caffau, E. 2007, *A&A*, 473, L37

Charbonnel, C. & Zahn, J.-P. 2007, *A&A*, 467, 15

Coc, A., Vangioni-Flam, E., Descouvemont, P., Adahchour, A., & Angulo, C. 2004, *ApJ*, 600, 544

† If you are willing to bet a bottle of wine that the remaining 0.15-0.22 dex are due to new physics, contact the author.

Deliyannis, C. P., Boesgaard, A. M., & King, J. R. 1995, *ApJ*, 452, L13

Dunkley, J., Komatsu, E., Nolta, M. R., Spergel, D. N., Larson, D., Hinshaw, G., Page, L., Bennett, C. L., Gold, B., Jarosik, N., Weiland, J. L., Halpern, M., Hill, R. S., Kogut, A., Limon, M., Meyer, S. S., Tucker, G. S., Wollack, E., & Wright, E. L. 2009, *ApJS*, 180, 306

González Hernández, J. I., Bonifacio, P., Caffau, E., Steffen, M., Ludwig, H.-G., Behara, N. T., Sbordone, L., Cayrel, R., & Zaggia, S. 2009, *A&A*, 505, L13

Jedamzik, K., Choi, K.-Y., Roszkowski, L., & Ruiz de Austri, R. 2006, *JCAP*, 07, 007

Korn, A. J., Grundahl, F., Richard, O., Barklem, P. S., Mashonkina, L., Collet, R., Piskunov, N., & Gustafsson, B. 2006, *Nature*, 442, 657

Korn, A. J., Grundahl, F., Richard, O., Mashonkina, L., Barklem, P. S., Collet, R., Gustafsson, B., & Piskunov, N. 2007, *ApJ*, 671, 402

Lee, J.-W., Kang, Y.-W., Lee, J., & Lee, Y.-W. 2009, *Nature*, 462, 480

Lind, K., Korn, A. J., Barklem, P. S., & Grundahl, F. 2008, *A&A*, 490, 777

Lind, K., Primas, F., Charbonnel, C., Grundahl, F., & Asplund, M. 2009, *A&A*, 503, 545

Michaud, G., Fontaine, G., & Beaudet, G. 1984, *ApJ*, 282, 206

Pasquini, L. & Molaro, P. 1996, *A&A*, 307, 761

Pasquini, L., Bonifacio, P., Molaro, P., Francois, P., Spite, F., Gratton, R. G., Carretta, E., & Wolff, B. 2005, *A&A*, 441, 549

Piau, L., Beers, T. C., Balsara, D. S., Sivarani, T., Truran, J. W., & Ferguson, J. W. 2006, *ApJ*, 653, 300

Piotto, G., Bedin, L. R., Anderson, J., King, I. R., Cassisi, S., Milone, A. P., Villanova, S., Pietrinferni, A., & Renzini, A. 2007, *ApJ*, 661, L53

Richard, O., Michaud, G., & Richer, J. 2001, *ApJ*, 558, 377

Spite, M. & Spite, F. 1982, *Nature*, 297, 483

Light Elements in the Universe
Proceedings IAU Symposium No. 268, 2009
Corinne Charbonnel, Monica Tosi, Francesca Primas
& Cristina Chiappini, eds.

© International Astronomical Union 2010
doi:10.1017/S1743921310004217

Main-sequence and sub-giant stars in the globular cluster NGC 6397: The complex evolution of the lithium abundance

J. I. González Hernández[1,2]†, P. Bonifacio[1,2,3], E. Caffau[1], M. Steffen[4], H.-G. Ludwig[1,2], N. Behara[1,2], L. Sbordone[1,2], R. Cayrel[1], and S. Zaggia[5]

[1] GEPI, Observatoire de Paris, CNRS, Université Paris Diderot;

Place Jules Janssen 92190 Meudon, France
email: Jonay.Gonzalez-Hernandez@obspm.fr

[2] Cosmological Impact of the First STars (CIFIST) Marie Curie Excellence Team

[3] Istituto Nazionale di Astrofisica - Observatorio

Astronomico di Trieste, Italy

[4] Astrophysikalisches Institut Potsdam, An der Sternwarte 16,

D-14482 Potsdam, Germany

[5] INAF - Osservatorio Astronomico di Padova,

Vicolo dell'Osservatorio 5, Padua 35122, Italy

Abstract. Thanks to the high multiplex and efficiency of Giraffe at the VLT we have been able for the first time to observe the Li I doublet in the Main Sequence stars of a globular cluster. At the same time we observed Li in a sample of Sub-Giant stars of the same B-V colour.

Our final sample is composed of 84 SG stars and 79 MS stars. In spite of the fact that SG and MS span the same temperature range we find that the equivalent widths of the Li I doublet in SG stars are systematically larger than those in MS stars, suggesting a higher Li content among SG stars. This is confirmed by our quantitative analysis carried out making use of 1D hydrostatic plane-parallel models and 3D hydrodynamical simulations of the stellar atmospheres.

We derived the effective temperatures of stars in our the sample from Hα fitting. Theoretical profiles were computed using 3D hydrodynamical simulations and 1D ATLAS models. Therefore, we are able to determined 1D and 3D-based effective temperatures. We then infer Li abundances taking into account non-local thermodynamical equilibrium effects when using both 1D and 3D models.

We find that SG stars have a mean Li abundance higher by 0.1 dex than MS stars. This result is obtained using both 1D and 3D models. We also detect a positive slope of Li abundance with effective temperature, the higher the temperature the higher the Li abundance, both for SG and MS stars, although the slope is slightly steeper for MS stars. These results provide an unambiguous evidence that the Li abundance changes with evolutionary status.

The physical mechanisms responsible for this behaviour are not yet clear, and none of the existing models seems to describe accurately these observations. Based on these conclusions, we believe that the cosmological lithium problem still remains an open question.

Keywords. Stars: abundances, fundamental parameters, Population II – Galaxy: globular clusters: individual: NGC 6397

---

† Present address: Dpto. de Astrofísica y Ciencias de la Atmósfera, Facultad de Física, Universidad Complutense de Madrid, E-28040 Madrid, Spain. Email: jonay@astrax.fis.ucm.es

1. Introduction

The determination of the baryonic density from the fluctuations of the cosmic microwave background (CMB) by the WMAP satellite (Spergel *et al.* 2007, Cyburt *et al.* 2008) implies a primordial Li abundance which is $\log(\mathrm{Li/H}) + 12 = 2.72 \pm 0.06$, at least 0.3–0.5 dex higher than the Li abundance determined in metal-poor stars of the Galactic halo (Spite & Spite 1982).

Many different models of Li depletion have been proposed to explain discrepancy: (a) Piau *et al.* (2006) proposed that the first generation of stars, Population III stars, could have processed some fraction of the halo gas, lowering the lithium abundance; (b) other authors suggest that the primordial Li abundance has been uniformly depleted in the atmospheres of metal-poor dwarfs by some physical mechanism (e.g. turbulent diffusion as in Richard *et al.* (2005), Korn *et al.* (2006); gravitational waves as in Charbonnel & Talon (2005), etc.); and (c) finally, it has been also suggested that the standard Big Bang nucleosynthesis (SBBN) calculations should be revised, possibly with the introduction of new physics as in e.g. Jedamzik (2004), Jedamzik (2006), Jittoh *et al.* (2008), Hisano *et al.* (2009).

Here we present the determination of Li abundances of subgiant (SG) and main-sequence (MS) stars of the cluster NGC 6397. This work provides the first observations of the Li doublet in MS stars of a globular cluster.

2. Observations

We performed spectroscopic observations of the globular cluster NGC 6397 with the multi-object spectrograph FLAMES-GIRAFFE at the VLT on 2007 April, May, June and July, covering the spectral range $\lambda\lambda 6400$–6800 Å at resolving power $\lambda/\delta\lambda \sim 17,000$.

We selected subgiant and dwarf stars in the colour range $B - V = 0.60 \pm 0.03$, which ensures that both set of stars fall in a similar and narrow effective temperature range (see Fig. 3 online in González Hernández *et al.* 2009a).

The spectra were reduced with the ESO pipeline and later on treated within MIDAS. We correct the spectra for sky lines and, barycentric and radial velocity. We typically combine 17 spectra of dwarfs and 4 spectra of subgiants to achieve a similar S/N ratio between 80 and 130 in both sets of stars. The mean radial velocity of the cluster stars is $v_r = 18.5$ km s^{-1}.

3. Stellar parameters

We derived the effective temperature by fitting the observed Hα line profiles with synthetic profiles, using 3D hydrodynamical model atmospheres computed with the CO^{5}BOLD code. The details of this code are provided in Freytag *et al.* (2002) and Wedemeyer *et al.* (2004). The ability of 3D models to reproduce Balmer line profiles has been shown in Behara *et al.* (2009). In that work, the Hα profiles of the Sun as well as the metal-poor stars HD 84937, HD 74000 and HD 140283 are studied. Ludwig *et al.* (2009) also quantified, from a purely theoretical point of view, the difference between the effective temperatures derived from Hα fitting using 1D and 3D models.

We also derived the effective temperatures of MS and SG stars using 1D ATLAS 9 model atmospheres (see Kurucz 2005) and the same fitting procedure. In Fig. 1 we display the histograms of the effective temperatures derived for MS and SG stars using both 1D and 3D models. In the 1D case we got similar effective temperatures for both sets of stars. However, using 3D hydrodynamical models we obtained hotter temperatures by

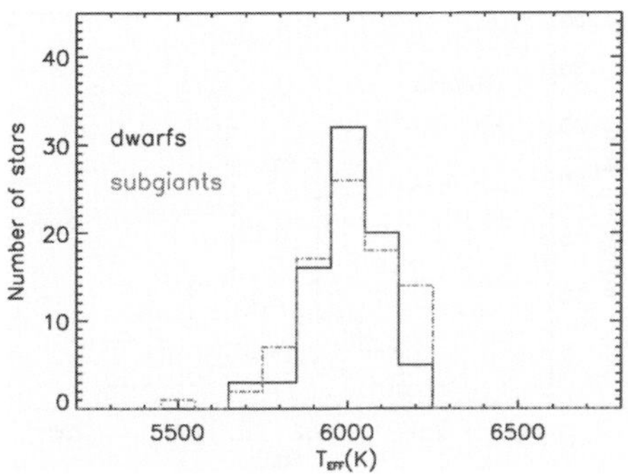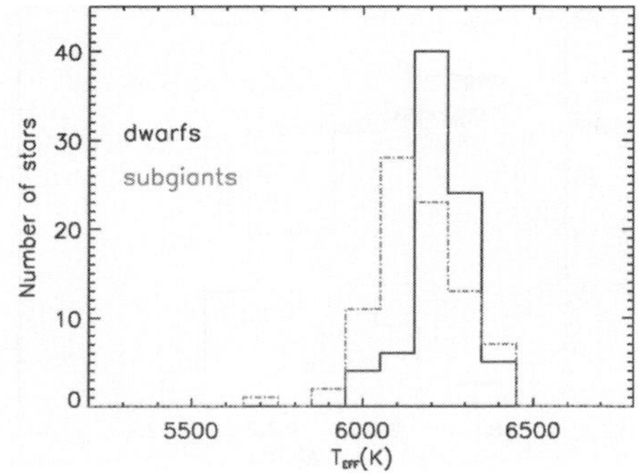

Figure 1. Histograms of 1D (left panel) and 3D (right panel) effective temperatures derived by fitting the observed Hα profiles with theoretical profiles computed using 1D (left panel) and 3D (right panel) model atmospheres, in bins of 100 K, in MS (solid line) and SG (dashed-dotted line) stars in the globular cluster NGC 6397.

approximately 250 K in MS stars and 150 K for SG stars. In the 3D case the histogram of SG stars is slightly shifted with respect to the histogram of MS stars to cooler temperatures, as expected from the difference in surface gravity between MS and SG stars and the sensitivity of the $B - V$ colour to the surface gravity.

Fixed values for the surface gravity were adopted for both subgiant and dwarf stars in the sample, according to the values that best match the position of the stars on a 12 Gyr isochrone (Straniero *et al.* 1997). The adopted values were $\log(g/\mathrm{cm\ s}^2) = 4.40$ and 3.85 for MS and SG stars, respectively.

4. Li abundances

We measure the equivalent width (EW) of the Li I 6708 Å line in SG and MS stars by fitting synthetic spectra of known EW to the observed Li profiles. González Hernández *et al.* (2009a) showed the histograms of the EWs measured in SG and MS stars of this cluster (see their Fig. 1). In that figure it is clearly seen that the EWs of SG stars are larger than those of MS stars. They also estimate the weighted mean EW of the SG stars, being $\sim$ 1.1 pm larger than the weighted mean EW of MS stars. Although the colour $B - V$ is sensitive to surface gravity, a priori, this result was not expected, and clearly suggests that subgiants in this cluster have actually higher Li abundances than dwarfs.

We derived Li abundances using 3D model atmospheres. The line formation of Li was treated in non-local thermodynamical equilibrium (NLTE) using the same code and model atom used in Cayrel *et al.* (2007). The analysis was also done using 1D model atmospheres providing essentially the same picture, even when T_{eff} in 1D are lower (see also Fig. 6 online in González Hernández *et al.* 2009a). In the 1D case we used the Carlsson *et al.* (1994) NLTE corrections.

5. Discussion and conclusions

In Fig. 2 we display the histograms of the derived 1D-NLTE and 3D-NLTE Li abundances of dwarf and subgiant stars of the globular cluster NGC 6397. In both 1D and 3D cases, the SG stars have on average larger amounts of Li content in their atmospheres than MS stars. The difference in the mean Li abundance of dwarfs and subgiants is $\sim$ 0.14 dex in the 1D case and $\sim$ 0.07 dex in the 3D case. This difference between the 1D and the 3D cases is due to the effective temperatures derived using 1D and 3D models,

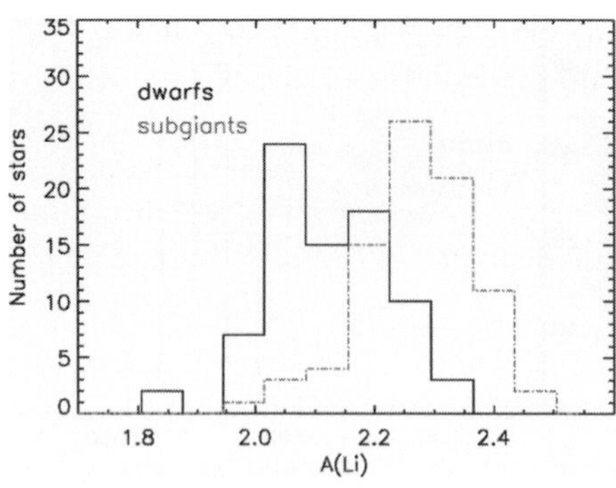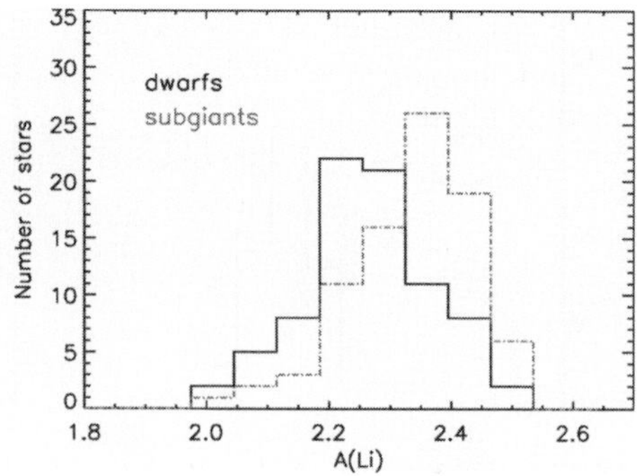

Figure 2. Histograms of 1D (left panel) and 3D (right panel) non-LTE Li abundances, in bins of 0.07 dex, in MS (solid line) and SG (dashed-dotted line) stars in the globular cluster NGC 6397.

and is not related to the NLTE corrections which were independently computed in the 1D and 3D cases. In fact, due to the cooler temperatures derived in 1D with respect to the 3D case, the mean 1D Li abundance lower by ~ 0.13 dex for the MS stars ~ 0.06 dex for the SG stars than the mean 3D abundance.

Lind *et al.* (2009) also find different mean Li abundances, using 1D models, in MSs and SGs, but only by 0.03 dex although still significant at 1σ. However, their result is partially affected by the very narrow range of $T_{\rm eff}$ for MSs deduced by Lind *et al.* (2009) (~ 80 K) compared to the wide range (~ 450 K) for the SGs (see Fig. 7 online in González Hernández *et al.* 2009a).

González Hernández *et al.* (2009a) showed, in their Fig. 2, the 3D-NLTE abundances of SG and MS stars versus 3D effective temperatures (see Fig. 1 in González Hernández *et al.* (2009b) for a similar picture but with $T_{\rm eff}$ and Li abundances computed using 1D models). The points in that figure display a decreasing trend of Li abundance with decreasing temperature. This lithium abundance pattern is different from what is found among field stars (see e.g. Meléndez & Ramírez 2004, Bonifacio *et al.* 2007, González Hernández *et al.* 2008).

Our results imply that the Li surface abundance depends on the evolutionary status of the star. In Fig. 2 of González Hernández *et al.* (2009a), the Li isochrones are shown for different turbulent diffusion models (Richard *et al.* 2005). These models were shifted up by 0.14 dex in Li abundance to make the initial abundance of the models, $\log({\rm Li/H}) = 2.58$, coincide with the primordial Li abundance predicted from fluctuations of the microwave background measured by the WMAP satellite (Cyburt *et al.* 2008). The models assuming pure atomic diffusion, and, among those including low levels of turbulent mixing, in particular the model T6.0, preferred model in Korn *et al.* (2006) and Lind *et al.* (2009) are ruled out by our observations, since they predict the Li content in MS stars to be higher than in SG stars. The only model that predicts a Li pattern which is qualitatively similar to that observed, is the T6.25 model. This model shows a decreasing trend of Li abundances with decreasing temperatures and also predicts higher abundances for SG stars than for MS stars although at temperatures lower than 6000 K, which is not consistent with the observed trend even in the 1D case (see Fig. 1 in González Hernández *et al.* 2009b).

Models including atomic diffusion and tachocline mixing (Piau 2008) do not seem to reproduce our observations either, since they provide a constant Li abundance up to 5500 K. More sophisticated models are required, for instance, those models that besides including diffusion and rotation also take into account the effect of internal gravity waves

(Talon & Charbonnel 2004), seem to predict accurately the Li abundance pattern in solar-type stars, at solar metallicity (Charbonnel & Talon 2005), but models at low metallicity are still needed.

The cosmological lithium discrepancy still needs to be solved. Given that none of the existing models of Li evolution in stellar atmospheres matches our observations in the globular cluster NGC 6397, we hope our results will prompt further new theoretical investigations.

References

Behara, N. T., Ludwig, H.-G., Steffen, M., & Bonifacio, P. 2009, *American Institute of Physics Conference Series*, 1094, 784

Bonifacio, P., Molaro, P., Sivarani, T., Cayrel, R., & Spite, M. 2007, *A&A*, 462, 851

Carlsson, M., Rutten, R. J., Bruls, J. H. M. J., & Shchukina, N. G. 1994, *A&A*, 288, 860

Cayrel, R., Steffen, M., Chand, H., Bonifacio, P., Spite, M., Spite, F., Petitjean, P., Ludwig, H. G., & Caffau, E. 2007, *A&A* (Letters), 473, 37

Charbonnel, C. & Talon, S. 2005, *Science*, 309, 2189

Cyburt, R. H., Fields, B. D., & Olive, K. A. 2008, *Journal of Cosmology and Astro-Particle Physics*, 11, 12

Freytag, B., Steffen, M., & Dorch, B. 2002, *Astronomische Nachrichten*, 323, 213

González Hernández, J. I. *et al.* 2008, *A&A*, 480, 233

González Hernández, J. I., Bonifacio, P., Caffau, E., Steffen, M., Ludwig, H.-G., Behara, N. T., Sbordone, L., Cayrel, R., & Zaggia, S. 2009a, *A&A* (Letters), 505, L13

González Hernández, J. I. *et al.* 2009b, Star Clusters: Basic Galactic Building Blocks Throughout Time And Space. Edited by R. de Grijs and J. Lepine. Proceedings of IAU Symposium #266, held 10-14 August, 2009 in Brazil, Rio de Janeiro. Cambridge, UK: *Cambridge University Press*, in press

Hisano, J., Kawasaki, M., Kohri, K., & Nakayama, K. 2009, *Phys. Rev. D.*, 79, 063514

Jedamzik, K. 2004, *Phys. Rev. D.*, 70, 083510

Jedamzik, K. 2006, *Phys. Rev. D.*, 74, 103509

Jittoh, T., Kohri, K., Koike, M., Sato, J., Shimomura, T., & Yamanaka, M. 2008, *Phys. Rev. D.*, 78, 055007

Kurucz, R. L. 2005, *Memorie della Società Astronomica Italiana Supplement*, 8, 14

Korn, A. J., Grundahl, F., Richard, O., Barklem, P. S., Mashonkina, L., Collet, R., Piskunov, N., & Gustafsson, B. 2006, *Nature*, 442, 657

Lind, K., Primas, F., Charbonnel, C., Grundahl, F., & Asplund, M. 2009, *A&A*, 503, 545

Ludwig, H.-G., Behara, N. T., Steffen, M., & Bonifacio, P. 2009, *A&A* (Letters), 502, 1

Meléndez, J. & Ramírez, I. 2004, *ApJ* (Letters), 615, 33

Piau, L., Beers, T. C., Balsara, D. S., Sivarani, T., Truran, J. W., & Ferguson, J. W. 2006, *ApJ*, 653, 300

Piau, L. 2008, *ApJ*, 689, 1279

Richard, O., Michaud, G., & Richer, J. 2005, *ApJ*, 619, 538

Spergel, D. N., Bean, R. Doré, O., Nolta, M. R., Bennett, C. L. *et al.* 2007 *ApJS*, 170, 377

Spite, M. & Spite, F. 1982, *Nature*, 297, 483

Straniero, O., Chieffi, A., & Limongi, M. 1997, *ApJ*, 490, 425

Talon, S. & Charbonnel, C. 2004, *A&A*, 418, 1051

Wedemeyer, S., Freytag, B., Steffen, M., Ludwig, H.-G., & Holweger, H. 2004, *A&A*, 414, 1121

Monique Spite & Katia Cunha

Corinne Charbonnel & Francesca Primas

Light Elements in the Universe
Proceedings IAU Symposium No. 268, 2009
C. Charbonnel, M. Tosi, F. Primas & C. Chiappini, eds.

© International Astronomical Union 2010
doi:10.1017/S1743921310004229

Observational signatures of lithium depletion in the metal-poor globular cluster NGC6397

Karin Lind[1], Francesca Primas[1], Corinne Charbonnel[2,3], Frank Grundahl[4], and Martin Asplund[5]

[1] European Southern Observatory, Karl-Schwarzschild-Strasse 2, 857 48 Garching
bei München, Germany
email: klind@eso.org

[2] Geneva Observatory, 51 chemin des Mailettes, 1290 Versoix, Switzerland

[3] Laboratoire d'Astrophysique de Toulouse-Tarbes, CNRS UMR 5572, Université de Tou louse,
14, Av. E. Belin, F-31400 Toulouse, France

[4] Department of Physics & Astronomy, Aarhus University, Ny Munkegade,
8000 Aarhus C, Denmark

[5] Max-Planck-Institut für Astrophysik, Karl-Schwarzschild-Strasse 1, 857 41 Garching
bei München, Germany

Abstract. The "stellar" solution to the cosmological lithium problem proposes that surface depletion of lithium in low-mass, metal-poor stars can reconcile the lower abundances found for Galactic halo stars with the primordial prediction. Globular clusters are ideal environments for studies of the surface evolution of lithium, with large number statistics possible to obtain for main sequence stars as well as giants. We discuss the Li abundances measured for >450 stars in the globular cluster NGC 6397, focusing on the evidence for lithium depletion and especially highlighting how the inferred abundances and interpretations are affected by early cluster self-enrichment and systematic uncertainties in the effective temperature determination.

Keywords. stars: abundances, atmospheres, evolution, globular clusters: individual (NGC 6397)

---

1. Introduction

Through the detailed mapping of the cosmic microwave background performed by the *Wilkinson Microwave Anisotropy Probe* (WMAP), a high-precision estimate of the baryon density of the Universe is nowadays possible. When the most up-to-date prediction, $\Omega_b h^2 = 0.02273 \pm 0.00062$ (Dunkley *et al.* 2009) is entered in standard Big Bang nucleosynthesis (SBBN), the primordial abundance ratios of D, ^{3}He, ^{4}He, and ^{7}Li with respect to H, are tightly constrained. Cyburt *et al.* (2008) thus determine a primordial lithium abundance of $N(^7\text{Li})/N(\text{H}) = 5.24^{+0.71}_{-0.67} \times 10^{-10}$ or $A(\text{Li}) = 2.72 \pm 0.06$†, which can be directly compared to the Li abundances inferred for old, metal-poor Galactic halo stars. Numerous high-resolution studies of Li in these stars have indeed revealed a close-to-constant abundance over a wide range in metallicities (forming the "Spite plateau", originally found by Spite & Spite 1982). However, the observationally inferred values are typically a factor of 3–5 lower than the primordial prediction. Assuming that both the WMAP+SBBN value and the Spite plateau abundances are correct, this implies that the Population II (Pop II) stars residing in the halo have undergone significant surface depletion of lithium during their life time.

That lithium depletion occurs in solar-type main sequence stars is well-known, e.g. since

$$\dagger \; A(\text{Li}) = \log\left(\tfrac{N(\text{Li})}{N(\text{H})}\right) + 12$$

the solar photospheric value is at least two orders of magnitudes lower than what is found in meteorites. Also in Pop I stars that are somewhat hotter than the Sun, characteristic depletion signatures like the famous 'Li dip' are seen in open clusters (first identified in the Hyades by Wallerstein *et al.* 1965). If indeed lithium depletion takes place also in Pop II stars, the process has left little traces behind, since the Spite Plateau is very homogeneous over a wide range in metallicities and effective temperatures, and shows small star-to-star variation. Richard *et al.* (2005) illustrated clearly how atomic diffusion (gravitational settling) of lithium can qualitatively account for the surface drainage, but needs to be moderated by a mixing process below the outer convective envelope to avoid too much Li depletion in stars lying on the hot end of the Spite Plateau. In a series of papers (Talon & Charbonnel 1998, 2003, 2004), it has been demonstrated that rotation-induced mixing is a plausible source of lithium destruction in both Pop I and II stars, given that the rotation of stars cooler than $\sim 6300\,\mathrm{K}$ also is efficiently stabilised by gravity waves. This stabilising effect is necessary to avoid a strong dependence of the lithium depletion on the initial angular momentum of the stars, which would result in a non-negligible scatter.

A next generation of stellar evolution models, accounting self-consistently for atomic diffusion, shear turbulence, and meridional circulation, should be confronted with the detailed lithium abundance patterns found for metal-poor stars. Of fundamental importance for constraining the models are Li abundance trends with effective temperature (which for main sequence stars also is indicative of the stellar mass), with metallicity (which constrains the contribution of post-primordial Li), and with evolutionary phase from the main sequence to the very end of the giant stages of evolution. In Lind *et al.* (2009b) we used FLAMES on VLT/UT2 to investigate the lithium abundances of a very large number of stars in a metal-poor globular cluster (NGC 6397, [Fe/H]=−2.0), with the primary goals to establish in detail how the abundances depend on evolutionary phase and investigate if there is any significant scatter in abundance for stars in the same phase. In the following we outline our main findings, with focus on the implications and robustness of the observed abundance trends.

Fig. 1 shows how the mean lithium abundance varies with stellar luminosity in NGC6397, from the dwarfs on the main sequence, over the turn-off point, and on the subgiant branch. This figure highlights the need for smooth turbulence, since the pure diffusive models show large gradients that disagree with the observations (models provided by O. Richard, *priv. comm*). On the other hand, a non-diffusive model is completely flat, which also is a poor match to the observed trend. A moderate degree of turbulence preserves the flat behaviour on the main sequence, while at least qualitatively reproducing the small upturn located prior to the point of strong depletion, although the location of maximum lithium abundance is not matched. However, the predicted level of depletion is only $\sim 0.2\,\mathrm{dex}$, which is not enough to reconcile the initial abundance with the cosmological prediction.

2. Li deficient un-evolved stars

Stars in globular clusters are known to show a larger spread in light elements (up to Al) than their counterparts in the halo field. This extra spread is believed to be a result of cluster self-enrichment of the rest products of high-temperature hydrogen burning from an early generation of more massive stars, made possible by the dense intra-cluster environment. Especially N, Na, and Al abundances are enhanced in the most polluted stars, whereas O and Mg levels are depleted. Further, since the stellar ejecta that

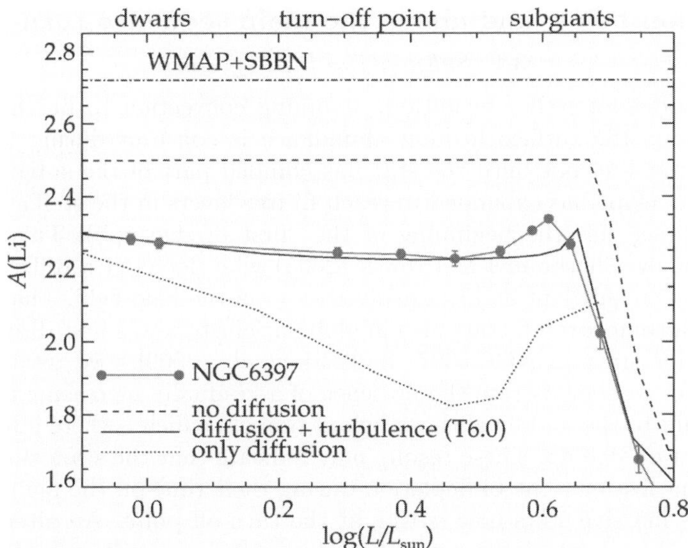

Figure 1. The red bullets represent bin-averaged non-LTE, lithium abundances against stellar luminosity in NGC 6397 (Lind *et al.* 2009a, 2009b). The dashed line shows the prediction from a model without atomic diffusion (Richard *et al.* 2005), the dotted line corresponds to a model with only diffusion, and the solid line a model with diffusion moderated by turbulence. The initial abundances of the models are shifted by -0.08 dex to $A(\mathrm{Li}) = 2.50$. The top line shows the prediction from WMAP+SBBN ($A(\mathrm{Li}) = 2.72$, Cyburt *et al.* 2008). The difference of 0.22 dex in Li abundance is still unaccounted for.

polluted the cluster gas are presumably void of Li (see e.g. Decressin *et al.* 2007), the Li abundances are theoretically expected to be lower in more polluted stars. Observational evidence of a Li-Na anti-correlation in the globular cluster NGC 6752 (Pasquini *et al.* 2005) supports this view.

In NGC 6397 we measured for the first time the Li and Na abundances for a large number (> 100) of turn-off stars and subgiants in a globular cluster. We found indeed, that a handful of rare lithium-deficient stars ($A(\mathrm{Li}) < 2.0$) also are most strongly enhanced in sodium. However, we also realised that a high degree of pollution is actually necessary to significantly affect the lithium abundances, and that the mean Li abundance trends are essentially unbiased by self-enrichment in NGC6397. We caution however, that this cannot be generalised to other globular clusters, which may have undergone more dramatic enrichment. Lithium abundances in globular clusters should thus preferably be studied in parallel with other light elements.

Limiting the sample to stars with $A(\mathrm{Na}) < 3.9$, having suffered a low degree of pollution, we found that the spread in lithium abundance is very low (0.09 dex), in accordance with what may be expected from observational uncertainty (see Lind *et al.* 2009b). We therefore expect the real star-to-star scatter in Li abundance in the first generation of cluster stars to be minimal (below 0.05 dex).

Note that both lithium and sodium abundances of metal-poor turn-off stars are sensitive to effective temperature, with $\pm100\,\mathrm{K}$ corresponding to $\pm0.07\,\mathrm{dex}$ in Li abundance and $\pm0.04\,\mathrm{dex}$ in Na abundance. Uncertainties in effective temperature therefore tend to correlate, rather than anti-correlate the abundances. The pollution signature we have identified in NGC6397 is thus robust in this respect.

3. Li depletion below and above the main sequence turn-off

3.1. *Subgiants vs turn-off stars*

In canonical models of stellar evolution, assuming convection to be the only mean of particle transport, the surface lithium abundance is constant during the whole main sequence life time, and not until the star has climbed part of the subgiant branch and its convective envelope has expanded to reach Li free layers in the stellar interior, strong surface dilution sets in (the beginning of the "first dredge up"). This view has been challenged, first by Charbonnel & Primas (2005) who detected a difference in lithium abundance between subgiant and less evolved stars in the halo field. That subgiant stars are more Li-rich than turn-off stars also in globular clusters was first discovered by Korn *et al.* (2007), for stars in NGC 6397. In Lind *et al.* (2009b) we used unprecedented number statistics to demonstrate the presence of a gradually increasing trend of Li from the turn-off point to the middle of the subgiant branch, which is there interrupted by the onset of the first dredge-up. These results may indicate that the stars that are subgiants today underwent less efficient Li depletion during their time on the main sequence than the slightly less massive stars now sitting at the turn-off point. An alternative scenario would be that the gravitational settling of Li formed a small over-abundance below the convective envelope which is dredged-up in the subgiants by the penetration of the convective envelope.

We found that the subgiants are at most 0.1 dex more Li-rich than the turn-off stars, an estimate which is not suffering from any significant statistical uncertainty, due to the large number of objects (> 200 subgiants and turn-off stars). However, as is always the case with lithium abundance determination, the systematic uncertainties stemming primarily from the effective temperature determination, are dominant. Since errors in effective temperature and lithium abundance are positively correlated, uncertainties in effective temperatures tend to create artificial trends of higher abundance for stars with higher effective temperature and vice versa. Lithium trends with effective temperature should thus be viewed with great caution. In particular, the true range in T_{eff} over which a change in abundance is discussed must be considerably larger than the uncertainty in relative effective temperature determination.

In Lind *et al.* (2009b) we relied on Strömgren photometry to infer effective temperatures, and based the final scale on the temperature sensitive $b - y$ index, translated to effective temperatures using synthetic colours computed from MARCS model atmospheres (Gustafsson *et al.* 2008, Önehag *et al.* 2009). To suppress artificial spread in stellar parameters, caused mainly by photometric uncertainty, we adopted bin-averaged colour-magnitude sequences, constructed from the full photometric catalogue of ~ 5000 stars. The mean locus of the main sequence turn-off point was thus determined to be 6428 K ($(b - y)_0 = 0.303$). In comparison, the new infra-red flux method (IRFM) calibration for dwarfs and subgiants of Casagrande *et al.* (2010), predicts a turn-off point effective temperature of 6435 K, based on $b - y$. Furthermore, broadband BVI photometry give very similar temperatures when adopting the IRFM calibrations: 6422 K based on $(B - V)_0 = 0.386$, and 6447 K based on $(V - I)_0 = 0.555$. The photometric evidence thus gives a very consistent view, especially when bearing in mind the different sensitivity to reddening of the colour indices, with $b - y$ the least affected, and $V - I$ the most.

The Li abundance difference between subgiants and turn-off stars depends primarily on the temperature difference between the stars. Measuring from the mean locus of the turn-off point ($M_V = 4.07$) to the point where the Li abundance reaches its maximum ($M_V = 3.4$), the different photometric indices and calibrations listed above predict $\Delta T_{\mathrm{eff}} = 312 \pm 40$ K (mean standard deviation). Taking 40 K to be an estimate of the uncertainty,

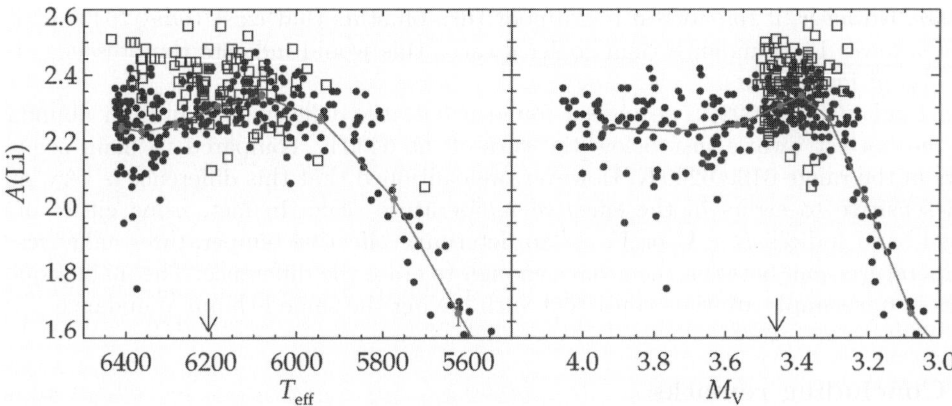

Figure 2. The left hand panel shows lithium abundances vs effective temperature for the post turn-off stars of Lind *et al.* (2009b, black bullets) and Gonzalez-Hernandez *et al.* (2009, open squares). In the right hand panel the same abundances are shown vs visual magnitude. The red, bullets connected with a solid line represent the bin-averaged abundances of Lind *et al.* The turn-off point is located at $M_V \approx 4.07$.

the 0.1 dex difference in A(Li) between turn-off stars and subgiants is significant at the 3σ level.

Spectroscopic effective temperatures have been determined by Hα profile fitting for stars around the turn-off point in NGC 6397 in other studies, with resulting scales that are typically $\sim$80 K cooler than our photometric scale (Korn *et al.* 2007), and $\sim$80 K hotter (Gonzalez-Hernandez *et al.* 2009). The systematic offsets originate in differences in the adopted model atmosphere and likely also in the quality of the spectroscopic observations. It is important to stress that the study by Gonzalez-Hernandez *et al.* only covered stars above and below but not actually at the turn-off point. The negative slope of A(Li) with decreasing $T_{\rm eff}$ inferred in that study for subgiant branch stars is thus not contradicting the positive increase we identified from the turn-off to the middle of the subgiants branch, contrary to what is claimed by the authors. This is displayed in Fig 2, where the left hand panel shows lithium abundance against effective temperature for post turn-off stars in Lind *et al.* and Gonzalez Hernandez *et al.*, and the right hand panel shows the same abundances, but with absolute visual magnitude as reference scale. The discrepancies seen in the left panel can partly be explained by the fact that their temperature scale is somewhat hotter. Note that the visual magnitudes listed by Gonzalez Hernandez *et al.* have been adjusted to agree with our absolute scale as described in Lind *et al.* (2009b).

3.2. *Hot vs cool dwarf stars*

In Pop I dwarfs of a certain age, very lithium-deficient stars are found in the range $T_{\rm eff} = 6400 - 6800$ K, with a gradual increase on the hot as well as the cool end, shaping the so called Li dip. Talon & Charbonnel (2004, see also references therein) describe how the shape and location of the dip can be explained when simultaneously accounting for how the efficiencies of convection, rotational-induced mixing, and gravity waves change with effective temperature. On the cool side of the dip, gravity waves dominate the transport of angular momentum, which reduces the differential rotation that otherwise would give rise to large lithium destruction. In Pop II stars, an analogous dip is theoretically predicted at the same location, but it has not been detected observationally, because the metal-poor stars that had such hot temperatures on the main sequence have evolved to the giant phases. However, if the hottest metal-poor turn-off stars that exist today (6400-6500 K) show a lower Li abundance than cooler dwarfs, this would indicate the presence of the dip also in Pop II stars.

In Lind *et al.* (2009b) we indeed found a 0.04 dex deficiency in lithium abundance for the hottest dwarfs just below the turn-off at 6430 K, compared to main sequence stars in the range 6100-6200 K. However, we cautioned that this difference is very small, and sensitive to errors in the effective temperature scale. In fact, using either of the broad-band indices $B - V$ or $V - I$ to determine effective temperatures enhances the temperature span between the groups enough to erase the difference. The main sequence stars in our sample are thus consistent with having the same lithium abundance.

4. Concluding remarks

Stellar lithium depletion remains a promising solution to the cosmological lithium problem, but there are several open issues. Non-standard mixing below the convective envelope is obviously of key importance for the surface evolution of lithium in Pop II stars. Progress on the modelling side as well as the observational side are needed to establish the underlying physical mechanisms and its correct temperature and metallicity dependence. From the point of view of inferring accurate Li abundances in stars, the effective temperature issue is the most crucial to settle, with the development of hydrodynamical model atmospheres for metal-poor stars being an important step in the right direction.

References

Casagrande, L., Ramírez, I., Meléndez, J., Bessel, M., & Asplund, M. 2010, *A&A*, *e-print:* *http://arxiv.org/abs/1001.3142*

Charbonnel, C. & Primas, F. 2005, *A&A*, 442, 961

Cyburt, R. H., Fields, B. D., & Olive, K. A. 2008, *JCAP*, 11, 12

Decressin, T., Charbonnel, C., & Meynet, G. 2007, *A&A*, 475, 859

Dunkley, J., Komatsu, E., Nolta, M. R. *et al.* 2009, *ApJs*, 180, 306

González Hernández, J. I., Bonifacio, P., Caffau, E. *et al.* 2009, *A&A*, 505, 13

Gustafsson, B., Edvardsson, B., Eriksson, K. *et al.* 2008, *A&A*, 486, 951

Korn, A. J., Grundahl, F., Richard, O. *et al.* 2007, *ApJ*, 671, 402

Lind, K., Asplund, M., & Barklem, P. S. 2009a, *A&A*, 503, 545

Lind, K., Primas, F., Charbonnel, C., Grundahl, F., & Asplund, M. 2009b, *A&A*, 503, 545

Önehag, A., Gustafsson, B., Eriksson, K., & Edvardsson, B. 2009, *A&A*, 498, 527

Pasquini, L., Bonifacio, P., Molaro, P. *et al.* 2005, *A&A*, 441, 549

Richard, O., Michaud, G., & Richer, J. 2005, *ApJ*, 619, 538

Spite, M. & Spite, F. 1982, *Nature*, 297, 483

Talon, S. & Charbonnel, C. 1998, *A&A*, 335, 959

Talon, S. & Charbonnel, C. 2003, *A&A*, 405, 1025

Talon, S. & Charbonnel, C. 2004, *A&A*, 418, 1051

Wallerstein, G., Herbig, G. H., & Conti, P. S. 1965, *ApJ*, 141, 610

Light Elements in the Universe
Proceedings IAU Symposium No. 268, 2009
C. Charbonnel, M. Tosi, F. Primas & C. Chiappini, eds.

© International Astronomical Union 2010
doi:10.1017/S1743921310004230

Lithium in a metal-poor external galaxy: ω Centauri

P. Bonifacio[1,2]**, L. Monaco**[3,4]**. L. Sbordone**[5]**, S. Villanova**[3]**, and E. Pancino**[6]

[1] GEPI, Observatoire de Paris, CNRS, Université Paris Diderot;
Place Jules Janssen, 92190 Meudon, France
email: `Piercarlo.Bonifacio@obspm.fr`

[2] Istituto Nazionale di Astrofisica, Osservatorio Astronomico di Trieste,
Via Tiepolo 11, I-34143 Trieste, Italy

[3] Universidad de Concepción, Casilla 160-C, Concepción, Chile

[4] European Southern Observatory, Casilla 19001, Santiago, Chile

[5] Max Planck Institut for Astrophysics Karl-Schwarzschild-Str. 1 85741 Garching, Germany

[6] Istituto Nazionale di Astrofisica, Osservatorio Astronomico di Bologna,
Via Ranzani 1, 40127, Bologna, Italy

Abstract. ω Centauri is a massive stellar system which is currently going through the Galactic Halo. Its compact aspect and spheroidal shape have for a long time led to it being classified as a Globular Cluster. However the fact that its stars cover a wide metallicity range ($-0.6 <$[Fe/H]< -2.1), points to this object as an external galaxy, satellite of the Milky Way. Lithium among warm metal-poor stars shows a roughly constant abundance, the "Spite Plateau". This has been interpreted as evidence for a primordial origin of the lithium nucleus, at the time of nucleosynthesis. After the physical conditions under which nucleosynthesis occurred, have been constrained by the observations of the fluctuations of the Cosmic Microwave Background, we are facing a "cosmological lithium problem", namely the primordial lithium was a factor of three to four higher than what observed in the Spite plateau. Several avenues may be taken to solve this conundrum, either relying on fundamental physics or on stellar physics, however the realm of possibilities may be considerably narrowed by observing stellar populations in different galaxies, which have experienced different evolutionary histories. Some of the proposed "solutions" may be clearly ruled out, depending on the observation of lithium in the metal-poor populations of external galaxies. ω Centauri is the only external galaxy amenable to such an investigation in the era of 8m telescopes. We have pushed to its limits FLAMES at the ESO 8.2m telescope to obtain high resolution spectra of the Li I doublet in 91 Turn-Off and Sub-Giant stars at V$\sim$ 18 in ω Centauri. We present our preliminary results on this data which suggest that the Li content in ω Centauri warm stars is comparable to that observed in Galactic Halo field stars of similar metallicities and temperatures. This may effectively rule out a whole class of models which invoke a severe Li depletion through processing of material in an early generation of massive stars.

Keywords. Nuclear reactions, nucleosynthesis, abundances – stars: abundances, Population II – Galaxy: globular clusters: individual (ω Cen) – galaxies: abundances, Local Group – cosmology: observations

1. Introduction

The Spite plateau is the constant Li abundance, observed in warm metal-poor stars of different effective temperature and metallicity. This remarkable feature in the abundance pattern of metal-poor stars was discovered by Monique and François Spite in 1982

(1982a,1982b) and was immediately interpreted as a signature of Big Bang Nucleosynthesis (BBN) and a mean to measure the baryonic density of the Universe. See the review of Spite & Spite (2010) on lithium and that of Steigman (2010) on nucleosynthesis in this volume, for an updated view of the problem. The most striking development came from the accurate measurement of the baryonic density from the WMAP satellite (Dunkley *et al.* 2009) which, coupled with standard BBN, implied a primordial Li abundance a factor of three to five higher than the Spite plateau. This discrepancy is often referred to as the "cosmological lithium problem".

Many solutions for this discrepancy have been proposed, including new physics at the time of the Big Bang (see for example Jedamzik 2004, 2006, Jittoh *et al.* 2008, Hisano *et al.* 2009), astration in the pristine Galaxy (Piau *et al.* 2006) and turbulent diffusion to deplete lithium in the stellar atmospheres (Richard *et al.* 2005). A fresh look at the problem can be afforded by the study of lithium in metal-poor populations of external galaxies. Theories like that of Piau *et al.* (2006) can be immediately tested and also theories which invoke stellar phenomena can be seriously constrained by the observation of stellar populations with different star formation histories. Unfortunately even nearby galaxies, like the Magellanic Clouds or the Sagittarius dwarf spheroidal are too far to allow such a study with existing telescopes, although they will be within the reach of the next generation of 40m class telescopes.

There is, however, one external galaxy which is, just about, within reach of our telescopes: ω Cen. The complexity of its colour magnitude diagram clearly testifies the existence of multiple stellar populations with a range of metallicities. It is currently generally accepted that ω Cen is not a globular cluster, but a satellite galaxy of the Milky Way. It is more massive than all other globular clusters (about 2.5×10^5 M$_\odot$, Van de Ven *et al.* 2006), but was probably more massive in the past and has lost mass due to tidal interaction with the Galaxy. In a galaxy which has such a different mass and history with respect to the Milky Way, one should expect a very different lithium content of the metal-poor populations, if the "cosmological lithium problem" is due to astration by an early population. Also in the case of stellar phenomena, such as diffusion, one may expect different lithium content, if the stars show a consistent age spread. Finally we might be able to capture the results of Li production by super-AGB stars, with A(Li) up to 4 (Ventura & D'Antona 2010, D'Antona & Ventura 2010).

2. Observations

We have selected targets on the turn-off and sub-giant branch of ω Cen, mainly from the high precision FORS/VLT photometry of Sollima *et al.* (2005) and from the spectroscopic survey of Villanova *et al.* (2007). Ten targets were selected to trace the faint subgiant branch, called SGB-a in Sollima *et al.* (2005). Our targets are shown in Fig. 1, where we have used the wide field photometry of Bellini *et al.* (2009).

The targets were observed on three nights from April 27th to 29th 2007 at ESO Paranal with FLAMES at the Kueyen 8.2m telescope. The fibres in Medusa mode fed the GIRAFFE spectrograph, configured in the HR15n setting, which covers both Hα and the Li$\scriptstyle\rm I$ resonance doublet at 670.8 nm at a resolution of 17 000. The same plate configuration was observed for all the three nights, with integration times between one hour and slightly over two hours. Both plates were used alternatively and configured in such a way as to minimise the light loss due to atmospheric refraction. One further plate configuration was observed, with a partial overlap with the main plate configuration. We thus obtained a total integration time between 17 and 19 hours for 91 stars on the MS/SGB

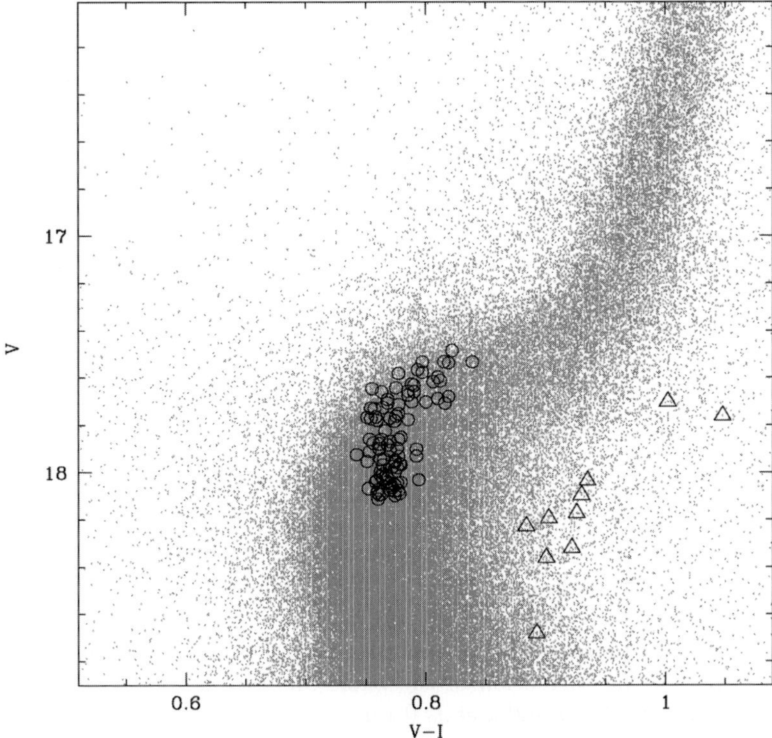

Figure 1. Colour-magnitude diagram of ω Cen from the WFI/2.2m photometry of Bellini *et al.* (2009) Our targets are shown as open circles, except for the targets on the SGB-a of Sollima *et al.* (2005), which are shown as open triangles.

and 10 stars on the SGB-a. After data reduction the spectra achieved S/N ratios in the range 30 to 90 with a mean around 60.

3. Analysis

Our analysis is based on one dimensional model atmospheres computed with version 9 of the ATLAS code (Kurucz 2005) in its Linux version (Sbordone 2005, Sbordone *et al.* 2004). The effective temperature of the stars has been determined by fitting the wings of Hα. The theoretical profiles were computed with a modified version of the BALMER code † which uses the Barklem *et al.* (2000a,b) self broadening theory and Stehlé & King (1999) Stark broadening. At this stage we assumed log g = 4.0 and a metallicity of –1.5 for all stars, thus ignoring the dependence of the Balmer line profiles on metallicity and surface gravity. The equivalent widths of the Li I resonance doublet were measured by fitting synthetic profiles, as done in Bonifacio *et al.* (2002). When we could not detect the Li line we estimated an upper limit as $2\sigma_{EW}$, where σ_{EW} was estimated from the Cayrel formula (Cayrel 1988). A model atmosphere with the appropriate effective temperature, log g = 4.0 and metallicity –1.5 was computed for each star and synthetic profiles were iteratively computed with SYNTHE until the equivalent width of the Li doublet matched the measured equivalent width. A microturbulent velocity of 1 kms^{-1} was assumed, this, like the assumed surface gravity and metallicity, have effects of a few hundredths of

† The original version of R.L. Kurucz is available at `http://kurucz.harvard.edu/`

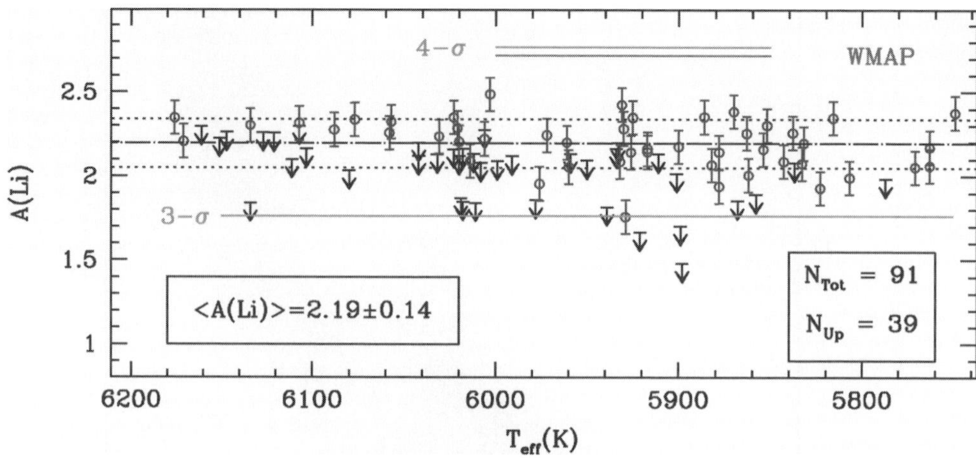

Figure 2. Li abundances in sub-giant and turn-off stars of ω Cen as a function of effective temperature.

dex on the derived Li abundance, this is totally negligible in the current context. Our S/N ratios are high enough that the error on Li abundances is totally dominated by the uncertainty in effective temperatures. The latter is of the order of 150 K and is dominated by the uncertainty in the correction of the blaze function of GIRAFFE. Our estimated total uncertainty on the Li abundance is 0.1 dex.

4. Results

For none of the stars of the SGB-a did we detect the Li doublet. For the other stars our results are summarised in Fig. 2, out of a total of 91 MS/SGB stars for 39 we could not detect any lithium. For the stars with measured Li the abundance appears to be uniform without any trend with effective temperature, the mean value is A(Li)=2.19 with a dispersion of 0.14 dex. We verified that if we use effective temperatures based on the V–I colour the mean Li abundance is similar (2.21) and the dispersion is a little smaller (0.11).

The spectral region covered by our observations is not rich of metallic lines, nevertheless we used the Fe I and Ca I to derive the metallicity of the stars, assuming [Ca/Fe]=+0.4. For the stars in common we get a rather good agreement with the results of Villanova *et al.* (2007) although we note a small offset of the order of 0.1 dex. Although this is, perhaps, not surprising, given the different spectral range, resolution, model atmospheres and spectrum synthesis codes used, we consider these metallicities yet preliminary and we plan to further investigate these differences in the future. However, even with these preliminary metallicities it is clear that there are no obvious trends of Li abundance with metallicity. Also for the upper limits, there does not appear to be present any clustering of upper limits at a particular range of metallicities.

5. Discussion

It appears that the stars of ω Cen lie on the Spite plateau. Comparison with the Galactic field stars, on the same effective temperature scale (see Sbordone *et al.* 2010) shows that the stars of ω Cen occupy the same zone of the Galactic stars, both in the $A(Li)$–T_{eff} and in the $A(Li)$–metallicity planes. This brings us to two robust conclusions:

- the Spite plateau exists also in other galaxies;
- the mechanism(s) which cause the "cosmological lithium problem" are the same in the Milky Way and in other galaxies.

Although this has only been established for ω Cen, it is simpler to assume that this is true for all galaxies than to assume that all galaxies behave differently and that only ω Cen behaves like the Milky Way. Until Li is measured in further external galaxies, this is an acceptable working hypothesis.

These facts immediately tell us that explanations of the "cosmological lithium problem" which require a special evolution for the Milky Way Halo are immediately ruled out. This is the case of the model by Piau *et al.* (2006), which requires that from one third to one half of the Galactic Halo ($\sim 10^9 M_\odot$) has been processed through massive stars which effectively depleted lithium from the primordial value to the current observed Spite plateau. It would be extremely contrived to assume that another galaxy, of current mass $2.5 \times 10^6 M_\odot$, thus of very different type and evolution from the Milky Way, has undergone a similar process with a fine tuning of the mass fraction processed through massive stars so that the Spite plateau results identical to that of the Milky Way. Occam's razor requires that this theory be discarded.

Among other solutions of the "cosmological lithium problem" all which invoke stellar atmospheric phenomena, such as diffusion might also be tightly constrained by our observations. All such phenomena are time dependent and they may produce a uniform Spite plateau in the Galactic Halo, only because the age-spread in the Halo is very small. So the question is: what is the age spread among the stars observed by us in ω Cen? For the stars in common with Villanova *et al.* (2007) we may use their age estimates, based on theoretical isochrones, their metallicity estimates and the requirement that, in the colour-magnitude diagram, each sub-giant branch be associated to a Main Sequence which contains the same relative number of stars. Such ages are given in their table 2 and for the stars in common with our study we find the age spread is 5.6 Gyr. If the age spread in ω Cen is indeed so large, then all theories which invoke time-dependent phenomena, such as diffusion, would be ruled out, since it would be impossible for stars of such different ages, which have started their lives with the same (primordial) Li abundance, to still show the same Li abundance.

However the actual age spread in ω Cen is still a matter of debate and the relative ages of Villanova *et al.* (2007) stand out in the literature for providing the largest spread. Our sample of stars captures essentially the metal–poor population of ω Cen. According to the vast majority of studies, the intrinsic age spread of this population is consistent with zero, from the first photometric estimates (e.g., Hughes *et al.* 2004, Hilker *et al.* 2004), to the most recent spectro-photometric studies based on high precision HST CMDs and low resolution spectroscopy of vast samples of SGB and TO stars (such as Sollima *et al.* 2005, Stanford *et al.* 2006, Kayser *et al.* 2006). Theoretical work supports these findings, implying that a first, coeval generation of stars (the metal–poor population) is responsible for at least part of the pollution of the subsequent generations (see e.g., Norris 2004, Lee *et al.* 2005, Romano *et al.* 2009). The uncertainties of the experimental age spread determination procedures still allow to accommodate for a maximum spread – within the metal–population – of about 1 Gyr. Besides Villanova *et al.* (2007), also the spectroscopic study by Johnson *et al.* (2009), supports a consistent age spread. In this study the metal-poor stars show different degrees of s-process enrichment (see their Figure 13), implying some 0.1–3 Gyrs (Schaller *et al.* 1992) for intermediate-mass AGB stars to pollute part of the metal-poor group, depending on the actual mean mass of the AGB population.

It is thus clear that at the present state of understanding of the age spread in ω Cen our observations do not provide a strong constraint on the viability of diffusion-like mechanisms for Li depletion. Future spectroscopic and photometric observations are likely to better pinpoint this issue. It is possible that even a spread of 1 Gyr would prove a strong constraint on possible Li depletion mechanisms

References

Barklem, P. S., Piskunov, N., & O'Mara, B. J. 2000a, *A&A*(Letters), 355, 5

Barklem, P. S., Piskunov, N., & O'Mara, B. J. 2000b, *A&A*, 363, 1091

Bellini, A., Piotto, G., Bedin, L.R., Anderson, J., Platais, I. *et al.* 2009, *A&A*, 493, 959

Bonifacio, P. *et al.* 2002, *A&A*, 390, 91

Cayrel, R. 1988, in The Impact of Very High S/N Spectroscopy on Stellar Physics, G. Cayrel de Strobel and M. Spite eds., IAU Symp. 132, p. 345

D'Antona, F. & Ventura, P. 2010, this volume

Dunkley, J. *et al.* 2009, *ApJS*, 180, 306

Hilker, M., Kayser, A., Richtler, T., & Willemsen, P. 2004, *A&A*(Letters), 422, 9

Hisano, J., Kawasaki, M., Kohri, K., & Nakayama, K. 2009, *Phys. Rev. D*, 79, 063514

Hughes, J., Wallerstein, G., van Leeuwen, F., & Hilker, M. 2004, *AJ*, 127, 980

Jedamzik, K. 2004, *Phys. Rev. D*, 70, 083510

Jedamzik, K. 2006, *Phys. Rev. D*, 74, 103509

Jittoh, T. *et al.* 2008, *Phys. Rev. D*, 78, 055007

Johnson, C. I., Pilachowski, C. A., Michael Rich, R., & Fulbright, J. P. 2009, *ApJ*, 698, 2048

Kayser, A., Hilker, M., Richtler, T., & Willemsen, P. G. 2006, *A&A*, 458, 777

Kurucz, R. L. 2005, Memorie della Società Astronomica Italiana Supplementi, 8, 14

Lee, Y.-W. *et al.* 2005, *ApJ*(Letters), 621, 57

Norris, J. E. 2004, *ApJ*(Letters), 612, 25

Piau, L. *et al.* 2006, *ApJ*, 653, 300

Richard, O., Michaud, G., & Richer, J. 2005, *ApJ*, 619, 538

Romano, D., Tosi, M., Cignoni, M., Matteucci, F., Pancino, E., & Bellazzini, M. 2009, *MNRAS*, 1604

Sbordone, L. 2005, Memorie della Societa Astronomica Italiana Supplementi, 8, 61

Sbordone, L., Bonifacio, P., Castelli, F., & Kurucz, R. L. 2004, Memorie della Societá Astronomica Italiana Supplementi, 5, 93

Sbordone *et al.* 2010, A&A submitted

Schaller, G., Schaerer, D., Meynet, G., & Maeder, A. 1992, *A&AS*, 96, 269

Sollima, A., Ferraro, F. R., Pancino, E., & Bellazzini, M. 2005, *MNRAS*, 357, 265

Spite, M. & Spite, F. 1982a, *Nature*, 297, 483

Spite, F. & Spite, M. 1982b, *A&A*, 115, 357

Spite, M. & Spite F. 2010, IAU Symposium 268: "Light elements in the Universe", C. Charbonnel, M. Tosi, F. Primas & C. Chiappini, eds., this volume

Steigman, G. 2010, IAU Symposium 268: "Light elements in the Universe", C. Charbonnel, M. Tosi, F. Primas & C. Chiappini, eds., this volume

Stanford, L. M., Da Costa, G. S., Norris, J. E., & Cannon, R. D. 2006, *ApJ*, 647, 1075

Stehle, R. & King, A. R. 1999, *MNRAS*, 304, 698

van de Ven, G., van den Bosch, R. C. E., Verolme, E. K., & de Zeeuw, P. T. 2006, *A&A*, 445, 513

Ventura, P. & D'Antona, F. 2010, *MNRAS*, in press, arXiv:0912.4399

Villanova, S. *et al.* 2007, *ApJ*, 663, 296

Light Elements in the Universe
Proceedings IAU Symposium No. 268, 2009
C. Charbonnel, M. Tosi, F. Primas & C. Chiappini, eds.

© International Astronomical Union 2010
doi:10.1017/S1743921310004242

Lithium and beryllium
in Population I dwarf stars

Sofia Randich[1]

[1]INAF-Osservatorio Astrofisico di Arcetri
Largo E. Fermi, 5, I-50125, Firenze, Italy
email: `randich@arcetri.astro.it`

Abstract. In the last few years a variety of lithium and beryllium surveys have been carried out among Pop. I stars in the field and open clusters, with the goal to trace the dependence of these element abundances on stellar mass, age, metallicity, and to understand the physical processes that lead to their depletion. I summarize here the most recent results, focusing on stars with temperatures similar to the Sun. In particular, I will discuss Li measurements in solar-type members of old open clusters, which definitively show that the low solar lithium is not the standard for a star of that age and mass.

Keywords. Stars: abundances, interiors – open clusters and associations: general

1. Introduction

Several papers in this conference have addressed the question whether metal poor Pop. II stars deplete some lithium (Li) during their lifetime and the issue is indeed still open and highly debated. On the contrary, a large number of observational evidences indicate that more metal rich Pop. I stars do deplete their initial Li content; this is indeed expected on theoretical grounds, due their thicker convective envelopes. However, understanding when and how Li depletion occurs in Pop. I stars of different masses is still a major challenge and several unsolved problems exist.

Among the most puzzling ones, I mention the so-called "Lithium-dip", a sharp decrease of Li abundance in a very narrow temperature range around 6500 K, first discovered in the Hyades (Boesgaard & Tripicco 1986) and then found in several other clusters older than a few hundred Myr; the dispersion in Li seen among stars cooler than $\sim$ 5000 K in the Pleiades and other young clusters (e.g., Jeffries 2006 and references therein); the main sequence (MS) Li depletion observed in the Sun and similar stars. In this paper I shall focus on the latter issue.

Li and Be are depleted from the stellar atmosphere when a mechanism exist which is able to transport surface material down in the stellar interior, where the temperature is high enough for Li/Be reactions. Standard models of stellar evolution (those including convection only as mixing mechanism) predict that solar-type stars should undergo a considerable amount of Li depletion during the pre-main sequence phases; on the contrary, these stars are not predicted to destroy any Li during the MS, since the base of the convective zone does not reach the layer where the temperature is high enough for Li burning to occur. Also, the amount of Li depletion should tightly depend on mass, age, and metallicity.

It is now well established on observational grounds that the Li pattern in solar-type stars contradicts standard model predictions. The Sun has undergone a factor of about 160 Li depletion with respect to its initial abundance witnessed by meteorites, and it is now well ascertained that solar depletion occurs during the MS, rather than in the

PMS phase. Indeed early observations of solar analogs in the Pleiades (100 Myr) and Hyades (600 Myr) clusters already suggested almost forty years ago that solar-analogs undergo less PMS depletion than predicted by models and that Li depletion must take place during the MS (Zappalà 1972).

Furthermore, as first found by Spite *et al.* (1987) and confirmed by several other studies (García López *et al.* 1988; Pasquini *et al.* 1997; Jones *et al.* 1999), otherwise similar members of the solar-age, solar-metallicity M 67 cluster are characterized by different Li abundances: part of them are similar to the Sun, while another fraction have a factor of ~ 10 larger Li, suggesting that the amount of MS depletion is not necessarily the same for stars with the same mass, metallicity, and age. Similar results were achieved for field stars: different studies reported both a dispersion in Li and relatively high Li abundances in old (as indicated by chromospheric activity or rotation) solar analogs (Duncan 1981; Spite & Spite 1982; Pallavicini *et al.* 1987; Pasquini *et al.* 1994).

All these findings support the idea that Li depletion in solar-type stars is not driven by age and mass only, but that some additional parameter likely plays a role. Several models including non standard physics (rotation, magnetic fields, gravity waves, termohaline instability) have indeed been developed, but so far no consensus has been reached neither on the extra-mixing mechanism, nor on the additional parameter(s).

In the last decade several new and more modern measurements of Li in solar-type stars have been performed both in the field and in clusters, allowing us to draw more solid conclusions on the Li depletion pattern in solar-type stars. I will summarize those results by addressing a few specific questions; Namely, **i.** Is the solar Li content typical for a star of that age and mass? **ii.** Is the dispersion seen among solar analogs in M 67 also observed in other open clusters? **iii.** What are the timescales of Li depletion for solar-type stars? **iv.** Does depletion depend on metallicity? **v.** Do solar analogs deplete Be along with Li?

I finally mention that, whereas I will follow here an empirical approach and I will not attempt any direct comparison between theoretical predictions and observed patterns, I encourage theoreticians to use the latter to put constraints on the different extra-mixing models and/or to discern among them.

2. Lithium in solar analogs in the field

New measurements of Li in field stars similar to the Sun were recently carried out by Lambert & Reddy (2009) and Meléndez *et al.* (2009), who reached discrepant conclusions.

The first authors compared the Sun and stars with mass around one solar mass and found that none of their sample stars was as Li-poor as the Sun; actually all of them showed a factor of ~ 10 larger abundance. Based on this, Lambert & Reddy concluded that the solar Li might not represent the standard. On the other hand, Meléndez *et al.* measured Li in a very small sample of solar twins (defined as stars with mass within 3% and [Fe/H] within ± 0.1 dex solar). Stars in their sample showed a decline of Li with age and, in particular, all the old solar twins were characterized by a low Li abundance. This led them to the conclusion that the solar Li is not abnormal and to the claim that previous results (i.e., the finding of Li-rich old solar-type stars) were due to biases in the samples and to considering young stars and/or stars that are not strict solar-twins.

Most obviously, ages of field stars are highly uncertain. A definitive answer to the question whether the solar Li is normal can only come from Li measurements in stars in old open clusters whose age is much better constrained.

3. Open clusters

Since the early study of Zappalá (1972), a huge amount of observational effort has been dedicated to Li measurements in open cluster stars. Sestito & Randich (2005) reanalyzed in a homogeneous fashion modern observations, putting together a sample of 22 open clusters from the literature, in order to derive the timescales of Li evolution in stars in different temperature intervals. They confirmed that solar-type stars deplete lithium during the MS phases; however, they evidenced that Li depletion slows down after the Hyades age and eventually stops for most of the stars after ~ 1 Gyr. Noticeably, solar-type members of the very old (6 Gyr) cluster NGC 188 showed a less than a factor of 2 lower abundance than similar stars in the ten times younger Hyades (see also Randich *et al.* 2003). At that time M 67 was the only known old cluster characterized by a dispersion and showing the presence of severely depleted stars. However, the sample of old clusters was small and only a few stars (typically less than 10) per clusters had Li measurements available.

Taking advantage of available multiplex facilities on 8m class telescopes, in the last five years new, high quality Li measurements in open clusters were obtained. Some of them focused on low- or very-low mass stars (e.g., Manzi *et al.* 2008; Jeffries *et al.* 2009) or on F-type stars (Anthony-Twarog *et al.* 2009), while others concentrated on solar-type stars (Prisinzano & Randich 2007; Pasquini *et al.* 2008; Randich *et al.* 2009; Pace *et al.* 2010, these Proceedings). In the following, the latter ones will be discussed in more detail.

3.1. *A new analysis of M 67*

Pasquini *et al.* (2008) carried out VLT/FLAMES observations of a large sample of MS stars in M 67 in order to identify true solar analogs. At variance with previous studies, effective temperature were derived by spectroscopic means, rather than from photometry. The Li vs. $T_{\rm eff}$ distribution based on these observations still show a dispersion, which however is more evident for stars in the temperature range $5850\,\mathrm{K} \leqslant T_{\rm eff} \leqslant 6000\,\mathrm{K}$, rather than for members within ± 50 K from the solar temperature. Most of these stars appear indeed Li-poor and only one of them has a high Li content. This apparently supports the claim of Meléndez *et al.* (2009) that the solar Li abundance is normal. However, we note that the sample of Pasquini et. al. includes only one of the high-lithium solar-type cluster members analyzed by Jones *et al.* (1999); hence, the inferred Li distribution for solar analogs is based on an incomplete sample and Li-rich stars might be missing. Indeed Randich *et al.* (2006) spectroscopically confirmed the solar-like temperature of star S969 ($T_{\rm eff} = 5800$ K); this star has a high Li ($\log n(\mathrm{Li}) = 2.06$) and is not included in the sample of Pasquini *et al.* (2008).

3.2. *The VLT/FLAMES survey*

We have very recently carried out a VLT/FLAMES spectroscopic survey of a large sample of Galactic open clusters (Randich *et al.* 2005; Pallavicini *et al.* 2006). UVES fibers were allocated to evolved cluster members in order to obtain their chemical composition and investigate the issue of the radial metallicity gradient. The Giraffe spectrograph was instead used to obtain high resolution spectra ($R\sim 20,000$) of unevolved cluster candidates; our primary goals were membership determination and Li measurements among confirmed members. A total of 11 clusters were observed, nine of which are close enough to allow us to obtain good quality spectra of their solar-type members. In Table 1 I list the seven sample clusters whose Li analysis has been completed. Analysis for the remaining two clusters (NGC 2324 and To 2) is in progress. As the table shows, the clusters cover the age interval between ~ 0.7 and 8 Gyr and the metallicity range $[\mathrm{Fe/H}] = -0.38 - +0.35$. After excluding radial velocity non members, each cluster sample typically consists of

Table 1. The sample clusters. I list ages, metallicities, and number of members for which we derived Li abundances.

Cluster	age (Gyr)	[Fe/H]	N_{stars}
NGC 3960	0.7	0.02 ± 0.04	36
NGC 2477	1.0	0.07 ± 0.04	73
NGC 2506	2.2	-0.20 ± 0.02	71
NGC 6253	3.0	0.36 ± 0.07	54
Melotte 66	4.0	-0.33 ± 0.03	53
Be 32	6.0	-0.29 ± 0.04	57
Cr 261	8.0	0.13 ± 0.05	135

$\sim$40-140 stars. The analysis of the data was homogeneously carried out as following Sestito & Randich (2005) and Randich *et al.* (2009). Briefly, effective temperatures ($T_{\rm eff}$) were derived employing the $T_{\rm eff}$ vs. B–V calibration of Soderblom *et al.* (1993a) for solar metallicity stars or that of Alonso *et al.* (1996) for stars with non-solar [Fe/H] content. Li abundances were then derived from equivalent widths and using curves of growths. For more details on the Li analysis (in particular on the estimate of the contribution of the Fe I 670.74 nm line to the Li I 670.8 nm feature) I refer to Randich *et al.* (2009).

I mention that Pace *et al.* (2010, these Proceedings) warn that effective temperatures estimated from colors might be in error, due to uncertain knowledge of reddening. However, I note that for all the FLAMES clusters reddening is known relatively well (in several cases has been estimated from spectroscopy). Also, whereas one cannot exclude that $T_{\rm eff}$ values of individual stars might be somewhat in error, we can exclude that the overall temperature distribution within a given cluster is shifted towards lower/higher temperatures. If this was indeed the case, there would be an inconsistency with cluster ages derived from turn-off stars. In conclusion, use of photometric temperatures, should not greatly affect our results and conclusions.

3.2.1. *Li vs. temperature distribution*

In Fig. 1, I compare the log n(Li) vs. $T_{\rm eff}$ distributions of M 67 and NGC 188 with those of four FLAMES clusters (Be 32 –6 Gyr, [Fe/H] $= -0.29$; Collinder 261 –6 Gyr, [Fe/H] $= 0.13$; Melotte 66 –4 Gyr, [Fe/H] $= -0.31$, and NGC 6253 –3 Gyr, [Fe/H] $= +0.36$). The comparison of the four panels in the figure clearly shows that the upper envelope of the Li vs. $T_{\rm eff}$ distribution is very similar for the six clusters; however, each of them behaves in a different way as far as the dispersion and the fraction of Li-poor and Li-rich stars around the solar temperature is concerned. The distribution of the metal poor Be 32 is similar to that of NGC 188: at variance with M 67, it is characterized by virtually no dispersion and all stars have a Li abundance more than a factor of 10 larger than the Sun. Cr 261 shows a larger amount of dispersion, but none of its members is as Li depleted as the Sun or the lower envelope of M 67. Viceversa, most members of Mel 66 are heavily Li depleted and only five stars (out of 53) have log n(Li)> 2. Finally, the very metal rich NGC 6253 exhibits an intermediate behaviour, more similar to the pattern of M 67: a fraction of stars shares the same low Li as the Sun, while another fraction has a much higher Li. As to the other three (younger) FLAMES clusters (NGC 3960, NGC 2477 and NGC 2506), all of them show almost no dispersion and all their members are Li-rich (log n(Li)$\sim 2.2 - 2.4$). Note that all these would hold true even if considering a very narrow (±50 K) temperature interval around the solar value.

I believe that these results confirm on solid and statistically significant grounds that old stars with temperatures close to the Sun are not necessarily as Li poor as the Sun.

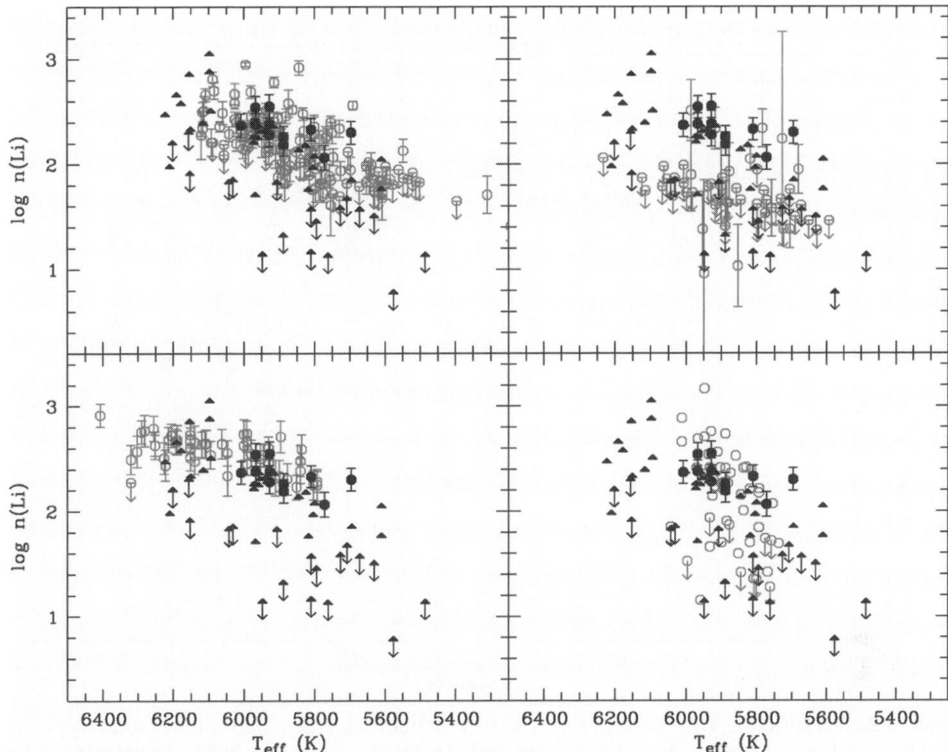

Figure 1. Li abundances (log n(Li) –in the usual logarithmic scale where log n(H) = 12) vs. effective temperature distributions of different FLAMES clusters (red open circles) are compared to M 67 (black filled triangles) and to NGC 188 (black filled circles). Namely, in the top panels the distributions of Collinder 261 (left-hand) and Melotte 66 (right-hand) are plotted, while in the bottom panels Berkeley 32 (left-hand) and NGC 6253 (right-hand) are shown. Li abundances for NGC 188 and M 67 have been retrieved from Sestito & Randich (2005) and had been derived in the same fashion as for the FLAMES sample clusters.

Old open cluster members show a variety of Li patterns and there is not a 'standard'. In a few clusters Li-rich and Li-poor stars are both present, while in others only the Li rich population exist. This provides strong support to the idea that the solar Li is not representative of the Li content of solar-type stars at the solar age and that other parameters, besides age and mass, drive Li depletion during the MS. Also note that the Li pattern of the clusters shown in Fig. 1 does not apparently depend on the cluster metallicity, since clusters with similar age and [Fe/H] (e.g., Be 32 and Mel 66) are characterized by completely different distributions, while clusters with significantly different metallicity (e.g. Be 32 and NGC 188) have similar Li patterns.

Our conclusion is that some initial cluster condition or property might have affected the following evolution of Li. Unfortunately, at the old ages of the clusters memory of that is lost, but Li might represent an important fossil record.

3.2.2. *Evolution of Li with age*

In Fig. 2 I show the evolution of Li with age as inferred from available open clusters measurements. For each cluster I consider stars in the temperature interval 5750 K $\leqslant$ $T_{\rm eff}$ $\leqslant$ 6050 K, i.e. solar-type stars and slightly warmer ones. Note that the 300 K temperature interval was chosen as to have relatively large samples of stars. Fig. 1 however

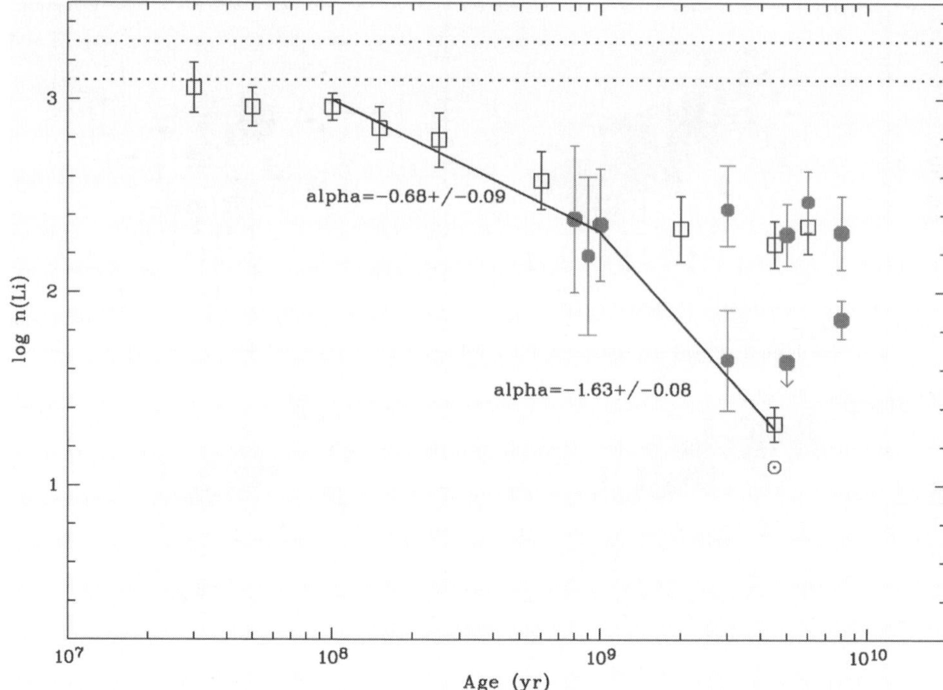

Figure 2. Average Li abundance as a function of age for solar-type stars (5750 K $\leq$ T$_{\text{eff}}$ $\leq$ 6050 K). Open symbols indicate clusters from Sestito & Randich (2005), while filled circles denote the new FLAMES clusters. Error bars indicate the 1σ deviation from the average. Symbols at the same age indicate the average of the upper and lower envelopes of the Li vs. T$_{\text{eff}}$ distribution of clusters characterized by a dispersion. The two dotted lines limit the range of initial Li abundances for Pop. I stars. The Sun is also plotted. The best fit exponential decays (log n(Li)$\propto$ t$^{-\alpha}$) between 100 Myr and 1 Gyr, and between 1 and 4.5 Gyr are also shown.

indicates that the results would not change very much if considering a narrower temperature range around the solar value.

Fig. 2 confirms, based on a much larger sample of clusters, the results of Sestito & Randich: very little depletion (less than a factor of 2) occurs up to 100 Myr. Then, a phase of continuous depletion follows up to an age of about 1 Gyr; the decay of Li abundance is well fitted with an exponential law log n(Li)$\propto$ t$^{-\alpha}$ with $\alpha = -0.68 \pm 0.09$. After 1 Gyr depletion becomes bimodal: part of the stars do not undergo any additional depletion and Li abundances converge towards a plateau value. I mention in passing that this plateau value is surprisingly similar to the Spite plateau of Pop. II stars. Another fraction of the stars, including the Sun, instead continue depleting Li at a very fast rate ($\alpha = -1.63 \pm 0.08$).

4. Beryllium

Be is destroyed at a higher temperature than Li (3.0 vs 2.5 MK) and thus simultaneous observations of Li and Be allow us to understand how deep in the stellar interior the mixing process extends and to put tighter constraints on models. Different models indeed predict different patterns of Be vs. Li depletion. Measurements of beryllium however rely on the resonant lines of Be II which are located in the near-UV spectral region at $\lambda\lambda = 3130.420, 3131.064$. Hence, the capability of observing these lines and measuring

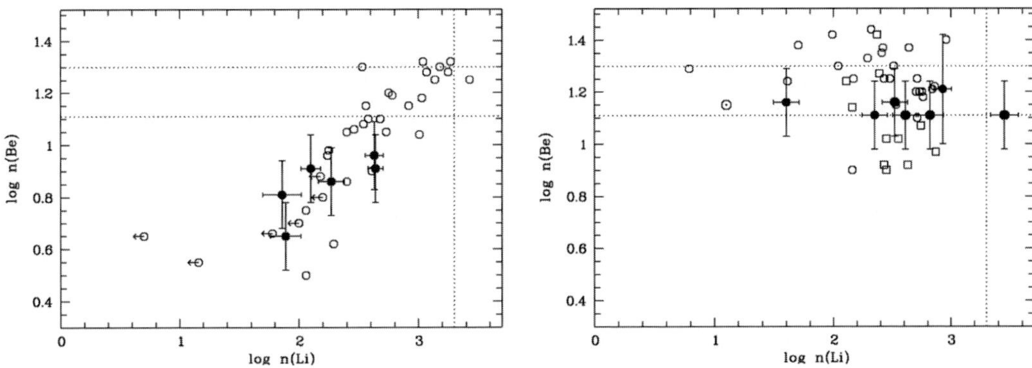

Figure 3. Be vs. Li abundances for stars with $T_{eff} > 6000$ K (left-hand panel) and for stars with $5700 \leqslant T_{eff} \leqslant 6000$ K (right-hand panel). Open circles are Hyades, Praesepe, Coma members from Boesgaard *et al.* (2002, 2003, 2004b), while filled circles denote M 67, IC 4651, IC 2391 and NGC 2516 members from Randich *et al.* (2007). Finally, open squares are solar mass field stars from Boesgaard *et al.* (2009). The position of the Sun is also shown. The horizontal lines bound the range of initial Be abundances in the different scales, while the vertical line indicates the initial Li abundance.

abundances is lower than for Li (see also Primas, these Proceedings). For this reason, until very recently much fewer surveys of Be have been performed, mostly focusing on bright F-type stars in the field and in the closest clusters (Hyades, Come Ber, Ursa Major, Pleiades). These measurements have shown that such stars undergo Be depletion (a Be dip has indeed been identified) and show a tight Be vs. Li correlation holds (Boesgaard *et al.* 2004a and references therein).

4.1. *Beryllium in solar-type stars*

As mentioned, the Sun has suffered a large amount of Li depletion. Until 10 years ago it was commonly believed that it had also undergone some (a factor of two) Be depletion. Balachandran & Bell (1998) performed a new analysis of the near-UV solar spectrum, correctly taking into account continuous opacity. They showed that the Sun has instead not depleted any Be, implying that mixing in the solar photosphere is more superficial than previously supposed.

The availability of state-of-the art high resolution spectrographs with high near-UV efficiency has made it possible to measure Be not only in bright stars in close-by clusters, but also in fainter members of more distant, old clusters (Randich *et al.* 2002, 2007; Smiljanic *et al.* 2009). In particular, Be has been measured in solar-analogs in two clusters older than the Hyades: IC 4651 (2 Gyr) and M 67. Furthermore, in a very recent paper Boesgaard & Krugler (2009) present Be measurements for the one solar mass stars of the sample of Lambert & Reddy (2004). All these new data give us the possibility to study the behaviour of Be depletion in old stars similar to our Sun.

In Fig. 3 I show Li vs. Be abundance for members of different clusters in two temperature regimes, as indicated in the caption. The figure evidences distinct behaviors for stars in the two subsamples. Namely, stars warmer than 6000 K in both young and old clusters deplete some amount of Be and a clear correlation between Li and Be abundances is present, in agreement with the results of Boesgaard and collaborators. I mention that Be depletion was also found by Smiljanic *et al.* (2009) among F-type stars in IC 4651. On the other hand, virtually all the stars in the range $5700 \leqslant T_{eff} \leqslant 6000$ K, including the Sun, show, within the errors, the same Be abundance. While some star-to-star dispersion is present, possibly due to the adoption of different abundance scales and to non-uniform initial abundance, the stars in the right-hand panel of the figure do not follow any Be vs. Li correlation. These stars span two orders of magnitude in Li abundances, they have different ages and metallicities, but they share the same Be content. The Sun also has a very similar Be abundance. I stress that stars in the old M 67 have virtually the same Be abundance as members of the 50 Myr old IC 2391 and the 100 Myr old NGC 2516 clusters. Also, stars in M 67 showing different amount of Li depletion have the same Be content (see also Randich *et al.* 2002, 2007). This in turn implies that *i.* solar-type stars do not deplete any beryllium at least up to the solar and M 67 age, in spite of the fact that they do deplete lithium; *ii.* at variance with the warmer star regime, where Be depletion is correlated to Li depletion, the mixing mechanism responsible for MS Li depletion in this temperature range is rather shallow and does not extend deep enough to cause also Be depletion.

5. Concluding remarks

I conclude by answering the five questions raised in the introduction.

Observations of solar-type stars in open clusters strongly suggest that the solar Li content is not typical. Whereas stars showing the same severe Li depletion as the Sun are found in the field and in clusters, otherwise similar old stars are present with a factor of at least ten higher Li. Part of the old open clusters so far observed show a dispersion similar to M 67, while the Li distribution in other clusters is very narrow. Solar-type stars deplete Li on the MS up to an age of about 1 Gyr; afterward, depletion either stops or becomes considerably more efficient, like in the Sun. The reasons why otherwise similar stars deplete Li at different rates after 1 Gyr and why similar cluster behave differently is so far not understood. The hypothesis that the Li behaviour might be related to the presence of a planetary system (Israelian, these Proceedings) is very tempting. In any case, this empirical evidence suggests that Li is a good age tracer for these stars up to about 1 Gyr. After that age , a 'low' solar-like Li abundance is indicative of an old age, plus, possibly, a peculiar evolution. Viceversa, a 'high' Li content ($\sim$ 10 times the solar value) only allows inferring a lower limit to the age.

Li vs. T_{eff} patterns and the presence of severely depleted stars do not depend on the cluster metallicity. I mention however that metallicity affects stellar structure and a given effective temperature corresponds to lower masses for lower metallicities (and viceversa). Therefore, similar Li vs. T_{teff} distributions do not correspond to similar Li vs. mass patterns, if the metallicity of the clusters is not the same (see Randich *et al.* 2009).

Finally, Be observations in solar-type stars in old clusters show that, like the Sun, they do not deplete any Be, in spite of the fact that they deplete [different amount of] Li. This indicates that the mixing process must be a shallow one.

References

Alonso, A., Arribas, S., & Martinez-Roger, C. 1996, *A&A*, 313, 873

Anthony-Twarog, B. J., Deliyannis, C. P., Twarog, B., Croxall, K. V., & Cummings, J. D. 2009, *AJ*, 138, 1171

Balachandran, S. C. & Bell, R. A. 1998, *Nature*, 392, 791

Boesgaard, A. M. & Tripicco, M. J. 1986, *ApJ*, 302, L49

Boesgaard, A. M. & King, J. R. 2002, *ApJ*, 565, 587

Boesgaard, A. M., Armengaud, E., & King, J. R. 2003, *ApJ*, 583, 955

Boesgaard, A. M., Armengaud, E., King., J. R., Deliyannis, C. P., & Stephens, A. 2004a, *ApJ*, 613, 1202

Boesgaard, A. M., Armengaud, E., & King, J. R. 2004b, *ApJ*, 605, 864

Boesgaard, A. M. & Krugler, H. J. 2009, *ApJ*, 691, 1412
Duncan, D. K. 1981, *ApJ*, 248, 651
García López, R. J., Rebolo, R., & Beckmann, J. E. 1988, *PASP*, 100, 1489
Jeffries, R. 2006, in *Chemical Abundances and Mixing in Stars in the Milky Way and it Satellites*,
    ESO Astrophysics Symposia, S. Randich, L. Pasquini (eds), (Springer-Verlag), p. 163
Jeffries, R., Jackson, R. J., James, D. J., & Cargile, P. A. 2009, *MNRAS*, 400, 317
Jones, B. F., Fisher, D., & Soderblom, D. R. 1999, *AJ*, 117, 330
Lambert, D. & Reddy, B. 2004, *MNRAS*, 349, 757
Manzi, S., Randich, S., de Wit, W. J., & Palla, F. 2009, *A&A*, 479, 141
Meléndez, J., Ramírez, I., Casagrande, L. *et al.* 2009, *Ap&SS*, in press
Pallavicini, R., Cerruti-Sola, M., & Duncan, D. K. 1987, *ApJ*, 174, 116
Pallavicini, R. *et al.* 2006, *In Chemical Abundances and Mixing in the Milky Way and its Satel-
    lites*, ESO Astrophysics Symposia, S. Randich, L. Pasquini (eds), Springer-Verlag, p. 181
Pasquini, L., Liu, Q., & Pallavicini, R. 1994, *A&A*, 287, cw191535
Pasquini, L., Randich, S., & Pallavicini, R. 1997, *A&A*, 325, 535
Pasquini, L., Biazzo, K., Bonifacio, P., Randich, S., & Bedin, L. R. 2008, *A&A*, 489, 677
Prisinzano, L. & Randich, S. 2007, *A&A*, 475, 535
Randich, S., Primas, F., Pasquini, L., & Pallavicini, R. 2002, *A&A*, 387, 222
Randich, S., Sestito, P., & Pallavicini, R. 2003, *A&A*, 399, 133
Randich, S. *et al.* 2005, *ESO Messenger*, 121, 18
Randich, S., Sestito, P., Primas, F., Pallavicini, R., & Pasquini, L. 2006, *A&A*, 450, 557
Randich, S., Primas, F., Pasquini, L., Sestito, P., & Pallavicini, R. 2007, *A&A*, 469, 163
Randich, S., Pace, G., Pastori, L., & Bragaglia, A. 2009, *A&A*, 496, 441
Sestito, P. & Randich, S. 2005, *A&A*, 442, 615
Smiljanic, R., Pasquini, L., Charbonnel, C., & Lagarde, N. 2009, *A&A*, in press
Soderblom, D. R., Stauffer, J. R., Hudon, J. D., & Jones, B. F. 1993, *ApJS*, 85, 113
Spite, M. & Spite, F. 1982, *A&A*, 115, 351
Spite, F., Spite, M., Peterson, R. C., & Chaffee, F. H., Jr. 1987, *A&A* 171, L8
Zappalà, R. R. 1972, *ApJ*, 72, 57

Andreas Kaufer

Garik Isrealian

Light Elements in the Universe
Proceedings IAU Symposium No. 268, 2009
C. Charbonnel, M. Tosi, F. Primas & C. Chiappini, eds.

© International Astronomical Union 2010
doi:10.1017/S1743921310004254

Lithium in stars with exoplanets

Garik Israelian

Instituto de Astrofisica de Canarias,
Calle Via Lactea s/n, E38200, La Laguna, Tenerife, Canary Islands, Spain
email: gil@iac.es

Abstract. Our recent study of solar-type stars from the HARPS GTO sample provides highly accurate information with regard to Lithium abundances in stars with and without detected planets (Israelian *et al.* 2009) . When the Li abundances of planet bearing stars are compared with the "single" stars, we find an excess of Li depletion in planet hosts with effective temperatures in the range 5700-5850 K. We also found that small amounts of Li have survived in the atmospheres of some planet-host solar analogs. Enhanced Li depletion in planet host stars puts constraints on mixing processes responsible for this phenomenon. We show that neither age nor metallicity are responsible for this observational fact.

Keywords. Stars: abundances, late-type, planetary systems

1. Introduction

The enhanced depletion of lithium in the Sun discovered more than 60 years ago remains the epitome of the Li puzzle. The base of the surface convective layer of the Sun is not hot enough for nuclear reactions to destroy Li, and yet the surface Li abundance is about 140 times less than the initial protosolar abundance which is the meteoritic value (Anders & Grevesse 1989). A large dispersion in Li abundance observed in solar-type stars of the same age, mass and metallicity is inconsistent with classical models of stellar evolution (DAntona & Mazzitelli 1994) and has reinforced the idea that the presence of planets may be responsible for this effect (King *et al.* 1997, Gonzalez & Laws 2000, Israelian *et al.* 2004).

King *et al.* (1997) were first to propose that the low Li abundances of the Sun and 16 Cyg B with respect to 16 Cyg A may be related to the presence of a planetary companion. Many studies have attempted to separate the effects of planets on Li abundance. Gonzalez & Laws (2000) corrected Li abundances in planet hosts for linear trends with age, metallicity, and $T_{\rm eff}$, and concluded that planet hosts contain less Li than field stars. This result was debated by Ryan (2000) who proposed that these differences were not significant. Many authors have revisited this topic since then (Israelian *et al.* 2004, Chen & Zhao 2006, Luck & Heiter 2006, Takeda *et al.* 2007, Gonzalez 2008) and all of them, except Luck & Heiter (2006) have concluded that stars with planets tend to have smaller Li abundance. However, stellar samples used in these studies were not homogeneous. It was possible that the planet host stars are Li poor because they are more metal rich. Moreover, neither of these authors could provide a homogeneous comparison sample of "single" stars without detected giant planets. These tasks have been undertaken by our group (Israelian *et al.* 2009) who recently has demonstrated that, indeed, stars with giant exoplanets contain less Li than "single" stars without (so far) detected planets in the HARPS GTO sample.

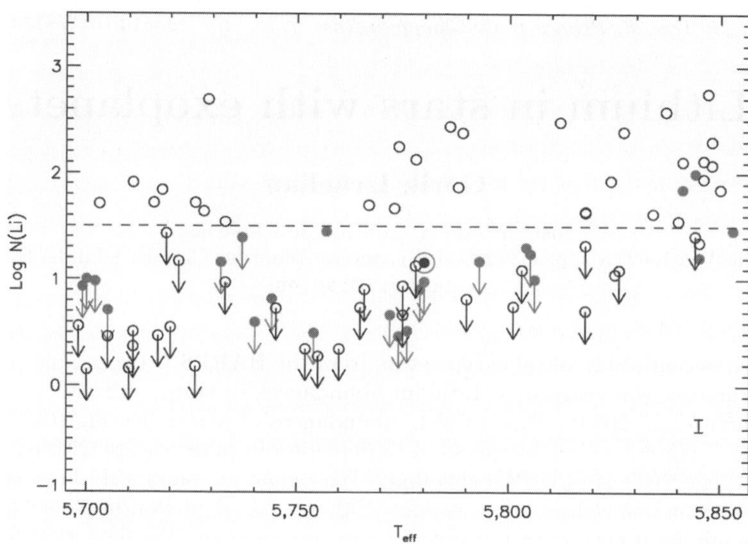

Figure 1. Lithium abundance against effective temperature in solar-analogue stars with and without detected planets. The planet-hosts are red filled circles. The minimum detectable Li abundance varies among the stars used in this study because their spectra have different signal–to-noise ratios. The straight line log N(Li) = 1.5 matches the upper envelope of the lower limits corresponding to a minimum S/N = 200 in a typical solar twin. We employ this line as a cut-off for selecting Li-depleted stars in our sample. The mean statistical errors (1?) for the log N(Li) and $T_{\rm eff}$ averaged over all stars are 0.06 dex and 30 K, respectively (Sousa *et al.* 2008). Errors in log N(Li) include uncertainties in $T_{\rm eff}$ and equivalent width measurements.

2. New analysis

To establish a definitive relation between the presence of planets and enhanced Li depletion in solar-type stars requires high quality observations and unbiased abundance analysis of Li for a large sample of stars with and without planets. We obtained Li abundances from high resolution, high S/N spectra for a sample of 451 stars in the HARPS high precision radial velocity survey (precision better than 1 m/s; Mayor *et al.* 2003) spanning the effective temperature range between 4900 and 6500 K. These are unevolved, slowly rotating non-active stars from a CORALIE catalogue (Mayor *et al.* 2003). Of these 451 stars, 70 are reported to host planets and the rest (often we call them single stars) have no detection so far. Our abundance analysis, which followed standard prescriptions for stellar models, spectral synthesis code and stellar parameter determination (Sousa *et al.* 2008), confirm a peculiar behavior of Li in the effective temperature range 5600-5900 K. To put this in a more solid statistical basis these two samples in the $T_{\rm eff}$ window 5600-5900 K were extended by adding 16 and 13 planet host and "single" stars, respectively, with Li abundances from our previous work where the same spectral synthesis tools have been employed (Israelian *et al.* 2004). It is remarkable that the immense majority of planet host stars have severely depleted lithium while in the comparison sample large fraction partially inhibited depletion. The Li abundance of 20 % of stars with exoplanets in the temperature range 5600-5900 K has log N(Li)$\geqslant$1.5 while for the 116 comparison stars the Li abundance shows a rather large dispersion with some 43 % of the stars displaying Li abundances log N(Li)> 1.5. This result becomes more obvious in solar analog stars where some 50 % of 60 single stars in narrow window of $T_{sun}\pm$80 K appear with Log N(Li)$\geqslant$1.5 while only one planet host, out of 24, has log N(Li) $\geqslant$1.5 (Fig 1). Lithium survival at $T_{eff} > 5850$ K is explained by the fact that the convective layers of stars more massive than the Sun are shallow and too far to reach the Li-burning layers. On the other hand, lower mass stars with $T_{eff} < 5700$ K have deeper convective layers and destroy Li more efficiently. We note that subgiants were not included in this study because they undergo dramatic changes in their internal structure that alters surface abundance of Li. The Li over-depletion in planet bearing main sequence stars is a generic feature over T_{eff} restricted range from 5700 to 5850 K. Let us now investigate if Li abundance in solar analog stars is determined by their age and/or metallicity.

3. Ages, metallicity, and rotation

3.1. *Metallicities*

Most of the planet-host stars discovered to date are metal-rich (Santos, Israelian & Mayor 2004). The metallicity excess could result from either the accretion of planets/planetesimals on to the star or the protostellar molecular cloud. This metallicity excess is also present in the solar analogue planet-bearing stars (see Fig. 2c). It is very important to investigate if a high metallicity is responsible for enhanced Li depletion in these stars. The increase of metal opacities in solar-type stars is responsible for the transition between radiative and convective energy transport. The main contributors to the total opacity at the base of the convective zone are oxygen and iron (Piau & Turck-Chièze 2002). Our data (Fig 2c) show that the fraction of single stars with log N(Li) >1.5 is 50% at [Fe/H]<0 and [Fe/H] > 0. This suggests that the Li depletion mechanism does not depend on the metallicity in the range $0.5 < [\mathrm{Fe/H}] < +0.5$. Apart from this, we have investigated the dependence of log N(Li) on [O/Fe] (Piau & Turck-Chièze 2002) for planet-host stars, using oxygen abundances in planet-host stars from the literature (Ecuvillon *et al.* 2006), and again found no correlation. We conclude that the metallicity or [O/Fe] ratio is not responsible for an enhanced Li depletion in metal-rich planet-host stars.

3.2. *Chromospheric ages and rotation*

It is often stated that the lithium abundance of solar-type main sequence stars decreases progressively with age (Sestito & Randich 2005). If that were the case, we should expect a correlation between lithium and stellar age indicators. Chromospheric activity is a reliable age indicator for solar-type stars from young ages to about 1 Gyr (Pace *et al.* 2009), or perhaps even to the age of the Sun (Wright *et al.* 2004). Abundances of Li versus chromospheric activity indices, R_{HK}, for the solar analogue stars with and without detected planets are shown in Fig. 2a. We find no correlation between Li and the activity index and conclude that the solar analogue stars with and without planets considered in this work have similar ages. This conclusion is valid as long as the chromospheric activity index R_{HK} can be used as an age indicator. It means our stars are older than 1 Gyr (Pace *et al.* 2009) or perhaps as old as 4.5 Gyr (Wright *et al.* 2004).

It is known (Cutispoto *et al.* 2003) that chromospheric activity correlates with stellar rotation (vsini). If the planet hosts were older than the comparison sample, their rotational velocities would be smaller than in the comparison sample. This is not observed either (Fig 2b), adding support to our conclusion that the ages of planet hosts and "single" stars in our sample exceed 1 Gyr.

We have compiled previously published isochrone ages for Li-rich (log N(Li) > 1.9) stars shown in Fig 1. One may think that these main sequence solar analogue stars are young since their atmospheres contain a lot of Lithium. However, the average isochrone ages of our planet-hosts and "single" stars are found in the window 5 to 9 Gyr (Holmberg

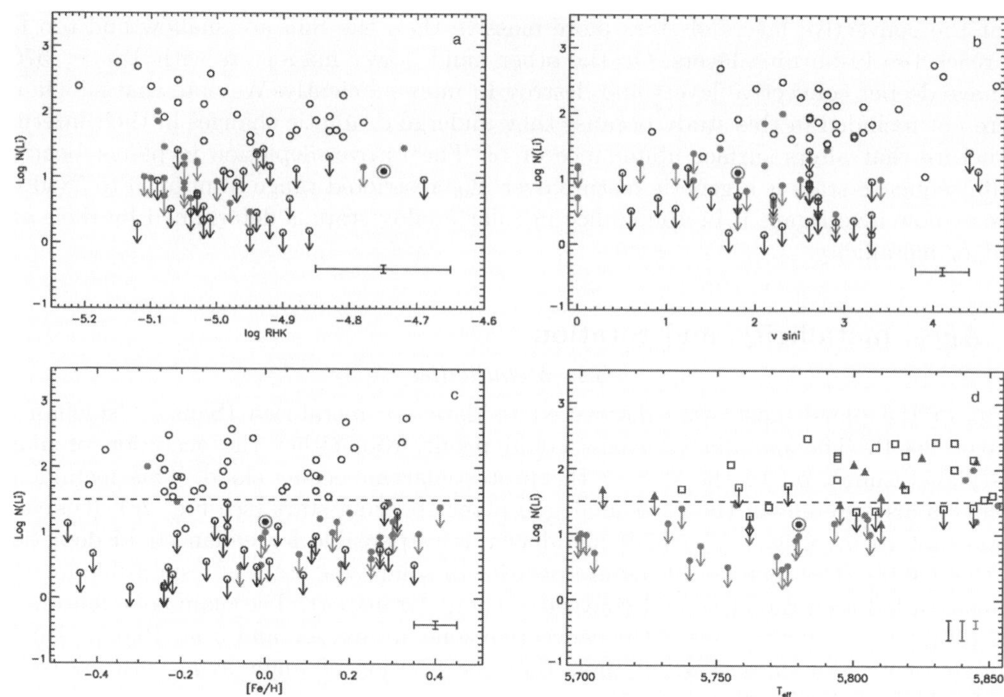

Figure 2. Panels (a) and (b). Lithium abundances in solar-analogue stars with and without detected planets versus chromospheric activity indices (log R_{HK}) and rotational velocity (vsini). R_{HK} values were taken from the literature (Wright *et al.* 2004, Saffe *et al.* 2005, Gray *et al.* 2006) while rotational velocities of the comparison sample stars and many planet hosts were measured from CORALIE and HARPS spectra using a cross correlation function (Santos *et al.* 2002). Typical 1σ uncertainties for log R_{HK} and vsini are 0.1 dex and 0.3 km/sec, respectively (Wright *et al.* 2004, Santos *et al.* 2002). Rotational velocities of several planet-hosts were taken from the literature (Valenti & Fischer 2005). Panel (c). Li abundances for the same stars versus metallicity. The latter was measured (Sousa *et al.* 2008) with a precision of 0.03 dex (1σ). Panel (d). Li abundances in planet hosts and stars of the open clusters M67 and NGC 6253. In this panel we plot Li abundances versus effective temperature in planet hosts (red filled circles), and stars of the open clusters M67 (blue triangles) and NGC 6253 (open squares). The data for M67 were taken from the literature. Li abundances in NGC 6253 have been derived from VLT/Giraffe spectra using standard methods (Randich *et al.*, in preparation). Typical 1σ error bars for cluster stars are 0.15 dex and 100 K for log N(Li) and T_{eff}, respectively.

et al. 2009). Nevertheless, stellar ages derived from isochrones cannot be used since they are very uncertain with the dispersion as large as 4 Gyr (Saffe *et al.* 2005).

3.3. *Lithium and ages in planet hosts and open clusters*

Observations of solar-type stars in the temperature range 5700-6100 K in open clusters (Sestito & Randich 2005, Randich 2008) show that the Li depletion continuously occurs from the Zero Age Main Sequence (ZAMS) up to ∼1 Gyr, with a time scale of 1.4 Gyr. It becomes bimodal at >1 Gyr as the fraction of stars continue depleting Li at higher rate. Our Sun is perhaps the best representative of this group. The Li depletion completely stops for the majority of stars and all cluster average abundances converge to a plateau value close to the Spite plateau of Pop II stars (see Sestito & Randich 2005, Randich 2008, Randich, S. this proceedings).

We have seen (Fig. 2) that the comparison with field stars leads to the conclusion that neither age nor metallicity is responsible for the excess Li depletion. This is reinforced by observations of Li in solar-type stars in old solar metallicity and/or metal-rich open clusters, which indeed show a wide dispersion of Li abundances with values ranging from log N(Li) = 2.5 down to 1.0 and lower (Sestito *et al.* 2007, Randich 2008, Pasquini *et al.* 2008). This is the case for M67 (age 3.5–4.8 Gyr and [Fe/H] = 0.06)(Pasquini *et al.* 2008) and NGC 6253 (age 3 Gyr and [Fe/H] = 0.35) (Yadav *et al.* 2008, Sestito *et al.* 2007), as is clearly seen in Fig 2d. These two clusters offer a homogeneous sample of solar analogues in terms of age and metallicity. Both high and low Li abundance solar analogues are present in these two clusters. The high Li abundance in a large fraction of old metal-rich stars in NGC 6253 and M67 supports our conclusion that high metallicity and/or age are not the main cause for the systematic low Li abundances in solar analogue planet-host stars. Two other clusters, Cr261 (age 6 Gyr and [Fe/H] = 0.13) and NGC 188 (age 8 Gyr and [Fe/H]~0) show significant Li dispersion (Sestito & Randich 2005, Randich 2008, Randich, S., this conference). Slow mixing models of solar-type stars predict Li abundances log N(Li)$\leqslant$1 and cannot explain a large fraction of Li-rich solar analogues in these clusters with log N(Li)~2.4. The analysis of several old open clusters observed with FLAMES confirm that the Li dispersion does not depend on age or metallicity. (S. Randich, this conference).

4. Conclusions

We propose that the low Li abundance of planet-host solar analogue stars is associated with the presence of planets. The presence of a planetary system may affect the angular momentum evolution of the star and the surface convective mixing, thereby causing enhanced lithium depletion. Planet migration could possibly trigger angular momentum transfer in the convective zone, leading to additional mixing below this zone. Theoretical models (Pinsonneault *et al.* 1989) show how magnetic braking scales with rotational velocity leading to turbulent diffusion mixing and enhanced lithium depletion. In this case we would expect severely Li-depleted stars to host planets with shorter orbital periods. On the other hand, long-lasting star–disc interaction during the pre-main sequence may cause planet-host stars to be slow rotators and develop a high degree of differential rotation between the radiative core and the convective envelope. This process may lead to enhanced lithium depletion too (Bouvier 2008).

We know that all stars used in the HARPS GTO program are non-active, slow rotators. Thus, from the RV point of view they are all equally good planet-host candidates. It is impossible to explain why HARPS was unable to detect the relatively young planetary systems with ages between (for example) 1 and 3 Gyr. This fact alone suggests that planet-host and "single" stars in our sample come from groups with the same age distribution. Thus, our finding presented in Fig. 1 can be considered as an independent confirmation of the well-known Li puzzle discovered in open clusters (Randich 2008). We confirm that factors other than age, mass, and metallicity, such as angular momentum (which planets may alter), can affect stellar lithium abundance.

Studying young stars to identify the effect that planets have on their rotation is the next step to explain the correlation between lithium abundance and planets. Young stars and planets are more difficult to observe, but they reveal more than old stars do about how fast they rotated in their infancy. In the future, it may be possible to test the hypothesis that altering stars' rotation affects lithium depletion. Discovery of planets in open clusters will certainly help to understand the Li puzzle.

References

Anders, E. & Grevesse, N. 1989, *Geochim. Cosmochim. Acta*, 53, 197

Bouvier, J. 2008, *A&A* (Letters), 489, 53

Chen, Y. Q. & Zhao, G. 2006, *AJ*, 131, 1861

Cutispoto, G., Tagliaferri, G., de Medeiros, J. R., Pastori, L., Pasquini, L. & Andersen, J. 2003, *A&A*, 397, 987

DAntona, F. & Mazzitelli, I. 1994, *ApJS*, 90, 467

Ecuvillon, A., Israelian, G., Santos, N. C., Shchukina, N. G., Mayor, M., & Rebolo, R. 2006, *A&A*, 445, 633

Gonzalez, G. 2008, *MNRAS*, 386, 928

Gonzalez, G. & Laws, C. 2000, *AJ*, 119, 390

Gray, R. O., Corbally, C. J., Garrison, R. F., McFadden, T., Bubar, E. J., McGahee, C. E., O'Donoghue, A. A., & Knox, E. R. 2006, *AJ*, 132, 161

Holmberg, J., Nordström, B., & Andersen, J. 2009, *A&A*, 501, 941

Israelian, G., Santos, N., Mayor, M., & Rebolo, R. 2004, *A&A*, 414, 601

Israelian, G., Delgado Mena, E., Santos, N., Sousa, S., Mayor, M., Udry, S., Dominguez Cerdena, C, Rebolo, R., & Randich, S. 2009, *Nature*, 462, 189

King, J. R., Deliyannis, C. P., Hiltgen, D. S., Stephen, A., Cunha, K., & Boesgaard, A. M. 1997, *AJ*, 113, 1871

Luck, E. & Heiter, U. 2006, 131, 3069

Mayor, M. *et al.* 2003, *Messenger*, 114, 20

Pace, G., Melendez, J., Pasquini, L., Carraro, G., Danziger, J., François, P., Matteucci, F., & Santos, N. C. 2009, *A&A* (Letters), 499, 9

Pasquini, L., Biazzo, K., Bonifacio, P., Randich, S., & Bedin, L. R. 2008, *A&A*, 489, 677

Piau, L. & Turck-Chièze, S. 2002, *A&A*, 566, 419

Pinsonneault, M., Kawaler, S. D., Sofia, S., & Demarque, P. 1989, *ApJ*, 338, 424

Randich, S. 2008, *Mem. Soc. Astr. It.*, 79, 516

Ryan, S. 2000, *MNRAS* (Letters), 316, 35

Saffe, C., Gomez, M., & Chavero, C. 2005, *A&A*, 443, 609

Santos, N. C., Mayor, M., Naef, D., Pepe, F., Queloz, D. *et al.* 2002, *A&A*, 392, 215

Santos, N. C., Israelian, G., & Mayor, M. 2004, *A&A*, 415, 1153

Sestito, P. & Randich, S. 2005, *A&A*, 442, 615

Sestito, P., Randich, S., & Bragaglia, A. 2007, *A&A*, 465, 185

Sousa, S. G., Santos, N. C., Mayor, M., Udry, S., Casagrande, L. *et al.* 2008, *A&A*, 487, 373

Takeda, Y., Kawanomoto, S., Honda, S., Ando, H., & Sakurai, T. 2007, *A&A*, 468, 663

Valenti, J. & Fischer, D. 2005, *ApJS*, 159, 141

Wright, J., Marcy, G. W., Butler, R. P., & Vogt, S. S. 2004, *ApJS*, 152, 261

Yadav, R. K. S., Bedin, L. R., Piotto, G., Anderson, J., Cassisi, S. *et al.* 2008, *A&A*, 484, 609

Light Elements in the Universe
Proceedings IAU Symposium No. 268, 2009
C. Charbonnel, M. Tosi, F. Primas & C. Chiappini, eds.
© International Astronomical Union 2010
doi:10.1017/S1743921310004266

Light elements in stars with exoplanets

N. C. Santos[1], E. Delgado Mena[2], G. Israelian[2], J. I. González-Hernández[3], M. C. Gálvez-Ortiz[4], M. Mayor[5], S. Udry[5], R. Rebolo[2,6], S. Sousa[1] and S. Randich[7]

[1] Centro de Astrofísica, Universidade do Porto, Rua das Estrelas, 4150-762 Porto, Portugal.
email: Nuno.Santos@astro.up.pt

[2] Instituto de Astrofísica de Canarias, E-38200 La Laguna, Tenerife, Spain.

[3] Departamento de Astrofísica, Facultad de Ciencias Físicas, Universidad Complutense de Madrid, E-28040, Spain.

[4] Centre for Astrophysics Research, Science and Technology Research Institute, University of Hertfordshire, Hatfield AL10 9AB, UK.

[5] Observatoire de Genève, 51 ch. des Maillettes, CH-1290 Sauverny, Switzerland.

[6] Consejo Superior de Investigaciones Científicas, E-28006, Madrid, Spain.

[7] INAF/Osservatorio Astrofisico di Arcetri, Largo Enrico Fermi 5, I-50125 Firenze, Italia.

Abstract. It is well known that stars orbited by giant planets have higher abundances of heavy elements when compared with average field dwarfs. A number of studies have also addressed the possibility that light element abundances are different in these stars. In this paper we will review the present status of these studies. The most significant trends will be discussed.

Keywords. Stars: abundances, fundamental parameters, atmospheres – planetary systems: formation

1. Introduction

Since the discovery of the first exoplanet around 51 Peg (Mayor & Queloz 1995) the number of known exoplanets orbiting solar type stars did not stop growing, being currently of more than 400 (including more than 30 multi-planetary systems)†. In addition, more than 50 planets are now known to transit their host stars. Finally, in the last few years about 30 planets with masses between 2 and 20 M_{Earth} have been discovered. Present results strongly suggest that planets are common around solar-type stars.

One remarkable characteristic of planet host stars is that they are considerably metal-rich when compared with single field dwarfs (Gonzalez 1998; Santos et al. 2000; Santos et al. 2004a; Fischer & Valenti 2005). As we can perfectly see in Fig. 1, the probability of finding a giant planet depends strongly on the metallicity of the star. Interestingly, as discussed in Santos et al. (2004a), there seems to be two regimes in this distribution. For super solar metallicities the presence of planets is a steep rising function of metallicity, while for metal poor stars there is no significant dependence with metallicity.

Two main explanations have been suggested to explain the observed metallicity "excess". In the first one it is considered that its origin is primordial, so the more metals you have in the proto-planetary disk, the higher should be the probability of forming a planet. Alternatively, this excess could be produced by accretion of rocky material by the star sometime after it reached the main-sequence.

In the right panel of Fig.1 metallicity is shown as a function of the stellar convective envelope mass. If pollution were the main responsible for the enhanced metallicity of

† See tables at http://www.exoplanet.eu

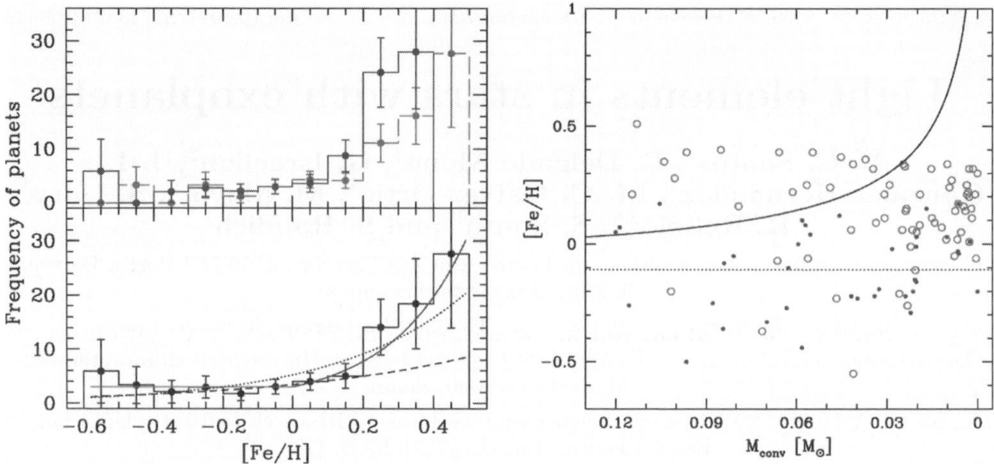

Figure 1. *Left panel:* Frequency of planet hosts as a function of stellar metallicity. Blue points are from Santos *et al.* (2004a) and red points are from Fischer & Valenti (2005). From Udry & Santos (2007). *Right panel:* Metallicity as a function of convective envelope mass for stars with planets (open symbols) and field stars (points). The [Fe/H] = constant line represents the mean [Fe/H] for the non-planet hosts stars from Santos *et al.* (2001). The curved line represents the result of adding 8 earth masses of iron to the convective envelope of stars having an initial metallicity equal to the non-planet hosts mean [Fe/H]. The resulting trend has no relation with the distribution of the stars with planets. From Santos *et al.* (2003).

planet hosts, we would expect to find higher metallicities as the convective envelope mass decreases. We do not find such a trend. Observations of stars arriving at the main sequence in open clusters do not show this correlation either (Shen *et al.* 2005). Another point against pollution is that we would need too high accretion rates to explain the metallicity observed in K-dwarfs. In addition, transit detections showed that the mass of heavy elements in the planets appears to be correlated with the metallicity of their parent stars (Guillot *et al.* 2006). The stars that are the most metal-rich host the most metal-rich planets. Finally, a recent work by Mordasini *et al.* (2009) finds that the distributions of planetary properties are well reproduced using core-accretion models (see below), which are dependent on dust content of the disk, thus supporting the primordial origin of overmetallicity in stars with planets.

These results have important implications for the models of giant planet formation and evolution. Two major giant planet formation models have been proposed. The core accretion model (Pollack *et al.* 1996) and the disk instability model (Boss 1997). In the former case, planets are formed by the collisional acumulation of planetesimals by a growing solid core, followed by accretion of a gaseous envelope onto the core. In the second one, a gravitationally unstable region in a protoplanetary disk forms self-gravitating clumps of gas and dust (Boss 1997). In the core accretion model, planet formation is critically dependent of the dust content of the disk (Pollack *et al.* 1996) while in the disk instability model it is not (Boss 2002). Present observations are thus more compatible with core accretion model, though do not exclude disk instability.

Although most data suggest that pollution is not the major source of the high metallicity levels in planet host stars, this issue is still not settled. In contrast with main sequence stars, planet-hosting giants do not show a tendency of being more metal-rich. Pasquini *et al.* (2008a) proposed that the lack of a metallicity-planet connection among giant stars is due to pollution of the star while on the main sequence, followed by dilution during the giant phase. Moreover, if hydrogen-poor material is accreted in the early phases of stellar evolution, during planet formation, the metal excess in the convective zone could be diluted, first by dynamical convection, then by thermohaline mixing, on a timescale much shorter than the stellar lifetime (Vauclair 2004).

Although the primordial origin of metallicity seems to be the more compatible explanation, pollution might have been able to alter more or less significantly the global metallicity of stars. In fact, though at low levels, cases of [Fe/H] pollution exist (e.g. Laws & Gonzalez 2001). In a different context, some Li-rich giants have been found (Brown *et al.* 1989), stars which should have depleted their lithium due to their deep convective envelopes, and that perhaps might have accreted metal-rich material. However, none of the Li-rich giants studied by Melo *et al.* (2005) were found to have detectable Be, something incompatible with an engulfment scenario (see also review by V.V. Smith in this book).

The key to understand these issues may be in the study of light element abundances. Light element should be particularly sensitive, since they are normally depleted in solar-type stars. If they are present in large quantities, a external origin might be the best explanation, rather than stellar evolution. Furthermore, light elements are important tracers of internal structure and mixing in stars and they can give us information about the rotational history of the star. It is indeed plausible that the planet formation process is able to alter the rotational history of a star, thus inducing changes in the observed abundances of light elements.

In this paper we will review the major results of the study of the light elements Li and Be in stars with planets.

2. ^{6}Li: Tracing pollution events.

The lithium isotope ^{6}Li is produced by spallation reactions in the insterstellar medium while it is easily destroyed in stellar interiors at a temperature of 2 million K. According to standard models (Forestini 1994, Montalbán & Rebolo 2002), at a given metallicity there is a mass range, where ^{6}Li but not ^{7}Li is being destroyed. These models predict that no ^{6}Li can survive pre-MS mixing in metal-rich solar-type stars. We note, however, that the mass and the depth of the convection zone also depend on the metal content of the star, and for this reason several old metal-poor stars have preserved a fraction of their initial ^{6}Li nuclei (Cayrel *et al.* 1999). In any case, these results suggests that any detected ^{6}Li in a metal-rich solar-type star would most probably be a signal of an external source, that is, accretion of planetary-like material (Sandquist *et al.* 2002). It is worth mentioning that ^{6}Li cannot be produced in large quantities in stellar flares, although some is produced (Ramaty *et al.* 2000).

The first ^{6}Li detection in a planet host star was reported by Israelian *et al.* (2001, 2003) in HD82943, a star which hosts two planets. This is an old and non active G0 dwarf with an effective temperature of 6010 K, [Fe/H]=+0.32 and log N(Li) $\sim$ 2.5. They found an isotopic ratio ^{6}Li/^{7}Li $\sim$ 0.05 (see Fig. 2), which may be explained by infall of 1 M$_J$ or equivalent, a value that would not significantly alter the stellar metallicity.

Observations of ^{6}Li are very difficult because it is a weak component of the much stronger doublet of ^{7}Li. Moreover, in metal-rich stars, blending with other weak absortions becomes important. This makes the detection of this isotope a controversial fact. For instance, Reddy *et al.* (2002) did not find signatures of ^{6}Li in HD82943, although they were using a blend of TiI in the Li region, instead of the SiI line used by Israelian *et al.* (2003). Other authors have also not found similar detections in other planet host stars (Mandell *et al.* 2004, Ghezzi *et al.* 2009). Uncertainties in the line lists can indeed lead to wrong determinations of ^{6}Li abundance. Perhaps most importantly, the use of

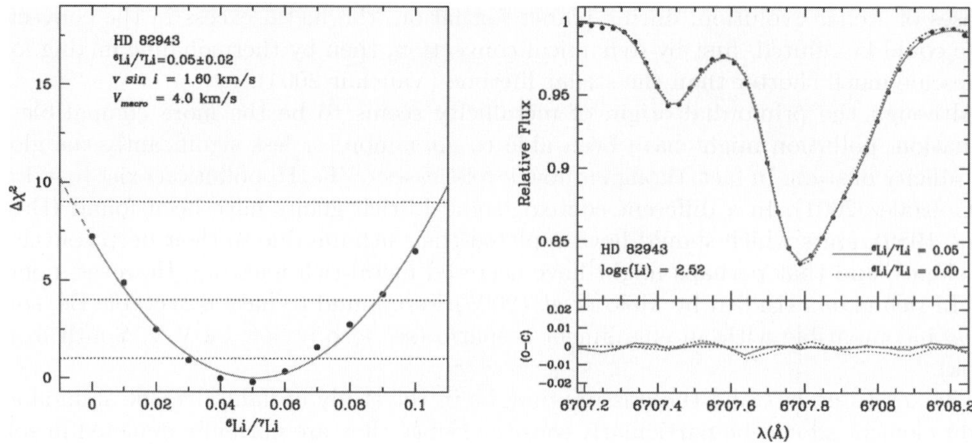

Figure 2. *Left panel:* Results from the χ^2 analysis for the ^{6}Li/^{7}Li ratio. $\Delta\chi^2 = 1$, 4, and 9 correspond to 1, 2 and 3σ confidence limit (dotted lines). Continuum adjustments were allowed within 0.2%. *Right panel:* Comparison of the observed (filled large dots) and synthetic spectra of HD 82943 corresponding to ^{6}Li/^{7}Li = 0.05 (continuous) and ^{6}Li/^{7}Li = 0 (dotted) isotopic ratios. They correspond to the best fits with f(^{6}Li) = 0 and a wavelength offset of -0.65 km s^{-1} (small dots) and to the f(^{6}Li) = 0.05 with a wavelength offset of -0.44 km s^{-1} (continuous line). Fits to the blue wing of the Li profile can be improved if we adopt the wavelengths of CN lines from Brault & Mueller (1975). The CN lines observed in the arc spectrum by these authors appear at 6707.55 Åwhile Reddy *et al.* (2002) and Lambert *et al.* (1993) list them between 6707.464 and 6707.529 Å. The residuals (O-C) of the observations after subtraction of the synthetic spectra are shown. Both plots are from Israelian *et al.* (2003).

3D models may change this panorama, since convection will produce an assymetry in the ^{7}Li line similar to the one produced by the presence of ^{6}Li (Cayrel *et al.* 2008, Ghezzi *et al.* 2009). This problem may affect the determination of ^{6}Li abundances not only for metal-rich planet host stars but also for their metal poor counterparts (see discussion in reviews by Asplund, Spite, Melendez, and Steffen, this Volume).

In any case, present results suggest that signs of "massive" pollution are not generalized in planet host stars.

3. ^{7}Li

The more common lithium isotope ^{7}Li was produced during Big Bang nucleosynthesis and it can be produced in stellar interiors during AGB phase. It is depleted at a temperature of 2.5 million K primarily during the PMS in solar-type stars but it can also be destroyed in stellar envelopes during MS if any mixing process exists.

The light element ^{7}Li can also be used to trace pollution events. For instance, Deliyannis *et al.* (2002) discovered an extremely Li-rich dwarf, J37, a F-star in the open cluster NGC6633 (see Fig. 3). Firstly they suggested upward, radiatively driven diffusion as the best explanation for this Li overabundance. However, later studies demonstrated a high Be abundance too (Ashwell *et al.* 2005), a result that is in contradiction with radiative diffusion models. Furthermore, Laws & Gonzalez (2003) showed that refractory elements were also overabundant, arriving at the conclusion that this star had accreted volatile-depleted material.

A study of the differences of Li abundances in stars with and without planets was first carried out by King *et al.* (1997). In that work they measured Li abundances for the binary system formed by 16CygA, a star without planets and a detectable Li abundance,

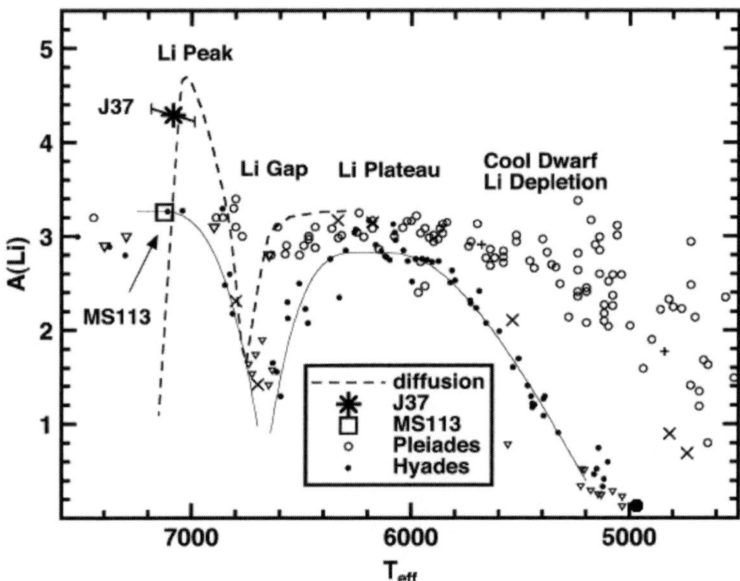

Figure 3. *Left panel:* Li abundances in the Hyades (detections: filled circles; upper limits: small inverted triangles; short-period binaries: multi crosses) and Pleiades (detections: open circles; upper limits: large inverted triangles; short-period binaries: plus signs) open clusters, in J37, and in MS113. From Deliyannis *et al.* (2002).

and 16CygB, a planet-host that is Li depleted. This work was followed by several studies claiming that stars with planets have different Li abundances when compared to stars without detected companions (Cochran *et al.* 1997; Gonzalez & Laws 2000; Takeda & Kawanomoto 2005; Chen & Zhao 2006; Israelian *et al.* 2004, 2009), though these results are not consensual (e.g. Ryan 2000; Luck & Heiter 2006). A recent uniform study by Israelian *et al.* (2009) seems to confirm, however, that planet host stars have lower Li abundances in the solar temperature region (see Fig. 4) and exclude metallicity, age, $v \sin i$, or activity as the cause of this anomaly (for more details see review by Israelian in this volume).

To explain the observed difference several possibilities exist. Pollution should be ruled out because it would have the opposite effect. On the other hand, it seems that stars with planets might have a different evolution. Extra mixing due to planet-star interaction, like migration, could take place (Castro *et al.* 2009). The infall of planets might also affect the mixing processes of those stars (Theado *et al.*, *in prep.*) . Finally a long-lasting star-disc interaction during PMS may cause planet hosts to be slow rotators and develop a high degree of differential rotation between the radiative core and the convective envelope, also leading to enhanced lithium depletion (Bouvier 2008).

4. Be

Be is mainly produced by spallation reactions in the interstellar medium while it is burned in the hot stellar furnaces at a temperature of 3.5 million K. Li is depleted at much lower temperatures than Be. Thus, by measuring Li in stars where Be is not depleted (early-G and late-F) and Be in stars where Li is depleted (late-G and K) we can obtain crucial information about the mixing, diffusion and angular momentum history of exoplanets hosts (Santos *et al.* 2002). Measurements of the abundance of this element

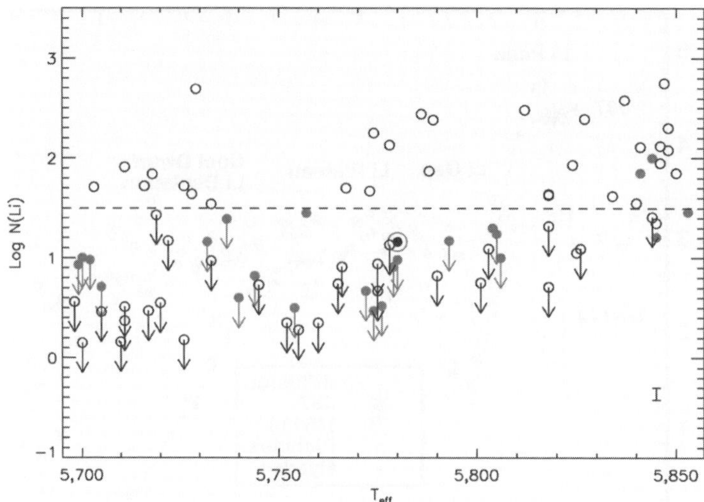

Figure 4. Li abundances as a function of effective temperature for stars with (red filled circles) and without known planets (open circles), from Israelian *et al.* (2009).

are however difficulted by the fact that the only available lines are in the near-UV, a very blended region in metal-rich stars (Fig. 5).

There are not many works about Be abundances in planet host stars in the literature. Some of the first works made (García López & Pérez de Taoro 1998; Deliyannis *et al.* 2000), had a very small number of objects. In Fig. 6 the largest samples made at the moment are plotted together, showing Be abundances as a function of effective temperature (Santos *et al.* 2002, 2004b; Gálvez-Ortiz *et al.* 2009; Delgado Mena *et al., in prep.*).

At a first sight, there are not clear differences between stars with and without planets. Globally, Be abundances decrease from a maximum at 6100 K towards both higher and lower temperatures, in a similar way as Li abundances behave. In the high temperature domain, the steep decrease with increasing temperature resembles the well-known Be gap for F-stars (Boesgaard *et al.* 1999). The decrease of the Be content towards lower temperatures is smoother and may show evidence for continuous Be-burning during the main-sequence evolution of these stars.

In the temperature range where Li abundances are different in stars with and without planets, we do not observe any difference in Be abundances. This is something we can expect because for those temperatures, convective envelopes are not deep enough to bring the material of the convective envelope down to the layers where the temperature is high enough to burn Be. On the other hand, for the lowest temperatures of the range it seems that stars with planets have lower abundances when compared with stars for which no planet has been discovered. Unfortunately, the number of comparison stars in that temperature regime is still small to allow us to take a strong conclusion.

In addition, in the low temperature region we find very low abundances of Be, in contradiction with models of Be depletion. In Fig. 6 , models with different initial rotation rates have been overplotted (Pinsonneault *et al.* 1990). As already noticed in Santos *et al.* (2004b,c), these models agree with the observations above roughly 5600 K, but while the observed Be abundance decreases towards lower temperatures when $T_{eff} < 5600$, these models predict either constant or increasing Be abundances. Even taking into account mixing by internal waves (Montalbán & Schatzman 2000), Be depletion is still lower than observed. Although uncertainties in Be abundances for the coolest stars are large

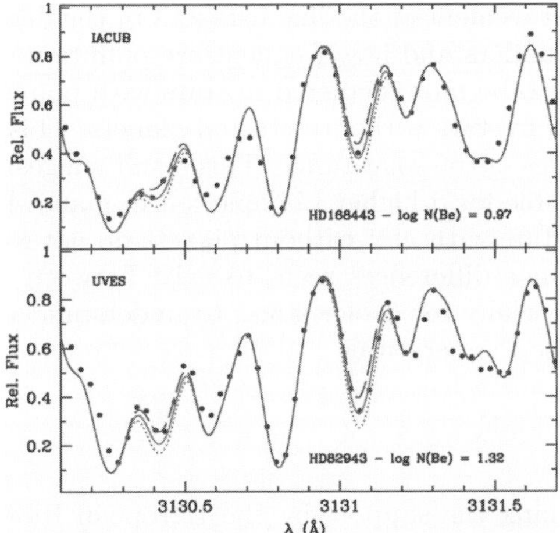

Figure 5. Spectra in the BeII region (dots) for HD168443 and HD82943, and three spectral synthesis with different abundances, corresponding to the best fit (solid line) and to changes of ± 0.2 dex, respectively. From Santos *et al.* (2002).

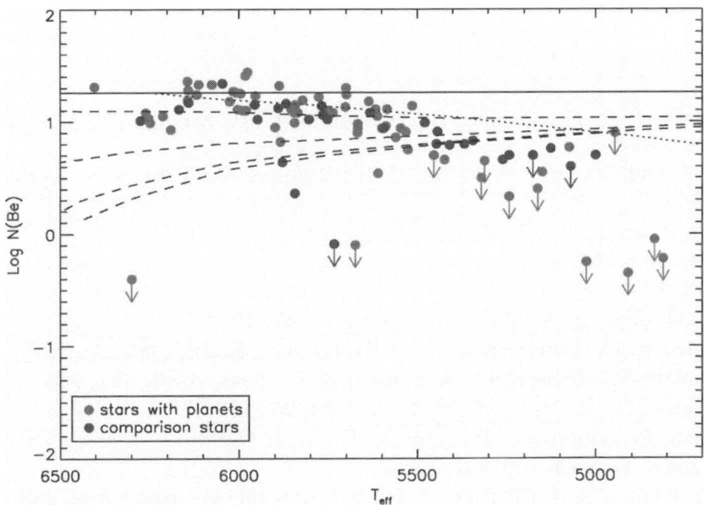

Figure 6. Be abundances as a function of effective temperature for stars with (red circles) and without planets (blue circles). The dashed lines represent 4 Be depletion models of Pinsonneault *et al.* (1990) (Case A) with different initial angular momentum for solar metallicity and an age of 1.7 Gyr. The solid line represents the initial Be abundance of 1.26. The dotted line represents the Be depletion isochrone for 4.6 Gyr taken from the models including mixing by internal waves of Montalbán & Schatzman (2000).

(significant systematic errors are not excluded), Be abundances are still overestimated by models.

5. Conclusions

In this paper we reviewed the main results concerning the study of light element abundances in stars with planets. The main conclusions can be listed as follows.

- ^{6}Li shows evidence that stars with planets may suffer (isolated) pollution events. However, the difficulty in deriving reliable abundances of this isotope makes this a very debatable issue. An improvement of the line lists and in the determination of convective blue shifts is needed to understand how frequent are pollution events.

- ^{7}Li has been found to be more depleted in stars with planetary companions in contrast with what happens in stars without detected planets. This trend is only observable in stars with temperatures in the solar range. This result suggests that some mechanisms is acting that is responsible for a higher Li depletion in planet-host stars.

- Be abundances in stars with and without planets do not follow the trend found for Li. No clear Be abundance differences seem to exist between the two groups of stars. More data is needed, specially for cooler stars, to understand the actual disagreements with the models for those temperatures.

Acknowledgements

NCS would like to thank the support by the European Research Council/European Community under the FP7 through a Starting Grant, as well from Fundação para a Ciência e a Tecnologia (FCT), Portugal, through a Ciência 2007 contract funded by FCT/M CTES (Portugal) and POPH/FSE (EC), and in the form of grants reference PTDC/CTE-AST/098528/2008 from FCT/MCTES.

References

Ashwell, J. F., Jeffries, R. D., Smalley, B., Deliyannis, C. P., Steinhauer, A., & King, J. R. 2005, *MNRAS*, 363, 81

Boesgaard, A. M., Deliyannis, C. P., King, J. R., Ryan, S. G., Vogt, S. S., & Beers, T. C. 1999, *AJ*, 117, 1549

Boss, A. P. 1997, *Science*, 276, 1836

Boss, A. P. 2002, *ApJ*, 567, 149

Bouvier, J. 2008, *A&A*, 489, 53

Brault, J. W. & Mueller, E. A. 1975, *Solar Physics*, 41, 43

Brown, J. A., Sneden, C., Lambert, D. L., & Dutchover, E. Jr. 1989, *ApJS*, 71, 293

Castro, M., Vauclair, S., Richard, O., & Santos, N. C. 2009, *A&A*, 494, 663

Cayrel, R., Lebreton, Y., & Morel, P. 1999, *Ap&SS*, 265, 87

Cayrel, R., Steffen, M., Bonifacio, P., Ludwig, H.-G., & Caffau, E. 2008, *nuco.conf*, E, 2C

Chen, Y. Q. & Zhao, G. 2006, *Aj*, 131, 1816

Cochran, W. D., Hatzes, A. P., Butler, R. P., & Marcy, G. W. 1997, *ApJ*, 483, 457

Deliyannis, C. P., Cunha, K., King, J. R., & Boesgaard, A. M. 2000, *AJ*, 119, 2437

Deliyannis, C. P., Steinhauer, A., & Jeffries, R. D. 2002, *ApJ*, 577, 39

Gálvez-Ortiz, M. C., Delgado Mena, E., González Hernández J. I., Israelian, G., Santos, N. C., & Rebolo. R. 2009, *A&A*, submitted

García López, R. J. & Pérez de Taoro, M. R. 1998, *A&A*, 334, 599

Fischer, D. A. & Valenti, J. 2005, *AJ*, 622, 1102

Forestini, M. 1994, *A&A*, 285, 473

Ghezzi, L., Cunha, K., Smith, V. V., Margheim, S., Schuler, S., de Arajo, F. X., & de la Reza, R. 2009, *ApJ*, 698, 451

Gonzalez, G. 1998, *A&A*, 334, 221

Gonzalez, G. & Laws, C. 2000, *AJ*, 119, 390

Guillot, T., Santos, N. C., Pont, F., Iro, N., Melo, C., & Ribas, I. 2006 *A&A* (Letters), 453, 21

Israelian, G., Santos, N. C., Mayor, M., & Rebolo, R. 2001, *Nature*, 411, 163

Israelian, G., Santos, N. C., Mayor, M., & Rebolo, R. 2003, *A&A*, 405, 753

Israelian, G., Santos, N. C., Mayor, M., & Rebolo, R. 2004, *A&A*, 414, 601

Israelian, G., Delgado Mena, E. Santos, N. C., Sousa, S. G., Mayor, M., Udry, S., Domnguez Cerdea, C., Rebolo, R., & Randich, S. 2009, *Nature*, 462, 189

King, J. R., Deliyannis, C. P., Hiltgen, D. D., Stephens, A., Cunha, K., & Boesgaard, A. M. 1997 *AJ*, 113, 1871

Laws, C. & Gonzalez, G. 2001, *ApJ*, 553, 405

Laws, C. & Gonzalez, G. 2003, *ApJ*, 595, 1148

Luck, R. E. & Heiter, U. 2006, *AJ*, 131, 3069

Mandell, A. M., Ge, J., & Murray, N. 2005, *AJ*, 127, 1147

Mayor, M. & Queloz, D. 1995, *Nature*, 378, 355

Melo, C. H. F., de Laverny, P., Santos, N. C., Israelian, G., Randich, S., Do Nascimento, J. D., Jr., & de Medeiros, J. R. 2005, *A&A*, 439, 227

Montalbán, J. & Schatzman, E. 2000, *A&A*, 354, 943

Montalbán, J. & Rebolo, R. 2002, *A&A*, 386, 1039

Mordasini, C., Alibert, Y., Benz, W., & Naef, D. 2009, *A&A*, 501, 1161

Pasquini, L., Dllinger, M. P., Hatzes, A., Setiawan, J., Girardi, L., da Silva, L., de Medeiros, J. R., & Weiss, A. 2008a, *IAUS*, 249, 209

Pasquini, L., Biazzo, K., Bonifacio, P., Randich, S., & Bedin, L. R. 2008b, *A&A*, 489, 677

Pinsonneault, M. H., Kawaler, S. D., & Demarque, P. 1990, *ApJS*, 74, 501

Pollack, J. B., Hubickyj, O., Bodenheimer, P., Lissauer, J. J., Podolak, M., & Greenzweig, Y. 1996, *Icarus*, 124, 62

Ramaty, R., Tatischeff, V., Thibaud, J. P., Kozlovsky, B., & Mandzhavidze, N., 2000, *ApJ* (Letters), 534, 207

Reddy, B. E., Lambert, D. L., Laws, C., Gonzalez, G., & Covey, K. 2002, *MNRAS*, 335, 1005

Ryan, S. G. 2000, *MNRAS*, 316, 35

Sandquist, E. L., Dokter, J. J., Lin, D. N. C., & Mardling, R. A. 2002, *ApJ*, 572, 1012

Santos, N. C., Israelian, G., & Mayor, M. 2000, *A&A*, 363, 228

Santos, N. C., Israelian, G., & Mayor, M. 2001a, *A&A*, 373, 1019

Santos, N. C., García López, R. J., Israelian, G., Mayor, M., Rebolo, R., García-Gil, A., Pérez de Taoro, M. R., & Randich, S. 2002, *A&A*, 386, 1028

Santos, N. C., Israelian, G., Mayor, M., Rebolo, R., & Udry, S. 2003, *A&A*, 398, 363

Santos, N. C., Israelian, G., & Mayor, M. 2004a, *A&A*, 415, 1153

Santos, N. C., Israelian, G., Randich, S., García López, R. J., & Rebolo, R. 2004b, *A&A*, 425, 1013

Santos, N. C., Israelian, G., García López, R. J., Mayor, M., Rebolo, R., Randich, S., Ecuvillon, A., & Domnguez Cerdea, C. 2004c, *A&A*, 427, 1085

Shen, Z.-X., Jones, B., Lin, D. N. C., Liu, X.-W., & Li, S.-L. 2005, *ApJ*, 635, 608

Takeda, Y. & Kawanomoto, S. 2005, *PASJ*, 57, 45

Udry, S. & Santos, N. C. 2007, *ARAA*, 45, 397

Vauclair, S. 2004, *ApJ*, 605, 874

George Wallerstein

Verne Smith and Evan Skillman

Light Elements in the Universe
Proceedings IAU Symposium No. 268, 2009
C. Charbonnel, M. Tosi, F. Primas & C. Chiappini, eds.

© International Astronomical Union 2010
doi:10.1017/S1743921310004278

Observations of Lithium in red giant stars

Verne V. Smith[1]

[1]National Optical Astronomy Observatory
950 N. Cherry Ave., Tucson, AZ 85719 USA
email: vsmith@noao.edu

Abstract. Connections between observations of the lithium abundance in various types of red giants and stellar evolution are discussed here. The emphasis is on three main topics; 1) the depletion of Li as stars ascend the red giant branch for the first time, 2) the synthesis of ^{7}Li in luminous and massive asymptotic giant branch stars via the mechanism of hot-bottom burning, and 3) the possible multiple sources of excess Li abundances found in a tiny fraction of various types of G and K giants.

Keywords. Stars: abundances, late-type

1. Introduction

The connection between lithium and evolution along the red giant branch has a long history, going back to early work by Bonsack (1959), who noted that lithium abundances declined towards later spectral types, which was interpreted as astration of Li in deepening convective envelopes in cooler stars. A quantitative analysis and comparison with models of the effect of red giant evolution on lithium abundances was carried out by Wallerstein (1966), who was able to identify the Li I λ6707Å line in both components of the double-lined spectroscopic binary Capella. Wallerstein found that the more luminous G–giant in the system had 60 times less Li than the less-evolved F–star component: the larger convective envelope of the G–giant had diluted its Li. The factor of 60 in dilution was in excellent agreement with predictions from stellar models by Iben (1965). Lithium had now become an important quantitative test of stellar evolution.

The simple evolution of the dilution of lithium from main sequence to red giant branch is not the whole story of "Lithium in Red Giants", however, due to the complex evolution of stars, initially up the first ascent of the red giant branch (which is referred to here as RGB), followed by evolution along the asymptotic giant branch (AGB), with both destruction and possible synthesis of lithium taking place during different evolutionary phases.

This short summary of lithium in red giants is divided into three sections:
- observations of Li in predominantly RGB stars, where mostly dilution/destruction occurs.
- the behavior of Li in AGB stars, where both destruction, as well as synthesis of ^{7}Li (via so-called "hot bottom burning") can take place.
- the nature of Li–rich G– and K–giant stars, where only dilution is nominally expected.

2. Lithium in first-ascent red giants

Due to the deepening convective envelope that occurs as stars evolve from the main sequence to the RGB, it is expected that the lithium abundance in RGB stars, in general, will be much lower than that observed in main sequence stars that have not destroyed their surface Li. By and large, this simple prediction has been born out by surveys of red

302 V. V. Smith

giants; two early quantitative abundance surveys contained 81 stars of spectral types G, K, and M by Lambert, Dominy, & Sivertson (1980) and Luck & Lambert (1982).

A more extensive survey of 644 G and K giants was presented by Brown *et al.* (1989) and their abundance results are summarized in histogram form in Figure 1. In Figure 1 the number of stars is plotted versus the lithium abundance, where A(Li) is the standard spectroscopic definition of $A(x) = Log[N(x)/N(H)] + 12.0$. The arrow indicates the solar system meteoritic abundance (A(Li) = 3.25) and the shift of the red giants to lower abundances is clear. Below abundances of about +0.50 in A(Li), the histogram is dominated by upper limits (that is, non-detections of the Li I λ6707Å line), so the real abundance distribution tails off to even lower Li abundances. There are two main points to take from Figure 1, the first of which is that the overall abundance distribution contains a large fraction of red giants with lower Li abundances than were predicted from standard convective dilution, thus suggesting that extra-mixing processes on the RGB removed more Li from the surface. The second point is that Brown *et al.* (1989) were able to quantify the percentage of G and K giants which contain more Li than predicted (giants with A(Li) $\geqslant$ 1.5 are considered to be "Li rich"): this small number of Li-rich giants can be seen in Figure 1. The number of Li-rich giants is roughly $\sim$1% of the total (this topic will be discussed further in Section 4).

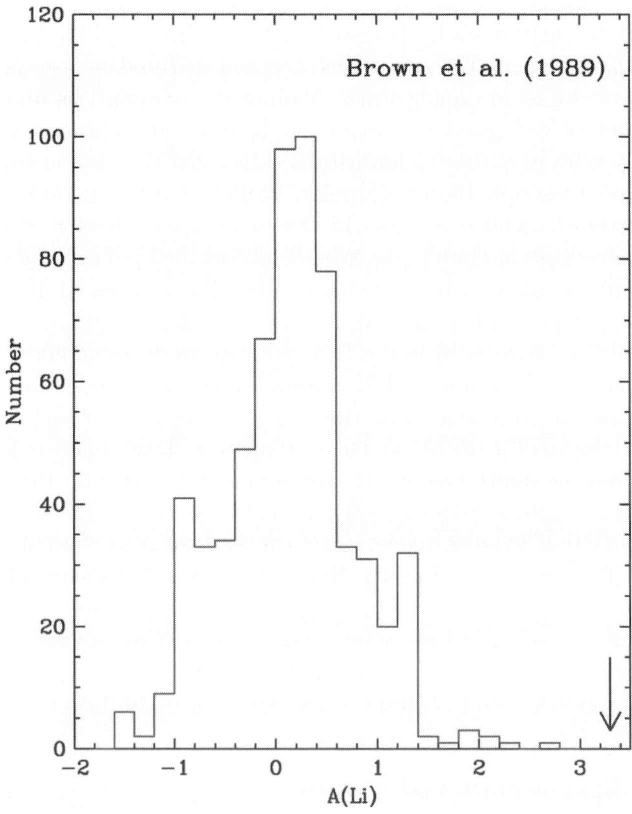

Figure 1. The distribution of lithium abundances in 644 G and K giants from Brown *et al.* (1989). Note that non-detections dominate the statistics below A(Li)$\sim$+0.5. The distribution of abundances suggests that extra-mixing and more astration of Li occurs on the RGB, with about $\sim$1% of the giants being "Li rich". The arrow indicates the solar system value of A(Li).

Gratton *et al.* (2000) conducted a survey of field metal-poor giants to study mixing, with lithium being included as one of their tests. One of the advantages in the Gratton *et al.* study, was that the luminosities of their sample giants could be estimated reasonably well, so that A(Li) could be investigated as a function of position on the RGB. The top panel of Figure 2 summarizes the results from Gratton *et al.* (2000), with A(Li) shown versus luminosity. The vertical dashed lines indicate the approximate luminosity of two important locations in an HR diagram; the first one shows the approximate location of the base of the RGB (at Log $(L/L_\odot)\sim+0.8$), while the second illustrates roughly where the RGB luminosity "bump" falls in low-mass low-metallicity giants. The dramatic drop in the Li abundance at the beginning of evolution up the RGB stands out clearly, with the abundance of lithium then remaining roughly constant, until another drop at the RGB bump. Displayed as in Figure 2, the evolution of the lithium abundance in most RGB stars seems reasonably well-defined.

Recent work on Li depletion along the RGB has focused on the globular cluster NGC6397, with studies by Korn *et al.* (2007) or Lind *et al.* (2009). In the case of a globular cluster, such as NGC6397, the advantage is that the luminosity (or put another way, the position on the RGB) is very well determined. The bottom panel of Figure 2 shows results from the Korn *et al.* (2007) study, where their results for 18 stars are averaged into 4 distinct evolutionary points: in increasing luminosity, the subgiant branch, the base of the RGB, the lower RGB, and the upper RGB. The overall behavior of A(Li) in the NGC6397 stars closely matches that for the metal-poor low-mass field giants from Gratton *et al.* (2000–the top panel of Figure 2). The recent work on NGC6397 (Korn *et al.* 2007; Lind *et al.* 2009) includes diffusion in their modelling and the results between Li depletion predicted by theory and what is observed is quite good.

Progress continues to be made using lithium to probe RGB evolution, e.g. Lagarde *et al.* (2009, this symposium) presents Li abundances in a large sample of field red giants, with luminosities derived from Hipparcos parallaxes, which were not available to Brown *et al.* (1989). It can be stated that our understanding of the behavior of Li abundances along the RGB is much better than it was compared to just a few years ago.

3. Lithium in hot-bottom burning asymptotic giant branch stars

Asymptotic Giant Branch stars are well-known producers of ^{12}C, with this primary carbon-12 produced during phases of shell-^{4}He-burning in thermal pulses, which drive convective mixing of ^{12}C-rich material to the stellar surface. This mixing results in the evolutionary sequence running from the oxygen-rich (C/O $\leqslant 1$) M-type giants to the carbon stars (C/O $\geqslant 1$). In addition to the production of ^{12}C, the mixing between He-burning shell and H-rich envelope results in the release of free neutrons via the chain ^{12}C(p,γ)^{13}N(β^+,ν)^{13}C(α,n)^{16}O. These neutrons drive the s-process and synthesize the neutron-rich nuclei heavier than Fe (such as Zr or Ba).

The synthesis of ^{12}C and the s-process elements, followed by dredge-up (the so-called "third dredge-up") to produce C- and s-process-rich AGB stars, occurs in a large fraction of AGB stars from M$\sim$1-8M$_\odot$, with the exact mass limits depending on stellar metallicity. Another process, referred to as "hot-bottom burning" (HBB), first studied in model stars by Scalo, Despain, & Ulrich (1975), occurs in the more massive AGB stars (say M $\geqslant 4$M$_\odot$) and can result in the production of ^{7}Li via the Cameron-Fowler mechanism (Cameron & Fowler 1971).

As a result of HBB, very Li-rich AGB stars can be created and observations pointed in this direction long before any understanding of internal nucleosynthesis in AGB stars. McKellar (1940) observed a very strong Li I λ6707Å line in the carbon star WZ Cas and suggested that this star had "an unusually high abundance of lithium". A survey of 30 disk carbon stars was conducted by Torres-Peimbert & Wallerstein (1966), who found some 16 of these stars to exhibit substantial Li I lines. The phenomenon of Li-rich AGB stars was extended to the S-stars (AGB stars with enhanced s-process and carbon-12 abundances, but with C/O still less than 1.0) with Keenan's (1967) detection of a strong Li I line in T Sgr.

Quantitative lithium abundance analyses of S stars began with Boesgaard (1970), who found very large abundances in a few S stars (including T Sgr), with values as large as A(Li)~4.0 (~10x the solar system abundance). Observationally measured luminosities of HBB Li-rich AGB stars was placed on a firmer footing by observations of such stars in the Magellanic Clouds (Smith & Lambert 1989; 1990; Plez *et al.* 1993; Smith *et al.* 1995), since the distances to the LMC and SMC are quite well known. More recently, attention has turned back to the Milky Way, with work by Garcia-Hernandez *et al.* (2007),

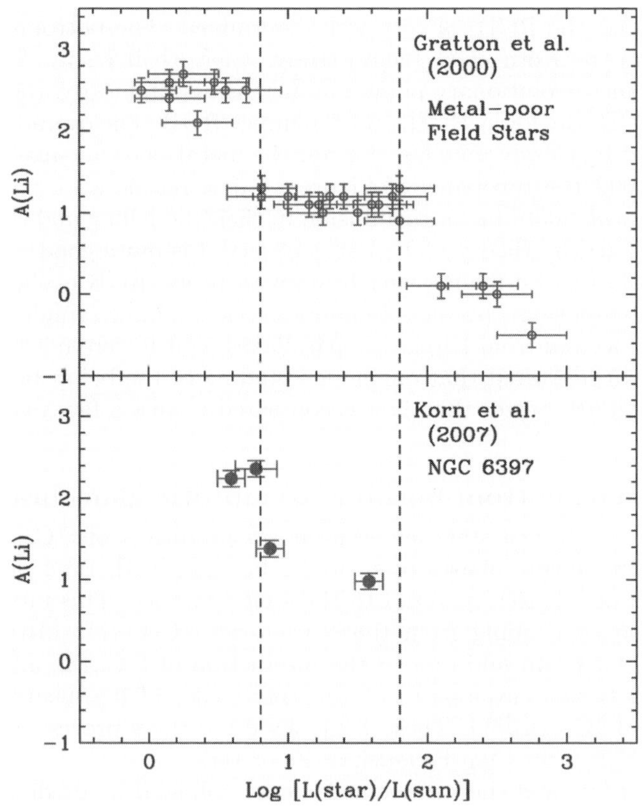

Figure 2. The top panel shows results from Gratton *et al.* (2000) lithium abundances in metal-poor low-mass field stars plotted versus luminosity. The vertical dashed lines mark the approximate luminosities of the base of the RGB (left line) and the RGB luminosity bump (right line). The precipitous drop on A(Li) as stars evolve onto the RGB is clear, with another decline near the location of the RGB bump. The bottom panel illustrates lithium abundances in stellar members of the globular cluster NGC6397 from Korn *et al.* (2007). The 4 points are averages from a sample of 18 stars, with the points representing stars on the subgiant branch, the base of the RGB, the lower RGB, and upper RGB. The behavior of the lithium abundance in the low−mass low-metallicity stars in NGC6397 is similar to those in the field-star sample from Gratton *et al.* (2000).

who determined Li abundances (along with other elements, including the s-process) in luminous Galactic AGB stars. A convenient way to view the combined lithium abundances in Galactic, LMC, and SMC luminous AGB stars is illustrated in Figure 3, where A(Li) is plotted versus pulsational period (all of these cool stars are long period variables). The overall trend is for the period to increase with increasing luminosity, so this figure shows the somewhat abrupt increase in lithium for periods greater than about 300-400 days. This figure also demonstrates that although ^{7}Li can be produced in amounts larger than that found in the current disk ISM, it can also be destroyed, so there is a wide distribution of lithium abundances in the luminous AGB stars; the exact net amount of ^{7}Li that might be contributed by these types of stars to chemical evolution remains uncertain, primarily due to uncertainties in how mass loss occurs along the AGB. Stellar models of HBB along the AGB by Sackmann & Boothroyd (1992) can produce the upper envelope of A(Li) observed in Figure 3 for stellar masses of 4-6$M_\odot$ at $M_{\rm Bol}\sim$ -6.2 to -6.8.

One interesting consequence of HBB in AGB stars is the conversion of primary ^{12}C (sometimes lots of it) into ^{14}N via the CN-cycle taking place within the deep convective envelope itself. The drop in ^{12}C abundance can result in the surface C/O ratio falling below 1.0, and the transformation of a carbon star back into an oxygen-rich star (in this case, since it would be s-process enriched star, it would be an S-star). This effect

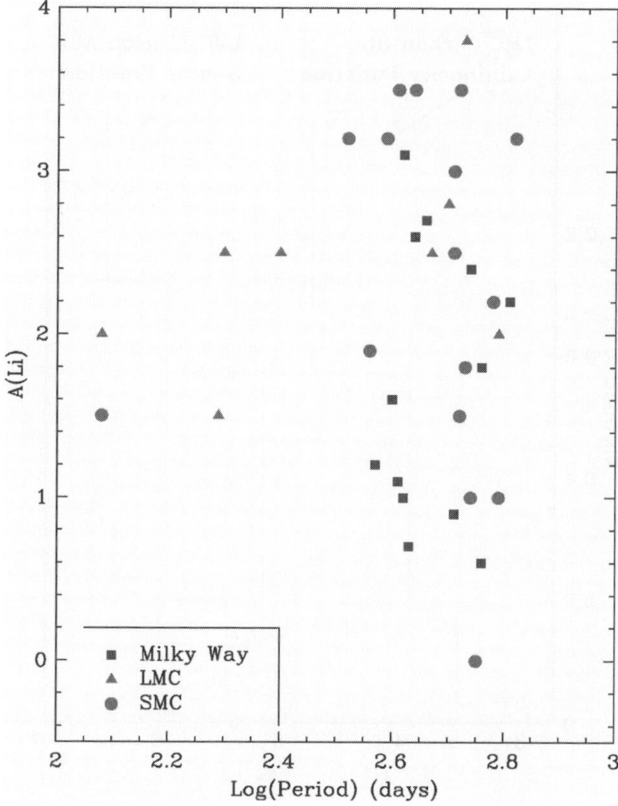

Figure 3. Lithium abundances versus pulsational period for luminous Galactic AGB stars from Garcia-Hernandez *et al.* (2007; filled blue squares) and for luminous SMC and LMC AGB stars from Smith *et al.* (1995); filled red circles for the SMC and filled magenta triangles for the LMC. The upper envelope of A(Li) is predicted by models of HBB in AGB stars by Sackmann & Boothroyd (1992).

is illustrated in Figure 4 for the LMC, where the C-star luminosity function from Iben (1981) is shown compared to the location, in luminosity space (with luminosity indicated by M_{Bol}), of the Li-rich S-stars. At the bolometric luminosity at which the number of C-stars drops to zero, the appearance of Li-rich S-stars takes place, with the lithium indicating the onset of HBB. With enough conversion of ^{12}C to ^{14}N in the hot envelopes of these stars, the C/O ratio is driven back down below unity and former carbon stars transform to S-stars.

4. The mysterious Lithium-rich red giants

As discussed in Section 2, stellar evolutionary models predict that lithium will be depleted during the first ascent of the RGB, with the amount of this depletion being 1 to 2 orders of magnitude. It was not expected that there would be significant sources of Li in stars evolving along the RGB, until the possible onset of HBB in massive, luminous AGB stars (the subject of Section 3). However, the discovery of an unexpectedly high lithium abundance (A(Li) = 2.7) in the K-giant HD112127 by Wallerstein & Sneden

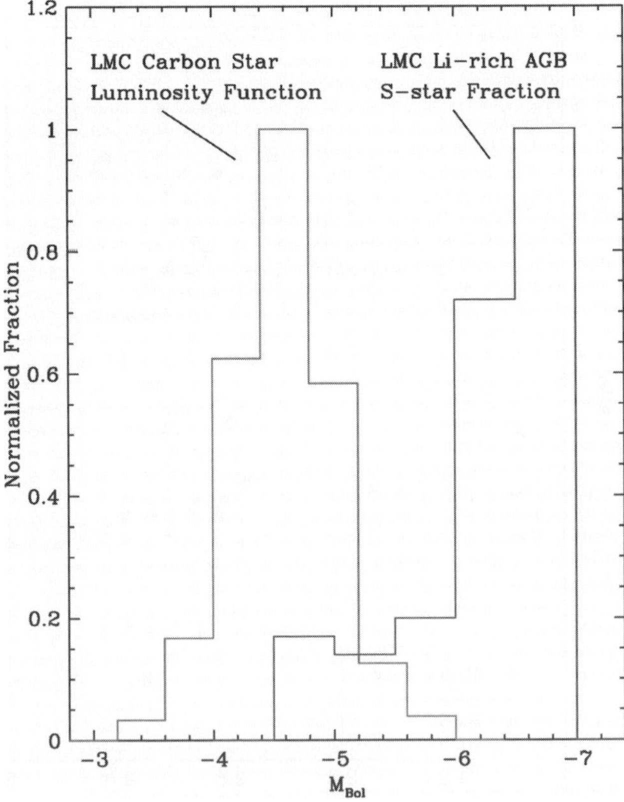

Figure 4. The carbon-star luminosity function for the LMC from Iben (1981) is shown as the lefthand histogram. On the right is plotted the fraction of AGB S-stars that are Li-rich in the LMC (Smith *et al.* 1995). Note that the fraction of Li-rich AGB stars with C/O $\geqslant$ 1.0 increases at about the same luminosity that the C-star luminosity function falls to zero; this transition between the two distributions represents the onset of hot bottom burning, the production of lithium and the conversion of carbon stars to luminous S-stars.

(1982) demonstrated that the behavior of Li in RGB stars was not simple. The subsequent survey by Brown *et al.* (1989–see Figure 1) showed that about 1% of G and K giants exhibit excess Li abundances (i.e, $A(Li) \geqslant 1.5$).

Continuing studies found additional examples of Li-rich RGB stars, with the interesting observation by Fekel & Balachandran (1993) that a significant fraction of the Li-rich giants are rapidly rotating (with vsini $\geqslant 8$ km-s^{-1}). In addition, de la Reza *et al.* (1996) found that these peculiar giants were surrounded by spatially extended, faint dust shells, as revealed in IRAS colors. Charbonnel & Balachandran (2000) found that the so-called "super Li-rich" giants (with $A(Li) \geqslant 3.3$) tended to be found in a specific location of the H-R Diagram, near the luminosity bump at $\log(L/L_\odot) \sim 1.4$–1.9 and $T_{eff} \sim 4450$–4600K. The combination of rapid rotation, dust shells which suggest mass-loss episodes, and location in a single part of the HR Diagram has led to a picture in which the excess Li has been produced within the RGB star (via the Cameron Fowler mechanism) and carried to the surface. This dredge-up episode also carried angular momentum from the more rapidly rotating interior and drove a mass loss event. The exact physical mechanism remains elusive, but may be the only viable hypothesis to explain the location of the super Li-rich giants near the luminosity bump.

In addition to the Li-rich RGB stars found in the disk, 3 Li-rich giants have been found in globular clusters; one each in the clusters M5 (Carney *et al.* 1998), NGC362 (Smith *et al.* 1999), and M3 (Kraft *et al.* 1999). Since the globular clusters have well-defined stellar populations and distances, it is straightforward to place these stars in an H-R Diagram. Figure 5 summarizes the locations of the Li-rich G and K giants in a Log L – T$_{eff}$ plane and it is immediately clear that the super Li-rich giants (filled circles) are in a distinctly different evolutionary state than the Li-rich globular cluster giants, which turn out to be AGB, or even post-AGB stars. Since the globular cluster AGB stars are not massive enough to have undergone HBB, the source of the lithium must be from a different mechanism. It is also worth noting that there is no evidence of rapid rotation in the 3 Li-rich globular cluster giants. One is tempted to conclude that different physical mechanisms are driving the mixing (if it is internal mixing) in these two classes of Li-rich giants.

Finally in Figure 5, we note the filled squares which represent rapidly rotating Li-rich G and K giants found by Carlberg *et al.* (2009b), and which extend to luminosities well below the luminosity bump. Carlberg *et al.* (2009b) speculate that these objects may result from the ingestion of closely orbiting large planets or brown dwarfs as the parent star evolves up the RGB. Simulations of the future evolution of known star-hot Jupiter systems by Carlberg *et al.* (2009a) suggest that such ingestion will occur, typically near the base of the RGB, and would result in an increase in the stellar surface Li abundance, as well as spin-up of the convective envelope.

With such a variety of low-mass giants exhibiting enhanced lithium abundances spread across the H-R Diagram, it may well be that the phenomenon of the Li–rich G and K giants is caused by a number of different mechanisms. Observers and modelers have much to sort out in this particular area.

Acknowledgements

I wish to thank the meeting organizers for inviting me to present this review of a truly fascinating and rich topic, involving lithium and stellar evolution. I would also like to note that of the 3 main issues that were covered here, George Wallerstein (one of the meeting attendees) played key roles in the early stages of each one of them: he is an observational pioneer in this area. Since GW was my thesis advisor, I would also like to

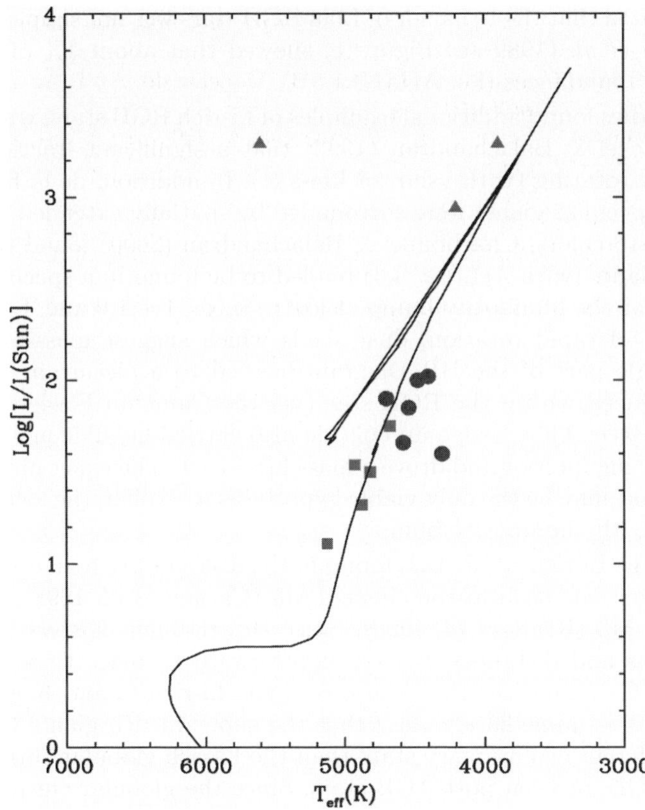

Figure 5. Luminosities and effective temperatures for samples of K-giants with enhanced lithium abundances. The filled triangles are stars in globular clusters from Carney *et al.* (1998) for M5, Smith *et al.* (1999) for NGC362, and Kraft *et al.* (1999) for M3. Filled circles are field K–giants discussed in Kumar & Reddy (2009), while the filled squares are rapidly-rotating Li–rich giants noted in Carlberg *et al.* (2009b). This type of diagram suggests that the "lithium-rich K-giant" phenomenon occurs over a wide evolutionary range and may be due to different processes, depending on the evolutionary state of the red giant. The solid line is a sample isochrone for a metallicity of -0.5 and an age of 9 Gyr (or a giant mass of about $1.3M_\odot$), not very different from the plotted giants; this stellar evolutionary track indicates that the Li-rich giants are found from near the base of the red giant branch to well along and past the AGB.

take this opportunity to thank him for introducing me, in his own unique style, to each of these topics and providing a background that stressed the key role of spectroscopy.

References

Boesgaard, A. M. 1970, *ApJ*, 161, 1003

Bonsack, W. K. 1959, *ApJ*, 130, 843

Brown, J. A., Sneden, C., Lambert, D. L., & Dutchover, E., Jr. 1989, *ApJS*, 71, 293

Cameron, A. G. W. & Fowler, W. A. 1971, *ApJ*, 164, 111

Carlberg, J. K., Majewski, S. R., & Arras, P. 2009a, *ApJ*, 700, 832

Carlberg, J. K., Majewski, S. R., Smith, V. V., Cunha, K., Patterson, R. J., Bizyaev, D., Arras, P., & Rood, R. T. 2009b, in: K. Cunha, M. Spite, & B. Barbuy (eds.), *Chemical Abundances in the Universe: Connecting First Stars to Planets* (Cambridge: Cambridge University Press), p. 408

Carney, B. W., Fry, A. M., & Gonzalez, G. 1998, *AJ*, 116, 2984

Charbonnel, C. & Balachandran, S. 2000, *A&A*, 359, 563

de La Reza, R., Drake, N. A., & da Silva, L. 1996, *ApJ* (Letters), 456, 456

Fekel, F. C. & Balachandran, S. 1993, *ApJ*, 403, 708

Garcia-Hernandez, D. A., Garcia-Lario, P., Plez, B., Manchando, A., D'Antona, F., Lub, J., & Habing, H. 2007, *A&A*, 462, 711

Gratton, R. G., Sneden, C., Carretta, E., & Bragaglia, A. 2000, *A&A*, 354, 169

Iben, I., Jr. 1965, *ApJ*, 142, 1447

Iben, I., Jr. 1981, *ApJ*, 243, 987

Keenan, P. C. 1967, *AJ*, 72, 808

Korn, A. J., Grundahl, F., Richard, O., Mashonkina, L., Barklem, P. S., Collet, R., Gustafsson, B., & Piskunov, N. 2007, *ApJ*, 671, 402

Kraft, R. P., Peterson, R. C., Guhathakurta, P., Sneden, C., Fulbright, J. P., & Langer, G. E. 1999, *ApJ*, 518, L53

Kumar, Y. B. & Reddy, B. E. 2009, *ApJ* (Letters), 703, 46

Lambert, D. L., Dominy, J. F., & Sivertsen, S. 1980, *ApJ*, 235, 114

Lind, K., Primas, F., Charbonnel, C., Grundahl, F., & Asplund, M. 2009, *A&A*, 503, 545

Luck, R. E. & Lambert, D. L. 1982, *ApJ*, 256, 189

McKellar, A. 1940, *PASP*, 52, 407

Plez, B., Smith, V. V., & Lambert, D. L. 1993, *ApJ*, 418, 812

Sackmann, I.-J. & Boothroyd, A. I. 1992, *ApJ*, 392, 71

Scalo, J. M., Despain, K. H., & Ulrich, R. K. 1975, *ApJ*, 196, 805

Smith, V. V. & Lambert, D. L. 1989, *ApJ* (Letters), 345, 75

Smith, V. V. & Lambert, D. L. 1990, *ApJ* (Letters), 361, 69

Smith, V. V., Plez, B., Lambert, D. L., & Lubowich, D. A. 1995, *ApJ*, 441, 735

Smith, V. V., Shetrone, M. D., & Keane, M. J. 1999, *ApJ* (Letters), 516, 73

Torres-Peimbert, S. & Wallerstein, G. 1966, *ApJ*, 146, 724

Wallerstein, G. 1966, *ApJ*, 143, 823

Wallerstein, G. & Sneden, C. 1982, *ApJ*, 255, 577

Sofia Randich

Andreas Korn & Jeffrey Linsky

Light Elements in the Universe
Proceedings IAU Symposium No. 268, 2009
C. Charbonnel, M. Tosi, F. Primas & C. Chiappini, eds.

© International Astronomical Union 2010
doi:10.1017/S174392131000428X

Mass loss and luminosities of S and C AGB stars with and without Li

Roald Guandalini[1], Sara Palmerini[1,2], Maurizio Busso[1,2], Enrico Maiorca[1,2], and Stefan Uttenthaler[3]

[1]Dipartimento di Fisica, Universitá degli Studi di Perugia,
via Pascoli, 06123, Perugia, Italy
email: `guandalini@fisica.unipg.it`

[2]I.N.F.N. sezione di Perugia

[3]Instituut voor Sterrenkunde, K. U. Leuven,
Celestijnenlaan 200D, 3000 Leuven, Belgium

Abstract. We present the preliminary results of an analysis performed on two samples of thermally pulsing Asymptotic Giant Branch stars from our Galaxy, the first made of carbon-rich sources and the second of S-type stars. We have estimated their absolute luminosities and updated rates of the stellar winds through methods based on their infrared spectrophotometry and on updated estimates of their variability and distance.

We then focus on those sources in our database showing Li in their spectra looking for correlations between the Li abundance and the other physical parameters, in the aim of establishing observational criteria for understanding the conditions for the occurrence of the deep mixing phenomena to which the production of Li is currently attributed.

Keywords. Stars: evolution, AGB, post-AGB – infrared: stars

1. Introduction on Li and evolved giant stars

Stars evolving along the red giant branch are on average depleted in lithium. We remind here that, for low-metallicity objects, an upper limit on the Li content was set by Gratton *et al.* (2000), stating that low-mass stars on the upper RGB have log ϵ(Li) $\leqslant 0$. Mixing processes on the Main Sequence and up to the first dredge-up imply that Li be gradually carried down from the photosphere to regions of high temperature and destroyed; at the same time new Li is not produced, because any mixing mechanism occurring is so slow that the parent nucleus ^{7}Be has time to capture protons on the path. This explains why red giants are mainly Li-poor. It was further established (Sweigart & Mengel 1979) and confirmed by observations (Gilroy 1989, Gilroy & Brown 1991, Kraft 1994) that there must be slow mixing episodes in low-mass stars also after the first dredge-up. They gradually carry to the surface a considerable amount of ^{13}C, so that the ^{12}C/^{13}C ratio decreases. These mixing phenomena can happen because after dredge-up, the H-burning shell advances into a homogeneous region, so that the natural barrier opposed to mass transport by a chemical stratification is not present. This happens in the so-called "L-bump phase". Slow mixing processes occurring at or after the L-bump would further reduce the Li abundance. However, a few red giants (about 2%) show enhanced Li abundances at levels sometimes higher than in the present Interstellar Medium (log ϵ(Li) $\simeq 3.3$). Amongst the various explanations attempted, the most common assumption is that some form of fast extra-mixing might transport ^{7}Be from above the H-burning shell to the envelope at a speed sufficient to overcome the rate of p-captures. This is the so-called Cameron-Fowler mechanism (Cameron & Fowler 1971). The attempts at involving stellar rotation as source for these extra-mixing processes were frustrated by the understanding that the stellar structure reacts to rotational distortions too quickly to induce significant mixing on long time scales (Palacios *et al.* 2006).

It has been suggested (Busso *et al.* 2007a, Wasserburg & Busso 2008, Nordhaus *et al.* 2008) that the processes that cause the mass transport might be linked to the buoyancy of magnetized H-burning ashes, based on the fact that magnetic bubbles are lighter than the stellar environment, due to the unbalance generated by magnetic pressure ($B^2/8\pi$). This requires a magnetic dynamo to occur in red giants, as indeed observed (Andrews *et al.* 1988). Magnetic buoyancy is an interesting possibility especially because it can occur at different speeds, depending on the amount of heat exchanged with the environment (Denissenkov *et al.* 2009, Palmerini & Busso 2008). Therefore, Li could experience production or destruction according to different mixing velocities.

In order to perform an accurate analysis on Li abundances and to attribute them to the correct evolutionary phases, we need reliable estimates of absolute luminosities and mass loss rates. Therefore, we here apply bolometric corrections for evolved stars, as derived in our recent work, to infer reliable estimates of these parameters (and of the distance) for stars having measurements of Li abundances in their photosphere. This helps in setting constraints on where extra-mixing is active. In Section 2 we discuss the analysis we performed and link it with our models. Instead, in Section 3 preliminary conclusions are illustrated.

2. Luminosities and mass loss rates of red giants showing Li

2.1. *The AGB sub-sample*

In order to improve our understanding of the evolutionary properties of Asymptotic Giant Branch stars (hereafter AGB) we need to study in an accurate way two crucial parameters: luminosities and mass loss rates. For both, determinations are hampered by the large distance uncertainties and by the fact that most of the radiation emitted by these sources can be observed only at infrared wavelengths and therefore depends on space-borne facilities, or on the exploitation of special ground-based locations.

In the last years we performed an extensive analysis of the photometric properties of AGB stars, using available infrared data from the ISO, MSX and IRAS-LRS experiments and looking for relations between their luminosities and their main chemical and physical parameters. In this way we are trying to find crucial tools that could help us in the improvement of the evolutionary models for AGB sources.

In the particular case of Li-bearing stars the formation of our sub-sample starts from the previous work by Uttenthaler *et al.* (2007) on AGB stars in the Galactic bulge with Li. We extend it to the Galactic AGB stars of the disk, already examined in their photometrical properties in our previous works (Guandalini *et al.* 2006, Guandalini & Busso 2008), with the aim of creating a homogeneous sub-sample with good estimates for mass and Li-abundances, good measurements of the distance and reliable near-to-mid infrared photometry. The resulting list of sources and all the details on the analysis performed are presented in Guandalini *et al.* (2009; Table 1). All sources come from large samples of C-rich or S-type AGB stars; the photometric data used and the techniques adopted are from Guandalini & Busso (2008), Busso *et al.* (2007b), and Guandalini *et al.* (2006).

2.2. *Luminosities and mass loss rates*

Figure 1 presents the selected sample of AGB stars showing Li in their spectra as compared with the luminosities obtained through the bolometric corrections from Guandalini *et al.* (2006) [C-rich] and Guandalini & Busso (2008) [S-type]. The sources richest in Li are all of CJ type: we believe, also thanks to our analysis performed on C-rich galactic sources in Guandalini *et al.* (2006), that they are all source with peculiar evolutionary properties. Their high Li abundance can be produced either by a relatively fast extra-mixing in low-mass stars or by hot bottom burning in more massive AGB stars. Going down along the axis of the Li-abundance we find the four stars from the Galactic bulge discussed by Uttenthaler *et al.* (2007): they are all O-rich. We have marked by dashed lines the region occupied by "standard" C(N) sources (circles). They are divided into two groups according to the higher (full circles) or lower (empty circles) reliability of their distance estimate. For this group we can see that there seems to be a pattern for which more luminous sources are more depleted in Li and fainter ones are instead less depleted. This could be originated by two concurring phenomena: 1) more luminous Carbon stars are on average more evolved, so had more time to destroy Li; 2) more luminous Carbon stars could be the most massive ones (in a relatively small range of masses), therefore they could destroy Li more efficiently due to higher H-burning temperatures. If we look at the vertical dashed lines, they indicate the range of the luminosity function for Galactic C-rich sources from Guandalini *et al.* (2006). Our stars belong to the typical luminosity range of normal C-stars and cover all the interval, therefore we can conclude that our sample is a good representative for the standard galactic Carbon stars from the disk. We believe that also their Li abundances should represent typical trends. Finally, we see that the few S-type (or M) sources we have in our sample are more depleted in Li than the average standard Carbon star.

If we look at Figure 1 we see a pattern for which the more luminous C-rich AGBs are the more depleted in Li. In order to study this property we have to consider in the picture also Red Giant Branch stars (hereafter RGB). Therefore, we add in the analysis a

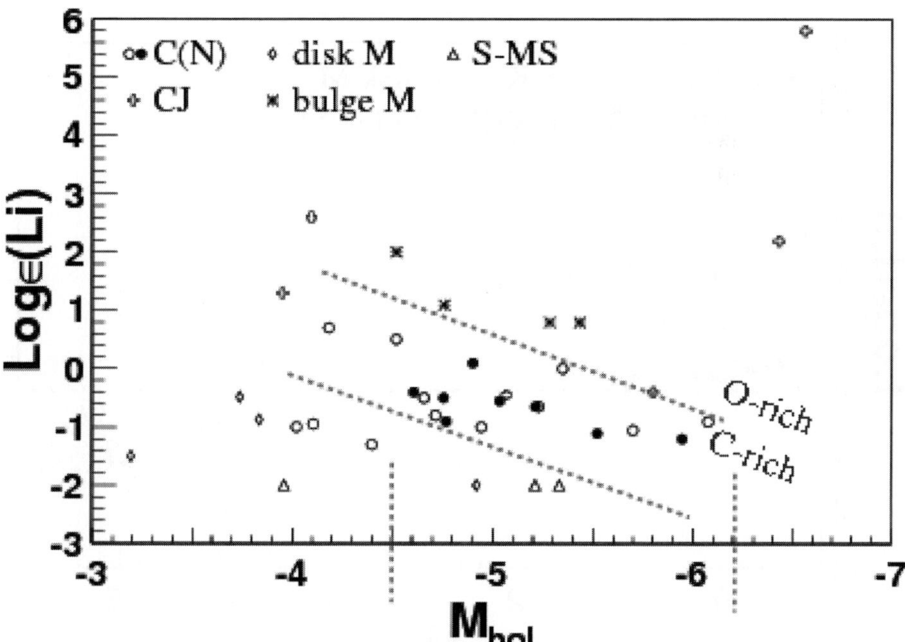

Figure 1. The sample of AGB stars showing Li in their spectra. Dashed lines roughly limit the region occupied by C(N) giants and their range of luminosities as inferred from the empirical Luminosity Function of Guandalini *et al.* (2006) (vertical lines).

314 R. Guandalini *et al.*

second sub-sample of RGB stars quite close to us, whose distances and Li-abundances are well-studied. The data for this group of K- and G-type sources are reported in Table 2 of Guandalini *et al.* (2009). In Figure 2 we consider both sub-samples in the same plot: RGB stars are the less luminous on the left side. The stars at the left of the plot, with log ε(Li) $\simeq 2$, are K giants with luminosities typical of the L-bump on the RGB. We consider them as being stars newly enriched in Li in agreement with Charbonnel & Balachandran (2000). After the extensive depletion of Li on the Main Sequence and on the early RGB phases, their envelopes must have seen some form of rapid mixing implementing a Cameron-Fowler mechanism. The second moderately Li-rich group is considered by us as being in the upper part of the RGB itself. Instead, Charbonnel & Balachandran (2000) interpreted this group as formed by early-AGB stars, producing Li (after some destruction in the late RGB phases). Inspection of stellar models, however, reveals that the early-AGB stages suitable to reproduce the luminosities and temperatures of these K giants fall in a temporal phase where the H-shell is extinguished, so that no ^{7}Be survives to be carried to the surface. We must therefore tentatively conclude that a new production of Li on the early-AGB at temperatures and luminosities compatible with observations does not occur and therefore the observed Li should be a relic of the previous production on the RGB. G-type stars show values of Li-abundance ranging between 0 and 1. They are not totally depleted but don't show the high values of K-type Li-rich sources. Finally, K-type sources without Li could be intermediate-mass stars with no hot bottom burning or low-mass stars with strong phenomena of depletion.

We searched also for possible relations between the rates of the stellar winds and Li-abundances for AGB stars. In order to estimate the rates, we adopted the methods explained in detail in Guandalini *et al.* (2006) and Guandalini (2010). Unfortunately, we didn't find any apparent relation between these two parameters: the data have a too large scatter to draw any conclusion.

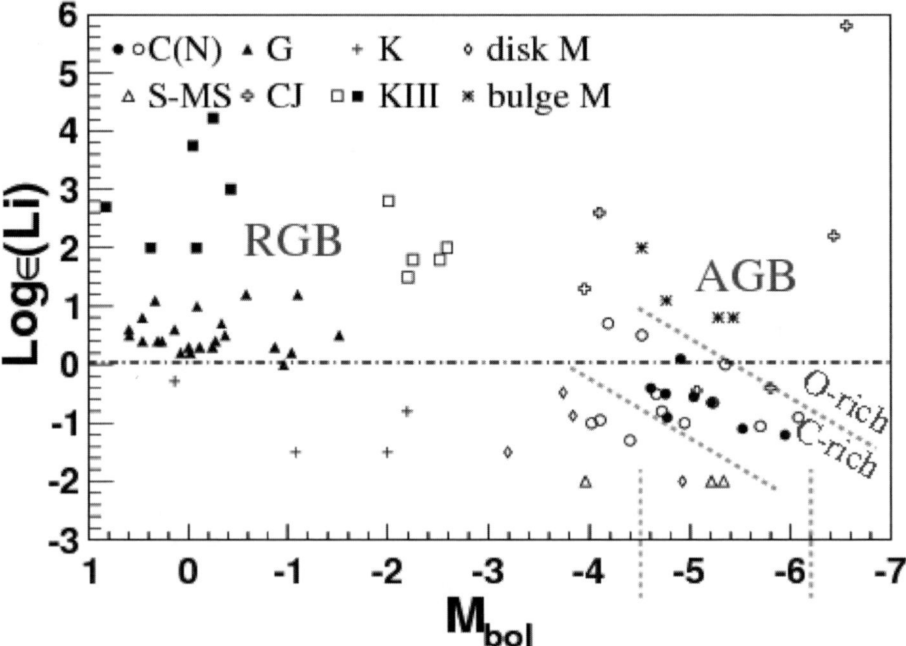

Figure 2. The same plot already presented in Figure 1, but including also the RGB sub-sample. See text and Guandalini *et al.* (2009) for details.

2.3. *Link with the models*

Our interpretation of the above results can be summarized in the following points [see Palmerini *et al.* (2010), this conference, for more details on this part]:

1) We observe in Figures 1 and 2 that Li is produced at the Luminosity bump and then is depleted with different rates during the successive evolution of the sources along the RGB and AGB stages.

2) We have so far no clear evidence for another phase of production after the L-bump.

3) This could be explained by the model presented by Palmerini *et al.* (2009) that suggests two types of extra-mixing. i) Fast Mixing, due to the buoyancy of small magnetized parcels, generated by detached magnetic instabilities. This causes ^{7}Be transport at rates faster than its decay lifetime (53 days) and therefore produces Li. This appears to happen at the bump of the Luminosity function following the first dredge-up. ii) Slow Mixing, due to the slower buoyancy of larger magnetized domains, which efficiently exchange heat with the environment, hence slowing down their motion. This happens during the remaining evolution and causes depletion of Li.

3. Conclusions

We have examined a sub-sample of Galactic AGB stars of moderate luminosity (Bolometric Magnitude fainter than -6) having phenomena of extra-mixing for Li. The study of this sub-sample gives us the chance to refine our previous studies made on larger and more general samples thanks to their higher homogeneity and to the presence of another constraint: the Li-abundance.

The relation between the Li abundance and the luminosity, examined for different classes of stars along the RGB and the AGB branches, shows us that, for stars of moderate luminosity, Li is produced at the Bump of the L-function and then is depleted at different rates during the successive evolution of the sources.

No clear correlation of Li abundances with mass loss rates emerged so far.

The behaviour of Li in evolved stars seems to be explained by the model presented in Palmerini *et al.* (2010), this conference, thanks to two different phenomena of extra-mixing:

- Fast Mixing, at the bump of the L-function, causing production of Li (buoyancy, at speeds close to the Alfvén Velocity, of small bubbles formed by magnetic instabilities).
- Slow Mixing, during the remaining evolution, which causes depletion of Li (buoyancy of larger magnetized structures exchancing heat efficiently with the environment).

Acknowledgments

This research was supported by the Italian Ministry of Research under contract PRIN2006-022731 and by INFN (sezione di Perugia) through the funds of Research Group n. 3. The idea of mixing by magnetic buoyancy was developed in common with G. J. Wasserburg and K. M. Nollett. R. G. wishes to thank the organizers of IAU Symposium 268 for the chance of giving a talk and for the wonderful conference.

References

Andrews, A. D., Rodonó, M., Lensky, J. L. *et al.* 1988, *A&A*, 204, 177
Busso, M., Wasserburg, G. J., Nollett, K. M., & Calandra, A. 2007a, *ApJ*, 671, 802
Busso, M., Guandalini, R., Persi, P., Corcione, L., & Ferrari-Toniolo, M. 2007b, *AJ*, 133, 2310
Cameron, A. G. W. & Fowler, W. A. 1971, *ApJ*, 164, 111

Charbonnel, C. & Balachandran, S. C. 2000, *A&A*, 359, 563

Denissenkov, P. A., Pinsonneault, M., & Mac Gregor, K. B. 2009, *ApJ*, 696, 1823

Gilroy, K. K. 1989, *ApJ*, 347, 835

Gilroy, K. K. & Brown, J. A. 1991, *ApJ*, 371, 578

Gratton, R. G., Carretta, E., Matteucci, F., & Sneden, C. 2000, *A&A*, 358, 671

Guandalini, R., Busso, M., Ciprini, S., Silvestro, G., & Persi, P. 2006, *A&A*, 445, 1069

Guandalini, R. & Busso, M. 2008, *A&A*, 488, 657

Guandalini, R., Palmerini, S., Busso, M., & Uttenthaler, S. 2009, *PASA*, 26, 168

Guandalini, R. 2010, *A&A*, in press

Kraft, R. P. 1994, *PASP*, 106, 553

Nordhaus, J., Busso, M., Wasserburg, G. J., Blackman, E. G., & Palmerini, S. 2008, *ApJ* (Letters), 684, 29

Palacios, A., Charbonnel, C., Talon, S., & Siess, L. 2006, *A&A*, 453, 261

Palmerini, S. & Busso, M. 2008, *New AR*, 52, 412

Palmerini,S., Busso, M., Maiorca, E., & Guandalini, R. 2009, *PASA*, 26, 161

Palmerini, S. *et al.* 2010, this conference

Sweigart, A. V. & Mengel, J. G. 1979, *ApJ*, 229, 624

Uttenthaler, S., Lebzelter, T., Palmerini, S., Busso, M., Aringer, B., & Lederer, M. T. 2007, *A&A* (Letters), 471, 41

Wasserburg, G. J. & Busso, M. 2008, *AIPC*, 1001, 295

Light Elements in the Universe
Proceedings IAU Symposium No. 268, 2009
C. Charbonnel, M. Tosi, F. Primas & C. Chiappini, eds.
© International Astronomical Union 2010
doi:10.1017/S1743921310004291

Observations of light elements
in massive stars

A. Kaufer

European Southern Observatory, Alonso de Cordova 3107,
Casilla 19001, Santiago de Chile
email: **akaufer@eso.org**

Abstract. Observations of light elements in hot massive stars are limited to few transitions of boron in the satellite-ultraviolet; lithium and beryllium are not observable at all. But because of its high sensitivity to the effects of rotational mixing, boron abundance determinations in massive stars have excelled as the definite test for evolutionary models with rotation. In this paper the observational evidence for rotational mixing in massive stars is reviewed and alternative interpretations are discussed.

Keywords. Stars: abundances, massive, rotation

1. Introduction

Despite their small number (only three stars out of a thousand have a mass larger than $8\,M_\odot$) and their small mass fraction (only 14% of the mass of all stars is found in massive stars with $M > 8\,M_\odot$), massive stars are important because they inject into the ISM large amounts of radiation, mass, and mechanical energy in shortest time $(3-30\,\mathrm{Myr})$ (Meynet 2008). Massive stars hence drive the chemical and dynamical evolution of the galaxies through nucleosynthesis and (re-)distribution of elements. Consequently, accurate galaxy evolution models require a detailed understanding of the internal and core properties of massive stars.

The observation of surface abundances of non-radioactive elements is an efficient probe for studying stellar evolution and verifying internal nuclear processes. The consideration of rotation in stellar evolution models predicts significant effects on the evolutionary tracks and surface abundances of massive stars (cf. Heger & Langer 2000 and Maeder & Meynet 2000). In particular the light trace elements lithium, beryllium, and boron are strongly affected by effects of rotationally induced mixing. Hence their observations and detailed surface abundance analyses are critical to test effects of rotational mixing. In the following an overview of the available observations of light elements in massive stars is given and the methods and achievable accuracies of abundance determinations from the observations are described. Then the observed light-element abundance patterns and their interpretation in context of stellar evolution models with rotation are discussed.

2. Observations

Lithium has been crucial to test models of cool stars at all possible stages of evolution, from pre-main sequence to the post-AGB stage and is conveniently observed through the resonance doublet at 6706 Å. Beryllium is observable in cool stars through the Be II $\lambda3130$ resonance doublet but of limited use because of the considerably more difficult access from the ground to this spectral region in the near-UV. Boron is the most difficult light element to observe since all transitions of B I, B II, and B III are below the atmospheric cut-off at 300 nm in the satellite ultraviolet. Unfortunately, boron is at the same time the only light element observable in hot massive stars. Boron can be observed either through the B II λ1362 resonance line or the B III λ2066 resonance doublet line.

After a 5-year forced hiatus, STIS onboard *HST* is today and the mid-term future the only observational resource to study boron in hot massive stars. In the past boron abundances have been determined using the B II λ1362 line by Boesgaard & Heacox (1978) using *Copernicus* observations of 16 'normal' A and B stars, by Venn, Lambert, & Lemke (1996) using *IUE* observations of 6 B and A sub/giants and supergiants, and by Cunha *et al.* (1997) using GHRS data from *HST* of 4 B stars in Orion.

The B II λ1362 line turns out to be rather problematic for accurate abundance determinations: the line is blended with Si III, Ni II, V II, Zn III, and Fe III and the line is sensitive to NLTE corrections as reported by Cunha *et al.* (1997) and Venn *et al.* (2002). For the temperature range of mid-B type stars and earlier, where B III is the dominant ion above a temperature of 18 kK, the B III λ2066 doublet line appears better suited: the lines are not blended and NLTE effects are small compared to the B II line.

Proffitt *et al.* (1999) were the first to move to the B III λ2066 line to determine boron abundances and ^{11}B/^{10}B isotopic ratios for 3 B stars using very high-SNR GHRS spectra. Proffitt *et al.* (1999) find isotopic ratios of the 3 B stars of ^{11}B/^{10}B$\sim$ 4 consistent with the solar-system value (Anders & Grevesse 1989). Proffitt & Quigley (2001) explored the *IUE* archive for high-resolution spectra of B-type stars and determined boron abundances (and upper limits) for 45 early B-type stars. Venn *et al.* (2002) determine 4 boron upper limits from 7 B-type main sequence stars using STIS onboard *HST*. Mendel *et al.* (2006) observe 7 additional B-type main sequence stars with STIS and find most of them boron depleted. Brooks *et al.* (2002) add to this handful of boron abundances of hot massive galactic B-type stars upper limits for boron abundances for two B-type stars in the Small Magellanic Cloud (SMC) obtained with STIS and the *HST* ($V \approx 15$ mag).

3. Abundance analysis of hot stars

To convert an observation of a stellar spectrum into an abundance measurement requires to carry out a detailed abundance analysis using atmosphere models and line-formation codes. For the early- to mid-B type main sequence stars, which are primarily observed in the context of light element abundances, the stellar atmospheres can be assumed to be plane-parallel and non-expanding, i.e., stellar winds are considered to be negligible. Then the stellar atmospheres can be described by 1D line-blanketed LTE models. Information on the gravity is obtained from the shape of the Balmer-line wings, which are broadened by the gravity-sensitive Stark effect. The line wings are fit with 1D (N)LTE Balmer line formation models. Effective temperature is derived from the so-called *ionisation equilibrium* of different ions of the same element (like Si II/III/IV) to reproduce the observed line profiles and equivalent widths for a given element abundance. Depending of the individual lines used for the analysis this requires 1D LTE line formation plus NLTE corrections or – if available – 1D NLTE line formation computations. Despite of sophisticated NLTE line-formation codes, for early-type B stars, usually a slight dependency of the computed line-by-line abundance on the strength of the lines remains, which is attributed to an additional line-broadening, the so-called microturbulence. It is determined from lines of ions that span a large range in equivalent widths, i.e., O or Fe in early B-type stars. Once effective temperature, gravity, and microturbulence have been determined in a quasi self-consistent way, the line-by-line element abundances can be determined using again 1D (N)LTE line formation.

Nieva & Przybilla (2007) and Nieva & Przybilla (2008) have pushed this hybrid NLTE approach for detailed abundance analyses of early B-type stars to a high level of perfection. Przybilla, Nieva, & Butler (2008) demonstrate that for a carefully selected sample of early-type B stars in the solar neighborhood with high-quality spectral observations, individual element abundances with 1 sigma errors clearly less than 0.1 dex can be obtained for all observable ions in the optical spectra of these stars.

A direct comparison of the proto-solar abundances of Asplund *et al.* (2009) with the B-star abundances of Przybilla, Nieva, & Butler (2008) shows nearly perfect agreement:

$$
\begin{array}{llllll}
\text{B stars} & X = 0.715 & Y = 0.271 & Z = 0.014 & Z/X = 0.020 & (\text{Przybilla } et\ al.\,2008) \\
\text{proto-solar} & X = 0.7154 & Y = 0.2703 & Z = 0.0142 & Z/X = 0.0198 & (\text{Asplund } et\ al.\,2009)
\end{array}
$$

Considering these advances and the improved accuracies that can be obtained with such state-of-the-art abundance determination techniques, it seems advisable to make the effort to reanalyse all available (but still few) spectra of early-type B stars for which satellite ultraviolet observations of light elements, i.e., of boron are available.

4. Boron abundances patterns in hot stars

Most of the boron abundance studies of hot massive stars aim to establish the present-day boron abundance to constrain the Galactic chemical evolution models. Cunha (2010) reviews the implications of these observations of cool and hot stars for chemical evolution models and constraints on the boron production mechanisms.

The latter is of particular interest since boron is the only light element whose production is neither due to nucleosynthesis in stellar interiors nor due to big-bang nucleosynthesis but is produced through cosmic ray spallation reactions in the interstellar gas. Boron abundance trends as function of oxygen provide constraints on the possible spallation reactions at work.

Cunha *et al.* (2000) and Smith, Cunha, & King (2001) find a linear trend for the boron abundances as function of oxygen abundances for galactic FG disk stars that show no beryllium depletion. The measured slope of 1.50 ± 0.08 (Cunha 2010) indicates a combination of primary source (expected slope 1) and secondary source (expected slope 2) production of boron in the Galaxy.

If the boron – oxygen abundances patterns determined for galactic hot stars by Proffitt & Quigley (2001) and Mendel *et al.* (2006) are compared with the galactic FG disk stars, a good agreement is found for the non-boron depleted hot stars.

Interestingly, an extrapolation of the boron – oxygen trend with a slope of 1.5 to low-metallicity environments is in excellent agreement with the abundance determinations (upper limits) of two B stars in the SMC by Brooks *et al.* (2002) (cf. their Fig. 6). The fact that boron in hot stars is frequently found to be severely depleted however complicates the rigorous interpretation of this interesting result. An increase in the sample size in the MCs would be the obvious demand to tighten the constraints on boron production by spallation in the MCs in comparison with the Galaxy but at the same time appears almost prohibitive considering the excessive amount of required STIS observing time (some 15 hours of *HST* integration time per star to obtain a S/N$\sim$ 50).

Additional constraints on cosmic ray spallation theory could be obtained through high resolution, high SNR observations of the B III $\lambda 2066$ line, which allows to determine ^{11}B/^{10}B isotopic ratios from the predicted isotopic shift of 42 mÅ. The solar-system value of ^{11}B/^{10}B$\sim$ 4 (Anders & Grevesse 1989) is larger than the prediction from current cosmic ray spallation theory (^{11}B/^{10}B$\sim$ 2.5, Meneguzzi, Audouze, & Reeves 1971). An observational determination of the isotopic ratio from hot massive stars would therefore provide strong constraints on the ^{11}B production processes. Again, such observations are possible with STIS but demanding in observing time.

5. Boron depletion in hot stars

The light elements lithium, beryllium, and boron are very sensitive to destruction by proton capture at temperatures much lower than those where the H-burning CN-cycle is effective. This makes the light elements a sensitive tracer of mixing of stellar surface layers to deeper layers and can be observed as apparent depletion of the respective element. Lithium and beryllium have been intensively used in this context to study stellar evolution processes in cool stars but cannot be observed in hot stars. For hotter stars, boron is the only observable light element as discussed before.

Boron is a very fragile element and destroyed by proton capture already at temperatures of $< 6 \times 10^6$ K, which corresponds to about $1\,M_\odot$ below the surface of a B-type main-sequence star. Due to this sensitivity of boron to moderate temperatures, any mild mixing of surface layers to deeper layers would already deplete boron while keeping less fragile elements like nitrogen unchanged. In more evolved stages of the star the same mixing process may enhance other elements like nitrogen (progressive mixing). Therefore, the determination of boron – nitrogen abundance patterns in hot stars of different evolutionary stage is a sensitive method to test stellar evolution models.

Venn, Lambert, & Lemke (1996) were the first to explore this possibility in a dedicated study of a sample of A and B stars of different evolutionary stage. Contrary to the expectation to find boron depletion for the more evolved stars showing at the same time nitrogen enrichment, the authors found severe boron depletion for all stars of their sample including for the non-evolved stars near the main sequence that show only little nitrogen enrichment. The finding of boron depletion without nitrogen enhancement could only be explained by invoking an additional mixing process. Based on their evolutionary tracks of massive stars including the effects of rotation, Fliegner, Langer, & Venn (1996) proposed rotationally induced mixing acting in the radiative zones of massive main star-sequence stars to explain the observations of Venn, Lambert, & Lemke (1996).

Figure 1 compiles all to date available boron and nitrogen abundance data of B-type stars from Venn *et al.* (2002), Lemke, Cunha, & Lambert (2000), Proffitt *et al.* (1999), and Proffitt & Quigley (2001). The dashed (blue) line shows an evolutionary track according to Heger & Langer (2000) for a fast rotating massive star of $12\,M_\odot$ and $v_{\mathrm{eq}} = 200$ km/s, which nicely demonstrate the rapid depletion of boron by a factor of ~ 50 while nitrogen enhances by a factor of ~ 2.5 only. The newer boron observations have added three measurement of the two stars HD 36591 and HD 30836 to the group of stars showing strong boron depletion with virtually no nitrogen enrichment (group II in the classification of Morel, Hubrig, & Briquet (2008)). These observations still remain difficult to explain by rotational mixing. However, new models of the Geneva stellar evolution code with rotational mixing and an extended reaction network including lithium, beryllium, and boron by Frischknecht (2010) promise to obtain an even faster boron depletion and therefore a better match to the observations.

It should be mentioned here that also other effects than rotational mixing can alter the stellar surface composition of hot massive stars: mass loss and mass transfer in binary systems are the primary candidates. While mass loss does not play a strong role in B-type main-sequence stars, the mass transfer in a close binary could produce considerable effects to the measured abundance patterns. For a detailed discussion of the latest results on element depletion and enhancements through binary interaction in massive stars see Langer (2010).

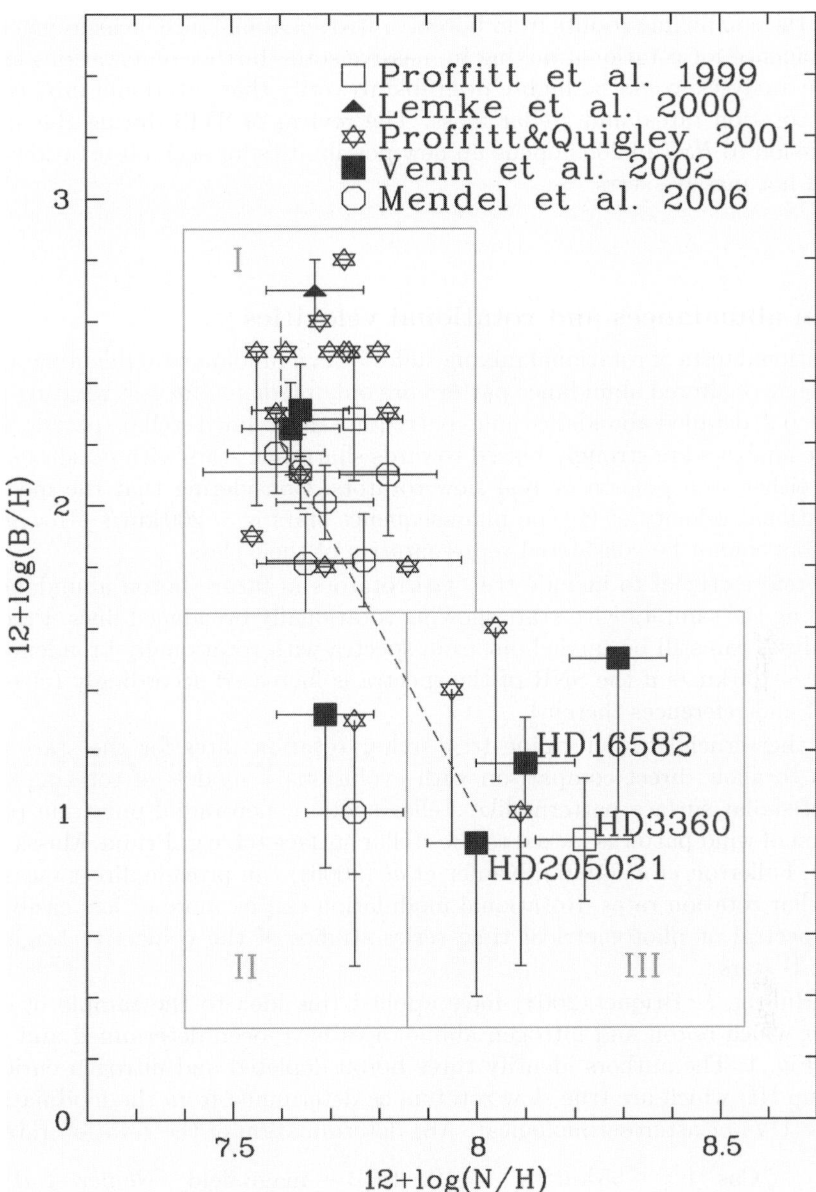

Figure 1. Boron and nitrogen abundances in B stars. Error bars of $+0/-0.5$ dex indicate upper limits. The dashed (blue) line shows an evolutionary track according to Heger & Langer (2000) for a rotating star of $12\,M_\odot$ and $v_{\mathrm{eq}} = 200\,\mathrm{km/s}$ and an initial nitrogen and boron abundance of 7.6 and 2.6, respectively. The (green) boxes indicate the different groups described in Morel, Hubrig, & Briquet (2008), i.e., I - unaltered, close to solar nitrogen and boron abundances, II - boron depleted but nitrogen unchanged, and III - boron depleted and nitrogen enhanced. Three stars are indicated with their HD number for which independent observations have established that these stars are slow rotators. See text for discussion.

Despite the compelling results from boron – nitrogen abundance measurements and the striking evidence for rotational mixing in massive stars, further observations and precise abundance analyses would be highly desirable to verify that rotational mixing increases with age, rotation rates, and stellar mass. The revival of STIS during the spectacular service mission to *HST* in 2009 opens up new possibilities for such observations on larger samples of hot massive stars.

6. Boron abundances and rotational velocities

Observational tests of rotational mixing suffer from a fundamental dilemma: while measurable effects of altered abundance pattern are only predicted for fast rotating stars with $v_{\rm eq}/v_{\rm crit} > 0.2$, detailed abundance analyses require sharp-lined stellar spectra. Therefore, abundance analyses are strongly biased towards sharp-line stars with $v_{\rm eq} \sin i < 20\,{\rm km/s}$, i.e., stars either seen pole-on or real slow rotators. Considering that the median equatorial rotational velocity of B-type main-sequence stars is $\sim 200\,{\rm km/s}$ (Howarth 2004), slow rotators cannot be considered representative of their class.

It is therefore crucial to include true *fast* rotators in future boron abundance studies by expanding the samples with stars showing rotationally broadened lines. Precise abundance analyses can still be carried out from spectra with rotationally broadened lines up to $v_{\rm eq} \sin i \sim 100\,{\rm km/s}$ if the SNR of the spectra is increased accordingly (cf. e.g. Kaufer *et al.* 1994 and references therein).

It is further crucial to determine true stellar rotation rates for the stars under examination to allow direct comparison with evolutionary models of rotating stars. The rotation of stellar surface patterns like stellar spots or non-radial pulsation patterns or the rotation of wind patterns locked to the stellar surface (cf. e.g. Prinja, Massa, & Fullerton (1995), Fullerton *et al.* (1997), Kaufer *et al.* (2006)) can provide direct measurements of true stellar rotation rates. Rotational modulation can be more or less easily observed through spectral or photometrical time series studies of the respective bright galactic early-type B stars.

Morel, Hubrig, & Briquet (2008) have applied this idea to the sample of early-type B stars for which boron and nitrogen abundances have been determined and which are shown in Fig. 1. The authors identify three boron depleted and nitrogen enriched stars (their group III) which are true slow rotators as determined from the modulation of UV wind lines (UV) or asteroseismological (AS) determination of the rotation rate:

HD 3360	ζ Cas	$v_{\rm eq} = 55\,{\rm km/s}$	UV	SPB + magn.field	Neiner *et al.* (2003)
HD 16582	δ Cet	$v_{\rm eq} = 14\,{\rm or}\,28\,{\rm km/s}$	AS	β Cep + magn.field	Aerts *et al.* (2006)
HD 205021	β Cep	$v_{\rm eq} = 26\,{\rm km/s}$	UV	β Cep + magn.field	Henrichs *et al.* (2000)

The fact that all three stars are either slowly pulsating B stars or β Cephei stars for which the presence of a weak magnetic field has been established leads Morel, Hubrig, & Briquet (2008) to propose that magnetic phenomena are important in altering the photospheric abundances of early B dwarfs, even for surface field strengths as low as at the one hundred Gauss level. The authors further argue that these stars only suffered a moderate amount of angular momentum loss along the main sequence and that they were likely slow rotators already on the ZAMS. At least in the cases of these three stars, the boron depletion can therefore not be explained by rotational mixing, which otherwise nicely reproduces the transition from group I to group III as illustrated in Fig. 1 by the evolutionary track from Heger & Langer (2000).

7. Conclusions

Undoubtedly the observational discovery of an additional mixing process to explain the boron depletion in the surface layers of non-evolved hot B stars on or near the main sequence by Venn, Lambert, & Lemke (1996) has been the most notable contribution by the studies of the light elements in massive stars. The subsequent identification of this mixing process to be due to rotationally induced mixing in the radiative layers of massive stars by Fliegner, Langer, & Venn (1996) has to be considered a major milestone in the development of stellar evolution models of massive stars. The understanding of the importance of stellar rotation for the evolution of massive stars has revolutionised the whole field of hot star research. Still, the number of actual measurements of boron in massive stars remains scarce due to the observational limitation imposed by the diagnostic lines in the satellite ultraviolet. The recent return of STIS onboard *HST* into science operation presents a rather unique opportunity to intensify the studies of rotational mixing as function of age, mass, and rotational rates of massive stars. Large and carefully selected samples of early-type stars should be examined to establish statistically significant dependencies on the relevant evolution parameters. The Galactic and MC samples already studied by the VLT-FLAMES survey of massive stars (Hunter el al. 2009) appear to be best suited for such future studies since they cover a large range of rotational velocities up to 300 km/s and a wide range of ages and masses.

References

Aerts, C. *et al.* 2006, *ApJ*, 642, 470

Anders, E. & Grevesse, N. 1989, *GeCoA*, 53, 197

Asplund, M., Grevesse, N., Sauval, A. J., & Scott, P. 2009, *ARA&A*, 47, 481

Boesgaard, A. M. & Heacox, W. D. 1978, *ApJ*, 226, 888

Brooks, A. M., Venn, K. A., Lambert, D. L., Lemke, M., Cunha, K., & Smith, V. V. 2002, *ApJ*, 573, 584

Cunha, K., Lambert, D. L., Lemke, M., Gies, D. R., & Roberts, L. C. 1997, *ApJ*, 478, 211

Cunha, K., Smith, V. V., Parizot, E., & Lambert, D. L. 2000, *ApJ*, 543, 850

Cunha, K. 2010, *these proceedings*

Fliegner, J., Langer, N., & Venn, K. A. 1996, *A&A*, 308, L13

Frischknecht, U. 2010, *these proceedings*

Fullerton, A. W., Massa, D. L., Prinja, R. K., Owocki, S. P., & Cranmer, S. R. 1997, *A&A*, 327, 699

Heger, A. & Langer, N. 2000, *ApJ*, 544, 1016

Henrichs, H. F. *et al.* 2000, *ASPC*, 214, 324

Howarth, I. D. 2004, *IAUS*, 215, 33

Hunter, I. *et al.* 2009, *A&A*, 496, 841

Kaufer, A., Szeifert, T., Krenzin, R., Baschek, B., & Wolf, B. 1994, *A&A*, 289, 740

Kaufer, A., Stahl, O., Prinja, R. K., & Witherick, D. 2006, *A&A*, 447, 325

Langer, N. 2010, *these proceedings*

Lemke, M., Cunha, K., & Lambert, D. L. 2000, *LIACo*, 35, 223

Maeder, A. & Meynet, G. 2000, *A&A*, 361, 159

Mendel, J. T., Venn, K. A., Proffitt, C. R., Brooks, A. M., & Lambert, D. L. 2006, *ApJ*, 640, 1039

Meneguzzi, M., Audouze, J., & Reeves, H. 1971, *A&A*, 15, 337

Meynet, G. 2008, *EAS Publication Series*, 32, 187

Morel, T., Hubrig, S., & Briquet, M. 2008, *A&A*, 481, 453

Neiner, C., Geers, V. C., Henrichs, H. F., Floquet, M., Frémat, Y., Hubert, A.-M., Preuss, O., & Wiersema, K. 2003, *A&A*, 406, 1019

Nieva, M. F. & Przybilla, N. 2008, *A&A*, 481, 199

Nieva, M. F. & Przybilla, N. 2007, *A&A*, 467, 295
Prinja, R. K., Massa, D., & Fullerton, A. W. 1995, *ApJ*, 452, L61
Proffitt, C. R., Jönsson, P., Litzén, U., Pickering, J. C., & Wahlgren, G. M. 1999, *ApJ*, 516, 342
Proffitt, C. R. & Quigley, M. F. 2001, *ApJ*, 548, 429
Przybilla, N., Nieva, M.-F., & Butler, K. 2008, *ApJ*, 688, L103
Smith, V. V., Cunha, K., & King, J. R. 2001, *AJ*, 122, 370
Venn, K. A., Lambert, D. L., & Lemke, M. 1996, *A&A*, 307, 849
Venn, K. A., Brooks, A. M., Lambert, D. L., Lemke, M., Langer, N., Lennon, D. J., & Keenan,
    F. P. 2002, *ApJ*, 565, 571

Light Elements in the Universe
Proceedings IAU Symposium No. 268, 2009
C. Charbonnel, M. Tosi, F. Primas & C. Chiappini, eds.

© International Astronomical Union 2010
doi:10.1017/S1743921310004308

Lithium abundances in Bulge-like SMR stars

Beatriz Barbuy[1], M. Trevisan[1], B. Gustafsson[2], K. Eriksson[2], M. Grenon[3], and L. Pompéia[4]

[1]Universidade de São Paulo, Brazil
email: barbuy@astro.iag.usp.br

[2]Uppsala Universitet, Sweden

[3]Observatoire de Genève, Switzerland

[4]Universidade do Vale do Paraíba, Brazil

Abstract. We analyze a sample of 21 super-metal-rich (SMR) stars, using high-resolution échelle spectra obtained with the FEROS Spectrograph at the 1.5m ESO telescope. The metallicities are in the range $0.15 <$ [Fe/H] < 0.5, 3 of them in common with Pompéia *et al.* (2002). Geneva photometry, astrometric data from *Hipparcos*, and radial velocities from CORAVEL are available for these stars. The peculiar kinematics suggests the thin disk close to the bulge as the probable birthplace of these stars (Grenon 1999). From *Hipparcos* data, it appears that the turnoff of this population indicates an age of 10-11 Gyr (Grenon 1999). Detailed analysis of the sample stars is carried out. Lithium abundances of these stars were derived, and their behaviour with effective temperature is shown.

Keywords. Stars: abundances, formation - Galaxy: formation

1. Introduction

Grenon (1989, 1999) selected a sample of about 6000 dwarf stars from the NLTT catalogue (see Grenon 1999), and studied them by means of Geneva photometry, radial velocity and Hipparcos astrometry. A sub-sample of super-metal-rich (SMR) stars was revealed (Grenon 1972, 1989, 1990, 1998, 2000). Optical spectra were obtained for 21 among the most metal-rich dwarfs of the sample, with the FEROS Spectrograph at the 1.52m telescope at ESO, La Silla. The total wavelength coverage is 3560–9200 Å with a resolving power R = 48,000.

The surface gravities $\log g$ were derived using *Hipparcos* parallaxes π with bolometric correction relations given in Alonso *et al.* (1995). The parameters [$T_{\rm eff}$, $v_{\rm t}$] were obtained by fixing trigonometric surface gravities and imposing excitation equilibrium for Fe I lines and ionization equilibrium for Fe I and Fe II. Independence between equivalent widths and the abundances of Fe I lines was imposed to determine $v_{\rm t}$. Derivations of metallicity were carried out through the Meudon code ABON2, and the code by the Uppsala group BSYN/EQWI. We are going through iterative checks, by computing FeI and FeII in the Sun, with both codes, and then for the stars, using the same tailored MARCS models. The codes are found to generate similar results, although some minor differences, which are not expected to affect the present results, are still being explored. The comparison, and more abundance results will be presented by Trevisan *et al.* (2009, in prep.). Lithium abundances were derived from the LiI 6707.8 Å line for the sample stars, by fitting the synthetic to the observed spectra. The spectrum synthesis code is described in Cayrel *et al.* (1991) and Barbuy *et al.* (2003). Photospheric 1D models for the sample were extracted from the MARCS grid (Gustafsson *et al.* 2008).

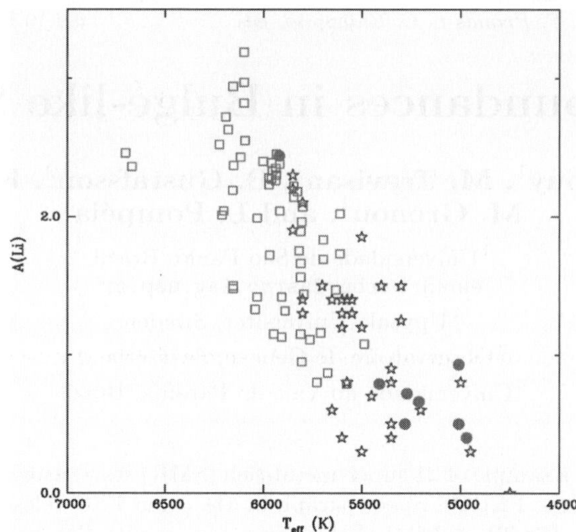

Figure 1. Log (Li/H) vs. effective temperature for the sample stars (red filled circles), compared with data for another sub-sample of bulgelike SMR stars by Pompéia *et al.* (2002) (open stars), and the metal-rich open cluster M67 (open squares, with results from Jones *et al.* (1999).

Comparisons are presented in Fig. 1 with the results by Pompéia *et al.* (2002), obtained for another sub-sample of bulgelike dwarfs, therefore compatible in terms of age and kinematics with the present sample, and with M67, a metal-rich open cluster of 4.5 Gyr, with data from Jones *et al.* (1999). Jones *et al.* had also shown that older stars deplete their Li with time, and it is more efficient in the cooler stars.

In conclusion, lithium abundances in our sample of very metal-rich stars are somewhat higher than in M67, and those found by Pompéia *et al.* (2002) for similar stars. One star IID 90054 shows a high Li abundance, compatible with being the hottest star of the sample.

Acknowledgements

The observations were carried out within brazilian time in a ESO-ON agreement, and within an IAG-ON agreement funded by FAPESP project n° 1998/10138-8.

References

Alonso, A., Arribas, S., & Martinez-Roger, C. 1995, *A&A*, 297, 197
Barbuy, B., Perrin, M.-N., Katz, D. *et al.* 2003, *A&A*, 404, 661.
Cayrel, R., Perrin, M.-N., Barbuy, B., & Buser, R. 1991, *A&A*, 247, 108.
Grenon, M. 1972, in IAU Colloq. 17, Age des Etoiles, eds. G. Cayrel de Strobel & A. M. Deplace (Paris : Obs. de Paris-Meudon), 55.
Grenon, M. 1989, *Ap&SS*, 156, 29.
Grenon, M. 1990, in ESO Workshop and Conf. Proc. 35, Bulges of Galaxies,1 st ESO / CTIO Workshop, ed. B. Jarvis & D. Terndrup (Garching : ESO), (Sunspot : NSO), 143.
Grenon, M. 1998,*Highlights of Astronomy*, Vol.11A, ed. J. Andersen (Dordrecht:Kluwer), 560.
Grenon, M. 1999, *Ap&SS*, 265, 331.
Grenon, M. 2000, in The Evolution of the Milky Way, ed. F. Matteucci & F. Giovannelli (Dordrecht : Kluwer), 47.
Gustafsson, B., Edvardsson, B., Eriksson, K. *et al.* 2008, *A&A*, 486, 951
Jones, B. F., Fischer, D., & Soderblom, D. R. 1999, *AJ*, 117, 330
Pompéia, L., Barbuy, B., Grenon, M., & Castilho, B. V. 2002, *ApJ*, 570, 820

Light Elements in the Universe
Proceedings IAU Symposium No. 268, 2009
C. Charbonnel, M. Tosi, F. Primas & C. Chiappini, eds.
© International Astronomical Union 2010
doi:10.1017/S174392131000431X

Survey for Li-rich K giants

Y. Bharat Kumar and Bacham E. Reddy

Indian Institute of Astrophysics, Bengaluru.
email: `bharat@iiap.res.in`, `ereddy@iiap.res.in`

Abstract. We present results from an ongoing survey of searching Li-rich K giants among low mass giants along the Red Giant Branch (RGB). A sample of 2500 stars with accurate astrometry have been selected from Hipparcos catalogue covering both the RGB luminosity bump and the red clump regions on the HR diagram. Lithium abundances have been determined for half of the sample from low resolution spectra using line depth ratio method. Results confirm the rarity of Li-rich K giants, just under 1%, in the solar neighbourhood. This study increased the total number of known Li-rich K giants by a factor of two. The analysis of high resolution spectra of candidate Li-rich K giants showed that the K giant HD 77361 is highly enriched in lithium (log ϵ(Li) = 3.82) and at the same time has anomalously low carbon isotopic ratio (^{12}C/^{13}C = 4.3). The results put important constraints on the theoretical modelling of the stellar structure and the mixing process, particularly, of the K giants.

Keywords. stars: late-type, abundances, individual (HD 77361), statistics – surveys

1. Li-rich K giants

Lithium is an important indicator of mixing processes in the stars on the Red Giant Branch (RGB). The expansion of convective envelope alters the surface chemical composition of lighter elements in the red giants. According to the classical stellar evolution theory (Iben 1967), any K giant with a surface lithium abundance exceeding log ϵ(Li) $\sim$ 1.4 can be characterized as Li-rich K giant (hereafter LRKG). The discovery of first LRKG (Wallerstein & Sneden 1982) has challenged the concept of gradual depletion of lithium on RGB. Since then, a dozen K giants were identified as LRKG. Charbonnel & Balachandran (2000) located all the known Li-rich stars on the Hertzsprung-Russell diagram using the Hipparcos parallaxes (Perryman *et al.* 1997). Most of them are fast rotators and show infrared excess. The origin of the anomalous lithium in these stars is not well understood. We have initiated a systematic survey to search for LRKG along the RGB, to pin down the source of lithium by studying the correlations between high lithium and other stellar parameters, and to estimate the fractional content of lithium contribution from the K giants to the Galaxy.

2. Samples and observations

We have selected 2500 sample giants along the RGB, sourced from Hipparcos catalog (Perryman *et al.* 1997, van Leeuwen 2007). We have restricted the survey to bright (mv $\leqslant$ 8) and nearby stars (d $\leqslant$ 200 pc), with accurate parallaxes ($\leqslant$ 15% error). Further, the sample stars were subjected to the following criteria: declination ($-60 \leqslant \delta \leqslant +80$), spectral type (K or late G) or T_{eff} ($4200 \leqslant T_{\mathrm{eff}} \leqslant 5000$ K) and luminosity ($1.0 \leqslant$ log $(L/L_\odot) \leqslant 2.5$). The last two criteria (luminosity and temperature) ensure that the sample covers the luminosity bump and/or red clump region on the HR diagram.

High quality low resolution spectra of sample stars were obtained using 2 m Himalaya Chandra Telescope (HCT), Hanle, 1 m Carl Zeiss telescope, and 2.34 m Vainu Bappu Telescope (VBT), Kavalur with resoultion R∼3500, R∼6000, R∼1500, respectively.

Along with the sample stars, we also made observations of known Li-rich giants, for which Li abundance was independently determined from high resolution spectra. An empirical relation is obtained between known Li-abundance and the ratios of line depths between Li I 6707 Å and Ca I 6717 Å., The derived relation is good to use for K giants with Li abundance $\log \epsilon(\mathrm{Li}) \geqslant 1.0$. The method is useful to efficiently eliminate K giants with low Li abundance and select LRKG based on line depth ratios.

3. Results & discussion

Preliminary results from the analysis of 1100 spectra are presented here. We could detect Li I line at 6707Å for Li abundances exceeding 0.6 dex. Li I line is seen in only 39% of the sample stars. The remaining 60% of stars in the sample are considered normal with $\log \epsilon(\mathrm{Li}) < 0.6$. The corresponding uncertainty in the derived abundance from the empirical relation is significantly large owing to the uncertainty in the line depth ratios. We found a dozen new K giants with Li significantly above the expected value and considered them as Li-rich K giants, which confirm the rarity of LRKG.

Detailed analysis from high resloution (R∼65000) spectra of one of the new LRKG HD 77361 at the RGB bump shows anomalous high lithium ($\log \epsilon(\mathrm{Li}) = 3.82$) and low carbon isotopic ratio ($^{12}\mathrm{C}/^{13}\mathrm{C} = 4.3$), which is different from the other known super Li-rich K giants. Results for HD 77361 do not fit with any of the explanations put forward for the source of enhanced Li in the photospheres of K giants: first dredge-up, extra deep mixing associated with cool bottom processing, lithium flash scenario, extra mixing triggered by spinning up the K giants with external angular momentum. The free mixing of material between hydrogen burning shell and the bottom of convective outer layer due to erasing the μ-barrier seems to be favored for the enrichment of products in the envelope, and hence the enhancement of surface lithium abundance and $^{13}\mathrm{C}$. We refer to Kumar & Reddy (2009) for a detail discussion.

4. Acknowledgement

We are grateful to the IAU, DST, and IIA for support to attend this meeting.

References

Charbonnel, C. & Balachandran, S. C. 2000, *A&A*, 359, 563
Iben, I. J. 1967, *ApJ*, 147, 624
Kumar, Y. B. & Reddy, B. E. 2009, *ApJ* (Letters), 703, L46
Perryman, M. A. C., Lindegren, L., Kovalevsky, J. *et al.* 1997, *A&A*, 323, L49
van Leeuwen, F. 2007, *A&A*, 474, 653
Wallerstein, G. & Sneden, C. 1982 *ApJ*, 255, 577

Light Elements in the Universe
Proceedings IAU Symposium No. 268, 2009
C. Charbonnel, M. Tosi, F. Primas & C. Chiappini, eds.

© International Astronomical Union 2010
doi:10.1017/S1743921310004321

A 3D-NLTE study of the 670 nm solar lithium feature

Elisabetta Caffau[1], Hans-Günter Ludwig[2,1], Matthias Steffen[3], and Piercarlo Bonifacio[1]

[1] GEPI, Observatoire de Paris, CNRS, Université Paris Diderot; 92195 Meudon Cedex, France
`Elisabetta.Caffau@obspm.fr`

[2] ZAH-Landessternwarte, Königstuhl 12, D-69117 Heidelberg, Germany

[3] Astrophysikalisches Institut Potsdam, An der Sternwarte 16, D-14482 Potsdam, Germany

Abstract. We derive the 3D-NLTE lithium abundance in the solar photosphere from the Li I line at 670 nm as measured in several solar atlases. The Li abundance is obtained from line profile fitting with 1D/3D-LTE/3D-NLTE synthetic spectra, considering several possibilities for the atomic parameters of the lines blending the Li feature. The 670 nm spectral region shows considerable differences in the two available disc-centre solar atlases, while the two integrated disc spectra are very similar. We obtain $A(\mathrm{Li})_{3D-NLTE} = 1.03$. The 1D-LTE abundance is 0.07 dex smaller. The line-lists giving the best fit for the Sun may fail for other stars, while some line-lists fail to reproduce the solar profile satisfactorily. We need a better knowledge of the atomic parameters of the lines blending the Li feature in order to be able to reproduce both the solar spectrum and the spectra of other stars. An improved line-list is also required to derive reliable estimates of the isotopic Li ratio in solar-metallicity stars.

Keywords. Sun: abundances – stars: abundances – hydrodynamics – line: formation

1. Introduction

For determining the lithium abundance in the solar photosphere one has to rely on the absorption feature of Li I at 670.7 nm. While the surrounding wavelength region is very clean in metal-poor stars making the abundance determination straightforward, this is not the case for the Sun and stars of comparable metallicity. The Li 670 nm feature is immersed in a forest of atomic and molecular lines whose atomic data are unfortunately not well known. Several line-lists for the blending components have been proposed. We applied them in 3D-NLTE spectral syntheses of the Li profile. The substantial differences among the resulting spectra indicate that for determining the Li abundance in stars of metallicity greater than 1/10 solar a major effort should be devoted to the search for data of the blending lines which are able to simultaneously reproduce the observed spectra over a wide range of effective temperatures and surface gravities.

Somewhat surprisingly, among the applied high-quality solar atlases we found noticeable differences in the Li range between the two disc-centre spectra of Neckel & Labs (1984) and Delbouille *et al.* (1973). While the lack of knowledge concerning the lines blending the Li 670 nm doublet is discussed in several papers, we could not find any information about this disagreement of the observed spectra in the literature.

2. Atomic and molecular data

In our analysis we assume that no ^{6}Li is present in the solar photosphere. For ^{7}Li we take into consideration ten line components. For the blending lines in the range, we

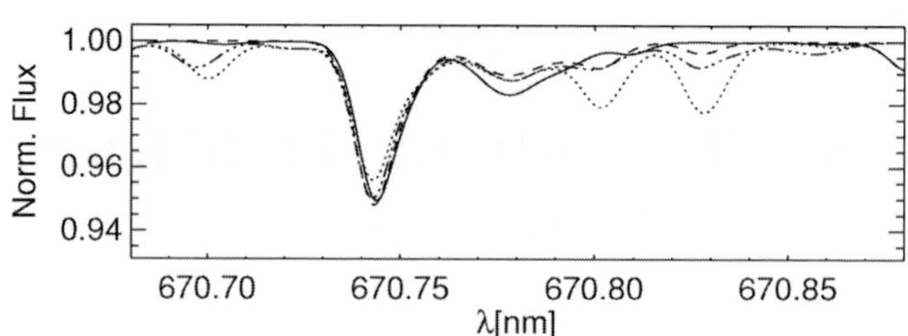

Figure 1. Synthetic, disc-integrated (flux) spectra for the Li I wavelength range and solar parameters: Li is modelled identically in 3D-NLTE in all syntheses while the blending components are computed in 3D-LTE, with line parameters from Hiltgen (1996) (solid), Reddy *et al.* (2002) (dash), Mandell *et al.* (2004) (dotted), and Ghezzi *et al.* (2009) (dash-dot).

used the line-lists of Hiltgen (1996), of Reddy *et al.* (2002), of Mandell *et al.* (2004), and the most recent work of Ghezzi *et al.* (2009). We computed the 3D spectral synthesis, taking into account the NLTE effects for the Li contribution. In Fig. 1, we compare the resulting synthetic profiles using the different line-lists. The disagreement between the profiles is evident, and can be explained by the fact that they have been optimised to analyse different stars.

3. The solar photospheric abundance of lithium

In the analysis of the solar Li abundance we only applied the line-lists for the blending components by Reddy *et al.* (2002) and Ghezzi *et al.* (2009). These authors applied and optimised their lists for the Sun, which naturally leads to a better agreement with the observed solar spectra. We further modified the line-lists slightly by changing the oscillator strength of the involved transitions to better reproduce the observations with our synthetic spectra based on 3D simulations. Performing a comprehensive fitting with synthetic 3D spectra is a computationally so demanding task that we had to restrict our approach to these small, manually introduced changes.

The abundances we obtain for the solar spectra of Neckel & Labs (1984) and of Kurucz (2005) are in close agreement. Despite that the parameters of the blending components in the two line-lists of Reddy *et al.* (2002) and Ghezzi *et al.* (2009) are different, the Li abundance does not change very much (within 0.04 dex). For the abundance determination we gave preference to the line-list of Ghezzi *et al.* (2009), and our result is $A(\mathrm{Li})_{\mathrm{3D-NLTE}} = 1.03 \pm 0.03$ where the uncertainty reflects the dispersion among the results for the different observed spectra.

The 1D-LTE abundance of lithium is 0.07 dex lower than the 3D-NLTE value. However, the synthetic 1D spectrum does not reproduce the observed line profile, in part because the parameters of the blending lines have been optimised for the 3D synthesis.

References

Delbouille, L., Roland, G., & Neven, L. 1973, Liege: Universite de Liege, Institut d'Astrophysique, 1973

Ghezzi, L., Cunha, K., Smith, V. V., Margheim, S., Schuler, S., de Araújo, F. X., & de la Reza, R. 2009, *ApJ*, 698, 451

Hiltgen, D. D. 1996, Ph. D. Thesis

Kurucz, R. L. 2005, *MemSAI* (Supplement), 8, 189

Mandell, A. M., Ge, J., & Murray, N. 2004, *AJ*, 127, 1147

Neckel, H. & Labs, D. 1984, *Solar Phys.*, 90, 205

Reddy, B. E., Lambert, D. L., Laws, C., Gonzalez, G., & Covey, K. 2002, *MNRAS*, 335, 1005

Light Elements in the Universe
Proceedings IAU Symposium No. 268, 2009
C. Charbonnel, M. Tosi, F. Primas & C. Chiappini, eds.

© International Astronomical Union 2010
doi:10.1017/S1743921310004333

Ultra-lithium-deficient halo stars

Lisa M. Elliott[1] and Sean G. Ryan[2]

[1] Centre for Stellar and Planetary Astrophysics, School of Mathematical Sciences
Building 28, Monash University, Victoria, 3800, Australia
email: `Lisa.Elliott@sci.monash.edu.au`

[2] School of Physics, Astronomy and Mathematics, and Centre for Astrophysics Research,
University of Hertfordshire,
College Lane Hatfield AL10 9AB, United Kingdom
email: `s.g.ryan@herts.ac.uk`

Abstract. While most warm halo dwarfs show lithium abundances at the level of the Spite Plateau, a small number ($\sim 5\%$) have undetectable lithium lines. The existence of these stars has long raised questions when interpreting the plateau abundances: are they an extreme example of a depletion mechanism that has affected the plateau stars, or do they have an entirely different history? We provide an overview of what is currently known about the lithium-poor halo stars and discuss a possible origin for the lithium deficiency in this unique group of stars.

Keywords. stars: abundances, blue stragglers – Galaxy: halo

1. Introduction

The existence of warm halo dwarfs with lithium abundances significantly below the Spite Plateau was first noted by Spite *et al.* (1984). These stars have undetectable lithium lines, with inferred upper limits for $A(\mathrm{Li})$ more than 0.4 dex below the level of the plateau (eg. Hobbs & Mathieu 1991; Thorburn 1992). Previous studies of lithium-deficient halo stars have revealed various abundance anomalies that affect some, although not all, stars (eg. Norris *et al.* 1997; Ryan *et al.* 1998). More recently, many of these objects were found to be deficient in Be (Boesgaard & Novicki 2005; Boesgaard 2007). Ryan *et al.* (2002) discovered that three out of four lithium-poor stars studied showed spectral line broadening and attributed this to rotation. Coupled with a high incidence of binarity, these findings have led to suggestions that the lithium-poor stars may be a product of the same mechanism that is responsible for the formation of field blue-stragglers. Understanding the origin of the lithium-poor stars is crucial for studies of the Spite Plateau: if they do have a different history to normal halo dwarfs they should be ruled out of future studies of lithium depletion in the general halo population.

Following the rotation study of Ryan *et al.* (2002), a further investigation of abundance trends and rotation properties in the lithium-poor stars was undertaken (Elliott & Ryan 2010; Ryan *et al.* 2010). Here we draw on results from these studies, along with previous results, to provide an overview of the observed characteristics of the lithium-poor halo stars.

2. Overview of results

A summary of observed properties for nine lithium-deficient halo stars is presented in Table 1. The key properties are as follows:

Abundances Two cool, metal-poor lithium-poor stars show abundance ratios that differ from the mean trends in the general halo population. This includes enhancements in the

Table 1. Overview of properties of lithium-deficient halo stars

Star	$T_{\rm eff}$	[Fe/H]	Abundance Anomalies?	Rotation	Binary[1]
HD97916	Warm[*]	High[*]	No	Yes	Yes
G202-65	Warm	High	No	Yes	Yes
G66-30	Warm	High	No	Yes	Yes
BD+51°1817	Warm	High	No	Yes[2]	Yes
BD+25°1981	Warm[3]	High[3]	No	Yes?[1]	Yes?
CD-31°19466	Cool	Med	No	No[2]	?
G122-69	Cool	Low	No	Yes	No
G139-8	Cool	Low	Yes	No	No
G186-26	Cool	Low	Yes	No	No

Notes:
[*] Warm refers to stars with $T_{\rm eff} \gtrsim 6300$ K, high refers to stars with [Fe/H] $\gtrsim -1.50$
[1] From Carney *et al.* 1994; Carney *et al.* 2001; Latham *et al.* 2002 [2] From Ryan *et al.* (2002). [3] From Ryan *et al.* 2001.

neutron-capture element ratios [Sr/Fe], [Y/Fe] and [Ba/Fe] in G186-26 and deficiencies in [Na/Fe], [Mg/Fe], [Al/Fe], [Sr/Fe] and [Ba/Fe] in G139-8.

Rotation At least five stars show evidence of line broadening indicative of rotation rates exceeding that expected in old halo dwarfs. Inferred projected rotation velocities for these five stars are between 4.7 and 10.4 km s^{-1}. At least four of the rotating stars are in binary systems, with periods ranging from 168 to 688 days (Carney *et al.* 1994; Carney *et al.* 2001; Latham *et al.* 2002). Rapid rotation is common in the warm, metal-rich subset of lithium-poor stars while only one of the cooler stars, G122-69, shows mildly enhanced rotation.

3. Conclusions

The high incidence of rotation among the lithium-poor stars suggests that these stars do have a different history to the lithium-normal halo population. Our results, particularly for the warm subset of lithium-poor stars, support a scenario in which mass and angular momentum have been transferred from a now evolved companion, similar to the mechanism that may be responsible for the formation of field blue-stragglers.

References

Boesgaard, A. M. & Novicki, M. C. 2005, *ApJ* (Letters), 633, L125

Boesgaard, A. M. 2007, *ApJ*, 667, 1196

Carney, B. W., Latham, D. W., Laird, J. B., & Aguilar, L. A. 1994, *AJ*, 107, 2240

Carney, B. W., Latham, D. W., Laird, J. B., Grant, C. E., & Morse, J. A. 2001, *AJ*, 122, 3419

Elliott, L. M. & Ryan S. G. 2010, *in prep.*

Hobbs, L. M. & Mathieu, R. D. 1991, *PASP*, 103, 431

Latham, D. W., Stefanik, R. P., Torres, G., Davis, R. J., Mazeh, T., Carney, B. W., Laird, J. B., & Morse, J. A. 2002, *AJ*, 124, 1144

Norris, J. E., Ryan, S. G., Beers, T. C., & Deliyannis, C. P. 1997, *ApJ*, 485, 370

Ryan, S. G., Norris, J. E., & Beers, T. C. 1998, *ApJ* 506, 892

Ryan, S. G., Gregory, S. G., Kolb, U., Beers, T. C., & Kajino, T. 2002, *ApJ* 571, 501

Ryan, S. G., Elliott, L. M., Ford, A., & Gregory, S. G. 2010, *in prep.*

Spite, M., Maillard, J. P., & Spite, F. 1984, *A&A*, 141, 56

Thorburn, J. A. 1992, *ApJ* (Letters), 399, L83

Light Elements in the Universe
Proceedings IAU Symposium No. 268, 2009
C. Charbonnel, M. Tosi, F. Primas & C. Chiappini, eds.

© International Astronomical Union 2010
doi:10.1017/S1743921310004345

Li-rich giants in the Galactic Bulge. Is Li linked only to evolutionary status?

Oscar A. Gonzalez[1]

[1]Southern Observatory, Karl-Schwarzschild-Strasse 2, D-85748 Garching, Germany
email: ogonzale@eso.org

Abstract. In our detailed study of chemical abundances in the Galactic bulge (see Zoccali *et al.* 2008 for a description of the entire project) we have measured Li abundances by fitting synthetic spectra to the ^{7}Li (6707.18Å) line for ~ 400 giants in Baade's Window and a field at b $= -6$ (Gonzalez *et al.* 2009). We have found 13 stars showing strong ^{7}Li lines in complete contrast to the rest of the sample for which only upper limits could be obtained. Our sample is at least 1.2 mag brighter than the expected RGB bump, therefore we interpreted our results as evidence for stars that might have avoided the observed extra-mixing process or undergoing a Li enrichment process not necessarily linked to the RGB bump.

Keywords. stars: abundances, late-type – Galaxy: bulge

1. Introduction

Even with the increasing number of low mass Li-rich giants found to date, we have not been able to fully understand their origin. The main reason is that is becoming difficult to relate Li-rich stars with a single evolutionary status. Charbonnel & Balachandran (2000) presented a first evidence for a connection between Li-rich stars and evolutionary status showing that a significant number of Li-rich giants fall close to the RGB bump, where an extra-mixing process is expected to act, affecting abundances of C, ^{12}C/^{13}C and ^{7}Li (Gratton *et al.* 2000). In particular, after the RGB bump, ^{7}Li abundances are expected to be close to A(Li) $= 0$ as the material from the convective envelope is brought to higher temperatures in which Li is expected to be destroyed. However, in a very short stage prior to its destruction, ^{7}Li could be produced by Cameron-Fowler mechanism (Cameron & Fowler 1971). In complete contrast, the number of Li-rich stars not fulfilling the RGB bump connection has been recently increased (Monaco *et al.* 2007; Uttenthaler *et al.* 2007; Gonzalez *et al.* 2009) requiring stellar models to explain a Li production phase almost at any instance along the giant evolutionary sequence.

2. Can we really constrain Li-rich giants to a single phenomena?

While 11 stars in our sample show A(Li) within 0.8 and 2.3 following a well defined trend with T$_{eff}$ (Fig. 1), 2 of them have A(Li) ~ 2.8 and fall away from this relation. From their location in the CMD (nearly 1 mag. above the red clump), Li-rich stars in our sample are not compatible with a Li production phase in the RGB bump, even considering the Bulge distance spread. However, the Li content in the remaining ~ 400 stars is in agreement with an extra-mixing process diluting Li at the RGB bump to a value of A(Li) ~ 0. Therefore, these Li-rich giants might be stars which have avoided the extra-mixing at the RGB bump and show the Li abundance expected from standard dilution (A(Li) ~ 1.5) which would explain the observed trend between ^{7}Li and T$_{eff}$. However, in this scenario, the problem remains for the two Li-rich stars showing A(Li) ~ 2.8 for

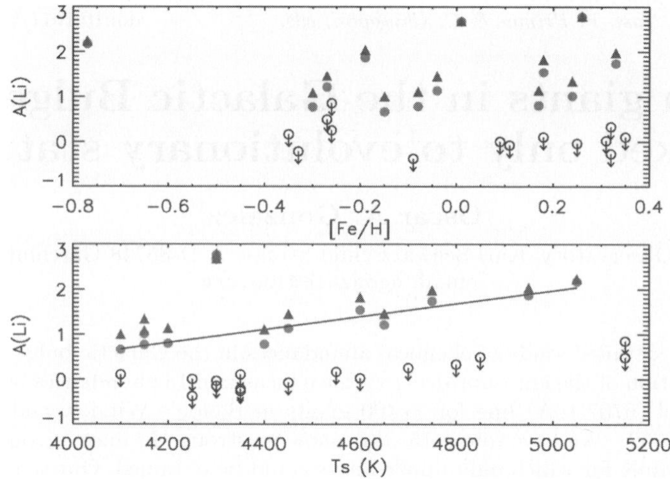

Figure 1. Relation between Li abundances and effective temperature (lower panel). Red filled circles are Li abundances measured under LTE and empty circles are upper limits. Blue triangles are A(Li) with NLTE corrections. No relation is observed between A(Li) and metallicity (upper panel).

which the high Li content could only be explained by an undergoing Li production phase. Models have shown that Li evolution in low-mass giants is highly dependent on magnetic field properties (Guandalini *et al.* 2009). By taking into account these magnetic properties, stars avoiding the extra-mixing process in the RGB bump might be explained (Charbonnel & Zahn 2007) as well as Li production in other instances of the giant evolution. A different approach must be considered as well, as Li has been probed to be affected by the presence of planets, at least in main sequence stars. In more evolved stars, this might also be the case. The engulfment of a planet would hardly contain enough material to enrich the surface of the star, but according to the models presented by Denissenkov & Herwig (2004), the addition of angular momentum might be enough to trigger a short Li production phase which might explain Li enriched giants at any moment during their giant evolution.

In conclusion, as the number of Li-rich stars has increased, it has became clear that their occurrence cannot be confined to a single evolutionary phase. Other parameters such as magnetic field properties or the presence of planets, might also be playing an important role.

References

Cameron, A. G. W. & Fowler, W. A. 1971, *ApJ*, 164, 111
Charbonnel, C. & Balachandran, S. C. 2000, *A&AS*, 359, 563
Charbonnel, C. & Zahn, J. P. 2007b, *A&A*, 476, L29
Denissenkov, P. A. & Herwig, F. 2004, *ApJ*, 612, 1081
Gonzalez, O. A., Zoccali, M., Monaco, L., Hill, V. *et al.* 2009, *A&A*, 508, 289
Gratton, R. G., Sneden, C., Carretta, E., & Bragaglia, A. 2000, *A&A*, 354, 169
Guandalini, R., Palmerini, M., Busso, M., & Uttenthaler, S. 2009, *PASA*, 26, 168
Monaco, L., Bellazzini, M., Bonifacio, P., Buzzoni, A., Ferraro, F. R., Marconi, G., Sbordone, L., & Zaggia, S. 2007, *A&A*, 464, 201
Monaco, L. & Bonifacio, P. 2008, *MemSAI*, 79, 1
Uttenthaler, S., Lebzelter, T., Palmerini *et al.* 2007, *A&A*, 471, L41
Zoccali, M., Lecureur, A., Hill, V. *et al.* 2008, *A&A*, 486, 177

Light Elements in the Universe
Proceedings IAU Symposium No. 268, 2009
C. Charbonnel, M. Tosi, F. Primas & C. Chiappini, eds.

© International Astronomical Union 2010
doi:10.1017/S1743921310004357

Interstellar Lithium as a probe of the primordial abundance

J. Christopher Howk[1]

[1]Department of Physics,
University of Notre Dame,
Notre Dame, IN, USA
email: jhowk@nd.edu

Abstract. The cosmic abundance of lithium continues to represent a conundrum, as predictions from BBN theory are inconsistent with measurements in the atmospheres of the lowest-metallicity stars. While there are worries that modifications of the stellar Li abundances may play a role in this discrepancy, no satisfactory solution has yet been found. We suggest an alternate approach to studying the cosmic abundance of Li: measurements of interstellar gas-phase Li in low-metallicity environments.

Keywords. ISM: abundances, clouds – galaxies: abundances – nuclear reactions, nucleosynthesis, abundances

1. Introduction

The primordial Li abundance predicted by standard Big Bang nucleosynthesis (BBN) is a factor of $\gtrsim 2-4$ above the best estimates of the Li abundance in halo star atmospheres (cf., summary in Cyburt *et al.* 2008). Several possibilities for this discrepancy exist, including destruction of Li within the stars themselves or the intriguing possibility of new physics (e.g., inhomogeneous nucleosynthesis, Li destruction in the first stars, non-thermal production by particle decays) as discussed elsewhere in these proceedings.

A new approach to estimating the primordial abundance of Li would be extremely useful. Here we reintroduce such an approach, the measurement of *interstellar* lithium abundances in low-metallicity galaxies. Estimating Li/H in the interstellar medium (ISM) of galaxies carries its own potential systematic uncertainties (see Steigman 1996), *but they are independent of those that may affect stellar abundance estimates*. In addition, the small velocity dispersions of interstellar gas makes it possible to measure the ^{6}Li/^{7}Li isotopic ratio (Kawanomoto *et al.* 2009), which can further constrain the origin of Li.

Previous searches for interstellar Li in external galaxies were limited to the SN 1987A sight line in the Large Magellanic Cloud as reported in Vidal-Madjar *et al.* (1987), ?, and Baade *et al.* (1991), which had somewhat limited usefulness for placing a limit on the absolute abundance measurements (Steigman 1996). More recently Prodanović & Fields (2004) suggested it may be possible to measure the gas-phase Li abundances in Galactic high velocity clouds, which are low-metallicity clouds that appear to be falling onto the Galaxy for the first time. While their calculations are overly optimistic (as described below), the rationale for such a measurement is clearly stated. With today's large aperture telescopes, however, it is possible to detect interstellar Li in other galaxies, and we will report on the first detection of interstellar Li in the Small Magellanic Cloud in an upcoming publication.

2. Interstellar lithium abundances

Measurements of Li in ISM clouds rely on the measurement of neutral lithium in absorption against background light sources using the Li I doublet near 6707 Å. However, the direct comparison of the column density of Li I with that of H I measured in the same way does not directly yield the interstellar Li abundance, because 1) Li^0 is not the dominant ionization stage of Li in the ISM (that being Li^+); and 2) Li may be incorporated into interstellar dust grains. These effects are such that Li I/H I $\ll$ Li/H. Thus, the abundance of Li in the ISM is given by

$$Li/H = N(Li\ I)N(H)^{-1}x(Li^0)^{-1}\delta_{Li}^{-1},\qquad(2.1)$$

where $N(H) \equiv N(H\ I) + 2N(H_2)$ is the total hydrogen column density, $x(Li^0)$ is the ionization fraction of neutral lithium, and δ_{Li} is the fraction of all Li present in the gas phase (the "depletion" due to dust). The ionization fraction in this case depends on the density of electrons (and perhaps of dust grains; see) and the strength of the radiation field.

These corrections can be quite large, with likely values $x(Li^0)^{-1} \gtrsim 100$ and $\delta_{Li}^{-1} \sim 4-5$ (Welty *et al.* 2003). Although the ionization corrections are typically derived in a relative sense using observed ratios of adjacent ions from other elements (e.g., Ca I/Ca II) so that the absolute strength of the radiation field is not important, these can still be quite uncertain. Steigman (1996) has argued that the ratio of Li I/K I can be a better approach to studying the interstellar abundance of Li, given the similar ionization characteristics of these two elements, although one is then in a position of estimating or assuming the K/H ratio if one is to estimate Li/H.

The recent work of Prodanović & Fields (2004) did not consider the strong effects of photoionization of Li I in their feasibility arguments when discussing measurements in high velocity clouds. The very small amounts of neutral lithium expected for such clouds, coupled with the relatively low column densities of these clouds makes the measurements extremely difficult. It will not be possible to probe Li in interstellar environments as metal poor as the halo star sample, but by probing gas in a range of metallicities, we will eventually be able to use the interstellar abundances to complement our understanding of the stellar results. Indeed, our first measurement beyond the Milky Way is in the gas of the Small Magellanic Cloud, which has a metallicity $\sim 0.25Z_\odot$. The next generation of very large aperture telescopes should allow us to push this approach to lower metallicity damped Lyman-α systems, although significant uncertainties in the absolute abundances derived through ionization analyses will likely persist.

References

Baade, D., Cristiani, S., Lanz, T., Malaney, R. A., Sahu, K. S., & Vladilo, G. 1991, *A&A*, 251, 253

Cyburt, R. H., Fields, B. D., & Olive, K. A. 2009, *J. Cosmology & Astroparticle Physics*, 11, 012

Kawanomoto, S. *et al.* 2009, *ApJ*, 701, 1506

Prodanović, T. & Fields, B. D. 2004, *ApJ* (Letters), 616, L115

Steigman, G. 1996, *ApJ*, 457, 737

Vidal-Madjar, A., Andreani, P., Cristiani, S., Ferlet, R., Lanz, T., & Vladilo, G. 1987, *A&A*, 177, L17

Welty, D. E., Hobbs, L. M., & Morton, D. C. 2003, *ApJS*, 147, 61

Light Elements in the Universe
Proceedings IAU Symposium No. 268, 2009
C. Charbonnel, M. Tosi, F. Primas & C. Chiappini, eds.

© International Astronomical Union 2010
doi:10.1017/S1743921310004369

A very low upper limit for a Be abundance of a carbon-enhanced metal-poor star

Hiroko Ito[1,2], Wako Aoki[1,2], Satoshi Honda[3],
Timothy C. Beers[4], and Nozomu Tominaga[5]

[1]Department of Astronomical Science, School of Physical Sciences, The Graduate University for Advanced Studies (SOKENDAI), 2-21-1, Osawa, Mitaka, Tokyo, 181-8588, Japan
email: hiroko.ito@nao.ac.jp
[2]National Astronomical Observatory of Japan, Mitaka, Tokyo, Japan
[3]Gumma Astronomical Observatory, Agatsuma, Gunma, Japan
[4]Michigan State University, East Lansing, MI 48824-1116, USA
[5]Konan University, Kobe, Hyogo, Japan

Abstract. We performed a 1D LTE chemical abundance analysis of an extremely metal-poor star BD+44°493 ([Fe/H] = −3.7), and set a very low upper limit for its Be abundance: $A(\mathrm{Be}) <$ −2.0. It may indicate that the decreasing trend of Be abundances with lower [Fe/H] still holds at [Fe/H] < −3.5, and demonstrate that high C and O abundances do not necessarily imply high Be abundances. However, since the star is a subgiant with $T_{\mathrm{eff}} \sim 5500$K, Be may be depleted.

Keywords. stars: abundances, individual (BD+44°493), Population II

1. Observation and analysis

High-resolution spectroscopy of BD+44°493 was carried out with Subaru/HDS. The atmospheric parameters that we adopt are $T_{\mathrm{eff}} = 5510$ K, and $\log g = 3.7$. Our 1D LTE abundance analysis derives [Fe/H] = −3.7, [C/Fe] = +1.3, and [O/Fe] = +1.6, indicating that this star is a carbon-enhanced metal-poor (CEMP) star. Its abundance pattern implies that a first-generation "faint" supernova (e.g., Tominaga *et al.* 2007) is the most likely origin of its carbon excess. See Ito *et al.* (2009) for detail.

2. Implications of its low beryllium abundance

We set a very low upper limit for its Be abundance ($A(\mathrm{Be}) <$ −2.0). This is the Be abundance reported at the lowest metallicity yet achieved, and is the lowest Be limit so far for metal-poor dwarfs or subgiants that have normal Li abundances. The result indicates that the decreasing trend of Be abundances with lower [Fe/H], which was revealed by previous studies (e.g., Boesgaard *et al.* 1999), still holds at [Fe/H] < −3.5 (Fig. 1).

Our analysis is the first attempt to measure a Be abundance for a CEMP star. Since Be is produced via the spallation of CNO nuclei, their abundances, especially O abundances, have been expected to correlate with Be abundances. However, our low Be upper limit shows that the high C and O abundances in BD+44°493 are irrelevant to its Be abundance (Fig. 1), which offers a new insight into the origin of CEMP stars.

3. Possibility of depletion

Previous studies of Be abundances in metal-poor subgiants indicate that Be is depleted in those with $T_{\mathrm{eff}} < 5500$K, so BD+44°493 is at the boundary (Fig. 2). In Ito *et al.* (2009), we adopt $T_{\mathrm{eff}} = 5510$ K determined by Carney *et al.* (2003), and assumed that Be in the

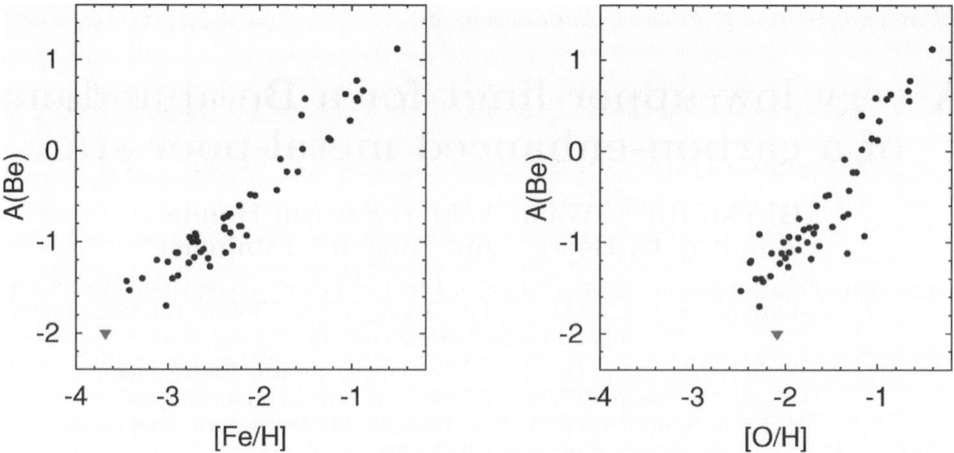

Figure 1. A(Be) vs. [Fe/H] and A(Be) vs. [O/H]. Our upper limit for BD+44°493 is shown by the red triangle. The filled circles indicate results of Rich & Boesgaard (2009).

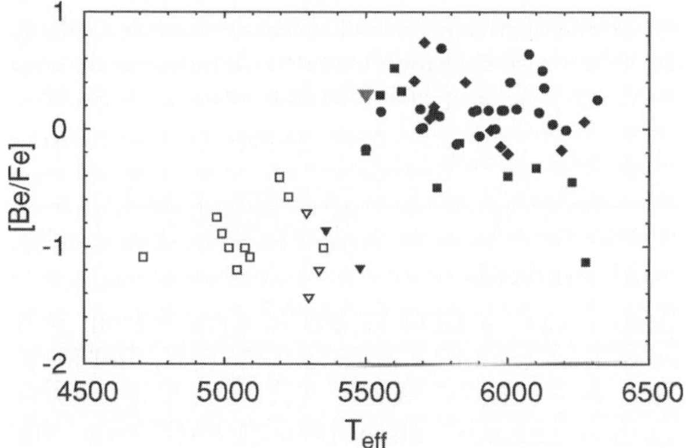

Figure 2. [Be/Fe] as a function of T_{eff}. Our upper limit for BD+44°493 is shown by the red (bigger) triangle above the centre of the figure. Filled circles and diamonds indicate results of Rich & Boesgaard (2009) and Tan *et al.* (2009), respectively. Filled squares and (smaller) triangles those of Smiljanic *et al.* (2009), and the open ones those of García Pérez & Primas (2006). All triangles represent upper limits. Only subgiants ($\log g < 4.0$) are plotted.

star is not depleted. However, Carney *et al.* (2003) seems to overestimate the reddening, and our re-estimate lowers its temperature by about 100K (Ito et al. in prep.), increasing the possibility of Be depletion. We cannot conclude whether its Be is depleted, but if it is, our interpretation of its low Be abundance needs to be revised.

References

Boesgaard, A. M., Deliyannis, C. P., King, J. R., Ryan, S. G. *et al.* 1999, *AJ*, 117, 1549
Carney, B. W., Latham, D. W., Stefanik, R. P., Laird, J. B., & Morse, J. A. 2003, *AJ*, 125, 293
García Pérez, A. E. & Primas, F. 2006, *A&A*, 447, 299
Ito, H., Aoki, W., Honda, S., & Beers, T. C. 2009, *ApJ* (Letters), 698, L37
Rich, J. A. & Boesgaard, A. M. 2009, *ApJ*, 701, 1519
Smiljanic, R., Pasquini, L., Bonifacio, P., Galli, D. *et al.* 2009, *A&A*, 499, 103
Tan, K. F., Shi, J. R., & Zhao, G. 2009, *MNRAS*, 392, 205
Tominaga, N., Maeda, K., Umeda, H., Nomoto, K. *et al.* 2007, *ApJ* (Letters), 657, L77

Light Elements in the Universe
Proceedings IAU Symposium No. 268, 2009
C. Charbonnel, M. Tosi, F. Primas & C. Chiappini, eds.
© International Astronomical Union 2010
doi:10.1017/S1743921310004370

Lithium abundances in the α Persei Cluster

Sushma V. Mallik[1], Suchitra C. Balachandran[2], and David L. Lambert[3]

[1] Indian Institute of Astrophysics, Bangalore 560034, India
email: sgvmlk@iiap.res.in

[2] Astronomy Department, University of Maryland, College Park, MD 20742-2421, USA
email: suchitra@astro.umd.edu

[3] W.J. McDonald Observatory, The University of Texas, Austin, TX 78712-0259, USA
email: dll@astro.as.utexas.edu

Abstract. As a sequel to the Li observations by Balachandran, Lambert & Stauffer (1988, 1996) in 35 stars of the 50 Myr old cluster α Persei, we have obtained and analyzed high resolution spectra of another 51 stars. Following a reconsideration of the cluster membership of the stars (Prosser 1992, Makarov 2006, Mermilliod *et al.* 2008, and Patience *et al.* 2002), we discuss the Li abundances for 70 stars. With our larger sample, we reexamine the question of whether the scatter in Li abundance at a given T_{eff} seen in young clusters at cool temperatures is real or not.

Keywords. Galaxy: open clusters – stars: abundances

1. Observations, abundances and their interpretation

High S/N echelle spectra were obtained for 30 stars at the 2.7m telescope at the McDonald Observatory at $R = 60,000$ and for 21 stars at the 4m telescope at KPNO at $R = 40,000$. The Li abundance as determined from the 6707.8 Å feature is singularly sensitive to the adopted T_{eff}. For an error of ± 200 K in T_{eff}, the uncertainty in log N(Li) varies from 0.28 to 0.14 over the temperature range from 4500 K to 6500 K. We choose (V$-$K) colour index as our principal indicator of T_{eff}. Our fresh estimates of reddening for stars with available Stromgren photometry yield E(b$-$y) in the range of 0.02 to 0.12 with an average of 0.075 ($\pm$ 0.05). We adopt this average that translates to E(V$-$K) = 0.284 and use the (V$-$K) - T_{eff} calibration of Alonso *et al.* (1986) to derive $T_{\text{eff}}(V - K)$ for all the stars. A mean Fe abundance of 7.40 $\pm$ 0.08 dex was obtained for the 26 slowly rotating stars from a fine spectroscopic analysis; the standard error comparable to the estimate of the precision of a single determination. As is apparent from Fig.1, Li abundance in the hottest stars approaches a constant value (taken as the initial value for young open clusters), close to the meteoritic value of 3.25 $\pm$ 0.06, as also to the abundance derived of T Tauri stars suggesting it has changed little in the last 4.5 Gyrs.

\# 1589, 1604, 56 and 93 (> 5300 K) and \# 1612, 1735 (< 5300 K) are outliers that do not define the mean relation. This could possibly be either because their assigned T_{eff} is too low or they are non-members that have experienced normal Li depletion for their age or an unusual amount of Li depletion has occurred. At $T_{\text{eff}} < 5300$ K, Li vs. T_{eff} relation plunges steeply and develops a scatter in Li abundance; it widens into a band with a lower envelope defined by low *vsini* stars from Li of 2.1 at 5000 K to -0.4 at 4500 K and an upper envelope defined by high *vsini* stars from Li of 2.5 at 4700 K to 0.6 at 4300 K; width of the band at $T_{\text{eff}} < 4700$ K is ~ 1.5 dex, as is observed in the Pleiades.

Pre-Main Sequence (PMS) depletion of Li is ineffective for the hotter stars of α Per. In cooler stars it is best mappable by looking at the youngest of clusters, for, stars in α Per have experienced PMS depletion that is essentially complete. A Li abundance

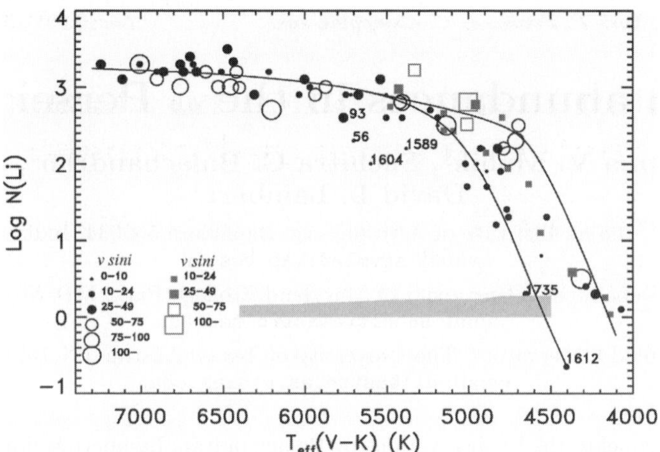

Figure 1. T_{eff} vs Li abundance for α Per with data (red squares) added from Randich *et al.* (1998). *vsini* is represented according to the legend of the figure. The shaded strip shows the Li spread resulting from an uncertainty of ±100K in T_{eff}. The labelled stars are possible outliers. Suggested upper and lower envelopes to the relation are indicated.

study of 5 associations with age ranging from 10 to 100 Myr (Mentuch *et al.* 2008) suggests increasing Li depletion with age as predicted for the PMS phase. The peak-to-peak scatter increases from ±0.3 (for β Pic) to $\pm0.5 - \pm0.8$ (for Tucanae-Horolagum and the AB Doradus), less than that seen for α Per. Clusters younger than α Per - IC 2602, IC 2391, NGC 4665, NGC 2547 also have smaller star-to-star scatter in Li. The older cluster, the Pleiades (70 Myr) has a similar scatter as α Per and in fact M 34 (250 Myr) has as much. These observations imply that that spread in Li among cooler stars is not well developed in the youngest clusters/associations and it may take 20 Myr or so to develop fully. Li depletion seems to be dominated by that occurring in the PMS phase for clusters up to the ages of 50-100 Myr. For older clusters, it is the Main Sequence depletion that begins to reduce Li in the coolest stars. Very apparent, *e.g.*, in M 34 where the star-to-star scatter remains similar to the Pleiades and α Per but the mean abundances are noticeably reduced.

The large scatter observed at cooler temperatures is far larger than could be explained by the standard sources of uncertainty. Several studies in the past report a strong correlation between the Li scatter and stellar activity. We have to have contemporaneous indicators of stellar activity for pairs of stars with maximum and minimum Li abundance but similar observed properties such as colour and rotation period. In the absence of these, the debate continues on whether the star-to-star spread is due to real differences in Li abundances or arises due to atmospheric effects.

References

Alonso, A., Arribas, S., & Martínez-Roger, C. 1996, *A&A*, 313, 873
Balachandran, S., Lambert, D. L., & Stauffer, J. R. 1988, *ApJ*, 333, 267
Balachandran, S., Lambert, D. L., & Stauffer, J. R. 1996, *ApJ*, 470, 1243
Makarov, V. V. 2006, *AJ*, 131, 2967
Mentuch, E. *et al.* 2008, *ApJ*, 689, 1127
Mermilliod, J.-C., Queloz, D., & Mayor, M. 2008, *A&A*, 488, 409
Patience, J., Ghez, A. M., Reid, I. N., & Matthews, K. 2002, *AJ*, 123, 1570
Prosser, C. F. 1992, *AJ*, 103, 488
Randich, S., Martín, E. L., García López, R. J., & Pallavicini, R. 1998, *A&A*, 333, 591

Light Elements in the Universe
Proceedings IAU Symposium No. 268, 2009
C. Charbonnel, M. Tosi, F. Primas & C. Chiappini, eds.
© International Astronomical Union 2010
doi:10.1017/S1743921310004382

Lithium in other Suns: no connection between stars and planets

Jorge Meléndez[1], Iván Ramírez[2], Martin Asplund[2], and Patrick Baumann[2]

[1] Centro de Astrofísica, Universidade do Porto, Rua das Estrelas, 4150-762 Porto, Portugal
email: `jorge@astro.up.pt`

[2] Max-Planck-Institut für Astrophysik, Germany

Abstract. An unbiased sample of solar twins shows that the Sun has a normal Li abundance for its age and that a low Li abundance does not imply the presence of planets. We find a tight correlation between Li and age, which holds for all stars analyzed in our sample: solar twins, stars with and without detected giant planets, and stars that may host terrestrial planets.

Keywords. Sun: abundances – stars: abundances, planetary systems

1. Lithium vs. age for solar twins in the field and clusters

The sample consists of more than one hundred stars similar to the Sun observed at McDonald (2.7m) and Las Campanas (6.5m Magellan Clay telescope). Most of them are from our solar twin survey (Meléndez *et al.* 2009a,b; Ramírez *et al.* 2009). We have also included ~ 20 solar analogs with and without detected giant planets.

Our high quality spectra (R = 60,000; S/N = 200-450) allows the determination of Li abundances in stars as Li-poor as the Sun. Due to the similarity between the Sun and the twins, accurate stellar parameters can be determined, thus reliable ages can be obtained from isochrones. For young stars ($<$1Gyr) whenever possible we use precise rotation periods to estimate ages. Both methods agree, but the rotational ages are more precise. In Fig. 1 we show only solar twins (open circles) with good ages (3-σ or better).

There is a clear correlation between Li and Age in our sample of solar twins (open circles). The one-solar-mass stars in solar-metallicity open clusters (filled triangles) also follow the same correlation. Note that we have only used clusters with reliable data, which are very few for old solar-metallicity open clusters. For example, the cluster Collinder 261 seems old but its age, reddening and metallicity are very uncertain (Spanò *et al.* 2005). Furthermore, only relatively high upper limits in the Li abundances of solar twins in Cr 261 are available (Spanò *et al.* 2005). NGC 188 is another potentially interesting old solar metallicity open cluster, but there is only one star that may be as cool as the Sun in the study of Randich *et al.* (2003, Fig. 2). Furthermore, according to Sestito & Randich (2005), the NGC 188 stars analyzed by Randich *et al.* (2003) have a S/N of only 20-35, which would be too low to detect the Li feature in a star like the Sun.

2. No planet connection

Our sample of solar twins without detected giant planets follows the same Li vs. Age as other solar twins. We have also studied 18 solar analogs included in radial velocity planet surveys; six of them with detected giant planets and 12 of them without detected giant planets. Both samples seem to follow the Li vs. Age correlation shown in Fig. 1.

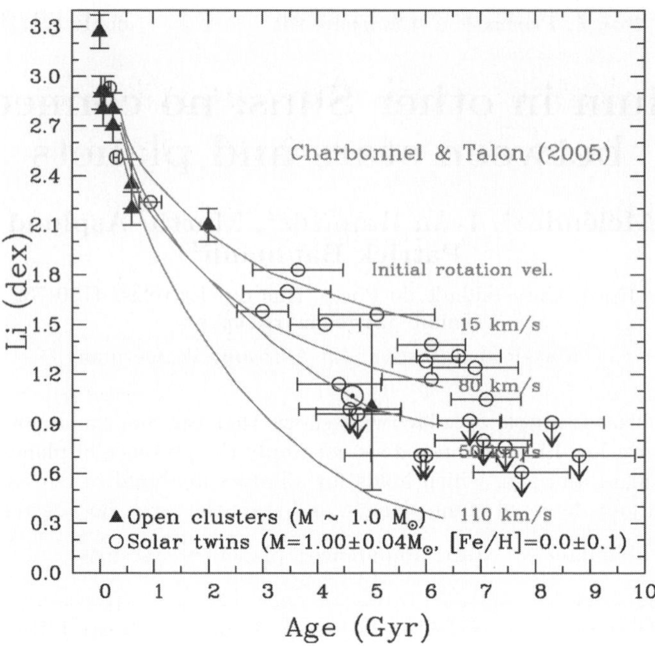

Figure 1. Li for our solar twins with one solar mass (±4%) and solar [Fe/H] (± 0.1 dex), and for one-solar- mass stars in solar metallicity (±0.15 dex) open clusters selected from Sestito & Randich (2005), although for M67 we used the sample of Pasquini *et al.* (2008). Field stars are shown as circles while open clusters with triangles. Figure updated from Meléndez *et al.* 2009b.

We have recently found a signature that may indicate the presence of terrestrial planets around other stars (Meléndez *et al.* 2009a; Ramírez *et al.* 2009). Solar twins showing the same terrestrial planet signature as our Sun also follow the same Li vs. Age relation as solar twins without the signature of other earths.

In conclusion, we find no relation between low levels of Li and the presence of planets. The low Li observed in the Sun is normal for a star of its age, mass, and metallicity.

As already suggested by other authors (see Meléndez *et al.* 2009b for a review), Israelian *et al.* (2009) have recently claimed that planet host stars in a narrow range around solar T$_{\rm eff}$, have enhanced depletion of lithium due to the presence of planets. Our work suggests as alternative explanation that the low Li in planet host stars around the solar T$_{\rm eff}$ may be due an age effect instead of the presence of planets.

References

Charbonnel, C. & Talon, S. 2005, *Science*, 309, 2189

Israelian, G. *et al.* 2009, *Nature*, 462, 189

Meléndez, J., Asplund, M., Gustafsson, B., & Yong, D. 2009a, *ApJ* (Letters), 704, L66

Meléndez, J. *et al.* 2009b, *Ap&SS*, 221, in press

Pasquini, L., Biazzo, K., Bonifacio, P., Randich, S., & Bedin, L. R. 2008, *A&A*, 489, 677

Ramírez, I., Meléndez, J., & Asplund, M. 2009, *A&A* (Letters), 508, L17

Randich, S., Sestito, P., & Pallavicini, R. 2003, *A&A*, 399, 133

Sestito, P. & Randich, S. 2005, *A&A*, 442, 615

Spanò, P., Pallavicini, R., & Randich, S. 2005, *IAU Symp.*, 228, 111

Light Elements in the Universe
Proceedings IAU Symposium No. 268, 2009
C. Charbonnel, M. Tosi, F. Primas & C. Chiappini, eds.
© International Astronomical Union 2010
doi:10.1017/S1743921310004394

Li abundances and chromospheric activity of BY Dra type stars

Tamara V. Mishenina[1,2] Caroline Soubiran[1], Valery V. Kovtyukh[2], and Stanislav I. Belik[2]

[1] Université de Bordeaux - CNRS - Laboratoire d'Astrophysique de Bordeaux,
BP 89, 33271 Floirac Cedex, France,
email: `caroline.soubiran@obs.u-bordeaux1.fr`

[2] Astronomical observatory, Odessa National University,
T.G.Shevchenko Park, Odessa 65014 Ukraine
email: `tamar@deneb1.odessa.ua`

Abstract. Atmospheric parameters and Li abundances have been determined for 162 stars observed at high resolution, high signal to noise ratio with the ELODIE echelle spectrograph (OHP, France). Among them, about 70 stars are active stars with a large fraction of BY Dra type stars. For all stars, rotational velocities were obtained with a calibration of the cross-correlation function, effective temperatures by the line depth ratio method, surface gravities by the parallaxe method and by the ionization balance of iron. The frequency of stars with observed lithium is significantly higher in active stars than in non active stars. Among active stars, no clear correlation has been found between different indicators of activity for our sample stars, but some correlation of an index R'_{HK} and $v\sin i$ is observed.

Keywords. Stars: abundances, fundamental parameters, activity

---

1. Introduction

Lithium abundances are of a particular interest, as reflecting various processes in stars and at the stellar surface and as a possible indicator of chromospheric activity. It is important also because the Sun is suspected to belong to the BY Dra type.

2. Observations, parameters, and Li determinations

The spectra of 162 stars were obtained using the 1.93 m telescope at Observatoire de Haute-Provence (OHP, France) equipped with the echelle spectrograph ELODIE (Baranne *et al.* 1996) which gives a resolving power of R = 42 000. The spectral processing was perfomed according to Katz *et al.* (1998). Rotational velocities $v\sin i$ were measured with a relation calibrated by Queloz *et al.* (1998).

The determination of $T_{\rm eff}$, $\log g$, [Fe/H] was performed following Mishenina *et al.* (2004, 2008) and Kovtyukh *et al.* (2004). Li abundances log A(Li) were determined by LTE spectral synthesis code STARSP (Tsymbal 1996). The list of lines in the region of Li I line 6707 Å was taken from Mishenina & Tsymbal (1997).

3. Results and discussion

Stars of BY Dra type are young stars. In their spectra the lines of lithium show different intensities not always correlating with other indicators of stellar activity. Sometimes Li lines are absent. The dependence of log A(Li) on $T_{\rm eff}$, is presented in Fig. 1 (the top value of an estimation of log A(Li) are marked by symbols with arrows).

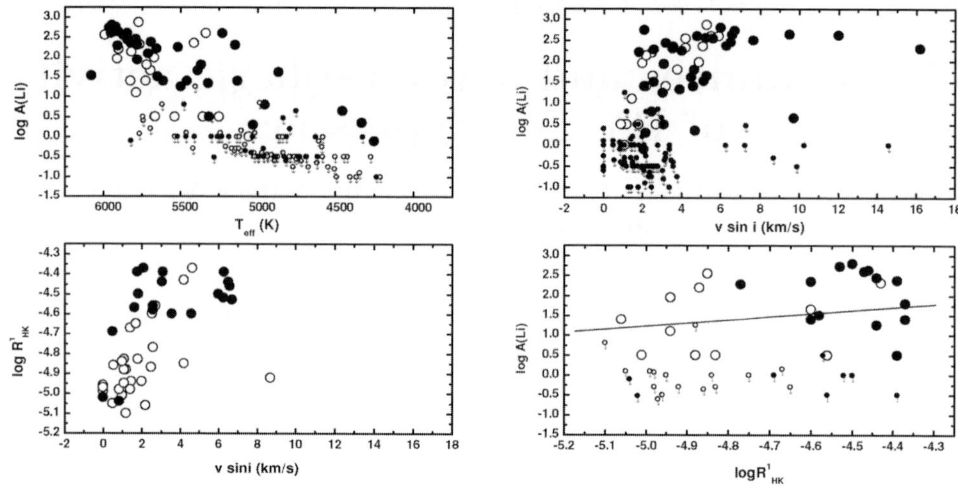

Figure 1. Connection between log A(Li), T_{eff}, $v\sin i$, and R'_{HK}.

We have detected the lithium in 19 stars among the 91 non active dwarfs and 39 stars among the 73 stars with chromospheric activity, correspondonding to about 20% and 54%, respectively. We confirm our earlier result (Mishenina *et al.* 2008) with almost doubled quantity of active stars. The beheavior of log A(Li) with $v\sin i$ is similar for active and non-active stars (see Fig. 1). To search for correlations between the lithium abundance and chromospheric activity we have used as an indicator of chromospheric activity the index R'_{HK} (Wright *et al.* 2004) for stars in a range of colours $0.4 < B - V < 0.9$. We have checked R'_{HK} versus rotation and lithium abundance versus R'_{HK} (see Fig. 1). We observe some correlation of R'_{HK} and $v\sin i$, but the correlation between log A(Li) and R'_{HK} is not so obvious.

4. Conclusions

I. The frequency of stars with observed lithium is significantly higher in active stars than in non active stars.

II. Active stars exhibit no clear correlation between log A(Li) and $v\sin i$ and the index R'_{HK}, but we observe some correlation of the index R'_{HK} and $v\sin i$.

References

Baranne, A., Queloz, D., Mayor, M. *et al.* 1996, *A&AS*, 119, 373

Katz, D., Soubiran, C., Cayrel, R. *et al.* 1998, *A&A*, 338, 151

Kovtyukh, V. V., Soubiran, C., & Belik, S. I. 2004, *A&A*, 427, 923

Mishenina, T. V. & Tsymbal, V. V. 1997, *Pis'ma v AZh*, 23, 693

Mishenina, T. V., Soubiran, C., Kovtyukh, V. V., & Korotin, S. A. 2004, *A&A*, 418, 551

Mishenina, T. V., Soubiran, C., Bienayme, O., Kovtyukh, V. V., & Korotin, S. A. 2008, *A&A*, 489, 923

Queloz, D., Allain, S., Mermilliod, J.-C. *et al.* 1998, *A&A*, 335, 183

Tsymbal, V. V. 1996, *ASP Conf. Ser.*, 108, 198

Wright, J. T., Marcy, G. W., Buter, R. P., & Vogt, S. S. 2004, *ApJS*, 152, 261

Light Elements in the Universe
Proceedings IAU Symposium No. 268, 2009
C. Charbonnel, M. Tosi, F. Primas & C. Chiappini, eds.

© International Astronomical Union 2010
doi:10.1017/S1743921310004400

Lithium abundances in dwarfs of intermediate age open clusters

Giancarlo Pace[1] and Jorge Mélendez[1]

[1]Centro de Astrofisica, Universidade do Porto,
Rua das Estrelas, 4150-762, Porto, Portugal
email: gpace@astro.up.pt

Abstract. Lithium abundance measurements in dwarf stars in open clusters are of crucial importance for our understanding of the mixing mechanism and have allowed us to achieve important conclusions on the matter. However, in order to further our understanding of what drives lithium depletion, lithium abundance measurements have to be coupled with accurate temperature determinations, which are best achieved when the analysis of iron lines is employed. Effective temperature estimations from photometry, on the contrary, can be affected by errors as large as several hundred kelvins due to uncertain open cluster reddening, especially when studying old open clusters, which tend to be more distant. We present lithium abundance in 12 dwarfs belonging to 4 open clusters at about 1 or 2 Gyr. The stellar effective temperatures, along with the other parameters, were estimated from the analysis of about 60 Fe I lines and 10 Fe II. Even though the few datapoints call for caution, we notice that stars in the open cluster IC 4651 seem to present a steep decline with temperature below 6000 K.

Keywords. Galaxy: open clusters – stars: low-mass, abundances

1. Data sample and analysis

The sample consists of 12 dwarf stars in the following 4 open clusters whose ages, as found by Salaris *et al.* (2004), are indicated in the legend of Figure 1. The spectra were taken with UVES @ VLT, they have a resolution of about R$\approx$ 100 000 and S/N of about 100 in the range between 4800 and 6800 Å.

Stellar parameters, namely temperature, gravity, metallicity, and microturbulence, were obtained (Pace *et al.* 2008, 2010) with the classical equivalent width analysis of the iron lines (about 60 for Fe I and about 10 for Fe II). Apart of the precision achievable, especially when, as in our case, spectra with high resolution and good S/N are available, this method has the great advantage of being independent of the cluster reddening.

We find differences of several hundred Kelvin between the temperatures as evaluated from the spectroscopic analysis and the temperatures evaluated by means of photometry and published calibration. Most of this difference is probably due to the error in the reddening of the cluster and in the zero point errors in the calibration of the photometry.

Lithium abundances were measured by comparing the observed spectrum of the lithium doublet at 6707.8 Å with the synthetic one, and changing the assumed abundance until a match between the two was obtained. The typical errors are between 0.05 and 0.15 dex. Using the same procedure, we also compared the synthetic spectrum of the Sun with the UVES-archive solar spectrum, and we obtained the best match by assuming A(Li) = 1 rather than the canonical A(Li) = 1.1.

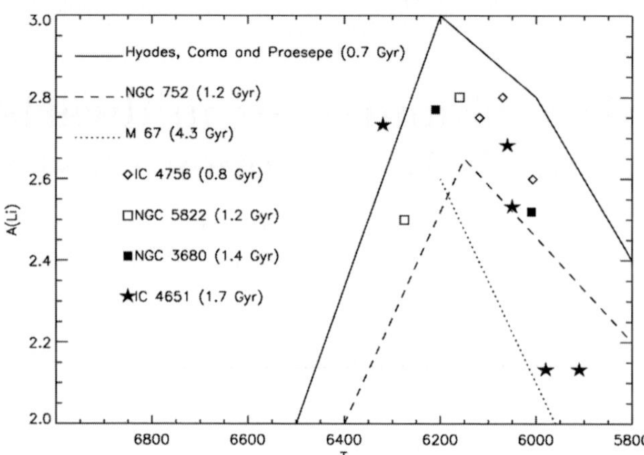

Figure 1. Our data on dwarfs in intermediate age open clusters compared with literature data about young open clusters and NGC 752 (also about 1 Gyr old) and the much older M 67.

2. Results and conclusions

In the Figure we show a temperature versus lithium abundance diagram, in which we compare the 12 data points from the present analysis with published open cluster data at 3 different ages. Hyades, Coma, and Praesepe data are depicted in one single curve, another represents NGC 752 and the third M 67. M 67 data are taken from Pasquini *et al.* (2008), the remainder from the compilation in Xiong & Deng (2009). The curves that represent the published cluster data at the 3 different ages, are fits by eye to the datapoints. The temperatures for the Hyades age clusters and NGC 752 are derived from photometry. For the former the uncertainty in the colour excess should not play a major role, since they are nearby clusters. The curve representing the clusters at the Hyades age form an upper envelope to the distribution of the other datapoints. The curve representing M 67 and, for temperatures higher than 6000 K, that of NGC 752, form, instead, a possible lower envelope. Only one datapoint falls slightly outside the region defined by the two envelopes. Even though the few datapoints call for caution, we notice that IC 4651 seems to present a steep decline below 6000K.

When studying the dependence of lithium abundances as a function of temperatures in stars in open clusters , the use of temperature measurements by spectroscopic analysis is essential to avoid errors due to uncertain reddening.

References

Pace, G., Danziger, J., Carraro, G., Melendez, J., François, P., Matteucci, F., & Santos, N. C. 2010 *A&A*, accepted

Pace, G., Pasquini, L., & François, P. 2008, *A&A* (Letters), 489, 403

Pasquini, L., Biazzo, K., Bonifacio, P., Randich, S., & Bedin, L. R. 2008, *A&A*, 489, 677

Salaris, M., Weiss, A., & Percival, S. M. 2004, *A&A*, 414, 163

Xiong, D. R. & Deng, L. 2009 *MNRAS*, 395, 2013

Light Elements in the Universe
Proceedings IAU Symposium No. 268, 2009
C. Charbonnel, M. Tosi, F. Primas & C. Chiappini, eds.
© International Astronomical Union 2010
doi:10.1017/S1743921310004412

HD 232 862 : a magnetic and lithium-rich giant star

A. Palacios[1], A. Lèbre[1], J. D. do Nascimento Jr[2], R. Konstantinova-Antova[3], D. Kolev[3], M. Aurière[4], P. de Laverny[5] and J. R. de Medeiros[2]

[1] GRAAL-CNRS/Université Montpellier II, France
email: palacios@graal.univ-montp2.fr

[2] DFTE - Universidade Federal do Rio Grande do Norte, Natal, Brazil

[3] Institute of Astronomy, Bulgarian Academy of Sciences, Sofia, Bulgaria

[4] LATT - Observatoire Midi-Pyrénées, Toulouse, France

[5] Cassiopée, Observatoire de la Côte d'Azur, Nice, France

Abstract. Using spectropolarimetric data acquired with the ESPaDOnS and NARVAL instruments at CFHT and at TBL, we present a detailed spectral synthesis analysis of HD 232 862, a field giant classified as a G8II star hosting a magnetic field. This star is the first lithium-rich field giant hosting a magnetic field. Stellar evolution models suggest that HD 232 862 should be a 1.5 to 2.0 $M_\odot$ star at the bottom of the red giant branch. Its unusually high lithium content (A(Li) = 2.45 ± 0.25 dex) is even more puzzling and challenges our understanding of the evolution of this star.

Keywords. Stars: spectroscopy, spectropolarimetry, abundances, evolution, magnetic fields

1. Observations and main features

HD 232 862 is classified as a G8II star in the SIMBAD database. This corresponds to a bright giant star in the mass range [2.5 $M_\odot$; 9 $M_\odot$]. With no Hipparcos parallax, a better determination of its mass and evolutionary status is not available in the literature. HD 232 862 presents several intriguing features : **(a)** rotational velocity v sini = 20.6 km/s, ten times larger than the mean value for luminosity class II objects (de Medeiros & Mayor 1999); **(b)** is an X-ray source in the ROSAT database and presents coronal and chromospheric actiity (IUE spectra); **(c)** is a visually tight binary (Couteau 1988).

The spectropolarimetric data acquired with ESPaDOnS (CFHT, Hawaii) in circular polarization mode between the 7 and 10 Dec. 2006, allowed us to characterize the Stokes V and I parameters. We could detect a complex and time-variable Stokes V profile, pointing to the existence of a magnetic field at the surface of this moderate rotating giant. On the other hand, thanks to the subarcsec conditions reached, we were able to separate the components of the binary and obtain the first high resolution and high S/N spectra for the main component alone. MARCS models atmospheres (Gustafsson *et al.* 2008) were used to derive the following fundamental parameters and Li abundance : $T_{eff} =$ 5000 $\pm$ 250 K, log g = 3.0 $\pm$ 0.5, [Fe/H] = –0.3 $\pm$0.1 dex, A(Li) = 2.45 $\pm$ 0.2 dex.

2. Evolutionary status

Lacking from any parallax, we use the gravity and temperature derived from the spectral synthesis to estimate the mass and evolutionary status of HD 232 862. To do so we use a grid of standard stellar evolution models computed with the STAREVOL code,

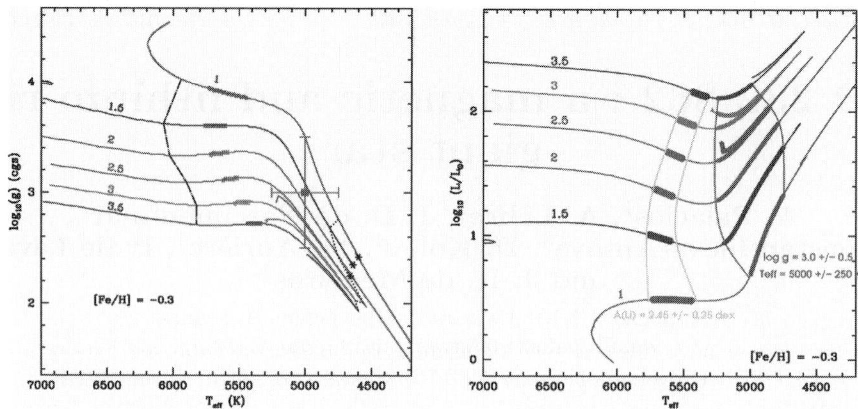

Figure 1. Evolutionary status of HD 232 862. *Left:* Dashed (left) and dotted (right) lines on the figure indicate the beginning and the end of the first dredge-up resp. The asterisks indicate the position of the bump. The bold parts on the tracks indicate the location where A(Li) $\in$ [2.7; 2.2] dex.

with masses ranging from 1 M$_\odot$ to 3.5 M$_\odot$. No diffusion, rotation nor magnetic fields are included in these models. We adopt [Fe/H] $= -0.3$, A(Li)$_{\text{init}} = 2.976$ dex and Grevesse & Sauval (1998) for the solar chemical composition. Figure 1 shows the (T$_{eff}$, log g) diagram. HD 232 862 appears to be undergoing the first dredge-up. During this phase, the deepening convective envelope reaches regions where lithium is depleted by nuclear processes, and in the temperature and gravity range estimated for HD 232 862, the surface abundance of this nuclide is predicted to have dropped by a factor of 60 to 100 (depending on the stellar mass), in contradiction with the determined abundance of A(Li) $=$ 2.45 $\pm$ 0.25 dex. Marking the regions in the theoretical HR diagram, where the surface Li abundance and (T$_{eff}$, log g) vary within the derived errorbars (Fig 1 right), we can see that there is no overlap of these regions : standard stellar evolution models cannot account for the high Li abundance found in HD 232862.

HD 232 862 appears to be a giant, more likely of luminosity class III, at the bottom of the red giant branch. Its surface gravity and effective temperature indicate a mass between 1 and 3.5 M$_\odot$. The high lithium abundance found at its surface is incompatible with the predictions of standard stellar evolution. It is even more intriguing since HD 232 862 should be at the end of the first dredge-up, a stage where the surface Li drops dramatically. As a comparison, the other known Li-rich giants are either at the beginning of the first dregde-up, or at the bump, an evolutionary point well beyond that of HD 232 862. HD 232 862 is also the first Li-rich giant with a detected surface magnetic field. This, together with its binary status, could be important to explain the unusual Li abundance of this star.

References

Grevesse, N. & Sauval, A. J. 1998, *Space Science Review*, 85, 161

de Medeiros, J. R. & Mayor, M. 1999, *A&A*, 139, 443

Couteau, P. 1988, *A&AS*, 75, 163

Gustafsson, B., Edvardsson, B., Eriksson, K. Jorgensen, U. G., Nordlund, A., & Plez, B. 2008, *A&A*, 486, 951

Light Elements in the Universe
Proceedings IAU Symposium No. 268, 2009
C. Charbonnel, M. Tosi, F. Primas & C. Chiappini, eds.

© International Astronomical Union 2010
doi:10.1017/S1743921310004424

Beryllium abundances in metal-rich stars

Ruth C. Peterson[1]

[1]UCO/Lick Observatories and Astrophysical Advances
email: peterson@ucolick.org

Abstract. In metal-rich stars as cool as the Sun, beryllium abundance determinations are difficult due to heavy line blanketing in the near-UV 3130 Å region where the accessible Be II lines reside. We can now attempt such determinations based on improved lists of atomic line identifications and gf-values in the near-UV. Here we report Be determinations for three metal-rich A, F, and G stars plus three solar-metallicity standards. All six stars have beryllium-to-hydrogen ratios at or below solar. More such determinations would provide stronger constraints on trends in Be abundance with temperature, metallicity, and age.

Keywords. Stars: abundances, individual (HD 61421, HD 72660, HD 165341, HD 179949, HD 217107)

Previous analyses of stellar beryllium abundances have shed light on both the production of beryllium in the interstellar medium, and its destruction in non-standard mechanisms within and among stars. However, except for members of the young Hyades open cluster (age 600 Myr), existing beryllium abundance determinations are for stars of solar metallicity and lower. Consequently little is known about beryllium production and depletion among old metal-rich stars.

The problem has been the extreme crowding by the rich assortment of spectral lines in the ultraviolet spectra of stars of type F and later. As reported by Peterson *et al.* (2004) we have empirically improved the lists of optical and near-UV atomic line parameters to the point where near-UV abundance analyses of such stars are tractable.

We first extracted from the ESO UVES and Keck HIRES archives the spectra for two slowly-rotating metal-rich stars from Takeda *et al.* (2005), HD 179949 (F8.5V) and HD 217107 (G8IV), and for the solar-metallicity standard HD 165341 (K0V; 70 Oph A). We analyzed these together with ground-based spectra of the Sun (G2V) and with space-based HST STIS E230H spectra of two Peterson *et al.* (2004) Hubble Treasury standards, HD 61421 (F5IV-V; Procyon) and the metal-rich HD 72660 (A1IV).

Beryllium abundances were determined by comparing the observed echelle spectra to calculations run with the Kurucz program SYNTHE and our updated atomic line list, using Castelli & Kurucz (2003) ODFNEW models. First, the stellar temperature T_{eff}, gravity log g, and metallicity [Fe/H] were found from the optical region of the spectra, except for HD 72660 where the near-UV continuum is seen. Our results confirm Takeda *et al.*'s parameters for the two cool metal-rich stars. We then matched the entire region near the Be II doublet with a small grid of calculations, to ensure that continuum placement and spectral resolution were determined as reliably as possible. Fig. 1 shows the results of these comparisons.

All stars are found to have a Be abundance of solar or less, including the two cool metal-rich stars. We are unaware of any previous Be determinations in metal-rich stars, nor in A stars. Before concluding that stellar beryllium never exceeds the solar value, more stars need to be analyzed, to understand what role may be played by the destruction of Be in cool or hot metal-rich stars.

This work was supported in part by NASA HST grants GO-9455 and AR-10970.

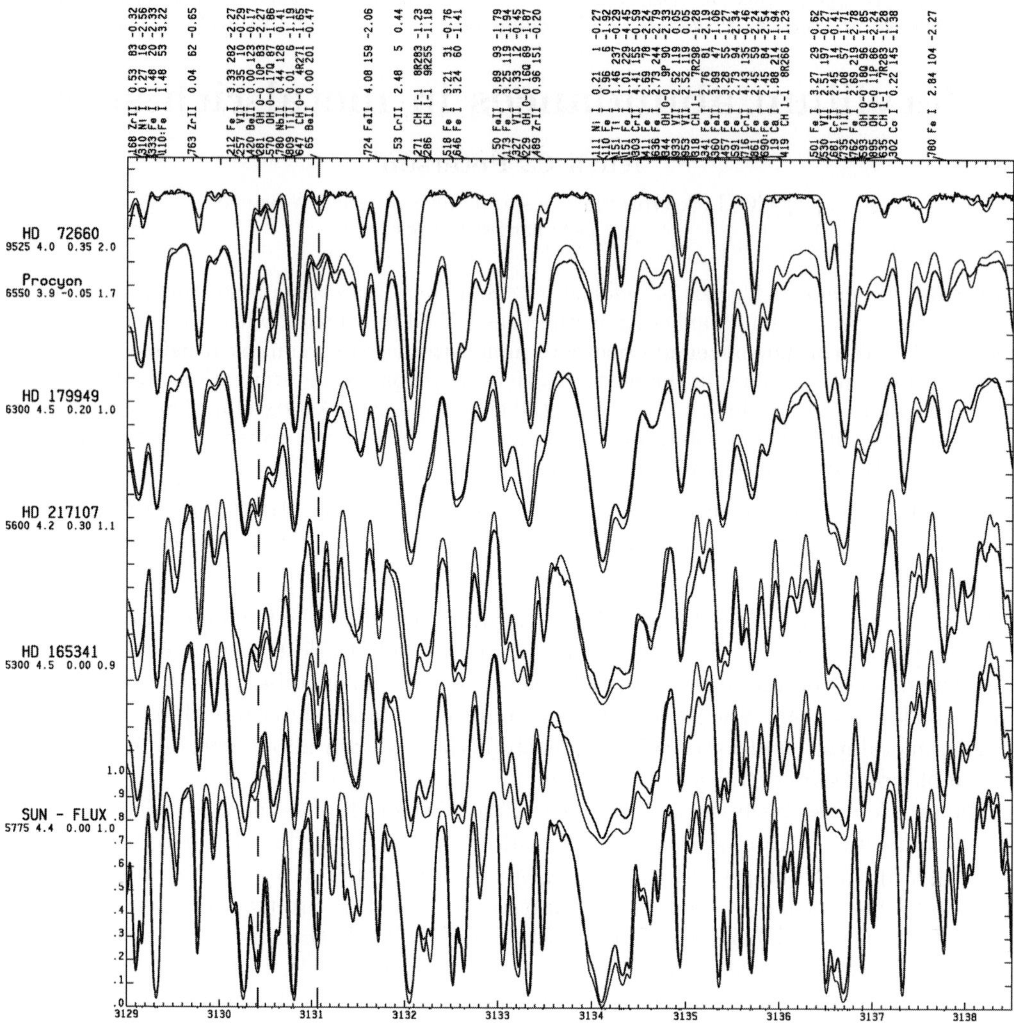

Figure 1. Comparisons between observed spectra (dark lines) and calculated spectra (light lines) are shown in an 11Å region including the Be II lines. Their positions are indicated by the vertical dashed lines. Each comparison is offset vertically for clarity; ticks denote 10% of the continuum level. Each star is identified on the left, with the T_{eff}, log g, and [Fe/H] values of the model used in its calculation. For each star, calculations are shown for two assumed Be abundances: one scaled to the stellar iron abundance, and one adopting the solar Be value itself (half this for HD 165341). For Procyon, which is severely Be-depleted (Boesgaard 1976; Stephens *et al.* 1997), three calculations show Be abundances of 1/10 and 1/40 solar as well as solar. Identifications of the strongest lines in the calculation appear at the top.

References

Boesgaard, A. M. 1976, *ApJ*, 210, 466

Castelli, F. & Kurucz, R. L. 2003, *IAU Symp. No. 210*, poster A20; models are at http://wwwuser.oat.ts.astro.it/castelli/grids.html

Peterson, R. C., Carney, B. W., Dorman, B., Green, E. M., Landsman, W., Liebert, J., O'Connell, R. W., Rood, R. T., & Schiavon, R. P. 2004, *STSCI Newsletter*, 21, no. 4, p. 1

Stephens, A., Boesgaards, A. M., King, J. R., & Deliyannis, C. P. 1997, *ApJ*, 491, 339

Takeda, Y., Ohkubo, M., Sato, B., Kambe, E., & Sadakane, K. 2005, *PASJ*, 57, 27

Light Elements in the Universe
Proceedings IAU Symposium No. 268, 2009
C. Charbonnel, M. Tosi, F. Primas & C. Chiappini, eds.

© International Astronomical Union 2010
doi:10.1017/S1743921310004436

New results of the spectral observations of CP stars

N. S. Polosukhina[1], A. V. Shavrina[2], N.A. Drake[3], D. O. Kudryavtsev[4], M. A. Smirnova[1]

[1]Crimean Astrophysical Observatory, Nauchnyi, Crimea, Ukraine – email:
`polo@crao.crimea.ua`

[2]Main Astronomical Observatory of NAS of Ukraine, Kyiv, Ukraine

[3]Sobolev Astronomical Institute, St. Petersburg State University, Russia

[4]Special Astrophysical Observatory, Russian Academy of Sciences, Russia

Abstract. The lithium problem in Ap-CP stars has been, for a long time, a subject of debate. Individual characteristics of CP stars, such as high abundance of the rare-earth elements presence of magnetic fields, complicate structure of the surface distribution of chemical elements, rapid oscillations of some CP-stars, make the detection of the lithium lines and the determination of the lithium abundance, a difficult task. During the International Meeting in Slovakia in 1996, the lithium problem in Ap-CP stars was discussed. The results of the Li study carried out in CrAO Polosukhina (1973–1976), the works of Hack & Faraggiana (1963), Wallerstein & Hack (1964), Faraggiana *et al.* (1992–1996) formed the basis of the International project 'Lithium in the cool CP-stars with magnetic fields'. The main goal of the project was, using systematical observations of Ap-CP stars with phase rotation in the spectral regions of the resonance doublet Li I 6708 $\mathring{A}$ and subordinate 6104 $\mathring{A}$ lithium lines with different telescopes, to create a database, which will permit to explain the physical origin of anomalous Li abundance in the atmospheres of these stars.

Keywords. Stars: magnetic fields, abundances

1. First results in the framework of the international Project: 'Lithium in cool CP stars with magnetic fields'

The first observations of roAp stars in a frame of the International Project, showed abnormally high Li abundances for some of these stars as well as different behavior of the Li doublet with stellar rotation. The most important result of the observations was the discovery of the profile variability of the Li I 6708 $\mathring{A}$ line with the rotation phase in the spectra of two southern roAp stars HD 83368 and HD 60435 (North *et al.* 1998)(see Figs. 1 and 2). The Doppler shift of the Li I line in the spectrum of HD 83368 is about 0.7 $\mathring{A}$ (v sin i = ±27.6 km/s) and is the result of the rotation-modulation of the spotted stellar surface (Fig. 3). We have shown also that Li spots are situated near the magnetic poles of dipole stellar magnetic field (Polosukhina *et al.* 1999).

1.1. *BTA program for Lithium observations and data treatment*

Observations have been carried out with the 6-m telescope and the echelle spectrometer 'NES' of the Special Astrophysical Observatory of RAS, Russia, (Panchuk and Klochkova, 1999) in the spectral region 6000–6800 $\mathring{A}$ with a signal-to-noise ratio S/N = 60–100. For the reduction of obtained spectra we used the package 'Reduce' (Piskunov and Valenti 2002).

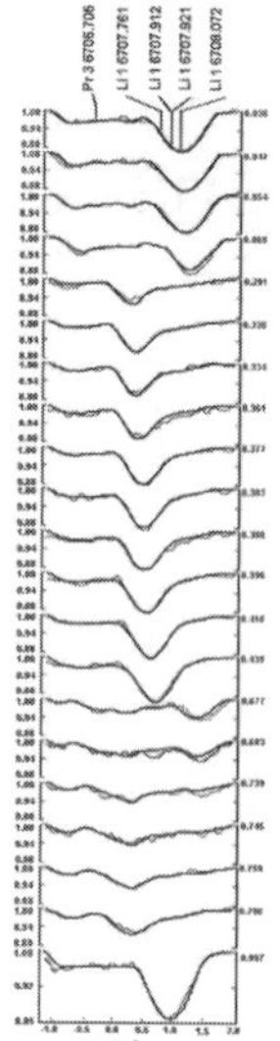

Figure 1. Observed and computed profiles of Li I 6708Å with rotational phases for HD 83368.

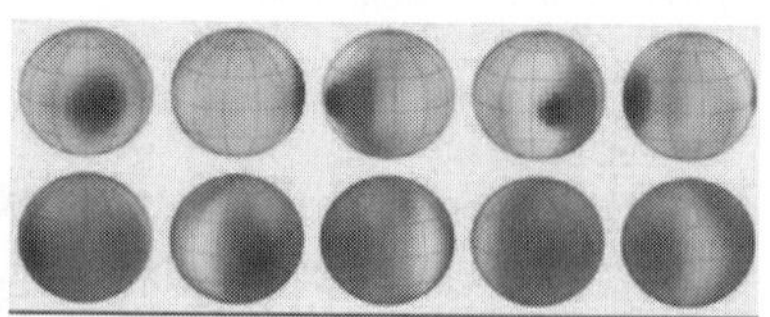

Figure 2. The test D.I. of Lithium 6708 Å blend (using CAT and FEROS telescopes) observations was made assuming that It is consist only of Li I and Pr III (Kochukhov *et al.* 2004)

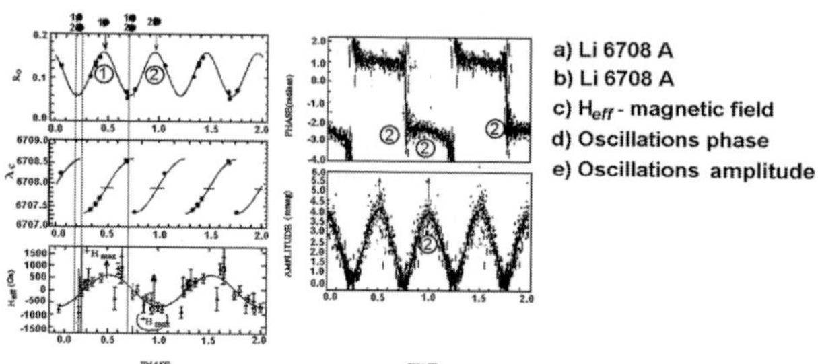

Figure 3. The observations 33Lib in CraO and in ESO

Test observations permitted us to discover some new stars with the detectable Li I 6708 Å line: HD 62140, HD 65339, HD 176232, HD 107612, HD 149822, HD 169842. Some of these stars have short rotational periods and we plan to carry out spectral monitoring of these objects in order to study the behavior of the Li I line with stellar rotation.

1) roAp-CP stars with sharp lines. The stars 33 Lib, HD134214, HD 166473 are low rotors (v sin i < 10 km/s). They have spectra rich REE lines and strong magnetic fields (1500 - 5000 G). Spectra of these stars confirmed the results obtained early and did not show any rotational variability of the strong Li I 6708 Å line.

2) Rapidly-rotating Ap-CP stars (v sin i >10 km/s): HD 65339, HD 169842, HD12098. Among these stars, HD 12098 deserves a special attention. In the spectrum of this star we detected a strong and variable Li I 6708 Å, indicating that Li spots must exist on the surface of this star, which is the first roAp star discovered on the northern hemisphere. Quantitative analysis of this star is presented in Shavrina *et al.* (2008).

1.2. *Variability spectra (dispersograms).*

We used the method of dispersograms for detection of the variable details in the stellar spectra. In order to present more clearly the variability of the spectrum, we calculated the spectrum of 'variability' (Malanushenco *et al.* 1992 as a value of the dispersion of intensity in each wavelength Ii from the mean intensity value I_{mean},

$$\sigma_{obs} = \frac{1}{I_{mean}} \sqrt{\frac{\sum (I_i - I_{mean})^2}{n-1}} \qquad (1.1)$$

where I_i - intensity of the spectrum in selected wavelengths (i), I_{mean} mean value of the intensity in corresponding wavelengths, n - the number of observed spectra (Fig. 5).

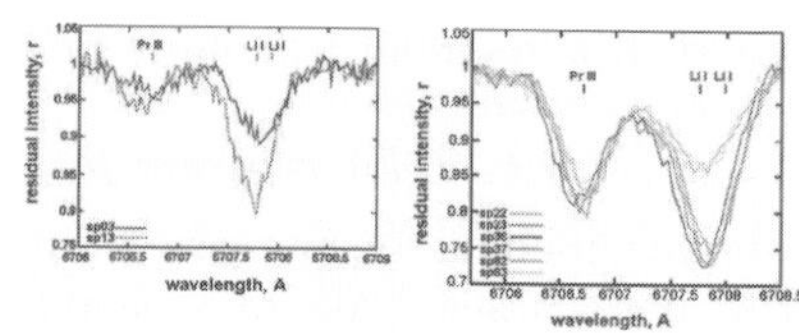

Figure 4. Original spectra of HD 12098 and HD 60435 (classical roAp-CP star with the lithium spots)

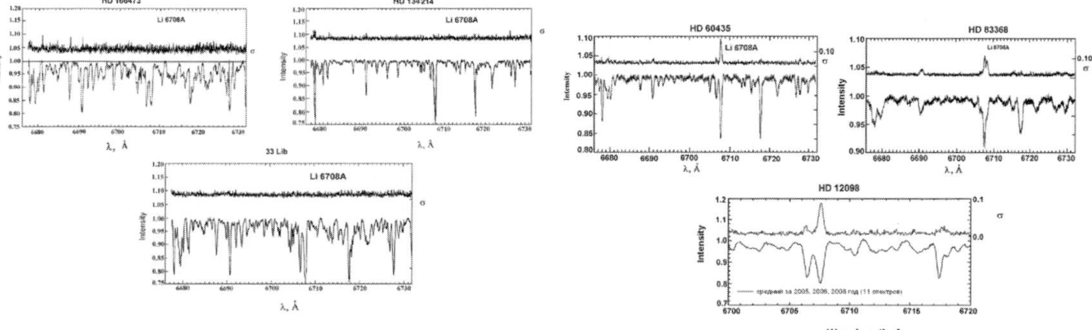

Figure 5. We present examples of the dispersograms for the sharp-lined stars with unvariable Li I 6708 Å line: HD134214, HD 166473, and 33 Lib.

Figure 6. Dispersograms for the stars HD 60435, HD 83368 and HD 12098 in the Li I 6708 A region.

Table 1. Lithium abundance sharp-lined stars roAp-CP stars.

	HD 101065	HD 134214	HD 137949	HD 137949	HD 166473	HD 201601
$T_{eff/logg}$	6600/4.2	7500/4.0	7750/4.5	7250/4.5	7750/4.0	7750/4.0
N(Li)6708Å	3.1	3.9	4.1	3.6	3.3	3.8
N(Li)6103Å	3.5	4.1	4.4	4.4	4.0	4.0
^{6}Li/^{7}Li	0.4:	0.3:	0.2:	0.3:	0.4:	0.5:

1.3. *Analysis of observed and synthetic spectra*

The stars with strong 6708 Å lithium doublets are very poorly studied. We study their spectra in detail in a narrow range near 6708 Å and 6103 Å, by the method of synthetic spectra, taking into account the Zeeman magnetic splitting and blending by REE lines. To calculate synthetic spectra we applied the magnetic spectrum synthesis code SYNTHM (Khan, 2004) which is similar to Piskunovs code SYNTHMAG. We also used the code STARSP of Tzymbal (1996).

Results of modeling observed and accounting profiles to lines 6708 Å Li I for stars HD 83368, HD 60435 and HD 12098 (Shavrina *et al.*, 2001, 2006):

- **HD 83368** for i $= 90^o$, $v_e = 35$ km/s, and a lithium content in the photoshpere $\log\varepsilon\varphi(\mathrm{Li}) = 1,8$
  - $\circ$ **Spots 1:** $l_1 = 173^o \pm 6^o$, $\varphi_1 = 0^o \pm 6^o$, $R_1 = 33^o \pm 6^o$, $\log \varepsilon_1(\mathrm{Li}) = 3.6 \pm 0.2$
  - $\circ$ **Spots 2:** $l_2 = 337^o \pm 6^o$, $\varphi_2 = 0^o \pm 6^o$, $R_2 = 35^o \pm 6^o$, $\log\varepsilon_2(\mathrm{Li}) = 3.5 \pm 0.2$
- **HD 60435** for i $= 47^o$ (133°), $v_e = 15$ km/s, and a lithium content in the photoshpere $\log \varepsilon_2(\mathrm{Li}) = 1,8$
  - $\circ$ **Spots 1:** $l_1 = 11^o \pm 6^o$, $\varphi_1 = -15^o \pm 6^o$, $R_1 = 44^o \pm 3^o$, $\log\varepsilon_1(\mathrm{Li}) = 3.8 \pm 0.2$
  - $\circ$ **Spots 2:** $l_2 = 205^o \pm 10^o$, $\varphi_1 = 15^o \pm 6^o$, $R_2 = 40^o \pm 7^o$, $\log \varepsilon_2(\mathrm{Li}) = 2.7 \pm 0.2$
- **HD 12098**
  - $\circ$ **Spot 1:** $l_1 = 30^o$, $\varphi = -20^o$, $R_1 = 40^o$, $\log \varepsilon_1 (\mathrm{Li}) = 5.0$
  - $\circ$ **Spot 2:** $l_2 = 180^o$, $\varphi = 25^o$, $R_2 = 70^o$, $\log\varepsilon_2 (\mathrm{Li}) = 4.2$
  - $\circ$ **Spot 3:** $l_3 = 290^o$, $\varphi = -20^o$, $R_3 = 40^o$, $\log\varepsilon_3 (\mathrm{Li}) = 4.4$

Analyses of the lithium lines 6708 Å and 6103 Å in spectra of these sharp-lined stars roAp stars were carried out All stars are characterize by strong overabundance of REE and by surface magnetic fields from 2 KG to 6.8 KG (Shavrina *et al.* 2004).

2. Conclusions

- The experience with spectra of roAp-CP stars in 6708 $\mathring{A}$ region shows that Li doublet is main contributor.
- Dispersograms clearly showed a different behavior of the lithium line in the spectra of slowly and rapidly rotating roAp-Cp stars. Dispersograms obtained for HD 12098 and the 'Li-spotted' stars HD 83368 and HD 60435, clearly show the variability of their spectra in the region of the Li I line 6708 $\mathring{A}$. The dispersograms show also that the amplitude of the intensity variations of the Li line is sufficiently higher than that of the lines of the rare-earth elements Nd III, Pr III, Ce II etc. This fact is an additional evidence that the blend at 6708 $\mathring{A}$ belongs to the lithium. - Observations of CP stars with the 6-m BTA telescope confirmed a non-variability of the Li I 6708 $\mathring{A}$ line in the spectra of the slowly rotating stars in comparison to the spectra observed at ESO in 1996. Analysis of the observed and synthetic spectra for the sharplined stars HD 134214, 33 Lib, HD 166473 showed table 1.
- The lithium abundance derived from the secondary Li I 6104 $\mathring{A}$ line is slightly higher than that derived from the resonance Li I 6708 $\mathring{A}$ line for all stars of this group.
- Lithium isotopic ratio (^{6}Li/^{7}Li) differs slightly from one star to another and is higher than ^{6}Li/^{7}Li ratio in the solar atmosphere and in the interstellar medium.
- Observations of HD 12098 with the BTA 6-m telescope in the Li I 6708 $\mathring{A}$ region revealed strong variations of this line testifying that one more Ap-Cp star with the Li spots was discovered. Figure 4 clearly shows the identical behavior of HD 12098 and HD 60435 spectra (HD 60435 is a classical roAp star with the Li spots on the magnetic poles).
- High Li abundance can be explained by means of physical processes which prevent the mixing in the stellar atmosphere and maintain its high initial abundance, suppression of convective motions by strong magnetic fields and action of the ambipolar diffusion. The observed lithium may also be produced in stellar atmospheres, for example by spallation reactions at the stellar surface in the regions of the magnetic poles, where lithium spots are found (Goriely 2007). The creation of a database of observational data in the region of the Li lines is very important for the CP-stars studies, and our new results on lithium abundances in Ap-CP stars and their interpretation, open new perspectives in the research of the physical nature of these stars.

References

Faraggiana, R. 1992-1996, *Mem. S. A. I.*
Goriely 2007, *A&A*, 466, 619
Hack, M. & Faraggina, R. 1963, *PASP*
Khan 2004, *JQSRT*, 88, N1-3, 71
Kochuknov, O., Hoppe, P., & Zinner, E. 2004, *A&A.*, 424, 935
Malanushenko, V. *et al.* 1992, *A&A.*, 259, 567
North *et al.* 1998, *A&A*, 333, 644
Panchuk & Klochkova, 1999.
Piskunov & Valenti 2002, *A&A.*, 385, 109
Polosukhina, N. 1973-1976, *Izv. CrAO*
Polosukhina, N. *et al.* 1999, *A&A*, 351, 283
Shavrina, A. V. *et al.* 2008, *Astrophysics*, Vol. 51, p. 517
Shavrina, A. V. *et al.* 2006 *in Physics of Magnetic Stars. Proceedings of the Int. Conf.*, p. 341
Shavrina, A. V. *et al.* 2004, *in The A-Star Puzzle, IAU Symp.*, 224, p. 711
Shavrina, A. V. *et al.* 2001 *A&A*, 372, 571
Tzymbal, V. 1996, *ASP Conf. Ser.*, 108, 198
Wallerstein, G. and Hack, M. 1964, *The Observatory*, Vol. 84, p. 160

Light Elements in the Universe
Proceedings IAU Symposium No. 268, 2009
C. Charbonnel, M. Tosi, F. Primas & C. Chiappini, eds.

© International Astronomical Union 2010
doi:10.1017/S1743921310004448

The metal–poor end of the Spite plateau: gravity sensitivity of the Hα wings fitting.

L. Sbordone[1,2,3], P. Bonifacio[1,2,4], E. Caffau[2], H.-G. Ludwig[1,2],
N. Behara[1,2], J. I. Gonzalez-Hernandez[1,2,5], M. Steffen[6], R. Cayrel[2],
B. Freytag[7], C. Van't Veer[2], P. Molaro[4], B. Plez[8], T. Sivarani[9],
M. Spite[2], F. Spite[2], T. C. Beers[10], N. Christlieb[11], P. François[2],
and V. Hill[2,12]

[1] CIFIST Marie Curie Excellence Team,
[2] GEPI – Observatoire de Paris – France
[3] Max-Planck Institut für Astrophysik, Garching – Germany
[4] INAF – Osservatorio Astronomico di Trieste – Italy
[5] Universidad Complutense de Madrid – Spain
[6] Astrophysikalische Institut Pottsdam – Germany
[7] Centre de Recherche Astrophisique de Lyon, UMR 5574 – France
[8] Université Montpellier 2 – France
[9] Indian Institute of Astrophysiscs, Bangalore – India
[10] Michigan State University and JINA, Lansing, MI – USA
[11] Landessternwarte Heidelberg – Germany
[12] Cassiopée, Observatoire de la Cote d'Azur, Nice – France

Abstract. We recently presented (Sbordone *et al.*, 2009a) the largest sample to date of lithium abundances in extremely metal-poor (EMP) Halo dwarf and Turn-Off (TO) stars. One of the most crucial aspects in estimating Li abundances is the $T_{\rm eff}$ determination, since the Li I 670.8 nm doublet is highly temperature sensitive. In this short contribution we concentrate on the $T_{\rm eff}$ determination based on Hα wings fitting, and on its sensitivity to the chosen stellar gravity.

Keywords. Nuclear reactions, nucleosynthesis – Galaxy: halo – stars: abundances, Population II

1. Introduction

In Sbordone *et al.* (2009a) we present lithium abundances for a sample of 28 stars in the $-3.6 <$[Fe/H]$<$-2.4 range, 10 of which have [Fe/H]$\leqslant -3.0$. We derived four different $T_{\rm eff}$ scales: Hα wings fitting against a grid of synthetic profiles from 1D models using Barklem *et al.* (2000) or Ali & Griem (1966) self-broadening theories (BA and ALI scales); Hα fitting against a 3D-hydrodynamical grid based on CO^{5}BOLD models (Freytag *et al.* 2002; Wedemeyer *et al.* 2004) with Barklem *et al.* (2000) self-broadening (3D scale); InfraRed Flux method (IRFM, González Hernández & Bonifacio 2009).

2. The Hα gravity sensitivity

Effective temperature ($T_{\rm eff}$) is the most crucial stellar atmosphere parameter influencing Li abundance determination: Li abundances derived from the Li I 670.75nm line are sensitive to $T_{\rm eff}$ at a level of about 0.03 dex for each 50 K variation in $T_{\rm eff}$. Being mostly temperature sensitive, the Hα wings are also influenced by the adopted surface gravity, to a level which might be significant in the present scope. In this short contribution we

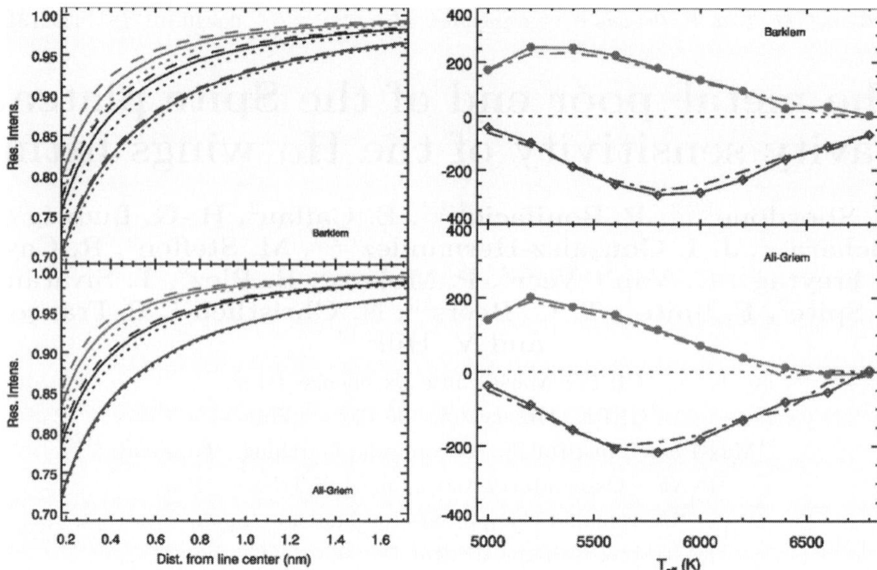

Figure 1. Left panel, for BA and ALI scales, 1D synthetic profiles of the Hα red wing are plotted for $T_{\rm eff} = 5400$ K, 6000 K, and 6600 K (red, black and blue respectively). Line type identify gravity, dashed for log g $= 3.5$, solid for log g $= 4.0$, dotted for log g $= 4.5$. In the right panel, the effect of gravity offsets on $T_{\rm eff}$ estimate are shown. Theoretical profiles computed with log g $= 4.0$ have been fitted with log g $= 3.5$ profiles (red lines, filled circles) and log g $= 4.5$ profiles (blue lines, open diamonds). The temperature difference (recovered-real) is plotted against the real temperature. Solid lines refer to [Fe/H] $= $ -3, dotted to [Fe/H] $= -2.5$.

concentrate on this effect. Barklem *et al.* (2002) already reported estimates of such a sensitivity down to [Fe/H]$=-2$. The effect is always in the sense of higher gravity leading to broader profiles, and appears generally stronger at lower metallicities, and for the BA profiles compared to the ALI profiles. The effect is shown in Fig. 1, where we show both the shape of the Hα red line wing and the temperature offset deriving from adopting the wrong gravity estimate in measuring effective temperature. As it can be seen, in the temperature range 6000-6500K, most significant for Li measurement in EMP TO and dwarf stars, significant ($>$100K) offsets can arise from neglecting the gravity dependence in fitting Hα profiles, which can appreciably skew the results when Li abundances are measured over a fairly broad temperature range (see Sbordone *et al.* 2009b).

References

Ali, A. W. & Griem, H. R. 1966, *Physical Review*, 144, 366

Barklem, P. S., Stempels, H. C., Allende Prieto, C., Kochukhov, O. P., Piskunov, N., & O'Mara, B. J. 2002, *A&A* , 385, 951

Barklem, P. S., Piskunov, N., & O'Mara, B. J. 2000, *A&A* (Letters), 355, 5

Freytag, B., Steffen, M., & Dorch, B. 2002, *Astronomische Nachrichten*, 323, 213

González Hernández, J. I., & Bonifacio, P. 2009, *A&A*, 497, 497

Sbordone, L., *et al.* 2009a, proceedings of IAU Symposium 265 "Chemical Abundances in the Universe: Connecting First Stars to Planets", K. Cunha, M. Spite & B. Barbuy, eds., in press.

Sbordone, L., *et al.* 2009b, *A&A*, submitted

Wedemeyer, S., Freytag, B., Steffen, M., Ludwig, H.-G., & Holweger, H., 2004, *A&A*, 414, 1121

Light Elements in the Universe
Proceedings IAU Symposium No. 268, 2009
C. Charbonnel, M. Tosi, F. Primas & C. Chiappini, eds.

© International Astronomical Union 2010
doi:10.1017/S174392131000445X

Beryllium abundances along the evolutionary sequence of the open cluster IC 4651

Rodolfo Smiljanic[1,2], L. Pasquini[2], C. Charbonnel[3,4], and N. Lagarde[3]

[1]IAG, University of São Paulo, Brazil, [2]ESO, Germany,
email: rsmiljan@eso.org

[3]Geneva Observatory, Switzerland, [4]LATT, CNRS, Université de Toulouse, France

Abstract. The simultaneous investigation of Li and Be in stars is a powerful tool in the study of the evolutionary mixing processes. Here, we present beryllium abundances in stars along the whole evolutionary sequence of the open cluster IC 4651. This cluster has a metallicity of [Fe/H] = +0.11 and an age of 1.2 or 1.7 Gyr. Abundances have been determined from high-resolution, high signal-to-noise UVES spectra using spectrum synthesis and model atmospheres. Lithium abundances for the same stars were determined in a previous work. Confirming previous results, we find that the Li dip is also a Be dip. For post-main-sequence stars, the Be dilution starts earlier within the Hertzsprung gap than expected from classical predictions, as does the Li dilution. Theoretical hydrodynamical models are able to reproduce well all the observed features.

Keywords. Stars: abundances, evolution, rotation – Open clusters and associations: individual: IC 4651

1. Introduction

In contradiction with standard stellar evolution models, where convection is the only mechanism driving mixing episodes, field and cluster F- and early G-type stars (including the Sun) deplete Li abundances during the main sequence (Lambert & Reddy 2004; Sestito & Randich 2005, and references therein). Different physical mechanisms have been proposed to explain these observations: atomic diffusion, mass loss, rotation-induced mixing, internal gravity waves, or combinations of these (see Charbonnel & Talon 2008 and references therein).

As Li and Be burn at different temperatures (2.5×10^6 K for Li and 3.5×10^6 K for Be), i.e. at different depths in the stellar interior, they help in constraining the transport mechanisms by performing a stellar tomography. In the study of mixing processes as a function of mass and evolutionary status cluster stars are ideal because they have well defined masses and share the same age and initial chemical composition. We derived Be abundances along the whole evolutionary sequence of the open cluster IC 4651, including solar-type, Li-dip, turn-off, subgiant, and red giant stars. With these new results, we investigate in detail the mixing processes in different stellar masses.

2. Discussion

The sample analyzed here is composed of 22 stars; 21 with atmospheric parameters and Li abundances from Pasquini *et al.* (2004) and 1 from Randich *et al.* (2002). The spectra have $40 \leqslant$ S/N $\leqslant 100$ (per resolution element) and R ~ 45000.

Be abundances were determined using synthetic spectra and the same codes and line lists used in Smiljanic *et al.* (2009a). As some of the sample stars are fast rotators, we first carefully modeled the slow-rotating stars and tested the effects in the abundances

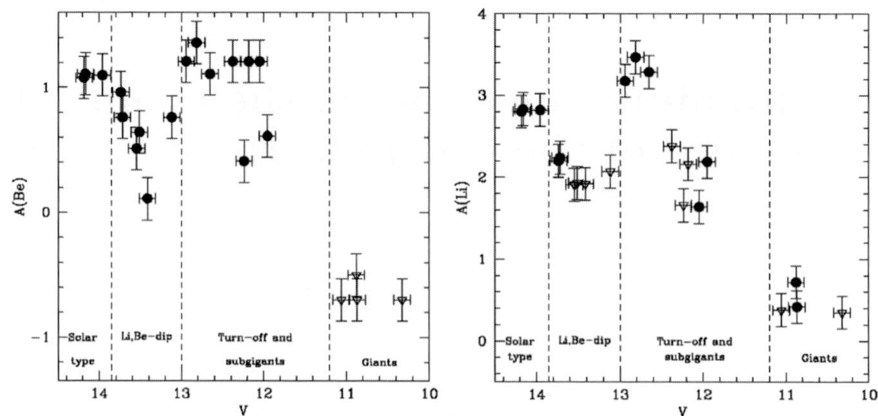

Figure 1. Abundances of Be (left panel) and Li (right panel) as a function of the V magnitude. Detections are shown as full circles and upper limits as open triangles.

of artificially broadening the spectra (see Smiljanic *et al.* 2009b for details). Beryllium was detected in all the sample stars except for the giants.

New evolutionary models for stars on the hot side of the dip, including atomic diffusion, meridional circulation, and shear turbulence, were calculated with STAREVOL V3.1 by Lagarde & Charbonnel (in preparation, see also Charbonnel & Lagarde this volume) for a range of stellar masses and initial rotation velocities. For stars on the cool side of the Li dip we use the 1.2 $M_\odot$ model computed by Talon & Charbonnel (2005) which has an initial rotation velocity of 50 km s^{-1}.

Beryllium abundances are found to follow closely the behavior of the Li abundances (Fig. 1). In a sequence of increasing mass we have first the coolest main-sequence stars that do not present a Be abundance dispersion. This is expected to be due to the impact of internal gravity waves. After that, a well-defined Be dip is seen. This confirms previous results that the Li dip is also a Be dip (Boesgaard & King 2002, Boesgaard *et al.* 2004). For post-main-sequence stars we confirm that Be dilution starts earlier than the expected classically. The Be abundances also present a significant dispersion.

The dispersion of Li and Be abundances on the blue side of the dip and in evolved stars is very well explained by the models when accounting for a dispersion in the initial values of the stellar rotational velocities. The models reproduce all the Li and Be features along the CMD of IC 4651. The success in explaining the Li and Be abundances along the whole evolutionary sequence shows that important steps have been taken towards the proper understanding of the physical mechanisms acting during the stellar evolution.

References

Boesgaard, A. M., Armengaud, E., & King, J. R. 2004, *ApJ*, 605, 864

Boesgaard, A. M. & King, J. R. 2002, *ApJ*, 565, 587

Charbonnel, C. & Talon, S. 2008, in Proceedings of IAUS 252, 163

Lambert, D. L. & Reddy, B. E. 2004, *MNRAS*, 349, 757

Pasquini, L., Randich, S., Zoccali, M. *et al.* 2004, *A&A*, 424, 951

Randich, S., Primas, F., Pasquini, L. *et al.* 2002, *A&A*, 387, 222

Sestito, P. & Randich, S. 2005, *A&A*, 442, 615

Smiljanic, R., Pasquini, L., Bonifacio, P., *et al.* 2009a, *A&A*, 499, 103

Smiljanic, R., Pasquini, L., Charbonnel, C., & Lagarde, N. 2009b, arXiv:0910.4399, *A&A*, in press

Talon, S. & Charbonnel, C. 2005, *A&A*, 440, 981

Light Elements in the Universe
Proceedings IAU Symposium No. 268, 2009
C. Charbonnel, M. Tosi, F. Primas & C. Chiappini, eds.

© International Astronomical Union 2010
doi:10.1017/S1743921310004461

Using lithium to estimate ages for solar-type stars

David R. Soderblom[1]

[1]Space Telescope Science Institute,
3700 San Martin Drive, Baltimore MD 21218 USA
email: drs@stsci.edu

Abstract. Can observations of Li in F, G, and K stars be used to derive an age for an individual star or a group? This is a brief progress report on using Li for age estimation in PMS and ZAMS stars.

Keywords. Stars:ages, solar-type, convection

The decline of the surface Li abundance with age is well known for stars near and below 1 $M_\odot$. Observations and models of this phenomenon should provide constraints on convection in these stars, the process believed to be the underlying cause of the Li depletion. This is a progress report on an effort to invert the problem to use observations of Li in F, G, and K dwarfs to estimate their ages.

A full understanding of Li depletion involves some complex and not well understood physics and requires accurate determinations of basic stellar quantities such as temperature and composition. Also, transforming an observation of Li to an abundance is very sensitive to temperature as well as non-LTE effects. However, when estimating ages these quantities are often only poorly constrained and I wished to address a simpler question: If observations of the Li equivalent width (W_λ) and color (($B - V$), say, or the equivalent derived from another color or from $T_{\rm eff}$) are at hand, how well can an age be estimated? Can an age be determined at all for a single star, or is some minimum number in a group required?

The data used will be described fully and illustrated in a forthcoming paper, and here I will summarize the key points:

- The youngest stars with good Li observations are T Tauri stars in Taurus-Auriga, the Orion Nebula Cluster, and NGC 2264, all with ages of 5 Myr or less. The scatter in W_λ among these stars is small and consistent with the observational errors. The consistency in Li among these different groups suggests stars in the present epoch all form with the same initial Li abundance, and the agreement with the primordial solar Li abundance (from meteorites) indicates that that value has not changed significantly in ~ 5 Gyr.

- By contrast, the Pleiades is a well-observed cluster with an age of 125 Myr and it exhibits an inherent spread in W_λ that grows in going to cooler stars, becoming as much as 1 dex for K dwarfs (see Fig. 1). This spread is much larger than the observing errors and also significantly larger than the scatter seen among T Tauri stars. The fact that T Tauris show little scatter implies that inhomogeneous stellar atmospheres (i.e., spots) are not the cause of the spread in Li in the Pleiades because the T Tauris are expected to be at least as spotted as ZAMS stars.

- The run of W_λ with color for the α Persei cluster and NGC 2451 at about 80 Myr is indistinguishable from that of the Pleiades (see Fig. 2). This implies that Li depletion – at least on average – goes through plateau stages where little or nothing happens for an extended time.

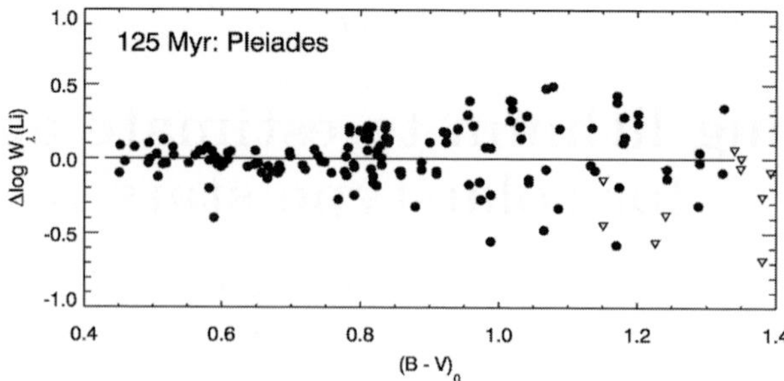

Figure 1. Lithium equivalent widths for F, G, and K stars in the Pleiades. A mean relation has been fitted and subtracted to show how the scatter in W_λ varies with color.

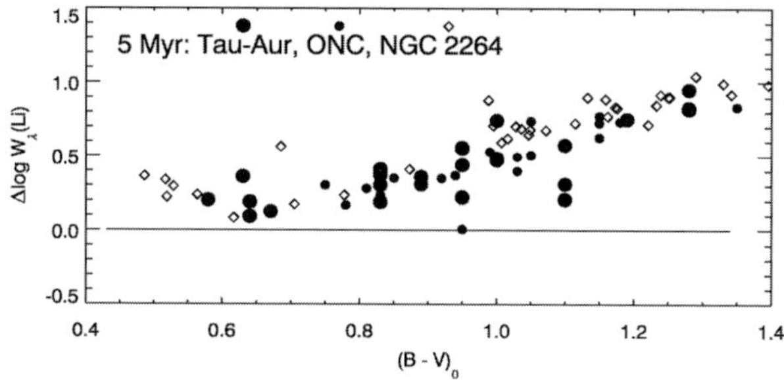

Figure 2. Li equivalent widths for T Tauri stars in Tau-Aur, the Orion Nebula Cluster, and NGC 2264, relative to the Pleiades.

- There are clusters with ages between those of the T Tauris and α Persei but they are sparse and so the data are less well defined. Nevertheless, W_λ declines in progression with age as expected, although there can again be plateaus where little difference in the average W_λ is seen despite a significant difference in age.

- Li continues to decline in stars after they reach the ZAMS, as seen in 200 Myr-old clusters (M35, NGC 2516 and Blanco 1), and at 300 Myr (M34 and NGC 6475).

- Stars in the Hyades and other clusters of that age (Praesepe, Coma, and NGC 6633) have well-defined runs of W_λ with color and with little scatter. This suggests that the large scatter seen in the Pleiades may later lessen, but there is no known mechanism to achieve that. Also, stars in the 4 Gyr-old cluster M67 show an appreciable spread in Li. The Hyades may possibly be misleading because it is metal-rich.

There are trends of declining average W_λ for the F and G stars but they are comparable to the scatter seen. For the K dwarfs, however, the decline in W_λ is large and rapid up to 200-300 Myr or so. The large scatter seen among stars of a single age (notably the Pleiades) means that for a single star one can only set limits to age, but if data for a group of 10-20 stars exist then a more precise age can be set. This makes it possible to use observations of Li in K dwarfs to at least set the age ordering of young kinematic groups such as β Pictoris, η Cha, TW Hya, etc.

Light Elements in the Universe
Proceedings IAU Symposium No. 268, 2009
C. Charbonnel, M. Tosi, F. Primas & C. Chiappini, eds.

© International Astronomical Union 2010
doi:10.1017/S1743921310004473

Lithium in metal-poor red giants

Laimons Začs and Arturs Barzdis

Faculty of Physics and Mathematics, University of Latvia,
Raiņa bulvāris 19, LV-1586 Rīga, Latvia
email: `zacs@latnet.lv`

Abstract. The lithium abundance was calculated for five metal-poor red giant stars from Li I doublet at 6707 Å by fitting the observed high-resolution spectra with synthetic spectra. The lithium abundance was found to be low in all stars, $\log \varepsilon(\mathrm{Li}) \leqslant 1.8$, confirming lithium depletion on the red giant and asymptotic giant branch.

Keywords. Stars: abundances, Population II, late-type, binaries: spectroscopic, stars: individual (HD 30443, HD 187216, HD 209621, HD 218732, HD 232078)

1. Observations and analysis

High resolution spectra for three analyzed stars (HD 209621, HD 218732, HD 232078) were observed with the cross-dispersed high-resolution échelle spectrograph FIES installed on NOT (Canary Islands) with a spectral resolution of R = 67 000. HD 187216 was observed using the fiber-fed échelle spectrograph BOES attached to the 1.8-m telescope of Bohyunsan Observatory (Korea) with a spectral resolution of R = 45 000. HD 30443 was observed using the coudé échelle spectrometer MAESTRO on the 2 m telescope at the Observatory on the Terskol Peak in Northern Caucasus (Russia) with a resolving power of 45 000. The basic data for all targets are provided in Table 1. Spectral classification was adopted from Yamashita (1975) and the SIMBAD database. All stars are first-ascent red giants or on the asymptotic giant branch. HD 30443 and HD 209621 are spectroscopic long period binaries (P = 2954 & 407.4 days; McClure & Woodsworth 1990). HD 218732 was suspected to be a long period binary (P = 567 days; Carney *et al.* 2003).

Synthetic spectra for some lithium abundances around the final value are calculated and convolved with the instrumental profile using spectrum synthesis technique which based on the R.L.Kurucz's code SYNTHE and the Uppsala atmospheric models. Figure 1 illustrates iterations (±0.1 dex for HD 30443, ±0.2 dex for HD 187216, HD 209621, and HD 232078, ±0.5 dex for HD 218732) around the accepted lithium value. The accepted lithium abundances are provided in Table 1. The line list was examined using the databases VALD and DREAM. Significant blending of Li doublet was expected because of low effective temperature and peculiar chemical composition of stellar atmospheres. C_2 and CN lines are strong in the spectra of HD 30443, HD 187216, and HD 209621 (see Figure 1) and lines of neutron capture elements are enhanced.

2. Conclusions

Five red giants in the metallicity range $-1 < [\mathrm{Fe/H}] < -2$ display $\log \varepsilon(\mathrm{Li}) \leqslant 1.8$, confirming a substantial lithium depletion on the red giant and asymptotic giant branch. Neutron capture elements are significantly enhanced in the amospheres of three analyzed giants.

Table 1. Basic data for the program stars. Accepted effective temperature (K), gravity (cgs), iron abundance relative to the solar value, lithium abundance and averaged abundance of neutron capture elements relative to the solar value are given.

Star	Sp.type	T_{eff}	$\log g$	[Fe I/H],[Fe II/H]	$\log\varepsilon$(Li)	[n/Fe II]	binary
HD30443	R, C4,3 CH	4100	1.5	$-1.1,-1.1$	1.8	$+1.6$	yes
HD187216	R, C3,3 CH	4000	0.75	$-2.6,-1.7$	0.45	$+1.2$	
HD209621	C, C1,2 CH	4300	-0.4	$-2.0,-2.0$	1.2	$+1.3$	yes
HD218732	G7 Ib	4200	0.50	$-1.5,-1.5$	-1.1	0.0	yes
HD232078	K3 IIp	4000	0.00	$-1.4,-1.4$	-0.55	$+0.3$	

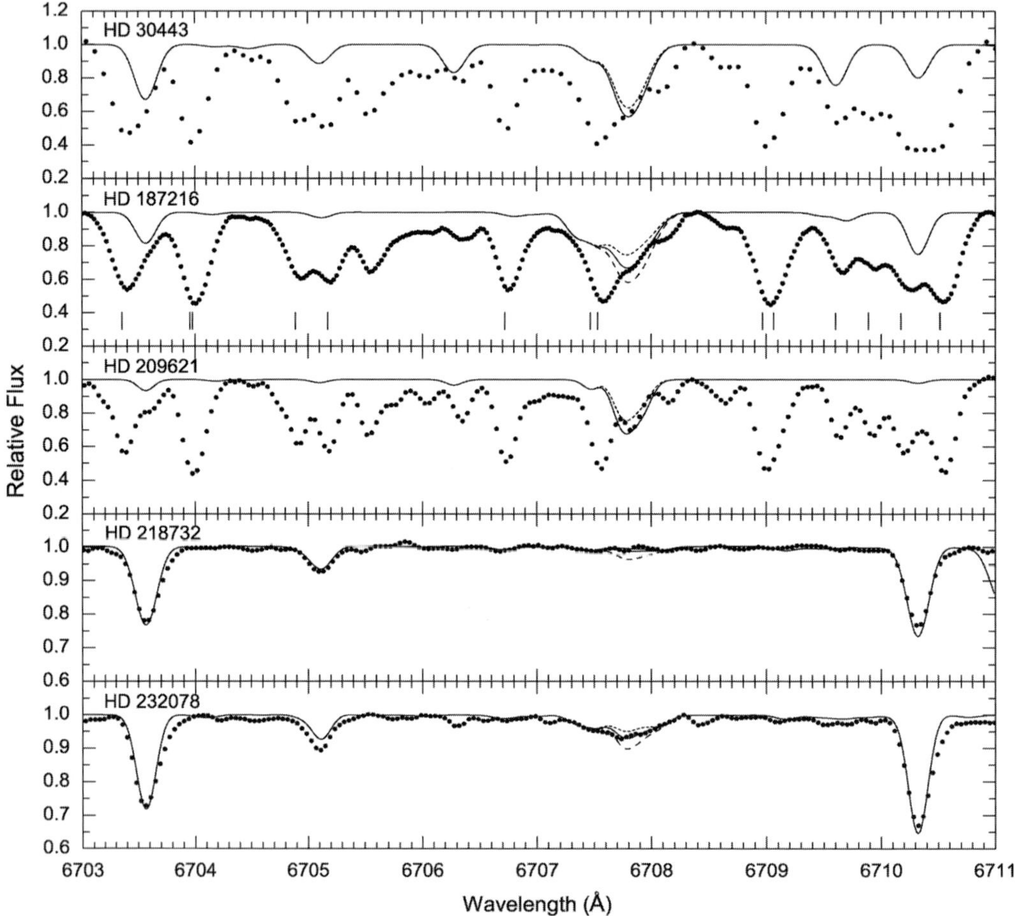

Figure 1. The observed spectra (dots) of red giants in spectral region around Li doublet at 6707 Å. The positions of CN Red system lines are marked for HD 187216. Synthesized atomic line spectra for the accepted lithium abundances are displayed by thick solid lines.

References

Carney, B. W., Latham, D. W., Stefanik, R. P., Laird, J. B., & Morse, J. A. 2003, *AJ*, 125, 293
McClure, R. D. & Woodsworth, A. W. 1990, *ApJ*, 352, 709
Yamashita, Y. 1975, *PASJ*, 27, 325

Session IV

Sources and sinks of light elements

Suzanne Talon

Takuji Tsujimoto

Light Elements in the Universe
Proceedings IAU Symposium No. 268, 2009
C. Charbonnel, M. Tosi, F. Primas & C. Chiappini, eds.

© International Astronomical Union 2010
doi:10.1017/S1743921310004485

Light elements as diagnostics on the structure and evolution of low-mass stars

Suzanne Talon[1] and Corinne Charbonnel[2,3]

[1]RQCHP, Département de physique, Université de Montréal, C.P. 6128, succ. centre-ville, Montréal, Québec, H3C 3J7
email: suzanne.talon@umontreal.ca

[2]Geneva Observatory, University of Geneva, ch. des Maillettes 51, 1290 Versoix, Switzerland
email: Corinne.Charbonnel@unige.ch

[3]Laboratoire d'Astrophysique de Toulouse-Tarbes, Université de Toulouse, CNRS UMR 5572, 14 Av. E. Belin, 31400 Toulouse, France

Abstract. Low-mass stars exhibit, at all stages of their evolution, the signatures of complex physical processes that require challenging modelling beyond standard stellar theory. In this review, we focus on lithium depletion in low-mass stars. After disecting the Li dip, we discuss how large scale mixing due to rotation and internal gravity waves may interact to explain this feature. We also briefly discuss the impact that is expected on Population II stars.

Keywords. Hydrodynamics – instabilities – turbulence – waves – stars: abundances, evolution, interiors, rotation

1. The lithium dip, atomic diffusion and rotation-induced mixing

During the last couple of decades, it became obvious that the art of modeling stars in the 21$^{\rm st}$ century strongly relies on the art of modelling transport processes. Observational evidences now give precise clues on the various processes that transport angular momentum and chemical elements in the radiative regions of low-mass stars, at various stages of their evolution. One of the most striking signatures of transport processes in low-mass stars is the so-called Li dip (see Fig. 1). This drop-off in the Li content of main-sequence F-stars in a range of $\sim 300\,\mathrm{K}$ centered around $6700\,\mathrm{K}$ was discovered in the Hyades by Wallerstein *et al.* (1965); its existence was later confirmed by Boesgaard & Tripicco (1986). This feature appears in all open clusters older than $\sim 200\,\mathrm{Myr}$, as well as in field stars (Balachandran 1995), an indication that it is a phenomenon occurring on the main sequence.

Michaud's (1986) original suggestion that this structure is shaped by gravitational settling (causing the drop in Li abundance on the cool side of the dip as the surface convection zone shrinks) and radiative levitation (which explains the rise on the warm side of the dip) suffers from two serious drawbacks:

• The expected concomitant underabundances of heavier elements (C, N, O, Mg, Si) are not observed in cluster stars (Takeda *et al.* 1998; Varenne & Monier 1999; Gebran *et al.* 2008).

• In this framework, Li is not destroyed; it settles and accumulates in a buffer zone below the convective envelope. Li should thus be dredged-up as a star enters the Hertzsprung gap and this is not observed, neither in the field nor in open cluster stars (Pilachowski *et al.* 1988; Deliyannis *et al.* 1997).

Boesgaard (1987) noticed that the effective temperature of the dip is also associated with a sharp drop in rotation velocities (see Fig. 1). Rotation was then suggested to play a dominant role in this mass range. To properly model rotation-induced mixing, one must follow the time evolution of the angular momentum distribution within a star, taking into account *all* relevant physical processes: contraction/expansion caused by stellar evolution, mass loss or accretion, tidal effects, as well as internal redistribution of angular momentum through meridional circulation, turbulence, magnetic torques and waves.

1.1. *Shaping the warm side of the dip with wind driven circulation*

The description of the internal physical processes related to stellar rotation greatly improved during the last two decades. Here, we review the results of the application of Zahn's (1992) model of rotation-induced mixing, using the same free parameters as those required to explain abundance anomalies in more massive stars (see Maeder & Meynet 2000). In this framework, transport of angular momentum is dominated by the Eddington-Sweet meridional circulation and shear instabilities.

In the dip region, the evolution of a star's angular momentum is influenced by various mechanisms depending on its effective temperature.

• Stars with $T_{\rm eff}$ higher than $\sim 6900\,$K have a shallow convective envelope and are not spun down by a magnetic torque. These stars soon reach a stationary regime where there is no net flux of angular momentum†, in which meridional circulation and shear-induced turbulence counterbalance each other. The associated weak mixing is just sufficient to counteract atomic diffusion. These rotating models account nicely for the observed constancy of Li and CNO in these stars, and they also explain the Li and Be behaviour in subgiant stars (Palacios *et al.* 2003; Pasquini *et al.* 2004; Smiljanic *et al.* 2009; see also Charbonnel & Lagarde, this Volume). Talon, Richard, & Michaud (2006) showed that the transport coefficients related to rotation-induced mixing lead to normal A stars for rotation velocities above $\sim 100\,\mathrm{km\,s^{-1}}$, and permit Am anomalies below with a mild correlation with rotation, provided a reduction of turbulent mixing by horizontal turbulence is taken into account. Fossati *et al.* (2008) have confirmed observationally this correlation with rotation.

• Between ~ 6900 and $6600\,$K, the convective envelope deepens and a weak magnetic torque, associated with the apparition of a surface dynamo, spins down the outer layers of the star. This creates internal shears and the transport of angular momentum by meridional circulation and shear turbulence increases, leading to a larger destruction of Li and Be, in agreement with the data. The rotating model thus perfectly fits the blue side of the Li and Be dips (Talon & Charbonnel 1998, Charbonnel & Talon 1999, Palacios *et al.* 2003, Pasquini *et al.* 2004; Smiljanic *et al.* 2009).

• Stars on the cool side of the Li dip ($T_{\rm eff} < 6600\,$K) have an even deeper convective envelope sustaining a very efficient dynamo, which produces a strong magnetic torque that spins down the outer layers very efficiently. If we assume that all the angular momentum transport is assured by the wind-driven circulation in these stars‡, we obtain too much Li depletion compared to the observations (see Fig. 2). *On the basis of these results, Talon & Charbonnel (1998) suggested that the Li dip corresponds to a transition region where the efficiency of another process for angular momentum transport rises.*

This signature of the existence of new physics can be drawn from studies performed by two other groups. In Théado & Vauclair (2003), where rotation-induced mixing is assumed to be linked to meridional circulation in a solid-body rotating star, the two free parameters associated with this mixing have to be varied by factors of 2 and 260 to fit

---

† In fact, there remains a small flux of angular momentum, just sufficient to counteract the effect of stellar contraction/expansion.

‡ Let us mention that in the case of large shears meridional circulation is far more efficient than shear for angular momentum transport.

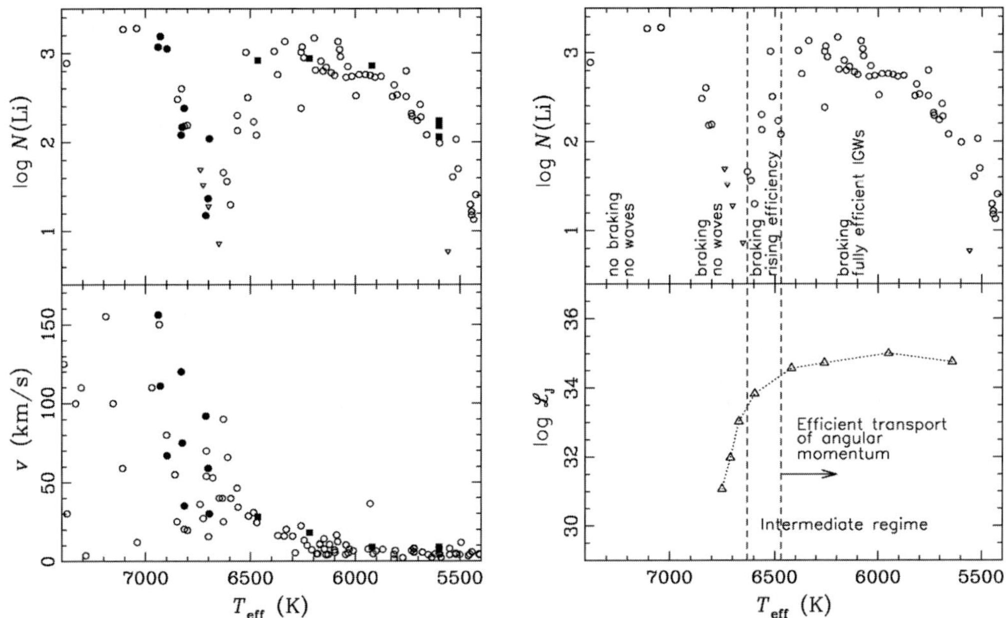

Figure 1. (**Left**) (*top*) Lithium abundance in the Hyades versus effective temperature (open symbols, data from Burkhart & Coupry 2000) plus model data (filled symbols). Circles: models where internal gravity waves (hereafter IGWs) are negligible. Squares: models where IGWs are significant (Charbonnel & Talon 2005; Talon & Charbonnel 2005). (*bottom*) Rotational velocity in the Hyades - unprojected (open symbols, data from Gaigé 1993) plus model data (filled symbols). (**Right**) (*top*) Lithium abundance plus approximate dependence of surface braking on effective temperature and the requirements for angular momentum transport for rotational mixing to lead to the formation of the lithium dip (Talon & Charbonnel 1998). (*bottom*) Filtered angular momentum luminosity of waves below the SLO measuring the efficiency of wave induced angular momentum transport. Adapted from Talon & Charbonnel (2003).

the Li dip. Similarly, Richard *et al.* (priv. comm.) have to change the ad-hoc turbulence they add to reduce atomic diffusion by an order of magnitude along along the Li dip. The strong variation of these *fudge factors* is actually an indication that the *physical processes* at play inside these stars change in that region.

1.2. *A link with the Sun's rotation*

The proposition by Talon & Charbonnel (1998) may be linked to another observation that fails to be reproduced by the pure hydrodynamic models, namely the flat solar rotation profile revealed by helioseismology (Brown *et al.* 1989; Kosovichev *et al.* 1997; Couvidat *et al.* 2003; García *et al.* 2007). At the solar age, models relying only on turbulence and meridional circulation for momentum transport predict large angular velocity gradients that are not present in the Sun (Pinsonneault *et al.* 1989; Chaboyer *et al.* 1995; Talon 1997; Matias & Zahn 1998). This indicates that another process participates to the transport of angular momentum in solar-type stars.

Two candidates have been examined. The first rests on the possible existence of a magnetic field within the radiation zone (Charbonneau & MacGregor 1993; Eggenberger *et al.* 2005). The second invokes traveling internal gravity waves (hereafter IGWs) generated at the base of the convection envelope (Schatzman 1993; Zahn *et al.* 1997; Kumar & Quataert 1997; Talon, Kumar & Zahn 2002). For either of these solutions to be convincing, they must be tested with numerical models coupling these processes with rotational instabilities and should explain all the aspects of the problem, including the lithium evolution with time.

The Li data suggest that the efficiency of the additional process is linked to the growth of the convective envelope in stars with effective temperatures around $T_{\rm eff} \simeq 6600\,{\rm K}$. As we shall see below, this is a characteristic of IGWs.

2. IGWs generation and momentum extraction in low-mass stars

In the Earth's atmosphere, wave-induced momentum transport is a key process in the understanding of several phenomena, the best known being the quasi-biennial oscillation of the stratosphere. In astrophysics, IGWs have initially been invoked as a source of mixing for chemicals (Press 1981; García Lopez & Spruit 1991; Schatzman 1993; Montalban 1994; Montalban & Schatzman 1996, 2000; Young et al. 2003). Ando (1986) studied the transport of angular momentum associated with standing gravity waves in Be stars. He was the first to clearly state, in the stellar context, that IGWs carry angular momentum from the region where they are excited to the region where they are dissipated. Traveling IGWs have since been invoked as an important source of angular momentum redistribution in single stars (Schatzman 1993; Kumar & Quataert 1997; Zahn, Talon & Matias 1997, Talon et al. 2002; Charbonnel & Talon 2005).

2.1. IGWs generation and wave spectrum

Two different processes contribute to IGWs excitation at the border of convective regions: convective overshooting in an adjacent stable region (García López & Spruit 1991; Kiraga et al. 2003; Rogers & Glatzmaier 2005a, 2005b), and Reynolds stresses in the convection zone (Goldreich & Kumar 1990; Balmforth 1992; Goldreich, Murray & Kumar 1994). In our studies, we shall use the second mechanism, which has been calibrated on solar p-modes and, thus, seems more reliable at this time. This gives a lower limit to the correct/total wave flux although neglecting the other excitation mechanism remains the weakest point of wave-induced transport models (see Charbonnel & Talon 2007 for a complete discussion on wave excitation and numerical simulations).

If we assume that prograde and retrograde waves are produced with the same efficiency and if they are damped in the same way as they propagate inside the star, then waves have no net impact on the angular momentum distribution. It is thus important to treat wave damping properly. In Talon et al. (2002), we showed that the development of a double-peaked shear layer (SLO, for Shear Layer Oscillation), acts as a filter for waves and that, when the core is rotating faster than the surface, the asymmetry of this filter produces momentum extraction.

In Talon & Charbonnel (2005), we developed a formalism to incorporate the contribution of IGWs to the transport of angular momentum and chemical elements in stellar models on secular time-scales. Using only this filtered flux, it is possible to follow the contribution of internal waves over long (evolutionary) time-scales. We use the formalism developed by Goldreich & Kumar (1990) and Goldreich et al. (1994) to estimate the angular momentum luminosity $\mathcal{L}_J$ below the convective envelope (see also Kumar & Quataert 1997). The local momentum luminosity of waves is then given by

$$\mathcal{L}_J(r) = \sum_{\sigma,\ell,m} \mathcal{L}_{J\,\ell,m}\left(r_{\rm cz}\right) \exp\left[-\tau(r,\sigma,\ell)\right] \qquad (2.1)$$

where 'cz' refers to the base of the convection zone. τ corresponds to the integration of the local damping rate, and takes into account the mean molecular weight stratification

$$\tau(r,\sigma,\ell) = [\ell(\ell+1)]^{\frac{3}{2}} \int_r^{r_c} (K_T + \nu_v) \, \frac{N N_T^2}{\sigma^4} \left(\frac{N^2}{N^2 - \sigma^2}\right)^{\frac{1}{2}} \frac{\mathrm{d}r}{r^3} \tag{2.2}$$

(Zahn *et al.* 1997). In this expression, N_T^2 is the thermal part of the Brunt-Väisälä frequency, K_T is the thermal diffusivity and ν_v the (vertical) turbulent viscosity. σ is the local, Doppler-shifted frequency

$$\sigma(r) = \omega - m \left[\Omega(r) - \Omega_{\mathrm{cz}}\right] \tag{2.3}$$

and ω is the wave frequency in the reference frame of the convection zone. Let us mention that, in this expression for damping, only the radial velocity gradients are taken into account. This is because angular momentum transport is dominated by the low frequency waves ($\sigma \ll N$) for which horizontal gradients are much smaller than vertical ones.

When meridional circulation, turbulence, and waves are all taken into account, the evolution of angular momentum follows

$$\rho \frac{\mathrm{d}}{\mathrm{d}t} \left[r^2 \Omega\right] = \frac{1}{5r^2} \frac{\partial}{\partial r} \left[\rho r^4 \Omega U\right] + \frac{1}{r^2} \frac{\partial}{\partial r} \left[\rho \nu_v r^4 \frac{\partial \Omega}{\partial r}\right] - \frac{3}{8\pi} \frac{1}{r^2} \frac{\partial}{\partial r} \mathcal{L}_J(r), \tag{2.4}$$

(Talon & Zahn 1998) where U is the radial meridional circulation velocity. This equation takes into account the advective nature of meridional circulation rather than modeling it as a diffusive process and assumes a "shellular" rotation (see Zahn 1992 for details). Horizontal averaging was performed, and meridional circulation is considered only at first order. When we calculate the fast SLO's dynamics, U is neglected in this equation. This is justified by the fact that when shears are large such as in the SLO, angular momentum redistribution is dominated by the (turbulent) diffusivity rather than by meridional circulation. However, the complete equation is used when we compute full evolution models as in Charbonnel & Talon (2005).

2.2. *Shear layer oscillation (SLO) and filtered angular momentum luminosity*

One key feature of the wave-mean flow interaction is that the dissipation of IGWs produces an increase in the local differential rotation: this is caused by the increased dissipation of waves that travel in the direction of the shear (see Eqs. 2.2 and 2.3). In conjunction with viscosity, this leads to the formation of an oscillating doubled-peak shear layer that oscillates on a short time-scale (Gough & McIntyre 1998; Ringot 1998; Kumar, Talon & Zahn 1999). This oscillation is similar to the Earth quasi-biennial oscillation that is also caused by the differential damping of internal waves in a shear region.

This SLO occurs if the deposition of angular momentum by IGWs is large enough when compared with (turbulent) viscosity (Kim & MacGregor 2001)†. To calculate the turbulence associated with this oscillation, we rely on a standard prescription for shear turbulence away from regions with mean molecular weight gradients

$$\nu_v = \frac{8}{5} Ri_{\mathrm{crit}} K \frac{(r\mathrm{d}\Omega/\mathrm{d}r)^2}{N_T^2} \tag{2.5}$$

which takes radiative losses into account (Townsend 1958; Maeder 1995). This coefficient is time-averaged over a complete oscillation cycle (for details, see Talon & Charbonnel 2005).

In the presence of differential rotation, the dissipation of prograde and retrograde waves in the SLO is not symmetric, and this leads to a finite amount of angular momentum

---

† If viscosity is large, a stationary state can be reached.

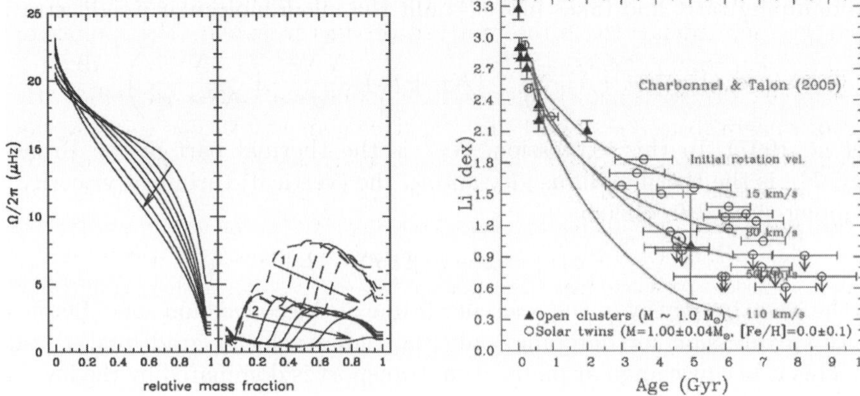

Figure 2. (Left) Evolution of the rotation profile in a solar-mass model with and without IGWs. The initial equatorial rotation velocity is $50\,\mathrm{km\,s^{-1}}$, and identical surface magnetic braking is applied. (*left*) Model without IGWs. Curves correspond to ages of 0.2, 0.5, 0.7, 1.0, 1.5, 3.0 and 4.6 Gy that increase in the direction of the arrow. (*right*) When IGWs are included, low-degree waves penetrate all the way to the core and deposit their negative angular momentum in the whole radiative region. Curves labelled 1 correspond to ages of 0.2, 0.21, 0.22, 0.23, 0.25 and 0.27 Gy and those labelled 2 to 0.5, 0.7, 1.0, 1.5, 3.0 and 4.6 Gy. From Charbonnel & Talon (2005). **(Right)** Evolution of surface lithium abundance with time for solar-mass models including gravity waves compared with observations in solar analogues. Updated from Meléndez *et al.* (2009).

being deposited in the radiative interior beyond the SLO. This is the filtered angular momentum luminosity $\mathcal{L}_J{}^{\mathrm{fil}}$. Let us mention that in fact, the existence of a SLO is not even required to obtain this differential damping between prograde and retrograde waves, and thus, as long as differential rotation exists at the base of the convection zone, waves will have a net impact of the rotation rate of the interior.

The SLO's dynamics is studied by solving Eq. (2.4) with small time-steps and using the whole wave spectrum while for the secular evolution of the star, one has to use instead the filtered angular momentum luminosity. Let us stress that in the case of the secular evolution we do not follow the SLO dynamics, because of its very short time scale. Rather, we only consider the filtered angular momentum luminosity beyond the SLO, and its effect on chemicals is given by a local turbulence calculated from a study of the SLO's dynamic over very short time-scales. Let us also mention here that, for both the SLO and the filtered angular momentum luminosity, differential damping is required. Since this relies on the Doppler shift of the frequency (see Eqs. 2.2 and 2.3), angular momentum redistribution will be dominated by the low frequency waves that experience a larger Doppler shift, but that are not so low that they will be immediately damped. Numerical tests indicate that this occurs around $\omega \simeq 1\ \mu$Hz.

3. The case of Population I low-mass stars, the Li dip and the Sun

An important property of IGWs is that their generation and efficiency in extracting angular momentum from stellar interiors depend on the structure of their convective envelope, which varies strongly with the effective temperature of the star. Figure 1 shows the T_{eff}-dependence of the filtered angular momentum luminosity of waves below the SLO, which directly measures the efficiency of wave-induced angular momentum extraction in zero-age main sequence stars around the Li dip. It appears that the net momentum luminosity slightly increases with increasing T_{eff}, presents a plateau, and suddenly drops at the T_{eff} of the dip. This clearly indicates that the momentum transport by IGWs has the proper T_{eff}-dependence to be the required process to explain the cool side of the Li dip (Talon & Charbonnel 2003).

Talon *et al.* (2002) have shown, in a static model, that waves can efficiently extract angular momentum from a star that has a surface convection zone rotating slower than the interior. Charbonnel & Talon (2005) then calculated the evolution of the internal rotation profile for a solar-mass star with surface spin-down. We showed that, in that case, waves tend to slow down the core, creating "slow" fronts that propagate from the core to the surface (Fig. 2). These calculations confirmed, in a complete evolution of solar-mass models evolved from the pre-main sequence to 4.6 Gy, that IGWs play a major role in braking the solar core. This momentum transport reduces rotational mixing in low-mass stars, leading to a theoretical surface lithium abundance in agreement with observations made in solar analogues of various ages (Fig. 2).

Figure 1 shows our predictions for rotation velocities and Li surface abundances together with the observed data at the age of the Hyades. On the left side of the dip, IGWs play no role and the predictions are taken from Charbonnel & Talon (1999). On the cool side of the dip IGWs are at act and lead to the rise of the surface Li. The model at ~ 5800 K corresponds to a $1.0\,M_\odot$ star. It was computed for 3 initial rotation velocities of 50, 80 and $110\,\text{km}\,\text{s}^{-1}$ (Charbonnel & Talon 2005). Models with IGWs are in perfect agreement with the observations, both regarding the amplitude of the Li depletion and the dispersion at a given effective temperature.

4. The case of Population II low-mass stars

In the context of primordial nucleosynthesis, it has long been debated whether Pop II stars could have depleted their surface Li abundance, just as their metal-rich counterpart did. Recent results on cosmic microwave background anisotropies, and especially those of the WMAP experiment, have firmly established that the primordial Li abundance is ~ 3 to 5 times higher than the measured Li value in dwarf stars along the so-called Spite plateau (Charbonnel & Primas 2005). The main theoretical difficulty to reproduce these data is that the Li abundance is remarkably constant in halo dwarfs, while it seems at first sight that depletion should lead to a larger dispersion.

A re-examination of Li data in halo stars available in the literature has led to a very surprising result: the mean Li value as well as its dispersion appear to be lower for the dwarfs than for the subgiant stars (Charbonnel & Primas 2005). In addition, all deviant stars, i.e. those with a strong Li deficiency or an abnormally high Li content, lie on or originate from the hot side of the Li plateau. These results indicate that halo stars that have now just passed the turnoff have experienced a Li history sightly different from that of their less massive counterparts.

This behaviour is the signature of a transport process for angular momentum whose efficiency changes on the extreme blue edge of the plateau and it corresponds to the generation and filtering of IGWs in Pop II stars (Fig. 3, Talon & Charbonnel 2004), just as it does in the case of Pop I stars.

Indeed and as discussed previously, the generation of IGWs and, consequently, their efficiency in transporting angular momentum, depend on the structure of the stellar convective envelope, which in turn depends on the effective temperature of the star as can be seen in Fig. 3. As in the case of Pop I stars on the red side of the Li dip, the net angular luminosity of IGWs is very high and constant in Pop II stars along the plateau up to $T_{\text{eff}} \sim 6300$ K. There, IGWs should dominate the transport of angular momentum and enforce quasi solid-body rotation of the stellar interior on very short timescale. As

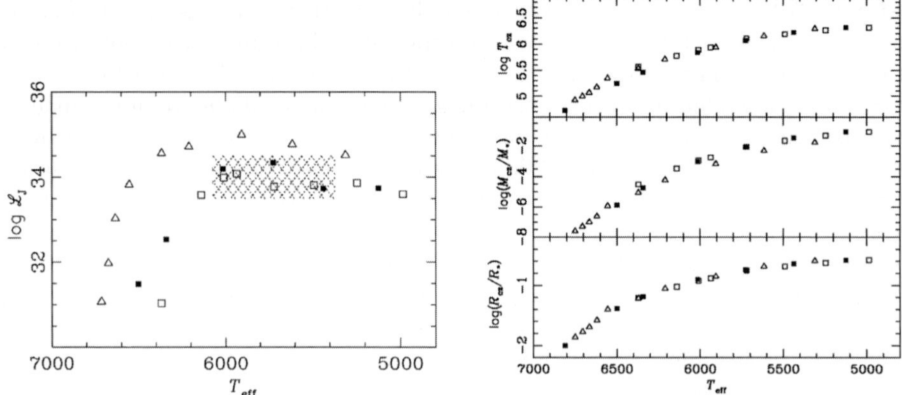

Figure 3. (Left) Net momentum luminosity at $0.03\,R_*$ below the surface convection zone as a function of T_{eff} for an initial differential rotation of $\delta\Omega = 0.01\,\mu$Hz over $0.05\,R_*$. The Li plateau region is dashed. From Charbonnel & Talon (2004). **(Right)** (*top*) Temperature at the base of the convection zone (T_{cz}), (*middle*) mass of the convection zone and (*bottom*) radius of the convection zone as a function of T_{eff}. Squares: Pop II stars on the zams (open squares) and at 10 Gyr (black squares); Triangles: Pop I stars on the zams. From Charbonnel & Talon (2004).

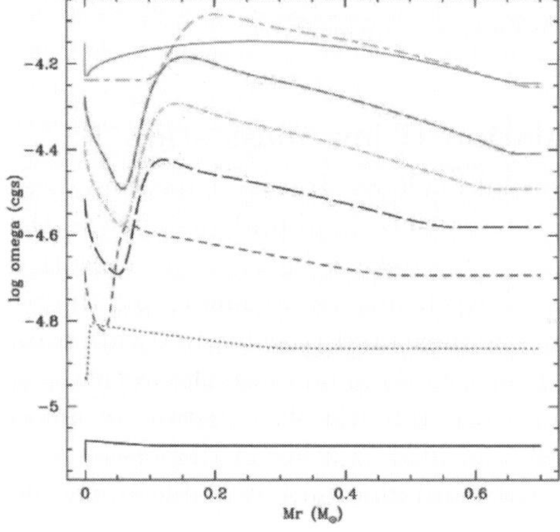

Figure 4. Evolution of the rotation profile inside a $0.7\,M_\odot$ Pop II star during the pre-main sequence. Curves correspond to 2, 5, 10, 15, 20, 25, 35 and 237 My, from bottom to top (acceleration is related to the star's overall contraction). The ZAMS velocity is $25\,\text{km}\,\text{s}^{-1}$.

a result, the surface Li depletion is expected to be independent of the initial angular momentum distribution, implying a very low dispersion of the Li abundance from star to star.

In more massive stars however the efficiency of IGWs decreases and internal differential rotation is expected to be maintained under the effect of meridional circulation and turbulence. Consistently, variations of the initial angular momentum from star to star would lead to more Li dispersion and to more frequent abnormalities in the case of the most massive stars where IGWs are not fully efficient, as required by the observations. We note that the mass-dependence of the IGWs efficiency also leads to a natural explanation of fast horizontal branch rotators.

Let us stress here that the absence of a Li dip in Pop II stars reflects the facts that Pop II dip stars already evolved past the main sequence.

We started the computation of complete stellar evolution models of Pop II stars from the pre-main sequence. Preliminary results show that the spin-down fronts that are observed in Pop I stars are also seen in Pop II stars during the PMS (see Fig. 4). Lithium depletion remains negligible during that phase. Calculation of the main sequence evolution is underway.

5. Conclusions

IGWs have a large impact on the evolution of low-mass stars, especially through their effect on the rotation profile, which then modifies meridional circulation and shear turbulence, and thus, the mixing of chemical elements. Within this framework, hydrodynamical models including the combined effects of meridional circulation, shear turbulence and internal gravity waves (using an excitation model that fits solar p-modes) successfully reproduce several observations:

- the time evolution of lithium in solar analogues (Fig. 2);
- the shape of the Li and Be dips (Fig. 1);
- the Li and Be behaviour in evolved stars (see Charbonnel & Lagarde, this Volume);
- the small amount of differential rotation measured by helioseismology (Fig. 2).

Up to now, no other theoretical model has achieved similar results. We also expect that IGWs can reduce the impact of the variety of initial rotation velocities during the spin-down phase of Pop II stars and thus, could be an important element in the understanding of the small amount of dispersion measured on the Li plateau.

Our comprehensive picture should have implications for other difficult unsolved problems related to the transport of chemicals and angular momentum in stars. We think in particular to the stars on the horizontal and asymptotic giant branches that exhibit unexplained abundance anomalies. No doubt that all these so-called "non-standard" physical processes must be part of the art of modelling stars in the 21st century. In this context, light elements such as Li and Be play a crucial role.

Acknowledgements

We acknowledge financial support from IAU, from the French "Programme National de Physique Stellaire" of CNRS/INSU, and from the Swiss National Science Foundation.

References

Ando, H. 1986, *A&A*, 163, 97

Balachandran, S. 1995, *ApJ* 446, 203

Balmforth, N. J. 1992, *MNRAS* 255, 639

Boesgaard, A. M. & Tripicco, M. J. 1986, *ApJ*, 302, L49

Boesgaard, A. M. 1987, *PASP* 99, 1067

Brown, T. M., Christensen-Dalsgaard, J., Dziembowski, W. A., Goode, P., Gough, D. O., & Morrow, C. A. 1989, *ApJ* 343, 526

Burkhart, C. & Coupry, M. F. 2000, *A&A* 354, 216

Chaboyer, N., Demarque, P., Guenther, D. B., & Pinsonneault, M. H. 1995, *ApJ* 446, 435

Charbonneau, P. & Mac Gregor, K. B. 1993, *ApJ* 417, 762

Charbonnel, C. & Primas, F. 2005, *A&A* 442, 961

Charbonnel, C. & Talon, S. 1999, *A&A* 351, 635

Charbonnel, C. & Talon, S. 2005, *Science* 309, 2189

Charbonnel, C. & Talon, S. 2007, AIP Conference Proceedings, Volume 948, pp. 15-26

Couvidat, S., García, R. A., Turck-Chièze, S., Corbard, T., Henney, C. J., & Jiménez-Reyes, S. 2003, *ApJ* 597, L77

Deliyannis, C. P., King, J. R., & Boesgaard, A. M. 1997, Kontikas E., *et al.* (eds), *"Wide-field spectroscopy"*, p. 201

Eggenberger, P., Maeder, A., & Meynet, G. 2005, *A&A*, 440, L9

Fossati, L., Bagnulo, S., Landstreet, J., Wade, G., Kochukhov, O., Monier, R., Weiss, W., & Gebran, M. 2008, *A&A*, 483, 891

Gaigé, Y. 1993, *A&A* 269, 267

García, R. A., Turck-Chièze, S., Jiménez-Reyes, S. J., Ballot, J., Pallé, P. L., Eff-Darwich, A., Mathur, S., & Provost, J. 2007, *Science*, 316, 1591

García López, R. J. & Spruit, H. C. 1991, *ApJ* 377, 268

Gebran, M., Monier, R., & Richard, O. 2008, *A&A*, 479, 189

Goldreich, P. & Kumar, P. 1990, *ApJ* 363, 694

Goldreich, P., Murray, N., & Kumar, P. 1994, *ApJ* 424, 466

Gough, D. O. & McIntyre, M. E. 1998, *Nature* 394, 755

Kim, E. & MacGregor, K. B. 2001, *ApJ*, 556, L117

Kiraga, M., Jahn, K., Stepien, K., & Zahn, J.-P. 2003, *Acta Astronomica* 53, 321

Kosovichev, A., *et al.* 1997, *Sol. Phys.* 170, 43

Kumar, P. & Quataert, E. J. 1997, *ApJ* 575, L143

Kumar, P., Talon, S., Zahn, J.-P. 1999, *ApJ* 520, 859

Maeder, A. 1995, *A&A*, 299, 84

Maeder, A. & Meynet, G. 2000, *ARAA* 38, 143

Matias, J. & Zahn, J.-P. 1998, Provost & Schmider (eds), *"Sounding solar and stellar interiors"*, IAU Symp. 181

Meléndez, J., Ramírez, I., Casagrande, L., Asplund, M., Gustafsson, B., Yong, D., Do Nascimento, J. D., Castro, M., & Bazot, M. 2009 *Ap&SS*, tmp, 221

Michaud, G. 1986, *ApJ* 302, 650

Montalban, J. 1994 *A&A*, 281, 421

Montalban, J. & Schatzman E. 1996 *A&A*, 305, 513

Montalban, J. & Schatzman E. 2000 *A&A*, 354, 943

Palacios, A., Talon, S., Charbonnel, C., & Forestini, M. 2003, *A&A* 399, 603

Pasquini, L., Randich, S., Zoccali, M., Hill, V., Charbonnel, C., & Nordström, B. 2004, *A&A* 424, 951

Pilachowski, C. A., Saha, A., & Hobbs, L. M. 1988, *PASP* 100, 474

Pinsonneault, M. H., Kawaler, S. D., Sofia, S., & Demarque, P. 1989 *ApJ* 338, 424

Press, W. H. 1981 *ApJ*, 245, 286

Ringot, O. 1998, *A&A* 335, 89

Rogers, T. M. & Glatzmaier, G. A. 2005a, *ApJ*, 620, 432

Rogers, T. M. & Glatzmaier, G. A. 2005b, *MNRAS*, 364, 1135

Schatzman, E. 1993 *A&A* 279, 431

Smiljanic, R., Pasquini, L., Charbonnel, C., & Lagarde, N., 2009 arXiv0910.4399, A&A, in press

Takeda, Y., Kawanomoto, S., Takada-Hidai, M., & Sadakane, K. 1998, *PASJ* 50, 509

Talon S. 1997 *PhD Thesis*, Université Paris VII

Talon, S., Kumar, P., Zahn, J.-P. 2002, *ApJL* 574, 175

Talon, S. & Charbonnel, C. 1998, *A&A* 335, 959

Talon, S. & Charbonnel, C. 2003, *A&A* 405, 1025

Talon, S. & Charbonnel, C. 2004, *A&A* 418, 1051

Talon, S. & Charbonnel, C. 2005, *A&A* 440, 981

Talon, S., Richard, O., & Michaud, G. 2006, *ApJ* 645, 634

Talon, S. & Zahn, J. P. 1998, *A&A*, 329, 315

Théado, S. & Vauclair, S. 2003, *ApJ*, 587, 795

Townsend A. A., 1958, *J. Fluid Mech.* 4, 361

Varenne, O. & Monier R. 1999 *A&A* 351, 247

Wallerstein, G., Herbig G. H., & Conti, P. S. 1965, *ApJ*, 141, 610

Young, P. A., Knierman, K. A., Rigby, J. R., & Arnett, D. 2003, *ApJ*, 585, 1114

Zahn, J. P. 1992, *A&A* 265, 115

Zahn, J. P., Talon, S., & Matias, J. 1997, *A&A* 322, 320

Light Elements in the Universe
Proceedings IAU Symposium No. 268, 2009
C. Charbonnel, M. Tosi, F. Primas & C. Chiappini, eds.

© International Astronomical Union 2010
doi:10.1017/S1743921310004497

Rotational mixing and Lithium depletion

M. H. Pinsonneault[1]

[1]Ohio State University, Dept. of Astronomy 140 W. 18th Ave. Columbus, OH 43210 USA
email: `pinsonneault.1@osu.edu`

Abstract. I review basic observational features in Population I stars which strongly implicate rotation as a mixing agent; these include dispersion at fixed temperature in coeval populations and main sequence lithium depletion for a range of masses at a rate which decays with time. New developments related to the possible suppression of mixing at late ages, close binary mergers and their lithium signature, and an alternate origin for dispersion in young cool stars tied to radius anomalies observed in active young stars are discussed. I highlight uncertainties in models of Population II lithium depletion and dispersion related to the treatment of angular momentum loss. Finally, the origins of rotation are tied to conditions in the pre-main sequence, and there is thus some evidence that environment and planet formation could impact stellar rotational properties. This may be related to recent observational evidence for cluster to cluster variations in lithium depletion and a connection between the presence of planets and stellar lithium depletion.

Keywords. Hydrodynamics – stars: abundances, rotation, spots

1. Introduction

Lithium is an extraordinarily sensitive diagnostic of stellar structure and evolution. The observed lithium abundances in stars, not surprisingly, reveal an extremely complex picture, and it can sometimes be difficult to remember why rotational mixing is a useful framework for interpreting this data. I therefore begin by briefly summarizing the case for rotation as the physical ingredient responsible for light element depletion in stars.

1.1. *Evidence for rotational mixing*

The first and most important point is that stellar rotation is capable of driving mild envelope mixing at the observationally required rates (Pinsonneault *et al.* 1989.) Rotation induces a departure from spherical symmetry which generates meridional circulation currents, and both structural evolution and angular momentum loss from magnetized winds generate shears which can drive mild turbulence. Lithium is easily destroyed in stellar interiors, and such mild mixing can therefore generate surface lithium depletion.

This leads directly to a second important feature of rotational mixing which is observationally required: namely, stars which rotate at different rates will have different mixing histories. Rapid rotators experience stronger torques and larger shears than slow rotators, and they also are less spherical. It is therefore a basic prediction of rotational mixing that there should be a dispersion in mixing rates which can manifest itself as a dispersion in lithium at fixed mass, composition, and age. Lithium is observed to have a significant dispersion in many clusters (see Pinsonneault 1997 for a theoretical review and Sestito & Randich 2005 for a more recent observational synthesis) while other elements in open clusters are very uniform (Paulson *et al.* 2003). Other mechanisms, such as gravity waves and microscopic diffusion, can generate depletion but not dispersion, so this observed feature allows us to discriminate between physical processes.

Finally, both the mass dependence and time dependence of the observed depletion pattern strongly implicate rotationally driven mixing as the culprit. Rotation declines with age, and so does lithium depletion. By contrast, processes such as gravitational settling tend to be more independent of age, or even increase in rate as stars get older. Rotational mixing also extends through stellar envelopes, and as a result it can simultaneously mix different elements and be present in stars with very different surface convection zone depths. We observe lithium depletion in all low mass open cluster stars, which would not be expected if lithium depletion were a phenomenon confined to the convection zone boundary. This does not rule out interesting interactions with other physics processes, such as magnetic or wave-driven angular momentum transport (see the contribution by Talon in these proceedings), but it does require rotation as a component of the solution.

However, the physics of stellar angular momentum evolution is extremely challenging, and it has proven difficult to develop a rigorous physical model. This has led to a sort of stasis in our understanding of phenomena such as rotational mixing. Fortunately, there have been positive developments, which I summarize below, which reveal a dynamic and more complete picture of stellar evolution. In Section 2 recent advances in our understanding of angular momentum evolution are reviewed; Section 3 then discusses three areas where there are either new observational or theoretical features in stellar lithium depletion. A discussion of some recent developments is given in section 4.

2. Angular momentum evolution

Stellar rotation is an initial value problem, and the initial conditions are set by the details of the star formation process. The angular momentum distribution is subsequently modified by angular momentum loss (via star-disk interactions) and internal angular momentum transport. The physics of the latter is vigorously debated in the literature, with three distinct mechanisms (hydrodynamic, wave-driven, and magnetic) all being in principle important. Rotational mixing is a natural byproduct of angular momentum transport in stellar radiative interiors, especially from hydrodynamic mechanisms. This is a rich field, so I will summarize the main developments relevant for rotational mixing.

Stars appear at the deuterium-burning birthline (Stahler 1988) with a range of rotation rates, typically well below that expected for accretion from a Keplerian disk. The currently favored explanation is that magnetic coupling between the protostar and the accretion disk regulates the rotation (Shu *et al.* 1994.) In this framework, the initial rotation rate can be thought of as related to the mass accretion rate in the early hydrodynamic stages of star formation. However, the predicted rotation rates on the main sequence are both too rapid and too uniform if models with the observed rotation rates are evolved to the main sequence, even if torques from solar-like winds are included.

However, if a coupling between protostars and their accretion disks exists, the initial spread of rotation rates can be amplified and stars can reach the main sequence as relatively slow rotators. Much observational work has also been invested in the question of star-disk coupling, with a diversity of results largely centered around the proper choice of disk proxies and disentangling evolutionary effects. However, recent Spitzer studies (Rebull *et al.* 2006) have provided strong evidence for a relationship between rotation and the presence of disks. This may reflect a coupling between the protostar and accretion disk similar to that operating at the earlier stages, or it could be induced by an enhanced stellar wind tied to accretion. In either case, the lifetime of accretion disks and their degree of coupling to the parent star is crucial for establishing the main sequence rotation. Rotation is therefore now perceived as a product of environment, and this raises the interesting possibility that rotational mixing may also depend on where a star was born or on how the accretion disk evolved.

There are also now very large databases of stellar rotation periods, ranging from star forming regions (Rodriguez-Ledesma *et al.* 2009) to extensive open cluster surveys such as the Monitor program(Irwin *et al.* 2009) and transit studies such as the one which yielded a large database of rotation periods in the 550 Myr system M37(Hartman *et al.* 2009). The latter study in particular indicates the ability of modern campaigns to infer rotation periods caused by spot modulation for large stellar samples at small amplitude.

These samples can in turn be used to reconstruct the angular momentum evolution of stellar populations, in particular the dependence of angular momentum loss on rotation rate and mass, as well as the coupling timescale between core and envelope (e.g. Irwin *et al.* 2007, Denissenkov *et al.* 2009). Different groups agree on the essential features. Angular momentum loss scales as the rotation rate cubed at low rotation rates, then saturates at a threshold which decreases as mass decreases. The net effect is that lower mass stars take longer to spin down and longer for their rotation rates to converge. The cores of the slowly rotating population couple to their envelopes with a timescale of order 100 Myr, while rapidly rotating stars appear to be more strongly coupled. These results are consistent with helioseismic data indicating that the rotation of the solar interior is strongly coupled to that of the surface convection zone.

This combination of theoretical advances and improved rotational data and empirical constraints therefore has significant promise for more robust rotational mixing predictions in the future.

3. Lithium depletion and rotational mixing revisited

The basic rotational mixing picture can be simply defined. Stars experience a mass-dependent pre-main sequence lithium burning epoch, which ends when they develop substantial radiative cores. They then experience rotational mixing on the main sequence, induced either by shears generated by angular momentum loss or their departure from spherical symmetry. As the stars spin down the rate of lithium depletion decreases. This overall picture is reasonable, but a number of phenomena defy easy categorization within it. This is in large part because of the interaction of rotation with other phenomena typically neglected in stellar models. Below are three examples.

3.1. *Structural effects of starspots*

There is a striking dispersion in lithium abundances among late-type stars in young open clusters; the Pleiades is the clearest example (Soderblom *et al.* 1993.) This trend is not expected from rotational mixing in such young stars, and the relative effect is also the opposite of the one expected: namely, the least depleted stars are the most active and heavily spotted. Much subsequent work has focused on whether the dispersion is real or induced by the heavily spotted nature of the stars in question; the model atmospheres used to interpret the data typically neglect the large changes in the strength of the lithium feature which would be associated with a substantial fraction of the surface covered with cool spots with ample neutral lithium. However, in recent work (King *et al.* 2009) we found that the scatter in K I was much less than the scatter in Li I, indicating that the bulk of the dispersion is real. The likely origin in our view is actually a different mechanism altogether, and it is motivated by recent data on radius anomalies in active stars from interferometric and eclipsing binary studies.

Eclipsing binaries such as YY Gem (Torres & Ribas 2002) were found to have radii significantly larger than those predicted by interiors theory. Subsequent work traced out a radius anomaly pattern. More recent interferometric data permits the measurement of radii for inactive field stars, which are found to be in accord with theoretical predictions (Demory *et al.* 2009.) High activity, such as that found in tidally synchronized short-period binaries, therefore appears to puff up stars. A similar effect during the fully convective protostellar phase would reduce the degree of pre-main sequence lithium depletion; if this varied from star to star it could generate a dispersion with characteristics remarkably like the data. This is illustrated in Fig. 1., where Pleiades data from Soderblom *et al.* (1993) is compared with standard stellar models (lower line) and models with a radius inflated by 10 percent, the level inferred in highly active stars (upper line). In addition to being an attractive solution for a longstanding problem, this leads to an interesting insight. Stellar activity is not a mere detail; it can impact the entire structure of a star and change its mixing history.

3.2. *Blue stragglers and halo Lithium depletion*

The lithium depletion pattern in metal-poor stars poses a different problem; the majority of stars exhibit little dispersion and the observed abundances appear to be nearly independent of surface temperature or metallicity. One striking counter-example is the presence of a small but real population of highly depleted stars (Thorburn 1994.) In recent work on blue stragglers we find that a population of sub-turnoff merger products, presumably highly lithiumn depleted, is predicted to arise from such mergers; the number expected is close to that observed in halo stars (Andronov *et al.* 2006). This confirms the suggestion in Ryan *et al.* (2001) that these highly depleted stars should not be regarded as the tail of a rotational mixing distribution, but rather that they have a distinct origin.

This does not, however, require that mixing be absent in halo stars. The nature of the dispersion predicted depends on both the distribution of initial conditions and the angular momentum loss history. At present we can only extrapolate Population I conditions to Population II stars. Their torques and initial conditions could very well have been different; for example, planet formation may be less common in them, and this could impact the distribution of accretion disk coupling timescales (see Section 4.) Future work on the activity properties of tidally synchronized halo stars may prove diagnostics of the braking law, while we may need rotation data in more metal-poor outer disk systems to test the metallicity dependence of the initial conditions.

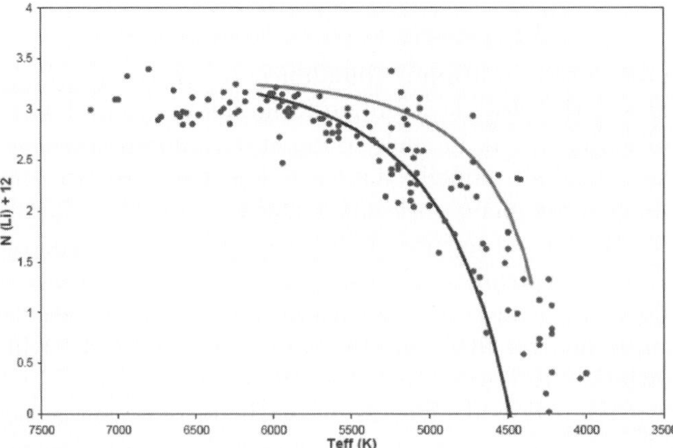

Figure 1. Lithium abundances as a function of effective temperature in the Pleiades cluster compared with standard models (lower line) and models which were inflated during the pre-main sequence lithium depletion epoch (upper line).

3.3. *Interaction of diffusion and rotational mixing*

Microscopic diffusion (or gravitational settling and thermal diffusion) is a basic physical process expected to occur in stars, typically over a very long timescale. The net effect is that heavy species tend to sink relative to light ones, although radiation pressure can drive some heavier elements upwards in sufficiently thin surface convection zones (Michaud 1970). There is clear evidence for diffusion in the Sun (Bahcall & Pinsonneault 1992), both in the sound speed profile and in the detection of a surface helium abundance lower than that initially required to reproduce the solar luminosity.

Diffusion can induce lithium depletion directly, but it also has interesting interactions with rotational mixing (see Richard *et al.* 1996 for a nice example in the solar context). Gravitational settling operates over shorter timescales for thinner convection zones, and it produces mean molecular weight gradients at the base of the surface convection zone. It is energetically unfavorfable to mix in the presence of a mu gradient, and mixing can erase composition gradients; there is thus a natural competition between the two processes. Furthermore, the timescale for mixing increases with age, while the timescale for settling changes very slowly (and tends to decrease as stars evolve to higher effective temperatures). One might therefore expect rotational mixing to predominate earlier while diffusion suppresses mixing at later ages, and for this interaction to depend on mass and composition. This may be related to the apparent stalling of lithium depletion in older open clusters (discussed by Randich in these proceedings), and could be an additional source of lithium depletion in halo stars as well. Recent evidence for settling in multiple elements (Korn, these proceedings) of globular cluster stars provides evidence that diffusion sets in for older stars; an earlier epoch of depletion is certainly permitted by theory, although establishing this observationally will require additional work as discussed in the angular momentum evolution section. Observations of multiple elements, as already done in globulars, could be used to establish the diffusion signature in old open clusters and the interaction between mixing and separation.

4. Future directions

In closing I'd like to note some other wrinkles which may prove important for understanding lithium depletion: differences in depletion patterns from cluster to cluster (see the presentation by Randich) and an apparent excess lithium depletion in stars which host planets (see the talks by Israelian and, for a contrary view, Melendez.) Both can be interpreted in the framework where stellar rotation properties are determined by interactions between protostars and accretion disks. In dense stellar environments the timescale for interactions can be comparable to the lifetimes inferred for accretion disks, raising the possibility that stars born in such regions might have a different distribution of disk lifetimes than stars born in loose associations. This hypothesis is testable in the measured rotation rates of young systems, and this is an important potential effect (especially if we use clusters as an evolutionary sequence!) which needs to be explored.

The recent report that stars with planets have excess lithium depletion (Israelian *et al.* 2009) may be a fascinating example of how the formation of planets can impact the properties of stars. Bouvier (2008) has proposed a linkage, arguing that systems with planets should have long-lived accretion disks. These in turn become slow rotators, with large relative shears, which in turn could drive excess mixing. He thus argued that there may be a connection between lithium overdepletion and planet formation. Such a link is certainly plausible, but the opposite correlation appears to be required by rotational mixing. More rapid rotators experience larger absolute torques (and in any case subsequently evolve to become slow rotators, thus in effect adding the mixing from the rapid to that of the slow phase). They also experience larger departures from spherical symmetry; both imply stronger mixing. However, the rotation is set primarily by the coupling between star and disk, not necessarily in the disk lifetime itself, and this may explain the apparent contradiction between "massive disk required for planets" and "weak star-disk interaction required for rapid rotation and lithium depletion." This avenue may prove promising to explore, and it would be a delightful turn of events if the planetary tail could wag the stellar dog.

References

Andronov, N., Pinsonneault, M. H., & Terndrup, D. M. 2006, *ApJ*, 646, 1160

Bahcall, J. N. & Pinsonneault, M. H. 1992, *RMP*, 64, 885

Bouvier, J. 2008, *A&A* (Letters), 489, 53

Demory, B.-O., Ségransan, D., Forveille, T., Queloz, D., Peuzit, J. L., *et al.* 2009, *A&A*, 505, 205

Denissenkov, P., A., Pinsonneault, M. H., Terndrup, D. M., & Newsham, G. 2009, submitted *ApJ* (astro-ph/0911.1121)

Hartman, J. D., Gaudi, B. S., Pinsonneault, M. H., Stanek, K. Z., Holman, M. J., McLeod, B. A., Meibom, S., Barranco, J. A., & Kalirai, J. S. 2009, *ApJ*, 691, 342

Irwin, J., Hodgkin, S., Aigrain, S., Hebb, L., Bouvier, J., Clarke, C., Moraux, E., & Bramich, D. M., 2007, *MNRAS*, 377, 741

Irwin, J., Aigrain, S., Bouvier, J., Hebb, L., Hodgkin, S., Irwein, M., & Moraux, E. 2009, *MNRAS*, 392, 1456

Israelian, G., Delgado Mena, E., Santos, N. C., Sousa, S. G., Mayor, M., Udry, S., Dominguez Cerdena, C., Rebolo, R., & Randich, S. 2009, *Nature*, 462, 189

King, J. R., Schuler, S. C., Hobbs, L. M., & Pinsonneault, M. H. 2009, *ApJ* in press (astro-ph/1001.2796)

Michaud, G. 1970, *ApJ*, 160, 641

Paulson, D. B., Sneden, C., & Cochran, W. D. 2003, *AJ*, 125, 3185

Pinsonneault, M. H. 1997, *ARAA*, 35, 557

Pinsonneault, M. H., Kawaler, S. D., Sofia, S., & Demarque, P. 1989, *ApJ*, 338, 424

Rebull, L. M., Stauffer, J. R., Megeath, S. T., Hora, J. L. & Hartmann, L. 2006, *ApJ*, 646, 297

Richard, O., Vauclair, S., Charbonnel, C., & Dziembowski, W. A.. 1996, *A&A*, 312, 1000

Rodriguez-Ledesma, M. V., Mundt, R., & Eisloffel, J. 2009 *A & A*, 502, 883

Ryan, S. G., Beers, T. C., Kajino, T., & Rosolankova, K. 2001, *ApJ*, 547, 231

Sestito, P. & Randich, S. 2005, *A&A*, 442, 615

Shu, F. H., Najita, J., Ostriker, E., Wilkin, F., Ruden, S., & Lizano, S. 1994, *ApJ*, 429, 781

Soderblom, D., Jones, B. F., Balachandran, S., Stauffer, J. R., Duncan, D. K., Fedele, S. B., & Hudon, J. D. 1993, *AJ*, 106, 1059

Stahler, S. W. 1988, *ApJ*, 332, 804

Thorburn, J. A. 1994, *ApJ*, 421, 318

Torres, G. & Ribas, I. 2002, *ApJ*, 567, 1140

Light Elements in the Universe
Proceedings IAU Symposium No. 268, 2009
C. Charbonnel, M. Tosi, F. Primas & C. Chiappini, eds.

© International Astronomical Union 2010
doi:10.1017/S1743921310004503

Effects of rotation and magnetic fields on the structure and surface abundances of solar-type stars

P. Eggenberger, A. Maeder, and G. Meynet

Observatoire de Genève, Université de Genève,
51 Ch. des Maillettes, CH-1290 Versoix, Suisse
email: [patrick.eggenberger;andre.maeder;georges.meynet]@unige.ch

Abstract. The effects of shellular rotation on the modelling of solar-type stars (in particular internal structure, evolutionary tracks in the HR diagram, lifetimes and surface abundances) are first examined. Then the effects of a dynamo possibly occuring in the internal stellar radiative zone by imposing nearly solid body rotation are studied. These results are finally discussed in the context of the rotational history of exoplanet host stars and the link between lithium depletion and the presence of exoplanets.

Keywords. Stars: rotation, magnetic fields, interiors, abundances

1. Introduction

Rotation is one of the key processes that changes the internal structure and global properties of stellar models with a peculiarly strong impact on the physics and evolution of massive stars (see e.g. Maeder 2009). In this work, we first focus on the effects of rotational mixing on the global properties and surface abundances of solar-type stars by comparing stellar models including shellular rotation to non-rotating models. We then investigate the effects of a dynamo in the radiative zone of a solar-type star on the efficiency of rotational mixing by computing models with shellular rotation and the Tayler-Spruit dynamo (Spruit 2002). A comparison between models of solar-type stars including rotation only and both rotation and magnetic fields is made in the framework of a scenario proposed by Bouvier (2008) that relates the enhanced lithium depletion in exoplanet host stars to their rotational history.

2. Effects of rotation

To investigate the effects of rotational mixing on the properties of solar-type stars, $1\,M_\odot$ models are computed with the Geneva stellar evolution code (Eggenberger *et al.* 2008). These models are computed with a solar chemical composition as given by Grevesse & Noels (1993) and a solar calibrated value for the mixing-length parameter. The main-sequence evolution of non-rotating models with and without atomic diffusion of helium and heavy elements is compared to the main-sequence evolution of a rotating model with an initial velocity of $50\,\mathrm{km\,s^{-1}}$ on the zero age main sequence (ZAMS). We adopt the braking law of Kawaler (1988) to account for the magnetic braking undergone by solar-type stars when arriving on the main sequence. Two parameters enter this braking law: the saturation velocity Ω_sat and the braking constant K. Following Bouvier *et al.* (1997), Ω_sat is fixed to $14\,\Omega_\odot$ and the braking constant K is calibrated so that a $1\,M_\odot$ star with an initial velocity of $50\,\mathrm{km\,s^{-1}}$ on the ZAMS reproduces the solar surface rotational velocity after $4.57\,\mathrm{Gyr}$. The rotating model also includes the effects of atomic diffusion

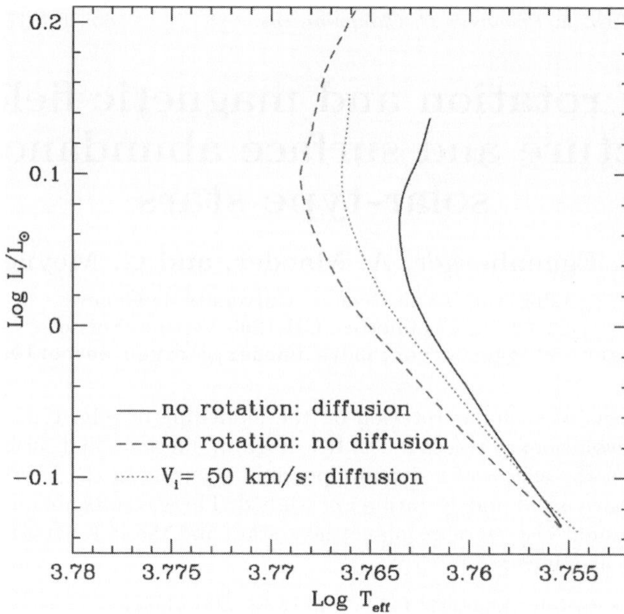

Figure 1. Main-sequence evolution of $1\,M_\odot$ models with and without rotation. The continuous and dashed lines correspond to non-rotating models computed with and without atomic diffusion, respectively. The dotted line indicates a rotating model computed with an initial velocity of $50\,\mathrm{km\,s^{-1}}$ on the ZAMS and with atomic diffusion.

of helium and heavy elements. All three $1\,M_\odot$ models share therefore exactly the same initial parameters except for the inclusion of shellular rotation and atomic diffusion.

The evolutionary tracks in the HR diagram corresponding to the main-sequence evolution of these models are shown in Fig. 1. By comparing the rotating model (dotted line) to the non-rotating model computed with the same initial parameters except for the inclusion of shellular rotation (continuous line), we see that the rotating model exhibits larger effective temperatures and slightly larger luminosities than the non-rotating one. This results in a shift of the evolutionary track to the blue part of the HR diagram when rotation is included in the computation. Concerning atomic diffusion, Fig. 1 shows that the non-rotating model including atomic diffusion (continuous line) is characterised by lower effective temperatures and luminosities compared to the non-rotating model without atomic diffusion (dashed line). The inclusion of atomic diffusion is thus found to shift the evolutionary track towards the red part of the HR diagram.

These changes observed in the HR diagram can be related to differences in the surface chemical composition of the models. Fig. 2 shows the helium surface mass fraction Y_s as a function of time during the main-sequence evolution for the three models shown in Fig. 1. The model computed without atomic diffusion and rotation (dashed line) exhibits a constant value of the helium surface abundance since no mixing mechanisms are at work in the radiative zone. The inclusion of atomic diffusion leads to a decrease of the helium mass fraction at the stellar surface. By comparing the models including atomic diffusion computed with and without rotation (dotted and continuous lines), we note a lower decrease of the helium surface abundance for the rotating model than for the non-rotating one. Rotational mixing is thus found to counteract the effects of atomic diffusion in the external layers of the star. As a result, rotating models exhibit larger values of helium abundances at the surface. This leads to a decrease of the opacity in

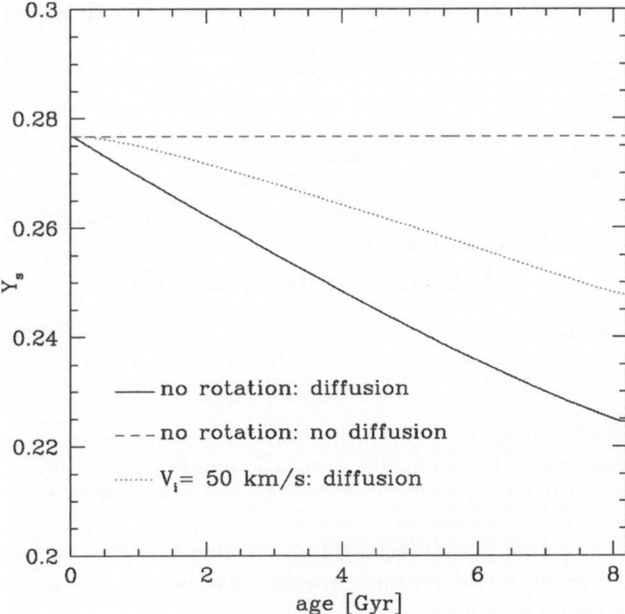

Figure 2. Surface helium mass fraction during the evolution on the main sequence for the $1\,M_\odot$ models shown in Fig. 1.

the external layers of the star and explains the shift towards the blue part of the HR diagram observed in Fig. 1.

The effects of rotation are of course not restricted to the external layers of the star. Rotational mixing has indeed a large impact on the properties of the central layers by bringing fresh hydrogen fuel to the central stellar core. As a result, the central hydrogen mass fraction at a given age is found to be larger for rotating models than for models without rotation. The inclusion of atomic diffusion leads to the opposite effect, since a non-rotating model computed with atomic diffusion exhibits a more rapid decrease of the central hydrogen abundance than a model without atomic diffusion. It is interesting to note that the efficiency of rotational mixing relative to atomic diffusion is found to be larger in the central layers of a solar-type star than in its external layers. In the external layers, rotation only reduces the efficiency of atomic diffusion but does not completely counteract these effects (see Fig. 2), while in the central layers the increase of the hydrogen abundance due to rotation is larger than the decrease related to atomic diffusion. Due to rotational mixing, the main-sequence lifetime is then larger for stellar models including rotation.

Rotational effects change the structure of a solar-type star and hence its asteroseismic properties. In particular, the change of the central chemical gradients and the increase of the central hydrogen abundance induced by rotational mixing lead to an increase of the values of the asteroseismic small separation for rotating models compared to non-rotating models (Eggenberger *et al.* 2006; Eggenberger & Carrier 2006). The effects of rotation on the external layers are of course reflected in the change of the surface abundances but can also be revealed by asteroseismic measurements. As discussed above, the inclusion of rotation results in a significant increase of the effective temperature. At a given luminosity, this leads to smaller radii for rotating models than for non-rotating models. Consequently, a rotating model will be characterised by a larger mean density and hence a larger value of the asteroseismic mean large separation than a non-rotating model. It is

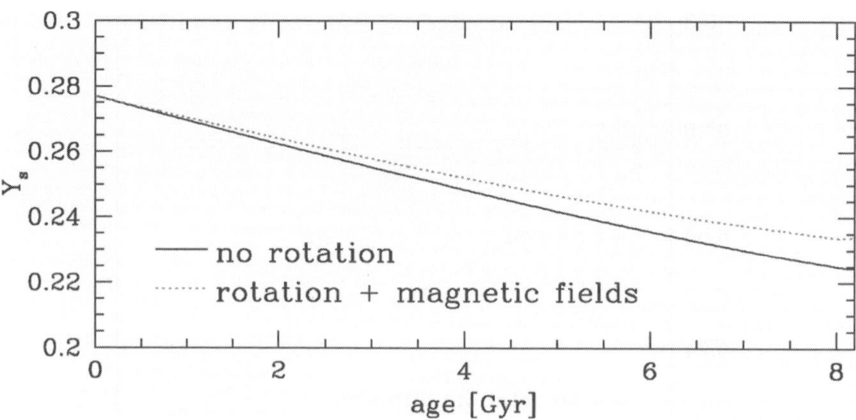

Figure 3. Surface helium mass fraction during the main-sequence evolution of a non-rotating model (continuous line) and a model including rotation and magnetic fields (dotted line). Both models are computed with atomic diffusion of helium and heavy elements.

thus interesting to note that spectroscopic measurements of the surface abundances and asteroseismic measurements can give us valuable insights about the effects of rotational mixing on the properties of solar-type stars.

3. Effects of magnetic fields

After the effects of rotation, the effects of magnetic fields and in particular of a dynamo possibly occuring in the radiative zone of the star are studied. These effects are particularly interesting to consider in the context of the internal rotation of the Sun, since models including only shellular rotation predict a rapidly rotating core in disagreement with helioseismic measurements. Models of solar-type stars including both shellular rotation and the Tayler-Spruit dynamo (Spruit 2002) are then computed. We recall here that the theoretical formulation of this dynamo is still a matter of debate (see e.g. Denissenkov & Pinsonneault 2007; Zahn *et al.* 2007; Rüdiger *et al.* 2009). It is however worth investigating the effects of such an efficient process for the transport of angular momentum on the properties of solar-type stars, since models including both the effects of shellular rotation and magnetic fields as prescribed by the Tayler-Spruit dynamo are found to correctly reproduce the helioseismic measurements of the solar rotation profile (Eggenberger *et al.* 2005).

Figure 3 shows the comparison between the helium surface abundance during the main sequence for a $1\,M_\odot$ model including both shellular rotation and the Tayler-Spruit dynamo computed with an initial velocity on the ZAMS of $50\,\mathrm{km\,s^{-1}}$ and the corresponding non-rotating model. Both models include atomic diffusion of helium and heavy elements. The inclusion of both rotation and magnetic fields results in a slight increase of the helium mass fraction at the stellar surface compared to a non-rotating model including only atomic diffusion. This increase is however much lower than for a model computed with rotation only (compare Fig. 2 and Fig. 3). The efficiency of rotational mixing is thus found to be strongly reduced when the effects of the Tayler-Spruit dynamo are taken into account. The near solid body rotation of models including this dynamo leads indeed to a low value of the diffusion coefficient associated to the shear turbulent mixing. The transport of chemicals by shear mixing is thus strongly reduced when the effects of the Tayler-Spruit are included in the computation. Compared to the case with rotation only, we also note a slight increase of the transport of chemicals by the meridional circulation for models with magnetic fields. This increase is however much smaller than the strong decrease of the shear turbulent mixing leading to a net decrease of the global efficiency of rotational mixing for a rotating model of a solar-type star computed with the Tayler-Spruit dynamo.

4. Rotational history of exoplanet host stars

Finally, the effects of rotation and magnetic fields are briefly discussed in the context of the rotational history of exoplanet host stars and in particular in the framework of a scenario suggested by Bouvier (2008). This scenario is first based on the fact that observations of rotational periodes of young solar-type stars suggest that slow rotators develop a high degree of differential rotation between the radiative core and the convective envelope, while solid-body rotation is favoured for fast rotators (see e.g. Irwin *et al.* 2007; Bouvier 2008). This result implies that slow rotators can be modelled with shellular rotation only, while fast rotators are modelled with both rotation and the Tayler-Spruit dynamo in order to produce a sufficient internal coupling to ensure solid body rotation. As discussed in the preceding section, this leads to different surface chemical compositions for slow and fast rotators, since the efficiency of rotational mixing is strongly reduced when the effects of magnetic fields are taken into account. Slow rotators will then exhibit lower values of surface lithium abundance than fast rotators.

As discussed by Bouvier (2008), the rotation of the star on the ZAMS depends on the initial velocity of the star and on the disk lifetime during the pre-main sequence. A longer disk lifetime enables the star to lose a larger amount of angular momentum during the pre-main sequence leading to a lower rotation rate on the ZAMS. As mentioned above, slow rotators are characterised by lower surface abundances of lithium than fast rotators. We thus obtain a relationship between the surface lithium abundance and the rotation rate on the ZAMS which seems to be in good agreement with observations in the Pleiades (Soderblom *et al.* 1993). Moreover, longer disk lifetimes may favor the formation and migration of giant exoplanets. This leads to an interesting relationship between the abundance of lithium and the presence of giant exoplanets. In this scenario, a longer disk lifetime leads indeed simultaneously to a lower lithium abundance (due to the smaller rotation rate on the ZAMS) and a higher probability to detect giant exoplanets. This seems to be in good agreeement with observations of lithium depletion in exoplanet host stars as reported by Israelian *et al.* (2009).

One may wonder whether other observational trends are predicted in the framework of this scenario. As mentioned in Sect. 2, the effects of rotation on the properties of the central layers of a solar-type star can be revealed by asteroseismic observations and in particular by changes of the small separation. In the scenario outlined here, the efficiency of rotational mixing is larger for slow rotators on the ZAMS than for fast rotators. Since the presence of giant exoplanets is favored for stars with slow rotation rates on the ZAMS, the efficiency of rotational mixing is then predicted to be larger in exoplanet host stars than in stars without planets. This explains the different lithium abundances of stars with and without planets but this also leads to changes in the structure and chemical composition of the central layers and hence to different values of the small separation. In the scenario proposed by Bouvier (2008) we thus expect larger values of the asteroseismic small separation for exoplanet host stars than for stars without planets. It will be particularly interesting to investigate this point in the light of new asteroseismic observations coming from space missions dedicated simultaneously to the search of exoplanets and asteroseismology like CoRoT and Kepler.

References

Bouvier, J. 2008, A&A, 489, L53

Bouvier, J., Forestini, M., & Allain, S. 1997, A&A, 326, 1023

Denissenkov, P. A. & Pinsonneault, M. 2007, ApJ, 655, 1157

Eggenberger, P. & Carrier, F. 2006, A&A, 449, 293

Eggenberger, P., Carrier, F., Maeder, A., & Meynet, G. 2006, Memorie della Societa Astronomica Italiana, 77, 309

Eggenberger, P., Maeder, A., & Meynet, G. 2005, A&A, 440, L9

Eggenberger, P., Meynet, G., Maeder, A., *et al.* 2008, Ap&SS, 316, 43

Grevesse, N. & Noels, A. 1993, in Origin and evolution of the elements: proceedings of a symposium in honour of H. Reeves, held in Paris, June 22-25, 1992. Edited by N. Prantzos, E. Vangioni-Flam and M. Casse. Published by Cambridge University Press, Cambridge, England, 1993, p.14, ed. N. Prantzos, E. Vangioni-Flam, & M. Casse, 14

Irwin, J., Hodgkin, S., Aigrain, S., *et al.* 2007, MNRAS, 377, 741

Israelian, G., Delgado Mena, E., Santos, N. C., *et al.* 2009, Nature, 462, 189

Kawaler, S. D. 1988, ApJ, 333, 236

Maeder, A. 2009, Physics, Formation and Evolution of Rotating Stars, ed. A. Maeder

Rüdiger, G., Gellert, M., & Schultz, M. 2009, MNRAS, 399, 996

Soderblom, D. R., Jones, B. F., Balachandran, S., *et al.* 1993, AJ, 106, 1059

Spruit, H. C. 2002, A&A, 381, 923

Zahn, J., Brun, A. S., & Mathis, S. 2007, A&A, 474, 145

Light Elements in the Universe
Proceedings IAU Symposium No. 268, 2009
C. Charbonnel, M. Tosi, F. Primas & C. Chiappini, eds.

© International Astronomical Union 2010
doi:10.1017/S1743921310004515

The light elements in a helio- and asteroseismic perspective

Sylvie Vauclair[1]

[1]Laboratoire d'Astronomie de Toulouse-Tarbes, Université de Toulouse,
14 Avenue Edouard Belin, 31400 Toulouse, France
email: sylvie.vauclair@ast.obs-mip.fr

Abstract. Asteroseismology is a powerful tool to derive stellar parameters, including the helium content and internal helium gradients, and the macroscopic motions which can lead to lithium, beryllium, and boron abundance variations. Precise determinations of these parameters need deep analyses for each individual stars. After a general introduction on helio and asteroseismology, I first discuss the solar case, the results which have been obtained in the past two decades, and the crisis induced by the new determination of the abundances of heavy elements. Then I discuss asteroseismology in relation with light element abundances, especially for the case of main sequence stars.

Keywords. Sun: abundances, heliosismology – stars: interiors

1. Introduction

The general study of stellar oscillations began long before the advent of the so-called helio and asteroseismology. In the past, astronomers only detected large amplitude oscillations and they spoke of "variable stars" or "pulsating stars". Nowadays, new techniques allow to detect very small amplitude oscillations, and variable stars are known all over the HR diagram. They can be classified according to:

- the type of waves which leads to their oscillations, either pressure of gravity waves or both
- their amplitudes
- their excitation mechanisms

Before the first discovery of solar oscillations, sola- type stars were not supposed to be variable, as the acoustic waves are damped in their interiors. We now know that stochastic excitation induced by convective motions leads to permanent destabilisation so that these stars behave like resonant cavities in spite of the waves damping.

The first report of a periodic solar velocity field was given by Leighton *et al.* (1962). Evidences of the five minute oscillations were later confirmed by Ulrich (1970) and Leibacher & Stein (1971). Some 10 millions p-modes are observed in the Sun, with frequencies around 2 to 4 mHz, velocity amplitudes about one cm.s^{-1} (max 20 cm.s^{-1}), relative variations of brilliance 10^{-7}, mode lifetimes of a few hours up to a few months.

All the solar-type stars which have been observed for seismology do oscillate with frequencies in the interval 0.1 to 10 mHz. However, only global modes can be detected, as stellar surfaces cannot be resolved. A few tens of modes only can typically be identified in such stars, so that the inversion techniques and the precision on the results are quite different from the solar case.

In this framework, light elements are related to helio- and asteroseismology in various ways. Helium 4, the second most abundant element in stars, is the only one which directly influences the oscillation scheme, as modifying its abundance changes the stellar structure. The helium abundance and abundance gradients may be derived from seismic studies. It can also be the cause for seismic destabilisation of the star through κ-mechanism.

On the other hand, the light elements lithium, beryllium, boron, and helium 3 are only indirectly related to asteroseismology, as they are destroyed by nuclear reactions in a way determined by macroscopic motions (convective zones, overshooting, internal mixing) which themselves have seismic signatures. As for deuterium, I do not think that any relation can be found between its abundance evolution and helio or asteroseismology.

2. The solar case

Helioseismology consists in analysing the sound waves that propagate throughout the Sun and using them to measure, by inversion procedures, the solar internal parameters like temperature, pressure, density, helium abundance, partial ionisation regions, zones of convection and macroscopic motions, internal rotation.

The oscillation modes are trapped in spherical-shell cavities extending between the surface and their "turning points", which are consequences of the refraction of the waves due to the inwardly increasing sound velocity. The angular component of the wave function of these oscillations is described by the spherical harmonics, characterized by their two quantum numbers: the degree l and the azimuthal-order m. The number l corresponds to the number of null circles around the sphere, and m to the number of these null circles which cross the poles (meridional circles). In the radial direction, the number of null spheres is characterized by the third quantum number n. The depth of a given cavity depends on both the oscillation frequency and the spherical-harmonic degree l of the associated mode. Modes with large l are confined near the surface, while modes with small l extend much deeper, those with $l = 0$ and $l = 1$ reaching the center of the Sun itself. Consequently, all these modes sample different, although overlapping, regions of the solar interior.

Various observational techniques have been developed to detect and precisely observe the solar oscillation frequencies. To obtain the needed precision, the observations must go on uninterrupted during at least one month (a solar rotation), which may be reached by three different solutions: instruments at the Earth's poles (Antarctica), space observations (e.g. SOHO: sohowww.nascom.nasa.gov) and networks of instruments dispatched in longitude all around the Earth (e.g. GONG: gong.nso.edu).

Inversion procedures have been developed to obtain very precisely the sound velocity throughout the Sun, and the individual internal parameters by comparison with models. Owing to the very large number of observed resonant modes, and to their propagations in different internal cavities, at different depths, it is possible to derive the solar internal parameters with an accuracy of 0.1 percent in most of the internal Sun (see Basu *et al.* 2009 and references therein). The first important success of helioseismology was that the seismic profile of the sound velocity inside the Sun clearly indicated the exact depth of the convective zone, at a fractional radius 0.713 +/−0.03 (Christensen-Dalsgaard *et al.*1996). It also showed that overshooting was not present or extended on a very small depth if any.

In the early phases of seismic inversion and comparison with solar models, around 1995, discrepancies of order one percent were found. Then it was realized that helium diffusion, which decreases the helium abundance in the convective zone by about 20 percent compared to the internal one, had to be added to obtain better agreement. Other physical processes were improved, motivated by these new constraints: opacities and equations of state (e.g. OPAL, Iglesias & Rogers 1996, Rogers & Nayfonov 2002) and nuclear reaction rates (Angulo *et al.* 1999). It was also found that a mild turbulence below the convective zone, which could smoothen the diffusion-induced helium gradient, was necessary to obtain a good fit between the seismic and model sound velocity in this special region. This mild turbulence was able to account for the lithium deficiency observed in the solar outer layers, as well as the constraints imposed by beryllium (Balachandran & Bell 1997) and helium 3 (Geiss & Gloecker 1998)(see Richard *et al.* 1996, Brun *et al.* 2002).

Precise analysis of the rotational splitting of the solar mode frequencies allowed to discover that, while the outer solar layers undergo a well-known differential rotation, the internal Sun, below the convective zone, rotates as a solid body. This offered an important challenge to hydrodynamicists and is not entirely solved. It may be related to internal gravity waves (Charbonnel & Talon 2005, Talon & Charbonnel 2008), or to the internal solar magnetic field (Turck-Chieze *et al.* 2005).

The picture of our Sun seemed to have well improved during about 20 years, with very good results, until a crisis came with the new determinations of the solar chemical composition by Asplund *et al.* (2005). Using 3D simulations of the atmospheric solar motions, they determined heavy element abundances significantly smaller than obtained before (Grevesse & Sauval 1998). Although these abundances have recently been revised and slighty increased (Asplund *et al.* 2009), the discrepancy between the models computed with the new abundances and the seismic inversions is unacceptable. As the new determinations of abundances seem convincing, there is a real challenge about solar physics (see Serenelli *et al.* 2009).

Several ideas have emerged to try and solve this discrepancy. None of them worked. The most promising one may be the idea of accretion by the young Sun of metal-poor material coming from the planetary disc gas (Castro *et al.* 2007). However, this leads to a steep abundance gradient below the convective zone, incompatible with the present sound velocity. New studies are under work, to see how this steep gradient could be smoothed, without modifying too much the internal Sun. Another solution could be that a systematic error occurred in the determinations of the new abundances, but at the present time it does not seem to be the case.

3. The stellar case

Asteroseismology can give much more precise values of the stellar parameters, including age, mass, radius, stellar gravity, effective temperature, metallicity, helium abundance value, depth of convective regions, than any other means. Such precise determinations need deep seismic analyses of individual stars, and cannot be obtained with approximate theories only, although these approximate theories are very useful for first insights.

The stellar oscillations may be observed either by spectroscopic observations, using Doppler effect on lines, or by photometric method, which gives access to the global luminosity variations of the star. For technical reasons, the first method is used on the ground (e.g. HARPS and SOPHIE spectrographs) while the second one is used on satellites, like COROT or KEPLER. In any case, only global modes can be detected, with l values ranging from 0 to 3. Typically a few tens of modes at most can be identified in stars, far from the ten millions observed in the Sun.

The inversion techniques used for the solar case cannot be applied to stars. In particular, the outer layers are not scanned by the different waves as it is in the Sun, as all the observed modes travel deep inside. However, if the waves encounter regions of rapid variation of the sound velocity on their way, they are partially reflected, creating secondary periods. This phenomenon is very helpful in the determination of internal stellar characteristics, as will be seen below.

Tassoul (1980) derived an asymptotic analytical expression for the mode frequencies in stellar cavities, which is widely used as a first approximation:

$$\nu_{n,\ell} \simeq 2\pi\Delta\nu_0\left(n + \frac{\ell}{2} + \epsilon\right) \tag{3.1}$$

with

$$\Delta\nu_0 = \left[2\int_0^R \frac{1}{r}dr\right]^{-1}, \tag{3.2}$$

which is the inverse of twice the time needed by the sound waves to travel from the stellar surface down to the center. This quantity is named the mean large separation.

One must be aware however that this theory is a crude approximation, valid only for small l and large n values, and only in stars in which the sound velocity varies smoothly as in the Sun. In a real star, there may be strong deviations from this theory and these deviations do provide important information about specific regions of the star (Soriano *et al.* 2007, see below).

In the first order approximation, two p-modes of successive radial order n, with the same degree ℓ are approximately separated by $\Delta\nu_0$. If we define the large separation by:

$$\Delta\nu = \nu_{n+1,\ell} - \nu_{n,\ell} \tag{3.3}$$

we have $\Delta\nu_{n,\ell}$ equal to $\Delta\nu_0$. But deviations from the asymptotic theory induce deviations of $\Delta\nu$ from $\Delta\nu_0$.

It is also useful to define the small separations as:

$$\delta\nu = \nu_{n,\ell} - \nu_{n-1,\ell+2} \tag{3.4}$$

At first order we clearly have $\delta\nu_{n,\ell} \simeq 0$. But in a real star, there are deviations from this theory and the small separations are not equal to zero. These quantities are very sensitive to the deep stellar interior (Gough 1986, Roxburgh 2009, and references therein) and can give us precious informations about the stellar core. These quantities, and in particular for the degrees $\ell = 0$ - $\ell = 2$, can even become negative (Soriano *et al.* 2007, Soriano & Vauclair 2008). This specific behaviour is related to the presence of a convective core or to a helium core with sharp edges. We can use this phenomenon to characterize helium-rich cores and to give strong constraints on the possible overshooting.

Finally, in regions with an important gradient of the sound velocity, like the boundary of the convective zone or the HeII ionization zone, there is partial reflexions of the waves that create modulations on the frequencies. These modulations clearly appear in the so-called second differences (Gough 1990, Vauclair & Theado 2004) which are defined by:

$$\delta_2\nu = \nu_{n+1,\ell} + \nu_{n-1,\ell} - 2\nu_{n,\ell} \tag{3.5}$$

The modulation period of the oscillations is equal to twice the acoustic depth, which is the time needed for the sound waves to travel from the considered region to the stellar surface:

$$t_s = \int_{r_s}^R \frac{dr}{c(r)} \tag{3.6}$$

$c(R)$ is the sound velocity at the radius r, and r_s the radius of the considered region.

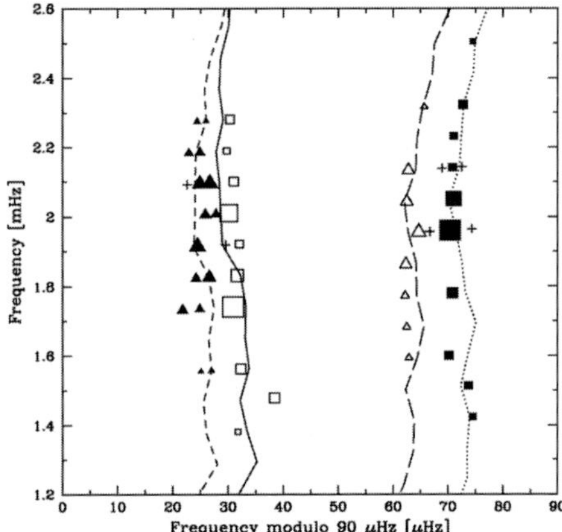

Figure 1. Example of echelle diagram: the star mu Arae. The symbols represent the observed modes, empty squares, $l = 0$, filled squares, $l = 1$, filled triangles, $l = 2$ and empty triangles, $l = 3$. The lines correspond to a stellar model. When two symbols are present, they correspond to two peaks due to rotational splitting.

A useful representation of the oscillation frequencies is the echelle diagram (Figure 1). In ordinates are plotted the frequencies and in abscissas, the same frequencies modulo the mean large separation. According to the asymptotic expressions, we should obtain straight lines corresponding to each value of the degree ℓ. In real stars, the specific features observed in the echelle diagramme structure are indicative of their internal structure.

4. Helium determinations from asteroseismology

4.1. *Helium in stellar outer layers and helium gradients*

Rapid variations of the sound velocity in the outer stellar layers lead to partial reflections of the sound waves, which appear as frequency modulations in the "second differences". The periods of these modulations are equal to $2t_s$ where t_s is the time needed for the acoustic waves to travel between the surface and the considered region (acoustic depth).

These rapid variations can be due to the boundary of the outer convective zone, to helium ionisation regions and to diffusion-induced helium gradients. In practice, when helium settling is taken into account, the importance of helium gradients is larger than those of the convective boundary and helium ionisation regions (Vauclair & Theado 2004, Castro & Vauclair 2006).

Similar computations can be done for evolutionary models in which helium diffusion is introduced or suppressed. The resulting second differences are quite different. When diffusion is suppressed, they show signatures of the helium ionization zones while the signatures of helium gradients have, of course, disappeared.

Houdek & Gough (2007) have shown in detail how to determine the helium abundance in stars from the observed helium-ionisation-induced features observed in the second differences. It gives interesting and precise results, but one must keep in mind that the obtained value corresponds to the helium abundance inside the convective zone, not the original one, as it has been decreased by gravitational settling.

4.2. *Determinations of the internal helium abundance from model comparisons*

Stars have to be observed during a sufficiently long time, typically eight nights or more with the HARPS spectrograph, several months for space observations, to allow precise comparisons with models. The Fourier analysis may lead to mode identification, as discussed in Bouchy *et al.* (2005), and frequency determinations. Meanwhile evolutionary tracks are computed, with various input parameters (mass, chemical composition, presence or not of overshooting, etc...). The computed mode frequencies are compared with the observed ones in a rigorous way. In this framework, departures from the "asymptotic theory" give fundamental information on the stellar parameters and internal structure.

For a given set of abundances (metallicity and helium value), only one model may reproduce the observed frequencies in a satisfying way (Soriano *et al.* 2007, Vauclair *et al.* 2008). The various models obtained in this way for different chemical composition have similar ages, gravities, and radii. The other parameters are then constrained with the help of the spectroscopic observational boxes.

In this way, it becomes possible to determine the internal helium abundance of the stars. Up to now, it has been done for two main-sequence exoplanet-host stars, which are both overmetallic compared to the Sun. The results for the helium abundance are quite different.

The exoplanet-host star μ Arae (HD160691) is a G5V star which has been observed for seismology during 8 nights in August 2004 with HARPS. The observations allowed to identify 43 oscillation modes of degrees $l = 0$ to $l = 3$ (Bouchy *et al.* 2005). They have been analysed by Bazot *et al.* (2005), and by Soriano *et al.* (2007) after revision of its Hipparcos parallax. The metallicity of this star, compared to the solar one, is $[Fe/H] = 0.30 \pm 0.01$, and the helium abundance is correspondingly high, $Y = 0.301 \pm 0.01$.

Solar-type oscillations of the exoplanet-host star ι Hor (HD17051) were detected with HARPS in November 2006. Up to 25 oscillation modes could be identified and compared with stellar models. The analysis is discussed in detail in Laymand & Vauclair (2007) and Vauclair *et al.* (2008). Contrary to μ Arae, which has a high helium value, as expected from the normal evolution of the chemical abundance of galaxies, according to its high metallicity, ι Hor has a very small helium value, comparable to that of the Hyades: $Y - 0.255 \pm 0.015$, for a metallicity $[Fe/H] = 0.16 \pm 0.02$. As we also know that this star has the same kinematics as the cluster in the Galaxy, this is an indication that it was formed with the cluster and evaporated.

5. Helium abundance variations and kappa mechanism

Helium is an important element in the framework of oscillation stars, as it may in some cases lead to wave amplification though kappa mechanism. Here I only discuss the Am - δ-Scuti case.

Among the main sequence stars which lie inside the instability strip, many chemically peculiar stars are found. The so-called Am stars are found in the H.R. diagram at the same place as the δ-Scuti stars. Generally speaking, the former ones show abundance peculiarities, namely a general overabundance of metals (except calcium and scandium), but no oscillations, while the later ones are pulsating but chemically normal. As discussed by Turcotte *et al.* (2000), in A-type stars, almost 70% of non chemically peculiar stars are δ-Scuti variables at current levels of sensitivity while most non-variable stars are Am stars. Furthermore, Am stars are slower rotators than δ-Scuti stars. In this region of the H.R. diagram, the stars display two different convective zones in their outer layers : the upper one due to the HI and HeI ionisations and the lower one to the HeII ionisation. The δ-Scuti stars pulsate due to a κ-mechanism which takes place in the second convective zone. When microscopic diffusion occurs, this convective zone disappears due to helium depletion and the κ-mechanism cannot take place anymore (Vauclair *et al.* 1974).

Some oscillating Am stars have been discovered (Kurtz 1989), which challenge the previously accepted theory. Richer *et al.* (2000) and Turcotte *et al.* (2000) computed models of Am stars in the framework of the Montreal models. They found that, due to the iron accumulation in the radiative zone below the H and He convective zone, a new convective region appears which increases the diffusion time scales compared to the previous models. In these new models, helium is still substantially present in the helium convective zone at the ages of the considered stars. They claim that it is possible to account for the existence of oscillating Am stars close to the cool boundary of the instability strip.

These computations ignored however an important process which occurs in case of μ-gradient inversion : thermohaline convection, or double-diffusive convection (Vauclair 2004, Charbonnel & Zahn 2007). Recently, Theado *et al.* (2009) have done precise computations of such diffusion-induced accumulation of elements, including thermohaline convection. The accumulation is much smaller than previously thought, but still present. This study will lead to many applications concerning element abundances in stars.

6. Conclusion

From the few examples already available, asteroseismology has proved to be a powerful tool for determining stellar parameters and constraints on their internal structure. However, tests have to be done for individual stars, observed during long periods. Usual approximate theories are not precise enough to obtain such results.

Only helium can be directly determined from asteroseismology. Lithium, beryllium, and boron, whose abundances are obtained from spectroscopy, are related to the macroscopic motions at work below the outer convective zones in solar-type stars. Asteroseismology can be used to constrain these macroscopic motions, in some cases. At least the depth of the convective zones can be derived somewhat precisely. Other kinds of mixing, like rotation-induced mixing, can influence the mode frequencies (e.g. Eggenberger 2009), but up to now the observed precision in the modes identifications and frequencies is not enough to measure it. Maybe in the future?

References

Angulo C., Arnould M., & Rayet M., (NACRE collaboration) 1999, *Nuclear Physics A*, 656, 1

Asplund, M., Grevesse, N., & Sauval, J. 2005, Cosmic Abundances as Records of Stellar Evolution and Nucleosynthesis, 336, 25

Asplund, M., Grevesse, N., Sauval, J., & Scott, P. 2009, *ARA&A*, 47, 481

Balachandran, S. & Bell, R.A. 1997, *AAS meeting* 29, 1325

Basu, S., Chaplin, W. J., Elsworth, Y., New, R., & Serenelli, A. M. 2009, *ApJ*, 699, 1403

Bazot, M., Vauclair, S., Bouchy, F., & Santos, N. C. 2005, *A&A*, 440, 615

Brun, A. S., Antia, H. M., Chitre, S. M., & Zahn, J.-P. 2002, *A&A*, 391, 725

Bouchy, F., Bazot, M., Santos, N. C., Vauclair, S., & Sosnowska, D. 2005, *A&A*, 440, 609

Castro, M. & Vauclair, S. 2006, *A&A*, 456, 611

Castro, M., Vauclair, S., & Richard, O. 2007, *A&A*, 463, 755

Charbonnel, C. & Talon, S. 2005, *Science*, 309, 2189

Charbonnel, C. & Zahn, J. P. 2007, *A&A* (Letters), 467, 15

Christensen-Dalsgaard J. *et al.* 1996, *Science*, 272, 1286

Eggenberger, P. 2009, this conference

Geiss, J. & Gloecker, G. 1998, *Space Science Reviews*, 84, 239

Gough, D. O. 1986, in Hydrodynamic and magnetohydrodynamic problems in the Sun and stars, ed. Y. Osaki (Uni. of Tokyo Press), p. 117

 S. Vauclair

Gough, D. O. 1990, In Progress of Seismology of the Sun and Stars, Proc. Oji International Seminar (Hakone) (Japan : Springer Verlag), *Lect. Notes Phys.*, 367, 283, eds Y. Osaki, H. Shibahashi

Grevesse, N. & Sauval, A. J. 1998, *Space Science Reviews*, 85, 161

Houdek, G. & Gough, D.O. 2007, *MNRAS*, 375, 861

Iglesias, C. A. & Rogers, F. J. 1996, *ApJ*, 464, 943

Kurtz, D. W. 1989, *MNRAS*, 238, 1077

Laymand, M. & Vauclair, S. 2007, *A&A*, 463, 657

Leibacher, J. W. & Stein, R. F. 1971, *APL*, 7, 191L

Leighton, R. B., Noyes, R. W., & Simon, G. W. 1962, *ApJ*, 135, 474

Richard, O., Vauclair, S., Charbonnel, C., & Dziembowski, W. A. 1996, *A&A*, 312, 1000

Richer, J., Michaud, G., & Turcotte, S. 2000, *ApJ*, 529, 338

Rogers, F. J. & Nayfonov, A. 2002, *ApJ*, 576, 1064

Roxburgh, I. W 2009, *A&A*, 493, 185

Serenelli, A. M., Basu, S., Ferguson, J. W., & Asplund, M. 2009, *ApJ*, 705, L123

Soriano, M., Vauclair, S., Vauclair, G., & Laymand, M. 2007, *A&A*, 471, 885

Soriano, M. & Vauclair, S. 2008, *A&A* 488, 975

Talon, S. & Charbonnel, C. 2008, *A&A*, 482, 597

Tassoul, M. 1980, *ApJS*, 43, 469

Theado, S., Vauclair, S., Alecian, G. & Le Blanc, F. 2009, *ApJ*, 704, 1262

Turck-Chieze, S., Appourchaux, T., Ballot, J., *et al.*, 2005, *ESASP*, 588, 193

Turcotte, S., Richer, J., Michaud, G., & Christensen-Dalsgaard, J. 2000, *A&A*, 360, 603

Ulrich, R. K. 1970, *ApJ*, 162, 993

Vauclair, G., Vauclair, S., & Pamjatnikh, A. 1974, *A&A*, 31, 63

Vauclair, S. 2004, *ApJ*, 605, 874

Vauclair, S. & Théado, S. 2004, *A&A*, 425, 179

Vauclair, S., Laymand, M., Bouchy, F., Vauclair, G., Hui Bon Hoa, A., Charpinet, & S., Bazot, M. 2008, *A&A* (Letters), 482, 5

Light Elements in the Universe
Proceedings IAU Symposium No. 268, 2009
C. Charbonnel, M. Tosi, F. Primas & C. Chiappini, eds.
© International Astronomical Union 2010
doi:10.1017/S1743921310004527

Lithium factories in the Galaxy: novae and AGB stars

Francesca D'Antona[1] **and Paolo Ventura**[1]

[1]INAF – Osservatorio di Roma,
via di Frascati 33, I-00040 Monteporzio, Italy
email: dantona@oa-roma.inaf.it ventura@oa-roma.inaf.it

Abstract. We review the state of the art in modelling lithium production, through the Cameron–Fowler mechanism, in two stellar sites: during nova explosions and in the envelopes of massive asymptotic giant branch (AGB) stars. We also show preliminary results concerning the computation of lithium yields from super–AGBs, and suggest that super–AGBs of metallicity close to solar may be the most important galactic lithium producers. Finally, we discuss how lithium abundances may help to understand the modalities of formation of the "second generation" stars in globular clusters.

Keywords. Stars: AGB and post-AGB, novae; convection, nuclear reactions, nucleosynthesis; Globular Clusters: general

1. Introduction

Although lithium is very fragile, its galactic abundance increases from $\log \epsilon(\mathrm{Li})$† ~ 2.2 at the surface of Population (Pop) II stars to $\log \epsilon(\mathrm{Li}) \sim 3.3$ or more in Pop I. Even if lithium is hidden in the atmospheres of Pop II, and its true primordial abundance is ~ 2.7, a galactic production by ~ 0.7dex is necessary.

The mechanism responsible for lithium production has been proposed by Cameron & Fowler (1971): ^{7}Be is produced by fusion of ^{3}He with ^{4}He, and rapidly transported to stellar regions where it can be converted into ^{7}Li by k–capture. Notice, then, that the lithium production may last only until there is ^{3}He available in the region of burning, and that the production ends when the ^{3}He is all consumed.

There are two main physical situations where this mechanism can produce enough lithium that it is important to investigate their role in the galactic production: the first one is the explosive hydrodynamical formation during the outbursts of novae (Arnould & Norgaard 1975, Starrfield *et al.* 1978), the second one is the hydrostatic, slow formation in the envelopes of asymptotic giant branch (AGB) stars, for which it was first proposed. In envelope models of AGB stars (Scalo *et al.* 1975), in which the bottom of the convective envelope reaches the hydrogen burning layers, and its temperature (T_{bce}) becomes as large as $T_{\mathrm{bce}} \sim 40$MK, the ^{3}He$(\alpha, \gamma)^7$Be chain acts. These models were able to explain the high lithium abundances found in some luminous red giants, and the process took the name of Hot Bottom Burning (HBB).

In Section 2 we will resume the state of the art of the modelling of lithium production during nova outbursts, and in Section 3 we will deal with the AGB models, to understand whether they can account for the lithium galactic evolution. In addition, we will show new models of lithium production in super–AGB stars (Ventura & D'Antona 2010) and speculate on the possible role of these stars as efficient lithium factories. Finally, in Section 4 we will shortly summarize the problem of lithium in the "second generation"

† we use the notation $\log \epsilon(\mathrm{Li}) = \log(N_{Li}/N_H) + 12$.

stars of globular clusters. We will not consider here the different, slow mixing process also based on the Cameron & Fowler (1971) mechanism and named "cool bottom burning" (e.g. Nollett *et al.* 2003). This process can explain the lithium abundances seen in lower luminosity red giants (e.g. Wasserburg *et al.* 1995, Sackmann & Boothroyd 1999), but its physical reasons are not well studied, while the nucleosynthesis in HBB is based on straightforward time–dependent mixing in standard convective regions.

2. Nova outbursts

Schatzman (1951) was the first to propose that the isotope ^{3}He could play a role in nova explosion, in the context of a theory of novae powered by thermonuclear detonations. Arnould & Noergaard (1975) proposed that the Cameron–Fowler mechanism, acting at the nova outburst, would produce a lithium abundance proportional to the ^{3}He abundance in the nova envelope. Starrfield *et al.* (1978) showed that the mechanism could be efficient for outburst temperatures >150MK, and the fast ejection of the ^{7}Be rich nova shell leads to ^{7}Li production; they quantified the expected linear relation, between lithium and the ^{3}He initial mass fraction X_{3i}, as:

$$[Li/H] \simeq 200 \times X_{3i}/X_{3\odot} \tag{2.1}$$

where $X_{3\odot}$ is the solar ^{3}He mass fraction. D'Antona & Matteucci (1991) modelled the galactic evolution of lithium, including the contribution of novae according to this result. The argument below their modelization was very simple: the nova explosion occurs when a critical hydrogen rich envelope is reached on the white dwarf component of the nova binary, by accretion from its low mass companion. By losing mass, low mass stars expose the stellar regions in which the hydrogen burning p–p chain is incomplete, and thus bring to the surface the ^{3}He accumulated in the envelope during the period preceding the mass transfer phase, and during the (slow) mass transfer phase itself (e.g. D'Antona & Mazzitelli 1982). Thus D'Antona & Matteucci (1991) linked the lithium abundance produced in the outburst to the "delay time" between the formation of the white dwarf and the occurrence of nova outbursts. As a result, mainly the novae containing an "old" white dwarf, and therefore an old and ^{3}He–rich low mass companion contribute to the galactic production of lithium, in agreement with the Li vs. [Fe/H] galactic relation.

Motivated by the D'Antona & Matteucci (1991) paper, Boffin *et al.* (1993) revisited the influence of ^{3}He on the nova outbursts with simple one–zone models, but found out that equation 2.1 was a large overestimate, due to two main reasons: 1) the neglect of the reaction $^8B(p,\gamma)^9C$ in the Starrfield *et al.* (1978) network, and 2) the increasing influence of the competitive reaction $^3He(^3He,2p)^4He$ when ^{3}He is enhanced in the nova envelope. Consequently, they found a milder dependence on the lithium production on the ^{3}He:

$$\frac{X(^7Li)}{X_0(^7Li)} \simeq 1 + 1.5 \log \frac{X(^3He)}{X_\odot(^3He)} \tag{2.2}$$

where X represents mass fractions, and $X_0(^7Li)$ is the lithium production when the solar ^{3}He abundance is adopted. Afterwards, Hernanz *et al.* (1996) and Jose & Hernanz (1998) re-examined the problem with an implicit hydro-code including a full reaction network, able to treat both the hydrostatic accretion phase and the explosion stage. They considered both the case of white dwarfs having a carbon–oxygen core and the case of oxygen– neon cores, showing that C–O cores are more efficient in the lithium production, as they have a shorter accretion phase, so that ^{3}He is not destroyed efficiently, and more ^{7}Be is produced. Overproduction of lithium is found, but its dependence on the initial ^{3}He abundance still follows Boffin *et al.* (1993) prescription. Including this revised lithium production in the galactic chemical evolution model, novae appear to be a modest lithium producer (Romano *et al.* 1999).

3. Luminous AGB stars

Above a luminosity of $\sim 2 \times 10^4 \, L_\odot$, the bottom of the convective envelope during the AGB evolution reaches the H–shell burning region, and the nuclear reaction products are transported to the surface by convection. This is the perfect site of lithium production through the Cameron–Fowler mechanism (Iben 1973, Sackmann *et al.* 1974). While we have seen that $T_{bce}\sim40$MK is sufficient to produce lithium by HBB, if T_{bce} becomes larger, other important reactions take place.

About 65MK are necessary to convert carbon to nitrogen. The very luminous ($M_{bol}<$ -6, that is L$> 2 \times 10^4 \, L_\odot$) lithium–rich giants of the Magellanic Clouds (Smith & Lambert 1989, 1990, Smith *et al.* 1995) are indeed M–stars, and not carbon stars. Their carbon star features may have been lost by CN processing in HBB. Carbon stars in the Clouds, in fact, populate only the region at $M_{bol}> -6$. Lithium rich – oxygen rich AGB stars embedded in thick circumstellar envelopes have also been discovered in a Galactic sample, in a survey by García-Hernández *et al.* (2007), aimed at obtaining spectroscopy of very massive AGB candidates.

A third possible processing occurs at even larger $T_{bce}(>80$MK), where H burns through the full CNO cycle. These very high temperatures are reached in low metallicity massive AGBs, and are possibly at the basis of the self– enrichment process in globular clusters (Ventura *et al.* 2001). The oxygen abundance in the envelopes of these AGB stars, and consequently in the matter ejected by wind or planetary nebula, is reduced, as we see in the "anomalous" stars of galactic globular clusters, see Sect.4.

Modelling of lithium rich AGB stars first of all requires to treat non– instantaneous mixing in the envelope, coupling the nuclear reaction network with the mixing process. This can be easily done by treating mixing as a diffusion. In Figure 1 we show the total phase of lithium production in a $5 \, M_\odot$ star of metallicity $Z = 10^{-3}$ (left side), and a zoom of the same figure between two thermal pulses (right side). We see that, when T_{bce}

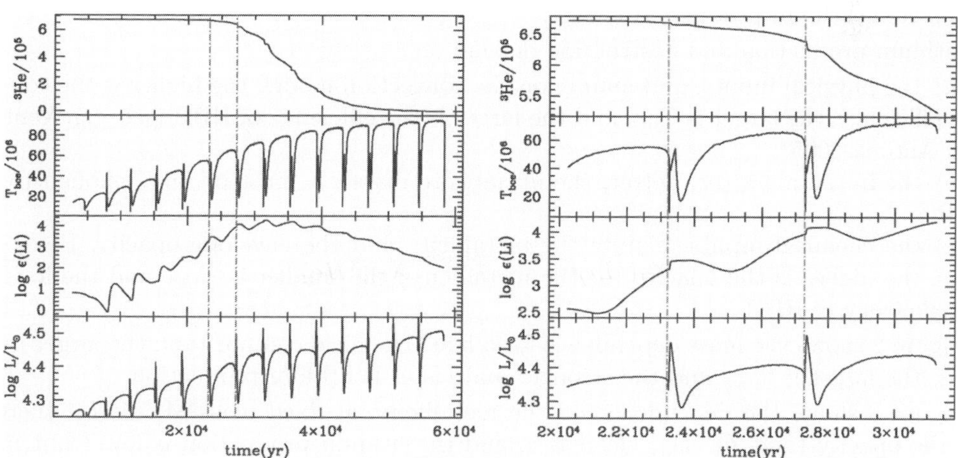

Figure 1. From bottom to top panel we plot the luminosity, surface lithium, HBB temperature and ^{3}He surface content along the AGB evolution of a star of $5 \, M_\odot$, metallicity $Z = 10^{-3}$. The total duration of the phase of the most important lithium production lasts $\sim 20 \times 10^3$yr. The two vertical lines delimit the time interval displayed in the right panel.

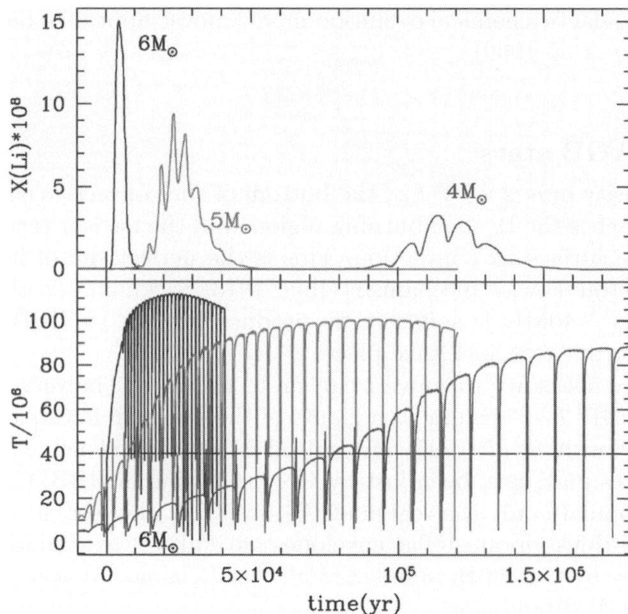

Figure 2. Temperature at the bottom of the convective layer (bottom) and lithium surface abundance (top) as a function of the time for the masses 6, 5 and $4\,M_\odot$, $Z = 10^{-3}$, from left to right. Time is computed from the beginning of the AGB phase, when the H–shell burning is reignited. The horizontal line at $T = 40$MK limits the temperature region for lithium production.

decreases, due to the ignition of the thermal pulse and the expansion of the envelope, the lithium abundance decreases. We can appreciate the delay time between the physical conditions in the burning region and the surface lithium, due to the non instantaneous mixing. The total phase of lithium production lasts more than 50×10^3 yr, but the phase in which $\log \epsilon(\mathrm{Li}) \gtrsim 3$ lasts only $\sim 20 \times 10^3$ yr. Once the initial ^{3}He present in the envelope is depleted, lithium production is over.

Lithium production and destruction depend on

(a) the physical inputs, and mainly on the convection model: the higher is the convection efficiency, the larger is T_{bce} and the larger is the efficiency of HBB (see, e.g. Ventura & D'Antona 2005);

(b) the initial mass (or, better, the initial core mass): it must be large enough to get HBB;

(c) the chemical inputs, mainly the metallicity and the envelope opacity. Fixed the mass, the higher is the opacity (or the metallicity) the smaller is T_{bce} and the lower is the efficiency of HBB.

Figure 2 shows the mass dependence for a fixed chemical composition: the larger is the mass, the larger is T_{bce} and the stronger and faster is lithium production.

Figure 3 shows the dependence on the metallicity, at fixed mass $M = 6\,M_\odot$. Increasing the opacity (and Z), T_{bce} decreases, and the lithium production is lower but more extended in time.

The computation of lithium production during the super–AGB evolution has been recently achieved by Ventura & D'Antona (2010) for $Z = 10^{-3}$. The results are very interesting, as we see in Fig. 4 for a mass of $7.5\,M_\odot$. Lithium achieves very large abundances,

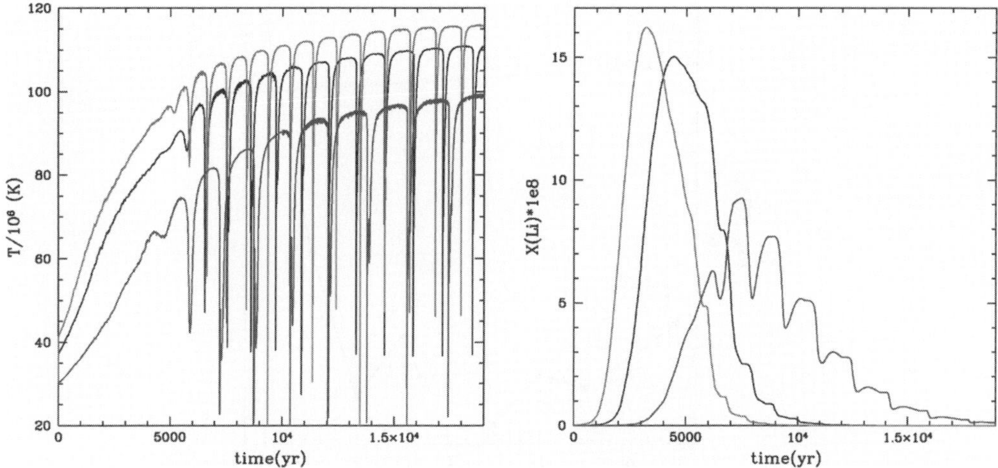

Figure 3. The evolution of 6 $M_\odot$ for metallicities Z $=0.0006$, Z $=0.001$ and Z $=0.004$ from top to bottom is displayed. On the left side, we plot T_{bce}, on the right side the lithium mass fraction X(Li).

due to the very high T_{bce}, and the Li–rich phase occurs even before the star begins the thermal pulse phase.

Of course 'production' does not mean 'yield': two ingredients are important: how much lithium is made, and how long it lasts, so that mass loss can recycle it into the interstellar medium. Consequently, the lithium yield is very dependent on the mass loss rate: larger rates during the phase of lithium production provide a higher lithium yield. Unfortunately, mass loss is another great uncertainty in the computation of stellar models. In Figure 6 we show as open (red) circles at [Fe/H] $=-1.3$ the average lithium abundance in the ejecta of models of 4, 5 and 6 $M_\odot$ with three different mass loss formulations: the middle points refer to the mass loss rate suggested by Blöcker (1995), who extends Reimers' recipe to describe the steep increase of mass loss with luminosity as the stars "climb" the AGB. The full expression, for Mira's periods exceeding 100d, is

$$\dot{M} = 4.83 \times 10^{-22}\eta_R M^{-3.1} L^{3.7} R \tag{3.1}$$

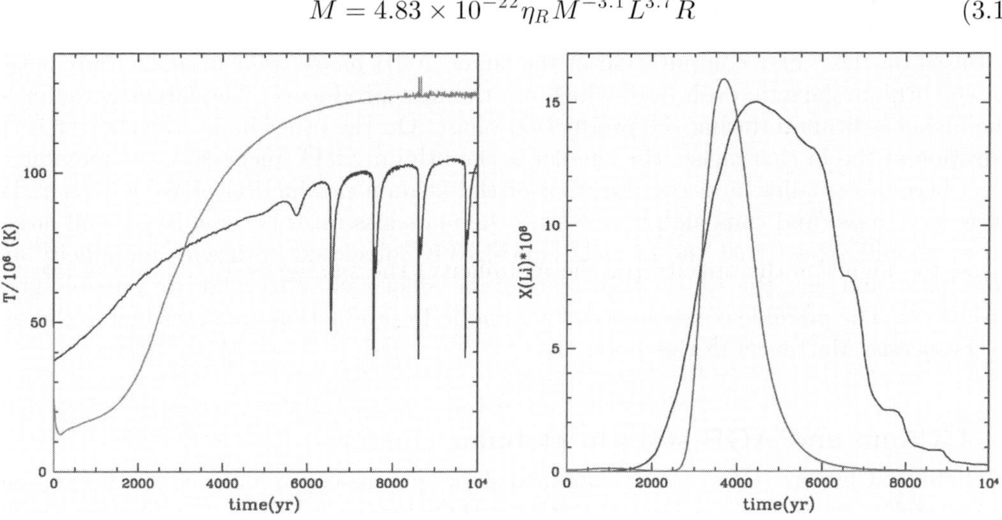

Figure 4. The same as Fig. 3 for Z $=0.001$ and masses 6 $M_\odot$ (lower curve) and a super–AGB model of 7.5 $M_\odot$.

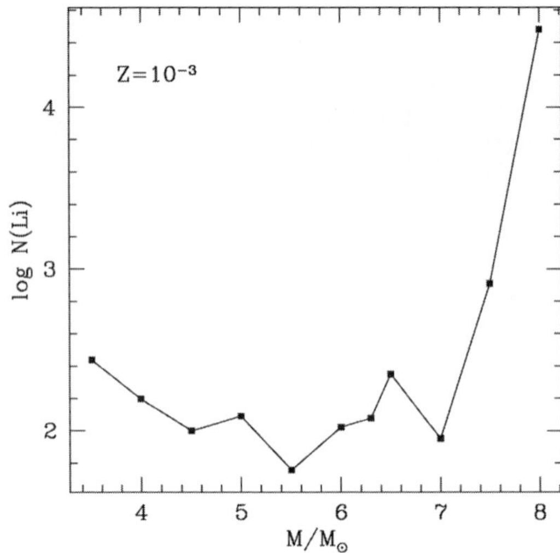

Figure 5. Lithium abundance averaged on the ejected envelope mass as a function of the total mass for $Z = 0.001$

where η_R is the free parameter entering the Reimers' (1977) prescription. In the "standard" models of Fig. 6 we adopt $\eta_R = 0.02$, according to a calibration based on the luminosity function of lithium rich stars in the Magellanic Clouds given in Ventura *et al.* (2000). The highest points in the figure are obtained for the extreme value of $\eta_R = 0.1$, while the models adopting the Vassiliadis & Wood (1993) mass loss rate are the lowest ones. We see then that the absolute values of the lithium yields must be considered highly uncertain (see also Ventura *et al.* 2002). The global behaviour of the average lithium abundance in the ejecta, as a function of the initial mass is given in Fig. 5 for the models computed with the standard mass loss ($\eta_R = 0.02$) prescription and $Z = 10^{-3}$. Increasing the mass, the average abundance first decreases, due to the faster consumption of ^{3}He, in spite of the larger abundances reached in the phase of production. For the super–AGB masses, the average abundance increases, and may also become very large, both due to the stronger production and to the huge mass loss rate achieved by the largest core masses.

Based on these first computation of the super–AGB phase, we can make a prediction on the lithium galactic evolution: which are the best producers? The larger is the mass, the higher is lithium during the production phase. On the other hand, the shorter is the duration of the Li rich phase, the smaller is the lithium yield. Increasing the metallicity, T_{bce} becomes smaller, and the duration of the lithium production phase is longer. For large core mass (and thus high luminosity) the mass loss rates become larger and larger. So we should expect that the Li yield is positively correlated both with metallicity and core mass, and that the super–AGB stars of metallicity close to solar are possibly great producers. The possible consequences for galactic Li production are described in the talk by Francesca Matteucci in this book.

4. Lithium and AGB stars in globular clusters

Globular Clusters (GCs) so far examined show spectroscopic evidence for the presence of two stellar generations: a First Generation (FG) having "normal" abundances, similar to those of halo stars of the same metallicity, and a Second Generation (SG) whose abundances are more spreaded, and bear the sign of hot CNO processing, with an often very significative oxygen reduction, evidence for the action of the Ne–Na cycle and sometimes of the Mg–Al cycle (see, e.g. Gratton *et al.* 2004). The SG contains at least 50% of the cluster stars (Carretta *et al.* 2009a,b). At low metallicity, in the most massive AGB stars, T_{bce} becomes larger than $\sim$80MK, and the ON chain of the CNO cycle becomes active. In these envelopes, oxygen is cycled to nitrogen, and its abundance can be dramatically reduced. Thus some models for the formation of the different populations attribute the presence of "anomalous" stars with low oxygen and high sodium, to a SG including matter processed by HBB (e.g. Ventura *et al.* 2001). Other models attribute the formation of the SG to the ejecta of fast rotating massive stars (FRMS, see e.g. Decressin *et al.* 2007a), or even to pollution from gas expelled during highly non conservative evolution of massive binaries (De Mink *et al.* 2009), although this latter model in particular can not explain the very high fraction of SG stars present in most of the GCs so far examined.

The lithium yield from AGB stars of different mass may contribute to understand the role (if any) of these stars in the formation of the SG in GCs. It is commonly believed that the polluting matter must be diluted with pristine matter to explain the abundance patterns, such as the Na–O anticorrelation (Prantzos & Carbonnel 2006, D'Antona & Ventura 2007). If the progenitors of the SG stars are massive stars, they have destroyed their original lithium, and the lithium in the SG must be due to the mixing with pristine gas. If instead the progenitors are massive AGB stars, they may have a non negligible lithium yield, that must be taken into account in the explanation of the SG abundances.

Figure 6 shows a compact summary of what we know about lithium abundances in the halo and in GCs in the plane $\log \epsilon$(Li) versus [Fe/H]. The halo stars are plotted as triangles, from Meléndez *et al.* (2009) (their non LTE abundances are plotted). The data for three clusters are added, at their [Fe/H] content, taken from Carretta *et al.* (2009c) scale. The references for the clusters data are in the figure label. Notice that the three open triangles of NGC 6397, at much lower ϵ(Li) than the other points, refer to subgiants, in which lithium can be reduced by mixing. Although the data analysis is not homogeneous among the different samples, the figure shows interesting trends. The lithium spread of the halo stars in the range of metallicities of the clusters NGC 6397 and NGC 6752 is very small around a plateau value $\log \epsilon$(Li)$\sim$2.2. In fact the full triangles at $\log \epsilon$(Li)$<$ 2 are lower mass stars for which depletion is expected (Meléndez *et al.* 2009). The WMAP – big bang nucleosynthesis "standard" abundance, $\log \epsilon$(Li)$=$2.72 (e.g. Cyburt *et al.* 2009) is much larger than the plateau abundance. The lithium spread in the clusters appears a bit larger, although Lind *et al.* (2009) point out that in NGC 6397 it is consistent with the observational error. We should expect a larger lithium spread among GC stars if there are SG stars, even if the pollutors' gas (AGB or massive stars envelopes) has been diluted with pristine gas (Decressin *et al.* 2007b, Prantzos *et al.* 2007). The dilution is very plausible if there is a direct correlation between lithium and sodium abundances, as convincingly shown in NGC 6752 (Pasquini *et al.* 2005). A similar correlation also appears in NGC 6397, but it is based only on the high sodium abundance of the three subgiants plotted as open triangles (Lind *et al.* 2009). A possible anticorrelation among the stars of 47 Tuc (Bonifacio *et al.* 2007) is not convincing, as these stars may be subject to lithium depletion mechanisms due to their larger iron content (D'Orazi, these proceedings). In addition, according to Pasquini *et al.* (2008), two stars in NGC 6397 differ by $\sim$0.6dex in oxygen, but have "normal" $\log \epsilon$(Li)$\sim$2.2: this is certainly not easily compatible with a simple dilution model, and may require that the pollutors are also important lithium producers. In fact, if the AGB pollutors produce enough lithium, a dilution model must take it into account.

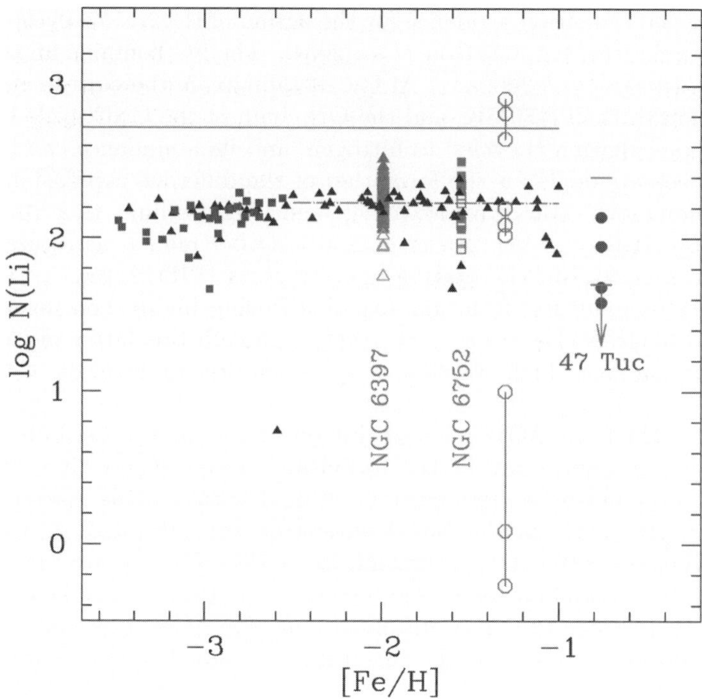

Figure 6. Lithium abundances as a function of [Fe/H] in halo stars and in scarcely evolved stars in three GCs. Halo data are from Meléndez *et al.* (2009), represented as black triangles (non LTE models). (Blue) full squares are from Sbordone *et al.* (2009), analyzed by 3D non LTE models. The top horizontal line represent a WMAP – standard Big Bang nucleosynthesis value $\log \epsilon(\mathrm{Li}) = 2.72$, the dot–dashed line represents an eye fit of the Meléndez *et al.* (2009) data in the range of the GC metallicities. Data for NGC 6397 are from Lind *et al.* (2009). The three open triangles are relative to the data for three subgiants, and may not represent the turnoff abundances in this cluster. Data for NGC 6752 are from Pasquini *et al.* (2005), plotted as open or full squares according to the two different temperature scales used in their work. The full circles are the data for 47 Tuc by Bonifacio *et al.* (2007). The limits of the lithium range in the 50 stars recently examined by D'Orazi (these proceedings) are also given. Open circles at $[\mathrm{Fe/H}] = -1.3$ represent the average abundances in the ejecta of models of 4, 5 an 6 $M_\odot$ for three different mass loss rate formulations (see text).

Notice that the dilution model is not so straightforward as we may think a priori: it will include a fraction α of matter with pristine Li, plus a fraction $(1-\alpha)$ having the Li of the ejecta (so, either the abundance of the AGB ejecta in the AGB mass range involved in the SG formation, or zero Li for the FRMS model). The dilution required to explain a given range in observed Li is different if we assign to the pristine Li the value $\log \epsilon(\mathrm{Li}) = 2.72$ (see above), or the atmospheric Pop II value ($\sim$2.2), or some intermediate value. In addition, if we are assuming that the uniform surface abundance of Li in Pop II is due to a depletion mechanism, also the abundance resulting from the dilution model must be decreased to take into account a similar depletion factor.

If we take our "standard mass loss" results of Fig. 6 at face value, ignoring the big question mark on mass loss, the yields can be used to predict the lithium expected in the SG, if the SG is a result of star formation from AGB ejecta diluted with pristine gas. The abundances will depend mainly on the mass range of the AGB progenitors: if the ejecta of masses in the range $4.5 - 6\,M_\odot$ are involved, their abundance is $\log \epsilon(\mathrm{Li}) \sim 2$, 0.7dex smaller than the Big Bang abundance. In order to explain the abundances observed in

NGC 6397 or in NGC 6752, a dilution model including the ejecta of these AGB stars will require a percentage of pristine matter only *slightly smaller* than in a model including the lithium free FRMS, and we will not be able to discriminate between the two models. The case is different if the Big Bang abundance is "non standard" and closer to the observed halo stars average value.

A different interesting problem is posed by the GCs in which a "blue" main sequence (MS) has been revealed from precise HST photometry, namely ω Cen (Bedin *et al.* 2004) and NGC 2808 (D'Antona *et al.* 2005, Piotto *et al.* 2007). The blue MS can only be interpreted as a very high helium MS (mass fraction Y$\sim$0.38) (Norris 2004, Piotto *et al.* 2005). Actually, in NGC 2808 three MS well separated each other in color are present (Piotto *et al.* 2007), corresponding to three main helium content values, and in agreement with the predictions made from the distribution of stars in the very extended and multimodal horizontal branch (see, e.g. D'Antona & Caloi 2004, D'Antona *et al.* 2005). Pumo *et al.* (2008) noticed that the helium abundances of super–AGB stars envelopes are within the small range $0.36<$Y<0.38 (Siess 2007) and D'Ercole *et al.* (2008) have shown that a full chemo–hydrodynamical model of the cluster can provide a reasonable interpretation of the three MSs of NGC 2808, *provided that the blue MS is formed directly by matter ejected from the super–AGB range, undiluted with pristine gas.* In the future, spectroscopic observations of the blue MS in ω Cen and NGC 2808 will provide a falsification of this hypothesis, e.g. by means of the oxygen and sodium abundance revealed. In particular lithium can be an important test too, as it could provide an independent calibration of the mass loss rate in the super–AGB phase. Already some observations of the turnoff stars in ω Cen are available (Bonifacio, in this book), but it is not clear whether stars belonging to the blue MS have been observed. The "standard mass–loss" super–AGB models shown in Fig. 5 predict that lithium in these stars may become very large if some blue MS stars are formed from the ejecta of the upper mass range of super–AGB stars. We need observations of the blue MS to falsify this prediction.

We thank Corinne Charbonnel and the organizing committees for the invitation and for the successful and intense meeting. We are grateful to J. Meléndez and L. Sbordone for allowing us to use their data in advance of publication, and to V. D'Orazi and D. Romano for useful information.

References

Arnould, M. & Norgaard, H. 1975, *A&A*, 42, 55

Bedin, L. R., Piotto, G., Anderson, J., Cassisi, S., King, I. R., Momany, Y., & Carraro, G. 2004, *ApJ*, 605, L125

Blöcker, T. 1995, *A&A*, 297, 727

Boffin, H. M. J., Paulus, G., Arnould, M., & Mowlavi, N. 1993, *A&A*, 279, 173

Bonifacio, P., *et al.* 2007, *A&A*, 470, 153

Cameron, A. G. W. & Fowler, W. A. 1971, *ApJ*, 164, 111

Carretta, E., *et al.* 2009a, *A&A*, 505, 117

Carretta, E., Bragaglia, A., Gratton, R., & Lucatello, S. 2009b, *A&A*, 505, 139

Carretta, E., Bragaglia, A., Gratton, R., D'Orazi, V., & Lucatello, S. 2009c, arXiv:0910.0675

Cyburt, R. H., Ellis, J., Fields, B. D., Luo, F., Olive, K. A., & Spanos, V. C. 2009, *Journal of Cosmology and Astro-Particle Physics*, 10, 21

D'Antona F., Caloi V., 2004, *ApJ*, 611, 871

D'Antona, F., Bellazzini, M., Caloi, V., Pecci, F. F., Galleti, S., & Rood, R. T. 2005b, *ApJ*, 631, 868

D'Antona, F. & Mazzitelli, I. 1982, *ApJ*, 260, 722

D'Antona, F. & Matteucci, F. 1991, *A&A*, 248, 62

D'Antona, F. & Ventura, P. 2007, *MNRAS*, 379, 1431
Decressin, T., Meynet, G., Charbonnel, C., Prantzos, N., & Ekström, S. 2007a, *A&A*, 464, 1029
Decressin, T., Charbonnel, C., & Meynet, G. 2007b, *A&A*, 475, 859
D'Ercole, A., Vesperini, E., D'Antona, F., McMillan, S. L. W., & Recchi, S. 2008, *MNRAS*, 391, 825
de Mink, S. E., Pols, O. R., Langer, N., & Izzard, R. G. 2009, *A&A*, 507, L1
García-Hernández, D. A., García-Lario, P., Plez, B., Manchado, A., D'Antona, F., Lub, J., & Habing, H. 2007, *A&A*, 462, 711
Gratton, R., Sneden, C., & Carretta, E. 2004, *ARA&A*, 42, 385
Hernanz, M., Jose, J., Coc, A., & Isern, J. 1996, *ApJ Letters*, 465, L27
Iben, I. J. 1973, *ApJ*, 185, 209
Jose, J. & Hernanz, M. 1998, *ApJ*, 494, 680
Lind, K., Primas, F., Charbonnel, C., Grundahl, F., & Asplund, M. 2009, *A&A*, 503, 545
Melèndez, J., Casagrande, L., Ramìrez, I., & Asplund, M., submitted
Nollett, K. M., Busso, M., & Wasserburg, G. J. 2003, *ApJ*, 582, 1036
Norris, J. E. 2004, *ApJ Letters*, 612, L25
Pasquini, L., Bonifacio, P., Molaro, P., Francois, P., Spite, F., Gratton, R. G., Carretta, E., & Wolff, B. 2005, *A &A*, 441, 549
Pasquini, L., Ecuvillon, A., Bonifacio, P., & Wolff, B. 2008, *A&A*, 489, 315
Piotto, G., et al. 2005, *ApJ*, 621, 777
Piotto, G., et al. 2007, *ApJ Letters*, 661, L53
Prantzos, N. & Charbonnel, C. 2006, *A&A*, 458, 135
Prantzos, N., Charbonnel, C., & Iliadis, C. 2007, *A&A*, 470, 179
Pumo, M. L., D'Antona, F., & Ventura, P. 2008, *ApJ Letters*, 672, L25
Reimers, D. 1977, *A&A*, 61, 217
Romano, D., Matteucci, F., Molaro, P., & Bonifacio, P. 1999, *A&A*, 352, 117
    & D'Antona, F. 2001, *A&A*, 374, 646
Sackmann, I.-J. & Boothroyd, A. I. 1999, *ApJ*, 510, 217
Sackmann, I.-J. Smith, R. L., & Despain, K. H. 1974, *ApJ*, 187, 555
Siess, L. 2007, *A&A*, 476, 893
Sbordone *et al.* 2009 *A&A* submitted
Scalo, J. M., Despain, K. H., & Ulrich, R. K. 1975, *ApJ*, 196, 805
Schatzman, E. 1951, *Annales d'Astrophysique*, 14, 294
Smith, V. V. & Lambert, D. L. 1989, *ApJ Letters*, 345, L75
Smith, V. V. & Lambert, D. L. 1990, *ApJ Letters*, 361, L69
Smith, V. V., Plez, B., Lambert, D. L., & Lubowich, D. A. 1995, *ApJ*, 441, 735
Starrfield, S., Truran, J. W., Sparks, W. M., & Arnould, M. 1978, *ApJ*, 222, 600
Vassiliadis, E. & Wood, P. R. 1993, *ApJ*, 413, 641
Ventura, P., D'Antona, F., & Mazzitelli, I. 2000, *A&A*, 363, 605
Ventura, P., D'Antona, F., Mazzitelli, I., & Gratton, R. 2001, *ApJ Letters*, 550, L65
Ventura, P., D'Antona, F., & Mazzitelli, I. 2002, *A&A*, 393, 215
Ventura, P. & D'Antona, F. 2005, *A&A*, 431, 279
Ventura, P. & D'Antona, F. 2010, *MNRAS Letters*, in press
Wasserburg, G. J., Boothroyd, A. I., & Sackmann, I.-J. 1995, *ApJ Letters*, 447, L37

Light Elements in the Universe
Proceedings IAU Symposium No. 268, 2009
C. Charbonnel, M. Tosi, F. Primas & C. Chiappini, eds.

© International Astronomical Union 2010
doi:10.1017/S1743921310004539

Lithium production by thermohaline mixing in low-mass, low-metallicity asymptotic giant branch stars

Richard J. Stancliffe, George C. Angelou and John C. Lattanzio

Centre for Stellar and Planetary Astrophysics, Monash University, VIC 3800, Australia
email: Richard.Stancliffe@sci.monash.edu.au

Abstract. We examine the effects of thermohaline mixing on the composition of the envelopes of low-metallicity asymptotic giant branch (AGB) stars. We have evolved models of 1, 1.5 and $2\,M_\odot$ and of metallicity $Z = 10^{-4}$ from the pre-main sequence to the end of the thermal pulsing asymptotic giant branch with thermohaline mixing applied throughout the simulations. We find that the small amount of ^{3}He that remains after the first giant branch is enough to drive thermohaline mixing on the AGB and that the mixing is most efficient in the early thermal pulses, with the efficiency dropping from pulse to pulse. We note a surprising increase in the ^{7}Li abundance, with $\log_{10} \epsilon(^7\text{Li})$ reaching values of over 2.5 in the $1.5\,M_\odot$ model. It is thus possible to get stars which are both C- and Li-rich at the same time. We compare our models to measurements of carbon and lithium in carbon-enhanced metal-poor stars which have not yet reached the giant branch. These models can simultaneously reproduced the observed C and Li abundances of carbon-enhanced metal-poor turn-off stars that are Li-rich.

Keywords. Stars: evolution, AGB and post-AGB, Population II, carbon

1. Introduction

It has long been known that models of asymptotic giant branch (AGB) stars that only include mixing in convective regions are incomplete. These canonical models cannot account for observations such as: the low ^{12}C/^{13}C ratios in low-mass AGB stars (Abia & Isern 1997; Lebzelter *et al.* 2008), Li and C-rich stars in our Galaxy (Abia & Isern 1997; Uttenthaler *et al.* 2007), isotopic ratios measured in pre-solar grains (e.g. Nollett *et al.* 2003, and references therein). It has therefore been suggested that material might circulate below the base of the convective envelope into regions where nuclear burning can happen. This process is often referred to as 'cool bottom processing'.

There have been detections of lithium in carbon-enhanced metal-poor (CEMP) stars (e.g. Thompson *et al.* 2008) and this is difficult to reconcile with standard models of AGB stars. Canonical AGB models produce Li via the Cameron-Fowler mechanism (Cameron & Fowler 1971), which involves the production of beryllium deep in the hydrogen burning shell via the reaction $^4\text{He}(^3\text{He},\gamma)^7\text{Be}$ and the immediate transport of this to cooler regions of the star where the ^{7}Li that forms (once the beryllium has undergone electron capture) is stable against proton captures. This takes place in the more massive stars which undergo hot bottom burning (HBB, where the base of the convective envelope lies in the top of the hydrogen-burning shell). Such stars would be rich in nitrogen, not carbon. These observations suggest that something is missing from the AGB models and an extra mixing mechanism must be at work.

Lithium is a particularly important element from the point of view of mixing processes in CEMP stars, especially those that are enriched in *s*-process elements. These CEMP-*s* stars are believed to have formed in binary systems where mass has been transferred from an AGB star which is no longer visible on to a companion that we now observe as carbon rich. Stancliffe *et al.* (2007) pointed out that material accreted on to a low-mass companion does not just remain at the surface of the companion star and that thermohaline mixing could efficiently mix this material deep into the stellar interior. However, detection of lithium in the CEMP binary system CS 22964-161 led Thompson *et al.* (2008) to suggest that the mixing efficiency could not be so high. Li is a fragile element and is easily destroyed at temperatures in excess of about 2.5×10^6 K. Even a modest depth of mixing can lead to efficient Li-depletion (Stancliffe 2009) and hence the measurement of Li in CEMP stars could be a good test of the efficiency of thermohaline mixing. It is therefore crucial that we understand the origin of this element.

Despite the apparent need for extra mixing on both the giant branches, the physical nature of the mechanism (or mechanisms) has proved illusive. Recently, Eggleton *et al.* (2006) showed that the lowering of the mean molecular weight by the reaction $^3\mathrm{He}(^3\mathrm{He},2\mathrm{p})^4\mathrm{He}$ could lead to mixing in red giants via the thermohaline instability. This can potentially explain the change in abundances seen in giants above the luminosity bump (Charbonnel & Zahn 2007; Eggleton *et al.* 2008). In this work, we wish to examine what the consequences of thermohaline mixing are for low-mass, low-metallicity AGB stars.

2. The stellar evolution code

Calculations in this work have been carried out using the STARS stellar evolution code (see Stancliffe & Eldridge 2009, for a detailed description of the code). Thermohaline mixing is included throughout all the evolutionary phases via the prescription of Kippen-hahn, Ruschenplatt & Thomas (1980), with the mixing coefficient being multiplied by a factor of 100 as suggested by the work of Charbonnel & Zahn (2007). These authors find that with a factor of this magnitude they are able to reproduced the abundance trends observed towards the tip of the red giant branch. Stancliffe *et al.* (2009) also showed that a coefficient of this magnitude could reproduced the observed mixing trends in both low-mass metal-poor stars and carbon-enhanced metal-poor stars on the upper part of the first giant branch.

We evolve stars of 1, 1.5 and $2\,\mathrm{M_\odot}$ from the pre-main sequence to the end of the thermally pulsing asymptotic giant branch (TP-AGB) using 999 mesh points. Reimers (1975) mass-loss prescription, with $\eta = 0.4$, is used from the Main Sequence up to the TP-AGB; the Vassiliadis & Wood (1993) mass-loss law is employed during the TP-AGB. A mixing length parameter of $\alpha = 2.0$ is employed. The metallicity of each model is $Z = 10^{-4}$ ([Fe/H]≈ -2.3) and the initial abundances are assumed to be solar-scaled according to Anders & Grevesse (1989), with the exception of $^7\mathrm{Li}$, for which we adopt a value of $X_{^7\mathrm{Li}} = 1.05 \times 10^{-9}$ which is equivalent to the Spite plateau value.

3. Lithium production on the AGB

While lithium is destroyed on the red giant branch, we find that it can be produced during the TP-AGB phase. Lithium production occurs in each of the models in the following way. First, the deepening of the convective envelope that occurs after every thermal pulse (referred to as third dredge-up) homogenises the envelope and flattens out the mean molecular weight above the burning shell (top left panel of Fig. 1). This is an essential precursor to thermohaline mixing. The reaction $^3\mathrm{He}(^3\mathrm{He},2\mathrm{p})^4\mathrm{He}$ can only produce a small reduction of the mean molecular weight and its effects will only be apparent in a region of zero mean molecular weight gradient.

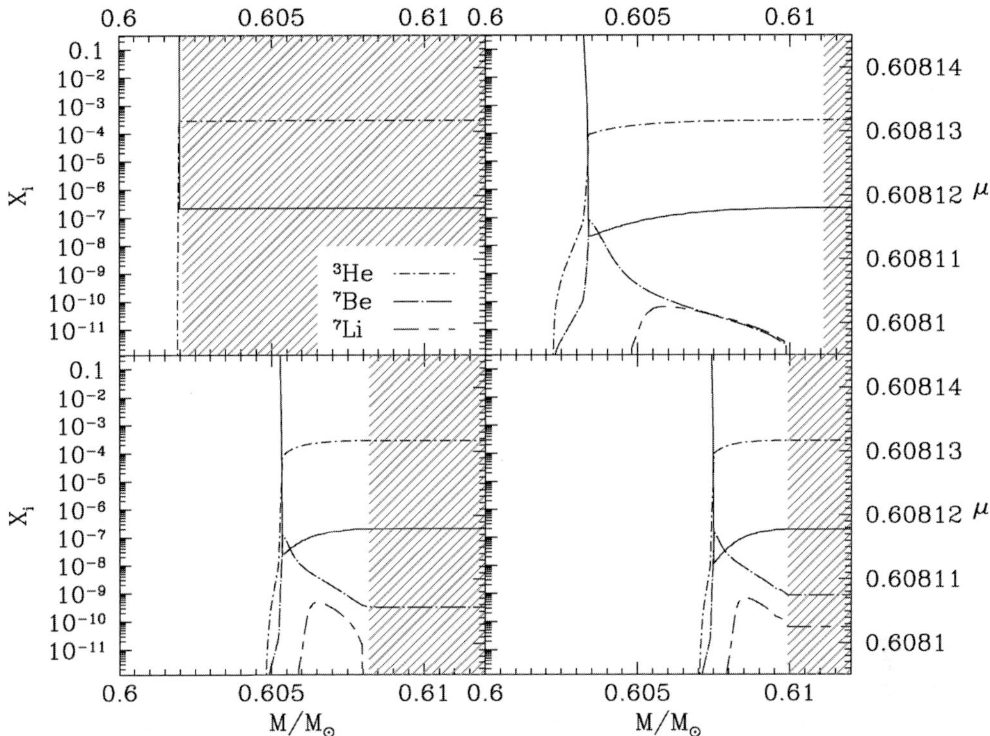

Figure 1. Abundance profiles of the $1.5\,M_\odot$ model after its second thermal pulse. In each panel, the abundances are: ^{3}He (dot-short dash line), ^{7}Be (dot-long dash line) and ^{7}Li (dashed line). The mean molecular weight, μ, is displayed by a solid line. The grey shading indicates convective regions. **Top left:** Just after the end of third dredge-up. **Top right:** Retreat of the convective envelope. Thermohaline mixing leads to the formation of a pocket of beryllium and lithium. **Bottom left:.** The hydrogen shell moves outward and the envelope comes back in again. **Bottom right** Just before the beginning of the next thermal pulse.

After the convective envelope has reached its maximum depth, it begins to retreat and hydrogen burning re-ignites. At the top of the hydrogen burning shell, two important reactions are occurring. The ^{3}He(^{3}He,2p)^{4}He lowers the mean molecular weight and thermohaline mixing starts to occur. In addition, the fusion of ^{3}He with ^{4}He produces ^{7}Be. Thermohaline mixing transports this newly synthesised beryllium toward the convective envelope. En route, ^{7}Be undergoes electron capture to produce ^{7}Li. While it remains deep in the star, this lithium is destroyed by proton captures. Thermohaline mixing is not efficient enough to transport the lithium to the envelope before it is destroyed. Pockets of ^{7}Be and ^{7}Li (top right panel of Fig. 1) are thus formed and their abundances depend on the rate at which fresh beryllium is transported up from hydrogen burning shell and the rate at which both species are being destroyed.

As the star settles into the interpulse phase, the convective envelope retreats inward again but not to the same extent as it did during third dredge-up. The envelope reaches into the lithium pocket that has been formed and a surface enrichment of lithium occurs. In addition the distance between the hydrogen burning shell and the convective envelope has now been reduced and thermohaline mixing is now able to transport lithium into the envelope before it is destroyed (bottom panels of Fig. 1). The lithium abundance thus continues to increase during the interpulse phase.

We find that the $1.5\,\mathrm{M_\odot}$ model obtains the greatest lithium enrichment of the three models we have run. By the time the star enters the superwind phase the surface lithium abundance has reached $\log_{10}\epsilon(^7\mathrm{Li}) = 2.51$. The $2\,\mathrm{M_\odot}$ model reaches a peak surface lithium enrichment of $\log_{10}\epsilon(^7\mathrm{Li}) = 1.44$ after 7 thermal pulses but then suffers a slight decline as the base of the convective envelope becomes sufficiently deep for some lithium destruction to take place. At the onset of the superwind phase, the surface lithium abundance is $\log_{10}\epsilon(^7\mathrm{Li}) = 1.38$. The $1\,\mathrm{M_\odot}$ model is the least enriched by this mechanism and only reaches $\log_{10}\epsilon(^7\mathrm{Li}) = 1.03$ by the end of the TP-AGB.

4. Discussion

The surprising outcome of these simulations is the high Li abundances that can be produced. It is usually supposed that only the higher mass AGB stars which undergo hot bottom burning are able to produce Li via the Cameron-Fowler mechanism (Cameron & Fowler 1971). This work shows that it is possible that *low-mass AGB stars could be producers of lithium-7*. We also note that the action of thermohaline mixing on the AGB is subtly different from its action of the RGB. On the RGB, it leads to a depletion of Li with the star leaving the RGB with virtually no lithium left. However, on the AGB thermohaline mixing can substantially increase the surface Li abundance above the Spite plateau value. It is therefore possible that low-mass AGB stars have contributed to the Galaxy's Li budget. The effect of a population of low-mass, low-metallicity lithium producers on Galactic chemical evolution models should be investigated.

These models do improve the agreement of the AGB models with observations of Li in carbon-enhanced metal-poor stars. We have extracted from the Stellar Abundances for Galactic Archaeology (SAGA) database (Suda *et al.* 2008) those CEMP stars that are both C-rich and have measured Li-abundances, and also that are still close to the main sequence turn-off (because first dredge-up will significantly reduce the surface Li abundance). We select only those stars in the metallicity range $-3 <[\mathrm{Fe/H}]< -2$ as the models presented herein may be expected to apply only over a limited range in metallicity. In particular, at very low metallicities, AGB stars can undergo additional mixing events not present in our models (see e.g. Fujimoto *et al.* 1990; Campbell & Lattanzio 2008; Lau *et al.* 2009, among many others for a discussion of these mixing events). SAGA lists 5 turn-off objects with measured lithium abundances, the properties of which are displayed in Table 1.

One caveat should be added to the following discussion. The scenario of mass transfer from an AGB primary star on to a lower mass secondary in a binary system is expected to apply to those stars belonging to the CEMP-*s* subclass, i.e. those stars which have $[\mathrm{Ba/Fe}]> 1$ and $[\mathrm{Ba/Eu}]>0.5$ according to the definitions given by Beers & Christlieb (2005). The origin of the $r + s$ subclass of CEMP stars, which have $0 <[\mathrm{Ba/Eu}]< 0.5$, is currently unknown. It is possible that their *s*-process enrichment has come from a binary mass transfer event, in which case the models presented herein would apply, but the enrichment may have another source entirely (see e.g. Lugaro *et al.* 2009, for a possible alternative formation scenario). This should be borne in mind throughout the following discussion.

We model CEMP stars by accreting material of the composition of the ejecta from our AGB models on to low-mass stars on the main sequence (see Stancliffe & Glebbeek 2008, for details). We accrete 0.001, 0.01 and $0.1\,\mathrm{M_\odot}$ of material on to a companion so that its final mass is $0.8\,\mathrm{M_\odot}$, equivalent to the turn-off mass of the Halo. The model is then evolved until the star reaches the main-sequence turn-off, i.e. the point at which the Li-rich CEMP stars are observed. An example of accretion from the $1.5\,\mathrm{M_\odot}$ model is shown in Fig. 2.

Object	[Fe/H]	[C/Fe]	$\log_{10} \epsilon(\mathrm{Li})$	log g	Refs.
CS 22964-161A	-2.41	1.35	2.09	3.7	1
CS 22964-161B	-2.39	1.15	2.09	4.1	1
HE 0024-2523	-2.7	2.6	1.5	4.3	2
CS 31080-095	-2.85	2.69	1.73	4.5	3
CS 31062-012	-2.53	2.14	2.3	4.3	4

Table 1. Properties of CEMP turn-off stars with measured Li-abundances, as extracted from the SAGA database. References: 1 – Thompson *et al.* (2008), 2 – Lucatello *et al.* (2003), 3 – Sivarani *et al.* (2006), 4 – Aoki *et al.* (2008)

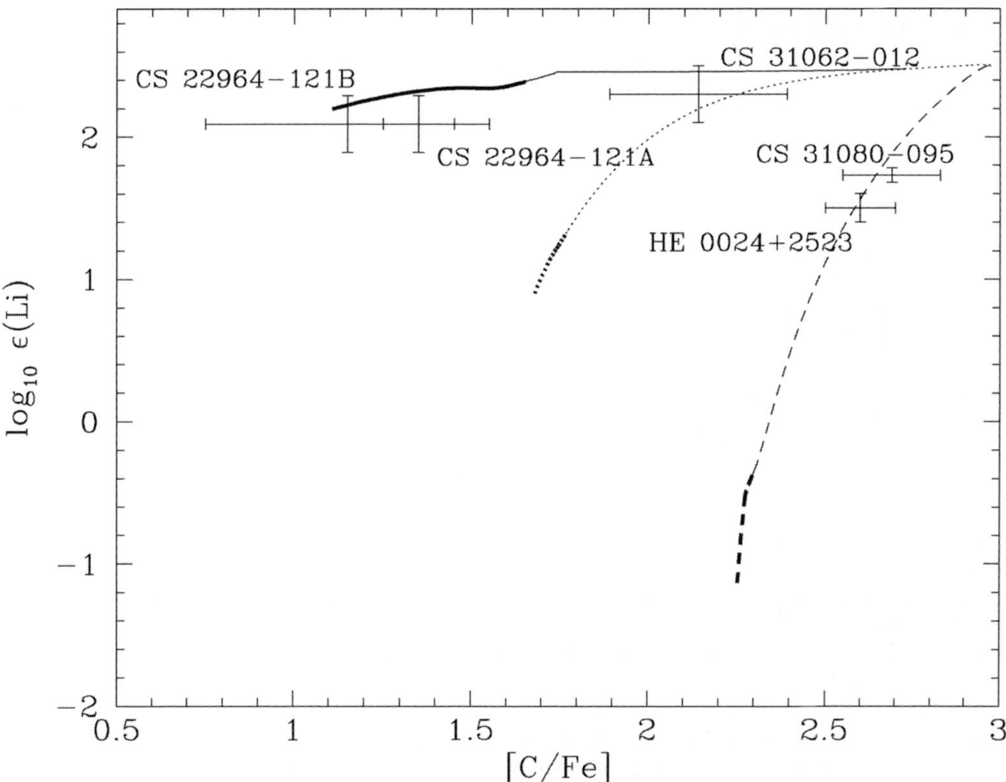

Figure 2. The evolution of $\log_{10} \epsilon(\mathrm{Li})$ with [C/Fe] when accreting material from a $1.5\,\mathrm{M}_\odot$ companion. The cases displayed are for when $0.001\,\mathrm{M}_\odot$ (solid line), $0.01\,\mathrm{M}_\odot$ (dotted line) and $0.1\,\mathrm{M}_\odot$ (dashed line) is accreted. In each case, the secondary is left with a total mass of $0.8\,\mathrm{M}_\odot$. Bold lines indicate where $\log g$ passes from 4.5 to 3.5 as the object evolves off the main sequence. The errorbars denote the locations of specific observed systems. The secondary is modelled including thermohaline mixing, gravitational settling and an extra turbulent process.

We find that a model that includes thermohaline mixing, gravitational settling and the *ad hoc* turbulent mixing of Richard *et al.* (2005) can reproduced the observed properties of CS 22964-121 if $0.001\,\mathrm{M}_\odot$ of material is accreted, while for CS 31062-012 around $0.002\,\mathrm{M}_\odot$ of material would have to be accreted. For CS 31080-095 and HE 0024+2523 a companion of between 1 and $1.5\,\mathrm{M}_\odot$ mass is most likely required, though based on the high carbon abundance in these objects, it seems unlikely that the accreted material would have undergone any mixing.

5. Conclusions

We have investigated the effect that thermohaline mixing has on the abundances of low-mass, low-metallicity AGB stars. We find that enough ^{3}He remains after the first giant branch that thermohaline mixing can still take place on the AGB. Thermohaline mixing can lead to substantial production of ^{7}Li – even up to abundances above the Spite plateau value. Thus it is possible to reconcile C- and Li-rich metal-poor stars with having come from a binary mass transfer scenario. The possibility that low-mass, low-metallicity stars could be producers of lithium is intriguing and their role in Galactic chemical evolution should be investigated.

6. Acknowledgements

RJS is funded by the Australian Research Council's Discovery Projects scheme under grant DP0879472. This work was supported by the NCI National Facility at the ANU.

References

Abia, C. & Isern, J. 1997, MNRAS, 289, L11

Anders, E. & Grevesse, N. 1989, Geo.Cosmo.Acta, 53, 197

Aoki, W., Beers, T. C., Sivarani, T., Marsteller, B., Lee, Y. S., Honda, S., Norris, J. E., Ryan, S. G., & Carollo, D. 2008, ApJ, 678, 1351

Beers, T. C. & Christlieb, N. 2005, ARA&A, 43, 531

Cameron, A. G. W. & Fowler, W. A. 1971, ApJ, 164, 111

Campbell, S. W. & Lattanzio, J. C. 2008, A&A, 490, 769

Charbonnel, C. & Zahn, J.-P. 2007, A&A, 467, L15

Eggleton, P. P., Dearborn, D. S. P., & Lattanzio, J. C. 2006, Science, 314, 1580

Eggleton, P. P., Dearborn, D. S. P., & Lattanzio, J. C. 2008, ApJ, 677, 581

Fujimoto M. Y., Iben I. J., & Hollowell, D. 1990, ApJ, 349, 580

Kippenhahn, R., Ruschenplatt, G., & Thomas, H.-C. 1980, A&A, 91, 175

Lau, H. H. B., Stancliffe, R. J. & Tout, C. A. 2009, MNRAS, 396, 1046

Lebzelter, T., Lederer, M. T., Cristallo, S., Hinkle, K. H., Straniero, O., & Aringer, B. 2008, A&A, 486, 511

Lucatello, S., Gratton, R., Cohen, J. G., Beers, T. C., Christlieb, N., Carretta, E., & Ramírez, S. 2003, AJ, 125, 875

Lugaro, M., Campbell, S. W., & de Mink, S. E. 2009, PASA, 26, 322

Nollett, K. M., Busso, M., & Wasserburg, G. J. 2003, ApJ, 582, 1036

Reimers, D. 1975, Memoires of the Societe Royale des Sciences de Liege, 8, 369

Richard, O., Michaud, G., & Richer, J. 2005, ApJ, 619, 538

Sivarani, T., Beers, T. C., Bonifacio, P., Molaro, P., Cayrel, R., Herwig, F., Spite, M., Spite, F., Plez, B., Andersen, J., Barbuy, B., Depagne, E., Hill, V., François, P., Nordström, B., & Primas, F. 2006, A&A, 459, 125

Stancliffe, R. J. 2009, MNRAS, 394, 1051

Stancliffe, R. J., Church, R. P., Angelou, G. C., & Lattanzio, J. C. 2009, MNRAS, 396, 2313

Stancliffe, R. J. & Eldridge, J. J. 2009, MNRAS, 396, 1699

Stancliffe, R. J. & Glebbeek, E. 2008, MNRAS, 389, 1828

Stancliffe, R. J., Glebbeek, E., Izzard, R. G., & Pols, O. R. 2007, A&A, 464, L57

Suda, T., Katsuta, Y., Yamada, S., Suwa, T., Ishizuka, C., Komiya, Y., Sorai, K., Aikawa, M., & Fujimoto, M. Y. 2008, PASJ, 60, 1159

Thompson, I. B., Ivans, I. I., Bisterzo, S., Sneden, C., Gallino, R., Vauclair, S., Burley, G. S., Shectman, S. A., & Preston, G. W. 2008, ApJ, 677, 556

Uttenthaler, S., Lebzelter, T., Palmerini, S., Busso, M., Aringer, B., & Lederer, M. T. 2007, A&A, 471, L41

Vassiliadis, E. & Wood, P. R. 1993, ApJ, 413, 641

Light Elements in the Universe
Proceedings IAU Symposium No. 268, 2009
C. Charbonnel, M. Tosi, F. Primas & C. Chiappini, eds.

© International Astronomical Union 2010
doi:10.1017/S1743921310004540

Light elements in massive single and binary stars

N. Langer[1,2], I. Brott[2], M. Cantiello[2], S. E. de Mink[2], R. G. Izzard[3], and S.-C. Yoon[1]

[1] Argelander-Institut für Astronomie, Universität Bonn,
Auf dem Hügel 71, Germany

[2] Sterrenkundig Instituut, University of Utrecht,
Postbus 80000, NL-3508TA, Utrecht, the Netherlands

[3] Institut d'Astronomie et d'Astrophysique, Université Libre de Bruxelles,
Boulevard du Triomphe, 1050 Brussels

Abstract. We highlight the role of the light elements (Li, Be, B) in the evolution of massive single and binary stars, which is largely restricted to a diagnostic value, and foremost so for the element boron. However, we show that the boron surface abundance in massive early type stars contains key information about their foregoing evolution which is not obtainable otherwise. In particular, it allows to constrain internal mixing processes and potential previous mass transfer event for binary stars (even if the companion has disappeared). It may also help solving the mystery of the slowly rotating nitrogen-rich massive main sequence stars.

Keywords. Stars: abundances, binaries

1. Introduction

A large effort has been undertaken in the last decades to measure and understand the surface chemical composition of massive main sequence stars. In particular, the detection of nitrogen enhancements in quite a number of such stars (e.g., Gies & Lambert 1992) has triggered the idea that internal mixing processes can bring material from the stellar core to the surface in rapid rotators (Meynet & Maeder 2000, Heger & Langer 2000).

The picture has become more complicated by the recent analysis of a large sample of early B type main sequence stars of Hunter *et al.* (2008b), who showed that the nitrogen-rich stars found by Gies & Lambert (1992), who restricted their analysis to objects with low projected rotational velocities, are likely part of a population of *intrinsically* slowly rotation main sequence stars. This view is supported by the work of Morel *et al.* (2006, 2008), who indeed identifies such a population in our Galaxy (see also Morel 2009). The origin of the nitrogen enrichment in these stars is not understood, but as they are slow rotators it appears difficult to reconcile them with the idea of rotational mixing.

On the other hand, Hunter *et al.* (2008b) also identified a nitrogen-rich population of rapidly rotating early B stars, which appears to be well in line with the predictions of theoretical models including rotational mixing (cf., Maeder *et al.* 2008). The caveat here is that evolutionary models of massive close binaries — whether or not they include rotationally induced chemical mixing (Langer *et al.* 2008) — appear to predict essentially the same trend of nitrogen enrichment with rotational velocity as the single star models (Meynet & Maeder 2000, Heger & Langer 2000).

The light elements lithium, beryllium and boron may play a key role to resolve this issue. They are so rare in the interstellar medium that they can not influence the course of stellar evolution, and thus are often neglected in massive star models. Furthermore, they are fragile nuclei which are generally not synthesised in stars, but rather destroyed, at least certainly on the main sequence. In the cool low-mass stars, lithium is most interesting, as it can be observed rather easily, and it is indeed used extensively to constrain internal mixing processes as can be seen in many contributions to this book. In massive stars this role can be played by boron as will be outlined below.

2. Single stars

Other than helium and the major CNO nuclei, the light elements Li, Be, and B are destroyed by proton capture relatively close to the stellar surface. For both stable boron isotopes, ^{10}B and ^{11}B, the life time against proton capture is equal to the main sequence life time of a $10 M_\odot$ star (10^7 yr) at a temperature of roughly $7\,10^6$ K. Fliegner *et al.* (1996) have computed the evolution of stars of $15 M_\odot$, and found this temperature to occur sufficiently deep inside the stellar envelope (i.e. roughly $1 M_\odot$ below the surface; cf. Fig. 1) that its surface abundance can not be altered due to mass loss alone on the main sequence in the B star regime. Thus, the boron abundance in B stars is a critical test of mixing processes in the upper stellar envelope, while CNO and helium abundances additionally trace the mixing in deeper layers.

Models which include rotational mixing show that boron depletion at the stellar surface is predicted for initial rotational velocities above 50 km/s. However, while in low-mass stars, the rotational velocity is a strong function of age, and so is the lithium abundance, a clear correlation of boron with stellar age is not expected in a population which contains initially fast and slow rotators (cf. Gies & Lambert 1992). However, a population synthesis simulation by Ines Brott shows (Fig. 2), that a clear correlation of the surface boron depletion with the stellar rotation rate can be expected in early B type single stars.

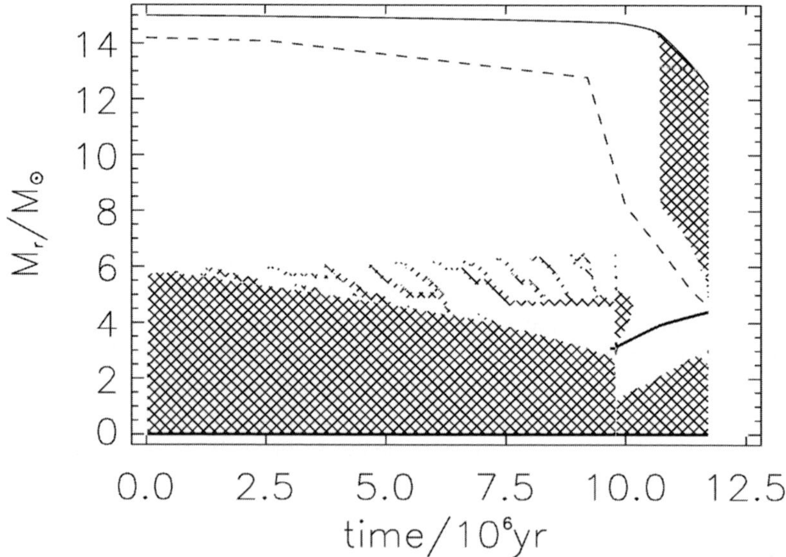

Figure 1. Internal structure of a $15 M_\odot$ star during core hydrogen and helium burning. The solid line on top indicates the total mass of the star as function of time. Hatched areas designate convectively unstable mass zones in the star. The full drawn line at $M_r \simeq 4 M_\odot$ and $t \gtrsim 10^7$ yr designates the location of the H-burning shell during core helium burning. The dashed line indicates the threshold temperature for boron destruction (cf. Fliegner *et al.* 1996).

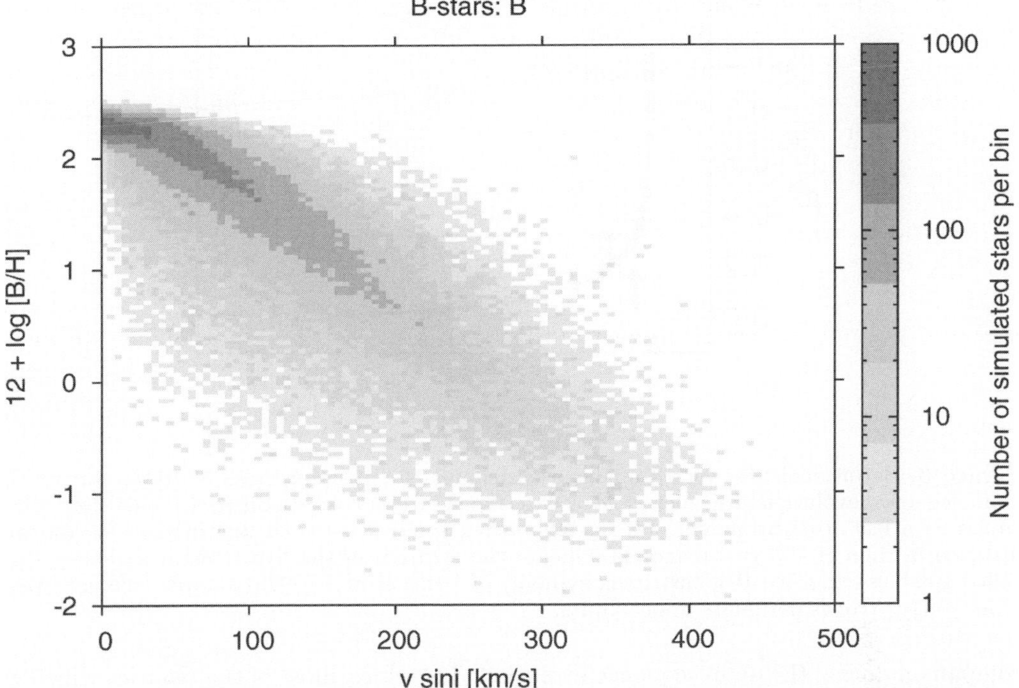

Figure 2. Result of a population synthesis calculation for massive main sequence stars (Brott *et al.*, in prep.), employing a Salpeter initial mass function, and a distribution of initial stellar rotation rates as derived by Hunter *et al.* (2008a), and a constant star formation rate, based on single star evolution models which include rotational mixing. A random Gaussian error of 0.2 dex was added to the predicted boron abundances.

Figure 3 shows the behaviour of the surface abundances of a rotating $15M_\odot$ in a plot of boron depletion versus nitrogen enhancement. Interestingly, it predicts that boron depletion happens essentially *before* nitrogen enrichment occurs. Thus, stars which are already depleted in boron, but which are still nitrogen normal are expected. Indeed, in the sample of Venn *et al.* (1996), several such stars seem to exist. The importance of this finding becomes more clear in the next section.

3. Binary stars

The binary fraction of massive stars is very high, and it may be futile to try to understand their surface abundances without considering effects of binarity (Langer *et al.* 2008). The issue of binarity is not easily resolved, because of two reasons.

Firstly, after a strong binary interaction, the object may not appear to be a binary any more. This is so since many mass transfer systems will produce a rejuvenated main sequence star which dominates the light of the system, with a faint helium star in a wide orbit. In fact, many such objects ought to exist, as we see many of their descendants, the Be-X-ray binaries. However, for the stage where the main sequence star has a helium star companion, we practically do not know any counterpart. Also, there will be many cases where the post-interaction system is in fact a single star. E.g., in many systems consisting of a main sequence star and a helium star companion, the supernova explosion of the helium star will disrupt the binary (instead of leading to a Be-X-ray binary stage). And furthermore, as many as 10% of all massive stars may actually merge with their close

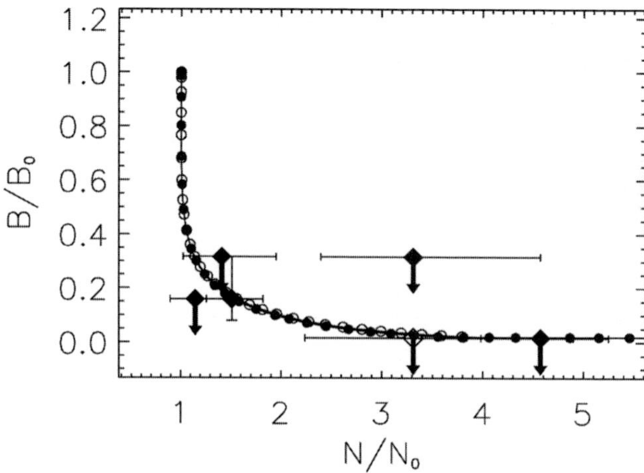

Figure 3. Boron depletion versus nitrogen enrichment for rotating stars of $10 M_\odot$ during the main sequence evolution (Fliegner *et al.* 1996). Open circles correspond to a moderately, filled circles to a fast rotating model. Two neighbouring symbols on each line indicate a constant distance in time of 10^6 yr. Diamonds indicate the location of the five B main sequence stars (filled symbols) and one B giant (open symbol) of Venn *et al.*'s (1996) sample (see also Venn *et al.* 2002). Arrows designate upper limits.

companion during the main sequence evolution. Therefore, most of the binaries which we detect in massive main sequence star populations may have in fact not yet interacted, while on the other hand, many apparent single stars may be the result of a strong binary interaction.

Secondly, binary interaction has many complex branches, and it is a big theoretical enterprise to even fully consider the most important ones in a population study. So far, no such study exists which takes the physics of rotation fully into account — which is what appears to be needed in order to demonstrate that rotational mixing works in Nature.

However, there are some binary evolution models available which include all the required physics, and while they will not allow to obtain a full view of the picture, they may give indications in one or the other direction. To that purpose, Fig. 4 shows the time evolution of the surface rotational velocity and of the surface boron abundance of the mass gainer (i.e. the star which will be visible after the mass transfer) of a close massive binary. The lower most panel also shows the boron depletion factor as function of the rotational velocity in this star.

The model shown in Fig. 4 does include rotational mixing, which produces the mild boron depletion before the first mass transfer event (at about $t = 8.5$ Myr). The mass transfer event itself, which occurs on a time scale of some 10^4 yr, puts boron depleted layers on the surface of the mass gainer, and subsequent thermohaline mixing brings the surface boron abundance back to a level of one per mille of the initial boron abundance.

Consecutively, the mass gainer is spun down by tidal interaction, while rotational mixing reduces the surface boron abundance slightly more. As a result of this phase, one obvious result from Fig. 4 is that in mass gainers of close binaries, one can *not* generally expect a correlation of the boron depletion factor with the surface rotation rate. Here, a slowly rotating main sequence star is produced which shows a surface boron depletion by 3 to 4 orders of magnitude. Fig. 2 shows that such a strong depletion is only expected in single stars with rotation rates above 300 km/s.

The model in Fig. 4 suffers from a second mass transfer at $t \simeq 10.2$ Myr, which leads to a second spin-up of the mass gainer and a further strong reduction of the surface boron

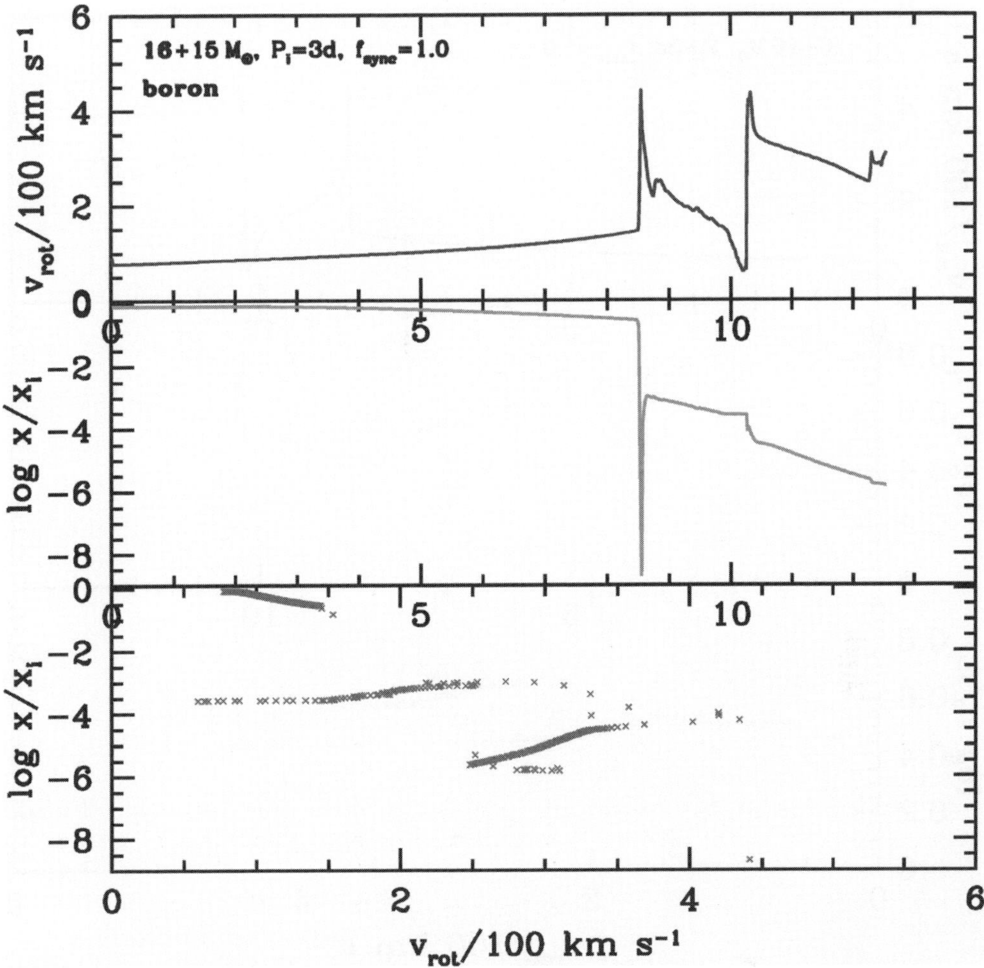

Figure 4. Equatorial rotational velocity (upper panel) and surface boron mass fraction relative to the initial value (middle panel) as function of time (in Myr), for the mass gainer in a solar metallicity $16M_\odot + 15M_\odot$ binary with an initial orbital period of 3 days. The computations include the physics of rotation for both components as in Heger *et al.* (2000), and Spin-Orbit coupling as in Detmers *et al.* (2008) with the nominal coupling parameter $f_{\mathrm{sync}} = 1$, and rotationally enhanced stellar wind mass loss (Langer 1998). Internal magnetic fields are not included. The bottom panel shows the evolution of the mass gainer in the boron depletion versus rotational velocity diagram, where each data point represents a duration of $20\,000\,\mathrm{yr}$. The spin-down of the star after the first accretion event ($t = 8.5...10\,\mathrm{Myr}$) is mostly due to tidal effects.

abundance. As the orbit widens strongly during the second mass transfer phase, no tidal spin-down occurs thereafter. The mass gainer is now a rapidly rotating main sequence star which is strongly boron depleted.

While rotational mixing is included in this model, we want to point out that the effects of rotational mixing and of mass transfer (and thermohaline mixing) are well separable in Fig. 4. The steps in the time evolution of the boron abundance are produced by the mass transfer, and the long time scale changes are due to rotational mixing. We conclude that the main features of the boron evolution of this model would not change if rotational mixing were switched off. From this consideration, we can argue that if rotational mixing would not operate in Nature, then perhaps binary models would not

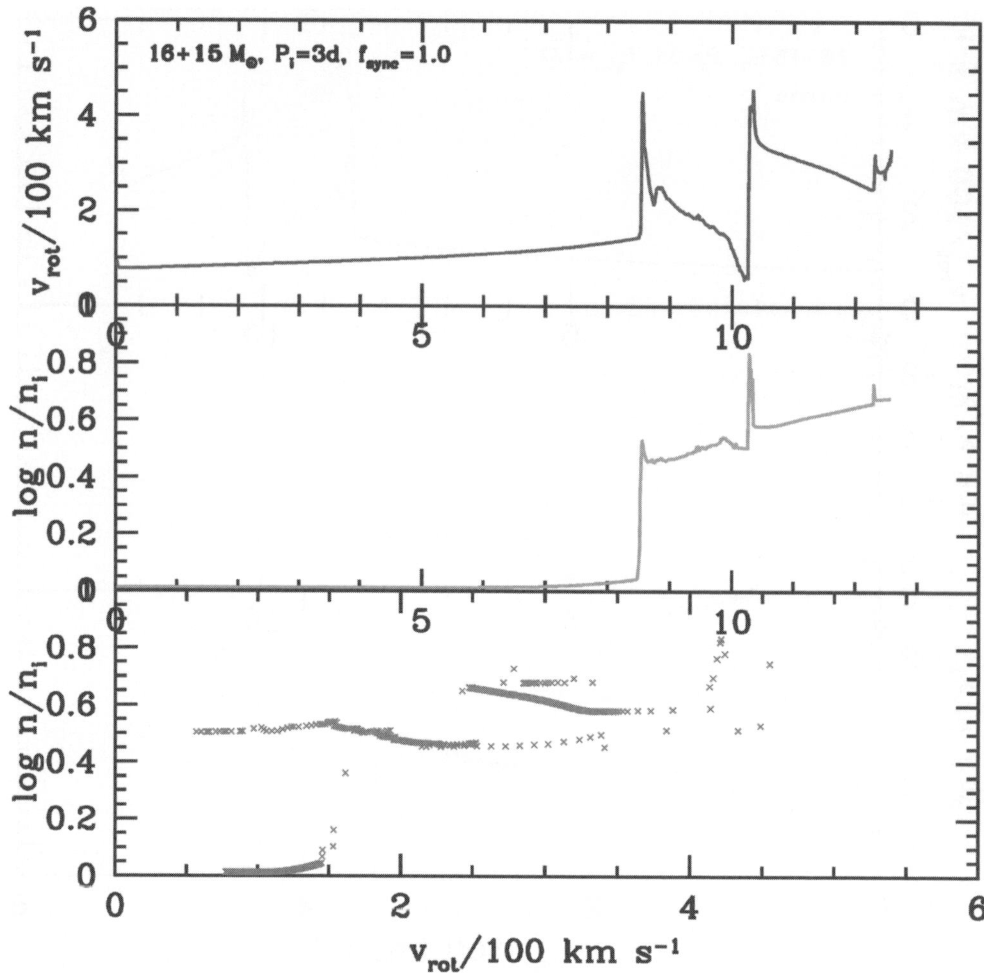

Figure 5. As Fig. 4, but showing the surface nitrogen abundance.

predict main sequence stars with only a mild boron depletion (unless one considered masses much higher than $15M_\odot$, where mass loss can gradually uncover boron-depleted layers).

The situation becomes more clear when nitrogen is considered at the same time. Fig. 5 shows the nitrogen surface abundances of the same binary model described above. The nitrogen surface abundance increases abruptly due to the mass transfer. Thus, from this sequence, one would not expect to observe stars which are boron depleted but not nitrogen enriched unless rotational mixing operates. The corresponding stars in Fig. 2 thus indicate that rotational mixing is perhaps operating as we expect.

However, this is not yet a definite conclusion. The example binary displayed in Figs. 4 and 5 evolves rather conservatively, i.e. about 70% of the transferred matter is actually accepted by the mass gainer (while the rest is ejected due to its excess angular momentum). We know from observations and predict theoretically that many mass transfer systems evolve rather non-conservatively (Petrovic *et al.* 2005ab). We can currently not exclude that some of these systems produce boron depletion with only mild or no nitrogen enhancement.

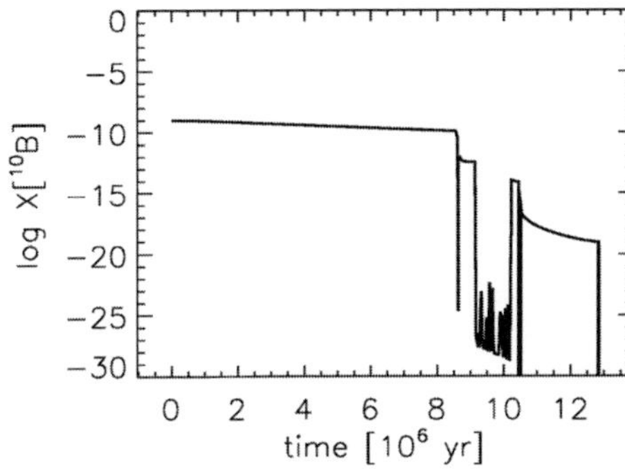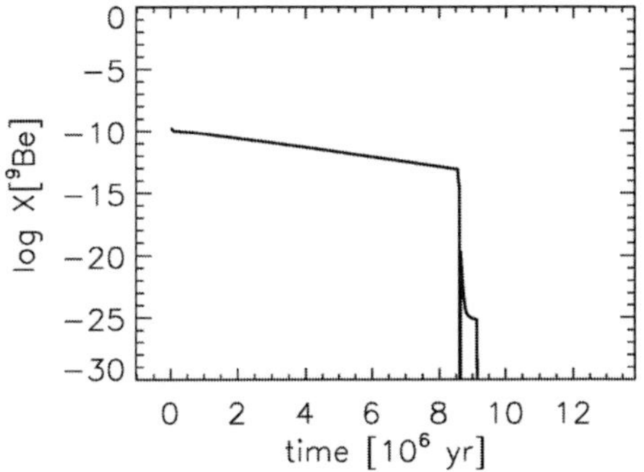

Figure 6. Surface abundance of boron 10 (left) and of beryllium (right) as function of time during core hydrogen burning, for the mass gainer in a solar metallicity $16M_\odot + 15M_\odot$ binary with an initial period of 3 days. In contrast to the models shown in Figs. 4 and 5, the stellar models shown here have been computed including angular momentum transport by internal magnetic fields (Yoon *et al.*, in prep.).

It is also interesting to consider boron as a test of rotational mixing in very close and very massive pre-mass transfer binaries, as explored by de Mink *et al.* (2009). In such very tight binaries, tidal synchronisation can enforce very rapid rotation of both stars. While de Mink *et al.* point out that generally the nitrogen surface abundance is the prime observable for such test, their results on boron depletion appear particularly interesting for Galactic binaries, since in those boron has a larger predicted relative change then nitrogen.

Finally, in Fig. 6, we show the surface abundances of boron and of beryllium in a similar binary as the one discussed above (even though some physics assumptions were different, which is not essential for our discussion here). Fig. 6 indicates that also beryllium is very interesting from the theoretical point of view. However, beryllium abundance determinations in hot main sequence stars appear to be difficult.

4. Unknown mixing processes

We have seen in the previous sections, that the surface abundances of massive main sequence stars are not yet fully understood, and that unambiguous evidence for the existence of rotationally induced mixing in massive stars is still lacking. Of particular worry is the solid evidence for a population of nitrogen-rich slowly rotating massive main sequence stars (Hunter *et al.* 2008b, 2009; Morel *et al.* 2006, 2008; Morel 2009). While the binary models discussed above do show a way to produce such stars (Langer *et al.*, 2008; cf. Fig. 5), the Galactic fraction of this population contains well investigated β Cephei pulsators (Morel *et al.* 2006, 2008) none of which seems to show any indication of binarity. Despite the warning above that binarity might be difficult to detect in post-mass transfer systems, the lack of any indication of a companion in well-studied nearby stars could imply that binarity is not the (only) answer to this question. It also remains to be seen whether binary evolution could produce enough of these objects, which may have a frequency of about 15% of all main sequence stars (Hunter *et al.* 2008b).

So we may face the situation that so far completely unaccounted mixing processes may operate in stars (see also Brott *et al.*, in preparation). Perhaps, they could be related to magnetic fields in the interior of massive stars. Also gravity waves could be excited in massive stars, as Talon & Charbonnel (2008) employ them for angular momentum transport in intermediate-mass stars. However, we want to end with a related possibility, for which recent observational evidence has accumulated.

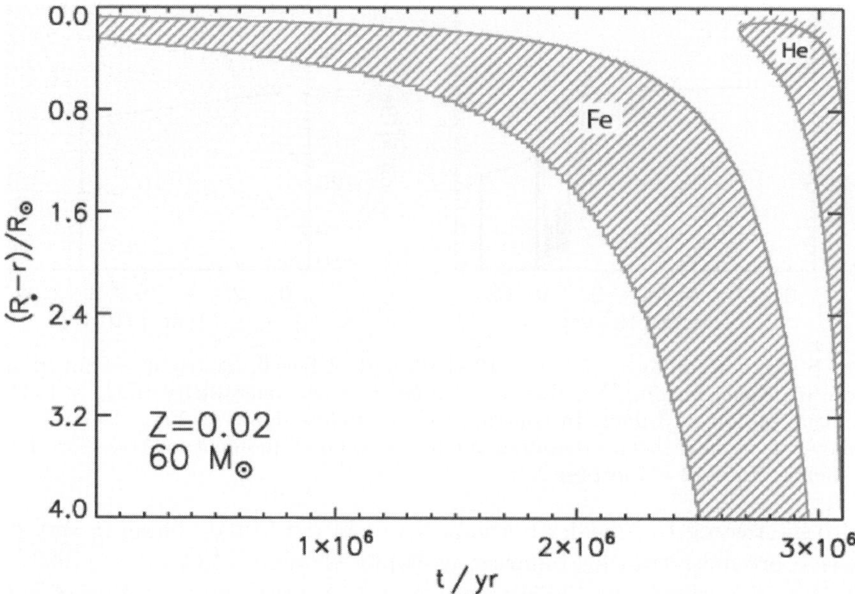

Figure 7. Evolution of the radial extent of the subsurface helium and iron convective regions (hatched) as function of time, from the zero age main sequence to roughly the end of core hydrogen burning, for a 60 $M_\odot$ star (Cantiello *et al.* 2009). The top of the plot represents the stellar surface. Only the upper 4 $R_\odot$ of the star are shown in the plot, while the stellar radius itself increases during the evolution. The star has a metallicity of $Z = 0.02$, and its effective temperature decreases from 48 000 K to 18 000 K during the main sequence phase.

Cantiello *et al.* (2009) investigated the subsurface convection zones which occur in the envclopes of hot massive stars due to the iron opacity peak (cf. Fig. 7). They found that the occurrence of strong subsurface convective motion predicted by the models correlates with observed large microturbulent velocities deduced from stellar spectroscopy. While this may not be sufficient to conclude that subsurface convection causes observable motion at the stellar surface, it appears to be a possibility. This could mean that the subsurface layers of massive stars, independent of their rotation, could be in motion, and perhaps lead to some mixing near the surface; this might produce a surface boron depletion without changing nitrogen.

5. Conclusions

It remains a major challenge to the theory of massive star evolution to explain the observed surface abundances of massive main sequence stars. While until recently, the incorporation of rotational mixing was thought to lead to a much better agreement, the discovery that a significant fraction of early B dwarfs are nitrogen-rich and intrinsically slowly rotating has cast some doubts on the previous ideas. We argue that boron observations of early type main sequence stars, as performed by Venn *et al.* (1996, 2002) and Morel *et al.* (2006, 2008), have the potential to move towards a solution. Clearly, binary evolution needs to be considered at the same time. Finally, we may be facing the situation that still not all mixing processes which can operate in massive main sequence stars have been described. Amongst possible candidates is mixing due to magnetic processes, and mixing induced by subsurface convection zones in hot massive stars.

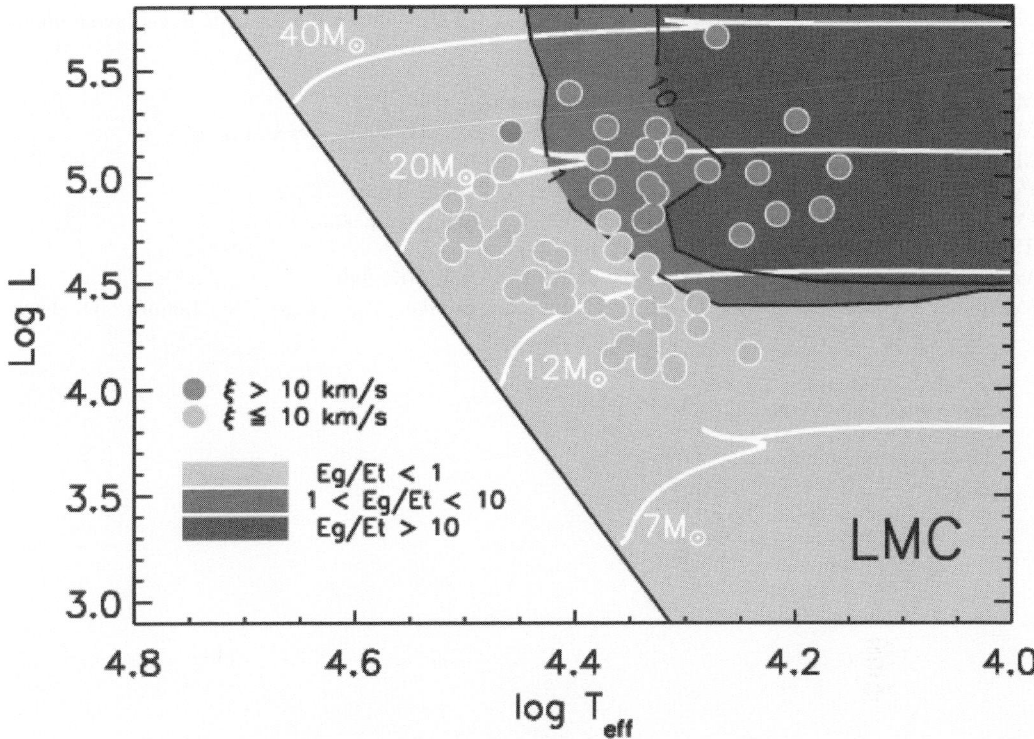

Figure 8. Values of the ratio $E_\mathrm{g}/E_\mathrm{s}$ of the kinetic energy in the form of gravity waves above the iron convection zone, to the kinetic energy of the surface velocity field, as a function of the location in the HR diagram (Cantiello *et al.* 2009). This plot is based on evolutionary models between 5 $M_\odot$ and 100 $M_\odot$ for LMC metallicity. The ratio $E_\mathrm{g}/E_\mathrm{s}$ (see Eq. (9) of Cantiello *et al.* 2009) is estimated using a value $v_\mathrm{s} = 10$ km s^{-1} for the surface velocity amplitude. Over–plotted as filled circles are stars which have photospheric microturbulent velocities ξ derived in a consistent way by Hunter *et al.* (2008a). Only data for stars with an apparent rotational velocity of $v \sin i < 80$ km s^{-1} is plotted. Solid white lines are reference evolutionary tracks, and the full drawn black line corresponds to the zero age main sequence.

References

Cantiello, M., Langer, N., Brott, I., de Koter, A., Shore, S. N., Vink, J. S., Voegler, A., Lennon, D. J., & Yoon, S.-C. 2009, *A&A*, 499, 279

de Mink, S. E., Cantiello, M., Langer, N., Pols, O. R., Brott, I., & Yoon, S.-Ch. 2009, *A&A*, 497, 243

Detmers, R. G., Langer, N., Podsiadlowski, Ph., & Izzard, R. G. 2008, *A&A*, 484, 831

Fliegner, J., Langer, N., & Venn, K. A. 1996, *A&A* (Letters), 308, 13

Gies D. R. & Lambert D.L 1992, *ApJ*, 387, 673

Heger, A. & Langer, N. 2000, *ApJ*, 544, 1016

Heger, A., Langer, N., & Woosley, S. E. 2000, *ApJ*, 528, 368

Hunter, I., Brott, I., Lennon, D. J., Langer, N., Dufton, P. L., Trundle, C., Smartt, S. J., de Koter, A., Evans, C. J., & Ryans, R. S. I. 2008b, *ApJ* (Letters), 676, 29

Hunter, I., Lennon, D. J., Dufton, P. L., Trundle, C., Simon-Diaz, S., Smartt, S. J., Ryans, R. S. I., & Evans, C. J. 2008a, *A&A*, 479, 541

Hunter, I., Brott, I., Langer, N., Lennon, D. J., Dufton, P. L., Howarth, I. D., Ryans, R. S. I., Trundle, C., Evans, C. J., de Koter, A., & Smartt, S. J. 2009, *A&A*, 496, 841

Langer, N. 1998, *A&A*, 329, 551

Langer, N., Cantiello, M., Yoon, S.-C., Hunter, I., Brott, I., Lennon, D., de Mink, S., & Verheijdt, M. 2008, IAU Symposium 250, p. 167

Maeder, A., Meynet, G., Ekström, S., & Georgy, C. 2009, *Communications in Asteroseismology*, 158, 72

Meynet, G. & Maeder, A. 2000, *A&A*, 361, 101

Morel, T. 2009, *Communications in Asteroseismology*, 158, 122

Morel, T., Butler, K., Aerts, C., Neiner, C., & Briquet, M. 2006, *A&A*, 457, 651

Morel, T., Hubrig, S., & Briquet, M. 2008, *A&A*, 481, 453

Petrovic, J., Langer, N., Yoon, S.-C., & Heger, A. 2005a, *A&A*, 435, 247

Petrovic, J., Langer, N., & van der Hucht, K. A. 2005b, *A&A*, 435, 1013

Talon, S. & Charbonnel, C. 2008, *A&A*, 482, 597

Venn, K. A., Lambert, D. L., & Lemke, M. 1996, *A&A*, 307, 849

Venn, K. A., Brooks, A. M., Lambert, David L., Lemke, M., Langer, N., Lennon, D. J., & Keenan, F. P. 2002, *ApJ*, 565, 571

Light Elements in the Universe
Proceedings IAU Symposium No. 268, 2009
C. Charbonnel, M. Tosi, F. Primas & C. Chiappini, eds.

© International Astronomical Union 2010
doi:10.1017/S1743921310004552

Boron depletion in 9 to 15 M$_\odot$ stars with rotation

U. Frischknecht[1], R. Hirschi[2], G. Meynet[3], S. Ekström[3], C. Georgy[3], T. Rauscher[1], C. Winteler[1] and F.-K. Thielemann[1]

[1]Department of Physics, University of Basel, CH-4056 Basel, Switzerland
email: urs.frischknecht@unibas.ch

[2]Astrophysics Group, Keele University, UK-ST5 5BG Keele and Institute for the Physics and Mathematics of the Universe, University of Tokyo, Japan

[3]Geneva Observatory, University of Geneva, CH-1290 Sauverny, Switzerland

Abstract. The treatment of mixing is still one of the major uncertainties in stellar evolution models. One open question is how well the prescriptions for rotational mixing describe the real effects. We tested the mixing prescriptions included in the Geneva stellar evolution code (GENEC) by following the evolution of surface abundances of light isotopes in massive stars, such as boron and nitrogen. We followed 9, 12 and 15 M$_\odot$ models with rotation from the zero age main sequence up to the end of He burning. The calculations show the expected behaviour with faster depletion of boron for faster rotating stars and more massive stars. The mixing at the surface is more efficient than predicted by prescriptions used in other codes and reproduces the majority of observations very well. However two observed stars with strong boron depletion but no nitrogen enrichment still can not be explained and let the question open whether additional mixing processes are acting in these massive stars.

Keywords. Stars: rotation, abundances, interiors

1. Introduction

Rotation is beside the stellar mass and the initial chemical composition the most important parameter in the evolution of single stars. It affects the physical and chemical structures of the stars and therefore quantities such as lifetime, luminosity, surface temperature etc. Recent models including rotation reproduce a wide range of observations better than those without (e.g. Meynet & Maeder 2003, 2005; Vázquez *et al.* 2007; Georgy *et al.* 2009). Still, the treatment of transport of angular momentum and chemical species is thought to be one of the main uncertainties in stellar evolution models.

Light elements and in particular boron and nitrogen can constrain the mixing induced by rotation and help distinguish between single stars and interacting binaries (Brott *et al.* 2009). Boron is destroyed at relatively low temperatures ($6 \cdot 10^6$ K) where the CNO-cycles are not yet efficient. Therefore shallow mixing due to rotation leads to a depletion of light elements such as Li, Be and B at the surface without considerable nitrogen enrichment. This effect can not be explained by mass transfer in a binary system, since there the accreted material is depleted in boron and enriched in nitrogen. An increasing number of boron surface abundances from O- and early B-type stars became available in the last few years (Proffitt & Quigley 2001; Venn *et al.* 2002; Mendel *et al.* 2006). The comparison in Mendel *et al.* (2006) of observational data with the models of Heger & Langer (2000) shows a good agreement with the exception of two stars. The strong boron depletion in these two young B-type stars raises the question if the efficiency of surface mixing due to rotation should be stronger or if there is another mixing effect which should be accounted for. We examined this question because GENEC includes the effect of rotation in a different way. The main difference comes from the fact that the transport of angular momentum is properly accounted for as an advection process and not as diffusion.

2. Models and comparison

We calculated 9, 12 and 15 $M_\odot$ models, each with different rotational velocities, to study the influence of rotational mixing on the light elements. A detailed description of the treatment of the transport of chemical species and angular momentum in GENEC can be found e.g. in Hirschi *et al.* (2004). In all our models shear turbulence is the dominant mixing process close to the stellar surface. In radiative zones, the gradient in angular momentum leading to shear mixing results from the concomitant effects of meridional currents, shear turbulence and from envelope expansion occurring on the main sequence.

In massive stars boron is only destroyed. The observations of boron in young massive stars in the solar vicinity show variations in $\log(B/H)$ from 2.9 down to unobservable quantities below 1. The large boron surface variations cannot be explained only by variation of initial composition (Venn *et al.* 2002). The boron vs nitrogen relation of the models is almost independent of the rotation velocity and initial stellar mass and therefore a good way to compare with observations since usually only $v \cdot \sin(i)$ is known from them. The early boron depletion and only subsequent enrichment of CN-cycle processed material is the main characteristic of rotational mixing. Most of the observations can be well reproduced. However the observed stars with boron depletion of about 1.5 dex or more and no enrichment of CNO-processed material at the surface are hard to explain with our and previous models. To reproduce these stars (HD 30836, HD 36591) the models would have to mix very efficiently the outermost envelope to destroy boron while at the same time not dredging nitrogen from the central regions up to the surface. Rotational mixing as it is treated in the current models does not seem to explain all the observations, meaning that additional physics like magnetic fields could play a role.

In comparison to the models of Heger & Langer (2000) ours show considerably more mixing for the outer part of the envelope at the end of central hydrogen burning, i.e. boron is in general more depleted at the end of the main sequence in our models with similar initial angular momentum. This means also that observed stars are better reproduced with our models since most of them are intrinsically slow rotators (Morel *et al.* 2008). The stronger surface mixing in our models originates in the meridional circulation, which is implemented as an advective process, whereas it is treated as a diffusive process in the models of Heger & Langer (2000). This allows meridional circulation to build stronger angular velocity gradients in the envelope instead of just diffusing the gradient away.

References

Brott, I., Hunter, I., de Koter, A., *et al.* 2009, Communications in Asteroseismology, 158, 55
Georgy, C., Meynet, G., Walder, R., Folini, D., & Maeder, A. 2009, *A&A*, 502, 611
Heger, A. & Langer, N. 2000, *ApJ*, 544, 1016
Hirschi, R., Meynet, G., & Maeder, A. 2004, *A&A* , 425, 649
Mendel, J. T., Venn, K. A., Proffitt, C. R., *et al.* 2006, *ApJ*, 640, 1039
Meynet, G. & Maeder, A. 2003, *A&A*, 404, 975
Meynet, G. & Maeder, A. 2005, *A&A*, 429, 581
Morel, T., Hubrig, S., & Briquet, M. 2008, *A&A*, 481, 453
Proffitt, C. R. & Quigley, M. F. 2001, *ApJ*, 548, 429
Vázquez, G. A., Leitherer, C., Schaerer, D., Meynet, G., & Maeder, A. 2007, *ApJ*, 663, 995
Venn, K. A., Brooks, A. M., Lambert, D. L., *et al.* 2002, *ApJ*, 565, 571

Light Elements in the Universe
Proceedings IAU Symposium No. 268, 2009
C. Charbonnel, M. Tosi, F. Primas & C. Chiappini, eds.

© International Astronomical Union 2010
doi:10.1017/S1743921310004564

Li survey in giant stars : probing non-standard stellar physics

N. Lagarde[1], C. Charbonnel[1,2], G. Jasniewicz[3], P. North[4], M. Shetrone[5], J. Hollek[5], and V. V. Smith[6]

[1] Geneva Observatory, University of Geneva , Switzerland
email: Nadege.Lagarde@unige.ch
[2] CNRS UMR 5572, Université de Toulouse, France
[3] Université de Montpellier II, CNRS/UM2 UMR 5024, France
[4] Laboratoire d'astrophysique, Ecole Polytechnique Fédérale de Lausanne, Switzerland
[5] University of Texas, McDonald Observatory, USA
[6] National Optical Astronomy Observatory, Tucson, USA

Abstract. Lithium has long been known to be a good tracer of non-standard mixing processes occurring in stellar interiors. Here we present the results of a large survey aimed at determining the surface Li abundance in a sample of about 800 giant (RGB and AGB) stars with accurate Hipparcos parallaxes. We compare the observed Li behaviour with that predicted by stellar models including rotation and thermohaline mixing.

Keywords. Hydrodynamics, instabilities, Stars: abundances, evolution, rotation

1. Introduction

Red giant stars present surface abundance anomalies that are not explained by classical stellar evolution models and that reveal the existence of extra-mixing (i.e., non convective) processes inside stellar interiors. Thermohaline mixing (Charbonnel & Zahn 2007) has been recently identified as the dominating process that governs the photospheric composition of low-mass bright giant stars, affecting in particular the surface Li abundances in red giants more luminous than the RGB bump. In order to test this assessment we present Li observations in a large sample of about 800 giant stars with Hipparcos parallaxes and compare our data for the solar-metallicity subsample with models computed with the evolutionary code STAREVOL and including thermohaline mixing and rotation-induced processes (see Charbonnel & Lagarde, this volume).

Observations were carried out with: (1) Standiford Cassegrain Echelle Spectrometer on the T2.1m at McDonald Observatory (2) Fiber-fed Extended Range Optical Spectrograph (FEROS) on the T1.52m at ESO ; (3) AURELIE spectrometer on the T1.52m at Haute-Provence Observatory.

2. Solar metallicity subsample

In Fig. 1 we compare our observations with the predictions of the solar metallicity models described in Charbonnel & Lagarde (this volume) and Lagarde & Charbonnel (in preparation) that were computed with the code STAREVOL taking into account (1) rotation-induced processes following the formalism by Zahn (92) and Maeder & Zahn (98), (2) thermohaline mixing as described by Charbonnel & Zahn (07), and (3) atomic diffusion. We also use these models to determine the mass and evolutionary status of each sample star.

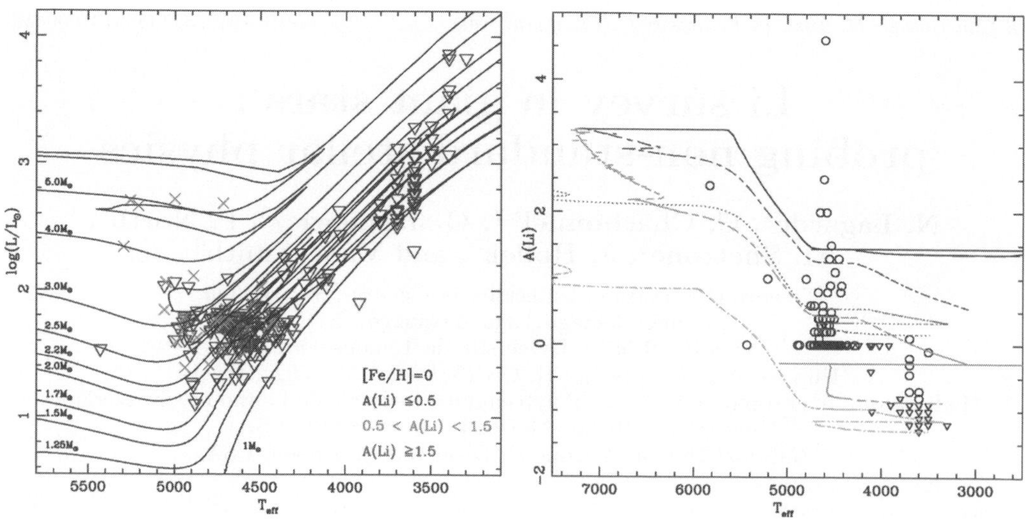

Figure 1. Comparison between models and data for the solar metallicity subsample. (Left:) HR diagram with indications on the surface Li abundance. Triangles : $A(Li) \leqslant 0.5$, square : $0.5 < A(Li) < 1.5$, crosses : $A(Li) \geqslant 1.5$. (Right:) Lithium as a function of Teff for stars with masses lower than $2M_\odot$ (Circles and triangles for actual determinations and upper limits respectively). Solid line, short dashed - long dashed line, and dashed line represent$1.5M_\odot$ stars in standard model, with thermohaline mixing and rotation $V_{ZAMS} = 50km/s$ and $V_{ZAMS} = 110km/s$ respectively. Dotted line and dot - short dashed line represent $2.0M_\odot$ with thermohaline mixing and rotation $V_{ZAMS} = 110km/s$ and $V_{ZAMS} = 250km/s$ respectively.

On the main sequence and early-RGB, rotation-induced processes lead to stronger Li depletion than in the standard case, in agreement with observations, and the observed Li dispersion reflects dispersion in the initial rotation velocity (see also Charbonnel & Talon 1999, Palacios *et al.* 2003, Smiljanic *et al.* 2009). After the end of the first dredge-up (Teff $\sim 4800K$), the Li abundance remains temporarily constant. When thermohaline mixing becomes efficient at the bump in the luminosity function (which corresponds here to Teff $\sim 4200K$), the Li abundance is predicted to drop again in drastic manner, explaining very well the Li upper limits obtained for the brightest RGB and AGB sample stars.

References

Charbonnel, C. & Zahn, J. P. 2007, *A&A*, 467, L15

Charbonnel, C. & Talon, S. 1999, *A&A*, 351, 635

Maeder, A. & Zahn, J. P. 1998, *A&A*, 334, 1000

Palacios, A., Talon, S., Charbonnel, C., & Forestini, M. 2003, *A&A*, 399, 603

Smiljanic, R., Pasquini, L., Charbonnel, C., & Lagarde, N. 2009, *A&A*,in press, astro-ph 0910.4399

Zahn, J. P. 1992, *A&A*, 265, 115

Light Elements in the Universe
Proceedings IAU Symposium No. 268, 2009
C. Charbonnel, M. Tosi, F. Primas & C. Chiappini, eds.

© International Astronomical Union 2010
doi:10.1017/S1743921310004576

Li and CNO isotopes from magnetically induced extra-mixing in evolved stars

Sara Palmerini[1,2], Maurizio Busso[1,2], Roald Guandalini[1] and Enrico Maiorca[1,2]

[1]Dipartimento di Fisica, Unoverstitá degli Studi di Perugia,
via Pascoli, 06125, Perugia, Italy
email: `sara.palmerini@fisica.unipg.it`

[2]I.N.F.N. sezione di Perugia, Italy

Abstract. Evolved low mass stars (LMS) contribute not only to the synthesis of s-process nuclei, but also to modifications in the isotopic mix of light elements (Li and CNO especially), induced by proton captures. In particular, RGB and AGB stars show a wide range of Li abundances. This spread is currently attributed to deep phenomena of non-convective mixing. These processes can, in principle, either produce or destroy Li, depending on their velocity. This is due to the fact that Li production requires preserving the unstable ^{7}Be, which has a half-life of only 53 days. Physical mechanisms devised so far to explain the existence of deep mixing in low mass stars generally fail in accounting for fast transport and in avoiding ^{7}Be destruction; on the contrary, this is easily obtained in Intermediate Mass Stars, where Hot Bottom Burning can occur. However, as Li-rich low-mass red giants do exist, we propose here a scenario where both production and destruction of Li are possible in LMS, thanks to the buoyancy of magnetized parcels of processed matter, traveling from the H shell to the envelope at different speeds (depending on their size). Consequences of this transport for CNO nuclei are also discussed.

Keywords. Nucleosynthesis – Stars: abundances, AGB, evolution, low-mass, magnetic fields

---

From the original poloidal field of a rotating star, a toroidal field of similar strength can be generated by the dynamo mechanism (Parker 1974). Indeed, the differential rotation of the external layers wraps the fields, aligning them to the equator (Spruit 1999). Toroidal flux tubes then develop, with their magneto hydrodynamics instabilities, and float to the surface (Spruit & van Ballegooijen 1982). This is so because a pressure term and a tension term (acting as an elastic force) are both created by the magnetic field. Torsion due to the Coriolis force then helps detaching portions of the tubes, or "bubbles". They move outward, with a relatively high speed, close to the Alfvén velocity (Case A; Busso *et al.* 2007). Due to their small size, the bubbles undergo minimal heat exchanges. When the average stellar magnetic field is strong enough to induce the buoyancy of a large portion of a flux tube, the resulting mixing velocity will be reduced (by heat exchanges) down to near the diffusion one (Case B; Palmerini & Busso 2008, Denissenkov *et al.* 2009). Hence magnetic buoyancy offers a scenario for the occurrence of both fast and slow mixing episodes.

A possible scenario for the evolution of the Li abundance in low mass red giants, as a function of their bolometric magnitude, on the basis of a new estimate for their distances and evolutionary status was suggested by Guandalini *et al.* (2009), to which we refer for details. These authors considered a phase of magnetic field growth, at the L-bump, where field strengths were assumed to be not strong enough to promote the buoyancy of entire magnetic flux tubes, but only to generate the intermittent detaching of fast bubbles (Case A). Hence, the induced fast mixing may induce Li enrichment, as shown

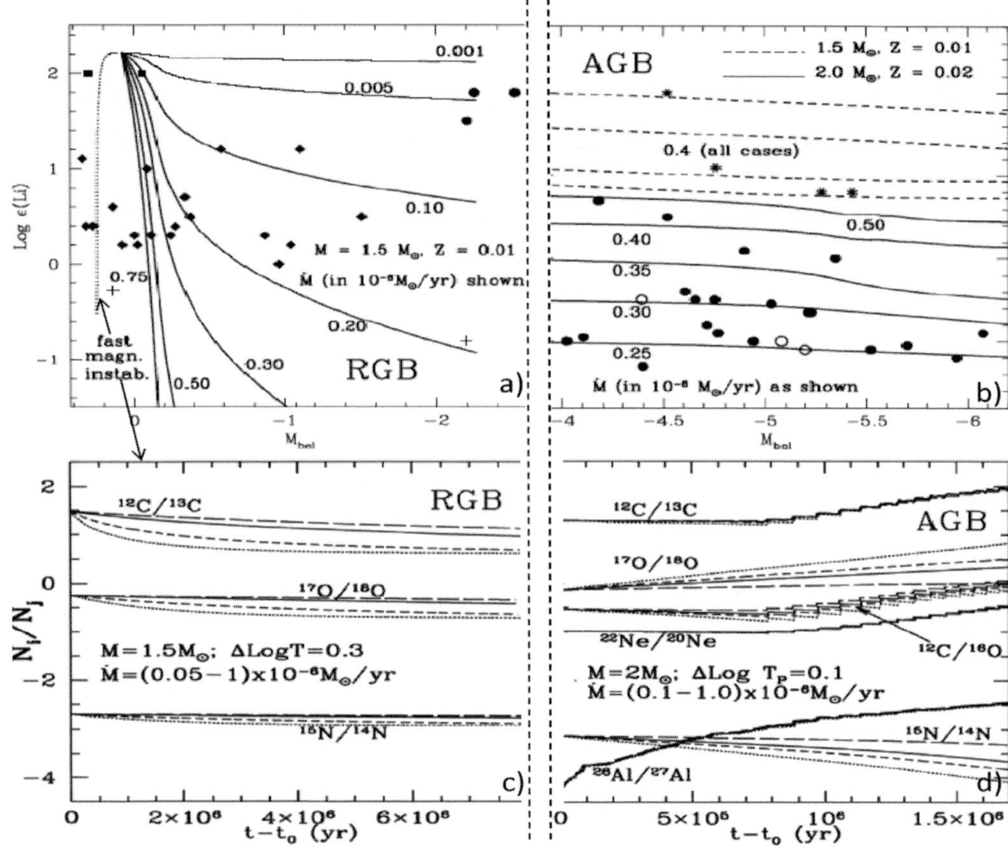

Figure 1. The evolution of Li (upper panels) and CNO isotopic ratios (lower panels) according to mixing cases A and B, see text for details.

by several red giants in this evolutionary phase (Fig.1 panel a). When the fields have grown sufficiently, larger magnetized structures can become buoyant, driving a slower mass circulation (case B), which depletes Li during the rest of the RGB phase and the whole AGB phase (Fig.1 panel a and b). The temporal evolution of Li is accompanied by modifications of CNO isotopic ratios during both fast and slow mixing episodes. Panel c of Fig.1 reports them at different values of the mixing rate ($\dot{M}$), for a $1.5M_\odot$, $Z = Z_\odot/2$ RGB model (Palmerini *et al.* 2009). Panel d instead shows temporal evolutions of CNO and Al isotopic ratios, in a $2M_\odot$, $Z = Z_\odot$ AGB model, due to both slow mixing (case B, at different $\dot{M}$) and convective third dredge–up, whose contribution results in the step-wise trend of ^{12}C/^{13}C, ^{12}C/^{16}O and ^{20}Ne/^{22}Ne ratios.

References

Busso, M., Wasserburg, G. J., Nollett, K. M., & Calandra, A., 2007, *ApJ*, 671, 802
Denissenkov, P. A., Pinsonneault, M., & Mac Gregor K. B. 2009, *ApJ*, 696,1823
Guandalini, R., Palmerini, S., Busso, M., & Uttenthaler, S. 2009, *PASA*, 26, 168
Palmerini, S. & Busso M., 2008, *New AR*, 52, 412
Palmerini, S., Busso, M., Maiorca, E., & Guandalini, R. 2009, *PASA*, 26, 161
Parker, E. N. 1974, *Ap&SS*, 31, 261
Spruit, H. C. 1999, *A&A*, 349, 189
Spruit H. C. & van Ballegooijen, A. 1982, *A&A*, 106, 58

Light Elements in the Universe
Proceedings IAU Symposium No. 268, 2009
C. Charbonnel, M. Tosi, F. Primas & C. Chiappini, eds.

© International Astronomical Union 2010
doi:10.1017/S1743921310004588

Lithium destruction induced by planetary accretion in solar-type stars

Sylvie Théado[1], Elise Bohuon [1], and Sylvie Vauclair[1]

[1]Laboratoire d'Astrophysique de Toulouse et Tarbes, CNRS, Université de Toulouse ; 14 avenue Edouard Belin, 31400 Toulouse, France
email: sylvie.vauclair@ast.obs-mip.fr

Abstract. Accretion of planetary (metal-rich) material onto a star in its early phases can produce episodes of thermohaline convection below the outer convective zone. These extra-mixing phases lead to rapid lithium destruction. The observed dispersion of lithium abundances in solar-type stars can be related to such events.

Keywords. Stars: abundances

Lithium depletion in solar-type stars remains a challenge for stellar models. Extra-mixing below the outer convective zone is needed to explain the observed abundances in the Sun, the solar analogs and the solar twins (e.g., Do Nascimento *et al.* 2009). Several processes have been invoked in the past to account for this lithium depletion, like rotational induced mixing, but they fail to account for all the observed features and observed abundance dispersion. Meanwhile recent detailed observations of heavy elements in the Sun and solar-type stars show systematic differences, which lead to the idea that accretion may play an important role in their early phases (e.g. Meléndez *et al.* 2009).

Although no precise statistics can yet be given, it seems possible that many solar-type stars were born with disks, some of them, but not all, leading to observable planets. According to the parameters of the disk and of the central star, matter with abundances different from those of the star can be accreted onto it. It may be either disk gaseous

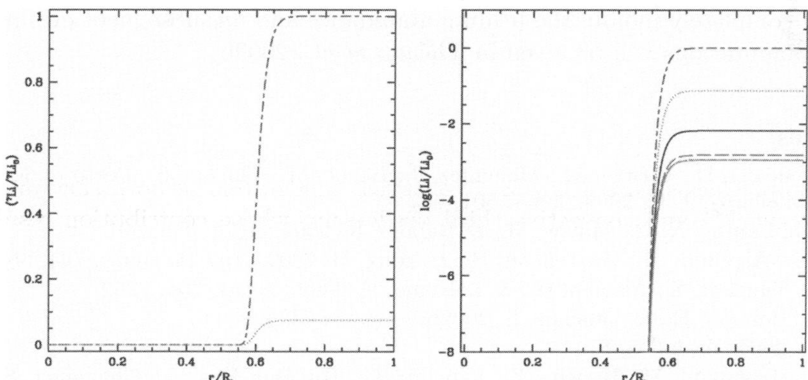

Figure 1. Lithium profiles in 1.10 $M_\odot$ stars with planetary accretion. The left panel displays the ratio of the lithium mass fraction to its initial value at 2 Myrs (evolution with atomic diffusion, blue line) and 4 Myrs (after the first accretion episode of $0.03M_{jup}$, orange line). The right panel presents the same ratio in logarithmic units at 2 Myrs (blue line) and after each accretion episode (at 4, 6, 8, 10 and 12 Myrs). Thermohaline mixing strongly decreases the lithium surface abundance. However the depletion saturates after 4 or 5 events. This occurs when the accreted amount of lithium becomes comparable to the depleted one.

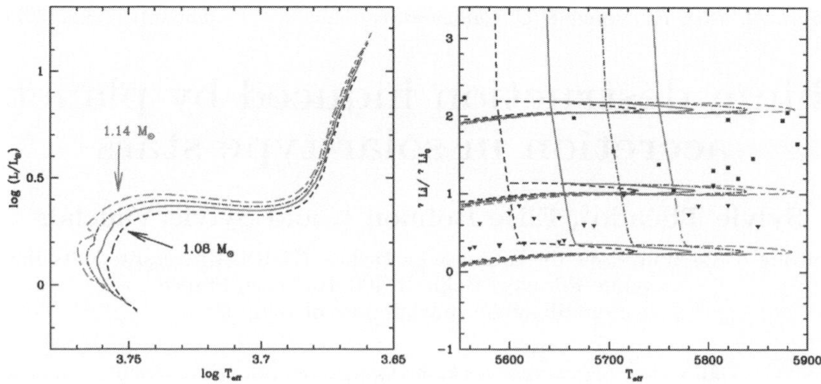

Figure 2. Lithium destruction in stars of various masses : 1.08 (dashed line), 1.10 (solid line), 1.12 (dashed-dotted line) and 1.14 $M_\odot$ (long dashed-dashed line) models. The left panel displays the evolutionary tracks. The right panel shows the lithium abundance variations along the evolutionary tracks, as a function of the effective temperature for the same models, including either one, two, or three accretion events of 0.03 M_{Jup} each. The points correspond to the lithium observations by Israelian *et al.* (2004).

matter (metal-poor) or matter already condensed in the form of planetesimals or planets (metal-rich). The idea that such metal rich-accretion could be the reason for the observed metal overabundances in exoplanet-host stars has now been ruled out (e.g. Vauclair *et al.* 2008). However, such accretion processes can happen, even if it does not lead to observable metal enrichment, and it may have important consequences on the lithium abundances.

Accretion of planetary matter onto a star in its early phases leads to mean molecular weight inversion below the outer convective zone, unstable against thermohaline convection (e.g. Vauclair 2004). Every time the star accretes planetary material, extra mixing induced by this hydrodynamical instability leads to lithium destruction.We first present computations for a 1.10 $M_\odot$ star with [Fe/H] = 0.20 suffering 1 to 5 accretion events of 0.03 M_{Jup} beginning at 2 Myrs and occurring every 2 Myrs afterwards. Then we show examples of other stellar masses and other accreted masses. In all these computations, thermohaline mixing is computed as described in Théado *et al.* (2009a). Clearly such events can completely modify the lithium abundance and its subsequent evolution. More detailed computations will be given in Théado *et al.* (2009b).

References

Do Nascimento, J. D., Castro, M., Meléndez, J., Bazot, M., Théado, S., Porto de Mello, G. F., & de Medeiros, J. R. 2009, *A&A*, 501, 687

Israelian, G., Santos, N. C., Mayor, M., & Rebolo, R. 2004, *A&A*, 414, 601

Meléndez, J., Asplund, M., Gustaffson, B., & Yong, D. 2009, *ApJ* (Letters), 704, 66

Théado, S., Vauclair, S., Alecian, G., & LeBlanc, F. 2009a , *ApJ*, 704, 1262

Théado, S., Bohuon, E., & Vauclair, S. 2009b, *in preparation*

Vauclair, S. 2004, *ApJ*, 605, 874

Vauclair, S., Laymand, M., Bouchy, F., Vauclair, G., Hui Bon Hoa, A., Charpinet, S. & Bazot, M. 2008, *A&A* (Letters), 482, 5

Session V

Evolution of light elements
in the Universe

Donatella Romano

Patrick Eggenberger

Light Elements in the Universe
Proceedings IAU Symposium No. 268, 2009
C. Charbonnel, M. Tosi, F. Primas & C. Chiappini, eds.

© International Astronomical Union 2010
doi:10.1017/S174392131000459X

Galactic evolution of D, ^{3}He and ^{4}He

Donatella Romano

Dept. of Astronomy, Bologna University,
Via Ranzani 1, I-40127, Bologna, Italy
and

INAF-Bologna Observatory,
Via Ranzani 1, I-40127, Bologna, Italy
email: `donatella.romano@oabo.inaf.it`

Abstract. The uncertainties which still plague our understanding of the evolution of the light nuclides D, ^{3}He and ^{4}He in the Galaxy are described. Measurements of the local abundance of deuterium range over a factor of 3. The observed dispersion can be reconciled with the predictions on deuterium evolution from standard Galactic chemical evolution models, if the true local abundance of deuterium proves to be high, but not too high, and lower observed values are due to depletion onto dust grains. The nearly constancy of the ^{3}He abundance with both time and position within the Galaxy implies a negligible production of this element in stars, at variance with predictions from standard stellar models which, however, do agree with the (few) measurements of ^{3}He in planetary nebulae. Thermohaline mixing, inhibited by magnetic fields in a small fraction of low-mass stars, could in principle explain the complexity of the overall scenario. However, complete grids of stellar yields taking this mechanism into account are not available for use in chemical evolution models yet. Much effort has been devoted to unravel the origin of the extreme helium-rich stars which seem to inhabit the most massive Galactic globular clusters. Yet, the issue of ^{4}He evolution is far from being fully settled even in the disc of the Milky Way.

Keywords. Galaxy: abundances, Galaxy: evolution, nuclear reactions, nucleosynthesis, abundances

1. Introduction

The discovery of the cosmic microwave background (CMB) by Penzias & Wilson (1965) in the mid sixties set the stage for a quantitative exploitation of the Big Bang nucleosynthesis theory. In their pioneer studies, Peebles (1966) and Wagoner *et al.* (1967) demonstrated that D, ^{3}He and ^{4}He could well have been produced in solar-system abundances in the 'primordial fireball'.

In the framework of the standard Big Bang nucleosynthesis (SBBN) theory, the baryon-to-photon ratio, η, is the only parameter regulating the amounts of D, ^{3}He and ^{4}He which emerge from the hot, early Universe (see Fig. 1 and contribution by G. Steigman, this volume). In the nineties, observations, seeking to constrain the primordial abundances of D, ^{3}He and ^{4}He – and, hence, the value of η – by probing the most metal-poor environments in the Universe, did not come up with consistent results. Both high and low values were suggested for the primordial deuterium abundance as measured in high-redshift, low-metallicity quasar absorption-line systems (QSOALS), $(D/H)_P = 2.5 \times 10^{-4}$ (Carswell *et al.* 1994, Songaila *et al.* 1994) or a few times 10^{-5} (Burles & Tytler 1998a,b). Similarly, both low and high values were suggested for the primordial ^{4}He abundance, e.g., $Y_P = 0.234 \pm 0.002$ (Olive *et al.* 1997) or 0.244 ± 0.002 (Izotov & Thuan 1998).

The difficulty to determine $(D/H)_P$ from observations led to turn the problem upside down and try to infer that quantity by using Galactic chemical evolution (GCE) models.

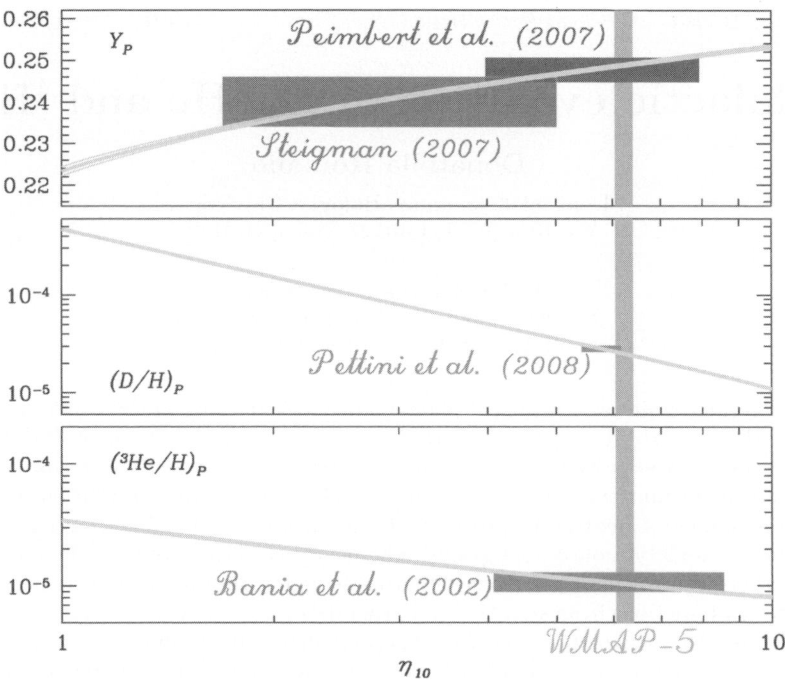

Figure 1. SBBN-predicted primordial mass fraction of ^{4}He (Y_P) and abundances of D and ^{3}He (relative to hydrogen by number), as functions of the η_{10} parameter, $\eta_{10} \equiv 10^{10}(n_B/n_\gamma)$. Theoretical predictions are from Hata *et al.* (1995), as updated by G. Steigman (courtesy of G. Steigman). The widths of the curves reflect the uncertainties in the nuclear and weak-interaction rates. The vertical band crossing all panels corresponds to the η_{10} value derived from analysis of five-years *WMAP* data on the CMB anisotropy (Dunkley *et al.* 2009). Also shown are the most recent estimates of the primordial abundances of D, ^{3}He and ^{4}He from observations, along with the allowed ranges of values for η_{10} (boxes; Bania *et al.* 2002, Peimbert *et al.* 2007, Steigman 2007, Pettini *et al.* 2008).

GCE models put stringent limits on the degree of astration suffered by deuterium in the solar vicinity over a Hubble time. Assuming that the local pre-solar (Geiss & Reeves 1972, Geiss & Gloeckler 1998) and current (Linsky 1998) D abundances are reasonably well known, they could settle tight limits to the primordial deuterium abundance, and definitively ruled out a high primordial deuterium.

Modelling the Galactic evolution of deuterium is a straightforward task. Since D is completely destroyed as gas cycles through stars and there are no known sources of substantial production other than BBN (Epstein *et al.* 1976, Prodanović & Fields 2003), its evolution is obtained for free from GCE models. Good models for the solar neighbourhood – i.e., models which satisfy the majority of the observational constraints available for the solar neighbourhood – have always predicted astration factors $f_D \equiv (D/H)_P/(D/H)_{LISM}$ not in excess of 2–3 for deuterium (Audouze & Tinsley 1974, Steigman & Tosi 1992, Edmunds 1994, Galli *et al.* 1995, Prantzos 1996, Tosi *et al.* 1998, Chiappini *et al.* 2002, Romano *et al.* 2003, 2006). Attempts to accommodate larger astration factors (Vangioni-Flam *et al.* 1994, Scully *et al.* 1997) have resulted in models which failed to reproduce important observational constraints. Moreover, the more D is burnt in the Galaxy, the more ^{3}He is produced. Since the abundance of ^{3}He is observed to stay rather constant with both time and position in the Milky Way (Bania *et al.* 2002), GCE models which overproduce this isotope must be discarded.

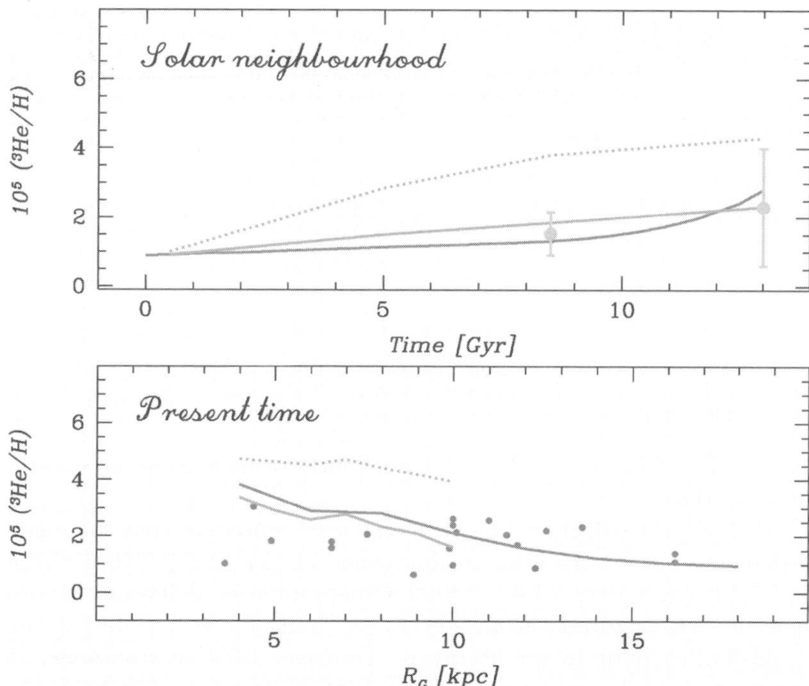

Figure 2. Evolution of ^{3}He/H in the solar neighbourhood (upper panel) and distribution of ^{3}He/H across the Galactic disc at the present time (lower panel) for different GCE models, assuming either $(D/H)_P = 2.5 \times 10^{-5}$ (solid lines) or $(D/H)_P = 20 \times 10^{-5}$ (dotted lines). All models assume zero net production of ^{3}He from 93% of low-mass stars (1–2 $M_\odot$) in order to fit the observations (filled circles; upper panel: local pre-solar and current values from Geiss & Gloeckler 1998; lower panel: H II region abundances from Bania *et al.* 2002). Figure adapted from Romano *et al.* (2003).

As first recognized by Truran & Cameron (1971), GCE models adopting standard prescriptions for the synthesis of ^{3}He in stars dramatically overestimate its abundance in the Milky Way (see also Rood *et al.* 1976). According to standard stellar models, ^{3}He is most efficiently produced on the main sequence (MS) of 1–2 $M_\odot$ stars through the action of the p-p chains. In order not to overproduce ^{3}He in the course of Galactic evolution, it has become customary to assume that some unknown ^{3}He-destruction mechanism is at work in more than 90% of low-mass stars (Dearborn *et al.* 1996, Galli *et al.* 1997, Chiappini *et al.* 2002, Romano *et al.* 2003). Hogan (1995) and Charbonnel (1995) have suggested 'extra mixing' during the red giant branch (RGB) phase of low-mass stars as a possible solution (see also Charbonnel & Do Nascimento 1998, Sackmann & Boothroyd 1999). In Fig. 2, we compare the predictions of two successful models for the chemical evolution of the Milky Way (the one by Chiappini *et al.* 2002 and Model 1 of Tosi 1988) to ^{3}He data for the solar neighbourhood (upper panel) and the Galactic disc (lower panel). Despite different assumptions about the infall law, star formation rate and stellar initial mass function (IMF), both models need to assume that at least 93% of low-mass stars burn the ^{3}He they have produced on the MS in later evolutionary phases in order to fit the observations. It is worth noticing that the good agreement between model predictions and observations depends also on the adopted value of the primordial deuterium abundance: if $(D/H)_P = 20 \times 10^{-5}$, rather than 2.5×10^{-5} (dotted versus solid lines in Fig. 2), both the local behaviour of ^{3}He with time and its present distribution across the Galactic disc

Table 1. Abundances of D, ^{3}He and ^{4}He at different epochs

Nuclide	Units	SBBN+$WMAP^a$ (13.7 Gyr ago)	Low-Z systems (10–13 Gyr ago)	Pre-solar matter (4.5 Gyr ago)	LISM (Today)
D	10^5 (D/H)	2.49 ± 0.17[1]	2.8 ± 0.2[2]	2.1 ± 0.5[3]	2.31 ± 0.24[4]
					0.98 ± 0.19[5]
					2.0 ± 0.1[6]
^{3}He	10^5 (^{3}He/H)	1.00 ± 0.07[1]	1.1 ± 0.2[7]	1.5 ± 0.2[3]	2.4 ± 0.7[8]
^{4}He	Y	0.2486 ± 0.0002[1]	0.2477 ± 0.0029[9]	0.2703[11]	
			0.240 ± 0.006[10]		

Notes:
[a] Using $\eta_{10} = 6.23 \pm 0.17$ from analysis of 5-years $WMAP$ data (Dunkley *et al.* 2009).
References:
(1) Cyburt *et al.* (2008); (2) Pettini *et al.* (2008); (3) Geiss & Gloeckler (1998); (4) Linsky *et al.* (2006); (5) Hébrard *et al.* (2005); (6) Prodanović *et al.* (2009); (7) Bania *et al.* (2002); (8) Gloeckler & Geiss (1996); (9) Peimbert *et al.* (2007); (10) Steigman (2007); (11) Asplund *et al.* (2009).

can not be reproduced by the models, independently of how many low-mass stars burn their ^{3}He on the RGB.

As far as ^{4}He is concerned, there has been a general consensus that the relative helium-to-metal enrichment ratio in the solar neighbourhood is $\Delta Y/\Delta Z \sim 2$, both from a theoretical (Chiosi & Matteucci 1982, Maeder 1992, Chiappini *et al.* 2003) and an observational point of view (e.g., Casagrande *et al.* 2007). Yet, hints for very different values of this ratio were reported early on in the literature (Danziger 1970, and references therein).

The determination of the parameter η from $WMAP$ data (see text by J. Dunkley, this volume) has allowed to fix, with unprecedented precision, the primordial abundances of the light elements in the framework of the SBBN model (see Cyburt *et al.* 2008 for recent work). The primordial abundances of D, ^{3}He and ^{4}He determined indirectly from the CMB anisotropies agree very well with those inferred from recent, direct observations (see Fig. 1 and Table 1), although one must be aware that the latter actually provide only lower/upper limits to the true primordial abundances. Above all, it is clear that the determination of the primordial abundance of deuterium is converging towards a low value, beautifully confirming earlier findings from GCE models.

In the following sections, the remaining (major) causes of uncertainty, which hamper our current understanding of the Galactic chemical evolution of the light elements D, ^{3}He and ^{4}He, are discussed, element by element.

2. Deuterium

The joint determinations of the primordial and pre-solar deuterium abundances (Table 1) point to a small depletion of deuterium from the Big Bang up to the solar system formation 4.5 Gyr ago. However, the present-day abundance of deuterium in the solar vicinity is currently under debate. The *FUSE* satellite has measured the deteurium abundance along the lines of sight to several stars in the Local Bubble as well as beyond it, up to 1–2 kpc away. The dispersion (by a factor of 3) which has been found in the measurements (Linsky *et al.* 2006) makes it hard to interpret the data in the context of standard GCE models. It has been suggested (Hébrard *et al.* 2005, Linsky *et al.* 2006) that either the lowest (see also text by G. Hébrard, this volume) or the highest (see also text by J. Linsky, this volume) observed abundances are indicative of the actual value of the deuterium abundance in the local ISM (LISM). However, neither of these values can be reproduced by GCE models in agreement with all the major observational constraints for the solar neighbourhood (Romano *et al.* 2006). Very recently, using a Bayesian analysis approach, Prodanović *et al.* (2009) have provided another estimate of the true LISM deuterium abundance, which places it very close to the D abundance at the time of the formation of the Sun (see Table 1).

In Fig. 3 we show the evolution of deuterium in the solar vicinity and the deuterium abundance profile across the Milky Way disc. The adopted GCE model, from Romano *et al.* (2006), is the one which allows for the lowest D astration factor in the solar vicinity ($f_\mathrm{D} = 1.39$). Here it has been recomputed assuming $(\mathrm{D/H})_\mathrm{P} = 2.8 \times 10^{-5}$ rather than 2.6×10^{-5} as in Romano *et al.* (2006). Extrapolation of the theoretical results towards the Galaxy center has been performed by taking into account the detailed results of a model for the Galactic bulge by Matteucci *et al.* (1999). Data are from Table 1 for local values (at a radius $R_\mathrm{G} = 8$ kpc), from Rogers *et al.* (2005) for the outer disc and from Lubowich *et al.* (2000) for a region at 10 pc distance from the Galactic center.

The agreement of the model predictions with the data is striking, especially if the true value for the local abundance of deuterium is the one suggested by Prodanović *et al.* (2009; see also contribution by T. Prodanović, this volume). In this context, lower observed local abundances of deuterium would be due to D depletion onto dust grains (Linsky *et al.* 2006, Steigman *et al.* 2007, and references therein). The highest observed local D values are marginally consistent with the estimate of the true D value by Prodanović *et al.* (2009).

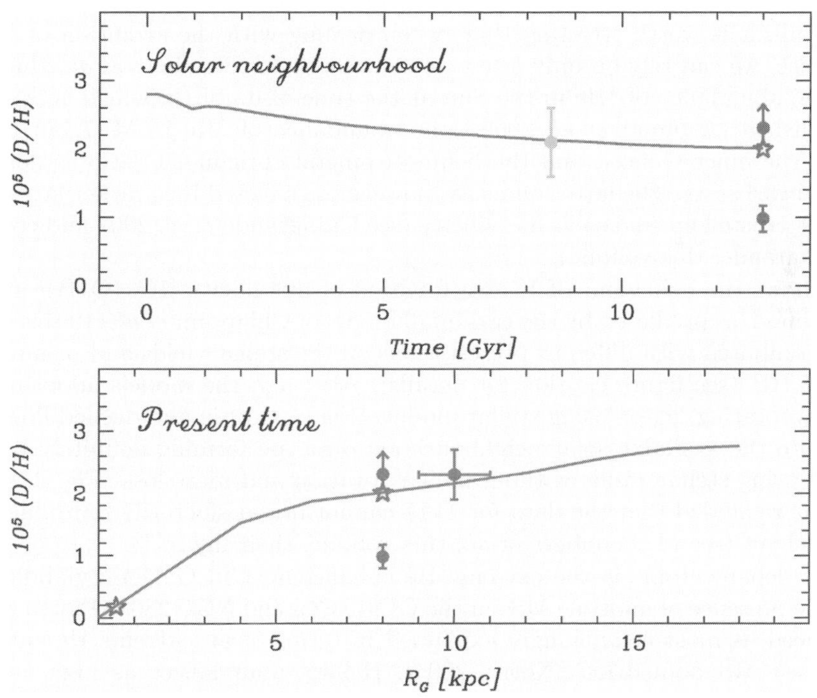

Figure 3. Evolution of D/H in the solar neighbourhood (upper panel) and distribution of D/H across the Galactic disc at the present time (lower panel) for the GCE model (solid lines in both panels) of Romano *et al.* (2006) with the lowest D astration factor, assuming $(\mathrm{D/H})_\mathrm{P} = 2.8 \times 10^{-5}$. Data are from Table 1 for local values, from Rogers *et al.* (2005) for the outer disc (filled circle at $R_\mathrm{G} = 10 \pm 1$ kpc, bottom panel) and from Lubowich *et al.* (2000) for the inner Galaxy (star at $R_\mathrm{G} = 10$ pc, bottom panel).

3. Helium-3

As far as ^{3}He is concerned, we must recognize that the problems raised in pioneering works by people such as Truran & Cameron (1971), Reeves *et al.* (1973) and Tinsley (1974) (to name a few) are still unsolved: we must postulate that some unknown ^{3}He-destruction mechanism is at work in not less than 90% of low-mass stars in order not to overproduce ^{3}He in the course of Galactic evolution (see discussion in Sect. 1). But which is the physical mechanism responsible for that? In a coherent picture, one must also be able to explain the existence of a few planetary nebulae (PNe) with high ^{3}He content, consistent with predictions from standard stellar models (e.g., Balser *et al.* 1999). Recently, the inclusion of thermohaline mixing in detailed stellar evolutionary models (Charbonnel & Zahn 2007) has shown that this is likely to be the 'extra mixing' mechanism we have been searching for years. Indeed, thermohaline mixing is able to efficiently destroy ^{3}He on the RGB of low-mass (1–2 $M_\odot$) stars, when not inhibited by magnetic fields. Details on this interesting process can be found in the contribution by N. Lagarde and C. Charbonnel to these conference proceedings (see also text by R. Stancliffe, this volume). The expected output of stellar models including thermohaline mixing are complete grids of yields. New GCE models computed with such new yields will hopefully lead the so-called ^{3}He problem to an end.

4. Helium-4

As far as ^{4}He is concerned, there are many open issues to be discussed.

First of all, it is worth stressing that, when dealing with the evolution of ^{4}He in the Galactic disc, we can rely on only a few data (see also M. Peimbert *et al.*, this volume), namely, the abundance of ^{4}He in the Sun at the time of its birth, which is by now quite well established (Asplund *et al.* 2009), the abundance of ^{4}He in M 17, an H II region located in the inner Galaxy, and the helium-to-metal enrichment ratio as derived from nearby K-dwarf stars. The latter quantity, however, is affected by a rather large error and can only be trusted around solar metallicity (see Casagrande *et al.* 2007 and contribution by L. Casagrande, this volume).

Fig. 4 shows the behaviour of Y as a function of metallicity [10^6 (O/H)] in the solar neighbourhood, as predicted by the two-infall model of Chiappini *et al.* (1997). The model has been computed with different prescriptions on the stellar yields and primordial mass fraction of ^{4}He (see figure caption for details). Although the models adopting the ^{4}He yields from rotating, mass-losing stellar models (Fig. 4, dotted and dashed lines) provide a good fit to the available solar neighbourhood data, performing definitely better than the model using stellar yields without stellar rotation and mass loss (Fig. 4, solid line), it has to be reminded that the data for M 17 cannot be satisfactorily reproduced by any GCE model yet (see M. Peimbert *et al.*, this volume, their figure 1).

Another debated topic is the extreme He enhancement in Galactic globular clusters (GCs). The presence of multiple MSs in the GCs ω Cen and NGC 2808 (Piotto *et al.* 2005, 2007), indeed, is most convincingly explained in terms of an extreme He enhancement of the bluest MS population (Norris 2004). Helium abundances as high as $Y \sim 0.4$ are suggested, which for ω Cen imply a helium-to-metal enrichment ratio $\Delta Y/\Delta Z \geqslant 70$ (Piotto *et al.* 2005). Such a value is outstandingly larger than that quoted for the solar neighbourhood around and above solar metallicity from a sample of nearby K-dwarf stars, $\Delta Y/\Delta Z = 2.1 \pm 0.9$ (Casagrande *et al.* 2007). Attemps have been made to explain such extreme ^{4}He abundances in the framework of two main competing scenarios, the so-called 'asymptotic giant branch (AGB) self-pollution scenario' (P. Ventura, this volume) and the so-called 'fast rotating massive star (FRMS) self-pollution scenario' (T. Decressin, this volume). Both scenarios have to reproduce other chemical peculiarities of GC stars besides the 'anomalous' ^{4}He abundances, and both have advantages and disadvantages. As a common limitation, they are presently able to deal only with two-population clusters. However, in ω Cen – the object for which the most compelling evidence for the need for a *huge* helium enrichment is found – there is clear-cut evidence of the presence of complex, multiple populations (e.g., Pancino *et al.* 2000). We (Romano *et al.* 2010) have recently proposed that the presence of extreme He-rich stars in ω Cen can be explained in the context of a model where the cluster is the remnant of a much more massive parent system, that evolved in isolation for a relatively long time – 3 Gyr, with the bulk of the stars forming during the first 1 Gyr. The system was then captured and partially disrupted by the Milky Way (see contribution by D. Romano *et al.*, this volume). The key ingredient in our model is the development of a differential galactic wind, which selectively removes from the progenitor galaxy mostly the elements restored to the ISM through fast polar winds from massive stars and supernova (SN) explosions. Elements restored to the ISM through gentle winds from both AGB and FRMSs are, instead, mostly retained in the cluster potential well, where they enter the formation of successive

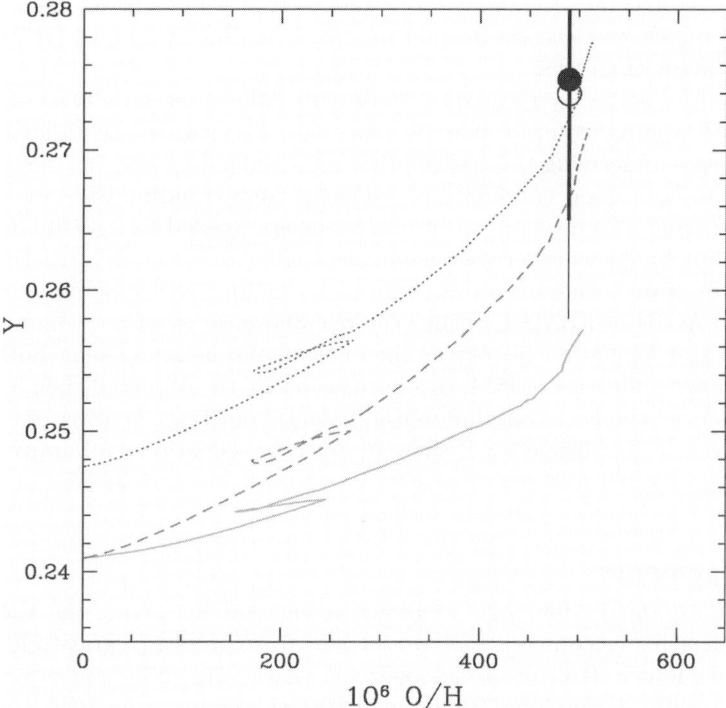

Figure 4. Y versus 10^6 (O/H) in the solar neighbourhood predicted by the *two-infall* model of Chiappini *et al.* (1997) with different prescriptions on the stellar nucleosynthesis. Solid line: the model adopts the van den Hoek & Groenewegen (1997) yields for low- and intermediate-mass stars and the Woosley & Weaver (1995) yields for massive stars; dashed line: the model is computed with the yields of Meynet & Maeder (2002), taking into account the effects of rotation on stellar evolution, for the whole range of stellar masses; dotted line: same model as the previous one, but starting from a higher primordial ^{4}He abundance, $Y_\mathrm{P} = 0.248$ rather than 0.241. The model predictions are compared with the solar value (oxygen from Allende-Prieto *et al.* 2001, Y from Anders & Grevesse 1989 – open circle with thin errorbar – and Grevesse & Sauval 1998 – filled circle with thick errorbar). Figure from Chiappini *et al.* (2003).

stellar generations. Since, according to the latest stellar evolutionary computations, ^{4}He is dispersed in the ISM by means of low-energy stellar winds by both AGBs and FRMSs, while metals are mostly expelled through SN explosions, a high $\Delta Y/\Delta Z$ is naturally obtained in the framework of our model. Other important observational constraints can also be satisfactorily reproduced.

Since differential galactic winds do, as a matter of fact, modify somewhat arbitrarily the true (effective) yields of the various elements, a better assessment of the stellar yields of ^{4}He is mandatory.

Finally, it is worth reminding that the usually quoted value of $Y \sim 0.4$ for the extreme He-rich GC stars could be revised downwards to as low as $Y \sim 0.3$ (L. Casagrande, this volume).

5. Conclusions

We have reviewed the evolution of the light elements D, ^{3}He and ^{4}He in the Milky Way, emphasizing recent developments and open problems. We summarize our conclusions as follows:

(a) GCE models are consistent with the relatively high value of $(D/H)_{LISM} = (2.0 \pm 0.1) \times 10^{-5}$ suggested by Prodanović $et~al.$ (2009) from their Bayesian analysis of $FUSE$ data. Standard GCE models are instead unable to explain values of $(D/H)_{LISM}$ significantly higher/lower than this.

(b) The need for some 'extra mixing' to destroy ^{3}He in most ($>$90%) of 1-2 $M_\odot$ stars came from GCE arguments more than 30 years ago. The way to the understanding of the physical processes underlying this assumption has been a long one, but now thermohaline mixing seems to be a good candidate to solve the long-standing issue of ^{3}He evolution. Stellar yields taking this mechanism into account are needed for use in GCE models.

(c) It is still debated whether the chemical peculiarities seen in a fraction of Galactic globular cluster stars – first of all an impressive helium enrichment – are due to self-pollution from AGBs or FRMSs. In the very peculiar case of ω Cen, which likely suffered a complicated star formation history as the nucleus of a larger system, both stellar categories should have polluted the ISM. In such a scenario, the observed chemical 'anomalies' would be driven by the action of differential galactic outflows, venting out preferentially metals. However, it is still unclear if such an extreme scenario could apply to other GCs as well.

Acknowledgements

I thank the organizers for their kind invitation and for giving me the opportunity to attend such a lively conference. I would like to express my gratitude to Francesca Matteucci and Monica Tosi for advice over the years, and to Johannes Geiss and Gary Steigman for helpful discussions. Generous financial support from IAU is also gratefully acknowledged. The author's research at Bologna University is supported by Italian MIUR under grant PRIN 2007, prot. 2007JJC53X_001.

References

Allende-Prieto, C., Lambert, D. L., & Asplund, M. 2001, ApJ, 556, L63
Anders, E. & Grevesse, N. 1989, $Geochim.~Cosmochim.~Acta$, 53, 197
Asplund, M., Grevesse, N., Sauval, A. J., & Scott, P. 2009, $ARA\&A$, 47, 481
Audouze, J. & Tinsley, B. M. 1974, ApJ, 192, 487

Balser, D., Rood, R. T., & Bania, T .M. 1999, *ApJ*, 522, L73

Bania, T. M., Rood, R. T., & Balser, D. S. 2002, *Nature*, 415, 54

Burles, S. & Tytler, D. 1998a, *ApJ*, 499, 699

Burles, S. & Tytler, D. 1998b, *ApJ*, 507, 732

Carswell, R. F., Rauch, M., Weymann, R. J., Cooke, A. J., & Webb, J. K. 1994, *MNRAS*, 268, L1

Casagrande, L., Flynn, C., Portinari, L., Girardi, L., & Jimenez, R. 2007, *MNRAS*, 382, 1516

Charbonnel, C. 1995, *ApJ*, 453, L41

Charbonnel, C. & Do Nascimento, J. D., Jr. 1998, *A&A*, 336, 915

Charbonnel, C. & Zahn, J.-P. 2007, *A&A*, 467, L15

Chiappini, C., Matteucci, F., & Gratton, R. 1997, *ApJ*, 477, 765

Chiappini, C., Matteucci, F., & Meynet, G. 2003, *A&A*, 410, 257

Chiappini, C., Renda, A., & Matteucci, F. 2002, *A&A*, 395, 789

Chiosi, C. & Matteucci, F. 1982, *A&A*, 105, 140

Cyburt, R. H., Fields, B. D., & Olive, K. A. 2008, *JCAP*, 11, 012

Danziger, I. J. 1970, *ARA&A*, 8, 161

Dearborn, D. S. P., Steigman, G., & Tosi, M. 1996, *ApJ*, 465, 887

Dunkley, J., *et al.* 2009, *ApJS*, 180, 306

Edmunds, M. G. 1994, *MNRAS*, 270, L37

Epstein, R. I., Lattimer, J. M., & Schramm, D. N. 1976, *Nature*, 263, 198

Galli, D., Palla, F., Ferrini, F., & Penco, U. 1995, *ApJ*, 443, 536

Galli, D., Stanghellini, L., Tosi, M., & Palla, F. 1997, *ApJ*, 477, 218

Geiss, J. & Gloeckler, G. 1998, *Space Sci. Rev.*, 84, 239

Geiss, J. & Reeves, H. 1972, *A&A*, 18, 126

Gloeckler, G. & Geiss, J. 1996, *Nature*, 381, 210

Grevesse, N. & Sauval, A. J. 1998, *Space Sci. Rev.*, 85, 161

Hata, N., Scherrer, R. J., Steigman, G., Thomas, D., Walker, T. P., Bludman, S., & Langacker, P. 1995, *Phys. Rev. Lett.*, 75, 3977

Hébrard, G., Tripp, T. M., Chayer, P., Friedman, S. D., Dupuis, J., Sonnentrucker, P., Williger, G. M., & Moos, M. W. 2005, *ApJ*, 635, 1136

Hogan, G. 1995, *ApJ*, 441, L17

Izotov, Y. I. & Thuan, T. X. 1998, *ApJ*, 500, 188

Linsky, J. L. 1998, *Space Sci. Rev.*, 84, 285

Linsky, J. L., *et al.* 2006, *ApJ*, 647, 1106

Lubowich, D. A., Pasachoff, J. M., Balonek, T. J., Millar, T. J., Tremonti, C., Roberts, H. & Galloway, R. P. 2000, *Nature*, 405, 1025

Maeder, A. 1992, *A&A*, 264, 105

Meynet, G. & Maeder, A. 2002, *A&A*, 390, 561

Matteucci, F., Romano, D., & Molaro, P. 1999, *A&A*, 341, 458

Norris, J. E. 2004, *ApJ*, 612, L25

Olive, K. A., Skillman, E., & Steigman, G. 1997, *ApJ*, 483, 788

Pancino, E., Ferraro, F. R., Bellazzini, M., Piotto, G., & Zoccali, M. 2000, *ApJ*, 534, L83

Peebles, P. J. E. 1966, *Phys. Rev. Lett.*, 16, 410

Peimbert, M., Luridiana, V., & Peimbert, A. 2007, *ApJ*, 666, 636

Penzias, A. A. & Wilson, R. W. 1965, *ApJ*, 142, 419

Pettini, M., Zych, B. J., Murphy, M. T., Lewis, A., & Steidel, C. C. 2008, *MNRAS*, 391, 1499

Piotto, G., *et al.* 2005, *ApJ*, 621, 777

Piotto, G., Bedin, L. R., Anderson, J., King, I. R., Cassisi, S., Milone, A. P., Villanova, S., Pietrinferni, A., & Renzini, A. 2007, ApJ, 661, L53

Prantzos, N. 1996, *A&A*, 310, 106

Prodanović, T. & Fields, B. D. 2003, *ApJ*, 597, 48

Prodanović, T., Steigman, G., & Fields, B. D. 2009, preprint (arXiv:0910.4961)

Reeves, H., Audouze, J., Fowler, W. A., & Schramm, D. N. 1973, *ApJ*, 179, 909

Rogers, A. E. E., Dudevoir, K. A., Carter, J. C., Fanous, B. J., Kratzenberg, E., & Bania, T. M. 2005, *ApJ*, 630, L41

Romano, D., Tosi, M., Matteucci, F., & Chiappini, C. 2003, *MNRAS*, 346, 295

Romano, D., Tosi, M., Chiappini, C., & Matteucci, F. 2006, *MNRAS*, 369, 295

Romano, D., Tosi, M., Cignoni, M., Matteucci, F., Pancino, E., & Bellazzini, M. 2010, *MNRAS*, in press (arXiv:0910.1299)

Rood, R. T., Steigman, G., & Tinsley, B. M. 1976, *ApJ*, 207, L57

Sackmann, I.-J. & Boothroyd, A. I. 1999, *ApJ*, 510, 217

Scully, S., Cassé, M., Olive, K. A., & Vangioni-Flam, E. 1997, *ApJ*, 476, 521

Songaila, A., Cowie, L. L., Hogan, C. J., & Rugers, M. 1994, *Nature*, 368, 599

Steigman, G. 2007, *Annu. Rev. Nucl. Part. Sci.*, 57, 463

Steigman, G. & Tosi, M. 1992, *ApJ*, 401, 150

Steigman, G., Romano, D., & Tosi, M. 2007, *MNRAS*, 378, 576

Tinsley, B. M. 1974, *ApJ*, 192, 629

Tosi, M. 1988, *A&A*, 197, 33

Tosi, M., Steigman, G., Matteucci, F., & Chiappini, C. 1998, *ApJ*, 498, 226

Truran, J. W. & Cameron, A. G. W. 1971, *Ap&SS*, 14, 179

van den Hoek, L. B. & Groenewegen, M. A. T. 1997, *A&AS*, 123, 305

Vangioni-Flam, E., Olive, K. A., Prantzos, N. 1994, *ApJ*, 427, 618

Wagoner, R. V., Fowler, W. A., & Hoyle, F. 1967, *ApJ*, 148, 3

Woosley, S. E. & Weaver, T. A. 1995, *ApJS*, 101, 181

Light Elements in the Universe
Proceedings IAU Symposium No. 268, 2009
C. Charbonnel, M. Tosi, F. Primas & C. Chiappini, eds.

© International Astronomical Union 2010
doi:10.1017/S1743921310004606

Thermohaline mixing in stars : solving the long-standing ^{3}He problem

Corinne Charbonnel[1,2] and Nadège Lagarde[1]

[1]Geneva Observatory, University of Geneva
Chemin des Maillettes 51, 1290 Versoix, Switzerland
email: `Corinne.Charbonnel@unige.ch`, `Nadege.Lagarde@unige.ch`

[2]CNRS UMR 5572, Toulouse University
14, av.E.Belin, 31400 Toulouse, France

Abstract. Thermohaline mixing has been recently identified as the dominating process that governs the photospheric composition of low-mass bright giant stars (Charbonnel & Zahn 2007a). Here we present the predictions of stellar models computed with the code STAREVOL that takes into account this mechanism together with rotational mixing and atomic diffusion. We compare our theorical predictions with recent observations and discuss how the corresponding yields for ^{3}He are compatible with the observed behaviour of this light element in our Galaxy.

Keywords. Hydrodynamics, instabilities, Stars: abundances, evolution, rotation, Galaxy: abundances

1. The "^{3}He problem"

The classical theory of stellar evolution predicts a very simple Galactic destiny to ^{3}He, dominated by a large production of this isotope by low-mass stars (Iben 1967; Rood 1972; Rood *et al.* 1976; Dearborn *et al.* 1996; Weiss *et al.* 1996). As a consequence, one expects a large increase of ^{3}He with time in the Galaxy with respect to its primordial abundance (e.g., Tosi 1996). However, the ^{3}He content of Galactic HII regions (Balser *et al.* 1994, 1999; Bania *et al.* 1997, 2002) is very similar to that of the Sun and solar system (Geiss & Reeves 1972; Geiss 1993, Mahaffy *et al.* 1998), and very close to the BBN value (Coc *et al.* 2004; Cyburt 2004; Serpico *et al.* 2004). This is the so-called "^{3}He problem" that could be resolved if only $\sim$10 % or less of the low-mass stars were releasing ^{3}He as predicted by classical stellar theory (Tosi 1998, 2000; Palla *et al.* 2000; Charbonnel 2002; Romano *et al.* 2003).

Charbonnel & Zahn (2007a) showed that thermohaline mixing drastically reduces the ^{3}He production in low-mass, low-metallicity stars. Simultaneously, this mechanism changes the surface carbon isotopic ratio as well as the abundances of lithium, carbon and nitrogen.

2. Stellar models including thermohaline convection, rotation-induced mixing, and atomic diffusion

Here we present the predictions of new stellar models computed with the code STAREVOL for solar metallicity and stellar masses between 1 and 4 M$_\odot$. Computations include the transport of chemical species in the radiative regions due to thermohaline instability, rotational mixing, and atomic diffusion. For thermohaline transport we use the diffusion coefficient advocated by Charbonnel & Zahn (2007a) based on Ulrich (1972) arguments for the aspect ratio of the salt fingers as supported by laboratory experiments

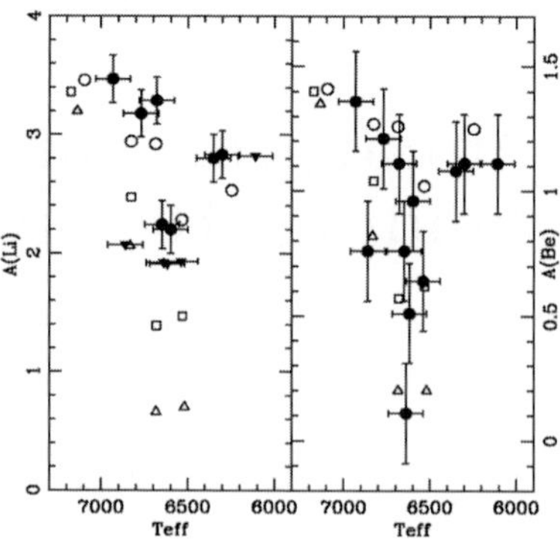

Figure 1. Li and Be abundances in IC 4651 main sequence and turnoff stars (black points and triangles for actual determinations and upper limits respectively). Open circles, squares, and triangles show model predictions for initial rotation velocities of 50, 80, and 110 km s^{-1} respectively. On the cool side of the Li and Be dip the model with Teff$\sim$6250 K is from Talon & Charbonnel (2005) and takes into account additional transport of angular momentum by internal gravity waves. Adapted from Smiljanic *et al.* (2009b)

(Krishnamurti 2003) and on Kippenhahn *et al.* (1980) extended expression for the case of a non-perfect gas. The evolution of the internal angular momentum profile and the associated transport of chemicals are accounted for with the complete formalism developed by Zahn (1992) and Maeder & Zahn (1998) that takes into account advection by meridional circulation and diffusion by shear turbulence (see Palacios *et al.* 2003, 2006, and Decressin *et al.* 2009 for a description of the implementation in STAREVOL). Typical initial (i.e., ZAMS) surface rotation velocities are chosen for all the models depending on the stellar mass. We assume magnetic braking on the early main sequence for the stars with Teff on the ZAMS lower than $\sim$ 6900 K that have relatively thick convective envelopes (Talon & Charbonnel 1998). The adopted braking law follows the description of Kawaler (1988). Rotational velocity further decreases when the stars evolve on the subgiant branch due to radius expansion. Atomic diffusion is included in the form of gravitational settling as well as that related to thermal gradients, using the formulation of Paquette *et al.* (1986).

3. Model predictions for the surface abundances

The model predictions for the evolution of the surface abundances of various species have been validated all along the evolutionary sequence. They reproduce for example very nicely the surface abundances of Li and Be along the colour-magnitude diagram of the open cluster IC 4651 as shown in Fig.1 and 2. Note that thermohaline mixing is efficient only when RGB stars reach the so-called bump in the luminosity function, which is located at Teff $\sim$ 4200 K in the present case. For stars less evolved than the bump as those shown in both figures, the Li and Be behaviours are thus dictated by rotation-induced mixing (see Smiljanic *et al.* 2009b for more details and Smiljanic *et al.*, this volume; see also Charbonnel & Talon 1999 and Palacios *et al.* 2003).

Predictions for the evolution of the surface carbon isotopic ratio are shown in Fig. 3 for models of 1.25 and 2 M$_\odot$ stars, and compared with observations in the open cluster M67 (turnoff mass $\sim$1.2 M$_\odot$). We note that rotation-induced mixing on the main sequence slightly lowers the post-dredge-up ^{12}C/^{13}C value compared to the classical case (i.e., no rotation, diffusion, nor thermohaline mixing). At the luminosity of the bump (log(L/L$_\odot$)$\sim$2 for the 1.25 M$_\odot$ star), thermohaline mixing leads to further decrease of

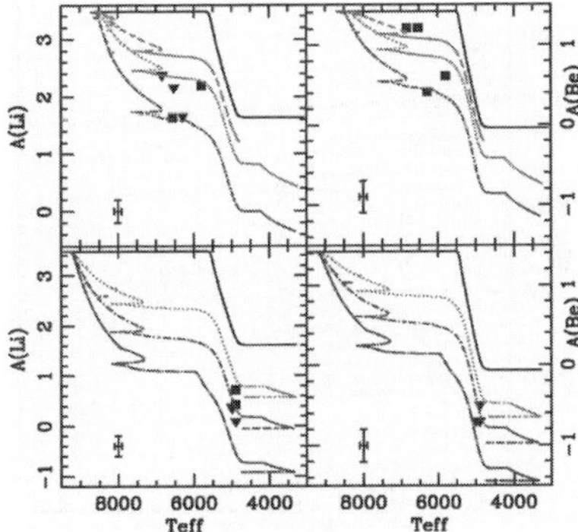

Figure 2. Li and Be abundances in IC 4651 evolved stars. Theoretical predictions for 1.8 and 2 $M_\odot$ models are compared to observations of subgiant and giant stars (upper and lower pannels resectively). Solid lines are for the classical case. Other lines correspond to different initial rotation velocities (80, 110, and 180 km s^{-1} for the 1.8 $M_\odot$ star; 110, 180, and 250 km s^{-1} for the 2.0 $M_\odot$ star). Adapted from Smiljanic *et al.* (2009b)

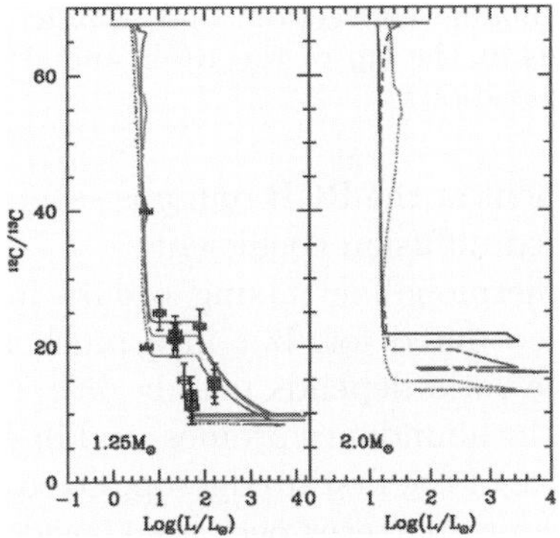

Figure 3. Evolution of the surface ^{12}C/^{13}C value as a function of stellar luminosity for models of 1.25 and 2 $M_\odot$ stars (left and right respectively). Different tracks are for different initial rotation velocities (50, 80, and 110 km.s^{-1} for the 1.25 $M_\odot$ star, and 0, 110, and 250 km.s^{-1} for the 2 $M_\odot$ star). Observations by Gilroy & Brown (1991) in evolved stars of the open cluster M67 (turnoff mass $\sim$1.2 $M_\odot$) are also shown (triangle, squares, and circles for subgiant, RGB, and clump stars respectively). Adapted from Lagarde & Charbonnel (in preparation)

the carbon isotopic ratio, in excellent agreement with M67 data. In the case of the 2 $M_\odot$ star, thermohaline mixing becomes efficient at the bump in the luminosity function only when rotation in earlier phases is accounted for. Importantly we note that at solar metallicity, the ^{12}C/^{13}C values reached when thermohaline mixing ceases are higher than in the case of metal-poor stars where the carbon isotopic ratios almost always reach the equilibrium value (see Fig. 3 of Charbonnel & Zahn 2007a). This metallicity-dependence is in perfect agreement with the observational behaviour (see Fig. 1 of Charbonnel & Do Nascimento 1998).

In Fig. 4 we show the predictions for the ^{12}C/^{13}C surface ratio at the tip of the RGB and of the AGB for our models over the whole considered mass range and compare them with observations in stars belonging to open clusters of various turnoff masses. We see that in models for stars with masses below $\sim$ 1.7 $M_\odot$, thermohaline mixing is the main physical process governing the photospheric composition of evolved giants, while rotation plays only a minor role on the red giant branch (see Palacios *et al.* 2006). In fact, the

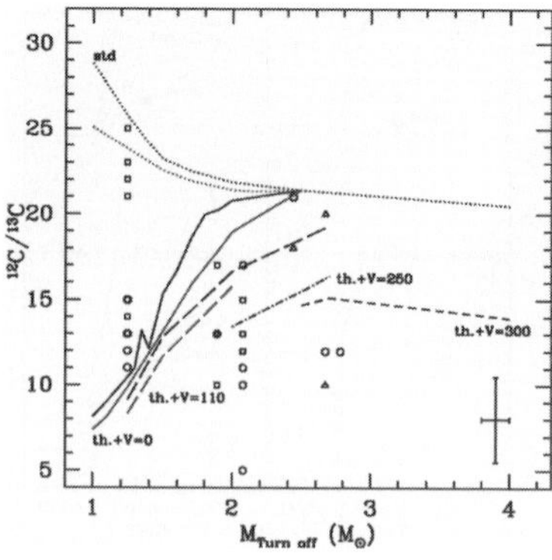

Figure 4. Theoretical predictions compared with observations of ^{12}C/^{13}C in open clusters spanning a large turnoff mass range. Data are from Smiljanic *et al.* (2009a) and Gilroy & Brown (1991). Squares, triangles and circles are for RGB, clump, and early-AGB stars respectively. Typical observational errors are indicated. Classical models (i.e., non-rotating and without thermohaline mixing) are shown as dotted lines. The solid lines are for models including thermohaline mixing only, while all the other models include rotation-induced mixing (with initial rotation velocities as indicated), thermohaline convection, and atomic diffusion. Black and blue lines correspond to model predictions at the tip of the RGB and AGB respectively. Adapted from Lagarde & Charbonnel (in preparation)

thermohaline diffusion coefficient at the RGB bump is several order of magnitudes higher than the total rotation-induced diffusion coefficient.

For more massive stars, thermohaline mixing occurs in the advanced phases when rotation-induced mixing is accounted for, but in a much less efficient manner. In this case, the final carbon isotopic ratio depends mainly on rotation-induced mixing on the main sequence that modifies the abundance profiles, and in particular the ^{13}C peak inside the stars, before the occurrence of the first dredge-up. Overal, the present models explain very well the observed abundance patterns over the considered mass range.

4. Model predictions for ^{3}He

On the main sequence, a ^{3}He peak builds up due to pp-reactions inside low-mass stars (Iben 1967), and is engulfed in the stellar envelope during the first dredge-up. As a consequence the surface abundance of ^{3}He strongly increases on the lower RGB as can be seen in Fig. 5 for various stellar masses. Its value reaches a maximum when the whole peak is engulfed. After the first dredge-up, the temperature at the base of the convective envelope is too low for ^{3}He to be nuclearly processed. As a result in canonical models this fresh ^{3}He is preserved until the ejection of the planetary nebula when it is released into the interstellar matter. This classical view is however contradicted by observations and chemical evolution models as discussed in § 1.

After the bump however, thermohaline mixing brings ^{3}He from the convective envelope down to the hydrogen-burning shell where it burns. This leads to a rapid decrease of the surface abundance (and thus of the corresponding yield) of this element as can be seen in Fig. 5, and as already shown by Charbonnel & Zahn (2007a) for low-metallicity stars. This confirms the early suggestion by Rood *et al.* (1984) that the variations of the carbon isotopic ratio and of ^{3}He are strongly connected (see also Charbonnel 1995; Charbonnel & Do Nascimento 1998; Eggleton *et al.* 2006). It is important to note that in the models presented here ^{3}He decreases by a large factor in the ejected material with respect to the canonical evolution predictions, but that low-mass stars remain net producers (while

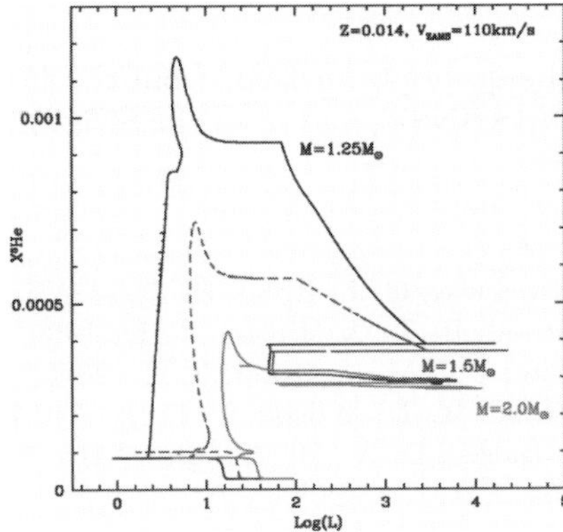

Figure 5. Evolution of the surface abundance of ^{3}He (in mass fraction) for stars of various initial masses and solar metallicity. Figure from Lagarde & Charbonnel (in preparation)

far much less efficient than in the canonical case) of ^{3}He. As already depicted by the ^{12}C/^{13}C behaviour that traces the dependance of the thermohaline mixing efficiency with metallicity, the destruction of fresh ^{3}He by this process is much more efficient in low-metallicity stars (see Fig. 4 of Charbonnel & Zahn 2007a).

Computations for a larger grid in stellar masses and metallicities are now being performed in order to quantify the actual contribution of low-mass stars to Galactic ^{3}He in the framework proposed here (Lagarde *et al.*, in preparation). We are confident that the corresponding ^{3}He yields will help reconciling the primordial nucleosynthesis with measurements of ^{3}He/H in Galactic HII regions (Charbonnel 2002).

5. The peculiar case of "thermohaline deviant stars": Ap star descendants?

However a couple of planetary nebulae, namely NGC 3242 and J320, have been found to behave "classically" (see Bania, this volume): slightly more massive than the Sun, they are currently returning fresh ^{3}He to the interstellar medium, in the amount predicted by classical stellar models (Rood *et al.* 1992; Galli *et al.* 1997; Balser *et al.* 1999, 2006).

To reconcile the ^{3}He/H measurements in Galactic HII regions with the high values of ^{3}He in NGC 3242 and J320, Charbonnel & Zahn (2007b) proposed that thermohaline mixing is inhibited by a fossil magnetic field in RGB stars that are descendants of Ap stars. They obtained a threshold for the magnetic field of 10^4-10^5 Gauss, above which it inhibits thermohaline mixing in red giant stars located at or above the L-bump. Fields of that order are expected in the descendants of Ap stars, taking into account the contraction of their core when they become red giants.

Charbonnel & Zahn (2007b) thus concluded that in a large fraction of descendants of Ap stars thermohaline mixing does not occur. As a consequence these objects should produce ^{3}He as predicted by the standard stellar theory and as observed in the planetary nebulae NGC 3242 and J320. The relative number of such stars with respect to non-magnetic objects that undergo thermohaline mixing is consistent with the statistical constraint coming from observations of the carbon isotopic ratio in red giant stars (Charbonnel & Do Nascimento 1998). It satisfies also the Galactic requirements for the evolution of the ^{3}He abundance.

Acknowledgements

We acknowledge financial support from IAU, from the French "Programme National de Physique Stellaire" of CNRS/INSU, and from the Swiss National Science Foundation.

References

Balser, D. A., Bania, T. M., Brockway, C. J., Rood, R. T., & Wilson, T. L., 1994, *ApJ*, 430, 667
Balser, D. A., Bania, T. M., Rood, R. T., & Wilson, T. L., 1999, *ApJ*, 510, 759
Balser, D. A., Goss, W. M., Bania, T. M., & Rood, R. T., 2006, *ApJ*, 640, 360
Bania, T. M., Balser, D. A., Rood, R. T., Wilson, T. L., & Wilson, T.J., 1997, *ApJS*, 113, 353
Bania, T. M., Rood, R. T., & Balser, D. A., 2002, *Nature*, 415, 54
Charbonnel, C. 1995, *ApJ*, 453, L41
Charbonnel, C. 2002, *Nature*, 415, 27
Charbonnel, C. & Do Nascimento, J. D. 1998, *A&A*, 336, 915
Charbonnel, C. & Talon, S. 1999, *A&A*, 351, 635
Charbonnel, C. & Zahn, J. P. 2007a, *A&A Letters*, 467, L15
Charbonnel, C. & Zahn, J. P. 2007b, *A&A Letters*, 476, L29
Coc, A., Vangioni-Flam, E., Descouvemont, P., Adahchour, A., & Angulo, C. 2004, *ApJ*, 600, 544
Cyburt, R. H. 2004, *Phys. Rev.D*, 70, 023 505
Dearborn, D. S. P., Steigman, G., & Tosi, M., 1996, *ApJ*, 465, 887
Decressin, T., Mathis, S., Palacios, A., *et al.* 2009, *A&A*, 495, 271
Eggleton, P. P., Dearborn, D. S. P., & Lattanzio, J. C 2006 *Science*, 314, 5805, 1580
Galli, D., Stanghellini, L., Tosi, M., & Palla, F. 1997, *ApJ*, 477, 218
Geiss, J. & Reeves, H., 1972, *A&A* 18, 126
Geiss, J., 1993, in Origin and evolution of the elements, eds. N. Prantzos *et al.*, p. 89
Gilroy, K. K. & Brown, J. A. 1991, *ApJ*, 371, 578
Iben, I., 1967, *ApJ*, 143, 642
Kawaler, S. D., 1988, *ApJ*, 333, 236
Kippenhahn, R., Ruschenplatt, G., & Thomas, H. C. 1980, *A&A*, 91, 175
Krishnamurti, R. 2003, *J. Fluid Mech.*, 483, 287
Maeder, A. & Zahn, J. P. 1998, *A&A*, 334, 1000
Palacios, A., Charbonnel, C., Talon, S., & Forestini, M. 2003, *A&A*, 399, 603
Palacios, A., Charbonnel, C., Talon, S., & Siess, L. 2006, *A&A*, 453, 261
Palla, F., Bachiller, R., Stanghellini, L., Tosi, M., Galli, D., 2000, *A&A*, 355, 69
Paquette, C., Pelletier, C., Fontaine, G., & Michaud, G., 1986, *ApJS*, 61,177
Rood, R. T., 1972, *ApJ*, 177, 681
Rood, R. T., Steigman, G., & Tinsley, B. M., 1976, *ApJ*, 207, L57
Rood, R. T., Bania, T. W., & Wilson, T. L. 1984, *ApJ*, 280, 629
Rood, R. T., Bania, T. W., & Wilson, T. L. 1992, *Nature*, 355, 618
Romano, D., Tosi, M., Matteucci, F., & Chiappini, C. 2003, *MNRAS*, 346, 295
Smiljanic, R., Gauderon, R., North, P., Barbuy, B., Charbonnel, C., & Mowlavi, N. 2009a, *A&A* 502, 267
Smiljanic, R., Pasquini, L., Charbonnel, C., & Lagarde, N. 2009b, *A&A*, in press, astro-ph 0910.4399
Talon, S. & Charbonnel, C 1998, *A&A*, 335, 959
Talon, S. & Charbonnel, C 2005, *A&A*, 440, 981
Tosi, M. 1996, *ASP Conference Series*, Vol. 98, 299
Tosi, M. 1998, *Space Science Reviews*, Vol. 84, 207
Tosi, M. 2000, IAUS 198, 525
Ulrich, R. K. 1972, *ApJ*, 172, 165
Weiss, A., Wagenhuber, J., & Denissenkov, P. A., 1996, *A&A*, 313, 581
Zahn, J. P. 1992, *A&A*, 265, 115

Light Elements in the Universe
Proceedings IAU Symposium No. 268, 2009
C. Charbonnel, M. Tosi, F. Primas & C. Chiappini, eds.

© International Astronomical Union 2010
doi:10.1017/S1743921310004618

Theoretical stellar $\Delta Y/\Delta O$ in the early Universe

Sylvia Ekström[1], Georges Meynet[1], André Maeder[1], Cristina Chiappini[1,2], Cyril Georgy[1] and Raphael Hirschi[3,4]

[1]Astronomical Observatory of the Geneva University
Maillettes 51 - Sauverny, 1290 Versoix GE, Switzerland
email: `Sylvia.Ekstrom@unige.ch`

[2]Osservatorio Astronomico di Trieste
OAT/INAF, Via G. B. Tiepolo 11, 34131 Trieste TS, Italy

[3]Astrophysics group, Keele University,
Lennard-Jones Lab., Keele, ST5 5BG, UK

[4]IPMU, University of Tokyo,
Kashiwa, Chiba 277-8582, Japan

Abstract. Population III stars initiated the chemical enrichment of the Universe. Chemical evolution models seem to favour fast rotators among the very low-metallicity population. When a star rotates fast, it ejects significant quantities of He and its nucleosynthesic products are modified compared to the case without rotation. The value of $\Delta Y/\Delta O$ is explored from a theoretical point of view through stellar models of zero- or very low-metallicity.

Keywords. nucleosynthesis, stars: rotation, early universe

1. Introduction

About a quarter of an hour after the Big Bang, the Universe has finished all possible nucleosynthesis leaving its chemical composition devoid of metals (see for example Iocco *et al.* 2007). Only when the first stars form does nucleosynthesis take place again, in their cores and envelopes, and the enrichment of the Universe in heavy elements can start.

In the very early Universe, the chemical enrichment follows the nucleosynthetic path of massive stars for the first few millions of years. Of course, in the galaxies we can observe now, the contribution of the first generations of stars has been overwhelmed by the following generations, where intermediate- or low-mass stars contribute actively to the nucleosynthesis. However, should we be able one day to observe galaxies with metallicities as low as $Z = 10^{-8}$, we would certainly observe a medium enriched only by massive stars and it is interesting to study how this enrichment would take place. Currently it is in the Milky Way halo that the most metal poor objects are observed. These low-mass, second generation stars retain the memory of the unique nucleosynthesis in the first generations of massive stars.

We have recently shown that only chemical evolution models adopting stellar yields from rotating star models can successfully explain some of the puzzling imprints of the first stellar generations, such as the production of primary nitrogen and ^{13}C in the early Universe (before intermediate-mass stars have had time to contribute to the chemical enrichment - see Chiappini *et al.* 2008, 2006). Here we address the following question: would fast rotators eject significant larger quantities of helium in the early chemical enrichment than present-day massive stars? Here, we explore the theoretical dY/dO value obtained by massive star models with rotation at extremely low or zero metallicities. We consider only stellar model results, since at those extreme metallicities only a few massive stars would have had time to contribute to the chemical enrichment.

2. Stellar models

In the present study, we use the same fast-rotating models of extremely-low metallicity ($Z = 0$ and $Z = 10^{-8}$) massive stars considered in our previous papers (Chiappini *et al.* 2008, 2006). The stellar models are described in Hirschi (2007) for $Z = 10^{-8}$ and in Ekström *et al.* (2008) for $Z = 0$. The mass range covered goes from 9 to 85 $M_\odot$. We let the interested reader refer to the original papers for the detailed physics of the models, we will just summarize here the main ingredients of the calculation:

• non-solid rotation is treated as in Hirschi *et al.* (2004), with the horizontal turbulence coefficient from Zahn (1992) and the shear diffusion coefficient from Talon & Zahn (1997);

• the radiative mass loss prescription is Kudritzki (2002) for the $Z = 0$ models, with the same adaptations as in Marigo *et al.* (2003), and Vink *et al.* (2001) for the $Z = 10^{-8}$ models;

• when (if) the star reaches the critical velocity†, a mechanical mass loss is applied as in Meynet *et al.* (2006), so the supercritical layers are removed;

• an important reaction rate for the results of $\Delta Y/\Delta O$ is the $^{12}C(\alpha,\gamma)^{16}O$ rate, which is still a subject of controversy. Here the models are computed with the NACRE recommanded rate;

• the convection criterion applied is the Schwarzschild (1958) criterion.

The models start on the ZAMS with a ratio $v/v_{\rm crit}$ around 0.5.

2.1. *Nucleosynthesis*

There are some peculiarities of the nucleosynthesis at extremely-low or zero metallicity. The stars are very compact, because of the lack of metals in their envelope, thus the hydrogen burning takes place at higher temperature in the core. Pop III stars even burn some helium during the main sequence. The compactness favours a mixing between the different burning zones, for example between the helium-burning convective core and the hydrogen burning radiative shell. This leads to the production of primordial nitrogen, for example (see Meynet & Maeder 2002). While the non-rotating models undergo this phenomenon only at specific mass domains (~ 25 and ~ 85 $M_\odot$), the rotating models present the primary nitrogen production at all masses. Later, some of this nitrogen can diffuse towards the core and increase the production of ^{22}Ne, which is an interesting source of neutrons for the $s-$process nucleosynthesis (Pignatari *et al.* 2008).

Rotation also leads to modifications in the yields of the elements that interest us here. At non-zero metallicity, the mass loss is enhanced, saving some helium from further burning. Fast rotators are thus expected to be strong helium producers (see the contribution of G. Meynet in this proceedings). When some C and O are diffused from the core to the H-burning shell, it drives a boost of the energy production in it (passing suddenly from the $pp-$chains to the CNO cycle). The boost of energy reduces the size of the core, and since the O yield is closely related to the core size, the O production is reduced in this case.

2.2. *Yields*

Figure 1 and Table 1 present the yields of O and He of all the models. The $Z = 0$ very massive (above 40 $M_\odot$) models produce more O and less He than the $Z = 10^{-8}$ models.

---

† *i.e.* the velocity at which the centrifugal force counterbalances exactly the gravitational force at the surface

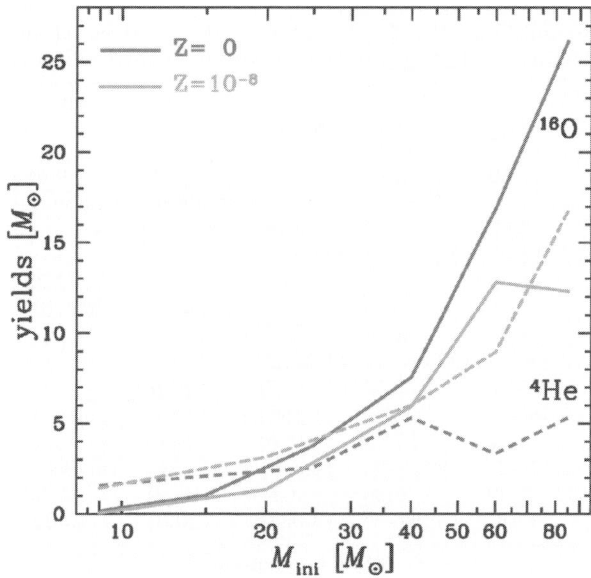

Figure 1. Yields in He (dashed lines) and O (solid lines): $Z = 0$ models (dark grey) and $Z = 10^{-8}$ (light grey).

This is mainly due to the difference in the mass loss rate: the Pop III models lose almost no mass at all while the $Z = 10^{-8}$ models experience a strong surface enrichment and the mass loss is largely increased thanks to the diffused metals. The evolution at almost constant mass and the extreme compactness of Pop III stars lead to very large CO cores, which favours a high O production at the expense of He.

As we mentioned previously, rotation enhances the mass loss. This is particularly interesting at very low metallicity, where radiative mass loss is supposed to be extremely weak. Also, according to Heger *et al.* (2003), some of the low-metallicity massive stars could end their life without the explosion of a supernova. The formation of a direct black hole would swallow the whole star without any ejection, except the winds the star lost during its life. In this case, the chemical contribution of the star would be of the 'wind-only' type. Table 1 shows the 'wind-only' contribution in parenthesis. We see that the wind is rich in He and very poor in O, as expected.

3. ΔY/ΔO

Figure 2 shows the dY/dO values obtained by stellar models. The more massive the model, the earlier the contribution is expected. Since high-mass models yield a low dY/dO value, the dY/dO ratio increases with time. The range of $\Delta Y/\Delta O$ values observed in the most metal-poor HII regions of the present-day Universe is indicated on the figure. Though metal-poor, those regions already bear the imprint of many stellar generations, including intermediate- and low-mass stars. The values are deduced from observations under the hypothesis that the IMF doesn't change at low metallicity (see Izotov *et al.* 2007). This hypothesis needs yet to be ascertained. In the very early Universe, it is thought that either the slope of the IMF is still valid but there is a mass-cut under $\simeq 10$ $M_\odot$ (Nakamura & Umemura 1999), either the IMF is flat, or it is doubled-peaked (Nakamura & Umemura 2001).

If we assume that all the stars die at the same moment (which is reasonnable given that the less massive stars live less than 30 million years), we can integrate their yields

Table 1. Yields in He and O for the $Z = 0$ and $Z = 10^{-8}$ models. All masses are in $M_\odot$. The values given in parenthesis are the 'wind-only' contribution (see text).

	$Z = 0$			$Z = 10^{-8}$		
Mass	^{4}He	^{16}O	dY/dO	^{4}He	^{16}O	dY/dO
9	1.59	0.17	9.35	1.43	0.06	24.2
				(2.80e-5)	(2.33e-8)	(1200)
15	2.10	1.05	2.00			
	(3.59e-4)	(1.79e-15)	(2.0e11)			
20				3.15	1.35	2.33
				(2.36e-4)	(2.54e-10)	(9.3e5)
25	2.57	3.76	0.68			
	(2.13e-3)	(2.06e-13)	(1.0e10)			
40	5.32	7.57	0.70	6.01	5.94	1.01
	(5.48e-2)	(4.09e-8)	(1.3e6)	(0.33)	(2.42e-3)	(136)
60	3.37	16.90	0.20	8.97	12.80	0.70
	(4.11e-2)	(8.88e-8)	(4.6e5)	(1.21)	(5.48e-5)	(2.2e4)
85	7.09	26.20	0.27	16.80	12.30	1.37
	(1.78)	(5.98e-5)	(3.0e4)	(20.0)	(3.02)	(6.62)

IMF integrated

	dY/dO	dY/dO
S55†	1.39	2.82
MS79‡	2.14	4.18

IMF integrated, with 'realistic' fate¶ for the stars

	dY/dO	dY/dO
S55	2.36	5.16
MS79	3.51	7.28

† Salpeter (1955)
‡ Miller & Scalo (1979)
¶ Heger et al. (2003)

with an IMF and get a mean value for a first burst of primordial stars. For this, we use the following formula:

$$\frac{\mathrm{dY}}{\mathrm{dO}} = \frac{\int_{M_{\mathrm{down}}}^{M_{\mathrm{up}}} \mathrm{dY}\, \Phi(M)\, \mathrm{d}M}{\int_{M_{\mathrm{down}}}^{M_{\mathrm{up}}} \mathrm{dO}\, \Phi(M)\, \mathrm{d}M}$$

where $\Phi(M) = AM^{-(1+x)}$ is the IMF. Since we consider only massive stars, we use $M_{\mathrm{down}} = 9\,M_\odot$ and $M_{\mathrm{up}} = 120\,M_\odot$.

The results are given in the middle panel of Table 1. Depending on the IMF used, the results change because different mass domains are favoured. Compared to Salpeter's, the slope $x = 2.30$ (for $M > 10\,M_\odot$) of Miller & Scalo is steeper and favours the lowest masses, leading to a higher value of dY/dO. The case of Pop III stars seems very peculiar, lower dY/dO values than the case of $Z = 10^{-8}$ stars. It reflects the peculiarities of the yields commented in section 2.2. When the metallicity grows slightly, the nucleosynthesis changes, and even with a metallicity as low as $Z = 10^{-8}$, the stars present a higher dY/dO value. The 'wind-only' case leads to extremely high dY/dO values.

Now let us take into account the 'realistic' fate of the star as determined by Heger *et al.* (2003). According to these authors, all stars with M_α† between 9 and 40 $M_\odot$ are supposed to end their life by collapsing into a black hole, without a supernova explosion. For such stars, the only contribution to the chemical enrichment of the Universe would

---

† M_α, the He core mass at the end of a star's evolution, is the mass coordinate at which the hydrogen abundance drops below 10^{-3}.

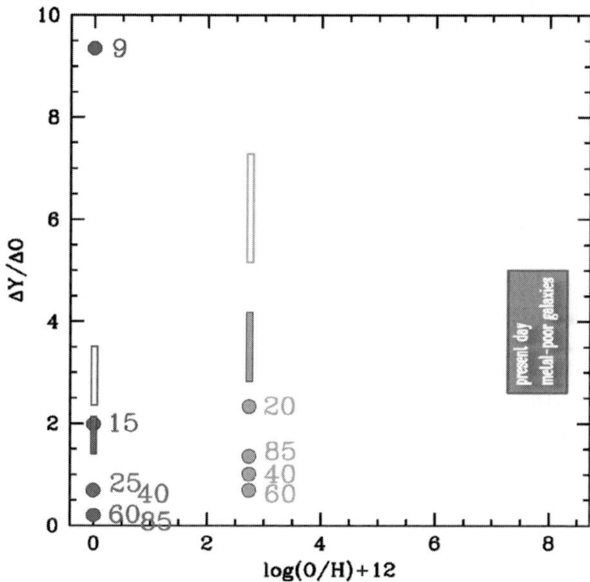

Figure 2. $\Delta Y/\Delta O$ values obtained with the total yields of the models as a function of the oxygen abundance: $Z = 0$ models (dark grey) and $Z = 10^{-8}$ (light grey). The IMF integrated values are given as a filled rectangle (range of values between S55 and MS79), and the ones obtained by taking into account the supposed fate of the models are shown with an empty rectangle. The range of values for the most metal-poor HII regions observed in the present-day Universe is also shown (data from Olive & Skillman 2004, Izotov *et al.* 2007 and Peimbert *et al.* 2007).

be made by the winds they experienced during their life. At $Z = 0$ the 25, 40, and 60 $M_\odot$ come into this category. At $Z = 10^{-8}$ the 40, 60, and 85 $M_\odot$ are concerned. Using these limits, we can compute more realistic dY/dO stellar ratios, which are given in the bottom panel of Table 1. Because of the high He and low O content of the winds, the dY/dO is much larger in this mixed case (i.e. winds+SN yields for stars with He core mass below $9M_\odot$ and only winds above that value).

4. Discussion

Here we computed the expected $\Delta Y/\Delta O$ ratio from a generation of Z=0 and ultra metal poor ($Z=10^{-8}$) massive stars. The stellar values presented here give the starting point for the further evolution of the $\Delta Y/\Delta O$ ratio. With time, the $\Delta Y/\Delta O$ ratio will increase thanks to the contribution to the helium of intermediate and low-mass stars. The evolution of $\Delta Y/\Delta O$ will depend on the star formation history, IMF and selective outflows of the particular galaxy under study.

We show how the initial $\Delta Y/\Delta O$ obtained from the first stars depend upon the fate of these stars. If the most massive stars do contribute to the chemical enrichment only via their stellar winds, the expected $\Delta Y/\Delta O$ ratio is larger due to the high helium and low oxygen content of the stellar winds. Under the hypothesis that all massive Z=0 stars ended up into black holes, contributing to the chemical enrichment only via their stellar winds, we get unrealistic high $\Delta Y/\Delta O$ ratios (due to a very low contribution to oxygen). A mixed scenario where, even at Z=0, only stars with helium core mass between 9 and 40 $M_\odot$ do implode directly as black holes seems more likely.

References

Bromm, V. & Larson, R. B. 2004, *ARA&A* 42, 79

Chiappini, C., Ekström, S., Meynet, G., Hirschi, R., Maeder, A., & Charbonnel, C. 2008, *A&A* 479, L9

Chiappini, C., Hirschi, R., Meynet, G., Ekström, S., Maeder, A., & Matteucci, F. 2006, *A&A* 449, L27

Ekström, S., Meynet, G., Chiappini, C., Hirschi, R., & Maeder, A. 2008, *A&A* 489, 685

Heger, A., Fryer, C. L., Woosley, S. E., Langer, N., & Hartmann, D. H. 2003, *ApJ* 591, 288

Hirschi, R. 2007, *A&A* 461, 571

Hirschi, R., Meynet, G., & Maeder, A. 2004, *A&A* 425, 649

Iocco, F., Mangano, G., Miele, G., Pisanti, O., & Serpico, P. D. 2007, *Phys. Rev. D* 75(8), 087304

Izotov, Y. I., Thuan, T. X., & Stasińska, G. 2007, *ApJ* 662, 15

Kudritzki, R. P. 2002, *ApJ* 577, 389

Maeder, A. 1992, *A&A* 264, 105

Marigo, P., Chiosi, C., & Kudritzki, R.-P. 2003, *A&A* 399, 617

Meynet, G., Ekström, S., & Maeder, A. 2006, *A&A* 447, 623

Meynet, G. & Maeder, A. 2002, *A&A* 390, 561

Miller, G. E. & Scalo, J. M. 1979, *ApJS* 41, 513

Nakamura, F. & Umemura, M. 1999, *ApJ* 515, 239

Nakamura, F. & Umemura, M. 2001, *ApJ* 548, 19

Olive, K. A. & Skillman, E. D. 2004, *ApJ* 617, 29

Peimbert, M., Luridiana, V., & Peimbert, A. 2007, *ApJ* 666, 636

Peimbert, M., Peimbert, A., Luridiana, V., & Ruiz, M. T. 2003, in E. Perez, R. M. Gonzalez Delgado, & G. Tenorio-Tagle (eds.), *Star Formation Through Time*, Vol. 297 of *Astronomical Society of the Pacific Conference Series*, p. 81

Pignatari, M., Gallino, R., Meynet, G., Hirschi, R., Herwig, F., & Wiescher, M. 2008, *ApJ* 687, L95

Salpeter, E. E. 1955, *ApJ* 121, 161

Schwarzschild, M. 1958, *Structure and evolution of the stars.*

Talon, S. & Zahn, J.-P. 1997, *A&A* 317, 749

Vink, J. S., de Koter, A., & Lamers, H. J. G. L. M. 2001, *A&A* 369, 574

Zahn, J.-P. 1992, *A&A* 265, 115

Light Elements in the Universe
Proceedings IAU Symposium No. 268, 2009
C. Charbonnel, M. Tosi, F. Primas & C. Chiappini, eds.

© International Astronomical Union 2010
doi:10.1017/S174392131000462X

Galactic evolution of ^7Li

Francesca Matteucci[1,2]

[1]Department of Physics, Astronomy Division, Trieste University, Italy
Via G. B. Tiepolo, 11 34134 Trieste, Italy
email: `matteucc@oats.inaf.it`

[2]INAF, Trieste
Via G. B. Tiepolo, 11 34134 Trieste, Italy

Abstract. Lithium represents a key element in cosmology, as it is one of the few nuclei synthesized during the Big Bang. The primordial abundance of ^7Li allows us to impose constraints on the primordial nucleosynthesis and on the baryon density of the universe. However, ^7Li is not only produced during the Big Bang but also during galactic evolution: measures of stellar Li in our Galaxy suggest an almost constant Li abundance (the so-called Spite plateau) at low metallicities and a subsequent increase in the disk stars, leading to a Li abundance in Population I stars higher by a factor of ten than in Population II stars. This means that there must exist several possible stellar sources of ^7Li: asymptotic giant branch stars, supernovae, novae, red giant stars. ^7Li is also partly produced in spallation processes while ^6Li is entirely produced by such processes. All of these sources have been included in galactic chemical evolution models and constraints have been derived on the primordial ^7Li and its evolution, as well on stellar models. I will review these models and their results and what we have learned about ^7Li evolution. Some still open problems, such as the disagreement between the primordial ^7Li abundance as derived by WMAP and as measured in Population II stars, and the uncertainties about the main sources of stellar ^7Li will be discussed.

Keywords. Stars: abundances – Galaxy: evolution

---

1. Introduction

Standard Big Bang nucleosynthesis predicts the abundances of ^2D, ^3He, ^4He and ^7Li and if these abundances are in agreement with the observationally derived primordial abundances, then we can derive the baryon to photon ratio and the baryonic density parameter Ω_b. This was the case until WMAP results became available (Spergel *et al.* 2003), suggesting directly a value for the baryon to photon ratio $\eta_{10} = 6.1^{+0.3}_{-0.2}$. More recently, the last release of WMAP (Hinshaw *et al.* 2009) suggests a revised value of $\eta_{10} = 6.23 \pm 0.17$. This value of η_{10} gives primordial abundances of ^2D, ^3He and ^4He in good agreement with observations except for ^7Li, for which the primordial abundance estimated from WMAP is higher by a factor of 3-4 than the abundance of ^7Li measured in low-metallicity halo stars, always interpreted as the primordial one. If the WMAP value is correct, we therefore should revise our interpretation of the ^7Li abundance in halo stars (Population II stars) and assume that some Li astration has taken place in these stars. In recent years some suggestions were put forward to explain the low Li abundance in metal-poor stars, such as diffusion and turbulence in the presence of weak turbulence (e.g. Richard *et al.* 2005; Melendez *et al.* 2009) and rotational mixing (Pinsonneault, this conference). Alternatively, this discrepancy could be solved by Li destruction during the Big Bang (Jedamzik, this conference). In this review we will focus on the interpretation of the plot $\mathrm{Log}\epsilon(Li)$ versus [Fe/H] in terms of Galactic production of ^7Li. We will review the main sources of ^7Li: stars, cosmic rays, and Big Bang. Then we will describe how to model the Galactic evolution of ^7Li by taking into account in detail all the various

^7Li sources. The most important results from 1990 up to now will be described, and the constraints derived from the comparison theory-observations as well as the still existing uncertainties will be summarized. Finally, we will discuss the ^7Li discrepancy originated by the difference between the primordial Li abundance derived from WMAP and the measured one.

2. ^7Li production in stars

There is only one way to produce ^7Li during normal stellar evolution and it is by means of the nuclear reaction $^3He(\alpha, \gamma)^7Be$. The fresh ^7Be should be then transported by convection into stellar regions of lower temperature, where it decays into ^7Li by k-capture. This mechanism was originally proposed by Cameron & Fowler (1971). However, ^7Li is also destroyed during stellar evolution, and in general it is assumed that all the original Li present in the star is destroyed in the course of the evolution. From the observational point of view, K giants and M supergiants are Li-rich (Smith & Lambert 1989, 1990) indicating that ^7Li is produced by asymptotic giant branch stars (AGBs) (see Gratton 1991, for an exhaustive review on the subject). Theoretical models of AGB stars (Sackmann & Boothroyd 1999) have suggested that stars in the mass range (4-6)$M_\odot$ can produce noticeable quantities of ^7Li ($Log\epsilon(Li) \sim 4 - 4.5$, with $Log\epsilon(Li)$ being the number density of Li in the unit system $Log(X/H) +12$, where 12 is the hydrogen). This result was later confirmed by Travaglio *et al.* (2001). On the other hand, Ventura *et al.* (1998) found a negligible ^7Li production from AGB stars, so the situation is still unclear. Classical novae could also in principle be ^7Li producers, although so far only one detection of ^7Li in novae has been reported (Della Valle *et al.* 2002). From the theoretical point of view, Starrfield *et al.* (1978) had suggested novae as possible ^7Li producers and that the amount of freshly produced Li would depend crucially on the amount of ^3He present in the matter accreted onto the white dwarf. Later on, José & Hernanz (1978) recomputed the ^7Li yields from novae and confirmed that these objects can be Li producers but their predicted Li was not as high as in Starrfield *et al.* (1978). In particular, José & Hernanz (1998) suggested an average mass of Li produced in a nova outburst $< M_{Li} >= (1.8 - 7.5) \cdot 10^{-7} M_\odot$, and considering that each nova has $\sim 10^4$ outbursts during its life, this leads to an average total mass of Li produced by a nova $< M_{Li} >= (1.8 - 7.5) \cdot 10^{-3} M_\odot$. Another possible stellar ^7Li source are supernovae (SN) II: neutrino-induced nucleosynthesis can produce ^7Li (Woosley *et al.* 1990). In particular, SNe II can produce ^7Li in the He-shell by excitation of He by μ and τ neutrinos, produced during the formation of the neutron star in the pre-explosion phase, followed by de-excitation with emission of a proton or a neutron which in turn reacts with He and forms ^7Li. Woosley & Weaver (1995) computed detailed ^7Li yields from SNe II as functions of the initial stellar metallicity. Unfortunately, no detection of ^7Li in SNe has been reported so far. Finally, low-mass giants could in principle produce ^7Li since in some of them it has been measured a high Li abundance. The production of ^7Li in this case should occur during the upper part of the red giant branch (RGB), between the first episode of dredge-up and the tip of the RGB, coupled with mass loss (see de la Reza *et al.* 2000). Unfortunately, only a very small fraction of these stars ($< 5\%$) show overabundances of ^7Li in their atmospheres ($Log \ \epsilon(Li) \sim 4$).

3. ^7Li production in the Big Bang and cosmic rays

During the Big Bang a tiny fraction of ^7Li was produced, whereas ^6Li has been entirely produced by cosmic rays (Reeves 1994). The isotope ^6Li has been detected in halo stars

(Asplund *et al.* 2006), thus suggesting that also some of the original ^7Li in the same stars originates from Galactic cosmic rays (GCRs). Calculations of the amount of ^7Li produced by GCRs can be found in Lemoine *et al.* (1988).

4. Galactic chemical evolution of ^7Li

The abundance of ^7Li measured in the stellar atmospheres of stars which have not yet depleted their original Li should provide, in principle, the history of the Galactic evolution of this element. In fact, the measured ^7Li abundance in very metal-poor halo stars is lower by a factor of ten than the ^7Li abundance measured in young stars, such as the Pleiades. This means that ^7Li has been produced by stars and cosmic rays during Galactic evolution. In Figure 1 we show the famous plot Log $\epsilon(Li)$ vs. [Fe/H], where the ^7Li evolution is given by the upper envelope of the data. The large spread in the disk stars is clearly due to the fact that Li is depleted in different ways in different stars. On the other hand, the halo stars ([Fe/H] <-1.0) show a quite remarkable plateau and a small spread. Such a plateau was discovered first by Spite & Spite (1982) and confirmed later on by other authors with some exceptions (Thornburn 1994; Ryan *et al.* 1999) and the primordial ^7Li value was fixed (Log $\epsilon(Li)$ = 2.1 − 2.3, e.g. Bonifacio *et al.* 2002). More recently, the newest data (28 halo dwarfs observed with UVES at the VLT) seem to indicate that the plateau does not exist any more for [Fe/H] < -3.0 (Sbordone *et al.* 2009, this conference). So the "Spite-plateau" does not exist any more and we should understand how Li has been depleted at an almost constant level in all stars with $-3.0 < [Fe/H] < -1.0$, and then how the depletion increases for [Fe/H] > -3.0. Anyway, the diagram of Figure 1 is the one that was adopted by all the chemical evolution modelists up to now and we will start this review by describing the history of the interpretation of this important diagram. In particular, chemical evolution models should aim at fitting the upper envelope of the diagram of Figure 1. Detailed chemical evolution models follow the evolution in time of the gas abundances in the Galaxy by taking into account the star formation history, the stellar yields and possible gas flows (infall/outflows) (see Matteucci 2001).

4.1. *The interpretation of the Log $\epsilon(Li)$ vs. [Fe/H]*

In principle, two possible interpretations can be suggested for the diagram of Figure 1: i) the ^7Li abundance of the plateau is the primordial Li value, whereas the Li abundance of young stars is the effect of Li production by stars and GCRs during Galactic evolution, ii) the primordial ^7Li abundance is the one measured in young stars, whereas that of the plateau is the consequence of Li depletion in stars. This second hypothesis, however, would require a non-standard Big Bang nucleosynthesis. Mathews *et al.* (1990) explored these two possibilities by means of a detailed chemical evolution model for the Milky Way and concluded that, on the basis of only chemical evolution models, it is difficult to distinguish which hypothesis works better. In Figure 2 we show the model predictions for the two cases: in case i) they started from the primordial abundance of the Spite plateau and assumed carbon stars (low-mass giants) and SNe II as possible ^7Li producers and explored different star formation rates. They found that the observed data could be reproduced by assuming ad hoc yields for both stellar sources, since at that time detailed Li yields were missing. In case ii) Mathews *et al.* (1990) started with a high ^7Li primordial abundance and then assumed a gradual exponential main-sequence destruction of this element. Also in this case the agreement with data is good. In the following years, case ii) was abandoned since it requires a non-standard Big Bang nucleosynthesis not supported by the primordial abundances of the other light elements, and since now on we will refer only to case i). D'Antona & Matteucci (1991) produced a chemical evolution model for the Milky Way, well reproducing the majority of observational constraints, and computing the Galactic evolution of the ^7Li abundance by considering Li production from novae and AGBs. The results are shown in Figure 3 where one can notice how the inclusion of novae as Li producers, reproduces very well the steep rise off the spite-plateau for [Fe/H] > -1.0. This is due to the fact that novae are long-living systems (white dwarfs in binary systems) and contribute to Galactic chemical enrichment with time delays as long as 1-5 Gyr. These authors computed the nova formation rate by assuming that it is a fraction of the rate of formation of the white dwarfs. Then they assumed that each nova has 10^4 bursts during its life and adopted the yields for Li as suggested by Starrfield $et\ al.$ (1978). Under these assumptions the evolution of the ^7Li abundance could be well reproduced just by assuming novae and AGBs as Li producers. In particular, each nova was assumed to produce from 10^{-8} to $10^{-5} M_\odot$ per outburst, thus contributing to more than 50% of the total Li production.

In Figure 4 we show the results of a more recent model from Romano $et\ al.$ (2001), who took into account novae, SNeII, low-mass giants, AGBs, and cosmic rays as ^7Li producers. Here, low-mass giants (M $= 1 - 2.5 M_\odot$) are assumed to produce a fair fraction of the total ^7Li: in fact, it is assumed that each star produces $Log\epsilon(Li) = 4.0$, whereas the AGB stars are producing a negligible amount of Li, as suggested by Ventura $et\ al.$ (1998). Novae produce a non negligible amount of Li, as suggested by José & Hernanz (1998). The Li yields from SNe II are from Woosley & Weaver (1995) and the ^7Li from GCRs is taken from Lemoine $et\ al.$ (1998). It should be noted that the yields of Woosley & Weaver (1995) have been halved in order not to overproduce boron, in agreement with results from Duncan $et\ al.$ (1997.) In Figure 4 the contributions from the different sources are separated. Again, we should note that assuming ^7Li sources with only short lifetimes (SNeII and AGBs) does not reproduce the steep rise off the Spite plateau which is instead obtained by long-living Li producers, such as novae and low-mass stars. In Figure 4 we

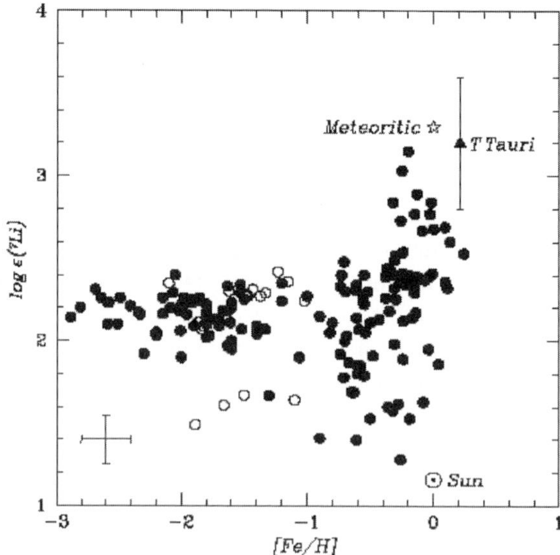

Figure 1. $Log\epsilon(Li)$ versus [Fe/H] for stars in the solar vicinity. The plateau value of $Log\epsilon(Li) \sim 2.2$ for [Fe/H] < -1.0 is interpreted as the primordial value, whereas the highest Li abundance relative to young stars and meteorites is $Log\epsilon(Li) = 3.3$. The value of the Li abundance in the Sun is indicated.

show also the predictions considering all sources together but the GCRs. The conclusions of Romano *et al.* (2001) were that most of Li in the solar system was produced by low-mass stars ($\sim 41\%$), that novae contribute by $\sim 18\%$, SNeII by $\sim 9\%$, AGB stars by only $\sim 0.5\%$ and finally that GCRs contribute for roughly 25%. The same conclusions were not shared by Travaglio *et al.* (2001,) who concluded that the major contributors to ^7Li in the Galaxy are AGB stars with minor contributions from low-mass stars, novae and SNe II. The different approach of Romano *et al.* (2001) and Travaglio *et al.* (2001) about ^7Li production from AGB stars was due to different assumptions about the mass loss in these stars made by these authors. However, not considering long-living Li producers as important, the Travaglio *et al.* model cannot reproduce the steep rise of the Spite plateau. In the following years, the advent of WMAP changed the situation depicted up to now. In particular, the primordial value suggested was Log $\epsilon(Li) = 2.6$ (Spergel *et al.* 2003), higher by a factor of ~ 2 than the value of the Spite plateau, thus implying that

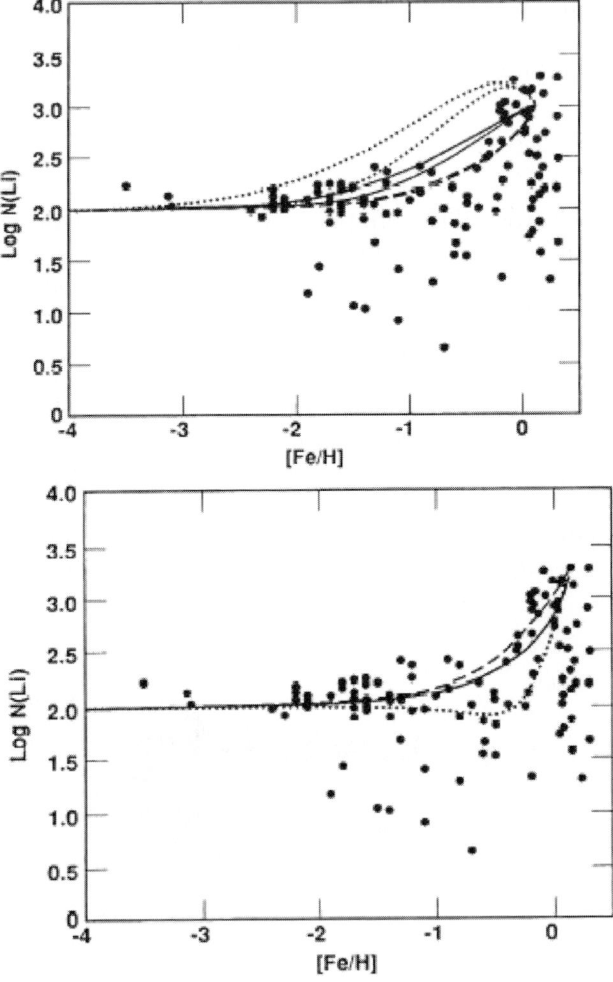

Figure 2. Log N(Li)(the equivalent of Log $\epsilon(Li)$) versus [Fe/H] for stars in the solar vicinity compared with chemical evolution models. Data are from Rebolo *et al.* (1988). In the upper panel are shown model results for case i) and different star formation histories, in the lower panel are shown model results for case ii) and the same star formation rates as for case i). Figure from Mathews *et al.* (1990).

the Li abundance in metal-poor stars is not the primordial one. This can be explained if the primordial Li is depleted in these stars, and some papers suggested that this is possible by means of gravitational settling and also that the amount of depletion is the same in all metal-poor stars, thus explaining the plateau (Richard *et al.* 2005). This was confirmed by observations of globular clusters where this depletion was measured (Korn *et al.* 2006). Unfortunately, more recently new data have appeared (Sbordone *et al.* 2009, this conference) and the situation has became even more complicated since the data have shown that the Spite plateau exists only down to [Fe/H] = –3.0, for lower metallicities a decreasing trend is evident. This finding implies that the Li depletion in very metal-poor stars must be a function of metallicity. In Figure 5 we show a more recent model (Romano *et al.* in preparation) where cosmic rays, novae, SNe II, super- AGBs (7-$11M_\odot$) but not low-mass stars are considered as Li producers. The prescriptions for Li production in GCRs, SNeII and novae are like in Romano *et al.* (2001, 2003) but the Li produced by super-AGBs is a function of stellar metallicity, and increases with metallicity. The yields for Li from super-AGBs are taken from D'Antona (this conference). As one can see, the steep rise off the Spite plateau is well reproduced even without low-mass stars because of novae and AGBs becoming important Li producers only for [Fe/H]> -1.0.

5. Conclusions

We have traced the history of the computation of the evolution of the abundance of ^7Li in our Galaxy. We started from the diagram tracing the evolution of the Li abundance as a function of [F/H] measured in stars of the solar neighbourhood, and interpreted the lower Li abundance in Pop II stars as the primordial Li abundance and that of Pop I stars as the effect of Li production during the galactic lifetime. Several models in the past eleven years have attempted to fit the upper envelope of the $\mathrm{Log}\epsilon(\mathrm{Li})$ vs. [Fe/H], and although the uncertainties still existing in the ^7Li yields from stars are preventing us from drawing firm conclusions, we have understood the following:

• one or more delayed stellar Li sources are necessary to reproduce the steep rise off the Spite plateau,

• possible candidates are classical novae and low-mass stars although these latter are unlikely since only a small fraction shows an overabundance of ^7Li in the atmosphere.

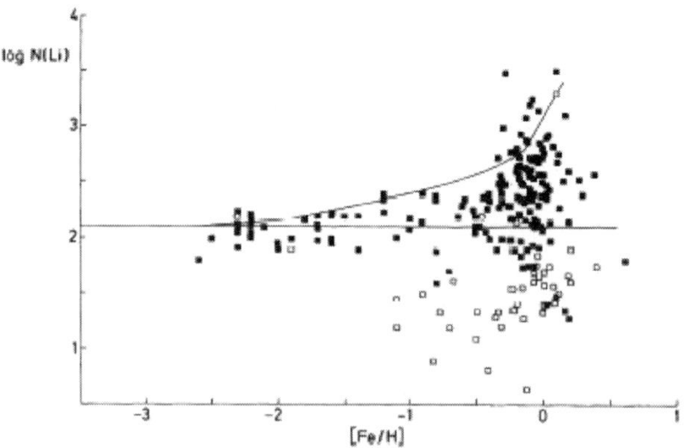

Figure 3. Log N(Li) versus [Fe/H] for stars in the solar vicinity compared with the chemical evolution model of D'Antona & Matteucci (1991) in the framework of case i). The assumed Li producers are novae and ABG stars.

- On the other hand, super-AGB stars (7-11$M_\odot$,) producing more and more ^7Li with increasing stellar metallicity, can be very promising candidates.
- The primordial Li abundance derived from WMAP is higher than the Li abundance of the Spite plateau, thus suggesting that the primordial Li abundance has been depleted in Pop II stars. Several mechanisms to explain such a depletion have been suggested but there is still a potential problem related to the possible detection of ^6Li in Pop II stars (Asplund *et al.* 2006), which should have been more heavily depleted than ^7Li.

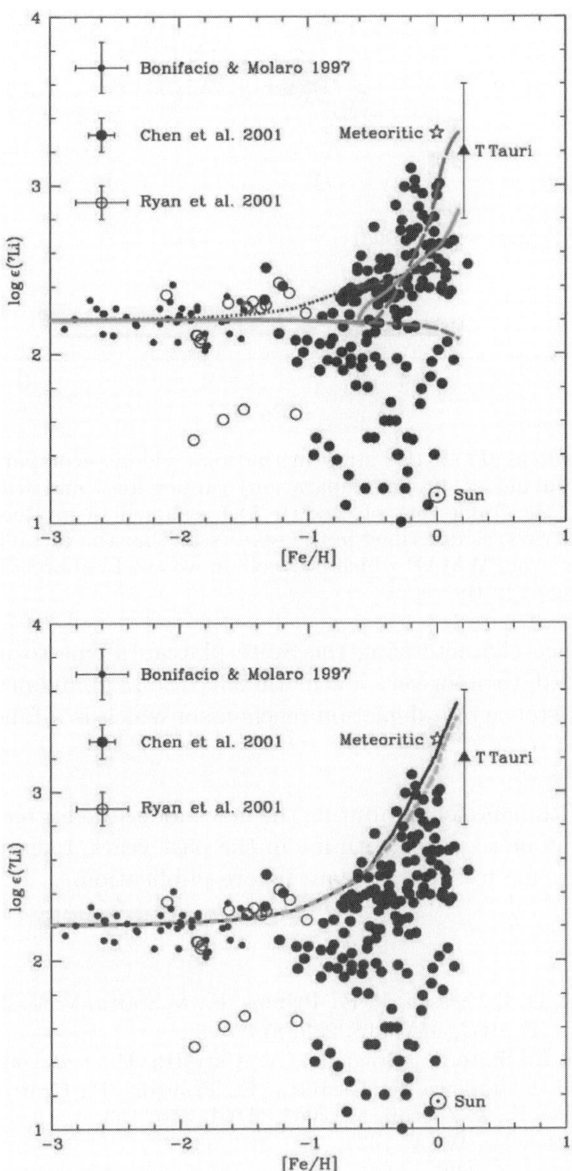

Figure 4. *Log$\epsilon(Li)$* versus [Fe/H] for stars in the solar vicinity compared with the chemical evolution model of Romano *et al.* (2001) in the framework of case i). The assumed Li producers are novae, SNeII, low-mass stars, AGB stars and cosmic rays. The data sources are indicated in the figure.

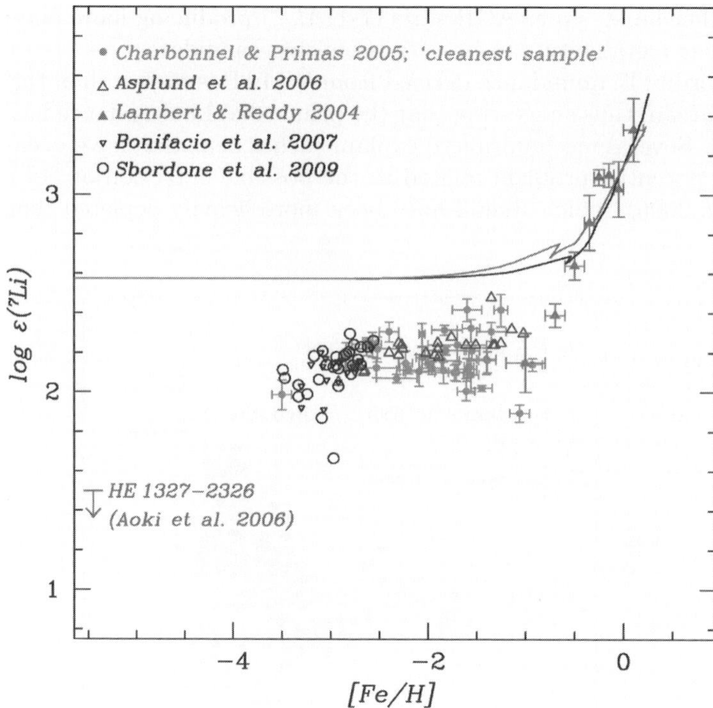

Figure 5. $Log\epsilon(Li)$ versus [Fe/H] for stars in the solar vicinity compared with the chemical evolution model of Romano *et al.* (in preparation) (upper line) and with the Romano *et al.* (2001) (lower line), in the framework of case i). The assumed Li producers are novae, SNeII, AGB stars, and cosmic rays, as described in the text. Note that the theoretical curve starts with the primordial Li value from WMAP which lies well above the Li abundances of halo stars. The data sources are indicated in the figure.

- The Li abundance characterizing the Spite plateau seems to be constant only for $-3.0 < [Fe/H] < -1.0$, then for very low metallicity the Li abundance seems to decrease. This suggests the existence of a depletion mechanism which is a function of metallicity.

Acknowledgements

I thank Donatella Romano for computing the new model and for reading the manuscript and for all the work done together with me in the past years. I also want to thank Luca Sbordone for allowing me to show his data before publication.

References

Asplund, M., Lambert, D. L., Nissen, P. E., Primas, F., & Smith, V. V. 2006, *ApJ*, 644, 229
Bonifacio, P. & Molaro, P. 1997, *MNRAS*, 285, 874
Bonifacio, P. Pasquini, L., Spite, F., Bragaglia, A., Carretta, E. *et al.* 2002, *A&A*, 390,91
Bonifacio, P., Pasquini, L., Molaro, P., Carretta, E., François, P., Gratton, R. G., James, G., Sbordone, L., Spite, F., & Zoccali, M. 2007, *A&A*, 470, 153
Cameron, A. G. W. & Fowler, W. A. 1971, *ApJ*, 164, 111
Charbonnel, C. & Primas, F. 2005, *A&A*, 442, 961
Chen, Y. Q., Nissen, P. E., Benoni, T., & Zhao, G. 2001, *A&A*, 371, 943
D'Antona, F. & Matteucci, F. 1991, *A&A*, 284, 62
de la Reza, R., da Silva, L., Drake, N. A., & Terra, M. A. 2000, *ApJ* (Letters), 535, 115
Della Valle, M., Pasquini, L., Daou, D., & William, R. E. 2002, *A&A*, 155,166

Duncan, D. K., Primas, F., Rebull, L. M., Boesgaard, A. M., Delyiannis, C. P., *et al.* 1997, *ApJ*, 488, 338

Gratton, R. 1991, *Mem. S. A. It.*, 62, 53

Hinshaw, G., Weiland, J. L., Hill, R. S., Odegard, N., Larson, D., *et al.* 2009, *ApJS*, 180, 225

José, J. & Hernanz, M. 1998, *ApJ*, 494, 680

Korn, A. Grundohl, F., Richard, O, Barklem, P. S., Mashonkina, L., Collet, R., Piskunov, N., & Gustafsson, B. 2006, *Nature*, 442, 657

Lambert, D. L. & Reddy, B. E. 2004, *MNRAS*, 349, 757

Lemoine, M., Vangioni-Flam, E., & Cassé, M. 1998, *ApJ*, 499,735

Mathews, G. J., Alcock, C. R., & Fuller, G. M. 1990, *ApJ*, 449, 457

Matteucci, F. 2001, "The chemical evolution of the Galaxy", *AASSL*, Kluwer Academic Publishers

Melendez J., Ramirez, I., Casagrande, L., Asplund, M., Gustafsson, B., Yong, D., do Nascimento, J. D., Jr. *et al.* 2009, astro-ph/090105845

Rebolo, R., Beckman, J. E., & Molaro, P. 1988, *A&A*, 192, 192

Reeves, H. 1994, *Rev. Mod. Phys*, 66, 193

Richard, O., Michaud, G., & Richer, J. 2005, *ApJ*, 619, 538

Romano, D., Matteucci, F., Ventura, P., & D'Antona, F. 2001, *A&A*, 374, 646

Romano, D., Tosi, M., Matteucci, F., & Chiappini, C. 2003, *MNRAS*, 346, 295

Ryan, S. G., Kajino, T., Beers, T. C., Suzuki, T. K., Romano, D., Matteucci, F., & Rosolankova, K. 2001, *ApJ*, 549, 55

Ryan, S. G., Norris, J. E., & Beers, T. C. 1999, *ApJ*, 523, 654

Sackmann, I.-J. & Boothroyd, A. I. 1999, *ApJ*, 510, 217

Smith, V. V. & Lambert, D. L. 1989, *ApJ* (Letters), 354, 75

Smith, V. V. & Lambert, D. L. 1990, *ApJ* (Letters), 361, 69

Spergel, D. N., Verde, L., Peiris, H. V., Komatsu, E., Nolta, M. R. *et al.* 2003, *ApJS*, 148, 175

Spite, F. & Spite, M. 1982, *A&A*, 115, 357

Starrfield, S., Truran, J. W., Sparks, W. M., & Arnould, M. 1978, *ApJ*, 222, 600

Thorburn, J. A. 1994, *ApJ*, 421, 318

Travaglio, C.,Randich, S., Galli, D., Lattanzio, J., Elliott, L. M., Forestini, M., & Ferrini, F. 2001, *ApJ*, 559, 909

Ventura, P., Zeppieri, A., Mazzitelli, I., & D'Antona, F. 1998, *A&A*, 334, 953

Woosley, S. E., Hartman, D. H., Hoffman, R. D., & Haxton, W. C. 1990, *ApJ*, 356, 272

Woosley, S. E. & Weaver, T. A. 1995, *ApJS*, 101, 181

Sylvia Ekström & Cyril Georgy

Ruth Peterson

Light Elements in the Universe
Proceedings IAU Symposium No. 268, 2009
C. Charbonnel, M. Tosi, F. Primas & C. Chiappini, eds.

© International Astronomical Union 2010
doi:10.1017/S1743921310004631

Lithium, beryllium, and boron production in core-collapse supernovae

Ko Nakamura[1], Takashi Yoshida[2], Toshikazu Shigeyama[2,3] and Toshitaka Kajino[1,2]

[1] National Astronomical Observatory of Japan,
Osawa 2-21-1, Mitaka, Tokyo, 181-8588 Japan
email: nakamura.ko@nao.ac.jp

[2] Department of Astronomy, Graduate School of Science, University of Tokyo,
Hongo 7-3-1, Bunkyo-ku, Tokyo, 113-0033 Japan

[3] Research Center for the Early Universe, Graduate School of Science, University of Tokyo

Abstract. Type Ic supernova (SN Ic) is the gravitational collapse of a massive star without H and He layers. It propels several solar masses of material to the typical velocity of 10,000 km/s, a very small fraction of the ejecta nearly to the speed of light. We investigate SNe Ic as production sites for the light elements Li, Be, and B, via the neutrino-process and spallations. As massive stars collapse, neutrinos are emitted in large numbers from the central remnants. Some of the neutrinos interact with nuclei in the exploding materials and mainly ^{7}Li and ^{11}B are produced. Subsequently, the ejected materials with very high energy impinge on the interstellar/circumstellar matter and spall into light elements. We find that the ν-process in the current SN Ic model produces a significant amount of ^{11}B, consistent with observations if combined with B isotopes from the following spallation production.

Keywords. Neutrinos, nuclear reactions, nucleosynthesis, abundances – Supernovae: general

1. Introduction

Most of metals up to iron are synthesized inside main sequence stars via nuclear reactions called nuclear burning. Newly-synthesized elements are ejected into interstellar space and pollute the interstellar gas primitively consisted of hydrogen and helium. However, light elements such as Li, Be, B and their isotopes (LiBeB) are hardly produced by nuclear burning partly because of Li fragile property and absence of stable nuclei with atomic mass number of eight. Some processes considered to mainly produce these elements are briefly summarized below.

Shortly after the Big Bang, protons and neutrons combine and produce hydrogen and helium isotopes. Subsequently, tritium and ^{3}He combine with α-particle to form ^{7}Li (and ^{7}Be eventually decaying to ^{7}Li), which make an almost constant Li abundance observed on the surfaces of metal-poor halo stars, the so-called Spite Plateau (Spite & Spite 1982).

After star formation and its gravitational contraction, hydrogen burning reaction is ignited at the center. Although light elements synthesized at the hot regions of stars are fated to be broken by surrounding hot protons, an escape route for them are known. Cameron proposed a mechanism (Cameron 1955) available for AGB stars, called Be-transport mechanism, where ^{7}Be in hydrogen burning shell are brought to cool envelopes by the circulation currents during a thermal pulse phase. A fraction of survived ^{7}Be (^{7}Li) capture α-particle and form ^{11}B, both of them are ejected into the interstellar space through stellar winds. This mechanism is effective for stars with masses in the range $4 \leqslant M/M_\odot \leqslant 6$. Some computational works, however, have revealed that lithium from AGB stars are not important for the Galactic chemical evolution (Romano *et al.* 2001; Ventura *et al.* 2002). Note that in some cases AGB stars show low, not high, lithium abundances caused by dilution during the previous evolutionary phases (Maceroni *et al.* 2002). For more massive stars, they are known to collapse and explode as supernovae at the end of their lives. Most of the gravitational energy released at the explosion of a massive star is carried away by neutrinos emitted from the central remnant. Their number is achieved to more than 10^{58}. Although cross sections of neutrino-nucleus reactions are very small, such a huge number of neutrinos enable to contribute the increase in the yields of some species of nuclei. This is called the ν-process (Woosley *et al.* 1990). The ν-process mainly contributes to the production of ^{7}Li and ^{11}B among light elements (e.g. Woosley *et al.* 1990; Heger *et al.* 2005; Yoshida *et al.* 2005, 2008).

The processes described above predominantly produce ^{7}Li and ^{11}B, while cosmic-ray (CR) nucleosynthesis produces all stable isotopes of LiBeB (e.g. Meneguzzi *et al.* 1971; Suzuki & Inoue 2002; Rollinde *et al.* 2008). In CR nucleosynthesis, low-energy CRs accelerated in shocks interact with the ambient medium and produce LiBeB via spallations (H,α+ C,N,O $\rightarrow$ LiBeB) or fusion reaction of α-particles ($\alpha + \alpha \rightarrow ^{6,7}$Li). Theoretical CR nucleosynthesis models predict strong correlations between BeB abundances and metallicity, consistent with observations although it is difficult to explain the observed linear dependences if the Galactic CRs consist of protons and α-particles.

Despite a great deal of efforts to investigate these processes, the origins of LiBeB are not fully understood and there remain some observational features conflicting with theoretical predictions. For instance, theoretical predictions for LiBeB production from Galactic CRs indicate quadratic relations between B(Be) abundances and metallicity, while observations clearly show linear relations. To consist with observational trend, CNO CRs from superbubbles (Higdon *et al.* 1998) and Type Ic supernovae (Fields *et al.* 2002; Nakamura & Shigeyama 2004) have been suggested. However, the CR spallations cannot be the only source of boron isotopes because the CR spallations do not reproduce the high ratio of ^{11}B to ^{10}B observed in meteorites. Therefore another ^{11}B source is necessary.

To solve these unsolved problems and consider the Galactic chemical evolution concerning with the light elements, it is necessary to accurately evaluate the contribution of each LiBeB productive process. Here we focus on the B isotopes from core-collapse SNe, in particular energetic Type Ic SNe (SNe Ic). The progenitor of an SN Ic is a C/O star and its H and He envelopes have been stripped during the stellar evolution. Although the explosion mechanism of SNe Ic has not clarified, neutrinos should be one of main carriers of the gravitational energy released from the collapsing core. The neutrino emission from a collapsing proto-neutron star evolved from a $\sim 40 M_\odot$ star has been investigated (Sumiyoshi *et al.* 2006). Even in the case that the central core of such a star becomes a black hole, a temporally formed proto-neutron star emits a huge amount of neutrinos before collapsing to the black hole. On the other hand, after the black hole formation, neutrinos are considered to be emitted from accretion disk surrounding the newly formed black hole (e.g. Kohri *et al.* 2005; Surman & McLaughlin 2005). Therefore, a huge amount of neutrinos are emitted from the central region of an SN Ic and the ν-process is expected to occur in the exploding supernova material. We investigate the ν-process in SN Ic for the first time. In addition, LiBeB production through spallations between the SN ejecta and the interstellar/circumstellar matter is considered. In §2, we present our calculations of supernova explosion (§2.1), nucleosynthesis by the ν-process (§2.2), and by spallations (§2.3). Section §3 discusses our results and clarifies the important role of SNe Ic in the light element synthesis.

2. Calculations and results

2.1. *Supernova explosion*

Here we consider a very energetic explosion of a 15 $M_\odot$ C/O star with the explosion energy $E_{\mathrm{ex}} = 3 \times 10^{52}$ ergs corresponding to SN 1998bw (Nakamura *et al.* 2001). The explosion energy is released at the center of the progenitor star as thermal energy. Resulting shock wave accelerates the stellar materials and explode the progenitor as a supernova. Time evolution of physical quantities in the progenitor are calculated with 1-dimensional hydrodynamic code taking the effects of special relativity into account. We solve the special relativistic hydrodynamic equations in Lagrangian coordinates with an ideal equation of state involving gas and radiation pressure. Adiabatic indices are treated as functions of pressure and gas density. Details on the numerical code are described in Nakamura & Shigeyama (2004).

2.2. *The Neutrino-Process*

Light element synthesis in the Type Ic supernova is calculated as a post process. We use a nuclear reaction network consisting of 291 species of nuclei and taking into account the ν-process (Yoshida *et al.* 2008).

Neutrinos are considered to be emitted from a collapsing proto-neutron star (e.g. Sumiyoshi *et al.* 2006) and/or the innermost region just above a black hole (e.g. Surman & McLaughlin 2005). However, properties of the neutrinos, particularly in SNe Ic, are still uncertain. So, we use the following neutrino model extended from a supernova neutrino model taken in Yoshida *et al.* (2008). We suppose that neutrino luminosity decreases exponentially with the decay time of 3 s while the neutrino temperature of each species does not change with time. The total energy carried out by neutrinos is assumed to be 3×10^{53} ergs.

We consider two models for these constant neutrino temperatures. In the first model, the temperatures of ν_e, $\bar{\nu}_e$, and $\nu_{\mu,\tau}$ and $\bar{\nu}_{\mu,\tau}$ are set to be $T_{\nu_e} = 3.2$ MeV, $T_{\bar{\nu}_e} = 5$ MeV, and $T_{\nu_{\mu,\tau}} = 6$ MeV, respectively. This temperature model is referred to as the "normal" $T_{\nu_{\mu,\tau}}$ model. The light element synthesis in an $\sim 20 M_\odot$ Type II supernova with the normal temperature model well reproduces the supernova contribution of the ^{11}B production during Galactic chemical evolution (Yoshida *et al.* 2005, 2008). The third peak of r-process elements is also achieved in neutrino-driven wind models with this temperature model. In the second model, the temperature of $\nu_{\mu,\tau}$ and $\bar{\nu}_{\mu,\tau}$ is set to be $T_{\nu_{\mu,\tau}} = 8$ MeV. The same values as the normal model are adopted for T_{ν_e} and $T_{\bar{\nu}_e}$. This temperature model is referred to as the high $T_{\nu_{\mu,\tau}}$ model. A newly forming proto-neutron star in an SN Ic should be more compact than that of an SN II. Therefore, the neutrino temperature of the collapsing proto-neutron star is larger than that of SNe II (e.g. Sumiyoshi *et al.* 2006).

The mass fraction distributions of ^{11}B, ^{11}C, ^{10}B, ^{7}Li, and ^{7}Be in the SN Ic are shown in Figure 1. The mass fractions are of the order of $\sim 10^{-8}$ for ^{11}B and ^{11}C and $\sim 10^{-11} - 10^{-10}$ for ^{10}B, ^{7}Li, and ^{7}Be. The yield of each species strongly depends on the corresponding branching ratio of the reactions. The light elements are mainly produced in the O/Ne layer ($M_r > 7.4 M_\odot$). The mass fractions in the O/Si layer ($4.6 M_\odot < M_r < 7.4 M_\odot$) are smaller than in the outer layer. The Si/S layer ($3.7 M_\odot < M_r < 4.6 M_\odot$) where carbon burned out indicates no light elements. The mass fractions of some light elements are also large in the innermost region where α-rich freeze out is achieved ($M_r < 2.6 M_\odot$) and complete Si-burning occurs ($2.6 M_\odot < M_r < 3.5 M_\odot$). There are no qualitative differences in the mass fraction distributions between the normal and high $T_{\nu_{\mu,\tau}}$ models.

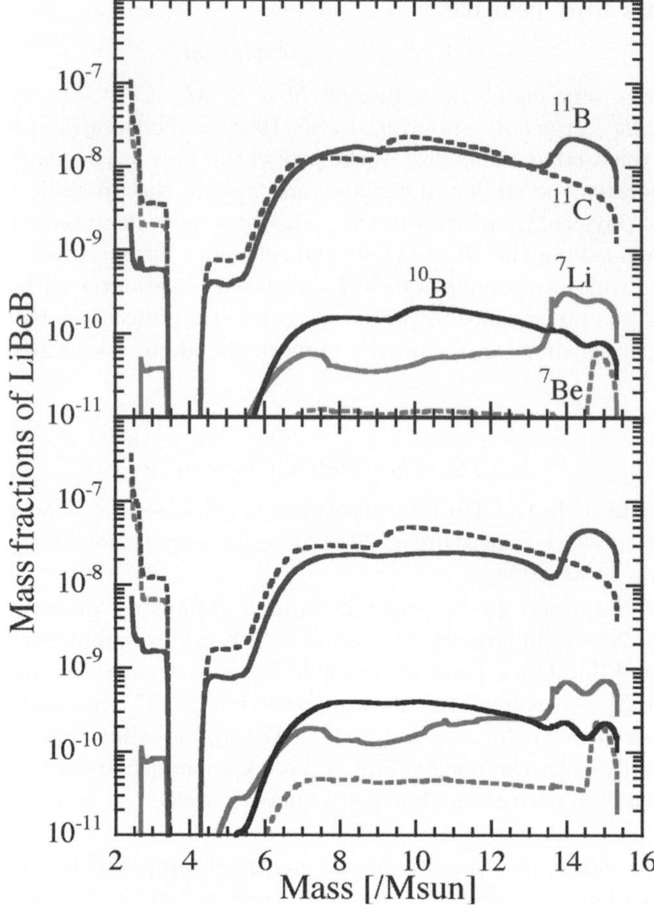

Figure 1. The mass fraction distributions of light elements produced via the ν-process as functions of the mass coordinate. Shown are the cases of the SN 1998bw model ($M_{\rm CO} = 15 M_\odot$, $E_{\rm ex} = 3 \times 10^{52}$ ergs) with $T_{\nu_e} = 3.2$ MeV, $T_{\bar\nu_e} = 5$ MeV, and $T_{\nu_{\mu\tau}} = 6$ MeV (*top*) and 8 MeV (*bottom*) at 75 seconds after the energy release.

The mass fraction of each species in the high $T_{\nu_{\mu,\tau}}$ model at a given mass coordinate is roughly twice as the corresponding one in the normal $T_{\nu_{\mu,\tau}}$ model.

Some of light elements produced through ^{12}C-neutrino reactions are decomposed by explosive nucleosynthesis occurring immediately after the shock arrival. ^{11}B, ^{11}C, and ^{10}B are decomposed by collisions with protons and α-particles. ^{7}Li and ^{7}Be are photo-disintegrated to ^{3}H and ^{3}He, respectively. After the shock passage, the exploding and cooling materials are still irradiated by neutrinos, so that light elements are produced through the ν-process again. We note that the mass fractions of ^{11}B, ^{7}Li, and ^{7}Be in the outermost region ($M_r > 13.5 M_\odot$) are larger than the corresponding ones in the inner region because they are not decomposed in explosive nucleosynthesis.

In the innermost region, light elements are produced after the termination of the nuclear statistical equilibrium. The main product in α-rich freeze out is ^{4}He. About 20% of ^{4}He by mass fraction is also produced in complete Si-burning. The ν-process of ^{4}He produces ^{3}H and ^{3}He in cooling materials. The α-capture reactions followed by the ν-process produce ^{7}Li and ^{7}Be. Furthermore, ^{11}B and ^{11}C are produced by α-captures and ^{10}B is produced through ^{7}Be$(\alpha, p)^{10}$B.

In both cases of $T_{\nu_{\mu,\tau}}$ a significant amount of ^{11}B is synthesized ($2.69 \times 10^{-7} M_\odot$ for the normal $T_{\nu_{\mu,\tau}}$ model and $5.46 \times 10^{-7} M_\odot$ for the high $T_{\nu_{\mu,\tau}}$ model). It should be noted that the amounts of ^{11}B produced through the ν-process include those of ^{11}C which decays to ^{11}B with a half life of 20 minutes. The yield of ^{7}Li is on the order of $10^{-9} - 10^{-8} M_\odot$ in the SN Ic model. This yield is much smaller than that produced in SNe II (e.g. Yoshida *et al.* 2008). Most of ^{7}Li in SNe II is produced through the ν-process of ^{4}He and the following α-capture reactions in the He-rich layer. On the other hand, almost all H and He layers of SN Ic progenitors have been stripped via stellar wind and/or binary effect before explosion. The yields of ^{6}Li, ^{9}Be, and ^{10}B are on the order of or below $10^{-9} M_\odot$. They are also much smaller than the ^{11}B yield. This is due to smaller branching ratios of neutrino-^{12}C reactions (Yoshida *et al.* 2008).

2.3. *Spallation reactions*

Supernova explosions accelerate and expel the stellar materials into the interstellar space, polluting the universe with newly synthesized elements. The progenitors of SNe Ic are so compact that a very small fraction of ejecta can be accelerated nearly to the speed of light, interact with the interstellar matter (Nakamura & Shigeyama 2004) or the circumstellar matter (Nakamura *et al.* 2006), and produce the light element isotopes via spallation reactions of CNO with protons or α-particles. The surface layers of SNe Ic, which are composed of C and O, impinge on the H or He nuclei in the ambient medium and spall into LiBeB. Here we use the result for a SN 1998bw model constructed by Nakamura & Shigeyama (2004). 0.3% ($0.04 M_\odot$) of the ejecta attain enough energy ($\gtrsim 10$ MeV/A) to undergo spallations (see eq. 2.2). They solved the transfer equation for each element i expressed as

$$\frac{\partial F_i(\epsilon, t)}{\partial t} = \frac{\partial[\omega_i(\epsilon)F_i(\epsilon, t)]}{\partial \epsilon} - \frac{F_i(\epsilon, t)}{\Lambda}\rho v_i(\epsilon), \tag{2.1}$$

where Λ is the loss lengths in g cm^{-2}, ρ denotes the mass density of the ISM, $v_i(\epsilon)$ the velocity of the element i with an energy per nucleon of ϵ. The initial condition for the mass of element i with an energy per nucleon ϵ at time $t = 0$, $F_i(\epsilon, t = 0)$, is derived from the numerical calculations of explosions or an empirical formula

$$\frac{M(> \epsilon)}{M_{\rm ej}} = A \left(\frac{E_{\rm ex}/10^{51}\,{\rm ergs}}{M_{\rm ej}/1\,M_\odot}\right)^{3.4} \times \left(\frac{\epsilon}{10\,{\rm MeV}}\right)^{-3.6}, \tag{2.2}$$

where $M_{\rm ej}$ is the mass of the ejecta and the constant A is equal to 1.9×10^{-4} for the current model. Then the yield of a light element l via the $i + j \to l + \cdots$ reaction is estimated from

$$\frac{dN_l}{dt} = n_j \int \sigma_{i,j}^l(\epsilon) \frac{F_i(\epsilon, t)}{A_i m_p} v_i(\epsilon) d\epsilon, \tag{2.3}$$

where N_l is the number of the produced light element l, n_j is the number density of an element j in the interstellar matter, $\sigma_{i,j}^l$ the cross section of $i + j \to l + \cdots$ reaction given by Read & Viola (1984), A_i the mass number of the element i. The interstellar matter is assumed to be composed of neutral H and He with the number densities of $n_{\rm H} = 1$ cm^{-3} and $n_{\rm He} = 0.1$ cm^{-3}.

Their results show that the mass of ^{11}B produced via spallation reactions is $1.34 \times 10^{-6} M_\odot$, which is larger than that synthesized via the ν-process even in the high $T_{\nu_{\mu,\tau}}$ model. The resultant isotopic ratio of ^{10}B/^{11}B ($\sim 1/3$) in this model is predominantly determined by the ratio of cross sections of the reaction p,α + O $\to^{10}$B to that of p,α + O $\to^{11}$B.

3. Conclusions

Nakamura & Shigeyama (2004) concluded that SNe Ic may not play an important role in B isotope productions because of the low isotope ratios $^{11}\mathrm{B}/^{10}\mathrm{B} \sim 2.8$ compared with observations in meteorites (4.05 ± 0.05), and other $^{11}\mathrm{B}$ sources like the ν-process in SNe II are necessary. We estimate the production of light elements including boron isotopes via the ν-process in an SN Ic and find that SNe Ic can be as powerful $^{11}\mathrm{B}$ producers as SNe II. The resulting number ratios of B isotopes from both the ν-process and spallations $^{11}\mathrm{B}(+^{11}\mathrm{C})/^{10}\mathrm{B} = 3.66 - 4.28$ well match with observations. It is reasonable to assume the small radii of neutrino spheres because of the compactness of SN Ic progenitors, leading to the high temperatures of emitted neutrinos and anti-neutrinos. High $T_{\nu_{\mu\tau}}$ results in high LiBeB yields through neutral current reactions in the ν-process, which raises the potential role of SNe Ic in the light element production.

The combined mechanism of light element production described here, the ν-process and spallations, may also be effective in the case of Type Ib supernovae where progenitor stars without H-rich envelope explode. In such a case, the outermost layers consist of He and the α-particle fusion reaction producing lithium isotopes become important. Furthermore, Meynet & Maeder (2002) suggested that the He layers of very metal-poor stars contain primary nitrogen produced during the He-burning phase by the CNO cycle in the H-burning shell stimulated by rotationally-induced diffusion of carbon into the H-burning shell. This nitrogen may enhance the light element production via spallations because the cross sections of spallation reactions involving nitrogen tend to have relatively low energy thresholds and high peak values. Observationally some low-metallicity halo stars have been found to have high abundances of $^{6}\mathrm{Li}$ or $^{9}\mathrm{Be}$ beyond the theoretical predictions , which engages our interest in SNe Ib even if they might not accelerate their envelopes effectively because their progenitors are not so compact as those of SNe Ic.

References

Cameron, A. G. W. 1955, *ApJ*, 121, 144

Fields, B. D., Daigne, F., Cassé, M., & Vangioni-Flam, E. 2002, *ApJ*, 581, 389

Heger, A., Kolbe, E., Haxton, W. C., Langanke, K., Martínez-Pinedo, G., & Woosley, S. E. 2005, Phys. Lett. B, 606, 258

Higdon, J. C., Lingenfelter, R. E., & Ramaty, R. 1998, *ApJl*, 509, L33

Kohri, K., Narayan, R., & Piran, T. 2005, *ApJ*, 629, 341

Maceroni, C., Testa, V., Plez, B., García Lario, P., & D'Antona, F. 2002, *A&A*, 395, 179

Meneguzzi, M., Audouze, J., & Reeves, H. 1971, *A&A*, 15, 337

Meynet, G. & Maeder, A. 2002, *A&A*, 390, 561

Nakamura, K. & Shigeyama, T. 2004, *ApJ*, 610, 888

Nakamura, K., Inoue, S., Wanajo, S., & Shigeyama, T. 2006, *ApJl*, 643, L115

Nakamura, T., Mazzali, P. A., Nomoto, K., & Iwamoto, K. 2001, *ApJ*, 550, 991

Read, S. M. & Viola, V. E. 1984, Atomic Data and Nuclear Data Tables, 31, 359

Rollinde, E., Maurin, D., Vangioni, E., Olive, K. A., & Inoue, S. 2008, *ApJ*, 673, 676

Romano, D., Matteucci, F., Ventura, P., & D'Antona, F. 2001, *A&A*, 374, 646

Spite, F. & Spite, M. 1982, *A&A*, 115, 357

Sumiyoshi, K., Yamada, S., Suzuki, H., & Chiba, S. 2006, Phys. Rev. Lett., 97, 091101

Surman, R. & McLaughlin, G. C. 2005, *ApJ*, 618, 397

Suzuki, T. K. & Inoue, S. 2002, *ApJ*, 573, 168

Ventura, P., D'Antona, F., & Mazzitelli, I. 2002, *A&A*, 393, 215

Woosley, S. E., Hartmann, D. H., Hoffman, R. D., & Haxton, W. C. 1990, *ApJ*, 356, 272

Yoshida, T., Kajino, T., & Hartmann, D. H. 2005, Phys. Rev. Lett., 94, 231101

Yoshida, T., Suzuki, T., Chiba, S., Kajino, T., Yokomakura, H., Kimura, K., Takamura, A., & Hartmann, D. H. 2008, *ApJ*, 686, 448

Light Elements in the Universe
Proceedings IAU Symposium No. 268, 2009
C. Charbonnel, M. Tosi, F. Primas & C. Chiappini, eds.

© International Astronomical Union 2010
doi:10.1017/S1743921310004643

The search for the origin of the light nuclei Li, Be, B

Hubert Reeves[1]

CNRS, France

Abstract. My aim is to show how the abundance ratios of the light elements (6 to 11) are related to the properties of the strong nuclear interaction and, in particular, to the major influence of closed shells of neutrons and protons, (the magic numbers : 2, 8, etc) on the binding energies of the nuclei.

Keywords. Nuclear reactions, nucleosynthesis, abundances – ISM: cosmic rays

1. Introduction

The α-particle has a very high binding energy, resulting from its doubly magic structure : two protons and two neutrons. This fact results, on the one hand, on the relatively high binding energies of the nuclei containing the equivalent of an integer number of α-particles (C-12, O-16, Ne-20, Mg-24, Si-28 : the so-called α-nuclei) but also on the very low binding energy of the light nuclei : lithium, beryllium, and boron, the most extreme case being the instability of the mass-5 and of the mass-8. Immediately after their formation these nuclei rapidly decay in an α-mix (a group of nuclei containing the maximum number of α-particles.)

As shown in Table 1, Be-9 is hardly bound, and B-11 in more bound than B-10. Lithium -6 and Lithium-7 are quite comparable to the Bs. These low binding energies with respect to the α-mix are the reasons why these elements can not withstand the high temperatures of the stellar interiors.

2. Where were they made?

There are several hints for the identification the formation mechanisms of lithium, beryllium, and boron.

Hint no 1: The amplitude of the nuclear cross sections for their formation by high energy protons on heavy nuclei (spallation) are closely related to their cosmic abundances.

We consider high energy processes in a cold region. For instance cosmic ray protons on C-12 or 0-16 on interstellar gas. The physics of these reactions is best described by the Enrico Fermi hot ball model. The target nucleus is first heated to very high temperatures by the energy of the incident bolid. It thermalizes and cools by evaporation of part of

Table 1. Table of binding energies of the light nuclei with respect to an α-mix The number in MeV gives the energy release by the break up of the nucleus in a group of light particles containing the maximum amount of α-particles (dubbed an α-mix)

Li-6	2.45 MeV
Li-7	1.24 MeV
Be-9	88 KeV
B-10	5.97 MeV
B-11	11.2 Mev

469

its nuclear constituents (neutrons, protons, alphas). The residual nucleus can be Li, Be, or B.

One of the tenets of quantum physics is that the probability of a reaction leading to a specific result is proportional to the number of equivalent ways of producing this result. This, in turn, is a measure of the volume of phase space available for the reactions, hence of the amount of kinetic energy released by the reactions. The products with low binding energies release a larger volume of space phase and have hence a larger formation probability. This is well illustrated when one considers the cross-sections for the formation of the light nuclei by high energy protons on C-12. As expected, Be has the smallest value, followed by Li and B. The cross-section for B-11 is larger than for B-10. The case of Li will be considered later.

Hint no 2: These nuclei are much more abundant in galactic cosmic rays than in nature. Note that boron is more abundant than nitrogen!

3. Nuclear physics data

In the 1960 when this model was presented, practically no data on the nuclear cross sections were available. In the following years I visited a number of nuclear laboratories (Los Alamos, CERN, Darmstad, Berkeley, Moscow, Leningrad, Erivan) to present this proposal. The reactions of the nuclear physicists was typically "Yes, this is interesting but because of the need of obtaining highly purified targets, it would take many years to complete this project. We can not get involved for such a long period of time."

Fortunately, in 1965, I met a group of physicists of Orsay, France led by René Bernas, who were already pursuing such experiments. We joined forces to interpret the results in the nucleosynthesis context.

Taking into account the fact that each accelerator works at a given energy, and that the cosmic ray particles span a large range of energies, from MeV to GeV and more, it became necessary for the nuclear physicists to bring their isotope separators to different accelerators in France, Germany, USSR, and Switzerland (CERN) in order to patiently build the full cross sections of the various reactions involved.

I want here to give credit to these scientists by giving their names : Marcelle Epherre, Elie Gradstajn, François Yiou, Robert Klapisch.

4. Comparison of abundances with the calculated values

Calculations were then performed of the abundances and isotopic ratios, taking into account the spectrum of energy in galactic cosmic rays and the nuclear cross sections obtained in the laboratories. The results and their astrophysical implications are discussed in the following.

The beryllium abundance can be taken as a measure of the integral activity of the bombardment of the cosmic rays throughout the life of our Galaxy. Indeed the product of the present formation rate of Be times the life of the Galaxy yields an abundance quite comparable to the observed abundance.

The relative probabilities of formation of the various isotopes can be used to evaluate the contribution of GCR to the other light isotopes. We define a term $F(obs/calc)(x/y)$ as the ratio of the observed abundances of two nuclei x and y over the calculated ratio of their formation rates in GCR.

Boron over beryllium

The observed ratio in the solar system SS (sun and meteorites) is 23. The calculated GCR ratio is 15. We have also observations of this ratio in the metal poor stars of ~ 20, constant with metallicities down to (-3) (one thousand solar).

The observed isotopic ratio in the solar system is $11/10 = 4.0$, while the formation ratio is 2.5. We probably need an extra source of B-11, which could come from stellar processes. One possibility is the C-12 desintegration by neutrino collisions in supernovae which would generate B-11 but no B-10. However the uncertainties involved in the calculations of the yields are so large that no meaningful comparison can be made.

We consider the case of B-10.

$$F = (10/9)obs/(10/9)calc = (116/22)/5 = 1.04$$

showing that this isotope is a pure GCR product.

Lithium 7/lithium-6

The GCR ratio is ~ 2 while the SS is 12.5. This was the first indication of a possible other source of Lithium-7. Thanks to the observations of François and Monique Spite, we associate this source to the Big Bang Nucleosynthesis. Studies of the fossil radiation (CMB) by the satellite WMAP have confirmed this association.

However, since the contribution of BBN to the formation of Li-6 is negligible, we may ask if this element is a pure product ot GCR. Computing the ratio of its abundance to the Be-9 abundance we find:

$$F(6/9)(obs)/calc) = 6.5/5 = 1.3$$

confirming the value of the hypothesis.

Some observations of lithium in interstellar gas, in particular around the star Rho Ophiucus, have locally given values much smaller than the solar system values. One possibility is the collision of fast alphas on Helium-4 which generates Li-6 and Li-7 with ratios around 2. We note that the respective cross-sections involve alphas of energies around ten to one hundred Mev per nucleon, appreciably smaller than the mean particle energy in the GCR (hundreds of Mev). This may involve special astrophysical contexts such as the activity of high energy emitting particles stars in these area.

To complete this operation we compute the ratio of B-10 to Li-6

$$F(10/6)obs/calc = (116/144)/1 = 0.8$$

confirming that the nuclei of Li-6, Be-9, and B-10 are pure products of the GCR.

5. Conclusions

It is of interest to note that the three nucleosynthetic processes : Big Bang, Galactic cosmic rays, and stellar nucleosynthesis cover suitably well the light elements abundances $(A<12)$.

The Big Bang is responsible for the atoms with mass number 2, 3, most of 4, and 7 (mostly in old stars).

The GCR is responsible for 6, 9, 10, and also a part of 7 and most of 11.

Stellar nucleosynthesis is responsible for most of 7, some 4, and perhaps some 3 and 11. And also of the elements from carbon to the heaviest elements : thorium and uranium.

References

The references are given in the paper of Nicolas Prantzos.

Hubert Reeves with some of his students and grand-students:
Thibaut Decressin, Ana Palacios, Nikos Prantzos, Sylvie Vauclair
Nadège Lagarde, Thierry Montmerle, Corinne Charbonnel

André Maeder & Sylvie Vauclair

Light Elements in the Universe
Proceedings IAU Symposium No. 268, 2009
C. Charbonnel, M. Tosi, F. Primas & C. Chiappini, eds.

© International Astronomical Union 2010
doi:10.1017/S1743921310004655

Origin of cosmic rays and evolution of spallogenic nuclides Li, Be and B

Nikos Prantzos

Institut d' Astrophysique de Paris
98bis, Bd. Arago, 75014 Paris
email: `prantzos@iap.fr`

Abstract. A short overview is presented of current issues concerning the production and evolution of Li, Be and B in the Milky Way. In particular, the observed "primary-like" evolution of Be is re-assessed in the light of a novel idea: it is argued that Galactic Cosmic Rays are accelerated from the wind material of *rotating* massive stars, hit by the forward shock of the subsequent supernova explosions. The pre-galactic levels of both Li isotopes remain controversial at present, making it difficult to predict their Galactic evolution. A quantitative estimate is provided of the contributions of various candidate sources to the solar abundance of Li.

1. Introduction

The idea that the light and fragile elements Li, Be and B are produced by the interaction of the energetic nuclei of galactic cosmic rays (CGR) with the nuclei of the interstellar medium (ISM) was introduced 40 years ago (Reeves *et al.* 1970, Meneguzzi *et al.* 1971, hereafter MAR). In those early works it was shown that, taking into account the relevant cross-sections and with plausible assumptions about the GCR properties - source composition, intensity and spectrum - one may reproduce reasonably well the abundances of those light elements observed in GCR and in meteorites (pre-solar).

Among the required ingredients for such a calculation, the relevant spallation cross sections of CNO nuclei are accurately measured in the laboratory. The source composition and the equilibrium energy spectrum of GCR are inferred from a combination of observations and models of GCR propagation in the Milky Way (e.g. in the framework of the so-called "leaky box" model). Once the equilibrium spectra of GCR in the ISM are established, the calculation of the resulting abundances of LiBeB is straightforward, at least to first order†. The production rate (s^{-1}) of the abundance $Y_L = N_L/N_H$ (by number) of LiBeB nuclei is given by

$$\frac{dY_L}{dt} = F_{p,a}^{GCR}\sigma_{pa+CNO}Y_{CNO}^{ISM} + F_{CNO}^{GCR}\sigma_{pa+CNO}Y_{p,a}^{ISM}P_L + F_a^{GCR}\sigma_{a+a}Y_a^{ISM}P_L \quad (1.1)$$

where: F (cm^{-2} s^{-1}) is the average GCR flux of protons, alphas or CNO, Y the abundances by number of those nuclei in the ISM, and σ (cm^2) is the average (over the equilibrium energy spectrum of GCR) cross-section for the corresponding spallation reactions producing LiBeB. The first term in the right hand member of this equation (fast protons and alphas hitting CNO nuclei of the ISM) is known as the "direct" term, the second one (fast CNO nuclei being fragmented on ISM protons and alphas) is the "reverse" term and the last one involves "spallation-fusion" reactions, concerning only the Li isotopes. P_L is the probability that nuclide L (produced at high energy) will be

---

† The full calculation should include production by spallation of other primary and secondary nuclides, such as ^{13}C; however, this has only second order effects.

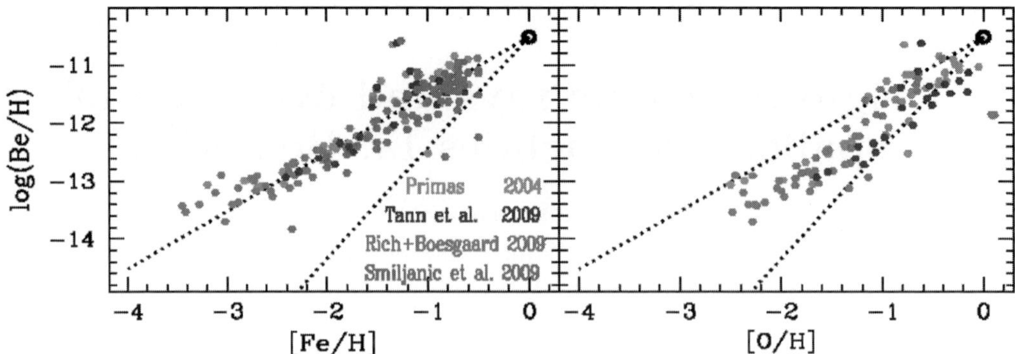

Figure 1. Observations of Be vs. Fe (*left*) and vs. O (*right*). In both panels, dotted lines indicate slopes of 1 (primary) and 2 (secondary). Be clearly behaves as a primary vs. Fe, whereas there is more scatter in the data vs. O.

thermalized and remain in the ISM (see, e.g. Prantzos 2006). Obviously, the GCR flux term $F_{CNO}^{GCR} \propto Y_{CNO}^{GCR}$ is proportional to the abundances of CNO nuclei in GCR, a fact of paramount importance for the evolution of Be and B (see next sections).

Substituting appropriate values for GCR fluxes ($F_p^{GCR} \sim 10$ p cm^{-2} s^{-1} for protons and scaled values for other GCR nuclei), for the corresponding cross sections (averaged over the GCR equilibrium spectrum $\sigma_{p,a+CNO \longrightarrow Be} \sim 10^{-26}$ cm^{-2}) and for ISM abundances $Y_{CNO} \sim 10^{-3}$, and integrating for $\Delta t \sim 10^{10}$ yr, one finds $Y_{Be} \sim 2\ 10^{-11}$, i.e. approximately the meteoritic Be value. Satisfactory results are also obtained for ^{6}Li and ^{10}B.

Two problems were identified with the GCR production, compared to meteoritic composition: the ^{7}Li/^{6}Li ratio (~ 2 in GCR, but ~ 12 in meteorites) and the ^{11}B/^{10}B ratio (~ 2.5 in GCR, but ~ 4 in meteorites). It was then suggested in MAR that supplementary sources are needed for ^{7}Li and ^{11}B. Modern solutions to those problems involve *stellar* production of $\sim 60\%$ of ^{7}Li (in the hot envelopes of AGB stars and/or novae, see Sec. 7) and of $\sim 40\%$ of ^{11}B (through ν-induced spallation of ^{12}C in SN, see Sec. 5). In both cases, however, uncertainties in the yields are such that observations are used to constrain the yields of the candidate sources rather than to confirm the validity of the scenario.

2. Primary Be: the problem

Observations of halo stars in the 90s revealed a linear relationship between Be/H and Fe/H (Gilmore *et al.* 1991, Ryan *et al.* 1992) as well as between B/H and Fe/H (Duncan *et al.* 1992). That was unexpected, since Be and B were thought to be produced as *secondaries*, by spallation of the increasingly abundant CNO nuclei. Indeed, the first two terms in Eq. 1.1 were thought to evolve in the same way with time (or metallicity), since the composition of GCR Y_{CNO}^{GCR} was supposed to evolve in step with the one of the ISM Y_{CNO}^{ISM}. Only the Li isotopes, produced at low metallicities mostly by $\alpha + \alpha$ reactions were thought to be produced as primaries (Steigman and Walker 1992) . The only way to produce primary Be is by assuming that GCR have always the same CNO content, as suggested in Duncan *et al.* (1992). Other efforts to enhance the early production of Be, by e.g. invoking a better confinement - and thus, higher fluxes - of GCR in the early Galaxy (Prantzos *et al.* 1993) failed. The reason for that failure was clearly revealed by the "energetics argument" put forward by Ramaty *et al.* (1997): if SN are the main source of GCR energy, there is a limit to the amount of light elements produced per SN, which depends on GCR and ISM composition. If the metal content of *both* ISM and GCR is

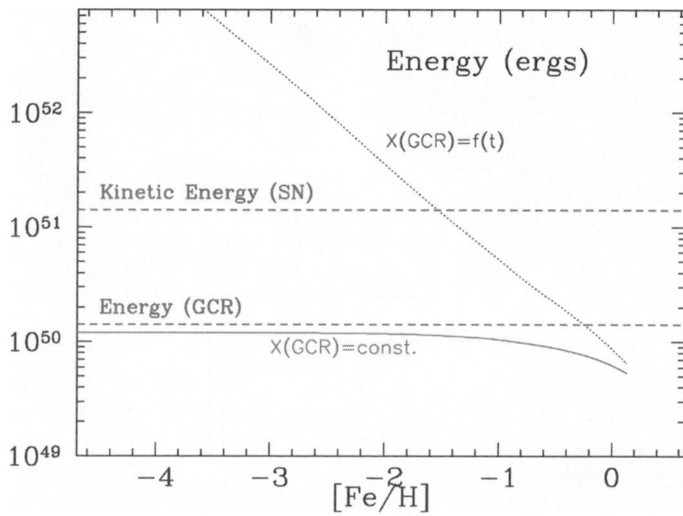

Figure 2. Energy input required from energetic particles accelerated by one CCSN in order to produce a given mass of Be, such as to have [Be/Fe] = 0 (solar), assuming that a core collapse SN produces, on average, 0.1 $M_\odot$ of Fe. *Solid* curve corresponds to the case of a constant composition for GCR, *dotted* curve corresponds to a time variable composition, following the one of the ISM. In the former case, the required energy is approximately equal to the energy imparted to energetic particles by supernovae, namely $\sim$0.1 of their kinetic energy of $\sim$1.5 10^{51} ergs; in the latter case, the energy required to keep [Be/Fe] = 0 becomes much larger than the total kinetic energy of a CCSN for metallicities [Fe/H]$\leqslant$-1.6.

low, there is simply not enough energy in GCR to keep the Be yields constant (Fig. 2)†. Since the ISM metallicity certainly increases with time, the "direct" component in Eq. 1.1 produces only secondary LiBeB. The only possibility to have $\sim$constant LiBeB yields is by assuming that the "reverse" component is primary, i.e. that GCR have a $\sim$constant metallicity. This has profound implications for our understanding of the GCR origin. It should be noted that before those Be and B observations, no one would have the idea to ask "what was the GCR composition in the early Galaxy?".

3. Origin of cosmic rays

For quite some time it was thought that GCR originate from the average ISM, where they are accelerated by the *forward shocks* of SN explosions (Fig. 3.A). However, this can only produce secondary Be.

A $\sim$constant abundance of C and O in GCR can "naturally" be understood if SN accelerate their own ejecta, trough their *reverse schock* (Ramaty *et al.* 1997, see Fig. 3.B). However, the absence of unstable ^{59}Ni (decaying through e^- capture within 10^5 yr) from observed GCR suggests that acceleration occurs $>10^5$ yr after the explosion (Wiedenbeck *et al.* 1999) when SN ejecta are presumably already diluted in the ISM. Furthermore, the reverse shock has only a small fraction of the SN kinetic energy, while observed GCR require a large fraction of it‡.

† For reasons unknown to the author, the energetics argument was obviously not understood by many prolific researchers in the field in the late 90ies.

‡ The power of GCR is estimated to $\sim10^{41}$ erg s^{-1} galaxywide, i.e. about 10% of the kinetic energy of SN, which is $\sim10^{42}$ erg s^{-1} (assuming 3 SN/century for the Milky Way, each one endowed with an average kinetic energy of 1.5 10^{51} ergs).

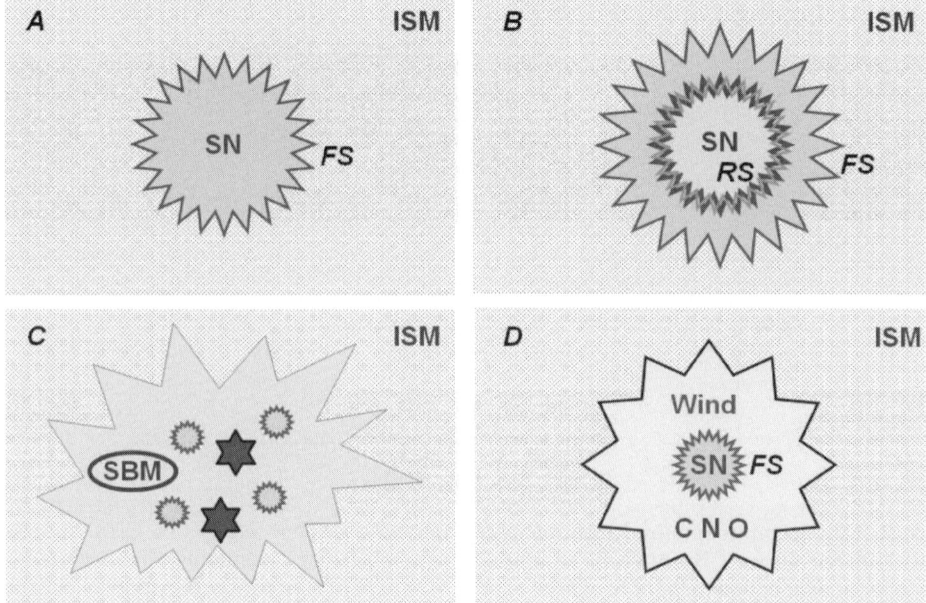

Figure 3. Scenarios for the origin of Galactic cosmic rays (GCR). *A*: GCR originate from the interstellar medium (ISM) and are accelerated from the forward shock (FS) of supernovae (SN). *B*: GCR originate from the interior of supernovae and are accelerated by the reverse shock (RS), propagating inwards. *C*: GCR originate from superbubble material (SBM), enriched by the metals ejected by supernovae and massive star winds; they are accelerated by the forward shocks of supernovae *and* stellar winds. *D*: GCR originate from the wind material of massive *rotating* stars, *always rich in CNO* (but not in heavier nuclei); they are accelerated by the forward shock of the SN explosion.

Higdon *et al.* (1998) suggested that GCR are accelerated out of *superbubbles* (SB) material (Fig. 3.C), enriched by the ejecta of many SN as to have a large and $\sim$constant metallicity. In this scenario, it is the forward shocks of SN that accelerate material ejected from other, previously exploded SN. Furthermore, it has been argued that in such an environment GCR could be accelerated to higher energies than in a single SN remnant (Parizot *et al.* 2004). That scenario has also been invoked in order to explain the present day source isotopic composition of GCR (Binns *et al.* 2005, Rauch *et al.* 2009). Notice that the main feature of that composition, namely a large ^{22}Ne/^{20}Ne ratio, is explained as due to the contribution of winds from Wolf-Rayet (WR) stars (e.g. Prantzos *et al.* 1987), and the SB scenario offers a plausible (but not unique) framework in bringing together contributions from both SN and WR stars.

However, the SB scenario suffers from (at least) two problems. First, core collapse SN are observationally associated to HII regions (van Dyk *et al.* 1996) and it is well known that the metallicity of HII regions reflects the one of the *ambient ISM* (i.e. it can be very low, as in IZw18) rather than the one of SN. Moreover, Higdon *et al.* (1998) evaluated the time interval Δt between SN explosions in a SB to a comfortable $\Delta t \sim 3\ 10^5$ yr, leaving enough time to ^{59}Ni to decay before the next SN explosion and subsequent acceleration. However, Prantzos (2005) noticed that SB are constantly powered not only by SN but also by the strong winds of massive stars (with integrated energy and acceleration efficiency similar to the SN one, e.g. Parizot *et al.* 2004), which should continuously accelerate ^{59}Ni, as soon as it is ejected from SN explosions. Binns *et al.* (2008) argued that the problem may be alleviated from the fact that only the most massive (and thus,

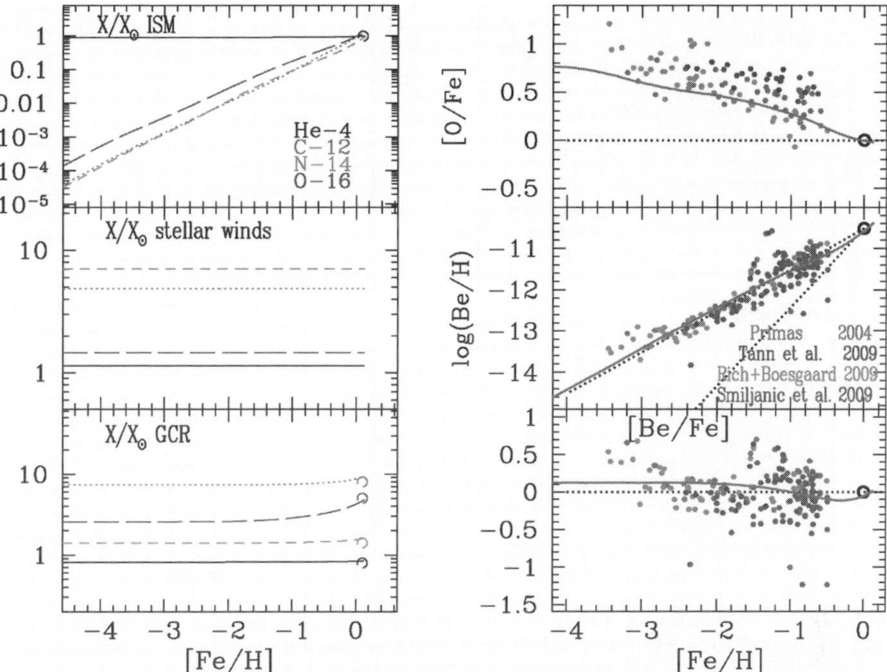

Figure 4. *Left:* Evolution of the chemical composition (in corresponding solar abundances) of He-4 (*solid*), C-12 (*dotted*), N (*short dashed*) and O (*long dashed*)in: ISM (*top*), massive star winds (*middle*) and GCR (*bottom*). *Dots* in lower panel indicate estimated GCR source composition (from Elison *et al.* 1997). *Right:* Evolution (*solid curves* of O/Fe (*top*), Be/H (*middle*) and Be/Fe (*bottom*); *dotted lines* indicate solar values in top and bottom panels, primary and secondary Be in middle panel.

short-lived) stars of an OB association emit strong winds; during the late (and longest) fraction of the lifetime of the SB (a few 10^7 years) particles are accelerated episodically (by SN explosions only) and no more continuously. Still, it is hard to imagine that superbubbles have always the same average metallicity, especially during the early Galaxy evolution, where metals were easily expelled out of the shallow potential wells of the small sub-units forming the Galactic halo (e.g. Prantzos 2008).

4. Cosmic rays from stellar winds and primary Be

In this work we propose a different explanation for the origin of GCR, which can also provide a satisfactory explanation for the primary nature of Be evolution. We first notice that there is now substantial evidence that GCR are indeed accelerated in SN remnants (e.g. Berezhko *et al.* 2009 and references therein). We then notice that, contrary to the case of non-rotating massive stars, which lose mass only at high metallicity, *rotating* massive stars display substantial mass loss down at very low (or even zero) metallicities (e.g. Meynet, this volume). The winds of those stars are enriched in CNO (products of H and He burning *within* the star itself) at all metallicities and at about the same level; it is precisely this enrichment of the WR winds at all metallicities that allows us to understand the observed primary behaviour of N down to the lowest metallicity halo stars (Chiappini *et al.* 2006). This gives some confidence in using the same model results to predict the composition of GCR over the history of the Milky Way.

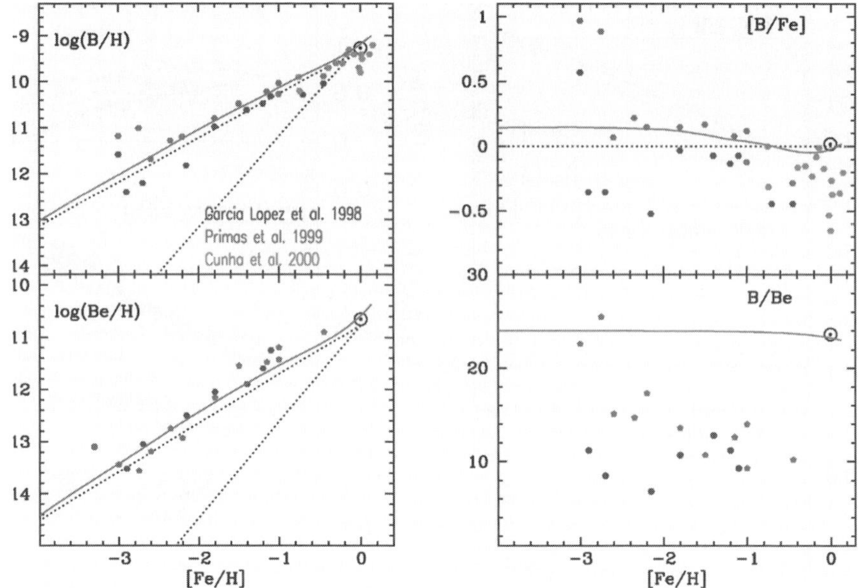

Figure 5. *Left:* Evolution of B (*top*) and Be (*bottom*); in both panels, *dotted lines* indicate primary and secondary evolution and *solid* curves indicate model evolution, including apropriately normalised ν-yields for ^{11}B. *Right:* Evolution of B/Fe (*top*) and B/Be (*bottom*). In the latter case, data indicate a subsolar mean value of B/Be$\sim$14, compatible with exclusively GCR production of both elements, but the uncertainties (not shown here) are too large to allow conclusions.

We assume then that GCR are accelerated when the forward shocks of SN propagate into the previously ejected envelopes of rotating massive stars, which have been partially mixed with the surrounding ISM. The calculation of the resulting GCR composition $Y^{GCR}(M)$ is far from trivial: it will be mostly $Y^{Wind}(M)$ in the case of SN with initial mass $M > 20$ M$_\odot$ (having lost a large fraction of their mass in the wind) and mostly Y^{ISM} in the case of $M=10$-20 M$_\odot$ stars, having suffered low mass losses. For ilustration purposes we adopt here, as a function of metallicity Z, $Y^{GCR}_{paCNO}(Z)=0.5\,[Y^{Wind}_{paCNO}(Z)+Y^{ISM}_{paCNO}(Z)]$, where $Y^{Wind}(Z)$ is provided by the Geneva models (G. Meynet, private communication) and is integrated over a stellar IMF, whereas $Y^{ISM}(Z)$ is provided by the chemical evolution model (left panels in Fig. 4).

The calculation of the Be evolution is then straightforward and nicely fits the data (right panels in Fig. 4); it is the first time that such a calculation is performed *not by assuming* a given $Y^{GCR}_{paCNO}(Z)$ but by *calculating* it in a (hopefully) realistic way.

5. Boron-11 from ν-nucleosynthesis?

As mentioned in Sec. 2, a supplementary source of ^{11}B is required in order to obtain the meteoritic ^{11}B/^{10}B $=4$ ratio. That source may be the ν-process in SN, extensively studied in Woosley *et al.* (1990): a fraction of the most energetic among the $\sim10^{59}$ neutrinos of a SN explosion spallate ^{12}C nuclei in the C-shell of the stellar envelope to provide ^{11}B (but no other light nuclide). Soon after the HST observations of the primary behaviour of B (Duncan *et al.* 1992) it was realised that the ν-process can provide just such a primary B (Olive *et al.* 1994). But, if Be is produced as primary by GCR (Sec. 5), then more than $\sim50\%$ of B is also produced as primary, leaving a rather small role to the ν-process. In fact, the large uncertainties in the ν yields of ^{11}B do not allow an accurate evaluation of

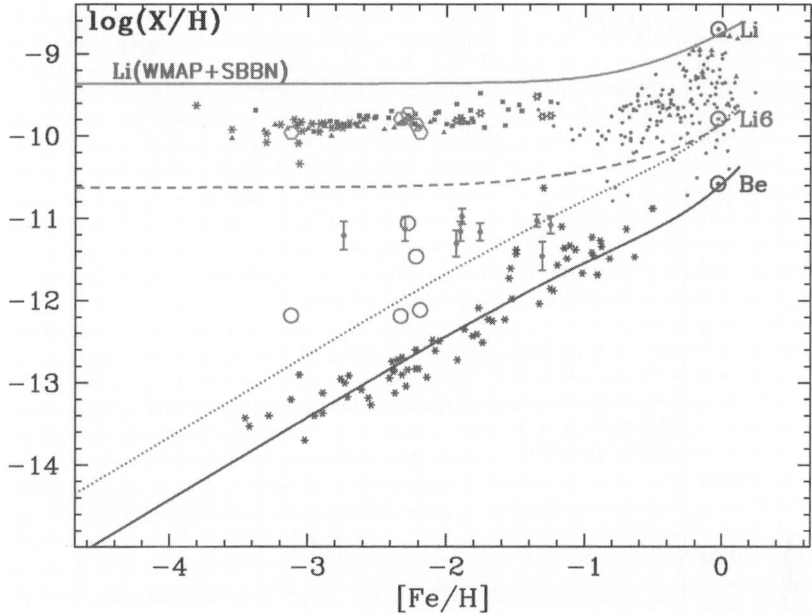

Figure 6. Evolution of total Li (*upper* set of data points and *solid* curve for model assuming high primordial ^{7}Li), Be (*lower* set of points and *solid* curve) and ^{6}Li (*intermediate* set of points and curves). ^{6}Li data are from Asplund *et al.* (2006, small filled circles with error bars) and Garcia-Perez *et al.* (2009, large open circles with - large - error bars not displayed), while model curves are for a canonical ("low") pre-galactic ^{6}Li (*dotted*) and a "high" pre-galactic ^{6}Li (*dashed*). In the latter case, a minimum amunt of depletion within stars (equal to that of ^{7}Li) has been conservatively assumed.

the B evolution: rather the B evolution (resulting from both GCR and ν-process) has to be used in order to constrain the B yields of SN.

The results of such an "exercise" appear in Fig. 5. In order to fit the observations, the ν yields of Woosley and Weaver (1995) had to be divided by a factor of $\sim$6, otherwise B/H and B/Fe would be overproduced. Notice the model B/Be ratio is always $\sim$24 (i.e. solar), substantially higher than the observed, but *highly uncertain*, B/Be$\sim$14 ratio in halo stars (which is consistent with pure GCR production of both elements!). Clearly, future observations with HST are required to clarify that important issue.

6. Early ^{7}Li and ^{6}Li: "high" or "low"?

For a long time, the Li "plateau" in low metallicity halo stars (discovered by Spite and Spite 1982) was considered to reflect the primordial abundance of ^{7}Li. However, the precise determination of baryonic density through observations of the cosmic microwave background, combined to results of standard Big bang nucleosynthesis (SBBN), suggests that the true value of primordial ^{7}Li should be 2-3 times higher. It is not yet clear whether this discrepancy is due to some problems with SBBN, whether non-standard particle physics might cure it, or whether primordial ^{7}Li is depleted in the surface convective zones of low metallicity stars with such an astonishing uniformity (see many contributions in this volume). Other suggestions, like e.g. astration by a pre-galactic Pop. III population of massive stars (Piau *et al.* 2006) face severe problems of metal overproduction (Prantzos 2006). This issue, one of the most important ones for our understanding of mixing in

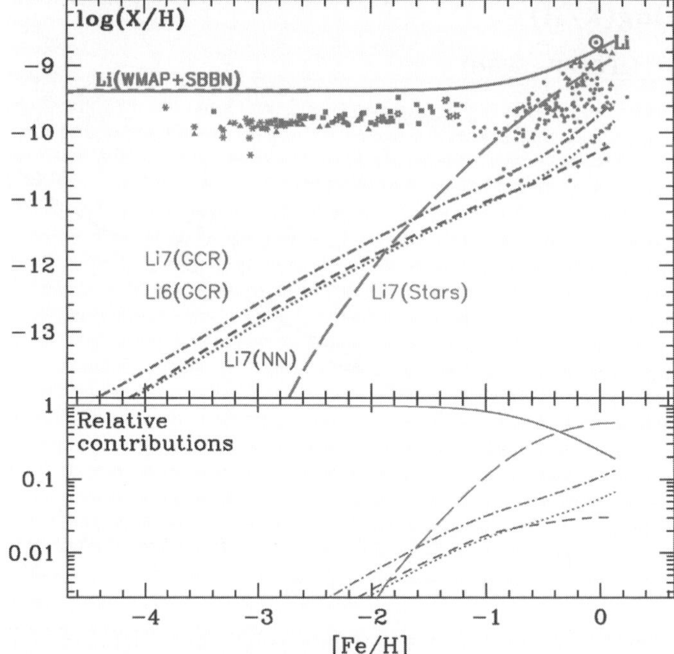

Figure 7. Evolution of total Li (*top*) and percentages of its various components (*bottom*): Li-7 from GCR (*dot-dashed*), Li-6 from GCR (*dotted*), Li-7 from ν-nucleosynthesis (NN, *dashed*) and Li-7 from a delayed stellar source (novae and/or AGB stars, *long dashed*). *Solid* curves indicate total Li (*upper* panel) and primordial ^{7}Li (*lower* panel).

stellar interiors, has also important implications for the chemical evolution of Li, as we shall see below.

The report of an "upper envelope" for ^{6}Li/H in low metallicity halo stars (Asplund *et al.* 2006) gave a new twist to the LiBeB saga. The reported ^{6}Li/H value at [Fe/H] = -2.7 is much larger (by a factor of 20-30) than expected if GCR are the only source of the observed ^{6}Li/H in that star, assuming that GCR can account for the observed evolution of Be (see Fig. 6). But, if it turns out that the true primordial Li is the one corresponding to the WMAP+SBBN value, then the initial ^{6}Li values in halo stars should be at least a factor of 3 higher than evaluated by Asplund *et al.* (2006, see Fig. 6). It should be noticed, however, that such high ^{6}Li values are not obtained in other investigations (Cayrel *et al.* 2008, Steffen *et al.* 2009).

In the past few years, the possibility of important pre-galactic production of ^{6}Li by non-standard GCR has drawn considerable attention from theoreticians, who proposed several scenarios:

1) Primordial, non-standard, production during Big Bang Nucleosynthesis: the decay/annihilation of some massive particle (e.g. neutralino) releases energetic nucleons/ photons which produce ^{3}He or ^{3}H by spallation/photodisintegration of ^{4}He, while subsequent fusion reactions between ^{4}He and ^{3}He or ^{3}H create ^{6}Li (e.g. Jedamzik 2004, and this meeting). Observations of ^{6}Li/H constrain then the masses/cross-sections/densities of the massive particle.

2) Pre-galactic, by fusion reactions of ^{4}He nuclei, accelerated by the energy released by massive stars (Reeves 2005) or by shocks induced during structure formation (Suzuki and Inoue 2002).

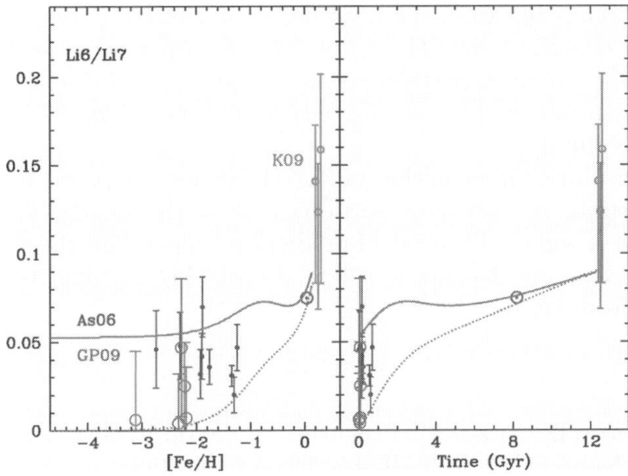

Figure 8. *Left:* Evolution of Li6/Li7 ratio as a function of [Fe/H] (*left*) and of time (*right*). Data are from Asplund *et al.* (2006, As06), Garcia-Perez *et al.* (2009, GP09) and Kawanomoto *et al.* (2009, K09). *Solid* curves corerspond to a "high" pre-galactic Li-6 and *dotted* curves to standard (low) pre-galactic Li-6.

3) In situ production by stellar flares, through ^{3}He$+^4$He reactions involving large amounts of accelerated ^{3}He (Tatischeff and Thibaud, 2007).

Prantzos (2006) showed that the energetics of ^{6}Li production by accelerated particles constrain severely any scenario proposed in category (2) above, including jets accelerated by massive black holes [this holds also for the "stellar flare" scenario, the parameters of which have to be pushed to their extreme values in order to obtain the "upper envelope" of the Asplund *et al.* (2006) observations]. This difficulty is confirmed by Evoli *et al.* (2008), who calculated pre-galactic ^{6}Li production by $\alpha + \alpha$ reactions with a semi-analytical model for the evolution of the early Milky Way; they found maximum values shorter by factors >10 (and plausible values shorter by 3 orders of magnitude) than the values reported by Asplund *et al.* (2006).

7. Evolution of Li and ^{6}Li/^{7}Li

Since GCR can only produce a ^{7}Li/^{6}Li ratio of $\sim$2, instead of the meteoritic (pre-solar) value of $\sim$12, another source of ^{7}Li had to be found. In the two decades following the original MAR paper, four such sources were identified: three possible stellar sources and the hot early Universe of the Big Bang. The latter has certainly operated, as testified by the observed Li "plateau" in low metallicity halo stars; depending on the true primordial value (see Fig. 7), it may contribute from 8 to 20% of the solar ^{7}Li. Among the stellar sources, observational evidence exists only for AGB stars, where high Li abundances have been detected in some cases. But the corresponding model yields (from ^{3}He$+^4$He in the bottom of the convective envelope) are highly uncertain, and this is also the case for the other two candidate sources of novae (from explosive H-burning) and core collapse SN (from ν-induced nucleosynthesis); notice that both novae and AGBs enter the Galactic scene with some time delay ("slow" ^{7}Li component), contrary to SN and GCR.

^{7}Li is thus the only isotope having three distintinctively different types of sources: stellar, BBN and GCR. *Assuming* that the ν-yields of ^{7}Li are well established (through the corresponding ^{11}B yields, see Sec. 6), one may try to estimate the evolution of the remaining "slow" stellar contribution to ^{7}Li, from the combined action of novae and

AGB stars, i.e. by removing from the observed evolutionary curve of Li/H vs Fe/H the BBN, GCR and ν contributions. The result of such an exercise is displayed in Fig. 7. The "slow" stellar component contributes from 50-65% of the solar ^{7}Li (depending on whether high or low primordial ^{7}Li is adopted); similar numbers are found in the analysis of Matteucci (this volume).

Finally, Fig. 8 displays the evolution of ^{6}Li/^{7}Li ratio, compared to data for the early halo (highly uncertain, see previous section) and in the nearby Galactic disk (along three different lines of sight). Theoretical predictions depend on the adopted pre-galactic ^{6}Li/^{7}Li ratio, but a generic feature is a late rise of ^{6}Li/^{7}Li, due to the late secondary production of ^{6}Li from GCR.

References

Asplund, M., Lambertm D., Nissen, P., Primas, F., & Smith, V. 2006, *ApJ* 644, 229

Berezhko, E., Ksenofontov, L., & Völk, H. J. 2009, *AA* 505, 169

Binns, W., Wiedenbeck, M., Arnould, M. *et al.* 2005, *ApJ* 634, 351

Binns, W., Wiedenbeck, M., Arnould, M. *et al.* 2008, *New Astronomy Reviews* 52, 427

Cayrel, R., Steffen, M., Bonifacio, P., Ludwig, H.-G., & Caffau, E. 2008, arXiv:0810.4290

Chiappini, C., Hirschi, R., Meynet, G. *et al.* 2006, *AA* 449, L27

Duncan, D., Lambert, D., & Lemke, M. 1992, *ApJ* 584, 595

Ellison, D., Drury, L., & Meyer, J.-P. 1997, *ApJ*487, 197

Evoli, C., Salvadori, S., & Ferrara, A. 2008, *MNRAS* 390, L14

Garca Prez, A. E., Aoki, W., & Inoue, S. *et al.* 2009, *AA* 504, 213

Gilmore, G., Gustafsson, B., Edvardsson, B., & Nissen, P. 1992, *Nature* 375, 379

Higdon, J., Lingenfelter, R., & Ramaty, R. 1998, *ApJ* 509, L33

Jedamzik, K. 2004, *PhysRevD* 70, 0603524

Kawanomoto, S., Kajino, T., Aoki, W. *et al.* 2009, *ApJ* 701, 1506

Meneguzzi, M., Audouze, J., & Reeves, H. 1971, *AA* 15, 337

Olive, K., Prantzos, N., Scully, S., & Vangioni-Flam, E. 1994, *ApJ* 424, 66

Parizot, E., Marcowith, A., van der Swaluw, E., Bykov, A., & Tatischeff, V. 2004, *AA* 424, 747

Piau, L., Beers, T., Balsara, D. *et al.* 2006, *ApJ* 653, 300

Prantzos, N. 2005, *NuPhA* 758, 249

Prantzos, N. 2006, *AA* 448, 665

Prantzos, N. 2008, *AA* 489, 525

Prantzos, N., Arnould, M., & Arcoragi, J. P. 1987, *ApJ* 315, 209

Prantzos, N., Casse, M., & Vangioni-Flam, E. 1993, *ApJ* 403, 630

Ramaty, R., Kozlovsky, B., Lingenfelter, R., & Reeves, H. 1997, *ApJ* 488, 730

Rauch, B. F., Link, J. T., Lodders, K. *et al.* 2009, *ApJ* 697, 2083

Reeves, H. 2005, *EAS Publications Series*, Vol. 17, p. 15

Reeves, H., Fowler, W., & Hoyle, F. 1970, *Nature* 226, 727

Ryan, S. Norris, J., Bessell, M., & Deliyannis, C. 1992, *ApJ* 388, 184

Spite, F. & Spite, M. 1982, *AA* 115, 357

Steffen, M., Cayrel, R., Bonifacio, P., Ludwig, H.-G., & Caffau, E. 2009, arXiv:0910.5917

Steigman, G. & Walker, T. 1992, *ApJ* 385, L13

Suzuki, T. & Inoue, S. 2002, *ApJ* 573, 168

Tatischeff, V. & Thibaud, J.-P. 2007, *AA* 467, 265

van Dyk, S., Hamuy, M., & Filippenko, A. 1996, AJ 111, 2017

Wiedenbeck, M. *et al.* 1999, *ApJ* 523, L61

Woosley, S., Hartmann, D., Hoffman, R., & Haxton, W. 1990, *ApJ* 356, 272

Woosley, S. & Weaver, T. 1995, *ApJS*, 101, 181

Light Elements in the Universe
Proceedings IAU Symposium No. 268, 2009
C. Charbonnel, M. Tosi, F. Primas & C. Chiappini, eds.

© International Astronomical Union 2010
doi:10.1017/S1743921310004667

Beryllium abundances and the formation of the halo and the thick disk

Rodolfo Smiljanic[1,2] L. Pasquini[2], P. Bonifacio[3,4,5], D. Galli[6], B. Barbuy[1], R. Gratton[7], and S. Randich[6]

[1]IAG, University of São Paulo, São Paulo, Brazil, [2]ESO, Garching bei München, Germany
email: `rsmiljan@eso.org`

[3]GEPI Observatoire de Paris - Meudon, France, [4]INAF, Osservatorio di Trieste, Trieste, Italy
[5]CIFIST Marie Curie Excellence Team, [6]INAF- Osservatorio di Arcetri, Firenze, Italy
[7]INAF-Osservatorio di Padova, Padova, Italy

Abstract. The single stable isotope of beryllium is a pure product of cosmic-ray spallation in the ISM. Assuming that the cosmic-rays are globally transported across the Galaxy, the beryllium production should be a widespread process and its abundance should be roughly homogeneous in the early-Galaxy at a given time. Thus, it could be useful as a tracer of time. In an investigation of the use of Be as a cosmochronometer and of its evolution in the Galaxy, we found evidence that in a $\log(\mathrm{Be/H})$ vs. $[\alpha/\mathrm{Fe}]$ diagram the halo stars separate into two components. One is consistent with predictions of evolutionary models while the other is chemically indistinguishable from the thick-disk stars. This is interpreted as a difference in the star formation history of the two components and suggests that the local halo is not a single uniform population where a clear age-metallicity relation can be defined. We also found evidence that the star formation rate was lower in the outer regions of the thick disk, pointing towards an inside-out formation.

Keywords. Stars: abundances, late-type – Galaxy: halo, thick disk

1. Introduction

The single stable isotope of beryllium, ^{9}Be, is a pure product of cosmic-ray spallation of heavy nuclei (mostly CNO) in the interstellar medium (Reeves *et al.* 1970). In this sense, the production of Be can occur in two ways. In the so-called direct process the cosmic rays are composed of protons and α-particles and these collide with CNO nuclei of the medium. In the so-called inverse process the cosmic rays are composed of accelerated CNO nuclei that collide with protons and α-particles of the medium.

Observational works on Be abundances in metal-poor stars (Rebolo *et al.* 1988; Gilmore *et al.* 1992; Molaro *et al.* 1997; Boesgaard *et al.* 1999; Smiljanic *et al.* 2009) find that $\log(\mathrm{Be/H})$ and [Fe/H] (or [O/H]) show a linear relation with slope close to one. Such slope argues that Be behaves as a primary element in the early Galaxy and its production mechanism is independent of the ISM metallicity. This means that the dominant production mechanism of Be is the inverse process (Duncan *et al.* 1992).

If one assumes that the cosmic-rays are globally transported across the Galaxy, than it follows that the Be production should be a widespread process. Beryllium can be produced anywhere in the Galaxy. One may thus expect that the Be abundances are rather homogeneous at a given time in the early Galaxy. It should have a smaller scatter than the products of stellar nucleosynthesis (Suzuki *et al.* 1999; Suzuki & Yoshii 2001) and could thus be employed as a cosmochronometer for the early stages of the Galaxy (Beers *et al.* 2000; Suzuki & Yoshii 2001).

This was tested by Pasquini *et al.* (2004, 2007) who calculated Be abundances in turn-off stars of two globular clusters, NGC 6397 and NGC 6752. These Be abundances

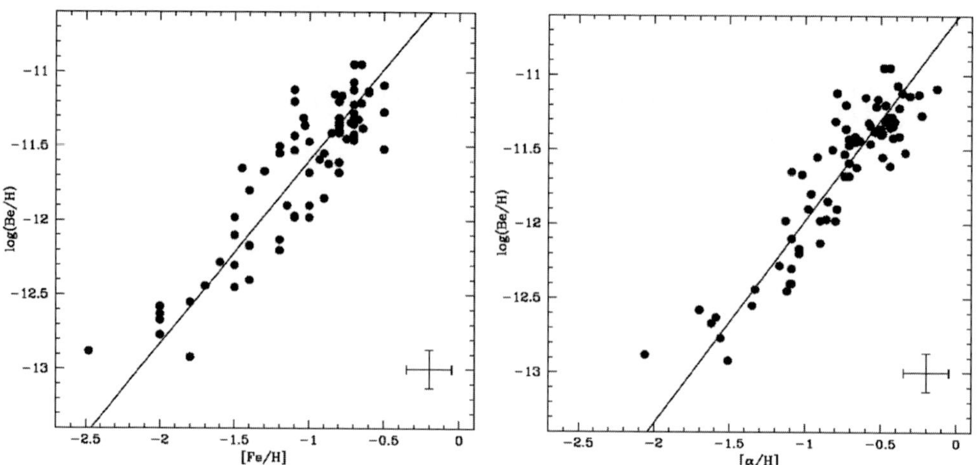

Figure 1. Abundances of Be as a function of [Fe/H] (left panel) and of [α/Fe] (right panel). This figure is adapted from Smiljanic *et al.* (2009).

were used to derive ages by means of a comparison with a model of the evolution of Be with time (Valle *et al.* 2002). These ages agree well with those derived from theoretical isochrones, supporting the use of Be as a cosmochronometer.

Pasquini *et al.* (2005) then used a sample of 20 halo and thick-disk stars, previously analyzed by Boesgaard *et al.* (1999), to investigate the evolution of star formation rate in these two Galactic components. The idea is to use a diagram of [O/Fe] vs. log(Be/H) where the abscissa can be considered as increasing time and the ordinate as the star formation rate. A possible separation between stars of the two components was found and interpreted as a difference in the time scales of star formation.

Smiljanic *et al.* (2009) analyzed the largest sample of halo and thick-disk stars to date, extending the investigation of Be as cosmochronometer and its role as a discriminator of the different stellar populations in the Galaxy. These results are discussed in more detail in the following sections.

More details about Be can also be found in the references cited above and in many contributions in this volume, e.g., A. Boesgaard, D. Lambert, F. Primas, and H. Reeves.

2. The relation of Be with [Fe/H] and [α/H]

The beryllium abundances calculated by Smiljanic *et al.* (2009) are shown in Fig. 1 as a function of both [Fe/H] and [α/H]. Linear relations with slopes close to one, as found in previous works, are seen in these plots (the slope is 1.23 for [Fe/H] and 1.37 for [α/H]).

The scatter of the abundances seen in these plots is larger than that found by previous works in the literature. Statistically, because of the size of the error bars, it is not possible to say whether the observed scatter is real. However, it was possible to find among the sample stars, stars that have similar atmospheric parameters, same metallicity, normal unaltered Li abundances, but different Be abundances. The direct comparison of the spectra of these stars is shown in Fig. 14 of Smiljanic *et al.* (2009). This comparison strongly argues that at least part of the observed scatter is real.

Two different explanations can be put forward to understand the scatter. One is that it is caused by local effects, such as the proximity to supernovae (or as suggested by Smiljanic *et al.* (2008) to explain the case of the Be-rich star HD 106038 the proximity to

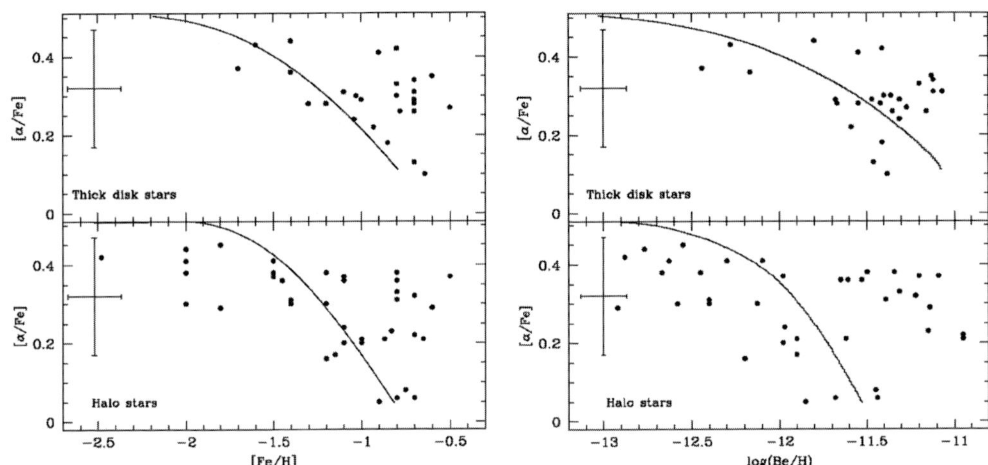

Figure 2. Diagram of [α/Fe] as a function of [Fe/H] (left panel) and of log(Be/H) (right panel). The thick-disk stars are shown in the upper panels and the halo stars in the lower ones. The thick-disk stars behave in the same way in both plots. The halo stars, however, clearly divide into two sequences when [α/Fe] is plotted against log(Be/H). The curves are the predictions of the models by Valle *et al.* (2002). This figure is adapted from Smiljanic *et al.* (2009).

a hypernova). The other is related to the stars belonging to different stellar populations. As we discuss below, this second explanation is the preferred one.

A figure similar to Fig. 1 with the stars divided according to its membership to the halo or the thick disk star does not show a obvious division. This may be caused in part by the narrow metallicity range of the thick-disk stars (metal-poor thick-disk stars are rare). However, if such a plot is made it becomes clear that the scatter of the abundances for the halo stars is larger than that for the thick-disk stars.

3. Stellar Populations

We further investigate the different stellar populations in Fig. 2, using a diagram of [α/Fe] vs. log(Be/H). In Smiljanic *et al.* (2009) oxygen abundances were not available, so mean abundances of α-elements were used instead. Again, the abscissa can be considered as increasing time and the ordinate as the star formation rate. Here we also present some new preliminary oxygen abundances determined from the infrared triplet at 777nm for a subsample of the halo stars (Fig. 3). The oxygen abundances were calculated using spectrum synthesis and the same codes and line lists used in Smiljanic *et al.* (2008, 2009). The oxygen abundances are corrected for NLTE effects following Fabbian *et al.* (2009). To calculate the oxygen abundances a new reduction of the red-arm UVES spectra, where the OI lines are located, was necessary. The red UVES spectra is usually strongly affected by fringing. This can, however, be corrected with a new reduction using the latest UVES pipeline.

When using the Be abundances instead of Fe, it becomes clear that, with either α or oxygen abundances, the halo stars define two clear distinct sequences (Figs. 2 and 3). One sequence is chemically similar to the thick disk, the other agrees with the behavior expected for the halo stars in the models of Valle *et al.* (2002). The thick disk stars behave the same no matter if using Be or Fe (Fig. 2).

Checking the kinematics of the stars, however, it is possible to notice that the group of low-α halo stars, that follows the expected behavior of the models, have similar kinematics. We define this group from the diagram of Fig. 2 as the stars with [α/Fe] ≤ 0.25

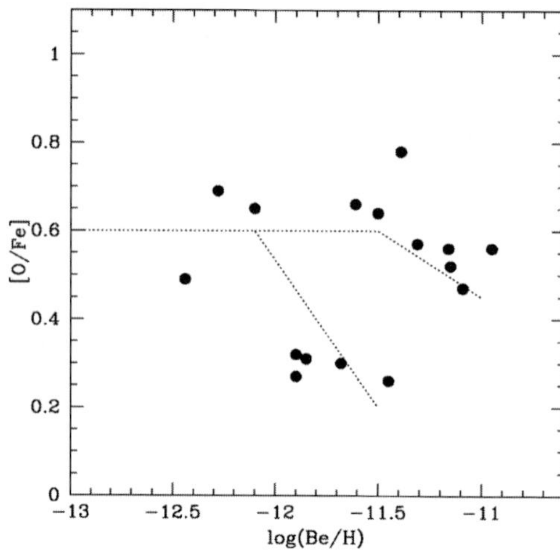

Figure 3. The [O/Fe] ratio as a function of beryllium abundances for a subsample of the halo stars analyzed in Smiljanic *et al.* (2009). These preliminary oxygen abundances were calculated using the OI triplet at 777nm. The abundances are corrected for NLTE (Fabbian *et al.* 2009). The lines are only traced to guide the eye.

and $\log(\mathrm{Be/H}) \leqslant -11.4$. In Fig. 4 we show diagrams of both $[\alpha/\mathrm{Fe}]$ and $\log(\mathrm{Be/H})$ as a function of V, the component of the space velocity of the star in the direction of the disk rotation, and of $\mathrm{R_{min}}$, the perigalactic distance of the stellar orbit. The low-α group (open symbols) have mostly $\mathrm{V} \sim 0$ and $\mathrm{R_{min}} \leqslant 1$ kpc. They seem to be a group of non-rotating stars going very close to the Galactic center, a behavior that might be expected to be shown by accreted stars that sink to the Galactic center by dynamical friction. They also form a very tight and well-defined sub-sequence in a diagram of [Fe/H] vs. log (Be/H) (Fig. 5). This fact helps to explain why the halo stars show a larger scatter in this kind of diagram.

The splitting into two components may be related to the accretion of external systems or to variations of star formation in different and initially independent regions of the early halo. The interpretation is still open, it is however clear that the halo is not a single uniform population with a single age-metallicity relation. A similar division was found by Nissen & Schuster (1997, 2009) but using Fe as a tracer of time. The division seems clearer when Be is used as a time scale.

For the thick disk, it is possible to see in the lower-right panel of Fig. 4 a significant anticorrelation of $[\alpha/\mathrm{Fe}]$ with perigalactic distance. This anticorrelation might be interpreted as evidence that the SFR was lower in the outer regions of the thick disk. A similar correlation however is not seen for Be. This lack of anticorrelation seems to indicate that Be is not affected by the local details of star formation, confirming that it can be used as a time scale. Although the range in Be abundances covered by the thick disk is very small, it is possible to see in Fig. 4 that the thick disk stars with smallest Be concentrate in the inner regions of the disk. As these are expected to be old stars, this result seems to point towards an inside-out formation of the thick disk.

One remaining question is how does this possible division of the halo stars compare to our current understanding of the formation of the Galactic halo? Given that we are analyzing a sample of local halo stars this division is likely not a consequence of the inner vs. outer halo dichotomy, as discussed for example by Carollo *et al.* (2007, 2009). According to Carollo *et al.* (2007), the outer halo has a metallicity distribution that peaks at $[\mathrm{Fe/H}] \sim -2.20$ and dominates the stellar population in distances beyond 15–20 Kpc

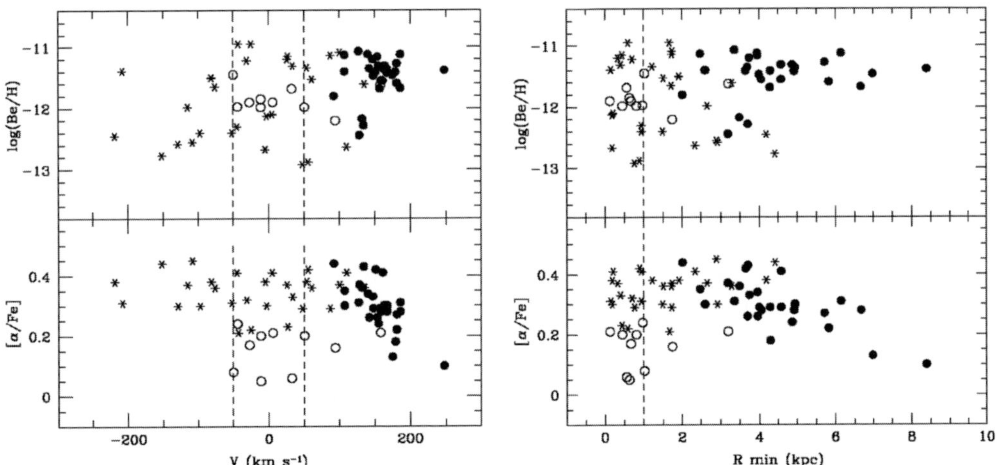

Figure 4. The $[\alpha/\mathrm{Fe}]$ ratio and the Be abundances as a function of V, the component of the space velocity of the star in the direction of the disk rotation (left panel), and of $\mathrm{R_{min}}$, the perigalactic distance of the stellar orbit (right panel). Thick disk stars are shown as filled circles, the low-α halo stars as open symbols, and the remaining halo stars as starred symbols. The low-α stars tend to have V close to zero and $\mathrm{R_{min}} \leqslant 1$ kpc. This figure is adapted from Smiljanic *et al.* (2009).

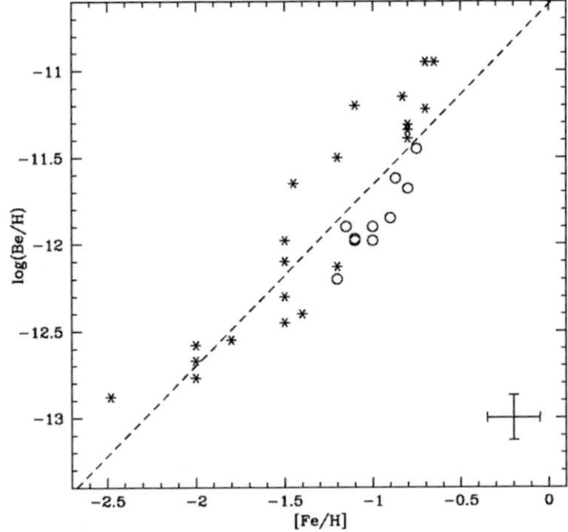

Figure 5. The beryllium abundances as a function of [Fe/H] only for the halo stars. The low-α stars seem to form a tight and well defined sub-sequence. This figure is adapted from Smiljanic *et al.* (2009).

from the Galactic center. The inner halo, on the other hand, has a metallicity distribution that peaks at $[\mathrm{Fe/H}] = \sim -1.60$ and dominates the stellar population in distances up to 10–15 Kpc. Although we did not check in detail to which of these halo components our sample stars belong, we note that they have metallicities between $[\mathrm{Fe/H}] = -2.00$ and -0.50. The low-α stars in particular have $-1.20 \leqslant [\mathrm{Fe/H}] \leqslant -0.70$. We thus believe most of our sample stars are definitely inner halo stars. This implies that our results suggest a dichotomy of the inner halo.

There are other literature results that seem to indicate a possible division of the inner halo, from both the observational and the theoretical point of view. On the observational side, Morrison *et al.* (2009), based on the kinematics of a sample of metal-poor stars, conclude that the local inner halo seems to divide into two components. One is moderately flattened, has no rotation, has a clumpy distribution in energy and angular momentum, and [Fe/H] < -1.50. The second is highly flattened, has a small prograde rotation, and $-1.50 <$ [Fe/H] < -1.00. This last component is distinct from the metal-weak thick disk.

On the theoretical side, Zolotov *et al.* (2009) present new simulations of the formation of disk galaxies in a ΛCDM universe. They allow for star formation both in the primary potential well of the galaxy being modelled and in dark matter subhalos that are later accreted. They show that the final stellar population in the inner halo of a galaxy formed like this has a dual nature. It is composed both by 'in situ stars', formed in the inner Galaxy and later displaced to the halo, and by accreted stars, formed in the subhalos.

Acknowledgements

This work has been supported by studentships and fellowships to R. S. by FAPESP (04/13667-4 and 08/55923-8) and CAPES (1521/06-3), and by financial support from the ESO DGDF. R. S. also acknowledges financial support from the organizers for his participation in this meeting.

References

Beers, T. C., Suzuki, T. K., & Yoshii, Y. 2000, in: Proc. IAU Symp. No. 198, p. 425
Boesgaard, A. M., Deliyannis, C. P., King, J. R. *et al.* 1999, *AJ*, 117, 1549
Carollo, D., Beers, T. C., Lee, Y.-S. *et al.* 2007, *Nature*, 450, 1020
Carollo, D., Beers, T. C., Chiba, M. *et al.* 2009, arXiv:0909.3019
Duncan, D. K., Lambert, D. L., & Lemke, M. 1992, *ApJ*, 401, 584
Fabbian, D., Asplund, M., Barklem, P. S. *et al.* 2009, *A&A*, 500, 1221
Gilmore, G., Gustafsson, B., Edvardsson, B., & Nissen, P. E. 1992, *Nature*, 357, 379
Molaro, P., Bonifacio, P., Castelli, F., & Pasquini, L. 1997, *A&A*, 319, 593
Morrison, H. L., Helmi, A., Sun, J. *et al.* 2009, *ApJ*, 694, 130
Nissen, P. E. & Schuster, W. J. 1997, *A&A*, 326, 751
Nissen, P. E. & Schuster, W. J. 2009, in: Proc. IAU Symp. No. 254, p. 103
Pasquini, L., Bonifacio, P., Randich, S., Galli, D., & Gratton, R. G. 2004, *A&A*, 426, 651
Pasquini, L., Bonifacio, P., Randich, S. *et al.* 2007, *A&A*, 464, 601
Pasquini, L., Galli, D., Gratton, R. G. *et al.* 2005, *A&A*, 436, L57
Rebolo, R., Abia, C., Beckman, J. E., & Molaro, P. 1988, *A&A*, 193, 193
Reeves, H., Fowler, W. A., & Hoyle, F. 1970, *Nature*, 226, 727
Smiljanic, R., Pasquini, L., Bonifacio, P. *et al.* 2009, *A&A*, 499, 103
Smiljanic, R., Pasquini, L., Primas, F. *et al.* 2008, *MNRAS*, 385, L93
Suzuki, T. K. & Yoshii, Y. 2001, *ApJ*, 549, 303
Suzuki, T. K., Yoshii, Y., & Kajino, T. 1999, *ApJ*, 522, L125
Valle, G., Ferrini, F., Galli, D., & Shore, S. N. 2002, *ApJ*, 566, 252
Zolotov, A., Willman, B., Brooks, A. M. *et al.* 2009, *ApJ*, 702, 1058

Light Elements in the Universe
Proceedings IAU Symposium No. 268, 2009
C. Charbonnel, M. Tosi, F. Primas & C. Chiappini, eds.

© International Astronomical Union 2010
doi:10.1017/S1743921310004679

DICUSSION C. The stellar yields in He-3, He-4 and Li-7: main sources, observational constraints, and problems

André Maeder

Geneva Observatory, University of Geneva, CH-1290 Versoix, Switzerland
email: andre.maeder@unige.ch

Keywords. Nucleosynthesis, abundances

I am pleased to recall that the first determination of the Li–abundance in the Sun was made at the Geneva Observatory in 1975 by Edith Müller, Eric Peytremann and Ramiro de la Reza on the basis of spectra taken at Kitt Peak. The first Be determination was also made in Geneva the same year by Y. Chmielewski, J. Brault (Kitt Peak National Observatory) and E. Müller. These two outstanding works opened the door for all further investigations on these light elements.

The yields of light elements are a matter where the shock of ideas frequently occurs, with new facts arising and different interpretations. This is particularly the case at present concerning the He abundance in globular clusters and the relation between the ^{7}Li abundances and planets. This made the discussions at IAU Symposium very lively and interesting. I will try my best to summarize the main points, restore the essence of the discussions and give a proper credit to the various contributors.

Normally, stellar evolution provides the chemical yields which are used as an input for the models of the chemical evolution of galaxies. The inverse path, i.e. using the observations of galactic trends in chemistry to infer the stellar properties is like trying to derive the properties of an egg from an omelette, as stated by Jean Audouze long time ago. Nevertheless, in the case of the very early stellar generations, the only direct infos we often have on the stellar properties come from the observed trends in the chemical evolution of galaxies. In this case, we can even say that we use a "fossil omelette" to reconstruct the original stellar properties of the very early stellar generations.

1. The yields in He-4, the globular clusters, and related questions

The relative helium to oxygen enrichment $\Delta Y / \Delta O$ obtained by Manuel Peimbert is 3.3 ±0.7, a ratio which typically applies to stars with solar abundances. We notice that the yields in He and O do not come from the same mass range, and that the above ratio depends heavily on mass loss, mixing, mass limit for black hole formation, cutoff mass in supernovae, etc... The question of the slope of the $\Delta Y / \Delta Z$ according to the metallicity Z is raised by Francesca Matteucci. Manuel Peimbert points out that $\Delta Y / \Delta Z$ may reach a value of about 4 at Z higher than solar.

The big question concerns the multiple sequences found in globular clusters, following the VLT study of ω Cen by Piotto and colleagues. The globular clusters have experienced successive starburst episodes, the recent one bearing the signature of strong He-enrichments. The red sequence corresponds to $Z = 10^{-3}$, $Y = 0.246$, while the blue sequence has $Z = 2 \cdot 10^{-3}$, $Y = 0.38$. Formally, this corresponds to a large He–enrichment for a minute one in heavy elements, i.e. $\Delta Y / \Delta Z \sim 10^2$. The discussions bear, not on the reality of the high He-content, but rather on its origin. Some authors in literature support the idea that the galactic winds driven by supernovae have removed the contributions in heavy elements, because they are ejected at high speed. Manuel Peimbert notes the C/O ratios in the gas may provide information on the importance of the ejecta of heavy elements in the galactic winds. The observations show low C/O ratios, which indicates that there is no important losses of O–rich material by the galactic winds. The material forming the ejected galactic winds is likely rather well mixed. Francesca Matteucci notes that the same remark seems to also apply to dwarf galaxies. Donatella Romano nevertheless mentions there are observations in a dwarf galaxy of galactic winds showing typical SN II ejecta.

One may wonder whether the galactic infall could not counterbalance the effects of selective ejecta. Monica Tosi emphasizes that all galaxies experience infalls, while only a few show galactic winds. She points out that dwarfs galaxies tend to develop galactic winds. She mentions that the above mentioned C observations are made in nearby galaxies, where there is likely no galactic winds.

Georges Meynet supports a different idea, i.e. that fast rotating massive stars at low Z lose a lot of mass as a result of rotational mixing and mass loss, thus making the He yields very large. He also relates these excesses to the anomalies of the CEMP (Carbon rich extremely metal poor stars). Some authors support losses from AGB stars or contributions from binaries. He emphasizes that massive and AGB stars do not produce exactly the same chemical anomalies and that this may be a way to disentangle the two types of contributions.

Why do these large He abundances do not affect the general $\Delta Y / \Delta Z$ ratio? Is it because the fraction of stars which form under the above peculiar conditions is too small? This seems likely, however this could be the source of some scatter in the Y vs. Z relation at low Z. Francesca Matteucci remarks that the peculiar solutions for the globular clusters do not necessarily apply to the Galaxy and she emphasizes the interest to perform some tests in the Galactic bulge.

Among other points, Marc Pinsonneault emphasizes the interest to get He abundances from asteroseismology in all possible objects. Nikos Prantzos remarks that the Sun places very good constraint on the $\Delta Y / \Delta Z$ ratio. This gives a value within the error bars of that given by Manuel Peimbert.

2. The surprising flat behaviors of the He-3/He ratio

In solar-type stars, a bump of ^{3}He forms at mass fraction 0.55 and the standard models predict that this ^{3}He should be ejected in the more advanced stages. The massive stars on the contrary destroy ^{3}He over most of their interior, both the initial cosmological one and the new amounts they may have synthesized, thus their yields are negative. The net result of the standard models integrated over the mass spectrum, as shown by models of the chemical evolution of galaxies made by Monica Tosi, is that the ^{3}He/He ratio should increase with the [O/H] abundance ratio. In the Galaxy, this ratio should decrease with the galactocentric distance.

However, the problem is that more than 90% of the solar-type stars show no ^{3}He enrichment or at least much less than predicted. Among the planetary nebulae, only a few (7 as discussed by Tom Bania) show ^{3}He enhancements in agreement with the current expectations. Also contrary to the model expectations, the ^{3}He/He ratio is flat with respect to [O/H] as discussed by Gary Steigmann and Tom Bania. The ^{3}He/He ratio is similar to that given by the standard Big Bang nucleosynthesis. Also, this ratio does not show the expected variations with the galactocentric radius or with the oxygen abundance. This is the so-called He-3 problem, the reality of which is not disputed.

Gary Steigman emphasizes that ^{3}He is both produced and destroyed, but in a way that the net production is zero, so that it keeps to the level of the standard Big–Bang synthesis. Corinne Charbonnel supports these views. She emphasizes that the low-mass stars have small positive yields (even with the thermohaline mixing), thus they do not destroy the cosmological ^{3}He. The planetary nebulae with the predicted ^{3}He abundances also have the classically predicted ^{12}C/^{13}C isotopic ratios. In relation with her models, Corinne Charbonnel mentions that it would be interesting to know the magnetic fields in RGB stars with the classical ^{12}C/^{13}C isotopic ratios.

Monica Tosi mentions that the present HII regions exhibit no ^{3}He/He gradient in the Milky Way, she emphasizes this confirms that there is no net overall production of ^{3}He at the present epoch. Manuel Peimbert questions whether there are biases in the studies of the planetary nebulae showing nitrogen enrichments. Tom Bania confirms that the sample of observed planetary nebulae is currently biased in favor of those showing nitrogen enrichments.

A possible solution for the ^{3}He problem is that proposed by Nadège Lagarde, Corinne Charbonnel, and Jean-Paul Zahn, it implies an instability due to thermohaline mixing. This instability, like salt fingers, occurs when matter with an higher mean molecular weight μ lies above matter with a lower μ. The reaction ^{3}He $+\,^3$He $\rightarrow\,^4$He $+$ 2 p results in the conversion of two particles into three ones at the basis of the ^{3}He bump. Thus, there it reduces the μ–value making a local inversion which creates an instability. One may note that this solution is reminiscent of the so-called "solar spoon" proposed decades ago by D. Gough.

However, the question arises why do some planetary nebulae still have the expected ^{3}He enrichments. The answer by Corinne Charbonnel is that this may be due to a magnetic field inhibiting the thermohaline instability. She proposes that these planetary nebulae are the descendants of the magnetic Ap stars. In both cases, the proportions of these objects are about 5%. Further investigations are needed to test these possibilities.

3. Questions regarding lithium: the Li plateau, Li in RGB and AGB stars, relation with planets

It is remarkable that thanks to the efforts of many people we now seem to have a consistent interpretation of the various physical effects shaping the Li abundance in the T_{eff} region surrounding the Li dip (see Fig. 1).

When one refers to the (in principle) unmodified Li abundance, it means the abundance on the left side of the Li dip as shown in the Figure. These Li abundances for stars covering a large range of [Fe/H] show an essentially flat distribution over the whole range of metallicities except their increase for the higher [Fe/H] corresponding to Pop. I stars. The almost flat part is the so–called "Spite plateau". The big problem is that this abundance plateau lies a factor of about 3 below the predictions of standard Big–Bang nucleosynthesis. In this respect, there is an interesting poster by Sbordone *et al.* on the low-metallicity end of the Spite plateau. Below [Fe/H] $= -3$, he notes that there is no plateau anymore, but a growth of A(Li) with [Fe/H] and a large scatter.

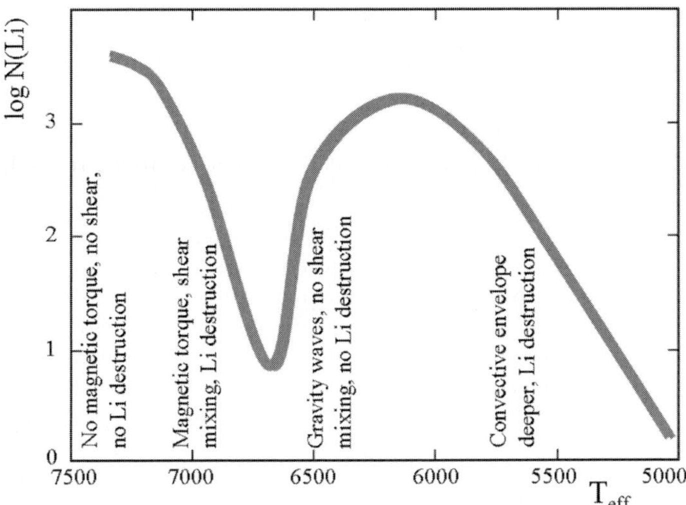

Figure 1. The physical effects shaping the Li dip as a function of the effective temperature. Adapted from S. Talon et C. Charbonnel.

The main lesson concerning the Li abundance $A(\text{Li})$ on the hot side of the dip is that it appears to be a multivariate function of the form

$$A(\text{Li}) \;=\; f(mass, Z, age, rotation, pre-MS, disk\; lifetime, ...) \,, \tag{3.1}$$

some effects being still disputed. The evolutionary effects along the tracks in the HR diagram are clearly predicted by the stellar models and they are also observed along the sequences of globular clusters. The T_{eff} scale is of particular importance along the subgiant branch, as emphasized in several interventions in the discussion.

There are clear differences of views on the age effect on the Li-abundances. For Sofia Randich, each cluster behaves in a different way and the relation with age is uncertain. Jorge Melendez however shows that the Li abundances for the available solar twins of the Hipparcos catalogue present a strong correlation with age in agreement with the Charbonnel & Talon (2005, Science, 309, 2189) models.

Andreas Korn wonders whether the Li dip is also observed in PopII stars as in Pop I stars. Suzanne Talon answers that Pop II stars are so old that the Pop II stars with masses at the level of the Li dip already occupy the subgiant branch, thus it may not be easy to see the it. Corinne Charbonnel further comments that the dip has nevertheless been found in halo stars (Charbonnel & Primas, 2005, A&A, 442, 961) and in NGC 6397 by Lind *et al.* (2009, A&A, 503, L545).

There are theoretical and observational evidences of some additional lithium depletion near the bump of the red giant branch (RGB) and various mechanisms are proposed to account for this depletion, in particular the hot-bottom burning and the thermohaline mixing. There is a variety of Li abundances in RGB and AGB stars. 1% of the GK giants are Li-rich, there are also Li rich AGB stars, which may be related to the Cameron-Fowler mechanism. To a question concerning the origin of the Li-depleted stars, Marc Pinsonneault suggests they are mergers.

Francesca Matteucci then makes some comments regarding ω Cen and the related problems. One may recall here that in his presentation Piercarlo Bonifacio has suggested, from the similarity of the Li content in ω Cen and in the Galactic halo field stars (which have experienced different evolutionary histories), that there is no evidence of a pre-galactic Li depletion by a generation of very massive stars as was suggested by Laurent Piau. Bonaficio has also noted the absence of difference in the Li content of stars in ω Cen over a range of 5 Gyr in ages, he has claimed that this in contradiction with the current explanations of the Li plateau based on diffusion processes. Francesca Matteucci wonders about the assumed 5 Gyr scatter and also on the sampling of data.

The last topic approached in this discussion concerns the relation between the enhanced Li-depletion and the occurrence of planets in solar-type stars. This interesting result was shown by Garik Israelian and also presented by Nuno Santos in his review. The point is however challenged in a poster by Jorge Melendez.

There are several possible interpretations of the above fact. The merging of a giant planet may bring additional angular momentum in the outer layers, thus driving mixing and Li-depletion, also the pollution resulting from the merging may give rise to thermohaline mixing as was suggested by Sylvie Theado and Sylvie Vauclair. Marc Pinsonneault makes comments on the importance of the disk and of rotation in the pre-Main Sequence evolution. He emphasizes the need to measure rotation rates. The key parameter (as firstly emphasized by Jerome Bouvier) is the disk lifetime. A longer disk lifetime means more disk locking in the pre-MS phase and thus a lower mean rotation on the ZAMS. The longer disk lifetime also leads to higher mixing due to more differential rotation in *slowly* rotating stars. Another consequence of the longer disk lifetime is that there is more possibilities for the formation of planets. This may establish a relation between the Li-depletion and the occurrence of planets via the longer disk lifetime.

This discussion was a beautiful illustration of the major open questions remaining in our understanding of the abundances of the light elements. I thank Patrick Eggenberger for his help in taking notes of the session. I apologize for any deviation from the exact meaning of the contributors, I hope nobody will bring me to the court for that.

Light Elements in the Universe
Proceedings IAU Symposium No. 268, 2009
C. Charbonnel, M. Tosi, F. Primas & C. Chiappini, eds.

© International Astronomical Union 2010
doi:10.1017/S1743921310004680

Discussion D: Observational problems with Li, Be and B

Poul Erik Nissen[1]

[1]Department of Physics and Astronomy, University of Aarhus, DK-8000 Aarhus C, Denmark
email: pen@phys.au.dk

Abstract. In Discussion D the following problems were addressed: Has ^{6}Li really been detected in the atmospheres of metal-poor halo stars? Is there a downward trend or increased scatter of Li abundances in stars on the 'Li-plateau' at metallicities [Fe/H] $\lesssim -2.5$? Are there significant differences of Li abundances in main-sequence, turn-off, and sub-giant stars in globular clusters? Is the Li abundance in solar-type stars related to the presence of planets? How does the Be abundance in dwarf stars increase with the heavy-element abundance, and is there a cosmic scatter in Be at a given [Fe/H]? The discussion of these problems is summarized and some suggestions for future observational and theoretical studies are mentioned.

Keywords. stars: abundances, atmospheres, interiors – planetary systems – ISM: abundances – early universe

1. Introduction

Observational problems with lithium and beryllium, which have attracted much attention in recent years, were taken up during Discussion D. Boron, on the other hand, was not included in the discussion, because observational progress for this element has been slow in recent years due to the 5-year hiatus in the UV high-resolution instrumentation at the HST. In the following, I briefly report on the main subjects of the discussion.

2. ^{6}Li in the atmospheres of metal-poor halo stars

The discussion focused on the claimed 1D LTE (> 2-sigma) detections of ^{6}Li in warm, metal-poor halo dwarf stars by Asplund *et al.* (2006) and Asplund & Meléndez (2008) including the very metal-poor stars G 64-12 with ^{6}Li/^{7}Li $= 0.059 \pm 0.021$ and G 64-37 with ^{6}Li/^{7}Li $= 0.111 \pm 0.032$. As emphasized by Martin Asplund, one may question a detection of ^{6}Li for a given star, but the collective distribution of ^{6}Li/^{7}Li can only be explained if ^{6}Li is present in the atmospheres of some of these metal-poor stars.

According to Martin Asplund (this volume), the ^{6}Li/^{7}Li ratio derived from the profile of the Li I 6708 Å resonance line does not change significantly if a 3D non-LTE analysis is applied instead of a 1D LTE analysis. Matthias Steffen (this volume), on the other hand, finds that the derived ^{6}Li/^{7}Li ratio is 0.01 to 0.02 smaller when using 3D non-LTE instead of 1D LTE. If such a zero-point shift in the derived ^{6}Li/^{7}Li values is included, then the distribution of ^{6}Li/^{7}Li is more compatible with the absence of ^{6}Li in the stellar atmospheres, although a few stars, such as HD 84937, still seem to have ^{6}Li detected at the 2-sigma level.

The difference in the estimation of the 3D (convective) effects on the derived value of ^{6}Li/^{7}Li may be connected to the way the analysis of the spectra is performed. Asplund *et al.* use other lines than Li I 6708 Å to determine the line broadening arising from the projected stellar rotation velocity, $V_{rot} \sin i$, whereas Steffen *et al.* assume a value

$V_{rot} \sin i = 0$ or 2 $\mathrm{km\,s}^{-1}$ and use only the Li line itself. Clearly, further 3D non-LTE studies are needed, and checks of line asymmetries induced by 3D convective motions should be performed for spectral lines that are formed in the same way and at the same atmospheric depth as the Li I line such as the K I 7698 Å line in the spectrum of HD 84937 (Smith *et al.* 2001) and the Na I D lines in the very metal-poor stars G 64-12 and G 64-37.

As emphasized by several participants, a clear detection of the ^{6}Li isotope at a level of ^{6}Li/^{7}Li ~ 0.05 in metal-poor halo stars with metallicities [Fe/H] $\lesssim -3$ would have profound effects on our understanding of the formation of Li. As shown by Nikos Prantzos (this volume) cosmic ray processes cannot explain the corresponding high ^{6}Li/Be ratio at [Fe/H] ~ -3. Pre-galactic production of ^{6}Li or non-standard Big-Bang nucleosynthesis have to be invoked (Karsten Jedamzik, this volume).

Concern was raised about the quality of the high-resolution spectra that have been used to determine ^{6}Li/^{7}Li. High resolution ($R \gtrsim 10^{5}$), and high signal-to-noise ($S/N \gtrsim 500$) are needed. In addition, one has to be very careful with the flat-fielding and rectification of the spectra to ensure that the continuum near the Li I 6708 Å line is accurate to better than 0.2 %. In this connection, it was mentioned that the non-detection of ^{6}Li in G 64-37 (^{6}Li/^{7}Li $= 0.01 \pm 0.04$) by García Pérez *et al.* (2009) (in contrast to the 3-sigma detection by Asplund & Meléndez quoted above) may be due to problems with residual fringes after flat-fielding of the Subaru/HDS spectra, as also discussed by García Pérez *et al.* themselves.

When discussing the Li isotope ratio, it should not be forgotten that ^{6}Li has been clearly detected in nearby interstellar clouds. As a new development in this field, Christopher Howk (this volume) presented the first measurements of the Li I 6708 Å line associated with gas in the Small Magellanic Cloud, which is known to be about a factor of four more metal-poor than the Sun. The high-resolution spectrum, which was taken with UVES and has a S/N of 250, suggests that ^{6}Li is present corresponding to ^{6}Li/^{7}Li ~ 0.12. A still higher S/N spectrum is needed to confirm this interesting detection of ^{6}Li in a metal-poor galaxy.

3. The 'Spite plateau' at the lowest metallicities

Sbordone *et al.* (this volume) presented evidence that halo stars with $T_{\mathrm{eff}} \gtrsim 5900\,\mathrm{K}$ and [Fe/H] < -2.5 show a decline and increasing scatter of Li abundances as a function of decreasing metallicity. At [Fe/H] ~ -3.5, the average Li abundance, $A(\mathrm{Li}) = \log(N_{\mathrm{Li}}/N_{\mathrm{H}}) + 12$, is 1.9 dex and the scatter is around 0.2 dex. In contrast, the Li abundance for halo stars with metallicities in the range $-2.5 < $ [Fe/H] < -1.5 is nearly constant at a level of $A(\mathrm{Li}) = 2.2$ with a scatter of 0.05 dex only. It was emphasized that this result does not depend on the effective temperature scale applied.

Meléndez *et al.* (this volume) find, however, that the increased scatter in Li abundance below [Fe/H] $= -2.5$ goes away if one selects the 'plateau' stars according to the condition $T_{\mathrm{eff}} > 5850 - 180\,[\mathrm{Fe/H}]\,\mathrm{K}$, i.e. with a strong metallicity dependence in the temperature limit; at [Fe/H] $= -3.0$ only stars with $T_{\mathrm{eff}} > 6390\,\mathrm{K}$ are included. Still, there is a tendency that stars with [Fe/H] < -2.5 have a 0.1 dex lower Li abundance than stars with [Fe/H] > -2.5.

Meléndez *et al.* (this volume) suggest that the turbulent diffusion models of Richard *et al.* (2005) can explain the large difference between the Li abundance predicted from WMAP + SBBN ($A(\mathrm{Li}) \sim 2.7$) and the abundance of the plateau stars ($A(\mathrm{Li}) \sim 2.2$), if models with a high parameter of turbulent mixing ($T6.25$) are adopted. The same models introduce a rather strong dependence of Li depletion on stellar mass, which may explain the scatter in $A(\mathrm{Li})$ at the lowest metallicities, but turbulent diffusion models with metallicities below [Fe/H] $= -2.0$ are still to be calculated. It would also be important to obtain a better understanding of the free parameter in the turbulent diffusion models from physical principles.

The derived stellar Li abundances depend critically on the adopted effective temperatures; if T_{eff} is increased by 100 K, $A(\mathrm{Li})$ increases by 0.07 dex. Over the years, there has been much discussion of the effective temperature scale for late-type stars especially for the metal-poor halo stars. Recent calibrations based on the IRFM method (González Hernández & Bonifacio 2009; Casagrande *et al.* 2009) agree, however, very well, and point to a fairly high temperature scale, i.e $T_{\mathrm{eff}} \sim 6500$ K for a very metal-poor star at the turn-off point. It is interesting that new 3D calculations of the profiles of Balmer lines (Ludwig *et al.* 2009) support such high temperatures. Further work on 3D non-LTE modelling of Balmer lines and comparison with observed profiles is, however, needed.

4. Lithium in globular clusters

Recent studies of Li abundance differences along the evolutionary sequence of stars in NGC 6397 were reviewed by Andreas Korn (this volume) and further discussed by Karin Lind (this volume) and Jonay González Hernández (this volume). Both high-resolution VLT/UVES spectra of 18 stars analyzed by Korn *et al.* (2007) and medium-resolution GIRAFFE spectra for 349 stars analyzed by Lind *et al.* (2009) suggest that the turn-off stars in NGC 6397 are more Li-poor, by about 0.1 dex, than subgiants which have not yet undergone dredge-up. Based on VLT/GIRAFFE spectra González Hernández *et al.* (2009) also find a difference of about 0.1 dex between 84 subgiants and 79 main-sequence stars in NGC 6397 selected in a narrow color range $0.57 < B - V < 0.63$. This suggests that Li sinks by diffusion during the MS/TO phase but to a depth low enough to prevent Li destruction such that some Li can be restored into the atmosphere, when a star evolves to the SG stage and the convection zone deepens.

It was noted that the two Li studies of NGC 6397 by Lind *et al.* and González Hernández *et al.* do not agree concerning the details of the trend of $A(\mathrm{Li})$ as a function of T_{eff}. The Lind *et al.* results suggest that $A(\mathrm{Li})$ decreases with increasing T_{eff} along the subgiant branch, whereas González Hernández *et al.* find the opposite trend. Karin Lind warned that these trends are somewhat uncertain, because of correlated errors in T_{eff} and $A(\mathrm{Li})$. Jonay González Hernández stressed that the turbulent diffusion model by Richard *et al.* (2005), which explains the difference between the WMAP based Li abundance, $A(\mathrm{Li}) \sim 2.7$, and the overall level of $A(\mathrm{Li}) \sim 2.3$ in the globular cluster stars, i.e. the $T6.25$ model, does not explain the observed difference between $A(\mathrm{Li})$ in SG and MS stars. Clearly, more work is needed. Li data for other globular clusters than NGC 6397 would be very important.

Piercarlo Bonifacio (this volume) presented a study of Li abundances in turnoff and subgiant stars belonging to ω Cen based on high-resolution spectra obtained with the ESO VLT. He considered ω Cen as the nucleus of an external galaxy that has been accreted by the Milky Way. According to Villanova *et al.* (2007), the stars observed span an age range of five Gyr. The preliminary results show that 30 stars with metallicities in the range $-1.9 < $ [Fe/H] $ < -1.5$ are distributed along a Li 'plateau' with $A(\mathrm{Li}) \sim 2.1$ and a scatter of the order of 0.1 dex. Given the five Gyr age-spread and the constancy of $A(\mathrm{Li})$ at 2.1 dex, Piercarlo Bonifacio pointed out that it is difficult to explain the difference in $A(\mathrm{Li})$ with respect to a primordial Li abundance of 2.7 dex by intrinsic stellar depletion. Li depletion in an early generation of massive stars, as suggested by Piau *et al.* (2006), was also considered an unlikely explanation by Piercarlo Bonifacio, because it seems surprising if this scenario would lead to the same depletion in the Milky

Way and in an external galaxy. Nikos Prantzos remarked that the suggestion of Piau *et al.* is excluded because Li astration in massive Pop III stars would be accompanied by heavy element production to a level much higher than we are observing in the very metal-poor halo stars.

5. Lithium in solar-type stars

Garik Israelian (this volume) described a new investigation of the Li abundance in Sun-like stars with and without detected planets based on spectra obtained with the HARPS spectrograph at the ESO 3.6 m telescope. For a limited range of ± 80 K around the solar effective temperature, only two out of 24 stars with detected planets have $A(\mathrm{Li})$ > 1.5, whereas 29 out of 60 stars without detected planets have $A(\mathrm{Li})$ > 1.5 (Israelian *et al.* 2009). Furthermore, there are no systematic differences in metallicity and age (as estimated from chromospheric activity) between stars with and without detected planets that could explain the difference in $A(\mathrm{Li})$. Hence, there seems to be evidence for enhanced Li depletion in Sun-like stars with orbiting planets. Israelian *et al.* suggest that the presence of a planetary system affects the angular momentum evolution of a star in a way that leads to a higher degree of rotational mixing of Li to interior regions, where it can be destroyed.

Meléndez *et al.* (this volume), on the other hand, find that solar-analogue stars with and without detected giant planets follow the same relation between atmospheric Li abundance and stellar age (determined from isochrones). This relation agrees well with that predicted by Charbonnel & Talon (2005) from models including the influence of gravity waves on the internal rotation of the Sun. Meléndez *et al.* also find that solar-twin stars (mass $M = (1.00 \pm 0.04)\, M_{\mathrm{Sun}}$ and metallicity [Fe/H] $= 0.0 \pm 0.1$) having the same *chemical* terrestrial planet signature as the Sun (Meléndez *et al.* 2009) follow the same Li vs. age relation as solar twins without the signature of terrestrial planets. These investigations suggest that one has to select stars with and and without planets in narrow mass, age and metallicity ranges before comparing their Li abundances. More work on larger samples of solar-twin stars is needed before final conclusions on Li differences between stars with and without planets can be made.

6. The Galactic evolution of beryllium

The formation and evolution of beryllium were discussed in talks by Ann Boesgaard, Francesca Primas and Rodolfo Smiljanic (this volume).

Beryllium abundances derived from Keck/HIRES spectra by Rich & Boesgaard (2009) for 49 stars ranging in metallicity from [Fe/H] $= -3.5$ to -0.5 and with oxygen abundances derived from OH lines in the UV suggest that the slope of $\log(\mathrm{Be})$ vs. $\log(\mathrm{O})$ may change from ~ 0.75 for stars with [O/H] < -1.6 to ~ 1.6 for stars with [O/H] > -1.6. As suggested by Ann Boesgaard, this could be due to a change in the dominant production mechanisms for Be. In the early Galaxy, Be formed by a primary process, i.e. high energy CNO atoms from supernovae hitting interstellar protons, whereas at later times the dominant (secondary) reaction is high energy protons from SNe hitting interstellar CNO.

Smiljanic *et al.* (2009) have used VLT/UVES spectra to determine Be abundances in a sample of 90 stars with metallicities mostly in the range $-2.0 <$ [Fe/H] < -0.5. Atmospheric parameters and abundances of the α-capture elements (Mg, Si, Ca, Ti) were adopted from literature. Li abundances are used to eliminate stars, possibly affected by depletion, from the sample. Interestingly, there is a significant scatter of Be at a given

[Fe/H]: the halo stars tend to split into two populations one with low Be and α/Fe, and another one with high Be and α/Fe like in the thick disk stars.

Francesca Primas (this volume) presented new Be data based on UVES spectra, which show that the two distinct populations of low-α and high-α halo stars discovered by Nissen & Schuster (2009) define different trends in the $\log$ (Be) - [Fe/H] diagram, thus confirming the results of Smiljanic *et al.* (2009). Furthermore, a hint of a flattening of the beryllium evolutionary trend at the lowest metallicities, [Fe/H] $\lesssim -2.5$, is found. As noted by Nikos Prantzos, this may be a signature of Be production by hypernovae.

From these works, it is clear that the evolution of Be in the Milky Way is more complicated than thought before, and that further studies of Be abundances in a larger sample of halo and disk stars may reveal interesting information about cosmic-ray processes and Galactic evolution. Investigations of non-LTE and 3D effects are, however, important in order to obtain accurate abundances of Be as well as Fe, O and the α-capture elements.

References

Asplund, M., Lambert, D. L., Nissen, P. E., Primas, F., & Smith, V. V. 2006, *ApJ*, 644, 229

Asplund, M. & Meléndez, J. 2008, in *First Stars III*, eds. B. W. O'Shea, A. Heger, and T. Abel, *AIP Conf. Proc.*, vol. 990, 342

Casagrande, L., Ramírez. I., Meléndez, J., Bessell, M., & Asplund, M. 2009, *A&A*, (submitted)

Charbonnel, C. & Talon, S. 2005, *Science*, 309, 2189

García Pérez, A. E., Aoki, W., Inoue, S. *et al.* 2009, *A&A*, 504, 213

González Hernández, J. I. & Bonifacio, P. 2009, *A&A*, 497, 497

González Hernández, J. I., Bonifacio, P., Caffau, E. *et al.* 2009, *A&A*, 505, L13

Israelian, G., Mena, E. D., Santos, N. C. *et al.* 2009, *Nature*, 462, 189

Korn, A. J., Grundahl, F., Richard, O. *et al.* 2007, *ApJ*, 671, 402

Lind, K., Primas, F., Charbonnel, C., Grundahl, F., & Asplund, M. 2009, *A&A*, 503, 545

Ludwig, H.-G., Behara, N. T., Steffen, M., & Bonifacio, P. 2009, *A&A*, 502, L1

Meléndez, J., Asplund, M., Gustafsson, B., & Yong, D. 2009, *ApJ*, 704, L66

Nissen, P. E. & Schuster, W. J. 2009, in *The Galaxy Disk in Cosmological Context (IAU Symp. 254)*, eds. J. Andersen, J. Bland-Hawthorn, and B. Nordström, (Cambridge Univ. Press), p. 103

Piau, L., Beers, T. C., Balsara, D. S. *et al.* 2006, *ApJ*, 653, 300

Rich, J. A. & Boesgaard, A. M. 2009, *ApJ*, 701, 1519

Richard, O., Michaud, G., & Richer, J. 2005, *ApJ*, 619, 538

Smiljanic, R., Pasquini, L., Bonifacio, P., Galli, D., Gratton, R. G., Randich, S., & Wolff, B. 2009, *A&A*, 499, 103

Smith, V. V., Vargas-Ferro, O., Lambert, D. L., & Olgin, J. G. 2001, *AJ*, 121, 453

Villanova, S., Piotto, G., King, I. R. *et al.* 2007, *ApJ*, 663, 296

Cristina Chiappini

Beatriz Barbuy

Light Elements in the Universe
Proceedings IAU Symposium No. 268, 2009
C. Charbonnel, M. Tosi, F. Primas & C. Chiappini, eds.

© International Astronomical Union 2010
doi:10.1017/S1743921310004692

Chemical evolution of D in the Local Disk

Takuji Tsujimoto[1] **and Joss-Bland-Hawthorn**[2]

[1] National Astronomical Observatory, Mitaka-shi, Tokyo 181-8588, Japan
email: `taku.tsujimoto@nao.ac.jp`
[2] Institute of Astronomy, School of Physics, University of Sydney, NSW 2006, Australia

Abstract. Chemical features of the local disk have firmly established the picture for the formation of the Galactic disk that the star formation has proceeded under the continuous accretion of low-metallicity gas from the halo. It sets two determinant processes for the evolution of deuterium (D), that is, the destruction of D in the interior of stars and the supply of new (nearly) primordial D associated with the gas infall. Conventional Galactic chemical evolution (GCE) models predict that this scheme leads to a monotonic decrease in D/H with time and ends up in the present-day D/H abundance $(D/H)_0$ which is severely lower than the recently observed estimates. These predicted features are the natural results of a construction of the metal-rich ($\sim$solar abundance) local star+gas system. Here we propose that the new GCE models, that incorporate large-scale winds form the Galactic bulge which entrain heavy elements and drop them on the disk with the recent tendency of star formation in tune with the observed implications, make the system rich in both metals and D. In addition, our finding of a gradual increase in D/H with time during the last several Gyr is observationally supported by the D/H abundance for the protosolar cloud lower than $(D/H)_0$.

Keywords. Galaxy: disk, evolution — Stars: abundances

Recent *Far-Ultraviolet Spectroscopic Explorer* (FUSE) observations reveal that the present-day D/H abundance $(D/H)_0$ for interstellar matter (ISM) in the Galactic disk is surprisingly as high as $\geqslant 2.31 \pm 0.24 \times 10^{-5}$ close to the primordial $(D/H)_p$ after correcting for dust depletion (Linsky *et al.* 2006). The high D/H abundance of $2.2^{+0.8}_{-0.6} \times 10^{-5}$ in the warm neutral medium of the lower Galactic halo, which originates in the disk and is elevated into the halo (Savage *et al.* 2007), strongly supports the claim by Linsky *et al.*

The difference between $(D/H)_p$ and $(D/H)_0$ is defined as a deuterium astration factor $f_d = (D/H)_p / (D/H)_0$. The value of f_d should absolutely be > 1, because it is believed that all D are produced during Big Bang nucleosynthesis, and only the destruction process of D occurs in the interior of stars in the post-Big Bang evolution. Thus, the value of f_d is determined by the fraction of matter which has never been cycled through stars in the present ISM. Accordingly, f_d is closely related to the enrichment degree by heavy elements in the Galaxy evolution through star formation that destroys D while fresh heavy elements are released into the ISM when the stars die. In the end, due to the fact that the D abundance in the ISM is anti-correlated with the metallicity in the system, the well-enriched ($\sim$solar metallicity) local disk demands a high f_d. Indeed, the standard GCE models predict $f_d = 1.39$–1.83 for the local disk (Romano *et al.* 2006). In other words, the local abundances of heavy elements such as Fe or α-elements in the ISM and long-lived stars imply $f_d \approx 1.4$ at the least, that is equivalent to $(D/H)_0 < 2.0 \times 10^{-5}$. On the other hand, the high $(D/H)_0$ implied by the recent observations means a smaller f_d. For the estimates of $(D/H)_p = 2.75 \times 10^{-5}$ (Cyburt *et al.* 2003) and $(D/H)_0 = 2.31 \times 10^{-5}$, we obtain $f_d = 1.19$, which is incompatible with the theoretical prediction.

The D/H abundances which bridge between the primordial $(D/H)_p$ and the present-day $(D/H)_0$ give a crucial information on the evolution of D. Viewing from this angle the

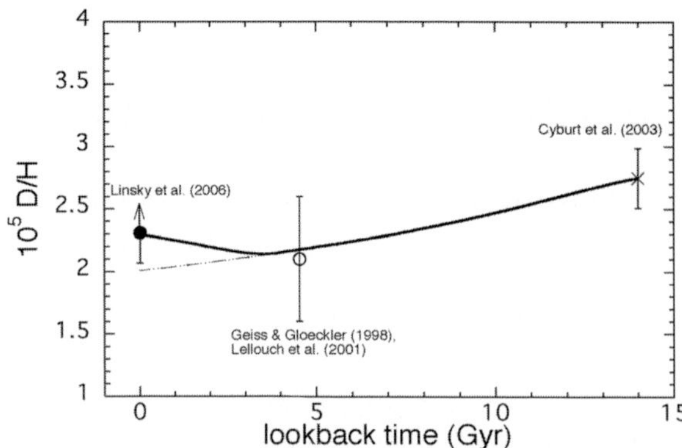

Figure 1. The evolution of D/H abundance in the local disk predicted by the wind+decreasing SFR model (solid line). For reference, the model case with a constant SFR coefficient is shown by the dashed line. Three observed points are for the primordial value (Cyburt *et al.* 2003), the value for the protosolar cloud (Geiss & Gloecker 1998, Lellouch *et al.* 2001), and the local ISM value (Linsky *et al.* 2006).

D/H in the protosolar cloud, which represents the abundance at the age of ~ 4.6 Gyr in the Galaxy, the obtained results are quite intriguing. The D/H abundances deduced from the atmosphere of jupiter (Lellouch *et al.* 2001) or from the solar wind (Geiss & Gloeckler 1998) are 2.1 ± 0.4 $(0.5) \times 10^{-5}$. If these estimates are correct, does it imply that D in the local disk has gradually increased during the last several Gyr?

Recent several works reveal the star formation history in the local disk, claiming that the star formation rate (SFR) has been clearly declining for the last several Gyr (Fuchs *et al.* 2009). In addition, recent new view on chemical evolution of the Galactic disk has suggested that large-scale winds from the Galactic bulge, which entrain a large amount of heavy elements, enrich the disk (Tsujimoto 2007). Theoretical framework that the Galaxy disk has evolved through an ongoing infall and the winds, together with the renewed constraint from the star formation history in the local disk, will open a new channel for the interpretation of the D evolution in the local disk.

The predicted time-evolution of D/H in this scheme is shown in Figure 1. Here the primordial $(D/H)_0$ is set to be 2.75×10^{-5}. From its initial value, the D/H abundance starts to decrease with time, and subsequently its change turns to a gradual increase around 4 Gyr ago. This turnover results from the supply of primordial D by an infall that overwhelms the destruction of D through the star formation owing to its low rate. Then, it finally enables to achieve a present-day high D/H abundance of 2.3×10^{-5}, fully consistent with the observed value, after passing D/H $\sim 2.1 \times 10^{-5}$ equivalent to the protosolar value around 4.6 Gyr ago. For reference, the case with a constant SFR over the full age is indicated by dashed line.

References

Cyburt, R. H., Fields, B. D., & Olive, K. A. 2003, *Phys. Lett. B*, 567, 227
Fuchs, B., Jahreiß, H., & Flynn, C. 2009, *AJ*, 137, 266
Geiss, J. & Gloeckler, G. 1998, *Space Sci. Rev.*, 84, 239
Lellouch, E. *et al.* 2001, *A&A*, 670, 610
Linsky, J. L. *et al.* 2006, *ApJ*, 647, 1106
Romano, D., Tosi, M., Chiappini, C., & Matteucci, F. 2006, *MNRAS*, 369, 295
Savage, B. D., Lehner, N., Fox, A., Wakker, B., & Sembach, K. 2007, *ApJ*, 659, 1222
Tsujimoto, T. 2007, *ApJ*, 665, L115

IAU 268 women in front of the Museum of Natural History

Tamara Mishenina, Corinne Charbonnel, Yuri Izotov

Anna Faviola Marino, Oscar Gonzalez,
Antonino Milone, Marcella di Criscienzo

Jean-Paul Zahn

Author Index